CRC HANDBOOK OF

Phase Equilibria and Thermodynamic Data of Aqueous Polymer Solutions

CRC HANDBOOK OF

Phase Equilibria and Thermodynamic Data of Aqueous Polymer Solutions

Christian Wohlfarth

CRC Press
Taylor & Francis Group
Boca Raton London New York

CRC Press is an imprint of the
Taylor & Francis Group, an **informa** business

CRC Press
Taylor & Francis Group
6000 Broken Sound Parkway NW, Suite 300
Boca Raton, FL 33487-2742

Printed in the United States of America on acid-free paper
Version Date: 20120530

International Standard Book Number: 978-1-4665-5438-2 (Hardback)

Library of Congress Cataloging-in-Publication Data

Wohlfarth, C.
CRC handbook of phase equilibria and thermodynamic data of aqueous polymer solutions / Christian Wohlfarth.
p. cm.
Includes bibliographical references and index.
ISBN 978-1-4665-5438-2 (hardback)
1. Polymer solutions--Thermal properties. 2. Phase rule and equilibrium. I. Title. II. Title: Handbook of phase equilibria and thermodynamic data of aqueous polymer solutions.

QD381.9.T54W64 2012
547'.704541--dc23 2012017889

Visit the Taylor & Francis Web site at
http://www.taylorandfrancis.com

and the CRC Press Web site at
http://www.crcpress.com

CONTENTS

PREFACE

Today, there is still a strong and continuing interest in thermodynamic properties of aqueous polymer solutions. Thus, about ten years after the *CRC Handbook of Thermodynamic Data of Aqueous Polymer Solutions* was published, necessity as well as desire arises for a supplementary book that includes and provides newly published experimental data from the last decade.

There are about **850 newly published references** containing about **150** new vapor-liquid equilibrium data sets and some new tables containing classical Henry's coefficients, about **600** new liquid-liquid equilibrium data sets and some new high-pressure fluid phase equilibrium data, **10** new enthalpic data sets, **20** new data sets describing PVT-properties of polymers, and **120** new data sets with densities or excess volumes. There are also new results on second osmotic virial coefficients of about **45** polymers in aqueous solution. So, in comparison to the original handbook, the new supplementary volume contains even a larger amount of data and will be a useful as well as necessary completion of the original handbook.

The *Supplement* will be divided again into the seven chapters as used before in the former *Handbook*: (1) Introduction, (2) Vapor-Liquid Equilibrium (VLE) Data of Aqueous Polymer Solutions, (3) Liquid-Liquid Equilibrium (LLE) Data of Aqueous Polymer Solutions, (4) High-Pressure Phase Equilibrium (HPPE) Data of Aqueous Polymer Solutions, (5) Enthalpy Changes for Aqueous Polymer Solutions, (6) PVT Data of Polymers and Aqueous Polymer Solutions, and (7) Second Virial Coefficients (A_2) of Aqueous Polymer Solutions. Finally, appendices quickly route the user to the desired data sets.

Additionally, tables of systems are provided where results were published only in graphical form in the original literature to lead the reader to further sources. Data are included only if numerical values were published or authors provided their numerical results by personal communication (and I wish to thank all those who did so). No digitized data have been included in this data collection.

The closing date for the data compilation was December 31, 2011. However, the user who is in need of new additional data sets is kindly invited to ask for new information beyond this book via e-mail at christian.wohlfarth@chemie.uni-halle.de. Additionally, the author will be grateful to all users who call his attention to mistakes and make suggestions for improvements.

The new *CRC Handbook of Phase Equilibria and Thermodynamic Data of Aqueous Polymer Solutions* will again be useful to researchers, specialists, and engineers working in the fields of polymer science, physical chemistry, chemical engineering, material science, biological science and technology, and those developing computerized predictive packages. The book should also be of use as a data source to Ph.D. students and faculty in chemistry, physics, chemical engineering, biotechnology, and materials science departments at universities.

Christian Wohlfarth

About the Author

Christian Wohlfarth is associate professor for physical chemistry at Martin-Luther University Halle-Wittenberg, Germany. He earned his degree in chemistry in 1974 and wrote his Ph.D. thesis in 1977 on investigations of the second dielectric virial coefficient and the intermolecular pair potential, both at Carl Schorlemmer Technical University Merseburg. In 1985, he wrote his habilitation thesis, *Phase Equilibria in Systems with Polymers and Copolymers*, at the Technical University Merseburg.

Since then, Dr. Wohlfarth's main research has been related to polymer systems. Currently, his research topics are molecular thermodynamics, continuous thermodynamics, phase equilibria in polymer mixtures and solutions, polymers in supercritical fluids, PVT behavior and equations of state, and sorption properties of polymers, about which he has published approximately 100 original papers. He has written the following books: *Vapor-Liquid Equilibria of Binary Polymer Solutions*, *CRC Handbook of Thermodynamic Data of Copolymer Solutions*, *CRC Handbook of Thermodynamic Data of Aqueous Polymer Solutions*, *CRC Handbook of Thermodynamic Data of Polymer Solutions at Elevated Pressures*, *CRC Handbook of Enthalpy Data of Polymer-Solvent Systems*, *CRC Handbook of Liquid-Liquid Equilibrium Data of Polymer Solutions*, and *CRC Handbook of Phase Equilibria and Thermodynamic Data of Copolymer Solutions*.

He is working on the evaluation, correlation, and calculation of thermophysical properties of pure compounds and binary mixtures resulting in eleven volumes of the *Landolt-Börnstein New Series*. He is a contributor to the *CRC Handbook of Chemistry and Physics*.

1. INTRODUCTION

1.1. Objectives of the handbook

Knowledge of thermodynamic data of aqueous polymer solutions is a necessity for industrial and laboratory processes. Furthermore, such data serve as essential tools for understanding the physical behavior of polymer solutions, for studying intermolecular interactions, and for gaining insights into the molecular nature of mixtures. They also provide the necessary basis for any developments of theoretical thermodynamic models. However, the database for aqueous polymer solutions is still modest in comparison with the enormous amount of data for low-molecular mixtures. On the other hand, there is still a strong and continuing interest in thermodynamic properties of aqueous polymer solutions, and during the last ten years after the former *CRC Handbook of Thermodynamic Data of Aqueous Polymer Solutions* was published, a large amount of new experimental data has been published, which is now compiled in this new *CRC Handbook of Phase Equilibria and Thermodynamic Data of Aqueous Polymer Solutions*.

Basic information on polymers can still be found in the *Polymer Handbook* (1999BRA), and there is also a chapter on properties of polymers and polymer solutions in the *CRC Handbook of Chemistry and Physics* (2011HAY). Older data for aqueous polymer solutions can be found in 1992WEN and 1993DAN, but also in the data books written by the author of this *Handbook* (1994WOH, 2001WOH, 2004WOH, 2005WOH, 2006WOH, 2008WOH, and 2011WOH). Older liquid-liquid equilibrium data for aqueous polymer systems were collected in 1985WAL, 1986ALB, and 1995ZAS. The *Handbook* does not present theories and models for polymer solution thermodynamics. Other publications (1971YAM, 1990FUJ, 1990KAM, 1999KLE, 1999PRA, and 2001KON) can serve as starting points for investigating those issues. Theories for aqueous two-phase systems are given by Cabezas (1996CAB). The state of the art for surfactants and polymers in aqueous solutions is summarized in 2003HOL. Thermodynamics of aqueous solutions of polyelectrolytes is reviewed by Maurer et al. (2011MAU). Aqueous solutions of non-ionic thermoresponsive polymers are reviewed in 2011ASE.

The data within this book are divided into six chapters:

- Vapor-liquid equilibrium (VLE) data of aqueous polymer solutions
- Liquid-liquid equilibrium (LLE) data of aqueous polymer solutions
- High-pressure phase equilibrium (HPPE) data of aqueous polymer solutions
- Enthalpy changes for aqueous polymer solutions
- PVT data of polymers and aqueous polymer solutions
- Second virial coefficients (A_2) of aqueous polymer solutions

Data from investigations applying to more than one chapter are divided and appear in the relevant chapters. Data are included only if numerical values were published or authors provided their results by personal communication. No digitized data have been included in this data collection. However, every data chapter is completed by a table that includes systems and references for data published only in graphical form as phase diagrams or related figures.

1.2. Experimental methods involved

Besides the common progress in instrumentation and computation, no remarkable new developments have been made with respect to the experimental methods involved here. So, a short summary of this chapter should be sufficient for the *Handbook*. The necessary equations are given together with some short explanations only.

Vapor-liquid equilibrium (VLE) measurements

Investigations on vapor-liquid equilibrium of polymer solutions can be made by various methods:

1. Absolute vapor pressure measurement
2. Differential vapor pressure measurement
3. Isopiestic sorption/desorption methods, i.e., gravimetric sorption, piezoelectric sorption, or isothermal distillation
4. Inverse gas-liquid chromatography (IGC) at infinite dilution, IGC at finite concentrations, and headspace gas chromatography (HSGC)
5. Steady-state vapor-pressure osmometry (VPO)

Experimental techniques for vapor pressure measurements were reviewed in 1975BON and 2000WOH. Methods and results of the application of IGC to polymers and polymer solutions were reviewed in 1976NES, 1988NES, 1989LLO, 1989VIL, and 1991MU1. Reviews on ebulliometry and/or vapor-pressure osmometry can be found in 1974TOM, 1975GLO, 1987COO, 1991MAY, and 1999PET.

VLE-experiments lead either to pressure vs. composition data or to activity vs. composition data. In the low-pressure region, solvent activities, a_A, are commonly determined from measured vapor pressures by:

$$a_A = (P_A / P_A^s)\exp\left[\frac{(B_{AA} - V_A^L)(P - P_A^s)}{RT}\right] \tag{1}$$

where:

a_A	activity of solvent A
B_{AA}	second virial coefficient of the pure solvent A at temperature T
P	system pressure
P_A	partial vapor pressure of the solvent A at temperature T
P_A^s	saturation vapor pressure of the pure liquid solvent A at temperature T
R	gas constant
T	measuring temperature
V_A^L	molar volume of the pure liquid solvent A at temperature T

Quite often, the exponential correction term is neglected, and only the ratio of partial pressure to pure solvent vapor pressure, P_A/P_A^s, is given.

IGC-experiments give specific retention volumes.

$$V_{net} = V_r - V_{dead} \tag{2}$$

where:

V_{net}	net retention volume
V_r	retention volume
V_{dead}	retention volume of the inert marker gas, dead retention, gas holdup

These net retention volumes are reduced to specific retention volumes, V_g^0, by division of equation (2) with the mass of the liquid (here the liquid is the molten polymer). They are corrected for the pressure difference between column inlet and outlet pressure, and reduced to a temperature $T_0 = 273.15$ K.

$$V_g^0 = \left(\frac{V_{net}}{m_B}\right)\left(\frac{T_0}{T}\right)\frac{3(P_{in}/P_{out})^2 - 1}{2(P_{in}/P_{out})^3 - 1} \tag{3}$$

where:

V_g^0	specific retention volume corrected to 0°C = 273.15 K
m_B	mass of the polymer in the liquid phase within the column
P_{in}	column inlet pressure
P_{out}	column outlet pressure
T_0	reference temperature = 273.15 K

Theory of GLC provides the relation between V_g^0 and thermodynamic data for the low-molecular component (solvent A) at infinite dilution (superscript ∞):

$$\left(\frac{P_A}{x_A^L}\right)^{\infty} = \frac{RT_0}{V_g^0 M_B} \quad \text{or} \quad \left(\frac{P_A}{w_A^L}\right)^{\infty} = \frac{RT_0}{V_g^0 M_A} \tag{4}$$

where:

M_A	molar mass of the solvent A
M_B	molar mass of the liquid (molten) polymer B (i.e., M_n)
P_A	partial vapor pressure of the solvent A at temperature T
x_A^L	mole fraction of solvent A in the liquid solution
w_A^L	mass fraction of solvent A in the liquid solution

The activity coefficients at infinite dilution read, if we neglect interactions to and between carrier gas molecules (which are normally helium):

$$\gamma_A^{\infty} = \left(\frac{RT_0}{V_g^0 M_B P_A^s}\right)\exp\left[\frac{P^s{}_A(V_A^L - B_{AA})}{RT}\right] \tag{5}$$

$$\Omega_A^{\infty} = \left(\frac{RT_0}{V_g^0 M_A P_A^s} \right) \exp \left[\frac{P_A^s (V_A^L - B_{AA})}{RT} \right] \tag{6}$$

where:

γ_A	activity coefficient of the solvent A in the liquid phase with activity $a_A = x_A \gamma_A$
Ω_A	mass fraction-based activity coefficient of the solvent A in the liquid phase with activity $a_A = w_A \Omega_A$

One should keep in mind that mole fraction-based activity coefficients, γ_A, become very small values for common polymer solutions and reach a value of zero for $M_B \rightarrow \infty$, which means a limited applicability at least to oligomer solutions. Therefore, the common literature provides only mass fraction-based activity coefficients for (high-molecular) polymer + (low-molecular) solvent pairs. The molar mass M_B of the polymeric liquid is an average value (M_n) according to the usual molar-mass distribution of polymers.

Furthermore, thermodynamic VLE data from GLC measurements are provided in the literature as values for $(P_A/w_A)^{\infty}$, i.e., classical mass fraction-based Henry's constants (if assuming ideal gas phase behavior):

$$H_{A,B} = \left(\frac{P_A}{w_A^L} \right)^{\infty} = \frac{RT_0}{V_g^0 M_A} \tag{7}$$

Since $V_{net} = V_r - V_{dead}$, the marker gas is assumed to not be retained by the polymer stationary phase and will elute at a retention time that is usually very small in comparison with those of the samples investigated. However, for small retention volumes, values for the mass fraction-based Henry's constants should be corrected for the solubility of the marker gas (1976LIU). The apparent Henry's constant is obtained from equation (7) above.

$$H_{A,B} = H_{A,B}^{app} \left[1 + \frac{M_A H_{A,B}^{app}}{M_{ref} H_{A,ref}} \right]^{-1} \tag{8}$$

M_{ref} is the molar mass of the marker gas. The Henry's constant of the marker gas itself, determined by an independent experiment, need not be known very accurately, as it is usually much larger than the apparent Henry's constant of the sample.

VPO-experiments give temperature differences for determining solvent activities:

$$\Delta T^{st} = -k_{VPO} \frac{RT^2}{\Delta_{LV} H_{0A}} \ln a_A \tag{9}$$

where:

k_{VPO}	VPO-specific constant (must be determined separately)
ΔT^{st}	temperature difference between solution and solvent drops in the steady state
$\Delta_{LV}H_{0A}$	molar enthalpy of vaporization of the pure solvent A at temperature T

The steady state must be sufficiently near the vapor-liquid equilibrium and linear non-equilibrium thermodynamics is valid then.

Liquid-liquid equilibrium (LLE) measurements

To understand the results of LLE experiments in polymer solutions, one has to take into account the strong influence of polymer distribution functions on LLE because fractionation occurs during demixing. Fractionation takes place with respect to molar mass distribution as well as to chemical distribution. Fractionation during demixing leads to some special effects by which the LLE phase behavior differs from that of an ordinary, strictly binary mixture because a common polymer solution is a multicomponent system. *Cloud-point curves* are measured instead of binodals; and per each individual feed concentration of the mixture, *two parts of a coexistence curve* occur below (for upper critical solution temperature, UCST, behavior) or above the cloud-point curve (for lower critical solution temperature, LCST, behavior), i.e., produce an infinite number of coexistence data. Distribution functions of the feed polymer belong only to cloud-point data. On the other hand, each pair of coexistence points is characterized by two new and different distribution functions in each coexisting phase. The critical concentration is the only feed concentration where both parts of the coexistence curve meet each other on the cloud-point curve at the critical point that belongs to the feed polymer distribution function. The threshold point (maximum or minimum corresponding to either UCST or LCST behavior) temperature (or pressure) is not equal to the critical point, since the critical point is to be found at a shoulder of the cloud-point curve. *Phase-volume-ratio method* (1968KON) or *coexistence concentration plot* (1969WOL) gives the critical point as the intersection of cloud-point curve and shadow curve. For more details, we refer to 1968KON, 1972KON, 2001KON, and 2008WOH.

Treating polymer solutions with distribution functions by continuous thermodynamics and procedures to measure and calculate liquid-liquid equilibria of such systems is reviewed in 1989RAE and 1990RAE.

High-pressure phase equilibrium (HPPE) measurements

Experimental methods for high-pressure fluid phase equilibria in polymer solutions follow the same lines as for VLE-, LLE-, or VLLE-experiments. The experimental equipment follows on the same techniques, however, extended to high pressure conditions. For more details, we refer to 1994MCH, 1997KIR, 1999KIR, and 2005WOH.

The solvents are in many cases supercritical fluids, i.e., gases/vapors above their critical temperature and pressure. Data were measured mainly for two kinds of solutions: solutions in supercritical CO_2 (and some other fluids) or solutions in supercritical monomers.

Measurement of enthalpy changes in polymer solutions

Experiments on enthalpy changes in binary polymer solutions can be made within common microcalorimeters by applying one of the following three methods:

1. Measurement of the enthalpy change caused by solving a given amount of the solute polymer in an (increasing) amount of solvent, i.e., the solution experiment
2. Measurement of the enthalpy change caused by mixing a given amount of a concentrated polymer solution with an amount of pure solvent, i.e., the dilution experiment
3. Measurement of the enthalpy change caused by mixing a given amount of a liquid/molten polymer with an amount of pure solvent, i.e., the mixing experiment

Care must be taken for polymer solutions with respect to the resolution of the instrument, which has to be higher than for common solutions with larger enthalpic effects. Usually employed calorimeters for such purposes are the Calvet-type calorimeters based on heat-flux principle. Details can be found in 1984HEM and 1994MAR.

The (integral) enthalpy of mixing or the (integral) enthalpy of solution of a binary system is the amount of heat which must be supplied when n_A mole of pure solvent A and n_B mole of pure polymer B are combined to form a homogeneous mixture/solution in order to keep the total system at constant temperature and pressure.

$$\Delta_M h = n_A H_A + n_B H_B - (n_A H_{0A} + n_B H_{0B}) \quad (10a)$$

$$\Delta_{sol} h = n_A H_A + n_B H_B - (n_A H_{0A} + n_B H_{0B}) \quad (10b)$$

where:

$\Delta_M h$	(integral) enthalpy of mixing
$\Delta_{sol} h$	(integral) enthalpy of solution
H_A	partial molar enthalpy of solvent A
H_B	partial molar enthalpy of polymer B
H_{0A}	molar enthalpy of pure solvent A
H_{0B}	molar enthalpy of pure polymer B
n_A	amount of substance of solvent A
n_B	amount of substance of polymer B

From thermodynamic reasons follows that the change $\Delta_M H$ of the molar (or specific or segment molar) enthalpy in an isothermal-isobaric mixing process is also the molar (or specific or segment molar) excess enthalpy, H^E, of the mixture. The dependence of H^E upon temperature, T, and pressure, P, permits the correlation of such data with excess heat capacities, C_p^E, and excess volumes, V^E.

$$\left(\partial H^E / \partial T\right)_P = C_P^E \quad (11)$$

$$\left(\partial H^E / \partial P\right)_T = V^E - T\left(\partial V^E / \partial T\right)_P \quad (12)$$

Partial molar enthalpies are given by:

$$\Delta_{sol}H_B = (\partial \Delta_{sol} h / \partial n_B)_{P,T,n_j} = H_B - H_{0B} \tag{13a}$$

$$\Delta_M H_B = (\partial \Delta_M h / \partial n_B)_{P,T,n_j} = H_B - H_{0B} \tag{13b}$$

Partial specific enthalpies are given by:

$$\Delta_{sol}H_B = (\partial \Delta_{sol} h / \partial m_B)_{P,T,m_j} \tag{13c}$$

$$\Delta_M H_B = (\partial \Delta_M h / \partial m_B)_{P,T,m_j} \tag{13d}$$

Intermediary enthalpy of dilution is often measured instead of excess enthalpies, i.e., the enthalpy effect obtained if solvent A is added to an existing homogeneous polymer solution. The extensive intermediary enthalpy of dilution is the difference between two values of the enthalpy of the polymer solution corresponding to the concentrations of the polymer solution at the beginning and at the end of the dilution process:

$$\Delta_{dil}H^{12} = H^{(2)} - H^{(1)} \tag{14}$$

with

$$H^{(1)} = n_A^{(1)} H_A^{(1)} + n_B H_B^{(1)} \tag{15a}$$

$$H^{(2)} = n_A^{(2)} H_A^{(2)} + n_B H_B^{(2)} \tag{15b}$$

and

$$n_A^{(2)} = n_A^{(1)} + \Delta n_A \tag{16}$$

where:

$\Delta_{dil}H^{12}$	(extensive) intermediary enthalpy of dilution
$H^{(1)}$, $H^{(2)}$	enthalpies of the polymer solution before and after the dilution step
$H_A^{(1)}$, $H_A^{(2)}$	partial molar enthalpies of solvent A before and after the dilution step
$H_B^{(1)}$, $H_B^{(2)}$	partial molar enthalpies of polymer B before and after the dilution step
$n_A^{(1)}$	amount of solvent in the solution before the dilution step
$n_A^{(2)}$	amount of solvent in the solution after the dilution step
Δn_A	amount of solvent added to solution (1)
n_B	amount of polymer in all solutions.

$\Delta_{dil}H^{12}$ is not directly related to $\Delta_M H$ but to $(\partial \Delta_M H / \partial n_A)_{P,T,n_j}$ by:

$$\Delta_{dil}H^{12} = \int_{n_A^{(1)}}^{n_A^{(2)}} (\partial \Delta_M H / \partial n_A)_{P,T,n_j} \, dn_A \tag{17}$$

Generally, it is known that partial molar enthalpies of mixing (or dilution) of the solvent can also be determined from the temperature dependence of the activity of the solvent, a_A:

$$\Delta_M H_A = R\,[\partial \ln a_A / \partial (1/T)]_P \tag{18}$$

Enthalpy data from light scattering, osmometry, vapor pressure or vapor sorption measurements, and demixing experiments can be found in the literature. However, agreement between enthalpy changes measured by calorimetry and results determined from the temperature dependence of solvent activity data is often of limited quality. In this *Handbook*, data for $\Delta_M H_A^\infty$ determined by inverse gas-liquid chromatography (IGC) have been included.

$$\Delta_M H_A^\infty = R\,[\partial \ln \Omega_A^\infty / \partial(1/T)]_P \tag{19a}$$

Additionally, the enthalpies of solution at infinite dilution determined by IGC, $\Delta_{sol} H_{A(vap)}^\infty$, of water vapor in molten polymers (with $\Delta_{sol} H_{A(vap)}^\infty = \Delta_M H_A^\infty - \Delta_{LV} H_{0A}$) have been included.

$$\Delta_{sol} H_{A(vap)}^\infty = -R\,[\partial \ln V_g^0 / \partial(1/T)]_P \tag{19b}$$

Measurement of PVT-behavior of polymer melt and of excess volume in solution

There are two widely practiced methods for the *PVT* measurement of polymers and polymers:

1. Piston-die technique
2. Confining fluid technique

which were described in detail by Zoller in papers and books (e.g., 1986ZOL, 1995ZOL). The tables in Chapter 6 provide specific volumes neither at or below the melting transition of semicrystalline materials nor at or below the glass transition of amorphous samples, since *PVT* data of solid polymer samples are non-equilibrium data and depend on sample history and experimental procedure (which will not be discussed here). Therefore, only equilibrium data for the liquid/molten state are tabulated. Their common accuracy (standard deviation) is about 0.001 cm^3/g in specific volume, 0.1 K in temperature, and $0.005{*}P$ in pressure (1995ZOL). Measurement of densities for polymer solutions is usually made today by U-tube vibrating densimeters.

Excess volumes at temperature T and pressure p are determined by:

$$V^E_{spec} = V_{spec} - (w_A V_{0A,\,spec} + w_B V_{0B,\,spec}) \tag{20a}$$
$$V^E = (x_A M_A + x_B M_B)/\rho - (x_A M_A/\rho_A + x_B M_B/\rho_B) \tag{20b}$$

where:

V_{spec}	specific volume of the polymer solution
V^E, V^E_{spec}	molar or specific excess volume
$V_{0A,\,spec}$, $V_{0B,\,spec}$	specific volume of pure polymer (B) and pure solvent (A)
w_A and w_B	mass fraction of polymer (B) and solvent (A) (definition see below)
x_A and x_B	mole fraction of polymer (B) and solvent (A) (definition see below)
ρ	density of the polymer solution
ρ_A, ρ_B	density of pure polymer (B) or pure solvent (A)

The second term in Eqs. (20a, b) corresponds to the *ideal volume*, V^{id}, of the polymer solution.

Determination of second virial coefficients A_2

There are a couple of methods for the experimental determination of the second virial coefficient: colligative properties (vapor pressure depression, freezing point depression, boiling point increase, membrane osmometry), scattering methods (classical light scattering, X-ray scattering, neutron scattering), sedimentation velocity, and sedimentation equilibrium. Details of the special experiments can be found in many textbooks and will not be repeated here (for example, 1972HUG, 1974TOM, 1975CAS, 1975FUJ, 1975GLO, 1987ADA, 1987BER, 1987COO, 1987KRA, 1987WIG, 1991CHU, 1991MAY, 1991MU2, 1992HAR, and 1999PET).

The *vapor pressure depression* of the solvent in a binary polymer solution, i.e., the difference between the saturation vapor pressure of the pure solvent and the corresponding partial pressure in the solution, $\Delta P_A = P_A^s - P_A$, is expressed as:

$$\frac{\Delta P_A}{P_A} = V_A^L c_B \left[\frac{1}{M_n} + A_2 c_B + A_3 c_B^2 + \ldots \right] \tag{21}$$

where:

$A_2, A_3, \ldots$	second, third, ... osmotic virial coefficients at temperature T
c_B	(mass/volume) concentration at temperature T
M_n	number-average relative molar mass of the polymer
ΔP_A	$P_A^s - P_A$, vapor pressure depression of the solvent A at temperature T
V_A^L	molar volume of the pure liquid solvent A at temperature T

The *freezing point depression*, $\Delta_{SL}T_A$, is:

$$\Delta_{SL} T_A = E_{SL} c_B \left[\frac{1}{M_n} + A_2 c_B + A_3 c_B^2 + \ldots \right] \tag{22}$$

and the *boiling point increase*, $\Delta_{LV}T_A$, is:

$$\Delta_{LV} T_A = E_{LV} c_B \left[\frac{1}{M_n} + A_2 c_B + A_3 c_B^2 + \ldots \right] \tag{23}$$

where:

E_{LV}	ebullioscopic constant
E_{SL}	cryoscopic constant
$\Delta_{SL}T_A$	freezing point temperature difference between pure solvent and solution, i.e., ${}_{SL}T_A^0 - {}_{SL}T_A$
$\Delta_{LV}T_A$	boiling point temperature difference between solution and pure solvent, i.e., ${}_{LV}T_A - {}_{LV}T_A^0$

The *osmotic pressure*, π, can be described as:

$$\frac{\pi}{c_B} = RT\left[\frac{1}{M_n} + A_2 c_B + A_3 c_B^{\ 2} + \ldots\right] \tag{24}$$

In the *dilute concentration region*, the virial equation is usually truncated after the second virial coefficient which leads to a linear relationship. A linearized relation over a wider concentration range can be constructed if the Stockmayer-Casassa relation between A_2 and A_3 is applied:

$$A_3 M_n = \left(\frac{A_2 M_n}{2}\right)^2 \tag{25}$$

$$\left(\frac{\pi}{c_2}\right)^{0.5} = \left(\frac{RT}{M_n}\right)^{0.5}\left[1 + \frac{A_2 M_n}{2} c_2\right] \tag{26}$$

Scattering methods enable the determination of A_2 via the common relation:

$$\frac{Kc_B}{R(q)} = \frac{1}{M_w P_z(q)} + 2A_2 Q(q) c_B + \ldots \tag{27}$$

with

$$q = \frac{4\pi}{\lambda}\sin\frac{\theta}{2} \tag{28}$$

where:

K	a constant that summarizes the optical parameters of a scattering experiment
M_w	mass-average relative molar mass of the polymer
$P_z(q)$	z-average of the scattering function
q	scattering vector
$Q(q)$	function for the q-dependence of A_2
$R(q)$	excess intensity of the scattered beam at the value q
λ	wavelength
θ	scattering angle

Depending on the chosen experiment (light, X-ray, or neutron scattering), the constant K is to be calculated from different relations. For details see the corresponding textbooks (1972HUG, 1975CAS, 1982GLA, 1986HIG, 1987BER, 1987KRA, 1987WIG, and 1991CHU).

1.3. Guide to the data tables

Characterization of the polymers

Polymers vary by a number of characterization variables. The molar mass and their distribution function are the most important variables, and also the chemical distribution and the average chemical composition have to be given. However, tacticity, sequence distribution, branching, and end groups determine their thermodynamic behavior in solutions too. Unfortunately, much less information is provided with respect to the polymers that were applied in most of the thermodynamic investigations in the original literature. In many cases, the samples are characterized only by one or two molar mass averages, the average chemical composition for copolymers, and some additional information (e.g., T_g, T_m, ρ_B, or how and where they were synthesized). Sometimes even this information is missed.

The molar mass averages are defined as follows:

number average M_n

$$M_n = \frac{\sum_i n_{B_i} M_{B_i}}{\sum_i n_{B_i}} = \frac{\sum_i w_{B_i}}{\sum_i w_{B_i} / M_{B_i}} \tag{29}$$

mass average M_w

$$M_w = \frac{\sum_i n_{B_i} M_{B_i}^2}{\sum_i n_{B_i} M_{B_i}} = \frac{\sum_i w_{B_i} M_{B_i}}{\sum_i w_{B_i}} \tag{30}$$

z-average M_z

$$M_z = \frac{\sum_i n_{B_i} M_{B_i}^3}{\sum_i n_{B_i} M_{B_i}^2} = \frac{\sum_i w_{B_i} M_{B_i}^2}{\sum_i w_{B_i} M_{B_i}} \tag{31}$$

viscosity average M_η

$$M_\eta = \left(\frac{\sum_i w_{B_i} M_{B_i}^a}{\sum_i w_{B_i}} \right)^{1/a} \tag{32}$$

where:

a	exponent in the viscosity-molar mass relationship
M_{Bi}	relative molar mass of the polymer species B_i
n_{Bi}	amount of substance of polymer species B_i
w_{Bi}	mass fraction of polymer species B_i

Measures for the polymer concentration

The following concentration measures are used in the tables of this *Handbook* (where B always denotes the main polymer, A denotes the solvent, and in ternary systems C denotes the third component):

mass/volume concentration

$$c_A = m_A/V \qquad c_B = m_B/V \qquad (33)$$

mass fraction

$$w_A = m_A/\Sigma\, m_i \qquad w_B = m_B/\Sigma\, m_i \qquad (34)$$

mole fraction

$$x_A = n_A/\Sigma\, n_i \qquad x_B = n_B/\Sigma\, n_i \qquad \text{with } n_i = m_i/M_i \text{ and } M_B = M_n \qquad (35)$$

volume fraction

$$\varphi_A = (m_A/\rho_A)/\Sigma\, (m_i/\rho_i) \qquad \varphi_B = (m_B/\rho_B)/\Sigma\, (m_i/\rho_i) \qquad (36)$$

segment fraction

$$\psi_A = x_A r_A/\Sigma\, x_i r_i \quad \psi_B = x_B r_B/\Sigma\, x_i r \qquad (37)$$

base mole fraction

$$z_A = x_A r_A/\Sigma\, x_i r_i \quad z_B = x_B r_B/\Sigma\, x_i r_i \quad \text{with } r_B = M_B/M_0 \text{ and } r_A = 1 \qquad (38)$$

where:

c_A	(mass/volume) concentration of solvent A
c_B	(mass/volume) concentration of polymer B
m_A	mass of solvent A
m_B	mass of polymer B
M_A	relative molar mass of the solvent A
M_B	relative molar mass of the polymer B
M_n	number-average relative molar mass of the polymer B
M_0	molar mass of a basic unit of the polymer B
n_A	amount of substance of solvent A
n_B	amount of substance of polymer B
r_A	segment number of the solvent A, usually $r_A = 1$
r_B	segment number of the polymer B
V	volume of the liquid solution at temperature T
w_A	mass fraction of solvent A
w_B	mass fraction of polymer B
x_A	mole fraction of solvent A
x_B	mole fraction of polymer B
z_A	base mole fraction of solvent A
z_B	base mole fraction of polymer B

φ_A	volume fraction of solvent A
φ_B	volume fraction of polymer B
ρ_A	density of solvent A
ρ_B	density of polymer B
ψ_A	segment fraction of solvent A
ψ_B	segment fraction of polymer B

For high-molecular polymers, a mole fraction is not an appropriate unit to characterize composition. However, for oligomeric products with rather low molar masses, mole fractions were sometimes used. In the common case of a distribution function for the molar mass, $M_B = M_n$ is to be chosen. Mass fraction and volume fraction can be considered as special cases of segment fractions depending on the way by which the segment size is actually determined: $r_i/r_A = M_i/M_A$ or $r_i/r_A = V_i/V_A = (M_i/\rho_i)/(M_A/\rho_A)$, respectively. Classical segment fractions are calculated by applying $r_i/r_A = V_i^{\text{vdW}}/V_A^{\text{vdW}}$ ratios where hard-core van der Waals volumes, V_i^{vdW}, are taken into account. Their special values depend on the chosen equation of state or simply some group contribution schemes (e.g., 1968BON, 1990KRE) and have to be specified. Volume fractions imply a temperature dependence and, as they are defined in the equations above, neglect excess volumes of mixing and, very often, the densities of the polymer in the state of the solution are not known correctly. However, volume fractions can be calculated without the exact knowledge of the polymer molar mass (or its averages). Base mole fractions are seldom applied for polymer systems today. The value for M_0, the molar mass of a basic unit of the polymer, has to be determined according to the corresponding chemical constitution and has to be specified.

Tables of experimental data

The data tables in each chapter are provided in order of the names of the polymers. In this *Handbook*, usually source-based polymer names are applied. These names are more common in use, and they are mostly given also in the original sources. For copolymers, their names were built by the two names of the comonomers which are connected by *-co-*, or more specifically by *-alt-* for alternating copolymers, by *-b-* for block copolymers, by *-g-* for graft copolymers, or by *-stat-* for statistical copolymers. Structure-based names, for which details about their nomenclature can be found in the *Polymer Handbook* (1999BRA), are chosen in some single cases only. CAS index names for polymers are usually not applied here. Latest *IUPAC Recommendations* for class names of polymers based on chemical structure and molecular architecture are given in (2009BAR). A list of the polymers in Appendix 1 utilizes the names as given in the chapters of this book.

Within types of polymers the individual samples are ordered by their increasing average molar mass. Subsequently, systems are ordered by their solvents. Solvents are listed alphabetically. When necessary, systems are ordered by increasing temperature. In ternary systems, ordering is additionally made subsequently according to the name of the third component in the system. Each data set begins with the lines for the solution components, e.g., in binary systems:

Polymer (B):	**poly(*N*,*N*-dimethylmethacrylamide)**		**2004ARC**
Characterization:	M_n/g.mol^{-1} = 4000, ρ = 1.15 g/cm^3, T_g/K = 378, Polymer Source		
Solvent (A):	**water**	**H_2O**	**7732-18-5**

where the polymer sample is given in the first line together with the reference. The second line provides then the characterization available for the polymer sample. The following line gives the solvent's chemical name, molecular formula, and CAS-registry number. In this data book, the solvent (A) is always water.

In ternary and quaternary systems, the following lines are either for a second solvent or a second polymer or a salt or another chemical compound, e.g., in ternary systems with two solvents

Polymer (B):	**polyesteramide (hyperbranched)**		**2003SE3**
Characterization:	M_n/g.mol^{-1} = 1200, M_w/g.mol^{-1} = 6000, 8 hydroxyl groups, Hybrane S1200, DSM Fine Chemicals, The Netherlands		
Solvent (A):	**water**	**H_2O**	**7732-18-5**
Solvent (C):	**ethanol**	**C_2H_6O**	**64-17-5**

or, e.g., in ternary systems with a second polymer

Polymer (B):	**poly(ethylene oxide-*b*-propylene oxide-*b*-ethylene oxide)**		**2010CAR**
Characterization:	M_n/g.mol^{-1} = 2900, 40 wt% ethylene oxide, Aldrich Chem. Co., Inc., Milwaukee, WI		
Solvent (A):	**water**	**H_2O**	**7732-18-5**
Polymer (C):	**maltodextrin**		
Characterization:	M_n/g.mol^{-1} = 990, M_w/g.mol^{-1} = 8800, Corn Products Brasil S/A, Mogi Guacu, SP, Brazil		

or, e.g., in ternary systems with a salt

Polymer (B):	**poly(1-vinyl-2-pyrrolidinone)**		**2005SA2**
Characterization:	M_w/g.mol^{-1} = 10000, PVP K15, Aldrich Chem. Co., Inc., Milwaukee, WI		
Solvent (A):	**water**	**H_2O**	**7732-18-5**
Salt (C):	**ammonium sulfate**	**$(NH_4)_2SO_4$**	**7783-20-2**

or, e.g., in quaternary (or higher) systems like

Polymer (B):	**poly(ethylene glycol)**		**2003HAG**
Characterization:	M_n/g.mol^{-1} = 1500, Merck KGaA, Darmstadt, Germany		
Solvent (A):	**water**	**H_2O**	**7732-18-5**
Salt (C):	**dipotassium phosphate**	**K_2HPO_4**	**7758-11-4**
Polymer (D):	**bovine serum albumin**		
Characterization:	M_n/g.mol^{-1} = 68000, Merck KGaA, Darmstadt, Germany		

There are some exceptions from this type of presentation within the tables for Henry's constants, A_2 values, UCST/LCST data, and PVT data of pure polymers. These tables are prepared in the forms as chosen in 2004WOH or 2008WOH.

The final date for including data into this *Handbook* was December 31, 2011.

1.4. List of symbols

a	exponent in the viscosity-molar mass relationship
a_A	activity of solvent A
B	parameter of the Tait equation
B_{AA}	second virial coefficient of the pure solvent A at temperature T
c_A	(mass/volume) concentration of solvent A
c_B	(mass/volume) concentration of polymer B
C	parameter of the Tait equation
E_{LV}	ebullioscopic constant
E_{SL}	cryoscopic constant
h_D	distance from the center of rotation
H^E	excess enthalpy = $\Delta_M H$ = enthalpy of mixing
H_A	partial molar enthalpy of solvent A
H_B	partial molar (or specific) enthalpy of polymer B
H_{0A}	molar enthalpy of pure solvent A
H_{0B}	molar (or specific) enthalpy of pure polymer B
$H_{A,B}$	classical mass fraction Henry's constant of solvent vapor A in the molten polymer B
$\Delta_{dil}H^{12}$	(integral) intermediary enthalpy of dilution (= $\Delta_M H^{(2)} - \Delta_M H^{(1)}$)
$\Delta_M H$	(integral) enthalpy of mixing
$\Delta_{sol}H$	(integral) enthalpy of solution
$^{int}\Delta_M H_A$	integral enthalpy of mixing of solvent A (= integral enthalpy of dilution)
$\Delta_M H_A$	partial molar enthalpy of mixing of the solvent A (= differential enthalpy of dilution)
$\Delta_M H_A^\infty$	partial molar enthalpy of mixing at infinite dilution of the solvent A
$^{int}\Delta_{sol}H_A$	integral enthalpy of solution of solvent A
$\Delta_{sol}H_A$	partial molar enthalpy of solution of the solvent A
$\Delta_{sol}H_A^\infty$	first integral enthalpy of solution of solvent A (= $\Delta_M H_A^\infty$ in the case of liquid/molten polymers <u>and</u> a liquid solvent, i.e., it is different from the values for solutions of solvent vapors or gases in a liquid/molten polymer $\Delta_{sol}H_{A(vap)}^\infty$)
$\Delta_{sol}H_{A(vap)}^\infty$	first integral enthalpy of solution of the vapor of solvent A (with $\Delta_{sol}H_{A(vap)}^\infty = \Delta_M H_A^\infty - \Delta_{LV}H_{0A}$)
$\Delta_{LV}H_{0A}$	molar enthalpy of vaporization of the pure solvent A at temperature T
$^{int}\Delta_M H_B$	integral enthalpy of mixing of polymer B
$\Delta_M H_B$	partial molar (or specific) enthalpy of mixing of polymer B
$\Delta_M H_B^\infty$	partial molar (or specific) enthalpy of mixing at infinite dilution of polymer B
$^{int}\Delta_{sol}H_B$	integral enthalpy of solution of polymer B
$\Delta_{sol}H_B$	partial molar (or specific) enthalpy of solution of polymer B
$\Delta_{sol}H_B^\infty$	first integral enthalpy of solution of polymer B ($\Delta_M H_B^\infty$ in the case of liquid/molten B)
k_{VPO}	VPO-specific constant (must be determined separately)
K	a constant that summarizes the optical parameters of a scattering experiment

m_A	mass of solvent A
m_B	mass of polymer B
M	relative molar mass
M_A	molar mass of the solvent A
M_B	molar mass of the polymer B
M_n	number-average relative molar mass of the polymer B
M_w	mass-average relative molar mass of the polymer B
M_η	viscosity-average relative molar mass of the polymer B
M_z	z-average relative molar mass of the polymer B
M_0	molar mass of a basic unit of the polymer B
MI	melting index of the polymer B
n_A	amount of substance of solvent A
n_B	amount of substance of polymer B
P	pressure
P_0	standard pressure (= 0.101325 MPa)
P_{crit}	critical pressure
P_A	partial vapor pressure of the solvent A at temperature T
P_A^s	saturation vapor pressure of the pure liquid solvent A at temperature T
ΔP_A	$P_A^s - P_A$, vapor pressure depression of the solvent A at temperature T
P_{in}	column inlet pressure in IGC
P_{out}	column outlet pressure in IGC
$P_z(q)$	z-average of the scattering function
q	scattering vector
$Q(q)$	function for the q-dependence of A_2
R	gas constant
$R(q)$	excess intensity of the scattered beam at the value q
r_A	segment number of the solvent A, usually $r_A = 1$
r_B	segment number of the polymer B
T	(measuring) temperature
T_{crit}	critical temperature
T_g	glass transition temperature
T_m	melting transition temperature
T_0	reference temperature (= 273.15 K)
ΔT^{st}	temperature difference between solution and solvent drops in VPO
$\Delta_{SL}T_A$	freezing point temperature difference between pure solvent and solution, i.e., ${}_{SL}T_A^0 - {}_{SL}T_A$
$\Delta_{LV}T_A$	boiling point temperature difference between solution and pure solvent, i.e., ${}_{LV}T_A - {}_{LV}T_A^0$
V, V_{spec}	volume or specific volume at temperature T
V_0	reference volume
V^E	excess volume at temperature T
V^{vdW}	hard-core van der Waals volume
V_A^L	molar volume of the pure liquid solvent A at temperature T
V_{net}	net retention volume in IGC
V_r	retention volume in IGC
V_{dead}	retention volume of the (inert) marker gas, dead retention, gas holdup in IGC
V_g^0	specific retention volume corrected to 0°C in IGC

w_A	mass fraction of solvent A
w_B	mass fraction of polymer B
$w_{B,\ crit}$	mass fraction of the polymer B at the critical point
x_A	mole fraction of solvent A
x_B	mole fraction of polymer B
z_A	base mole fraction of solvent A
z_B	base mole fraction of polymer B
α	critical exponent
γ_A	activity coefficient of the solvent A in the liquid phase with activity $a_A = x_A\gamma_A$
λ	wavelength
φ_A	volume fraction of solvent A
φ_B	volume fraction of polymer B
$\varphi_{B,\ crit}$	volume fraction of the polymer B at the critical point
ρ	density (of the mixture) at temperature T
ρ_A	density of solvent A at temperature T
ρ_B	density of polymer B at temperature T
ψ_A	segment fraction of solvent A
ψ_B	segment fraction of polymer B
π	osmotic pressure
θ	scattering angle
ω	angular velocity
Ω_A	mass fraction-based activity coefficient of the solvent A in the liquid phase with activity $a_A = w_A\Omega_A$
Ω_A^{∞}	mass fraction-based activity coefficient of the solvent A at infinite dilution

1.5. References

1968BON Bondi, A., *Physical Properties of Molecular Crystals, Liquids and Glasses*, J. Wiley & Sons, New York, 1968.

1968KON Koningsveld, R. and Staverman, A.J., Liquid-liquid phase separation in multicomponent polymer solutions I and II, *J. Polym. Sci.*, Pt. A-2, 6, 305, 325, 1968.

1969WOL Wolf, B.A., Zur Bestimmung der kritischen Konzentration von Polymerlösungen, *Makromol. Chem.*, 128, 284, 1969.

1971YAM Yamakawa, H., *Modern Theory of Polymer Solutions*, Harper & Row, New York, 1971.

1972HUG Huglin, M.B., Ed., *Light Scattering from Polymer Solutions*, Academic Press, New York, 1972.

1972KON Koningsveld, R., Polymer solutions and fractionation, in *Polymer Science*, Jenkins, E.D., Ed., North-Holland, Amsterdam, 1972, 1047.

1974TOM Tombs, M.P. and Peacock, A.R., *The Osmotic Pressure of Macromolecules*, Oxford University Press, London, 1974.

1975BON Bonner, D.C., Vapor-liquid equilibria in concentrated polymer solutions, *Macromol. Sci. Rev. Macromol. Chem.*, C13, 263, 1975.

1975CAS Casassa, E.F. and Berry, G.C., Light scattering from solutions of macromolecules, in *Polymer Molecular Weights*, Marcel Dekker, New York, 1975, Pt. 1, 161.

1975FUJ Fujita, H., *Foundations of Ultracentrifugal Analysis*, J. Wiley & Sons, New York, 1975.

1975GLO Glover, C.A., Absolute colligative property methods, in *Polymer Molecular Weights*, Marcel Dekker, New York, 1975, Pt. 1, 79.

1976LIU Liu, D.D. and Prausnitz, J.M, Solubilities of gases and volatile liquids in polyethylene and in ethylene-vinyl acetate copolymers in the region 125-225 ^{o}C, *Ind. Eng. Chem. Fundam.*, 15, 330, 1976.

1976NES Nesterov, A.E. and Lipatov, Yu.S., *Obrashchennaya Gasovaya Khromatografiya v Termo-dinamike Polimerov*, Naukova Dumka, Kiev, 1976.

1982GLA Glatter, O. and Kratky, O., Eds., *Small-Angle X-Ray Scattering*, Academic Press, London, 1982.

1984HEM Hemminger, W. and Höhne, G., *Calorimetry: Fundamentals and Practice*, Verlag Chemie, Weinheim, 1984.

1985WAL Walter, H., Brooks, D.E., and Fischer, D., *Partitioning in Aqueous Two-Phase Systems: Theory, Methods, Uses, and Applications to Biotechnology*, Academic Press, New York, 1985.

1986ALB Albertson, P.-A., *Partition of Cell Particles and Macromolecules*, 3rd ed., J. Wiley & Sons, New York, 1986.

1986HIG Higgins, J.S. and Macconachie, A., Neutron scattering from macromolecules in solution, in *Polymer Solutions*, Forsman, W.C., Ed., Plenum Press, New York, 1986, 183.

1986ZOL Zoller, P., Dilatometry, in *Encyclopedia of Polymer Science and Engineering*, Vol. 5, 2nd ed., Mark, H. et al., Eds., J. Wiley & Sons, New York, 1986, 69.

1987ADA Adams, E.T., Osmometry, in *Encyclopedia of Polymer Science and Engineering*, Vol. 10, 2nd ed., Mark, H. et al., Eds., J. Wiley & Sons, New York, 1987, 636.

1987BER Berry, G.C., Light scattering, in *Encyclopedia of Polymer Science and Engineering*, Vol. 8, 2nd ed., Mark, H. et al., Eds., J. Wiley & Sons, New York, 1987, 721.
1987COO Cooper, A.R., Molecular weight determination, in *Encyclopedia of Polymer Science and Engineering*, Vol. 10, 2nd ed., Mark, H. et al., Eds., J. Wiley & Sons, New York, 1987, 1.
1987KRA Kratochvil, P., *Classical Light Scattering from Polymer Solutions*, Elsevier, Amsterdam, 1987.
1987WIG Wignall, G.D., Neutron scattering, in *Encyclopedia of Polymer Science and Engineering*, Vol. 10, 2nd ed., Mark, H. et al., Eds., J. Wiley & Sons, New York, 1986, 112.
1988NES Nesterov, A.E., *Obrashchennaya Gasovaya Khromatografiya Polimerov*, Naukova Dumka, Kiev, 1988.
1989LLO Lloyd, D.R., Ward, T.C., Schreiber, H.P., and Pizana, C.C., Eds., *Inverse Gas Chromatography*, ACS Symposium Series 391, American Chemical Society, Washington, 1989.
1989RAE Rätzsch, M.T. and Kehlen, H., Continuous thermodynamics of polymer systems, *Prog. Polym. Sci.*, 14, 1, 1989.
1989VIL Vilcu, R. and Leca, M., *Polymer Thermodynamics by Gas Chromatography*, Elsevier, Amsterdam, 1989.
1990BAR Barton, A.F.M., *CRC Handbook of Polymer-Liquid Interaction Parameters and Solubility Parameters*, CRC Press, Boca Raton, 1990.
1990FUJ Fujita, H., *Polymer Solutions*, Elsevier, Amsterdam, 1990.
1990KAM Kamide, K., *Thermodynamics of Polymer Solutions*, Elsevier, Amsterdam, 1990.
1990KRE [Van] Krevelen, D.W., *Properties of Polymers*, 3rd ed., Elsevier, Amsterdam, 1990.
1990RAE Rätzsch, M.T. and Wohlfarth, C., Continuous thermodynamics of copolymer systems, *Adv. Polym. Sci.*, 98, 49, 1990.
1991CHU Chu, B., *Laser Light Scattering*, Academic Press, New York, 1991.
1991MAY Mays, J.W. and Hadjichristidis, N., Measurement of molecular weight of polymers by osmometry, in *Modern Methods of Polymer Characterization*, Barth, H.G. and Mays, J.W., Eds., J. Wiley & Sons, New York, 1991, 201.
1991MU1 Munk, P., Polymer characterization using inverse gas chromatography, in *Modern Methods of Polymer Characterization*, Barth, H.G. and Mays, J.W., Eds., J. Wiley & Sons, New York, 1991, 151.
1991MU2 Munk, P., Polymer characterization using the ultracentrifuge, in *Modern Methods of Polymer Characterization*, Barth, H.G. and Mays, J.W., Eds., J. Wiley & Sons, New York, 1991, 271.
1992HAR Harding, S.E., Rowe, A.J., and Horton, J.C., *Analytical Ultracentrifugation in Biochemistry and Polymer Science*, Royal Society of Chemistry, Cambridge, 1992.
1992WEN Wen, H., Elbro, H.S., and Alessi, P., *Polymer Solution Data Collection*, Chemistry Data Series, Vol. 15, DECHEMA, Frankfurt am Main, 1992.
1993DAN Danner, R.P. and High, M.S., *Handbook of Polymer Solution Thermodynamics*, American Institute of Chemical Engineers, New York, 1993.
1994MAR Marsh, K.N., Ed., *Experimental Thermodynamics, Volume 4, Solution Calorimetry*, Blackwell Science, Oxford, 1994.
1994MCH McHugh, M.A. and Krukonis, V.J., *Supercritical Fluid Extraction: Principles and Practice*, 2nd ed., Butterworth Publishing, Stoneham, 1994.
1994WOH Wohlfarth, C., *Vapour-Liquid Equilibrium Data of Binary Polymer Solutions: Physical Science Data*, 44, Elsevier, Amsterdam, 1994.

1995ZAS Zaslavsky, B.Y., *Aqueous Two-Phase Partitioning. Physical Chemistry and Bioanalytical Applications*, Marcel Dekker Inc., New York, 1995.

1995ZOL Zoller, P. and Walsh, D.J., *Standard Pressure-Volume-Temperature Data for Polymers*, Technomic Publishing, Lancaster, 1995.

1996CAB Cabezas, H., Jr., Theory of phase formation in aqueous two-phase systems, *J. Chromatogr. B*, 680, 3, 1996.

1997KIR Kiran, E. and Zhuang, W., *Miscibility and Phase Separation of Polymers in Near- and Supercritical Fluids*, ACS Symposium Series 670, 2, 1997.

1999BRA Brandrup, J., Immergut, E.H., and Grulke, E.A., Eds., *Polymer Handbook*, 4th ed., J. Wiley & Sons, New York, 1999.

1999KIR Kirby, C.F. and McHugh, M.A., Phase behavior of polymers in supercritical fluid solvents, *Chem. Rev.*, 99, 565, 1999.

1999KLE Klenin, V.J., *Thermodynamics of Systems Containing Flexible-Chain Polymers*, Elsevier, Amsterdam, 1999.

1999PET Pethrick, R.A. and Dawkins, J.V., Eds., *Modern Techniques for Polymer Characterization*, J. Wiley & Sons, Chichester, 1999.

1999PRA Prausnitz, J.M., Lichtenthaler, R.N., and de Azevedo, E.G., *Molecular Thermodynamics of Fluid Phase Equilibria*, 3rd ed., Prentice Hall, Upper Saddle River, NJ, 1999.

2000WOH Wohlfarth, C., Methods for the measurement of solvent activity of polymer solutions, in *Handbook of Solvents*, Wypych, G., Ed., ChemTec Publishing, Toronto, 2000, 146.

2001KON Koningsveld, R., Stockmayer, W.H., and Nies, E., *Polymer Phase Diagrams*, Oxford University Press, Oxford, 2001.

2001WOH Wohlfarth, C., *CRC Handbook of Thermodynamic Data of Copolymer Solutions*, CRC Press, Boca Raton, 2001.

2003HOL Holmberg, K., Jönsson, B., Kronberg, B., and Lindman, B., *Surfactants and Polymers in Aqueous Solutions*, 2nd ed., J. Wiley & Sons, Ltd., New York, 2003.

2004WOH Wohlfarth, C., *CRC Handbook of Thermodynamic Data of Aqueous Polymer Solutions*, CRC Press, Boca Raton, 2004.

2005WOH Wohlfarth, C., *CRC Handbook of Thermodynamic Data of Polymer Solutions at Elevated Pressures*, Taylor & Francis, CRC Press, Boca Raton, 2005.

2006WOH Wohlfarth, C., *CRC Handbook of Enthalpy Data of Polymer-Sovent Systems*, Taylor & Francis, CRC Press, Boca Raton, 2006.

2008WOH Wohlfarth, C., *CRC Handbook of Liquid-Liquid Equilibrium Data of Polymer Solutions*, Taylor & Francis, CRC Press, Boca Raton, 2008.

2009BAR Baron, M. et al., Glossary of class names of polymers based on chemical structure and molecular architecture (IUPAC Recommendations 2009), *Pure Appl. Chem.*, 81, 1131, 2008.

2011ASE Aseyev, V., Tenhu, H., and Winnik, F.M., Non-ionic thermoresponsive polymers in water, *Adv. Polym. Sci.*, 242, 29, 2011.

2011HAY Haynes, W.M., Ed., *CRC Handbook of Chemistry and Physics, Section 13: Polymer Properties*, 92nd ed., Taylor & Francis, CRC Press, Boca Raton, 2011.

2011MAU Maurer, G., Lammertz, S., and Schäfer, L.N., Aqueous solutions of polyelectrolytes: Vapor–liquid equilibrium and some related properties, *Adv. Polym. Sci.*, 238, 67, 2011.

2011WOH Wohlfarth, C., *CRC Handbook of Phase Equilibria and Thermodynamic Data of Copolymer Solutions*, CRC Press, Boca Raton, 2011.

2. VAPOR-LIQUID EQUILIBRIUM (VLE) DATA OF AQUEOUS POLYMER SOLUTIONS

2.1. Partial water vapor pressures or water activities for binary polymer solutions

Polymer (B): **bovine serum albumin** **2005SUG**
Characterization: M_w/g.mol^{-1} = 66000, Across Organics, New Jersey
Solvent (A): **water** **H_2O** **7732-18-5**

Type of data: vapor-liquid equilibrium

T/K = 298.15

φ_B	0.936	0.872	0.806	0.747	0.531	0.246
P_A/P_{0A}	0.120	0.370	0.610	0.760	0.960	0.990

Polymer (B): **dextran** **2005SUG**
Characterization: M_n/g.mol^{-1} = 27800, M_w/g.mol^{-1} = 59500, ρ (298 K) = 1.376 g/cm^3, Polymer Standard Service GmbH, Mainz, Germany
Solvent (A): **water** **H_2O** **7732-18-5**

Type of data: vapor-liquid equilibrium

T/K = 298.15

φ_B	0.931	0.848	0.781	0.713	0.594	0.387
P_A/P_{0A}	0.097	0.384	0.707	0.741	0.893	0.994

Polymer (B): **dextran** **2005SUG**
Characterization: M_n/g.mol^{-1} = 395000, M_w/g.mol^{-1} = 3420000, ρ (298 K) = 2.050 g/cm^3, Polymer Standard Service GmbH, Mainz, Germany
Solvent (A): **water** **H_2O** **7732-18-5**

Type of data: vapor-liquid equilibrium

T/K = 298.15

φ_B	0.957	0.928	0.842	0.778	0.718	0.595	0.213
P_A/P_{0A}	0.059	0.123	0.531	0.638	0.745	0.894	0.955

Polymer (B): **maltodextrin** **2010CAR**
Characterization: M_n/g.mol^{-1} = 990, M_w/g.mol^{-1} = 8800,
Corn Products Brasil S/A, Mogi Guacu, SP, Brazil
Solvent (A): **water** **H_2O** **7732-18-5**

Type of data: vapor-liquid equilibrium

T/K = 298.15

w_B	0.0471	0.0937	0.1543	0.2160	0.2781	0.3280	0.3939	0.4590	0.5284
a_A	0.998	0.996	0.995	0.995	0.992	0.987	0.984	0.978	0.969
w_B	0.5616								
a_A	0.956								

Polymer (B): **maltodextrin** **2005NIN**
Characterization: M_n/g.mol^{-1} = 1140, M_w/g.mol^{-1} = 8283,
Aldrich Chem. Co., Inc., Milwaukee, WI
Solvent (A): **water** **H_2O** **7732-18-5**

Type of data: vapor-liquid equilibrium

T/K = 298.15

w_B	0.2859	0.2644	0.2914	0.3167	0.3365	0.3399	0.3379	0.3488	0.3802
a_A	0.9967	0.9970	0.9966	0.9960	0.9953	0.9957	0.9957	0.9956	0.9945
w_B	0.4099	0.4590	0.4615						
a_A	0.9939	0.9919	0.9922						

T/K = 318.15

w_B	0.2683	0.3156	0.4327	0.4852
a_A	0.9927	0.9903	0.9821	0.9767

Polymer (B): **maltodextrin** **2005NIN**
Characterization: M_n/g.mol^{-1} = 1475, M_w/g.mol^{-1} = 13710,
Aldrich Chem. Co., Inc., Milwaukee, WI
Solvent (A): **water** **H_2O** **7732-18-5**

Type of data: vapor-liquid equilibrium

T/K = 298.15

w_B	0.1750	0.2234	0.2419	0.3029	0.3408	0.3796	0.4286	0.4408	0.4872
a_A	0.9962	0.9948	0.9942	0.9917	0.9898	0.9875	0.9838	0.9829	0.9779
w_B	0.4867	0.5307	0.5671	0.5898					
a_A	0.9781	0.9724	0.9664	0.9622					

T/K = 318.15

w_B	0.2767	0.3284	0.4452	0.4968
a_A	0.9927	0.9903	0.9821	0.9767

Polymer (B): **maltodextrin** **2010CAR**
Characterization: M_n/g.mol^{-1} = 1500, M_w/g.mol^{-1} = 20500,
Corn Products Brasil S/A, Mogi Guacu, SP, Brazil
Solvent (A): **water** **H_2O** **7732-18-5**

Type of data: vapor-liquid equilibrium

T/K = 298.15

w_B	0.0473	0.0943	0.1542	0.2177	0.2725	0.3289	0.3896	0.4534	0.5233
a_A	0.998	0.996	0.993	0.992	0.992	0.993	0.987	0.984	0.975
w_B	0.5656								
a_A	0.964								

Polymer (B): **maltodextrin** **2005NIN**
Characterization: M_n/g.mol^{-1} = 2683, M_w/g.mol^{-1} = 38360,
Aldrich Chem. Co., Inc., Milwaukee, WI
Solvent (A): **water** **H_2O** **7732-18-5**

Type of data: vapor-liquid equilibrium

T/K = 298.15

w_B	0.2598	0.2743	0.3434	0.3887	0.3844	0.4197	0.4492	0.4559	0.4824
a_A	0.9928	0.9922	0.9886	0.9854	0.9860	0.9831	0.9802	0.9797	0.9763
w_B	0.5357	0.5850							
a_A	0.9689	0.9594							

Polymer (B): **maltodextrin** **2010CAR**
Characterization: M_n/g.mol^{-1} = 3000, M_w/g.mol^{-1} = 44400,
Corn Products Brasil S/A, Mogi Guacu, SP, Brazil
Solvent (A): **water** **H_2O** **7732-18-5**

Type of data: vapor-liquid equilibrium

T/K = 298.15

w_B	0.0474	0.0946	0.1546	0.2183	0.2731	0.3327	0.3906	0.4545	0.5246
a_A	0.999	0.998	0.994	0.992	0.992	0.992	0.987	0.984	0.975
w_B	0.5675								
a_A	0.967								

Polymer (B): **poly(acrylic acid)** **2004ARC**
Characterization: M_n/g.mol^{-1} = 250000, ρ = 1.15 g/cm^3, T_g/K = 381
Solvent (A): **water** **H_2O** **7732-18-5**

Type of data: vapor sorption

T/K = 308.15

w_B	0.963	0.914	0.862	0.800	0.708	0.605
a_A	0.29	0.48	0.59	0.69	0.80	0.86

Polymer (B): **poly(ammonium acrylate)** **2004LAM, 2008LA1**
Characterization: M_n/g.mol^{-1} = 2300, M_w/g.mol^{-1} = 3900, M_z/g.mol^{-1} = 5800, Rohagit SL 135, Roehm GmbH, Darmstadt, Germany
Solvent (A): **water** **H_2O** **7732-18-5**

Type of data: vapor-liquid equilibrium

T/K = 298.15

w_B	0.1202	0.1536	0.1794	0.1871	0.2367	0.2407	0.2556	0.2781	0.2854
a_A	0.9887	0.9842	0.9798	0.9787	0.9690	0.9672	0.9646	0.9588	0.9545
w_B	0.2865	0.2989	0.3506						
a_A	0.9571	0.9523	0.9307						

T/K = 323.15

w_B	0.1250	0.1550	0.1864	0.2308	0.2793	0.3050	0.3428
a_A	0.9888	0.9847	0.9801	0.9706	0.9614	0.9550	0.9420

Polymer (B): **poly(ammonium acrylate)** **2004LAM, 2008LA1**
Characterization: M_n/g.mol^{-1} = 7300, M_w/g.mol^{-1} = 13600, M_z/g.mol^{-1} = 19200, Rohagit SL 252, Roehm GmbH, Darmstadt, Germany
Solvent (A): **water** **H_2O** **7732-18-5**

Type of data: vapor-liquid equilibrium

T/K = 298.15

w_B	0.1282	0.1615	0.1875	0.2054
a_A	0.9887	0.9842	0.9798	0.9766

T/K = 323.15

w_B	0.1344	0.1655	0.1957	0.2396
a_A	0.9888	0.9847	0.9801	0.9706

Polymer (B): **poly(ammonium acrylate)** **2004LAM, 2008LA1**
Characterization: M_n/g.mol^{-1} = 7700, M_w/g.mol^{-1} = 13300, M_z/g.mol^{-1} = 19400, Rohagit SL 159, Roehm GmbH, Darmstadt, Germany
Solvent (A): **water** **H_2O** **7732-18-5**

Type of data: vapor-liquid equilibrium

T/K = 298.15

w_B	0.1361	0.1715	0.1974	0.2051	0.2571	0.2592	0.2756	0.2975	0.3015
a_A	0.9887	0.9842	0.9798	0.9787	0.9690	0.9672	0.9646	0.9588	0.9545
w_B	0.3061	0.3158	0.3190	0.3207	0.3442	0.3678	0.3681		
a_A	0.9571	0.9526	0.9523	0.9529	0.9402	0.9299	0.9307		

T/K = 323.15

w_B	0.1423	0.1749	0.2058	0.2479	0.3016	0.3272	0.3623
a_A	0.9888	0.9847	0.9801	0.9706	0.9614	0.9550	0.9420

Polymer (B): **poly(*N,N*-dimethyl acrylamide-*co*-*tert*-butylacrylamide)** **2009FO1**
Characterization: M_n/g.mol^{-1} = 1350, unspecified comonomer content, synthesized in the laboratory
Solvent (A): **water** **H_2O** **7732-18-5**

Type of data: vapor-liquid equilibrium

T/K = 308.15

w_B	0.45	0.47	0.49	0.51	0.53	0.54	0.57	0.58	0.60
a_A	0.82	0.74	0.72	0.62	0.54	0.51	0.38	0.32	0.24
w_B	0.62	0.64	0.67						
a_A	0.17	0.15	0.11						

Polymer (B): **poly(*N,N*-dimethyl acrylamide-*co*-*tert*-butylacrylamide)** **2009FO1**
Characterization: M_n/g.mol^{-1} = 1700, unspecified comonomer content, synthesized in the laboratory
Solvent (A): **water** **H_2O** **7732-18-5**

Type of data: vapor-liquid equilibrium

T/K = 308.15

w_B	0.47	0.49	0.52	0.53	0.54	0.56	0.58	0.61	0.62
a_A	0.84	0.77	0.69	0.70	0.69	0.63	0.58	0.55	0.51
w_B	0.63	0.65	0.69						
a_A	0.39	0.30	0.22						

Polymer (B): **poly(*N,N*-dimethyl acrylamide-*co*-*tert*-butylacrylamide)** **2009FO1**
Characterization: M_n/g.mol^{-1} = 2520, unspecified comonomer content, synthesized in the laboratory
Solvent (A): **water** **H_2O** **7732-18-5**

Type of data: vapor-liquid equilibrium

T/K = 308.15

w_B	0.48	0.51	0.53	0.56	0.59	0.61	0.63	0.65	0.67
a_A	0.89	0.83	0.78	0.75	0.71	0.65	0.58	0.53	0.45
w_B	0.69	0.72	0.74						
a_A	0.38	0.35	0.30						

Polymer (B): **poly[2-(*N,N*-dimethylamino)ethyl methacrylate]** **2004ARC**
Characterization: M_n/g.mol^{-1} = 100000, ρ = 1.10 g/cm^3, T_g/K = 291, Scientific Polymer Products, Inc., Ontario, NY
Solvent (A): **water** **H_2O** **7732-18-5**

Type of data: vapor sorption

T/K = 308.15

w_B	0.996	0.992	0.981	0.968	0.952	0.924	0.888	0.826
a_A	0.14	0.24	0.38	0.49	0.58	0.68	0.76	0.83

Polymer (B): **poly(*N,N*-dimethylmethacrylamide)** **2004ARC**
Characterization: M_n/g.mol^{-1} = 4000, ρ = 1.15 g/cm^3, T_g/K = 378, Polymer Source, Inc., Quebec, Canada
Solvent (A): **water** **H_2O** **7732-18-5**

Type of data: vapor sorption

T/K = 308.15

w_B	0.975	0.947	0.905	0.864	0.822	0.765	0.707	0.628
a_A	0.14	0.24	0.38	0.49	0.58	0.68	0.76	0.83

Polymer (B): **poly(ethylene glycol)** **2011SA1**
Characterization: M_n/g.mol^{-1} = 200, Merck, Darmstadt, Germany
Solvent (A): **water** **H_2O** **7732-18-5**

Type of data: vapor-liquid equilibrium

T/K = 298.15

w_B	0.0615	0.0786	0.0851	0.0908	0.0984	0.1041	0.2389	0.2632	0.3064
a_A	0.9932	0.9915	0.9906	0.9903	0.9894	0.9880	0.9679	0.9628	0.9526
w_B	0.3215	0.3565	0.3962	0.4346	0.4429	0.5255	0.5499		
a_A	0.9484	0.9391	0.9258	0.9122	0.9088	0.8695	0.8556		

Comments: by isopiestic method

T/K = 298.15

w_B	0.0107	0.0304	0.0615	0.0877	0.0962	0.1094	0.1218
a_A	0.9990	0.9968	0.9925	0.9896	0.9884	0.9870	0.9843

Comments: by vapor-pressure osmometry (VPO)

T/K = 303.15

w_B	0.0036	0.0055	0.0082	0.0103	0.0313	0.0494	0.0648	0.0786	0.0992
a_A	0.9998	0.9996	0.9994	0.9992	0.9972	0.9959	0.9949	0.9934	0.9912
w_B	0.1065	0.1178	0.1330	0.1401	0.1502				
a_A	0.9905	0.9892	0.9863	0.9854	0.9839				

Comments: by vapor-pressure osmometry (VPO)

continued

continued

T/K = 308.15

w_B	0.0053	0.0082	0.0108	0.0129	0.0154	0.0245	0.0402	0.0550	0.0678
a_A	0.9998	0.9995	0.9993	0.9991	0.9989	0.9979	0.9963	0.9948	0.9937
w_B	0.0798	0.0963	0.1110	0.1241					
a_A	0.9927	0.9907	0.9889	0.9863					

Comments: by vapor-pressure osmometry (VPO)

Polymer (B): **poly(ethylene glycol)** **2006SA1**
Characterization: M_n/g.mol^{-1} = 400, Merck, Darmstadt, Germany
Solvent (A): **water** H_2O **7732-18-5**

Type of data: vapor-liquid equilibrium

T/K = 293.15

w_B	0.1533	0.1573	0.2220	0.2772	0.3423	0.3465	0.4349	0.5362	0.5709
a_A	0.9893	0.9880	0.9812	0.9740	0.9585	0.9582	0.9340	0.8830	0.8685
w_B	0.6152	0.6586							
a_A	0.8411	0.8055							

T/K = 298.15

w_B	0.1702	0.1797	0.1940	0.2428	0.2689	0.3657	0.4490	0.5088	0.5601
a_A	0.9884	0.9873	0.9860	0.9792	0.9753	0.9543	0.9299	0.9033	0.8784
w_B	0.6060	0.7173							
a_A	0.8526	0.7617							

T/K = 303.15

w_B	0.1734	0.2136	0.2752	0.3118	0.3367	0.3577	0.3984	0.4261	0.4459
a_A	0.9882	0.9836	0.9749	0.9693	0.9640	0.9580	0.9450	0.9384	0.9332
w_B	0.4652	0.4702	0.5020	0.5269	0.6176	0.6463	0.6920	0.7071	0.7204
a_A	0.9233	0.9230	0.9118	0.9011	0.8518	0.8342	0.7945	0.7806	0.7674

T/K = 308.15

w_B	0.1569	0.2169	0.2250	0.3437	0.3627	0.3797	0.4324	0.4780	0.5326
a_A	0.9899	0.9838	0.9823	0.9642	0.9568	0.9523	0.9399	0.9245	0.9007
w_B	0.5934	0.6593	0.6841	0.6962					
a_A	0.8682	0.8277	0.8079	0.8019					

T/K = 313.15

w_B	0.2055	0.2151	0.2310	0.2793	0.2885	0.3748	0.4518	0.5528	0.5983
a_A	0.9853	0.9842	0.9824	0.9763	0.9754	0.9590	0.9376	0.8966	0.8723
w_B	0.6920	0.7363	0.7396						
a_A	0.8105	0.7632	0.7654						

Polymer (B): **poly(ethylene glycol)** **2011SA2**
Characterization: M_n/g.mol^{-1} = 400, Merck, Darmstadt, Germany
Solvent (A): **water** H_2O **7732-18-5**

Type of data: vapor-liquid equilibrium

T/K = 318.15

w_B	0.0250	0.0277	0.0496	0.0604	0.0990	0.0991	0.1297	0.1493	0.1600
a_A	0.9992	0.9986	0.9976	0.9972	0.9946	0.9951	0.9927	0.9913	0.9910
w_B	0.1753	0.1835	0.1987	0.2268					
a_A	0.9898	0.9891	0.9870	0.9845					

Polymer (B): **poly(ethylene glycol)** **2008ZAF**
Characterization: M_n/g.mol^{-1} = 4230, Merck, Darmstadt, Germany
Solvent (A): **water** H_2O **7732-18-5**

Type of data: vapor-liquid equilibrium

T/K = 298.15

c_B/(mol/kg A)	0.0494	0.0603	0.0605	0.0690	0.0775	0.0846	0.0859	0.0933
a_A	0.9947	0.9933	0.9932	0.9917	0.9898	0.9884	0.9875	0.9863
c_B/(mol/kg A)	0.0983	0.1106	0.1264	0.1469	0.1641	0.1766	0.1936	0.2040
a_A	0.9841	0.9810	0.9754	0.9689	0.9629	0.9576	0.9510	0.9465
c_B/(mol/kg A)	0.2248	0.2370	0.2482	0.2703	0.2894	0.3006	0.3159	0.3326
a_A	0.9380	0.9330	0.9266	0.9182	0.9087	0.9022	0.8952	0.8836

T/K = 308.15

c_B/(mol/kg A)	0.0230	0.0266	0.0426	0.0494	0.0660	0.0768	0.0805	0.0913
a_A	0.9966	0.9961	0.9943	0.9934	0.9915	0.9900	0.9890	0.9867
c_B/(mol/kg A)	0.1089	0.1230	0.1257	0.1257	0.1351	0.1435	0.1565	0.1727
a_A	0.9830	0.9795	0.9789	0.9789	0.9765	0.9724	0.9695	0.9643
c_B/(mol/kg A)	0.1828	0.2002	0.2106	0.2179	0.2397	0.2470	0.2699	0.2978
a_A	0.9610	0.9561	0.9524	0.9492	0.9404	0.9390	0.9311	0.9227
c_B/(mol/kg A)	0.3148	0.3389	0.3559	0.3753	0.3908			
a_A	0.9167	0.9085	0.9041	0.8945	0.8873			

T/K = 318.15

c_B/(mol/kg A)	0.0238	0.0505	0.0715	0.0753	0.0806	0.0842	0.1111	0.1237
a_A	0.9950	0.9933	0.9903	0.9892	0.9882	0.9879	0.9832	0.9799
c_B/(mol/kg A)	0.1374	0.1496	0.1598	0.1710	0.1839	0.2020	0.2280	0.2562
a_A	0.9782	0.9753	0.9724	0.9698	0.9669	0.9617	0.9538	0.9472
c_B/(mol/kg A)	0.2791	0.3011	0.3322	0.3530	0.3895	0.4043	0.4419	
a_A	0.9379	0.9321	0.9230	0.9147	0.9027	0.8958	0.8816	

Polymer (B): **poly(ethylene glycol)** **2011SA1**
Characterization: M_n/g.mol^{-1} = 6000, Merck, Darmstadt, Germany
Solvent (A): **water** H_2O **7732-18-5**

Type of data: vapor-liquid equilibrium

T/K = 298.15

w_B	0.2041	0.2399	0.2643	0.2903	0.3055	0.3261	0.3297	0.3334	0.3549
a_A	0.9932	0.9915	0.9895	0.9869	0.9846	0.9818	0.9810	0.9805	0.9768
w_B	0.3592	0.3952	0.3958	0.4125	0.4319	0.4458	0.4514	0.4751	0.5010
a_A	0.9756	0.9678	0.9677	0.9628	0.9577	0.9526	0.9505	0.9413	0.9303
w_B	0.5101	0.5416							
a_A	0.9258	0.9122							

Comments: by isopiestic method

T/K = 298.15

w_B	0.0674	0.0897	0.1229	0.1554	0.1802	0.2089	0.2407	0.2946	0.3179
a_A	0.9991	0.9988	0.9982	0.9970	0.9958	0.9942	0.9922	0.9872	0.9833

Comments: by vapor-pressure osmometry (VPO)

T/K = 303.15

w_B	0.0654	0.0860	0.1024	0.1225	0.1470	0.1529	0.1693	0.1961	0.2102
a_A	0.9995	0.9993	0.9990	0.9987	0.9980	0.9979	0.9974	0.9961	0.9955
w_B	0.2162	0.2368	0.2484	0.2619	0.2787	0.2862	0.2987	0.3194	0.3352
a_A	0.9950	0.9940	0.9932	0.9924	0.9908	0.9900	0.9887	0.9869	0.9851

Comments: by vapor-pressure osmometry (VPO)

T/K = 308.15

w_B	0.0635	0.0880	0.1030	0.1152	0.1263	0.1418	0.1546	0.1635	0.1657
a_A	0.9996	0.9993	0.9991	0.9989	0.9986	0.9983	0.9979	0.9977	0.9976
w_B	0.1741	0.1844	0.1933	0.1985	0.2067	0.2185	0.2218	0.2436	0.2546
a_A	0.9973	0.9968	0.9964	0.9961	0.9956	0.9949	0.9948	0.9935	0.9926
w_B	0.2720	0.2837	0.3042	0.3150	0.3205				
a_A	0.9912	0.9900	0.9881	0.9867	0.9861				

Comments: by vapor-pressure osmometry (VPO)

Polymer (B): **poly(ethylene glycol)** **2010SA1**
Characterization: M_n/g.mol^{-1} = 6000, Merck, Darmstadt, Germany
Solvent (A): **water** **H_2O** **7732-18-5**

Type of data: vapor-liquid equilibrium

T/K = 298.15

w_B	0.3047	0.3271	0.3347	0.3624	0.3803	0.4153	0.4484	0.3153	0.3285
a_A	0.9847	0.9815	0.9797	0.9760	0.9712	0.9626	0.9525	0.9828	0.9814
w_B	0.3469	0.3718	0.3882	0.4236					
a_A	0.9786	0.9738	0.9703	0.9609					

Polymer (B): **poly(ethylene glycol)** **2010SA2**
Characterization: M_n/g.mol^{-1} = 6000, Merck, Darmstadt, Germany
Solvent (A): **water** **H_2O** **7732-18-5**

Type of data: vapor-liquid equilibrium

T/K = 298.15

w_B	0.2812	0.2903	0.3055	0.3261	0.3297	0.3334	0.3549	0.3592	0.3952
a_A	0.9880	0.9869	0.9846	0.9818	0.9810	0.9805	0.9768	0.9756	0.9673
w_B	0.3958	0.4319	0.4399	0.4514	0.4751	0.5137	0.5282		
a_A	0.9677	0.9577	0.9540	0.9505	0.9413	0.9259	0.9178		

Polymer (B): **poly(ethylene glycol) diacetate** **2007CHA**
Characterization: M_n/g.mol^{-1} = 300, LG Chem. Ltd., Seoul, South Korea
Solvent (A): **water** **H_2O** **7732-18-5**

Type of data: vapor-liquid equilibrium

T/K = 333.15

x_2	0.000	0.205	0.399	0.502	0.602	0.698	0.800
P/kPa	0.86	18.13	35.24	45.09	55.05	66.09	75.01

T/K = 343.15

x_2	0.000	0.205	0.399	0.502	0.602	0.698	0.800
P/kPa	1.25	24.46	50.68	64.62	76.94	94.84	100.59

T/K = 353.15

x_2	0.000	0.205	0.399	0.502	0.602	0.698	0.800
P/kPa	1.82	34.09	70.67	90.03	107.79	129.31	143.37

T/K = 363.15

x_2	0.000	0.205	0.399	0.502	0.602	0.698	0.800
P/kPa	2.45	46.05	95.78	120.42	149.97	179.59	201.84

continued

continued

T/K = 373.15

x_2	0.000	0.205	0.399	0.502	0.602	0.698	0.800
P/kPa	3.19	61.09	123.84	162.44	202.99	252.79	280.30

T/K = 383.15

x_2	0.000	0.205	0.399	0.502	0.602	0.698	0.800
P/kPa	4.02	79.34	167.91	209.95	275.83	317.18	377.45

T/K = 393.15

x_2	0.000	0.205	0.399	0.502	0.602
P/kPa	4.96	101.04	216.55	284.03	500.09

Comments: PEGDAE possesses a non-negligible vapor pressure, so that the pressure given in the table is the total vapor pressure of the system.

Polymer (B): **poly(ethylene glycol) dimethyl ether** **2011SA1**
Characterization: M_n/g.mol^{-1} = 250, Merck, Darmstadt, Germany
Solvent (A): **water** **H_2O** **7732-18-5**

Type of data: vapor-liquid equilibrium

T/K = 298.15

w_B	0.0844	0.1075	0.1164	0.1168	0.1313	0.1393	0.1924	0.2384	0.2678
a_A	0.9932	0.9915	0.9906	0.9903	0.9894	0.9880	0.9823	0.9750	0.9706
w_B	0.2829	0.2876	0.3065	0.3483	0.3561	0.3621	0.3977	0.4325	0.4579
a_A	0.9679	0.9664	0.9628	0.9526	0.9501	0.9484	0.9391	0.9258	0.9158
w_B	0.4936	0.5553	0.5639	0.5913					
a_A	0.9019	0.8741	0.8695	0.8556					

Comments: by isopiestic method

T/K = 298.15

w_B	0.0090	0.0143	0.0207	0.0404	0.0613	0.0806	0.0962	0.1112	0.1215
a_A	0.9996	0.9993	0.9987	0.9973	0.9955	0.9939	0.9928	0.9916	0.9903
w_B	0.1397	0.1589	0.1714	0.1793					
a_A	0.9890	0.9871	0.9852	0.9836					

Comments: by vapor-pressure osmometry (VPO)

T/K = 303.15

w_B	0.0017	0.0032	0.0066	0.0081	0.0107	0.0249	0.0344	0.0402	0.0611
a_A	0.9999	0.9998	0.9996	0.9996	0.9994	0.9986	0.9979	0.9974	0.9960
w_B	0.0753	0.0801	0.0891	0.1008	0.1166	0.1372	0.1601	0.1724	0.1878
a_A	0.9948	0.9946	0.9939	0.9930	0.9915	0.9895	0.9875	0.9862	0.9850

continued

continued

w_B	0.2015
a_A	0.9839

Comments: by vapor-pressure osmometry (VPO)

T/K = 308.15

w_B	0.0008	0.0049	0.0081	0.0100	0.0140	0.0255	0.0307	0.0323	0.0499
a_A	0.9999	0.9997	0.9996	0.9995	0.9993	0.9986	0.9982	0.9982	0.9972
w_B	0.0503	0.0665	0.0714	0.0797	0.0962	0.1026	0.1208	0.1325	0.1360
a_A	0.9971	0.9960	0.9956	0.9951	0.9939	0.9935	0.9923	0.9912	0.9907
w_B	0.1467	0.1515	0.1636	0.1732					
a_A	0.9897	0.9889	0.9874	0.9859					

Comments: by vapor-pressure osmometry (VPO)

Polymer (B): **poly(ethylene glycol) dimethyl ether** **2011SA2**
Characterization: M_n/g.mol^{-1} = 250, Merck, Darmstadt, Germany
Solvent (A): **water** H_2O **7732-18-5**

Type of data: vapor-liquid equilibrium

T/K = 318.15

w_B	0.0100	0.0251	0.0304	0.0496	0.0498	0.0721	0.0819	0.0954	0.0999
a_A	0.9994	0.9988	0.9983	0.9974	0.9972	0.9955	0.9950	0.9935	0.9932
w_B	0.1152	0.1355	0.1413	0.1461	0.1654	0.1973	0.2249	0.2430	
a_A	0.9918	0.9898	0.9897	0.9891	0.9873	0.9840	0.9809	0.9788	

Comments: by vapor-pressure osmometry (VPO)

Polymer (B): **poly(ethylene glycol) dimethyl ether** **2011SA1**
Characterization: M_n/g.mol^{-1} = 500, Merck, Darmstadt, Germany
Solvent (A): **water** H_2O **7732-18-5**

Type of data: vapor-liquid equilibrium

T/K = 298.15

w_B	0.0101	0.0395	0.0622	0.0792	0.0998	0.1169	0.1435	0.1576	0.1873
a_A	0.9998	0.9985	0.9973	0.9964	0.9950	0.9936	0.9915	0.9904	0.9875
w_B	0.2013	0.2184							
a_A	0.9861	0.9840							

Comments: by vapor-pressure osmometry (VPO)

T/K = 303.15

w_B	0.0057	0.0113	0.0161	0.0180	0.0220	0.0402	0.0629	0.0824	0.1021
a_A	0.9998	0.9996	0.9995	0.9994	0.9993	0.9985	0.9977	0.9968	0.9957

continued

continued

w_B	0.1250	0.1395	0.1571	0.1789	0.1913	0.2046	0.2213	0.2224
a_A	0.9944	0.9934	0.9920	0.9904	0.9895	0.9881	0.9866	0.9864

Comments: by vapor-pressure osmometry (VPO)

T/K = 308.15

w_B	0.0036	0.0088	0.0104	0.0287	0.0405	0.0697	0.0907	0.1026	0.1154
a_A	0.9999	0.9997	0.9997	0.9990	0.9985	0.9974	0.9965	0.9960	0.9954
w_B	0.1300	0.1419	0.1581	0.1713	0.1819	0.2009	0.2199	0.2378	0.2484
a_A	0.9948	0.9939	0.9928	0.9918	0.9911	0.9895	0.9878	0.9862	0.9852

Comments: by vapor-pressure osmometry (VPO)

Polymer (B): **poly(ethylene glycol) dimethyl ether** **2011SA2**
Characterization: M_n/g.mol^{-1} = 500, Merck, Darmstadt, Germany
Solvent (A): **water** H_2O **7732-18-5**

Type of data: vapor-liquid equilibrium

T/K = 318.15

w_B	0.0118	0.0301	0.0499	0.0503	0.0734	0.1007	0.1208	0.1488	0.1604
a_A	0.9991	0.9988	0.9985	0.9984	0.9972	0.9959	0.9951	0.9934	0.9927
w_B	0.1776	0.2053	0.2486	0.2700	0.2960	0.3165			
a_A	0.9908	0.9893	0.9855	0.9833	0.9806	0.9777			

Comments: by vapor-pressure osmometry (VPO)

Polymer (B): **poly(ethylene glycol) dimethyl ether** **2011SA1**
Characterization: M_n/g.mol^{-1} = 2000, Merck, Darmstadt, Germany
Solvent (A): **water** H_2O **7732-18-5**

Type of data: vapor-liquid equilibrium

T/K = 298.15

w_B	0.1753	0.2127	0.2222	0.2394	0.2504	0.2971	0.3370	0.3633	0.3773
a_A	0.9932	0.9915	0.9906	0.9894	0.9880	0.9823	0.9750	0.9706	0.9679
w_B	0.3841	0.3987	0.4148	0.4350	0.4499	0.4741	0.5097	0.6173	0.6422
a_A	0.9664	0.9628	0.9592	0.9526	0.9484	0.9391	0.9258	0.8695	0.8556

Comments: by isopiestic method

T/K = 298.15

w_B	0.0112	0.0216	0.0407	0.0613	0.0850	0.1036	0.1191	0.1341	0.1467
a_A	0.9998	0.9997	0.9994	0.9989	0.9978	0.9969	0.9961	0.9954	0.9948
w_B	0.1692	0.1997	0.2204	0.2367	0.2674				
a_A	0.9934	0.9913	0.9899	0.9887	0.9854				

Comments: by vapor-pressure osmometry (VPO)

continued

continued

T/K = 303.15

w_B	0.0052	0.0100	0.0131	0.0202	0.0253	0.0494	0.0788	0.0782	0.1181
a_A	0.9998	0.9998	0.9998	0.9997	0.9996	0.9993	0.9986	0.9985	0.9972
w_B	0.1593	0.1777	0.2004	0.2209	0.2464	0.2538	0.2936		
a_A	0.9956	0.9943	0.9924	0.9910	0.9888	0.9883	0.9844		

Comments: by vapor-pressure osmometry (VPO)

T/K = 308.15

w_B	0.0101	0.0159	0.0192	0.0414	0.0600	0.0844	0.0998	0.1212	0.1377
a_A	0.9999	0.9998	0.9998	0.9994	0.9991	0.9986	0.9984	0.9978	0.9972
w_B	0.1591	0.1817	0.2007	0.2179	0.2369	0.2595	0.2742		
a_A	0.9963	0.9952	0.9940	0.9932	0.9916	0.9891	0.9876		

Comments: by vapor-pressure osmometry (VPO)

Polymer (B): **poly(ethylene glycol) dimethyl ether** **2011SA2**
Characterization: M_n/g.mol^{-1} = 2000, Merck, Darmstadt, Germany
Solvent (A): **water** **H_2O** **7732-18-5**

Type of data: vapor-liquid equilibrium

T/K = 318.15

w_B	0.0301	0.0578	0.0901	0.0992	0.1197	0.1515	0.1791	0.1976	0.2109
a_A	0.9995	0.9991	0.9981	0.9977	0.9973	0.9962	0.9953	0.9940	0.9936
w_B	0.2454	0.2734	0.2869	0.2964	0.3298	0.3417	0.3605		
a_A	0.9914	0.9900	0.9891	0.9869	0.9831	0.9821	0.9796		

Comments: by vapor-pressure osmometry (VPO)

Polymer (B): **poly(ethylene glycol) dimethyl ether** **2011ZAF**
Characterization: M_n/g.mol^{-1} = 2305, Merck, Darmstadt, Germany
Solvent (A): **water** **H_2O** **7732-18-5**

Type of data: vapor-liquid equilibrium

T/K = 298.15

w_B	0.1306	0.1539	0.1978	0.2065	0.2183	0.2476	0.2653	0.2710	0.3032
a_A	0.9961	0.9944	0.9924	0.9909	0.9912	0.9881	0.9861	0.9863	0.9810
w_B	0.3172	0.3320	0.3582	0.4070	0.4096	0.4250	0.4789	0.4921	0.5353
a_A	0.9800	0.9774	0.9712	0.9609	0.9606	0.9568	0.9385	0.9306	0.9139
w_B	0.5459	0.5679	0.6341	0.6357					
a_A	0.9077	0.8959	0.8605	0.8583					

continued

continued

T/K = 308.15

w_B	0.1679	0.1982	0.2011	0.2281	0.2506	0.3122	0.3428	0.3688	0.4291
a_A	0.9954	0.9929	0.9933	0.9905	0.9899	0.9834	0.9798	0.9743	0.9608
w_B	0.4296	0.4518	0.4818	0.5689	0.7011				
a_A	0.9613	0.9557	0.9448	0.9104	0.8391				

T/K = 318.15

w_B	0.1704	0.1840	0.1907	0.2384	0.3126	0.3840	0.3953	0.4320	0.4432
a_A	0.9956	0.9939	0.9934	0.9909	0.9850	0.9761	0.9799	0.9626	0.9611
w_B	0.5196	0.5332	0.6110	0.6208	0.7201				
a_A	0.9434	0.9378	0.9078	0.9026	0.8499				

Polymer (B): **poly(ethylene glycol) monomethyl ether** **2011SA1**
Characterization: M_n/g.mol^{-1} = 350, Fluka AG, Buchs, Switzerland
Solvent (A): **water** **H_2O** **7732-18-5**

Type of data: vapor-liquid equilibrium

T/K = 298.15

w_B	0.0890	0.1211	0.1265	0.1280	0.1463	0.1514	0.2058	0.2514	0.2811
a_A	0.9932	0.9915	0.9906	0.9903	0.9894	0.9880	0.9823	0.9750	0.9706
w_B	0.2964	0.2983	0.3205	0.3618	0.3688	0.3756	0.4051	0.4442	0.4704
a_A	0.9679	0.9664	0.9628	0.9526	0.9501	0.9484	0.9391	0.9258	0.9158
w_B	0.4767	0.5023	0.5578	0.5642	0.5834				
a_A	0.9122	0.9019	0.8741	0.8695	0.8556				

Comments: by isopiestic method

T/K = 298.15

w_B	0.0112	0.0213	0.0412	0.0600	0.0806	0.1000	0.1187	0.1285	0.1483
a_A	0.9996	0.9990	0.9979	0.9967	0.9947	0.9930	0.9914	0.9905	0.9884
w_B	0.1563	0.1827							
a_A	0.9877	0.9847							

Comments: by vapor-pressure osmometry (VPO)

T/K = 303.15

w_B	0.0027	0.0039	0.0063	0.0082	0.0121	0.0154	0.0206	0.0415	0.0589
a_A	0.9999	0.9998	0.9997	0.9995	0.9993	0.9991	0.9989	0.9977	0.9966
w_B	0.0796	0.0935	0.1053	0.1223	0.1373	0.1583	0.1663	0.1816	0.1952
a_A	0.9955	0.9945	0.9937	0.9924	0.9909	0.9888	0.9881	0.9863	0.9847

Comments: by vapor-pressure osmometry (VPO)

continued

continued

T/K = 308.15

w_B	0.0064	0.0095	0.0127	0.0147	0.0202	0.0337	0.0509	0.0614	0.0725
a_A	0.9996	0.9995	0.9994	0.9993	0.9990	0.9985	0.9973	0.9967	0.9959
w_B	0.0815	0.1003	0.1191	0.1365	0.1479	0.1570	0.1694	0.1811	
a_A	0.9954	0.9938	0.9926	0.9914	0.9905	0.9894	0.9881	0.9864	

Comments: by vapor-pressure osmometry (VPO)

Polymer (B): **poly(ethylene oxide)** **2008SAD**
Characterization: M_n/g.mol^{-1} = 6000, Merck, Darmstadt, Germany
Solvent (A): **water** H_2O **7732-18-5**

Type of data: vapor-liquid equilibrium

T/K = 298.15

w_B	0.2604	0.2802	0.2839	0.2840	0.3152	0.3171	0.3172	0.3286	0.3312
a_A	0.9871	0.9865	0.9865	0.9868	0.9827	0.9820	0.9822	0.9815	0.9810
w_B	0.3379	0.3473	0.3662	0.3705					
a_A	0.9799	0.9764	0.9736	0.9713					

T/K = 308.15

w_B	0.2116	0.2171	0.2198	0.2214	0.2581	0.2606	0.2624	0.2702	0.2728
a_A	0.9944	0.9927	0.9912	0.9935	0.9910	0.9895	0.9896	0.9888	0.9888
w_B	0.2750	0.2875	0.2899	0.3023	0.3064	0.3070	0.3185	0.3385	0.3391
a_A	0.9885	0.9886	0.9877	0.9856	0.9869	0.9865	0.9821	0.9821	0.9820
w_B	0.3545	0.3597	0.3874	0.3971	0.4348				
a_A	0.9781	0.9784	0.9731	0.9695	0.9626				

Polymer (B): **poly(ethylene oxide)-poly(butylene terephthalate) multiblock copolymer** **2003MET**
Characterization: see comments, samples by Iso Tis BV, The Netherlands
Solvent (A): **water** H_2O **7732-18-5**

Type of data: sorption of water vapor into block copolymer films

T/K = 293.15

w_B	0.9943	0.9938	0.9916	0.9887	0.9818	0.9733	0.9730
P_A/P_{0A}	0.41	0.53	0.64	0.76	0.87	0.93	0.93

Comments: PEO: M_w/(g/mol) = 300, PBT: T_m/K = 451.45, crystallinity degree = 0.24, mass ratio of PEO/PBT = 32/68.

continued

continued

T/K = 293.15

w_B	0.9949	0.9932	0.9889	0.9883	0.9840	0.9823	0.9765	0.9714	0.9606
P_A/P_{0A}	0.30	0.40	0.50	0.53	0.60	0.64	0.70	0.76	0.80
w_B	0.9411	0.9045	0.8744						
P_A/P_{0A}	0.87	0.93	0.96						

Comments: PEO: M_w/(g/mol) = 600, T_g/K = 224.15, PBT: T_m/K = 451.95, crystallinity degree = 0.18, mass ratio of PEO/PBT = 41/49.

T/K = 293.15

w_B	0.9976	0.9939	0.9881	0.9880	0.9822	0.9807	0.9731	0.9678	0.9565
P_A/P_{0A}	0.18	0.30	0.40	0.41	0.50	0.53	0.60	0.64	0.70
w_B	0.9376	0.9173	0.9078	0.8654	0.8425	0.8114			
P_A/P_{0A}	0.76	0.80	0.81	0.87	0.90	0.93			

Comments: PEO: M_w/(g/mol) = 1000, T_g/K = 224.15, T_m/K = 262.75, crystallinity degree = 0.05, PBT: T_m/K = 436.15, crystallinity degree = 0.12, mass ratio of PEO/PBT = 52/48.

T/K = 293.15

w_B	0.9913	0.9865	0.9844	0.9794	0.9748	0.9676	0.9556	0.9427	0.9053
P_A/P_{0A}	0.30	0.40	0.41	0.50	0.53	0.60	0.64	0.70	0.76
w_B	0.8894	0.8067	0.7358						
P_A/P_{0A}	0.80	0.87	0.93						

Comments: PEO: M_w/(g/mol) = 2000, T_g/K = 220.15, T_m/K = 280.05, crystallinity degree = 0.20, PBT: T_m/K = 455.85, crystallinity degree = 0.15, mass ratio of PEO/PBT = 52/48.

T/K = 293.15

w_B	0.9900	0.9841	0.9846	0.9831	0.9748	0.9589	0.9275	0.9261	0.9264
P_A/P_{0A}	0.30	0.40	0.41	0.41	0.50	0.60	0.70	0.70	0.70
w_B	0.8607	0.8396	0.6969						
P_A/P_{0A}	0.80	0.81	0.93						

Comments: PEO: M_w/(g/mol) = 3000, T_g/K = 221.15, T_m/K = 292.65, crystallinity degree = 0.26, PBT: T_m/K = 466.65, crystallinity degree = 0.15, mass ratio of PEO/PBT = 52/48.

Polymer (B): **poly(ethylene oxide-*b*-propylene oxide-*b*-ethylene oxide)** **2010CAR**

Characterization: M_n/g.mol^{-1} = 2100, 20 wt% ethylene oxide, Oxiteno S/A, Maua, SP, Brazil

Solvent (A): **water** **H_2O** **7732-18-5**

continued

continued

Type of data: vapor-liquid equilibrium

T/K = 298.15

w_B	0.100	0.249	0.397	0.468	0.536	0.601	0.670	0.745	0.809
a_A	0.998	0.996	0.995	0.992	0.989	0.982	0.974	0.959	0.930
w_B	0.875	0.939	0.959	0.977	0.994				
a_A	0.875	0.766	0.694	0.547	0.247				

Polymer (B): **poly(ethylene oxide-*b*-propylene oxide-*b*-ethylene oxide)** **2010CAR**

Characterization: M_n/g.mol^{-1} = 2450, 40 wt% ethylene oxide, Oxiteno S/A, Maua, SP, Brazil

Solvent (A): **water** **H_2O** **7732-18-5**

Type of data: vapor-liquid equilibrium

T/K = 298.15

w_B	0.100	0.242	0.400	0.468	0.538	0.607	0.673	0.745	0.812
a_A	0.997	0.994	0.991	0.987	0.980	0.969	0.952	0.921	0.886
w_B	0.874	0.948	0.968	0.978	0.994				
a_A	0.832	0.684	0.571	0.484	0.253				

Polymer (B): **poly(ethylene oxide-*b*-propylene oxide-*b*-ethylene oxide)** **2010CAR**

Characterization: M_n/g.mol^{-1} = 2900, 40 wt% ethylene oxide, Aldrich Chem. Co., Inc., Milwaukee, WI

Solvent (A): **water** **H_2O** **7732-18-5**

Type of data: vapor-liquid equilibrium

T/K = 298.15

w_B	0.101	0.251	0.398	0.470	0.536	0.609	0.676	0.746	0.826
a_A	0.996	0.995	0.988	0.985	0.982	0.971	0.951	0.925	0.873
w_B	0.883	0.950	0.980	0.991	0.998				
a_A	0.818	0.675	0.464	0.306	0.184				

Polymer (B): **poly(ethylene oxide-*b*-propylene oxide-*b*-ethylene oxide)** **2010CAR**

Characterization: M_n/g.mol^{-1} = 1900, 50 wt% ethylene oxide, Aldrich Chem. Co., Inc., Milwaukee, WI

Solvent (A): **water** **H_2O** **7732-18-5**

Type of data: vapor-liquid equilibrium

continued

continued

T/K = 298.15

w_B	0.099	0.250	0.399	0.468	0.542	0.609	0.671	0.752	0.824
a_A	0.996	0.988	0.983	0.975	0.965	0.953	0.930	0.894	0.843
w_B	0.901	0.944	0.979	0.987	0.994				
a_A	0.752	0.634	0.373	0.250	0.160				

Polymer (B): **poly(ethylene oxide-*b*-propylene oxide-*b*-ethylene oxide)** **2010CAR**

Characterization: M_n/g.mol^{-1} = 8400, 80 wt% ethylene oxide, Aldrich Chem. Co., Inc., Milwaukee, WI

Solvent (A): **water** **H_2O** **7732-18-5**

Type of data: vapor-liquid equilibrium

T/K = 298.15

w_B	0.050	0.101	0.149	0.182	0.212	0.241	0.274	0.303	0.335
a_A	0.998	0.998	0.998	0.995	0.994	0.993	0.990	0.989	0.984
w_B	0.364	0.398							
a_A	0.983	0.976							

Polymer (B): **polyglycerol** **2004SEI**

Characterization: M_n/g.mol^{-1} = 1400, perfect linear analogue of the above hyperbranched sample, synthesized in the laboratory

Solvent (A): **water** **H_2O** **7732-18-5**

Type of data: vapor-liquid equilibrium

T/K = 363.15

w_B	0.999	0.996	0.992	0.988	0.987	0.978	0.965	0.933	0.923
P_A/kPa	4.80	11.31	16.95	24.73	25.00	36.65	50.68	82.29	92.92
w_B	0.917	0.790	0.903	0.892					
P_A/kPa	100.15	113.02	115.20	128.66					

Polymer (B): **polyglycerol (hyperbranched)** **2004SEI**

Characterization: M_n/g.mol^{-1} = 1400, M_w/g.mol^{-1} = 2100, 20 hydroxyl groups, synthesized in the laboratory

Solvent (A): **water** **H_2O** **7732-18-5**

Type of data: vapor-liquid equilibrium

T/K = 363.15

w_B	1.000	0.996	0.989	0.985	0.979	0.971	0.961	0.951	0.939
P_A/kPa	5.49	11.31	30.32	37.86	45.83	55.21	69.98	80.29	95.78
w_B	0.929	0.927	0.922	0.916					
P_A/kPa	108.86	110.45	115.08	123.07					

Polymer (B):	**poly(DL-lactic acid)**	**2004OLI**
Characterization:	M_w/g.mol^{-1} = 102800, T_g/K = 329.9, L:D ratio = 80:20, Cargill Dow Polymers LLC, Savage, MN	
Solvent (A):	**water** **H_2O**	**7732-18-5**

Type of data: vapor solubility

T/K = 293.2

P/bar	0.003	0.007	0.007	0.012	0.015	0.018	0.020
c_A/[cm^3 (STP)/cm^3 polymer]	2.18	5.21	4.80	7.87	10.30	12.95	15.96
P/bar	0.024						
c_A/[cm^3 (STP)/cm^3 polymer]	19.45						

Comments: The measuring temperature is below T_g.

T/K = 303.2

P/bar	0.005	0.011	0.011	0.012	0.014	0.020	0.022
c_A/[cm^3 (STP)/cm^3 polymer]	1.61	3.10	3.20	4.00	4.81	6.50	7.83
P/bar	0.026	0.032	0.035				
c_A/[cm^3 (STP)/cm^3 polymer]	9.67	11.85	14.47				

Comments: The measuring temperature is below T_g.

T/K = 313.2

P/bar	0.004	0.011	0.017	0.026	0.035	0.043	0.043
c_A/[cm^3 (STP)/cm^3 polymer]	0.67	1.60	3.33	5.25	7.03	9.81	9.73
P/bar	0.045	0.046					
c_A/[cm^3 (STP)/cm^3 polymer]	9.45	10.20					

Comments: The measuring temperature is below T_g.

Polymer (B):	**poly(L-lactic acid)**	**2006OLI**
Characterization:	T_g/K = 326.0, L:D ratio = 98:02, 10% crystallinity, Cargill Dow Polymers LLC, Savage, MN	
Solvent (A):	**water** **H_2O**	**7732-18-5**

Type of data: vapor solubility

T/K = 283.6

P/bar	0.00107	0.00108	0.00184	0.00263	0.00270	0.00407	0.00505
c_A/[cm^3 (STP)/cm^3 polymer]	1.03	1.18	2.12	2.67	2.63	3.93	4.91
c_A/[cm^3 (STP)/cm^3 polymer]*	1.14	1.30	2.33	2.94	2.89	4.32	5.40
P/bar	0.00657	0.00849	0.01007				
c_A/[cm^3 (STP)/cm^3 polymer]	6.22	8.08	9.90				
c_A/[cm^3 (STP)/cm^3 polymer]*	6.84	8.89	10.89				

Comments: * Values are calculated per cm^3 amorphous polymer considering a 10% crystallinity. The measuring temperature is below T_g.

continued

continued

T/K = 293.0

P/bar	0.00148	0.00197	0.00518	0.01036	0.01466	0.01966
c_A/[cm^3 (STP)/cm^3 polymer]	1.00	1.35	3.45	6.90	9.91	13.25
c_A/[cm^3 (STP)/cm^3 polymer]*	1.10	1.49	3.79	7.59	10.90	14.57

Comments: * Values are calculated per cm^3 amorphous polymer considering a 10% crystallinity. The measuring temperature is below T_g.

T/K = 303.0

P/bar	0.00190	0.00395	0.01236	0.01979	0.02650	0.03232	0.03732
c_A/[cm^3 (STP)/cm^3 polymer]	1.03	1.59	4.85	7.88	10.63	12.93	15.23
c_A/[cm^3 (STP)/cm^3 polymer]*	1.13	1.75	5.34	8.67	11.69	14.22	16.75

Comments: * Values are calculated per cm^3 amorphous polymer considering a 10% crystallinity. The measuring temperature is below T_g.

T/K = 313.0

P/bar	0.00130	0.00735	0.01541	0.02392	0.02913	0.03449	0.03974
c_A/[cm^3 (STP)/cm^3 polymer]	0.29	1.81	3.82	5.76	7.15	8.55	9.95
c_A/[cm^3 (STP)/cm^3 polymer]*	0.32	1.99	4.20	6.33	7.86	9.40	10.95

P/bar	0.04407
c_A/[cm^3 (STP)/cm^3 polymer]	11.46
c_A/[cm^3 (STP)/cm^3 polymer]*	12.61

Comments: * Values are calculated per cm^3 amorphous polymer considering a 10% crystallinity. The measuring temperature is below T_g.

Polymer (B): **poly(propylene glycol)** **2011ZAF**
Characterization: M_n/g.mol^{-1} = 400, Fluka AG, Buchs, Switzerland
Solvent (A): **water** H_2O **7732-18-5**

Type of data: vapor-liquid equilibrium

T/K = 298.15

w_B	0.1478	0.1664	0.2427	0.3128	0.3660	0.4209	0.5983	0.6134	0.6640
a_A	0.9924	0.9912	0.9861	0.9810	0.9774	0.9712	0.9611	0.9606	0.9568

w_B	0.7882	0.8448
a_A	0.9306	0.8959

T/K = 308.15

w_B	0.1443	0.2100	0.2735	0.4240	0.4701	0.5970	0.7500	0.7695	0.8065
a_A	0.9929	0.9905	0.9866	0.9813	0.9798	0.9724	0.9613	0.9557	0.9410

w_B	0.8064	0.8599	0.8723
a_A	0.9361	0.9025	0.8802

continued

continued

T/K = 318.15

w_B	0.0892	0.1253	0.1898	0.2597	0.4170	0.5134	0.5439	0.5784	0.6004
a_A	0.9945	0.9941	0.9924	0.9901	0.9874	0.9855	0.9852	0.9835	0.9817
w_B	0.6696	0.6944	0.7375	0.7696	0.8142	0.8271	0.8597	0.8733	
a_A	0.9777	0.9759	0.9690	0.9611	0.9434	0.9378	0.9078	0.9026	

Polymer (B): **poly(propylene glycol)** **2011SA1**
Characterization: M_n/g.mol^{-1} = 400, Fluka AG, Buchs, Switzerland
Solvent (A): **water** **H_2O** **7732-18-5**

Type of data: vapor-liquid equilibrium

T/K = 298.15

w_B	0.2126	0.3056	0.5421	0.5663	0.6121	0.6556	0.7112	0.7217	0.7328
a_A	0.9880	0.9823	0.9685	0.9671	0.9637	0.9596	0.9533	0.9503	0.9484
w_B	0.7788	0.8089	0.8382	0.8401	0.8760	0.8837			
a_A	0.9391	0.9258	0.9122	0.9088	0.8695	0.8556			

Comments: by isopiestic method

T/K = 298.15

w_B	0.0100	0.0205	0.0489	0.0815	0.1064	0.1260	0.1547	0.1935
a_A	0.9993	0.9988	0.9972	0.9957	0.9945	0.9933	0.9916	0.9896

Comments: by vapor-pressure osmometry (VPO)

T/K = 303.15

w_B	0.0100	0.0214	0.0400	0.0565	0.0720	0.0928	0.0954	0.1243	0.1626
a_A	0.9998	0.9991	0.9983	0.9975	0.9971	0.9958	0.9957	0.9943	0.9926
w_B	0.1867	0.2070	0.2255	0.2402	0.2522	0.2831	0.3001		
a_A	0.9916	0.9906	0.9894	0.9883	0.9878	0.9862	0.9855		

Comments: by vapor-pressure osmometry (VPO)

T/K = 308.15

w_B	0.0101	0.0200	0.0400	0.0517	0.0634	0.0745	0.0886	0.1030	0.1186
a_A	0.9996	0.9992	0.9983	0.9979	0.9974	0.9971	0.9965	0.9959	0.9951
w_B	0.1594	0.1787	0.1852	0.2002	0.2251	0.2327	0.2504	0.2864	0.2995
a_A	0.9932	0.9923	0.9920	0.9913	0.9904	0.9902	0.9894	0.9881	0.9876
w_B	0.3181								
a_A	0.9867								

Comments: by vapor-pressure osmometry (VPO)

Polymer (B): **poly(propylene glycol)** **2011SA2**
Characterization: M_n/g.mol^{-1} = 400, Fluka AG, Buchs, Switzerland
Solvent (A): **water** H_2O **7732-18-5**

Type of data: vapor-liquid equilibrium

T/K = 318.15

w_B	0.0481	0.1014	0.1505	0.1981	0.2023	0.3006	0.3010	0.3818	0.3990
a_A	0.9981	0.9963	0.9943	0.9930	0.9929	0.9910	0.9909	0.9900	0.9896
w_B	0.4879	0.5040	0.5803	0.5998	0.7069				
a_A	0.9882	0.9881	0.9874	0.9870	0.9847				

Comments: by vapor-pressure osmometry (VPO)

Polymer (B): **poly(propylene glycol)** **2010CAR**
Characterization: M_n/g.mol^{-1} = 400, Aldrich Chem. Co., Inc., Milwaukee, WI
Solvent (A): **water** H_2O **7732-18-5**

Type of data: vapor-liquid equilibrium

T/K = 298.15

w_B	0.100	0.249	0.400	0.470	0.525	0.607	0.676	0.745	0.814
a_A	0.985	0.981	0.975	0.968	0.967	0.962	0.957	0.943	0.920
w_B	0.884	0.949	0.982	0.994	0.999				
a_A	0.851	0.632	0.340	0.152	0.076				

Polymer (B): **poly(propylene glycol)** **1998ZAF**
Characterization: M_n/g.mol^{-1} = 404, Aldrich Chem. Co., Inc., Milwaukee, WI
Solvent (A): **water** H_2O **7732-18-5**

Type of data: vapor-liquid equilibrium

T/K = 298.15

c_B/(mol/kg A)	0.5538	0.7366	0.9127	1.0453	1.1931	1.3148	1.5521	1.8376
a_A	0.9897	0.9860	0.9833	0.9820	0.9799	0.9776	0.9743	0.9728

Polymer (B): **poly(propylene glycol)** **2004SAL**
Characterization: M_n/g.mol^{-1} = 405, Aldrich Chem. Co., Inc., Milwaukee, WI
Solvent (A): **water** H_2O **7732-18-5**

Type of data: vapor-liquid equilibrium

T/K = 298.15

w_B	0.1955	0.2180	0.2365	0.2661	0.2975	0.3106	0.3334	0.3727	0.3896
a_A	0.9895	0.9872	0.9855	0.9844	0.9804	0.9810	0.9773	0.9766	0.9729
w_B	0.4427	0.4508	0.4965	0.5201	0.5691	0.5753	0.6040	0.6505	0.6938
a_A	0.9706	0.9674	0.9659	0.9633	0.9564	0.9561	0.9542	0.9454	0.9423

Polymer (B): **poly(propylene oxide)** **2006SA4**
Characterization: M_n/g.mol^{-1} = 400, PPO 400, Fluka AG, Buchs, Switzerland
Solvent (A): **water** **H_2O** **7732-18-5**

Type of data: vapor-liquid equilibrium

T/K = 293.15

w_B	0.0864	0.1376	0.1749	0.2798	0.3052	0.3650	0.3740	0.4168	0.5501
a_A	0.9952	0.9920	0.9893	0.9817	0.9806	0.9751	0.9742	0.9704	0.9603
w_B	0.6439	0.6662	0.6979	0.7962	0.8056	0.8117	0.8169	0.8396	0.8681
a_A	0.9516	0.9482	0.9458	0.9226	0.9196	0.9154	0.9140	0.8956	0.8702
w_B	0.8777	0.8863	0.9008	0.9138	0.9149				
a_A	0.8608	0.8506	0.8144	0.7897	0.7768				

T/K = 298.15

w_B	0.1080	0.1761	0.1823	0.2165	0.2814	0.2932	0.3363	0.3432	0.3655
a_A	0.9942	0.9903	0.9900	0.9879	0.9839	0.9829	0.9803	0.9800	0.9781
w_B	0.3788	0.4282	0.4901	0.5213	0.5926	0.6676	0.6679	0.6864	0.7125
a_A	0.9775	0.9738	0.9705	0.9677	0.9634	0.9580	0.9592	0.9543	0.9513
w_B	0.7716	0.7770	0.7920	0.8393	0.8569	0.8716	0.8901	0.9038	
a_A	0.9372	0.9339	0.9323	0.9125	0.8948	0.8810	0.8443	0.8200	

T/K = 303.15

w_B	0.1113	0.1855	0.1886	0.2290	0.2401	0.3117	0.3249	0.4243	0.4583
a_A	0.9943	0.9907	0.9905	0.9881	0.9870	0.9833	0.9837	0.9786	0.9773
w_B	0.4690	0.5053	0.5458	0.5869	0.6721	0.7014	0.7317	0.7560	0.8098
a_A	0.9765	0.9749	0.9735	0.9708	0.9650	0.9615	0.9579	0.9524	0.9317
w_B	0.8339	0.8578	0.8668	0.8947	0.8993	0.9043	0.9232		
a_A	0.9221	0.9031	0.8932	0.8505	0.8382	0.8265	0.7697		

T/K = 308.15

w_B	0.1113	0.1712	0.1944	0.2023	0.2075	0.2132	0.2302	0.2519	0.2761
a_A	0.9945	0.9917	0.9906	0.9905	0.9904	0.9900	0.9898	0.9886	0.9875
w_B	0.3361	0.3887	0.4996	0.5208	0.5257	0.6334	0.7194	0.7205	0.7660
a_A	0.9848	0.9839	0.9790	0.9790	0.9786	0.9743	0.9678	0.9661	0.9571
w_B	0.7935	0.7996	0.8017	0.8343	0.8447	0.8549	0.8788	0.8906	0.9064
a_A	0.9497	0.9479	0.9446	0.9251	0.9217	0.9169	0.8803	0.8556	0.8272

T/K = 313.15

w_B	0.1297	0.1381	0.1439	0.1478	0.1765	0.2648	0.3242	0.5402	0.5443
a_A	0.9943	0.9941	0.9935	0.9933	0.9923	0.9893	0.9876	0.9836	0.9825
w_B	0.6556	0.6883	0.6960	0.7201	0.7319	0.7421	0.7524	0.7819	0.7967
a_A	0.9793	0.9743	0.9743	0.9705	0.9702	0.9696	0.9688	0.9624	0.9586

continued

continued

w_B	0.8690	0.8748	0.8797	0.8911	0.9050	0.9224
a_A	0.9057	0.8971	0.8901	0.8659	0.8426	0.7920

T/K = 318.15

w_B	0.1450	0.2110	0.2737	0.3571	0.5170	0.6647	0.7498	0.7548	0.7940
a_A	0.9941	0.9916	0.9906	0.9893	0.9865	0.9806	0.9766	0.9755	0.9637
w_B	0.8090	0.8229	0.8345	0.8461	0.8624	0.8810	0.8937	0.9136	0.9246
a_A	0.9594	0.9559	0.9502	0.9397	0.9180	0.8944	0.8700	0.8307	0.7975

Polymer (B): **poly(sodium acrylate)** **2004LAM, 2008LA1**
Characterization: M_n/g.mol^{-1} = 2600, M_w/g.mol^{-1} = 4300, M_z/g.mol^{-1} = 6100, Rohagit SL 137, Roehm GmbH, Darmstadt, Germany
Solvent (A): **water** H_2O **7732-18-5**

Type of data: vapor-liquid equilibrium

T/K = 298.15

w_B	0.1322	0.1469	0.1476	0.1680	0.1929	0.2000	0.2097	0.2402	0.2426
a_A	0.9887	0.9880	0.9881	0.9842	0.9798	0.9788	0.9777	0.9701	0.9690
w_B	0.2593	0.2756	0.2773	0.2932	0.2936	0.2962	0.3401		
a_A	0.9646	0.9617	0.9588	0.9526	0.9523	0.9531	0.9307		

Polymer (B): **poly(sodium acrylate)** **2004LAM, 2008LA1**
Characterization: M_n/g.mol^{-1} = 6900, M_w/g.mol^{-1} = 17300, M_z/g.mol^{-1} = 39700, Sokalan SA 40, BASF AG, Ludwigshafen, Germany
Solvent (A): **water** H_2O **7732-18-5**

Type of data: vapor-liquid equilibrium

T/K = 298.15

w_B	0.1406	0.1795	0.2090	0.2228	0.2620	0.2776	0.2783	0.2974	0.3123
a_A	0.9887	0.9842	0.9798	0.9778	0.9690	0.9646	0.9657	0.9588	0.9523
w_B	0.3149	0.3155	0.3412	0.3557	0.3607	0.3696	0.4144		
a_A	0.9529	0.9526	0.9402	0.9307	0.9299	0.9291	0.8958		

T/K = 323.15

w_B	0.1398	0.1807	0.2081	0.2587	0.2946	0.3149	0.3457
a_A	0.9888	0.9847	0.9801	0.9706	0.9614	0.9550	0.9420

Polymer (B): **poly(sodium ethylenesulfonate)** **2004LAM, 2008LA1**
Characterization: M_n/g.mol^{-1} = 1600, M_w/g.mol^{-1} = 2800, M_z/g.mol^{-1} = 5100, Polysciences Europe GmbH, Eppelheim, Germany
Solvent (A): **water** H_2O **7732-18-5**

Type of data: vapor-liquid equilibrium

continued

continued

T/K = 298.15

w_B	0.2551	0.3560	0.4497
a_A	0.9677	0.9431	0.9054

T/K = 323.15

w_B	0.1290	0.1704
a_A	0.9866	0.9809

Polymer (B): **poly(sodium ethylenesulfonate)** **2004LAM, 2008LA1**
Characterization: M_n/g.mol^{-1} = 6900, M_w/g.mol^{-1} = 11800, M_z/g.mol^{-1} = 16200, Natriumpolat PIPU 005, Hoechst GmbH, Frankfurt, Germany
Solvent (A): **water** **H_2O** **7732-18-5**

Type of data: vapor-liquid equilibrium

T/K = 298.15

w_B	0.0793	0.2113	0.3179	0.3911	0.4057	0.4171	0.4462	0.4777	0.4979
a_A	0.9974	0.9883	0.9803	0.9687	0.9667	0.9647	0.9559	0.9485	0.9345
w_B	0.5295								
a_A	0.9270								

T/K = 323.15

w_B	0.2438	0.3061	0.3139	0.3543	0.4031	0.4135	0.4253	0.4570
a_A	0.9866	0.9809	0.9800	0.9750	0.9651	0.9644	0.9595	0.9495

Polymer (B): **poly(sodium methacrylate)** **2004LAM, 2008LA1**
Characterization: M_n/g.mol^{-1} = 6100, M_w/g.mol^{-1} = 10000, M_z/g.mol^{-1} = 15400, Sigma-Aldrich Chemie GmbH, Steinheim, Germany
Solvent (A): **water** **H_2O** **7732-18-5**

Type of data: vapor-liquid equilibrium

T/K = 298.15

w_B	0.1077	0.1517	0.2054	0.2072	0.2153	0.2381	0.2588	0.2602	0.3006
a_A	0.9926	0.9876	0.9802	0.9799	0.9784	0.9728	0.9675	0.9674	0.9533
w_B	0.3015	0.3262	0.3280						
a_A	0.9528	0.9408	0.9396						

T/K = 323.15

w_B	0.1587	0.1958	0.2002	0.2248	0.2652	0.2660	0.2803	0.3052
a_A	0.9866	0.9809	0.9800	0.9750	0.9651	0.9644	0.9595	0.9495

Polymer (B): **poly(sodium methacrylate)** **2004LAM, 2008LA1**
Characterization: M_n/g.mol^{-1} = 14200, M_w/g.mol^{-1} = 20500, M_z/g.mol^{-1} = 27600, Polysciences Europe GmbH, Eppelheim, Germany
Solvent (A): **water** **H_2O** **7732-18-5**

Type of data: vapor-liquid equilibrium

T/K = 298.15

w_B	0.1435	0.1663	0.1977	0.2598	0.3017	0.3257
a_A	0.9893	0.9867	0.9823	0.9697	0.9519	0.9388

T/K = 323.15

w_B	0.1734	0.2084	0.2324	0.2689	0.2712	0.2850	0.3102
a_A	0.9866	0.9800	0.9750	0.9651	0.9644	0.9595	0.9495

Polymer (B): **poly(sodium styrenesulfonate)** **2004LAM, 2008LA1**
Characterization: M_n/g.mol^{-1} = 127000, M_w/g.mol^{-1} = 148000, M_z/g.mol^{-1} = 155000, Sigma-Aldrich Chemie GmbH, Steinheim, Germany
Solvent (A): **water** **H_2O** **7732-18-5**

Type of data: vapor-liquid equilibrium

T/K = 298.15

w_B	0.1082	0.1696	0.2475	0.2759	0.3013	0.3288	0.3851	0.3909	0.4460
a_A	0.9933	0.9884	0.9794	0.9749	0.9707	0.9674	0.9558	0.9551	0.9417
w_B	0.4867								
a_A	0.9292								

T/K = 323.15

w_B	0.1847	0.2389	0.2738	0.3240	0.3303
a_A	0.9866	0.9800	0.9750	0.9651	0.9644

Polymer (B): **poly(vinyl acetate)** **2004PAL**
Characterization: M_w/g.mol^{-1} = 113000, Aldrich Chem. Co., Inc., Milwaukee, WI
Solvent (A): **water** **H_2O** **7732-18-5**

Type of data: vapor-liquid equilibrium

T/K = 313.15

w_A	0.0038	0.0048	0.0050	0.0060	0.0061	0.0101	0.0153
P_A/P_{0A}	0.263	0.265	0.265	0.351	0.351	0.485	0.623

Polymer (B): **poly(vinyl acetate-*co*-vinyl alcohol)** **2004PAL**
Characterization: M_n/g.mol^{-1} = 57000, M_w/g.mol^{-1} = 103000, 12 mol% vinyl acetate, semicrystalline sample with 39% crystallinity (by MDSC), Air Products and Chemicals, Inc., Allentown, PA
Solvent (A): **water** **H_2O** **7732-18-5**

Type of data: vapor-liquid equilibrium

T/K = 363.15

w_A	0.0237	0.0273	0.0373	0.0523	0.0701	0.1170	0.1037	0.189	0.219
P_A/P_{0A}	0.222	0.224	0.295	0.461	0.467	0.623	0.701	0.891	0.909

T/K = 373.15

w_A	0.0228	0.0623	0.112	0.208
P_A/P_{0A}	0.221	0.490	0.675	0.849

T/K = 383.15

w_A	0.0222	0.0506	0.0930
P_A/P_{0A}	0.244	0.442	0.621

Comments: The concentration of solvent A is related to the 61% amorphous region of the polymer, i.e., the crystalline part is not influenced by sorption and is subtracted when determining w_B.

Polymer (B): **poly(vinyl alcohol)** **2004PAL**
Characterization: M_n/g.mol^{-1} = 64000, M_w/g.mol^{-1} = 116000, semicrystalline sample with 64% crystallinity (by MDSC), Air Products and Chemicals, Inc., Allentown, PA
Solvent (A): **water** **H_2O** **7732-18-5**

Type of data: vapor-liquid equilibrium

T/K = 363.15

w_A	0.0272	0.0404	0.0475	0.0828	0.0989	0.120	0.188	0.197	0.338
P_A/P_{0A}	0.152	0.215	0.272	0.425	0.483	0.548	0.700	0.714	0.869

w_A	0.373
P_A/P_{0A}	0.885

T/K = 373.15

w_A	0.0335	0.122	0.202	0.296
P_A/P_{0A}	0.193	0.551	0.717	0.876

T/K = 383.15

w_A	0.0409	0.107	0.228	0.336
P_A/P_{0A}	0.229	0.491	0.723	0.844

Comments: The concentration of solvent A is related to the 36% amorphous region of the polymer, i.e., the crystalline part is not influenced by sorption and is subtracted when determining w_B.

Polymer (B): **poly(vinyl alcohol)** **2005CSA**
Characterization: M_n/g.mol^{-1} = 75600, prepared and fractionated in the laboratory from Poval 420, Kuraray Co., Japan
Solvent (A): **water** H_2O **7732-18-5**

Type of data: vapor-liquid equilibrium

T/K = 298.15

φ_B	0.00548	0.01271	0.01623	0.02104	0.02461	0.03078	0.03808
$\ln(a_A)$	$-3.46\ 10^{-6}$	$-7.05\ 10^{-6}$	$-1.10\ 10^{-5}$	$-1.73\ 10^{-5}$	$-2.27\ 10^{-5}$	$-3.14\ 10^{-5}$	$-4.60\ 10^{-5}$
φ_B	0.03949	0.05426					
$\ln(a_A)$	$-4.63\ 10^{-5}$	$-8.07\ 10^{-5}$					

Polymer (B): **poly(vinyl alcohol-*co*-vinyl acetal)** **2005CSA**
Characterization: M_n/g.mol^{-1} = 75600, 7.0 mol% vinyl acetal, prepared and fractionated in the laboratory from Poval 420, Kuraray Co., Japan
Solvent (A): **water** H_2O **7732-18-5**

Type of data: vapor-liquid equilibrium

T/K = 298.15

φ_B	0.00574	0.01256	0.01563	0.02079	0.02461	0.03028	0.03538
$\ln(a_A)$	$-2.52\ 10^{-6}$	$-8.36\ 10^{-6}$	$-1.02\ 10^{-5}$	$-1.47\ 10^{-5}$	$-1.91\ 10^{-5}$	$-3.04\ 10^{-5}$	$-3.86\ 10^{-5}$
φ_B	0.04022	0.04962					
$\ln(a_A)$	$-4.68\ 10^{-5}$	$-6.70\ 10^{-5}$					

Polymer (B): **poly(vinyl alcohol-*co*-vinyl butyral)** **2005CSA**
Characterization: M_n/g.mol^{-1} = 77260, 7.0 mol% vinyl butyral, prepared and fractionated in the laboratory from Poval 420, Kuraray Co., Japan
Solvent (A): **water** H_2O **7732-18-5**

Type of data: vapor-liquid equilibrium

T/K = 298.15

φ_B	0.00586	0.00970	0.01527	0.02008	0.02416	0.02902	0.03089
$\ln(a_A)$	$-2.73\ 10^{-6}$	$-3.24\ 10^{-6}$	$-4.82\ 10^{-6}$	$-6.90\ 10^{-6}$	$-1.14\ 10^{-5}$	$-1.20\ 10^{-5}$	$-1.36\ 10^{-5}$
φ_B	0.03807	0.04490					
$\ln(a_A)$	$-1.69\ 10^{-5}$	$-2.22\ 10^{-5}$					

Polymer (B): **poly(vinyl alcohol-*co*-vinyl propional)** **2005CSA**
Characterization: M_n/g.mol^{-1} = 76430, 7.0 mol% vinyl propional, prepared and fractionated in the laboratory from Poval 420, Kuraray Co., Japan
Solvent (A): **water** **H_2O** **7732-18-5**

Type of data: vapor-liquid equilibrium

T/K = 298.15

φ_B	0.00409	0.01112	0.01383	0.02045	0.02209	0.03115	0.0350
$\ln(a_A)$	$-1.39\ 10^{-6}$	$-4.30\ 10^{-6}$	$-6.54\ 10^{-6}$	$-1.02\ 10^{-5}$	$-1.13\ 10^{-5}$	$-2.15\ 10^{-5}$	$-2.75\ 10^{-5}$
φ_B	0.04621						
$\ln(a_A)$	$-4.31\ 10^{-5}$						

Polymer (B): **poly(vinyl alcohol-*co*-vinyl propional)** **2005CSA**
Characterization: M_n/g.mol^{-1} = 77270, 9.5 mol% vinyl propional, prepared and fractionated in the laboratory from Poval 420, Kuraray Co., Japan
Solvent (A): **water** **H_2O** **7732-18-5**

Type of data: vapor-liquid equilibrium

T/K = 298.15

φ_B	0.00475	0.00806	0.01122	0.02801	0.02675	0.04273	0.04160
$\ln(a_A)$	$-1.76\ 10^{-6}$	$-3.48\ 10^{-6}$	$-4.46\ 10^{-6}$	$-1.56\ 10^{-5}$	$-1.42\ 10^{-5}$	$-3.19\ 10^{-5}$	$-3.20\ 10^{-5}$
φ_B	0.05267						
$\ln(a_A)$	$-4.49\ 10^{-5}$						

Polymer (B): **poly(*N*-vinylcaprolactam)** **2009FO2**
Characterization: M_n/g.mol^{-1} = 1150, synthesized in the laboratory
Solvent (A): **water** **H_2O** **7732-18-5**

Type of data: vapor-liquid equilibrium

T/K = 308.15

w_B	0.46	0.48	0.49	0.52	0.54	0.56	0.58	0.60	0.62
P_A/kPa	5.59	5.59	5.58	5.57	5.57	5.55	5.54	5.50	5.48
w_B	0.63	0.65	0.67						
P_A/kPa	5.46	5.43	5.39						

Polymer (B): **poly(*N*-vinylcaprolactam)** **2009FO2**
Characterization: M_n/g.mol^{-1} = 1500, synthesized in the laboratory
Solvent (A): **water** **H_2O** **7732-18-5**

Type of data: vapor-liquid equilibrium

continued

continued

T/K = 308.15

w_B	0.495	0.503	0.513	0.531	0.546	0.582	0.598	0.621	0.657
P_A/kPa	5.598	5.591	5.583	5.577	5.570	5.552	5.542	5.507	5.488
w_B	0.673	0.681	0.713						
P_A/kPa	5.463	5.432	5.398						

Polymer (B): **poly(*N*-vinylcaprolactam)** **2009FO2**
Characterization: M_n/g.mol^{-1} = 1760, synthesized in the laboratory
Solvent (A): **water** H_2O **7732-18-5**

Type of data: vapor-liquid equilibrium

T/K = 308.15

w_B	0.512	0.525	0.548	0.573	0.612	0.633	0.647	0.685	0.720
P_A/kPa	5.608	5.602	5.599	5.593	5.575	5.555	5.537	5.517	5.487
w_B	0.741	0.755	0.786						
P_A/kPa	5.462	5.411	5.407						

Polymer (B): **poly(vinyl chloride)** **2007SER**
Characterization: M_w/g.mol^{-1} = 175000,
Scientific Polymer Products, Inc., Ontario, NY
Solvent (A): **water** H_2O **7732-18-5**

Type of data: vapor-liquid equilibrium

T/K = 303.15

w_A	0.003	0.007	0.010	0.012	0.016	0.023	0.027	0.032	0.040
P_A/kPa	2.1	2.2	2.5	2.6	2.8	3.1	3.5	3.7	3.8

T/K = 313.15

w_A	0.001	0.004	0.008	0.015	0.027	0.031	0.045	0.053	0.055
P_A/kPa	2.5	2.8	3.0	3.6	4.1	4.5	5.3	6.0	6.2

Polymer (B): **poly(1-vinyl-2-pyrrolidinone)** **2004SAD**
Characterization: M_w/g.mol^{-1} = 10000, PVP K15,
Aldrich Chem. Co., Inc., Milwaukee, WI
Solvent (A): **water** H_2O **7732-18-5**

Type of data: vapor-liquid equilibrium

T/K = 298.15

w_B	0.2960	0.3192	0.3858	0.4070	0.4192	0.4362	0.4838	0.4904	0.4940
a_A	0.9868	0.9858	0.9787	0.9749	0.9724	0.9685	0.9576	0.9565	0.9563
w_B	0.5825	0.6328	0.6807	0.7250	0.7540				
a_A	0.9217	0.8900	0.8519	0.7998	0.7529				

continued

continued

T/K = 308.15

w_B	0.1852	0.2110	0.2401	0.2865	0.3083	0.3375	0.3633	0.3779	0.3879
a_A	0.9935	0.9930	0.9906	0.9888	0.9871	0.9849	0.9833	0.9814	0.9794
w_B	0.4214	0.4441	0.4514	0.4735	0.4838	0.4854	0.5003	0.5029	0.5333
a_A	0.9754	0.9718	0.9702	0.9651	0.9627	0.9622	0.9592	0.9573	0.9482
w_B	0.5500	0.5885	0.6001	0.6015	0.6159	0.6648	0.6956	0.7143	0.7655
a_A	0.9439	0.9263	0.9209	0.9207	0.9128	0.8804	0.8532	0.8333	0.7701

T/K = 318.15

w_B	0.2623	0.3526	0.3940	0.4519	0.5007	0.5513	0.6052	0.6316	0.6869
a_A	0.9914	0.9844	0.9795	0.9722	0.9638	0.9493	0.9265	0.9137	0.8749
w_B	0.7291	0.7761							
a_A	0.8346	0.7779							

T/K = 328.15

w_B	0.2961	0.3392	0.3687	0.4130	0.4584	0.4980	0.5089	0.5502	0.5558
a_A	0.9912	0.9897	0.9856	0.9810	0.9743	0.9675	0.9644	0.9532	0.9510
w_B	0.5616	0.6079	0.6245	0.6623	0.6727	0.7275	0.7450	0.7807	
a_A	0.9497	0.9336	0.9238	0.9034	0.8951	0.8512	0.8326	0.7965	

Polymer (B): **poly(1-vinyl-2-pyrrolidinone)** **2005SA1**
Characterization: M_w/g.mol^{-1} = 10000, PVP K15,
Aldrich Chem. Co., Inc., Milwaukee, WI
Solvent (A): **water** **H_2O** **7732-18-5**

Type of data: vapor-liquid equilibrium

T/K = 298.15

w_B	0.2821	0.2910	0.3858	0.4362	0.4530	0.4850	0.4944	0.5350	0.5941
a_A	0.9890	0.9884	0.9794	0.9685	0.9669	0.9588	0.9543	0.9417	0.9259

T/K = 308.15

w_B	0.3245	0.3985	0.4838	0.4977	0.5297	0.5702	0.5837	0.6273
a_A	0.9865	0.9774	0.9640	0.9603	0.9496	0.9329	0.9273	0.9051

T/K = 318.15

w_B	0.3705	0.4475	0.4604	0.5228	0.5310	0.5908	0.6095	0.6165
a_A	0.9830	0.9733	0.9713	0.9589	0.9563	0.9336	0.9243	0.9214

Polymer (B): **poly(1-vinyl-2-pyrrolidinone)** **2005SA2**
Characterization: M_w/g.mol^{-1} = 10000, PVP K15,
Aldrich Chem. Co., Inc., Milwaukee, WI
Solvent (A): **water** $\mathbf{H_2O}$ **7732-18-5**

Type of data: vapor-liquid equilibrium

T/K = 298.15

w_B	0.2970	0.3742	0.4180	0.4440	0.4821	0.4951	0.5000	0.5260
a_A	0.9863	0.9797	0.9733	0.9670	0.9590	0.9554	0.9540	0.9451

T/K = 308.15

w_B	0.3970	0.4445	0.4560	0.4835	0.4952	0.5081	0.5289	0.5635
a_A	0.9787	0.9714	0.9691	0.9629	0.9601	0.9560	0.9504	0.9371

T/K = 318.15

w_B	0.3330	0.4025	0.4520	0.4600	0.4850	0.5481	0.5700	0.5940
a_A	0.9864	0.9782	0.9724	0.9715	0.9669	0.9507	0.9439	0.9325

Polymer (B): **poly(1-vinyl-2-pyrrolidinone)** **2006SA2**
Characterization: M_w/g.mol^{-1} = 10000, PVP K15,
Aldrich Chem. Co., Inc., Milwaukee, WI
Solvent (A): **water** $\mathbf{H_2O}$ **7732-18-5**

Type of data: vapor-liquid equilibrium

T/K = 298.15

w_B	0.3085	0.3192	0.3193	0.3352	0.3858	0.3975	0.4723	0.4947
a_A	0.9872	0.9859	0.9857	0.9848	0.9787	0.9766	0.9614	0.9561

T/K = 308.15

w_B	0.3368	0.4000	0.4350	0.4672	0.4725	0.5085	0.5440	0.5508	0.5837
a_A	0.9855	0.9783	0.9730	0.9671	0.9658	0.9557	0.9453	0.9431	0.9273

T/K = 318.15

w_B	0.2963	0.3610	0.3940	0.4435	0.4440	0.4832	0.5330	0.5430
a_A	0.9891	0.9834	0.9794	0.9735	0.9741	0.9679	0.9558	0.9530

Polymer (B): **poly(1-vinyl-2-pyrrolidinone)** **2006SA3**
Characterization: M_w/g.mol^{-1} = 10000, PVP K15,
Aldrich Chem. Co., Inc., Milwaukee, WI
Solvent (A): **water** $\mathbf{H_2O}$ **7732-18-5**

Type of data: vapor-liquid equilibrium

T/K = 298.15

w_B	0.3200	0.3520	0.3780	0.4076	0.4190
a_A	0.9852	0.9828	0.9804	0.9740	0.9729

continued

continued

T/K = 308.15

w_B	0.3501	0.3785	0.4050	0.4150	0.4510	0.4780
a_A	0.9838	0.9807	0.9776	0.9763	0.9707	0.9640

T/K = 318.15

w_B	0.3350	0.3730	0.4110	0.4370	0.4520	0.4940
a_A	0.9863	0.9823	0.9785	0.9747	0.9721	0.9654

Polymer (B): **poly(1-vinyl-2-pyrrolidinone)** **2010SA2**
Characterization: M_w/g.mol^{-1} = 10000, PVP K15,
Aldrich Chem. Co., Inc., Milwaukee, WI
Solvent (A): **water** H_2O **7732-18-5**

Type of data: vapor-liquid equilibrium

T/K = 298.15

w_B	0.3837	0.4209	0.4513	0.4858	0.4990	0.5418	0.5829	0.5856	0.6237
a_A	0.9871	0.9820	0.9770	0.9697	0.9666	0.9542	0.9389	0.9373	0.9125
w_B	0.3855	0.4325	0.4623	0.4942	0.5161	0.5470	0.5835	0.5905	0.6299
a_A	0.9870	0.9808	0.9749	0.9691	0.9614	0.9537	0.9365	0.9316	0.9083

Polymer (B): **poly(1-vinyl-2-pyrrolidinone)** **2005CSA**
Characterization: M_n/g.mol^{-1} = 117000, fractionated in the laboratory from
Fluka K90 PVP, Fluka AG, Buchs, Switzerland
Solvent (A): **water** H_2O **7732-18-5**

Type of data: vapor-liquid equilibrium

T/K = 298.15

φ_B	0.00779	0.01562	0.02349	0.03139	0.03932	0.04729	0.05530
$\ln(a_A)$	$-2.85\ 10^{-6}$	$-8.83\ 10^{-6}$	$-1.87\ 10^{-5}$	$-3.33\ 10^{-5}$	$-5.34\ 10^{-5}$	$-7.99\ 10^{-5}$	$-1.14\ 10^{-4}$
φ_B	0.06334	0.07142					
$\ln(a_A)$	$-1.55\ 10^{-4}$	$-2.06\ 10^{-4}$					

2.2. Partial solvent vapor pressures or solvent activities for ternary aqueous polymer solutions

Polymer (B): **poly(ammonium acrylate)** **2004LAM, 2008LA2**
Characterization: M_n/g.mol^{-1} = 2300, M_w/g.mol^{-1} = 3900, M_z/g.mol^{-1} = 5800, Rohagit SL 135, Roehm GmbH, Darmstadt, Germany
Solvent (A): **water** **H_2O** **7732-18-5**
Salt (C): **sodium chloride** **NaCl** **7647-14-5**

Type of data: vapor-liquid equilibrium

T/K = 298.15

w_B	0.1871	0.1475	0.1099	0.0740	0.0000	0.2407	0.1925	0.1447	0.0944
w_C	0.0000	0.0085	0.0163	0.0234	0.0365	0.0000	0.0125	0.0242	0.0356
a_A	0.9787	0.9787	0.9787	0.9787	0.9787	0.9672	0.9672	0.9672	0.9672
w_B	0.0000	0.2854	0.2309	0.1750	0.1145	0.0000	0.3186	0.2600	0.2010
w_C	0.0547	0.0000	0.0162	0.0314	0.0470	0.0736	0.0000	0.0194	0.0377
a_A	0.9672	0.9545	0.9545	0.9545	0.9545	0.9545	0.9415	0.9415	0.9415
w_B	0.1341	0.0000							
w_C	0.0572	0.0917							
a_A	0.9415	0.9415							

Polymer (B): **poly(ammonium acrylate)** **2004LAM, 2008LA2**
Characterization: M_n/g.mol^{-1} = 7300, M_w/g.mol^{-1} = 13600, M_z/g.mol^{-1} = 19200, Rohagit SL 252, Roehm GmbH, Darmstadt, Germany
Solvent (A): **water** **H_2O** **7732-18-5**
Salt (C): **sodium chloride** **NaCl** **7647-14-5**

Type of data: vapor-liquid equilibrium

T/K = 298.15

w_B	0.2054	0.1665	0.1289	0.0904	0.0000	0.2984	0.2280	0.1728	0.1134
w_C	0.0000	0.0086	0.0168	0.0243	0.0398	0.0000	0.0148	0.0282	0.0410
a_A	0.9766	0.9766	0.9766	0.9766	0.9766	0.9618	0.9618	0.9618	0.9618
w_B	0.0000	0.3951	0.3196	0.2542	0.1880	0.0000			
w_C	0.0629	0.0000	0.0231	0.0425	0.0609	0.1090			
a_A	0.9618	0.9281	0.9281	0.9281	0.9281	0.9281			

Polymer (B): **poly(ammonium acrylate)** **2004LAM, 2008LA2**
Characterization: M_n/g.mol^{-1} = 7700, M_w/g.mol^{-1} = 13300, M_z/g.mol^{-1} = 19400, Rohagit SL 159, Roehm GmbH, Darmstadt, Germany
Solvent (A): **water** **H_2O** **7732-18-5**
Salt (C): **sodium chloride** **NaCl** **7647-14-5**

continued

continued

Type of data: vapor-liquid equilibrium

T/K = 298.15

w_B	0.2051	0.1650	0.1242	0.0851	0.0000	0.2592	0.2086	0.1584	0.1062
w_C	0.0000	0.0086	0.0160	0.0229	0.0365	0.0000	0.0123	0.0238	0.0346
a_A	0.9787	0.9787	0.9787	0.9787	0.9787	0.9672	0.9672	0.9672	0.9672
w_B	0.0000	0.3015	0.2464	0.1891	0.1277	0.0000	0.3321	0.2742	0.2150
w_C	0.0547	0.0000	0.0159	0.0308	0.0458	0.0736	0.0000	0.0189	0.0367
a_A	0.9672	0.9545	0.9545	0.9545	0.9545	0.9545	0.9415	0.9415	0.9415
w_B	0.1456	0.0000							
w_C	0.0557	0.0917							
a_A	0.9415	0.9415							

Polymer (B): **polyester (hyperbranched, aliphatic)** **2003SE2**

Characterization: M_n/g.mol^{-1} = 1620, M_w/g.mol^{-1} = 2100, 16 OH groups per macromolecule, hydroxyl no. = 490-520 mg KOH/g, acid no. = 5-9 mg KOH/g, hydroxyl functional hyperbranched polyesters produced from polyalcohol cores and hydroxy acids, Boltorn H20, Perstorp Specialty Chemicals AB, Sweden

Solvent (A): **water** H_2O **7732-18-5**

Solvent (C): **ethanol** C_2H_6O **64-17-5**

Type of data: vapor-liquid equilibrium

T/K = 363.15

w_B	0.20	0.20	0.20	0.20	0.20	0.20	0.50	0.50	0.50
x_C	0.200	0.300	0.416	0.599	0.802	0.902	0.200	0.400	0.600
y_C	0.523	0.571	0.618	0.705	0.827	0.905	0.504	0.637	0.731
w_B	0.50	0.50	0.60	0.60	0.60	0.60			
x_C	0.800	0.951	0.401	0.799	0.901	0.950			
y_C	0.857	0.954	0.631	0.862	0.912	0.959			

Polymer (B): **polyester (hyperbranched, aliphatic)** **2003SE3**

Characterization: M_n/g.mol^{-1} = 1620, M_w/g.mol^{-1} = 2100, 16 OH groups per macromolecule, hydroxyl no. = 490-520 mg KOH/g, acid no. = 5-9 mg KOH/g, hydroxyl functional hyperbranched polyesters produced from polyalcohol cores and hydroxy acids, Boltorn H20, Perstorp Specialty Chemicals AB, Sweden

Solvent (A): **water** H_2O **7732-18-5**

Solvent (C): **ethanol** C_2H_6O **64-17-5**

Type of data: vapor-liquid equilibrium

continued

continued

T/K = 363.15

w_B	0.20	0.20	0.20	0.20	0.20	0.20	0.50	0.50	0.50
x_C	0.1999	0.2998	0.4160	0.5987	0.8018	0.9021	0.2001	0.3999	0.5996
y_C	0.5225	0.5708	0.6179	0.7054	0.8274	0.9046	0.5043	0.6367	0.7306

w_B	0.50	0.50	0.50
x_C	0.8002	0.9010	0.9509
y_C	0.8565	0.9204	0.9541

Polymer (B): **polyester (hyperbranched, aliphatic)** **2003SE2**
Characterization: M_n/g.mol^{-1} = 2330, M_w/g.mol^{-1} = 3500, 32 OH groups per macromolecule, hydroxyl functional hyperbranched polyesters produced from polyalcohol cores and hydroxy acids, Boltorn H30, Perstorp Specialty Chemicals AB, Sweden
Solvent (A): **water** **H_2O** **7732-18-5**
Solvent (C): **ethanol** **C_2H_6O** **64-17-5**

Type of data: vapor-liquid equilibrium

T/K = 363.15

w_B	0.20	0.20	0.20	0.20	0.20
x_C	0.200	0.300	0.400	0.800	0.898
y_C	0.529	0.568	0.618	0.822	0.892

Polymer (B): **polyester (hyperbranched, aliphatic)** **2003SE2**
Characterization: M_n/g.mol^{-1} = 2830, M_w/g.mol^{-1} = 5100, 64 OH groups per macromolecule, hydroxyl no. = 470-500 mg KOH/g, acid no. = 7-11 mg KOH/g, hydroxyl functional hyperbranched polyesters produced from polyalcohol cores and hydroxy acids, Boltorn H40, Perstorp Specialty Chemicals AB, Sweden
Solvent (A): **water** **H_2O** **7732-18-5**
Solvent (C): **ethanol** **C_2H_6O** **64-17-5**

Type of data: vapor-liquid equilibrium

T/K = 363.15

w_B	0.20	0.20	0.20	0.20	0.20	0.20	0.50	0.50	0.50
x_C	0.200	0.300	0.400	0.600	0.798	0.901	0.800	0.900	0.951
y_C	0.517	0.565	0.621	0.709	0.825	0.905	0.845	0.908	0.942

Polymer (B): **polyesteramide (hyperbranched)** **2003SE3**
Characterization: M_n/g.mol^{-1} = 1200, M_w/g.mol^{-1} = 6000, 8 hydroxyl groups, Hybrane S1200, DSM Fine Chemicals, The Netherlands
Solvent (A): **water** **H_2O** **7732-18-5**
Solvent (C): **ethanol** **C_2H_6O** **64-17-5**

Type of data: vapor-liquid equilibrium

T/K = 363.15

w_B	0.40	0.40	0.40	0.40	0.40	0.40	0.40	0.60	0.60
x_C	0.2008	0.3993	0.6002	0.8013	0.8503	0.9000	0.9500	0.1994	0.4028
y_C	0.5191	0.6269	0.7261	0.8412	0.8776	0.9101	0.9507	0.4782	0.6289
w_B	0.60	0.60	0.60	0.60	0.60	0.60	0.60	0.70	0.70
x_C	0.6005	0.8001	0.8500	0.8999	0.9292	0.9515	0.9724	0.2028	0.4033
y_C	0.7428	0.8593	0.8874	0.9192	0.9575	0.9706	0.9832	0.4824	0.6391
w_B	0.70	0.70	0.70	0.70	0.70	0.70	0.70	0.80	0.80
x_C	0.6050	0.7020	0.8003	0.8498	0.9007	0.9482	0.9701	0.1492	0.3949
y_C	0.7629	0.8272	0.8917	0.9133	0.9447	0.9747	0.9860	0.4038	0.6392
w_B	0.80	0.80	0.80	0.80	0.80	0.80			
x_C	0.5489	0.6995	0.8332	0.9009	0.9511	0.9731			
y_C	0.7504	0.8440	0.9212	0.9560	0.9807	0.9913			

Polymer (B): **polyesteramide (hyperbranched)** **2003SE3**
Characterization: M_n/g.mol^{-1} = 1500, M_w/g.mol^{-1} = 7500, 10 hydroxyl groups, Hybrane H1500, DSM Fine Chemicals, The Netherlands
Solvent (A): **water** **H_2O** **7732-18-5**
Solvent (C): **ethanol** **C_2H_6O** **64-17-5**

Type of data: vapor-liquid equilibrium

T/K = 363.15

w_B	0.40	0.40	0.40	0.40	0.40	0.40	0.40	0.60	0.60
x_C	0.1997	0.4000	0.5999	0.7998	0.8499	0.9000	0.9499	0.1998	0.3999
y_C	0.5026	0.6120	0.7095	0.8323	0.8689	0.9082	0.9487	0.4621	0.5883
w_B	0.60	0.60	0.60	0.60	0.60	0.70	0.70	0.70	0.70
x_C	0.6000	0.8001	0.8500	0.8999	0.9501	0.1992	0.4004	0.6037	0.7999
y_C	0.7045	0.8390	0.8732	0.9120	0.9498	0.4303	0.5874	0.7142	0.8519
w_B	0.70	0.70	0.70						
x_C	0.8496	0.8999	0.9498						
y_C	0.8861	0.9196	0.9550						

Polymer (B): **polyesteramide (hyperbranched)** **2003SE3**
Characterization: M_n/g.mol^{-1} = 1200, M_w/g.mol^{-1} = 6000, 8 hydroxyl groups, Hybrane S1200, DSM Fine Chemicals, The Netherlands
Solvent (A): **water** H_2O **7732-18-5**
Solvent (C): **tetrahydrofuran** C_4H_8O **109-99-9**

Type of data: vapor-liquid equilibrium

T/K = 343.15

w_B	0.40	0.40	0.40	0.40	0.40	0.40	0.40	0.60	0.60
x_C	0.201	0.398	0.700	0.798	0.898	0.949	0.968	0.096	0.204
y_C	0.751	0.787	0.846	0.881	0.924	0.954	0.966	0.647	0.731
w_B	0.60	0.60	0.60	0.60	0.60	0.60	0.60	0.70	0.70
x_C	0.396	0.601	0.700	0.800	0.900	0.950	0.971	0.202	0.395
y_C	0.781	0.839	0.881	0.918	0.955	0.975	0.984	0.714	0.806
w_B	0.70	0.70	0.70	0.70	0.70	0.70	0.70		
x_C	0.597	0.796	0.860	0.903	0.923	0.945	0.979		
y_C	0.870	0.930	0.950	0.965	0.973	0.980	0.991		

Polymer (B): **poly(ethylene glycol)** **2003SAL**
Characterization: M_n/g.mol^{-1} = 1000, Merck, Darmstadt, Germany
Solvent (A): **water** H_2O **7732-18-5**
Salt (C): **ammonium sulfate** $(NH_4)_2SO_4$ **7783-20-2**

Type of data: vapor-liquid equilibrium

T/K = 298.15

w_A	0.8688	0.9077	0.9598	0.9868	0.8190	0.8578	0.9343	0.9767	0.7547
w_B	0.1312	0.0880	0.0298	0.0000	0.1810	0.1364	0.0486	0.0000	0.2453
w_C	0.0000	0.0043	0.0104	0.0132	0.0000	0.0058	0.0171	0.0233	0.0000
a_A	0.9957	0.9957	0.9957	0.9957	0.9925	0.9925	0.9925	0.9925	0.9868
w_A	0.7912	0.8902	0.9589	0.7107	0.7542	0.8613	0.9415	0.6662	0.7222
w_B	0.2026	0.0837	0.0000	0.2893	0.2369	0.1031	0.0000	0.3338	0.2651
w_C	0.0062	0.0261	0.0411	0.0000	0.0089	0.0356	0.0585	0.0000	0.0127
a_A	0.9868	0.9868	0.9868	0.9811	0.9811	0.9811	0.9811	0.9744	0.9744
w_A	0.8413	0.9228	0.6169	0.7097	0.8063	0.8925	0.5893	0.6789	0.7867
w_B	0.1102	0.0000	0.3831	0.2654	0.1322	0.0000	0.4107	0.2953	0.1447
w_C	0.0485	0.0772	0.0000	0.0249	0.0615	0.1075	0.0000	0.0258	0.0686
a_A	0.9744	0.9744	0.9647	0.9647	0.9647	0.9647	0.9581	0.9581	0.9581

Polymer (B): **poly(ethylene glycol)** **2003SAL**
Characterization: M_n/g.mol^{-1} = 6000, Merck, Darmstadt, Germany
Solvent (A): **water** H_2O **7732-18-5**
Salt (C): **ammonium sulfate** $(NH_4)_2SO_4$ **7783-20-2**

Type of data: vapor-liquid equilibrium

continued

continued

T/K = 298.15

w_A	0.8606	0.9095	0.9546	0.9933	0.8026	0.8489	0.9392	0.9863	0.7553
w_B	0.1394	0.0879	0.0407	0.0000	0.1974	0.1477	0.0510	0.0000	0.2447
w_C	0.0000	0.0026	0.0047	0.0067	0.0000	0.0034	0.0098	0.0137	0.0000
a_A	0.9977	0.9977	0.9977	0.9977	0.9955	0.9955	0.9955	0.9955	0.9925
w_A	0.8245	0.9216	0.9763	0.6897	0.7459	0.8935	0.9569	0.6369	0.6861
w_B	0.1682	0.0610	0.0000	0.3103	0.2460	0.0750	0.0000	0.3631	0.3047
w_C	0.0073	0.0174	0.0237	0.0000	0.0081	0.0315	0.0431	0.0000	0.0092
a_A	0.9925	0.9925	0.9925	0.9862	0.9862	0.9862	0.9862	0.9783	0.9783
w_A	0.8564	0.9327	0.5875	0.6393	0.8227	0.9021			
w_B	0.0976	0.0000	0.4125	0.3491	0.1136	0.0000			
w_C	0.0460	0.0673	0.0000	0.0116	0.0637	0.0979			
a_A	0.9783	0.9783	0.9680	0.9680	0.9680	0.9680			

Polymer (B): **poly(ethylene glycol)** **1996LIN**
Characterization: M_n/g.mol^{-1} = 9000-10000,
Shanghai Chemical Reagent Factory, P.R. China
Solvent (A): **water** H_2O **7732-18-5**
Salt (C): **ammonium sulfate** $(NH_4)_2SO_4$ **7783-20-2**

Type of data: vapor-liquid equilibrium

T/K = 298.15

w_A	0.8670	0.8962	0.9282	0.9666	0.9951	0.8209	0.8603	0.8909	0.9471
w_B	0.1330	0.1030	0.0697	0.0298	0.0000	0.1791	0.1386	0.1063	0.0474
w_C	0.0000	0.00083	0.0021	0.0036	0.0049	0.0000	0.0011	0.0028	0.0055
a_A	0.9984	0.9984	0.9984	0.9984	0.9984	0.9972	0.9972	0.9972	0.9972
w_A	0.9913	0.7086	0.7383	0.7729	0.8535	0.9638			
w_B	0.0000	0.2914	0.2596	0.2206	0.1295	0.0000			
w_C	0.0087	0.0000	0.0021	0.0065	0.0170	0.0362			
a_A	0.9972	0.9889	0.9889	0.9889	0.9889	0.9889			

Polymer (B): **poly(ethylene glycol)** **1996LIN**
Characterization: M_n/g.mol^{-1} = 990-1100,
Shanghai Chemical Reagent Factory, P.R. China
Solvent (A): **water** H_2O **7732-18-5**
Salt (C): **dipotassium phosphate** K_2HPO_4 **7758-11-4**

Type of data: vapor-liquid equilibrium

T/K = 298.15

w_A	0.8883	0.9176	0.9316	0.9511	0.9679	0.9843	0.8406	0.8790	0.9023
w_B	0.1117	0.0791	0.0623	0.0397	0.0196	0.0000	0.1594	0.1162	0.0890
w_C	0.0000	0.0033	0.0061	0.0092	0.0125	0.0157	0.0000	0.0048	0.0087
a_A	0.9961	0.9961	0.9961	0.9961	0.9961	0.9961	0.9938	0.9938	0.9938

continued

continued

w_A	0.9217	0.9506	0.9743	0.8150	0.8596	0.8810	0.9076	0.9361	0.9666
w_B	0.0636	0.0301	0.0000	0.1850	0.1349	0.1087	0.0754	0.0393	0.0000
w_C	0.0147	0.0193	0.0257	0.0000	0.0055	0.0103	0.0170	0.0246	0.0334
a_A	0.9938	0.9938	0.9938	0.9920	0.9920	0.9920	0.9920	0.9920	0.9920
w_A	0.7629	0.8034	0.8312	0.8654	0.9028	0.9442	0.6326	0.6707	0.7035
w_B	0.2371	0.1889	0.1542	0.1099	0.0598	0.0000	0.3674	0.3164	0.2708
w_C	0.0000	0.0077	0.0146	0.0247	0.0374	0.0558	0.0000	0.0129	0.0257
a_A	0.9868	0.9868	0.9868	0.9868	0.9868	0.9868	0.9655	0.9655	0.9655
w_A	0.7453	0.7968	0.8553						
w_B	0.2079	0.1250	0.0000						
w_C	0.0468	0.0782	0.1447						
a_A	0.9655	0.9655	0.9655						

Polymer (B): **poly(ethylene glycol)** **1996LIN**
Characterization: M_n/g.mol^{-1} = 3500-4500,
Shanghai Chemical Reagent Factory, P.R. China
Solvent (A): **water** **H_2O** **7732-18-5**
Salt (C): **dipotassium phosphate** **K_2HPO_4** **7758-11-4**

Type of data: vapor-liquid equilibrium

T/K = 298.15

w_A	0.8755	0.9001	0.9311	0.9509	0.9740	0.9904	0.8287	0.8655	0.9039
w_B	0.1245	0.0979	0.0649	0.0432	0.0178	0.0000	0.1713	0.1316	0.0906
w_C	0.0000	0.0020	0.0040	0.0059	0.0082	0.0096	0.0000	0.0029	0.0055
a_A	0.9974	0.9974	0.9974	0.9974	0.9974	0.9974	0.9962	0.9962	0.9962
w_A	0.9289	0.9621	0.9844	0.7782	0.8140	0.8561	0.8889	0.9333	0.9703
w_B	0.0625	0.0252	0.0000	0.2218	0.1823	0.1356	0.0977	0.0456	0.0000
w_C	0.0086	0.0127	0.0156	0.0000	0.0037	0.0083	0.0134	0.0211	0.0297
a_A	0.9962	0.9962	0.9962	0.9930	0.9930	0.9930	0.9930	0.9930	0.9930
w_A	0.6954	0.7341	0.7730	0.8130	0.8827	0.9370			
w_B	0.3046	0.2605	0.2139	0.1653	0.0797	0.0000			
w_C	0.0000	0.0054	0.0131	0.0217	0.0376	0.0630			
a_A	0.9848	0.9848	0.9848	0.9848	0.9848	0.9848			

Polymer (B): **poly(ethylene glycol)** **2003SE2**
Characterization: M_n/g.mol^{-1} = 365, M_w/g.mol^{-1} = 400,
Polysciences, Inc., Warrington, PA
Solvent (A): **water** **H_2O** **7732-18-5**
Solvent (C): **ethanol** **C_2H_6O** **64-17-5**

Type of data: vapor-liquid equilibrium

continued

continued

T/K = 363.15

w_B	0.20	0.20	0.20	0.20	0.20	0.20	0.20	0.40	0.40
x_C	0.202	0.400	0.600	0.798	0.850	0.895	0.950	0.202	0.602
y_C	0.535	0.628	0.719	0.835	0.872	0.905	0.948	0.531	0.748
w_B	0.40	0.40	0.40	0.40	0.70	0.70	0.70	0.70	0.70
x_C	0.801	0.851	0.900	0.950	0.400	0.699	0.800	0.851	0.902
y_C	0.860	0.893	0.924	0.958	0.648	0.794	0.862	0.897	0.929
w_B	0.70								
x_C	0.950								
y_C	0.959								

Polymer (B): **poly(ethylene glycol)** **2003SE3**
Characterization: M_n/g.mol^{-1} = 365, M_w/g.mol^{-1} = 400,
Polysciences, Inc., Warrington, PA
Solvent (A): **water** **H_2O** **7732-18-5**
Solvent (C): **ethanol** **C_2H_6O** **64-17-5**

Type of data: vapor-liquid equilibrium

T/K = 363.15

w_B	0.70	0.70	0.70	0.70	0.70	0.70
x_C	0.4001	0.6989	0.8002	0.8511	0.9018	0.9495
y_C	0.6480	0.7936	0.8622	0.8967	0.9285	0.9587

Polymer (B): **poly(ethylene glycol)** **2003SE2**
Characterization: M_n/g.mol^{-1} = 910, M_w/g.mol^{-1} = 1000,
Polysciences, Inc., Warrington, PA
Solvent (A): **water** **H_2O** **7732-18-5**
Solvent (C): **ethanol** **C_2H_6O** **64-17-5**

Type of data: vapor-liquid equilibrium

T/K = 363.15

w_B	0.40	0.40	0.40	0.40	0.40
x_C	0.200	0.400	0.599	0.900	0.952
y_C	0.529	0.624	0.717	0.924	0.953

Polymer (B): **poly(ethylene glycol)** **2003SE2**
Characterization: M_n/g.mol^{-1} = 2830, M_w/g.mol^{-1} = 3400,
Polysciences, Inc., Warrington, PA
Solvent (A): **water** **H_2O** **7732-18-5**
Solvent (C): **ethanol** **C_2H_6O** **64-17-5**

Type of data: vapor-liquid equilibrium

continued

continued

T/K = 363.15

w_B	0.40	0.40	0.40	0.40	0.40	0.40
x_C	0.189	0.400	0.600	0.800	0.900	0.950
y_C	0.510	0.628	0.720	0.842	0.913	0.951

Polymer (B): **poly(ethylene glycol)** **1996LIN**
Characterization: M_n/g.mol^{-1} = 9000-10000,
Shanghai Chemical Reagent Factory, P.R. China
Solvent (A): **water** H_2O **7732-18-5**
Salt (C): **magnesium sulfate** $MgSO_4$ **7487-88-9**

Type of data: vapor-liquid equilibrium

T/K = 298.15

w_A	0.8811	0.9026	0.9292	0.9546	0.9942	0.8400	0.8597	0.8868	0.9214
w_B	0.1189	0.0967	0.0692	0.0423	0.0000	0.1600	0.1393	0.1104	0.0731
w_C	0.0000	0.0007	0.0016	0.0031	0.0058	0.0000	0.0010	0.0028	0.0055
a_A	0.9987	0.9987	0.9987	0.9987	0.9987	0.9978	0.9978	0.9978	0.9978

Polymer (B): **poly(ethylene glycol)** **1996LIN**
Characterization: M_n/g.mol^{-1} = 990-1100,
Shanghai Chemical Reagent Factory, P.R. China
Solvent (A): **water** H_2O **7732-18-5**
Salt (C): **monopotassium phosphate** KH_2PO_4 **7778-77-0**

Type of data: vapor-liquid equilibrium

T/K = 298.15

w_A	0.8790	0.8969	0.9210	0.9405	0.9631	0.9828	0.8277	0.8501	0.8788
w_B	0.1210	0.1001	0.0723	0.0494	0.0233	0.0000	0.1723	0.1455	0.1110
w_C	0.0000	0.0030	0.0067	0.0101	0.0136	0.0172	0.0000	0.0044	0.0102
a_A	0.9960	0.9960	0.9960	0.9960	0.9960	0.9960	0.9928	0.9928	0.9928
w_A	0.9061	0.9380	0.9697	0.7238	0.7485	0.7824	0.8181	0.8661	0.9234
w_B	0.0780	0.0392	0.0000	0.2762	0.2442	0.1992	0.1511	0.0847	0.0000
w_C	0.0159	0.0228	0.0303	0.0000	0.0073	0.0184	0.0308	0.0492	0.0766
a_A	0.9928	0.9928	0.9928	0.9827	0.9827	0.9827	0.9827	0.9827	0.9827
w_A	0.6833	0.7071	0.7416	0.7784	0.8297	0.8941			
w_B	0.3167	0.2844	0.2366	0.1841	0.1077	0.0000			
w_C	0.0000	0.0085	0.0218	0.0375	0.0626	0.1059			
a_A	0.9766	0.9766	0.9766	0.9766	0.9766	0.9766			

Polymer (B): **poly(ethylene glycol)** **1996LIN**
Characterization: M_n/g.mol^{-1} = 3500-4500,
Shanghai Chemical Reagent Factory, P.R. China
Solvent (A): **water** H_2O **7732-18-5**
Salt (C): **monopotassium phosphate** KH_2PO_4 **7778-77-0**

Type of data: vapor-liquid equilibrium

T/K = 298.15

w_A	0.8703	0.8930	0.9205	0.9406	0.9681	0.9893	0.8087	0.8436	0.8704
w_B	0.1297	0.1045	0.0745	0.0526	0.0230	0.0000	0.1913	0.1528	0.1233
w_C	0.0000	0.0025	0.0050	0.0068	0.0089	0.0107	0.0000	0.0036	0.0063
a_A	0.9972	0.9972	0.9972	0.9972	0.9972	0.9972	0.9950	0.9950	0.9950
w_A	0.9073	0.9450	0.9796	0.7384	0.7680	0.8003	0.8323	0.8882	0.9535
w_B	0.0821	0.0397	0.0000	0.2616	0.2266	0.1871	0.1485	0.0807	0.0000
w_C	0.0106	0.0153	0.0204	0.0000	0.0054	0.0126	0.0192	0.0311	0.0465
a_A	0.9950	0.9950	0.9950	0.9896	0.9896	0.9896	0.9896	0.9896	0.9896
w_A	0.6788	0.7044	0.7357	0.7692	0.8323	0.9188			
w_B	0.3212	0.2888	0.2477	0.2043	0.1211	0.0000			
w_C	0.0000	0.0068	0.0166	0.0265	0.0466	0.0812			
a_A	0.9820	0.9820	0.9820	0.9820	0.9820	0.9820			

Polymer (B): **poly(ethylene glycol)** **2007KAZ**
Characterization: M_n/g.mol^{-1} = 6000, Merck, Darmstadt, Germany
Solvent (A): **water** H_2O **7732-18-5**
Salt (C): **potassium citrate** $C_6H_5K_3O_7$ **866-84-2**

Type of data: vapor-liquid equilibrium

T/K = 298.15

w_A	0.7637	0.8666	0.9382	0.9571	0.7419	0.8237	0.9148	0.9453	0.7159
w_B	0.2363	0.1106	0.0226	0.0000	0.2581	0.1550	0.0397	0.0000	0.2841
w_C	0.0000	0.0228	0.0392	0.0429	0.0000	0.0213	0.0455	0.0547	0.0000
a_A	0.9915	0.9915	0.9915	0.9915	0.9901	0.9901	0.9901	0.9901	0.9874
w_A	0.8256	0.8801	0.9282	0.6954	0.7755	0.8560	0.9174	0.6911	0.7811
w_B	0.1411	0.0678	0.0000	0.3046	0.1982	0.0883	0.0000	0.3089	0.1886
w_C	0.0333	0.0521	0.0718	0.0000	0.0263	0.0557	0.0826	0.0000	0.0303
a_A	0.9874	0.9874	0.9874	0.9852	0.9852	0.9852	0.9852	0.9844	0.9844
w_A	0.8447	0.9131	0.6682	0.7218	0.8468	0.8946	0.6620	0.7455	0.8286
w_B	0.0986	0.0000	0.3318	0.2581	0.0759	0.0000	0.3380	0.2211	0.0989
w_C	0.0567	0.0869	0.0000	0.0201	0.0773	0.1054	0.0000	0.0334	0.0725
a_A	0.9844	0.9844	0.9809	0.9809	0.9809	0.9809	0.9802	0.9802	0.9802
w_A	0.8910	0.6405	0.7256	0.8058	0.8725	0.6279	0.7127	0.7924	0.8609
w_B	0.0000	0.3595	0.2365	0.1126	0.0000	0.3721	0.2467	0.1203	0.0000
w_C	0.1090	0.0000	0.0379	0.0816	0.1275	0.0000	0.0406	0.0873	0.1391
a_A	0.9802	0.9762	0.9762	0.9762	0.9762	0.9736	0.9736	0.9736	0.9736

continued

continued

w_A	0.6113	0.6834	0.7855	0.8443	0.6092	0.6911	0.7707	0.8429	0.6031
w_B	0.3887	0.2787	0.1106	0.0000	0.3908	0.2656	0.1337	0.0000	0.3969
w_C	0.0000	0.0379	0.1039	0.1557	0.0000	0.0433	0.0956	0.1571	0.0000
a_A	0.9696	0.9696	0.9696	0.9696	0.9692	0.9692	0.9692	0.9692	0.9679
w_A	0.6755	0.7763	0.8369	0.5987	0.6724	0.7757	0.8333		
w_B	0.2857	0.1164	0.0000	0.4013	0.2887	0.1129	0.0000		
w_C	0.0388	0.1073	0.1631	0.0000	0.0389	0.1114	0.1667		
a_A	0.9679	0.9679	0.9679	0.9672	0.9672	0.9672	0.9672		

Polymer (B): **poly(ethylene glycol)** **1996TSU**
Characterization: M_w/g.mol^{-1} = 200, M_w/M_n = 1.085,
Wako Pure Chemical Industries Co. Ltd., Japan
Solvent (A): **water** **H_2O** **7732-18-5**
Solvent (C): **2-propanol** **C_3H_8O** **67-63-0**

Type of data: vapor-liquid equilibrium

Comments: The tables give the mass fraction of the polymer B in the liquid solution, w_B, the partial pressures of 2-propanol, P_C, and of water, P_A, and the mole fractions of 2-propanol in the liquid phase, x_C. To calculate the mole fraction in the liquid phase, a molar mass of 194.23 for PEG200 is used.

T/K = 298.15

w_B	0.00	0.00	0.00	0.00	0.00	0.00	0.00	0.00	0.00
x_C	0.0695	0.1200	0.2009	0.2995	0.3999	0.5003	0.6018	0.6985	0.7976
P_A/kPa	2.942	2.879	2.844	2.771	2.681	2.555	2.343	2.104	1.665
P_C/kPa	2.339	3.008	3.327	3.550	3.673	3.867	4.132	4.418	4.759
w_B	0.00	0.05	0.05	0.05	0.05	0.05	0.05	0.05	0.05
x_C	0.9002	0.0000	0.1125	0.2035	0.2951	0.3970	0.4993	0.6454	0.6895
P_A/kPa	0.964	3.140	2.786	2.762	2.680	2.620	2.480	2.163	1.990
P_C/kPa	5.214	0.000	2.858	3.236	3.418	3.646	3.856	4.245	4.367
w_B	0.05	0.05	0.05	0.20	0.20	0.20	0.20	0.20	0.20
x_C	0.7951	0.8863	0.9840	0.0000	0.0980	0.1934	0.2883	0.3826	0.4794
P_A/kPa	1.463	0.867	0.000	3.078	2.792	2.680	2.592	2.463	2.291
P_C/kPa	4.651	5.083	5.719	0.000	2.784	3.318	3.554	3.749	4.002
w_B	0.20	0.20	0.20	0.20	0.20	0.35	0.35	0.35	0.35
x_C	0.5695	0.6522	0.7530	0.8255	0.9282	0.0000	0.0938	0.1829	0.2711
P_A/kPa	2.019	1.678	1.194	0.736	0.000	2.974	2.706	2.615	2.438
P_C/kPa	4.247	4.439	4.764	5.018	5.493	0.000	2.589	3.197	3.497
w_B	0.35	0.35	0.35	0.35	0.35	0.35	0.35		
x_C	0.3696	0.4505	0.4709	0.5424	0.6529	0.7414	0.8572		
P_A/kPa	2.217	1.992	1.931	1.671	1.165	0.687	0.000		
P_C/kPa	3.759	3.985	4.040	4.211	4.520	4.816	5.184		

Polymer (B):	**poly(ethylene glycol)**		**1996TSU**
Characterization:	M_w/g.mol^{-1} = 1000, M_w/M_n = 1.176, Wako Pure Chemical Industries Co. Ltd., Japan		
Solvent (A):	**water**	**H_2O**	**7732-18-5**
Solvent (C):	**2-propanol**	**C_3H_8O**	**67-63-0**

Type of data: vapor-liquid equilibrium

Comments: The tables give the mass fraction of the polymer B in the liquid solution, w_B, the partial pressures of 2-propanol, P_C, and of water, P_A, and the mole fractions of 2-propanol in the liquid phase, x_C. To calculate the mole fraction in the liquid phase, a molar mass of 987.23 for PEG1000 is used.

T/K = 298.15

w_B	0.00	0.00	0.00	0.00	0.00	0.00	0.00	0.00	0.00
x_C	0.0695	0.1200	0.2009	0.2995	0.3999	0.5003	0.6018	0.6985	0.7976
P_A/kPa	2.942	2.879	2.844	2.771	2.681	2.555	2.343	2.104	1.665
P_C/kPa	2.339	3.008	3.327	3.550	3.673	3.867	4.132	4.418	4.759

w_B	0.00	0.05	0.05	0.05	0.05	0.05	0.05	0.05	0.05
x_C	0.9002	0.0000	0.1073	0.2061	0.3024	0.4000	0.5018	0.6001	0.7032
P_A/kPa	0.964	3.155	2.919	2.837	2.821	2.710	2.540	2.374	2.053
P_C/kPa	5.214	0.000	3.097	3.401	3.612	3.829	4.057	4.219	4.558

w_B	0.05	0.05	0.05	0.20	0.20	0.20	0.20	0.20	0.20
x_C	0.7501	0.7975	0.8923	0.0000	0.2018	0.2975	0.3919	0.5885	0.5007
P_A/kPa	1.831	1.586	0.932	3.140	2.794	2.685	2.549	2.110	2.346
P_C/kPa	4.772	4.885	5.183	0.000	3.408	3.656	3.886	4.368	4.140

w_B	0.20	0.20	0.20
x_C	0.6851	0.7857	0.8855
P_A/kPa	1.762	1.309	0.703
P_C/kPa	4.638	4.969	5.350

Polymer (B):	**poly(ethylene glycol)**		**1996TSU**
Characterization:	M_w/g.mol^{-1} = 20000, M_w/M_n = 1.108, Wako Pure Chemical Industries Co. Ltd., Japan		
Solvent (A):	**water**	**H_2O**	**7732-18-5**
Solvent (C):	**2-propanol**	**C_3H_8O**	**67-63-0**

Type of data: vapor-liquid equilibrium

Comments: The tables give the mass fraction of the polymer B in the liquid solution, w_B, the partial pressures of 2-propanol, P_C, and of water, P_A, and the mole fractions of 2-propanol in the liquid phase, x_C. To calculate the mole fraction in the liquid phase, a molar mass of 20018.15 for PEG20000 is used.

continued

continued

T/K = 298.15

w_B	0.00	0.00	0.00	0.00	0.00	0.00	0.00	0.00	0.00
x_C	0.0695	0.1200	0.2009	0.2995	0.3999	0.5003	0.6018	0.6985	0.7976
P_A/kPa	2.942	2.879	2.844	2.771	2.681	2.555	2.343	2.104	1.665
P_C/kPa	2.339	3.008	3.327	3.550	3.673	3.867	4.132	4.418	4.759
w_B	0.00	0.05	0.05	0.05	0.05	0.05	0.05	0.05	0.05
x_C	0.9002	0.0836	0.1239	0.1935	0.2471	0.2915	0.3563	0.4402	0.5199
P_A/kPa	0.964	2.903	2.897	2.859	2.828	2.783	2.749	2.643	2.549
P_C/kPa	5.214	2.767	3.160	3.469	3.598	3.664	3.780	3.915	4.118
w_B	0.05	0.05	0.05	0.05	0.05	0.20	0.20	0.20	0.20
x_C	0.5942	0.6291	0.6913	0.7426	0.7779	0.0842	0.1755	0.2998	0.4080
P_A/kPa	2.390	2.277	2.097	1.877	1.708	2.923	2.812	2.700	2.550
P_C/kPa	4.309	4.348	4.517	4.666	4.738	2.687	3.368	3.732	3.983
w_B	0.20	0.20	0.20						
x_C	0.4887	0.6030	0.6914						
P_A/kPa	2.389	2.105	1.799						
P_C/kPa	4.130	4.431	4.666						

Polymer (B): **poly(ethylene glycol)** **2006SA1**
Characterization: M_n/g.mol^{-1} = 400, Merck, Darmstadt, Germany
Solvent (A): **water** H_2O **7732-18-5**
Salt (C): **sodium chloride** **NaCl** **7647-14-5**

Type of data: vapor-liquid equilibrium

T/K = 293.15

w_A	0.8467	0.8977	0.9385	0.9814	0.8427	0.8845	0.9283	0.9793	0.7780
w_B	0.1533	0.0950	0.0483	0.0000	0.1573	0.1077	0.0584	0.0000	0.2220
w_C	0.0000	0.0073	0.0132	0.0186	0.0000	0.0078	0.0133	0.0207	0.0000
a_A	0.9893	0.9893	0.9893	0.9893	0.9880	0.9880	0.9880	0.9880	0.9812
w_A	0.8460	0.9031	0.9679	0.7228	0.8113	0.8790	0.9563	0.6577	0.7861
w_B	0.1425	0.0767	0.0000	0.2772	0.1719	0.0929	0.0000	0.3423	0.1823
w_C	0.0115	0.0202	0.0321	0.0000	0.0168	0.0281	0.0437	0.0000	0.0316
a_A	0.9812	0.9812	0.9812	0.9740	0.9740	0.9740	0.9740	0.9585	0.9585
w_A	0.8563	0.9327	0.6535	0.7286	0.8218	0.9323	0.5651	0.6857	0.7715
w_B	0.0959	0.0000	0.3465	0.2520	0.1365	0.0000	0.4349	0.2768	0.1685
w_C	0.0478	0.0673	0.0000	0.0194	0.0417	0.0677	0.0000	0.0375	0.0600
a_A	0.9585	0.9585	0.9582	0.9582	0.9582	0.9582	0.9340	0.9340	0.9340
w_A	0.8993	0.4638	0.5836	0.6740	0.8415	0.4291	0.5486	0.6590	0.8273
w_B	0.0000	0.5362	0.3611	0.2341	0.0000	0.5709	0.3948	0.2374	0.0000
w_C	0.1007	0.0000	0.0553	0.0919	0.1585	0.0000	0.0566	0.1036	0.1727
a_A	0.9340	0.8830	0.8830	0.8830	0.8830	0.8685	0.8685	0.8685	0.8685

continued

continued

w_A	0.3848	0.5090	0.6246	0.8025	0.3414	0.4581	0.5568	0.7736	
w_B	0.6152	0.4244	0.2566	0.0000	0.6586	0.4706	0.3234	0.0000	
w_C	0.0000	0.0666	0.1188	0.1975	0.0000	0.0713	0.1198	0.2264	
a_A	0.8411	0.8411	0.8411	0.8411	0.8055	0.8055	0.8055	0.8055	

T/K = 298.15

w_A	0.8298	0.8742	0.9269	0.9799	0.8203	0.8726	0.9122	0.9780	0.8060
w_B	0.1702	0.1185	0.0591	0.0000	0.1797	0.1196	0.0750	0.0000	0.1940
w_C	0.0000	0.0073	0.0140	0.0201	0.0000	0.0078	0.0128	0.0220	0.0000
a_A	0.9884	0.9884	0.9884	0.9884	0.9873	0.9873	0.9873	0.9873	0.9860
w_A	0.8512	0.9040	0.9758	0.7572	0.8148	0.8868	0.9646	0.7311	0.8036
w_B	0.1400	0.0808	0.0000	0.2428	0.1748	0.0905	0.0000	0.2689	0.1818
w_C	0.0088	0.0152	0.0242	0.0000	0.0104	0.0227	0.0354	0.0000	0.0146
a_A	0.9860	0.9860	0.9860	0.9792	0.9792	0.9792	0.9792	0.9753	0.9753
w_A	0.8798	0.9582	0.6343	0.7487	0.8345	0.9265	0.5510	0.6993	0.7832
w_B	0.0917	0.0000	0.3657	0.2216	0.1149	0.0000	0.4490	0.2526	0.1435
w_C	0.0285	0.0418	0.0000	0.0297	0.0506	0.0735	0.0000	0.0481	0.0733
a_A	0.9753	0.9753	0.9543	0.9543	0.9543	0.9543	0.9299	0.9299	0.9299
w_A	0.8939	0.4912	0.6081	0.7191	0.8627	0.4399	0.5651	0.6744	0.8367
w_B	0.0000	0.5088	0.3455	0.1950	0.0000	0.5601	0.3803	0.2266	0.0000
w_C	0.1061	0.0000	0.0464	0.0859	0.1373	0.0000	0.0546	0.0990	0.1633
a_A	0.9299	0.9033	0.9033	0.9033	0.9033	0.8784	0.8784	0.8784	0.8784
w_A	0.3940	0.5326	0.6390	0.8124	0.2827	0.4067	0.5017	0.7417	
w_B	0.6060	0.4040	0.2467	0.0000	0.7173	0.5124	0.3668	0.0000	
w_C	0.0000	0.0634	0.1143	0.1876	0.0000	0.0809	0.1315	0.2583	
a_A	0.8526	0.8526	0.8526	0.8526	0.7617	0.7617	0.7617	0.7617	

T/K = 303.15

w_A	0.8266	0.8850	0.9278	0.9796	0.7864	0.8655	0.9177	0.9718	0.7248
w_B	0.1734	0.1072	0.0588	0.0000	0.2136	0.1223	0.0626	0.0000	0.2752
w_C	0.0000	0.0078	0.0134	0.0204	0.0000	0.0122	0.0197	0.0282	0.0000
a_A	0.9882	0.9882	0.9882	0.9882	0.9836	0.9836	0.9836	0.9836	0.9749
w_A	0.7865	0.8722	0.9576	0.6882	0.7679	0.8564	0.9488	0.6633	0.7408
w_B	0.1998	0.0982	0.0000	0.3118	0.2155	0.1100	0.0000	0.3367	0.2417
w_C	0.0137	0.0296	0.0424	0.0000	0.0166	0.0336	0.0512	0.0000	0.0175
a_A	0.9749	0.9749	0.9749	0.9693	0.9693	0.9693	0.9693	0.9640	0.9640
w_A	0.8203	0.9406	0.6423	0.6919	0.7515	0.9318	0.6016	0.7388	0.8009
w_B	0.1459	0.0000	0.3577	0.2989	0.2243	0.0000	0.3984	0.2235	0.1441
w_C	0.0338	0.0594	0.0000	0.0092	0.0242	0.0682	0.0000	0.0377	0.0550
a_A	0.9640	0.9640	0.9580	0.9580	0.9580	0.9580	0.9450	0.9450	0.9450

continued

continued

w_A	0.9135	0.5739	0.7301	0.8336	0.9047	0.5541	0.7122	0.7985	0.8980
w_B	0.0000	0.4261	0.2245	0.0915	0.0000	0.4459	0.2412	0.1283	0.0000
w_C	0.0865	0.0000	0.0454	0.0749	0.0953	0.0000	0.0466	0.0732	0.1020
a_A	0.9450	0.9384	0.9384	0.9384	0.9384	0.9332	0.9332	0.9332	0.9332
w_A	0.5348	0.6662	0.7682	0.8857	0.5298	0.7324	0.7872	0.8853	0.4980
w_B	0.4652	0.2889	0.1552	0.0000	0.4702	0.2025	0.1282	0.0000	0.5020
w_C	0.0000	0.0449	0.0766	0.1143	0.0000	0.0651	0.0846	0.1147	0.0000
a_A	0.9233	0.9233	0.9233	0.9233	0.9230	0.9230	0.9230	0.9230	0.9118
w_A	0.6498	0.7246	0.8721	0.4731	0.6196	0.7201	0.8601	0.3824	0.5097
w_B	0.2960	0.1978	0.0000	0.5269	0.3249	0.1895	0.0000	0.6176	0.4323
w_C	0.0542	0.0776	0.1279	0.0000	0.0555	0.0904	0.1399	0.0000	0.0580
a_A	0.9118	0.9118	0.9118	0.9011	0.9011	0.9011	0.9011	0.8518	0.8518
w_A	0.6424	0.8114	0.3537	0.4818	0.6179	0.7962	0.3080	0.4295	0.5523
w_B	0.2437	0.0000	0.6463	0.4569	0.2604	0.0000	0.6920	0.5018	0.3193
w_C	0.1139	0.1886	0.0000	0.0613	0.1217	0.2038	0.0000	0.0687	0.1284
a_A	0.8518	0.8518	0.8342	0.8342	0.8342	0.8342	0.7945	0.7945	0.7945
w_A	0.7648	0.2929	0.3887	0.5027	0.7547	0.2796	0.3646	0.4681	0.7454
w_B	0.0000	0.7071	0.5526	0.3781	0.0000	0.7204	0.5823	0.4208	0.0000
w_C	0.2352	0.0000	0.0587	0.1192	0.2453	0.0000	0.0531	0.1111	0.2546
a_A	0.7945	0.7806	0.7806	0.7806	0.7806	0.7674	0.7674	0.7674	0.7674
T/K = 308.15									
w_A	0.8431	0.8843	0.9384	0.9824	0.7831	0.8299	0.9081	0.9721	0.7750
w_B	0.1569	0.1106	0.0499	0.0000	0.2169	0.1636	0.0737	0.0000	0.2250
w_C	0.0000	0.0051	0.0117	0.0176	0.0000	0.0065	0.0182	0.0279	0.0000
a_A	0.9899	0.9899	0.9899	0.9899	0.9838	0.9838	0.9838	0.9838	0.9823
w_A	0.8455	0.8922	0.9695	0.6563	0.7514	0.8476	0.9409	0.6373	0.7478
w_B	0.1432	0.0889	0.0000	0.3437	0.2287	0.1119	0.0000	0.3627	0.2270
w_C	0.0113	0.0189	0.0305	0.0000	0.0199	0.0405	0.0591	0.0000	0.0252
a_A	0.9823	0.9823	0.9823	0.9642	0.9642	0.9642	0.9642	0.9568	0.9568
w_A	0.8389	0.9299	0.6203	0.7170	0.8382	0.9235	0.5676	0.6701	0.7871
w_B	0.1135	0.0000	0.3797	0.2589	0.1086	0.0000	0.4324	0.3024	0.1526
w_C	0.0476	0.0701	0.0000	0.0241	0.0532	0.0765	0.0000	0.0275	0.0603
a_A	0.9568	0.9568	0.9523	0.9523	0.9523	0.9523	0.9399	0.9399	0.9399
w_A	0.9065	0.5220	0.6537	0.7537	0.8870	0.4674	0.5877	0.6765	0.8595
w_B	0.0000	0.4780	0.3044	0.1743	0.0000	0.5326	0.3665	0.2488	0.0000
w_C	0.0935	0.0000	0.0419	0.0720	0.1130	0.0000	0.0458	0.0747	0.1405
a_A	0.9399	0.9245	0.9245	0.9245	0.9245	0.9007	0.9007	0.9007	0.9007
w_A	0.4066	0.5455	0.6349	0.8263	0.3407	0.4732	0.5978	0.7905	0.3159
w_B	0.5934	0.3930	0.2647	0.0000	0.6593	0.4593	0.2790	0.0000	0.6841
w_C	0.0000	0.0615	0.1004	0.1737	0.0000	0.0675	0.1232	0.2095	0.0000
a_A	0.8682	0.8682	0.8682	0.8682	0.8277	0.8277	0.8277	0.8277	0.8079

continued

continued

w_A	0.4181	0.5305	0.7747	0.3038	0.4412	0.5695	0.7701
w_B	0.5268	0.3612	0.0000	0.6962	0.4830	0.2942	0.0000
w_C	0.0551	0.1083	0.2253	0.0000	0.0758	0.1363	0.2299
a_A	0.8079	0.8079	0.8079	0.8019	0.8019	0.8019	0.8019

T/K = 313.15

w_A	0.7945	0.8425	0.9209	0.9745	0.7849	0.8376	0.9139	0.9727	0.7690
w_B	0.2055	0.1514	0.0619	0.0000	0.2151	0.1540	0.0673	0.0000	0.2310
w_C	0.0000	0.0061	0.0172	0.0255	0.0000	0.0084	0.0188	0.0273	0.0000
a_A	0.9853	0.9853	0.9853	0.9853	0.9842	0.9842	0.9842	0.9842	0.9824
w_A	0.8291	0.8935	0.9696	0.7207	0.8202	0.8825	0.9597	0.7115	0.7827
w_B	0.1605	0.0875	0.0000	0.2793	0.1638	0.0902	0.0000	0.2885	0.2058
w_C	0.0104	0.0190	0.0304	0.0000	0.0160	0.0273	0.0403	0.0000	0.0115
a_A	0.9824	0.9824	0.9824	0.9763	0.9763	0.9763	0.9763	0.9754	0.9754
w_A	0.8841	0.9582	0.6252	0.7601	0.8408	0.9330	0.5482	0.6610	0.7819
w_B	0.0864	0.0000	0.3748	0.2110	0.1135	0.0000	0.4518	0.3081	0.1544
w_C	0.0295	0.0418	0.0000	0.0289	0.0457	0.0670	0.0000	0.0309	0.0637
a_A	0.9754	0.9754	0.9590	0.9590	0.9590	0.9590	0.9376	0.9376	0.9376
w_A	0.9034	0.4472	0.5685	0.7203	0.8548	0.4017	0.5310	0.6426	0.8301
w_B	0.0000	0.5528	0.3865	0.1829	0.0000	0.5983	0.4152	0.2608	0.0000
w_C	0.0966	0.0000	0.0450	0.0968	0.1452	0.0000	0.0538	0.0966	0.1699
a_A	0.9376	0.8966	0.8966	0.8966	0.8966	0.8723	0.8723	0.8723	0.8723
w_A	0.3080	0.4654	0.5744	0.7765	0.2637	0.3648	0.4887	0.7420	0.2604
w_B	0.6920	0.4556	0.2959	0.0000	0.7363	0.5806	0.3919	0.0000	0.7396
w_C	0.0000	0.0790	0.1297	0.2235	0.0000	0.0546	0.1194	0.2580	0.0000
a_A	0.8105	0.8105	0.8105	0.8105	0.7632	0.7632	0.7632	0.7632	0.7654
w_A	0.3820	0.4967	0.7435						
w_B	0.5483	0.3747	0.0000						
w_C	0.0697	0.1286	0.2565						
a_A	0.7654	0.7654	0.7654						

Polymer (B):	**poly(ethylene glycol)**		**2007KAZ**
Characterization:	M_n/g.mol^{-1} = 6000, Merck, Darmstadt, Germany		
Solvent (A):	**water**	**H_2O**	**7732-18-5**
Salt (C):	**sodium citrate**	**$C_6H_5Na_3O_7$**	**68-04-2**

Type of data: vapor-liquid equilibrium

T/K = 298.15

w_A	0.7255	0.8136	0.8873	0.9438	0.7111	0.8322	0.8792	0.9356	0.6968
w_B	0.2745	0.1664	0.0736	0.0000	0.2889	0.1373	0.0758	0.0000	0.3032
w_C	0.0000	0.0200	0.0391	0.0562	0.0000	0.0305	0.0450	0.0644	0.0000
a_A	0.9883	0.9883	0.9883	0.9883	0.9871	0.9871	0.9871	0.9871	0.9853

continued

continued

w_A	0.8043	0.8791	0.9266	0.6996	0.9105	0.9183	0.9223	0.6788	0.8862
w_B	0.1666	0.0661	0.0000	0.3004	0.0165	0.0052	0.0000	0.3212	0.0368
w_C	0.0291	0.0548	0.0734	0.0000	0.0730	0.0765	0.0777	0.0000	0.0770
a_A	0.9853	0.9853	0.9853	0.9840	0.9840	0.9840	0.9840	0.9818	0.9818
w_A	0.8986	0.9094	0.6647	0.7552	0.8590	0.9014	0.6594	0.7682	0.8324
w_B	0.0178	0.0000	0.3353	0.2162	0.0665	0.0000	0.3406	0.1962	0.1032
w_C	0.0836	0.0906	0.0000	0.0286	0.0745	0.0986	0.0000	0.0356	0.0644
a_A	0.9818	0.9818	0.9800	0.9800	0.9800	0.9800	0.9796	0.9796	0.9796
w_A	0.8983	0.6413	0.7641	0.7894	0.8845	0.6384	0.7471	0.8122	0.8804
w_B	0.0000	0.3587	0.1896	0.1517	0.0000	0.3616	0.2126	0.1142	0.0000
w_C	0.1017	0.0000	0.0463	0.0589	0.1155	0.0000	0.0403	0.0736	0.1196
a_A	0.9796	0.9759	0.9759	0.9759	0.9759	0.9758	0.9758	0.9758	0.9758
w_A	0.6460	0.7489	0.7702	0.8864	0.6350	0.7427	0.8100	0.8778	0.6232
w_B	0.3540	0.2148	0.1848	0.0000	0.3650	0.2179	0.1142	0.0000	0.3768
w_C	0.0000	0.0363	0.0450	0.1136	0.0000	0.0394	0.0758	0.1222	0.0000
a_A	0.9756	0.9756	0.9756	0.9756	0.9752	0.9752	0.9752	0.9752	0.9719
w_A	0.7543	0.7809	0.8651						
w_B	0.1893	0.1476	0.0000						
w_C	0.0564	0.0715	0.1349						
a_A	0.9719	0.9719	0.9719						

Polymer (B): **poly(ethylene glycol)** **2010SA2**
Characterization: M_n/g.mol^{-1} = 6000, Merck, Darmstadt, Germany
Solvent (A): **water** **H_2O** **7732-18-5**
Salt (C): **sodium 1-pentanesulfonate** **$C_5H_{11}NaO_3S$** **22767-49-3**

Type of data: vapor-liquid equilibrium

T/K = 298.15

w_B	0.2812	0.1660	0.0892	0.0439	0.0000	0.3055	0.1894	0.1057	0.0532
w_C	0.0000	0.0304	0.0454	0.0520	0.0570	0.0000	0.0347	0.0538	0.0630
a_A	0.9880	0.9880	0.9880	0.9880	0.9880	0.9846	0.9846	0.9846	0.9846
w_B	0.0000	0.3297	0.2144	0.1248	0.0643	0.0000	0.3549	0.2504	0.1618
w_C	0.0723	0.0000	0.0394	0.0636	0.0762	0.0867	0.0000	0.0403	0.0685
a_A	0.9846	0.9810	0.9810	0.9810	0.9810	0.9810	0.9768	0.9768	0.9768
w_B	0.0868	0.0000	0.3952	0.2814	0.1793	0.0987	0.0000	0.4319	0.3331
w_C	0.0878	0.1045	0.0000	0.0516	0.0913	0.1169	0.1402	0.0000	0.0510
a_A	0.9768	0.9768	0.9673	0.9673	0.9673	0.9673	0.9673	0.9577	0.9577
w_B	0.2382	0.1251	0.0000	0.4514	0.3509	0.2525	0.1527	0.0000	0.5137
w_C	0.0953	0.1418	0.1820	0.0000	0.0546	0.1070	0.1545	0.2238	0.0000
a_A	0.9577	0.9577	0.9577	0.9505	0.9505	0.9505	0.9505	0.9505	0.9259

continued

continued

w_B	0.4205	0.2889	0.1837	0.0000	0.2903	0.1849	0.1088	0.0545	0.0000
w_C	0.0647	0.1475	0.2219	0.3856	0.0000	0.0297	0.0461	0.0552	0.0623
a_A	0.9259	0.9259	0.9259	0.9259	0.9869	0.9869	0.9869	0.9869	0.9869
w_B	0.3261	0.2100	0.1214	0.0624	0.0000	0.3334	0.2287	0.1439	0.0758
w_C	0.0000	0.0385	0.0618	0.0739	0.0836	0.0000	0.0368	0.0609	10.0767
a_A	0.9818	0.9818	0.9818	0.9818	0.9818	0.9805	0.9805	0.9805	0.9805
w_B	0.0000	0.3592	0.2443	0.1483	0.0787	0.0000	0.3958	0.2927	0.1985
w_C	0.0896	0.0000	0.0448	0.0755	0.0932	0.1083	0.0000	0.0471	0.0841
a_A	0.9805	0.9756	0.9756	0.9756	0.9756	0.9756	0.9677	0.9677	0.9677
w_B	0.1114	0.0000	0.4399	0.3281	0.2208	0.1285	0.0000	0.4751	0.3793
w_C	0.1127	0.1401	0.0000	0.0602	0.1124	0.1522	0.1992	0.0000	0.0525
a_A	0.9677	0.9677	0.9540	0.9540	0.9540	0.9540	0.9540	0.9413	0.9413
w_B	0.2823	0.1609	0.0000	0.5282	0.4362	0.3038	0.1966	0.0000	
w_C	0.1129	0.1822	0.2978	0.0000	0.0671	0.1541	0.2373	0.4162	
a_A	0.9413	0.9413	0.9413	0.9178	0.9178	0.9178	0.9178	0.9178	

Polymer (B): **poly(ethylene glycol)** **2010SA1**
Characterization: M_n/g.mol^{-1} = 6000, Merck, Darmstadt, Germany
Solvent (A): **water** **H_2O** **7732-18-5**
Salt (C): **sodium tungstate** **Na_2WO_4** **13472-45-2**

Type of data: vapor-liquid equilibrium

T/K = 298.15

w_B	0.3047	0.1846	0.0999	0.0581	0.0000	0.3271	0.1613	0.0804	0.0501
w_C	0.0000	0.0341	0.0591	0.0721	0.0897	0.0000	0.0526	0.0803	0.0909
a_A	0.9847	0.9847	0.9847	0.9847	0.9847	0.9815	0.9815	0.9815	0.9815
w_B	0.0000	0.3347	0.2069	0.1188	0.0702	0.0000	0.3624	0.2289	0.1362
w_C	0.1089	0.0000	0.0422	0.0716	0.0889	0.1155	0.0000	0.0468	0.0821
a_A	0.9815	0.9797	0.9797	0.9797	0.9797	0.9797	0.9760	0.9760	0.9760
w_B	0.0820	0.0000	0.3803	0.2060	0.1098	0.0702	0.0000	0.4153	0.2878
w_C	0.1039	0.1393	0.0000	0.0676	0.1095	0.1273	0.1628	0.0000	0.0532
a_A	0.9760	0.9760	0.9712	0.9712	0.9712	0.9712	0.9712	0.9626	0.9626
w_B	0.1763	0.1111	0.0000	0.4484	0.3091	0.1986	0.1267	0.0000	0.3153
w_C	0.1043	0.1377	0.1993	0.0000	0.0632	0.1197	0.1604	0.2385	0.0000
a_A	0.9626	0.9626	0.9626	0.9525	0.9525	0.9525	0.9525	0.9525	0.9828
w_B	0.1947	0.1069	0.0630	0.0000	0.3285	0.2081	0.1165	0.0690	0.0000
w_C	0.0359	0.0632	0.0775	0.0994	0.0000	0.0384	0.0686	0.0856	0.1102
a_A	0.9828	0.9828	0.9828	0.9828	0.9814	0.9814	0.9814	0.9814	0.9814
w_B	0.3469	0.1761	0.0901	0.0567	0.0000	0.3718	0.2421	0.1481	0.0976
w_C	0.0000	0.0578	0.0899	0.1029	0.1254	0.0000	0.0471	0.0842	0.1056
a_A	0.9786	0.9786	0.9786	0.9786	0.9786	0.9738	0.9738	0.9738	0.9738

continued

continued

w_B	0.0000	0.3882	0.2527	0.1545	0.0947	0.0000	0.4236	0.2881	0.1845
w_C	0.1502	0.0000	0.0516	0.0932	0.1200	0.1669	0.0000	0.0561	0.1049
a_A	0.9738	0.9703	0.9703	0.9703	0.9703	0.9703	0.9609	0.9609	0.9609
w_B	0.1252	0.0000							
w_C	0.1354	0.2072							
a_A	0.9609	0.9609							

Polymer (B): **poly(ethylene glycol) dimethyl ether** **2011ZAF**
Characterization: M_n/g.mol^{-1} = 2305, Merck, Darmstadt, Germany
Solvent (A): **water** **H_2O** **7732-18-5**
Polymer (C): **poly(propylene glycol)**
Characterization: M_n/g.mol^{-1} = 400, Fluka AG, Buchs, Switzerland

Type of data: vapor-liquid equilibrium

T/K = 298.15

w_B	0.1978	0.1164	0.0586	0.0000	0.2183	0.1311	0.0499	0.0000	0.2653
w_C	0.0000	0.0642	0.1056	0.1478	0.0000	0.0642	0.1315	0.1664	0.0000
a_A	0.9924	0.9924	0.9924	0.9924	0.9912	0.9912	0.9912	0.9912	0.9861
w_B	0.1335	0.0610	0.0000	0.3032	0.1693	0.0761	0.0000	0.3320	0.2032
w_C	0.1199	0.1874	0.2427	0.0000	0.1340	0.2310	0.3128	0.0000	0.1248
a_A	0.9861	0.9861	0.9861	0.9810	0.9810	0.9810	0.9810	0.9774	0.9774
w_B	0.0782	0.0000	0.3582	0.2426	0.1037	0.0000	0.3984	0.2414	0.1234
w_C	0.2658	0.3660	0.0000	0.1213	0.2761	0.4209	0.0000	0.1835	0.3532
a_A	0.9774	0.9774	0.9712	0.9712	0.9712	0.9712	0.9611	0.9611	0.9611
w_B	0.0000	0.4096	0.2486	0.1266	0.0000	0.4250	0.2612	0.1325	0.0000
w_C	0.5983	0.0000	0.1867	0.3620	0.6134	0.0000	0.1852	0.3819	0.6640
a_A	0.9611	0.9606	0.9606	0.9606	0.9606	0.9568	0.9568	0.9568	0.9568
w_B	0.4921	0.3233	0.1701	0.0000	0.5679	0.3814	0.1854	0.0000	
w_C	0.0000	0.2251	0.4866	0.7882	0.0000	0.2582	0.5553	0.8448	
a_A	0.9306	0.9306	0.9306	0.9306	0.8959	0.8959	0.8959	0.8959	

T/K = 308.15

w_B	0.1982	0.1012	0.0417	0.0000	0.2281	0.1195	0.0531	0.0000	0.2676
w_C	0.0000	0.0734	0.1200	0.1443	0.0000	0.1039	0.1617	0.2100	0.0000
a_A	0.9929	0.9929	0.9929	0.9929	0.9905	0.9905	0.9905	0.9905	0.9866
w_B	0.1622	0.0710	0.0000	0.3149	0.2024	0.0968	0.0000	0.3428	0.2100
w_C	0.1022	0.2011	0.2735	0.0000	0.1201	0.2612	0.4240	0.0000	0.1394
a_A	0.9866	0.9866	0.9866	0.9813	0.9813	0.9813	0.9813	0.9798	0.9798
w_B	0.1095	0.0000	0.3539	0.2322	0.1179	0.0000	0.4296	0.2845	0.1502
w_C	0.2723	0.4701	0.0000	0.1466	0.3333	0.5970	0.0000	0.1902	0.4357
a_A	0.9798	0.9798	0.9724	0.9724	0.9724	0.9724	0.9613	0.9613	0.9613

continued

continued

w_B	0.0000	0.4518	0.2967	0.1553	0.0000	0.4973	0.3362	0.1690	0.0000
w_C	0.7500	0.0000	0.2033	0.4582	0.7695	0.0000	0.2236	0.4920	0.8009
a_A	0.9613	0.9557	0.9557	0.9557	0.9557	0.9410	0.9410	0.9410	0.9410
w_B	0.5072	0.3432	0.1726	0.0000	0.5882	0.4042	0.1912	0.0000	0.6166
w_C	0.0000	0.2266	0.5005	0.8064	0.0000	0.2613	0.5661	0.8599	0.0000
a_A	0.9361	0.9361	0.9361	0.9361	0.9025	0.9025	0.9025	0.9025	0.8802
w_B	0.4139	0.2055	0.0000						
w_C	0.2892	0.5947	0.8723						
a_A	0.8802	0.8802	0.8802						

T/K = 318.15

w_B	0.1460	0.0885	0.0508	0.0000	0.2079	0.1253	0.0529	0.0000	0.2156
w_C	0.0000	0.0547	0.0763	0.1253	0.0000	0.0602	0.1270	0.1819	0.0000
a_A	0.9945	0.9945	0.9945	0.9945	0.9941	0.9941	0.9941	0.9941	0.9924
w_B	0.1391	0.0571	0.0000	0.2282	0.1337	0.0581	0.0000	0.2420	0.1661
w_C	0.0639	0.1329	0.1898	0.0000	0.0802	0.1606	0.2597	0.0000	0.1047
a_A	0.9924	0.9924	0.9924	0.9901	0.9901	0.9901	0.9901	0.9874	0.9874
w_B	0.0759	0.0000	0.2932	0.1759	0.0889	0.0000	0.3132	0.1857	0.0937
w_C	0.2185	0.4170	0.0000	0.1229	0.2662	0.5134	0.0000	0.1346	0.2836
a_A	0.9874	0.9874	0.9855	0.9855	0.9855	0.9855	0.9852	0.9852	0.9852
w_B	0.0000	0.3180	0.1860	0.0948	0.0000	0.3248	0.1997	0.1054	0.0000
w_C	0.5439	0.0000	0.1550	0.2993	0.5784	0.0000	0.1447	0.3190	0.6004
a_A	0.9852	0.9835	0.9835	0.9835	0.9835	0.9817	0.9817	0.9817	0.9817
w_B	0.3286	0.2290	0.1077	0.0000	0.3549	0.2220	0.1154	0.0000	0.3684
w_C	0.0000	0.1267	0.3404	0.6242	0.0000	0.1609	0.3495	0.6458	0.0000
a_A	0.9796	0.9796	0.9796	0.9796	0.9777	0.9777	0.9777	0.9777	0.9732
w_B	0.2306	0.1377	0.0000	0.4432	0.2881	0.1559	0.0000	0.4515	0.2934
w_C	0.1890	0.3730	0.6587	0.0000	0.2088	0.4454	0.7458	0.0000	0.2157
a_A	0.9732	0.9732	0.9732	0.9611	0.9611	0.9611	0.9611	0.9558	0.9558
w_B	0.1556	0.0000	0.5196	0.3358	0.1764	0.0000	0.5332	0.3452	0.1813
w_C	0.4712	0.7553	0.0000	0.2433	0.5039	0.8142	0.0000	0.2433	0.5190
a_A	0.9558	0.9558	0.9434	0.9434	0.9434	0.9434	0.9378	0.9378	0.9378
w_B	0.0000	0.6110	0.4006	0.1975	0.0000	0.6208	0.4050	0.2033	0.0000
w_C	0.8271	0.0000	0.2784	0.5707	0.8597	0.0000	0.2935	0.5808	0.8733
a_A	0.9378	0.9078	0.9078	0.9078	0.9078	0.9026	0.9026	0.9026	0.9026

Polymer (B):	**poly(ethylene oxide)**		**2008SAD**
Characterization:	M_n/g.mol^{-1} = 6000, Merck, Darmstadt, Germany		
Solvent (A):	**water**	**H_2O**	**7732-18-5**
Salt (C):	**disodium phosphate**	**Na_2HPO_4**	**7558-79-4**

Type of data: vapor-liquid equilibrium

continued

continued

T/K = 298.15

w_B	0.2839	0.1525	0.1021	0.0529	0.0000	0.3152	0.1807	0.1247	0.0511
w_C	0.0000	0.0167	0.0248	0.0339	0.0446	0.0000	0.0188	0.0287	0.0456
a_A	0.9865	0.9865	0.9865	0.9865	0.9865	0.9827	0.9827	0.9827	0.9827
w_B	0.0000	0.3172	0.1824	0.1047	0.0710	0.0000	0.3312	0.1666	0.1197
w_C	0.0618	0.0000	0.0197	0.0333	0.0426	0.0612	0.0000	0.0272	0.0375
a_A	0.9827	0.9822	0.9822	0.9822	0.9822	0.9822	0.9810	0.9810	0.9810
w_B	0.0776	0.0000							
w_C	0.0498	0.0750							
a_A	0.9810	0.9810							

T/K = 308.15

w_B	0.2116	0.1116	0.0618	0.0296	0.0000	0.2214	0.1318	0.1180	0.0381
w_C	0.0000	0.0073	0.0108	0.0136	0.0172	0.0000	0.0078	0.0098	0.0170
a_A	0.9944	0.9944	0.9944	0.9944	0.9944	0.9935	0.9935	0.9935	0.9935
w_B	0.0000	0.2606	0.1677	0.1019	0.0522	0.0000	0.2750	0.1779	0.1140
w_C	0.0205	0.0000	0.0088	0.0150	0.0215	0.0298	0.0000	0.0098	0.0164
a_A	0.9935	0.9895	0.9895	0.9895	0.9895	0.9895	0.9885	0.9885	0.9885
w_B	0.0564	0.0000	0.2899	0.1948	0.1193	0.0619	0.0000	0.3070	0.1504
w_C	0.0237	0.0338	0.0000	0.0091	0.0184	0.0263	0.0373	0.0000	0.0195
a_A	0.9885	0.9885	0.9877	0.9877	0.9877	0.9877	0.9877	0.9865	0.9865
w_B	0.0827	0.0513	0.0000	0.3391	0.1797	0.1097	0.0630	0.0000	0.3597
w_C	0.0298	0.0352	0.0453	0.0000	0.0242	0.0376	0.0471	0.0638	0.0000
a_A	0.9865	0.9865	0.9865	0.9820	0.9820	0.9820	0.9820	0.9820	0.9784
w_B	0.2620	0.1713	0.1095	0.0000					
w_C	0.0143	0.0299	0.0448	0.0733					
a_A	0.9784	0.9784	0.9784	0.9784					

Polymer (B): **poly(ethylene oxide)** **2008SAD**
Characterization: M_n/g.mol^{-1} = 6000, Merck, Darmstadt, Germany
Solvent (A): **water** **H_2O** **7732-18-5**
Salt (C): **monosodium phosphate** **NaH_2PO_4** **7558-80-7**

Type of data: vapor-liquid equilibrium

T/K = 298.15

w_B	0.2802	0.1712	0.0983	0.0459	0.0000	0.3379	0.2185	0.1419	0.0719
w_C	0.0000	0.0170	0.0303	0.0410	0.0497	0.0000	0.0257	0.0432	0.0613
a_A	0.9865	0.9865	0.9865	0.9865	0.9865	0.9799	0.9799	0.9799	0.9799
w_B	0.0000	0.3473	0.2072	0.1266	0.0716	0.0000	0.3662	0.2405	0.1614
w_C	0.0813	0.0000	0.0310	0.0539	0.0664	0.0860	0.0000	0.0302	0.0518
a_A	0.9799	0.9764	0.9764	0.9764	0.9764	0.9764	0.9736	0.9736	0.9736

continued

continued

w_B	0.0890	0.0000	0.3705	0.2288	0.1440	0.0834	0.0000
w_C	0.0734	0.1033	0.0000	0.0342	0.0613	0.0773	0.1039
a_A	0.9736	0.9736	0.9713	0.9713	0.9713	0.9713	0.9713

T/K = 308.15

w_B	0.2581	0.1789	0.1048	0.0477	0.0000	0.2702	0.1634	0.0881	0.0302
w_C	0.0000	0.0103	0.0188	0.0251	0.0312	0.0000	0.0136	0.0224	0.0305
a_A	0.9910	0.9910	0.9910	0.9910	0.9910	0.9888	0.9888	0.9888	0.9888
w_B	0.0000	0.3023	0.2011	0.1202	0.0494	0.0000	0.3185	0.2423	0.1475
w_C	0.0353	0.0000	0.0163	0.0284	0.0390	0.0485	0.0000	0.0137	0.0304
a_A	0.9888	0.9856	0.9856	0.9856	0.9856	0.9856	0.9821	0.9821	0.9821
w_B	0.1014	0.0000	0.3545	0.2741	0.1886	0.1030	0.0000	0.3874	0.3031
w_C	0.0391	0.0612	0.0000	0.0153	0.0327	0.0521	0.0759	0.0000	0.0183
a_A	0.9821	0.9821	0.9781	0.9781	0.9781	0.9781	0.9781	0.9731	0.9731
w_B	0.2191	0.1233	0.0000	0.3971	0.2913	0.2175	0.1049	0.0000	0.4348
w_C	0.0379	0.0637	0.0992	0.0000	0.0258	0.0444	0.0776	0.1141	0.0000
a_A	0.9731	0.9731	0.9731	0.9695	0.9695	0.9695	0.9695	0.9695	0.9626
w_B	0.3334	0.2508	0.1287	0.0000					
w_C	0.0277	0.0522	0.0957	0.1486					
a_A	0.9626	0.9626	0.9626	0.9626					

Polymer (B): **poly(ethylene oxide)** **2008SAD**
Characterization: M_n/g.mol^{-1} = 6000, Merck, Darmstadt, Germany
Solvent (A): **water** H_2O **7732-18-5**
Salt (C): **sodium phosphate** Na_3PO_4 **7601-54-9**

Type of data: vapor-liquid equilibrium

T/K = 298.15

w_B	0.2604	0.1646	0.1104	0.0509	0.0000	0.2840	0.1882	0.1244	0.0763
w_C	0.0000	0.0085	0.0158	0.0239	0.0305	0.0000	0.0099	0.0188	0.0264
a_A	0.9871	0.9871	0.9871	0.9871	0.9871	0.9868	0.9868	0.9868	0.9868
w_B	0.0000	0.3171	0.2137	0.1618	0.0787	0.0000	0.3286	0.2306	0.1724
w_C	0.0373	0.0000	0.0127	0.0211	0.0364	0.0538	0.0000	0.0125	0.0219
a_A	0.9868	0.9820	0.9820	0.9820	0.9820	0.9820	0.9815	0.9815	0.9815
w_B	0.1172	0.0000							
w_C	0.0319	0.0584							
a_A	0.9815	0.9815							

continued

continued

T/K = 308.15

w_B	0.2171	0.1285	0.0695	0.0266	0.0000	0.2198	0.1274	0.0754	0.0308
w_C	0.0000	0.0051	0.0090	0.0124	0.0145	0.0000	0.0062	0.0100	0.0139
a_A	0.9927	0.9927	0.9927	0.9927	0.9927	0.9912	0.9912	0.9912	0.9912
w_B	0.0000	0.2624	0.1732	0.0998	0.0410	0.0000	0.2728	0.1826	0.1077
w_C	0.0173	0.0000	0.0068	0.0134	0.1930	0.0235	0.0000	0.0076	0.0147
a_A	0.9912	0.9896	0.9896	0.9896	0.9896	0.9896	0.9888	0.9888	0.9888
w_B	0.0464	0.0000	0.2875	0.1972	0.1221	0.0532	0.0000	0.3064	0.2392
w_C	0.0206	0.0257	0.0000	0.0082	0.0160	0.0243	0.0324	0.0000	0.0064
a_A	0.9888	0.9888	0.9886	0.9886	0.9886	0.9886	0.9886	0.9869	0.9869
w_B	0.1877	0.1172	0.0000	0.3385	0.2450	0.1656	0.0795	0.0000	
w_C	0.0126	0.0217	0.0394	0.0000	0.0097	0.0215	0.0363	0.0525	
a_A	0.9869	0.9869	0.9869	0.9821	0.9821	0.9821	0.9821	0.9821	

Polymer (B): **poly(ethylene oxide-*b*-propylene oxide-*b*-ethylene oxide)** **2010CAR**
Characterization: M_n/g.mol^{-1} = 2450, 40 wt% ethylene oxide, Oxiteno S/A, Maua, SP, Brazil
Solvent (A): **water** **H_2O** **7732-18-5**
Polymer (C): **maltodextrin**
Characterization: M_n/g.mol^{-1} = 990, M_w/g.mol^{-1} = 8800, Corn Products Brasil S/A, Mogi Guacu, SP, Brazil

Type of data: vapor-liquid equilibrium

T/K = 298.15

w_B	0.050	0.025	0.075
w_C	0.231	0.277	0.136
a_A	0.991	0.985	0.989

Polymer (B): **poly(ethylene oxide-*b*-propylene oxide-*b*-ethylene oxide)** **2010CAR**
Characterization: M_n/g.mol^{-1} = 2900, 40 wt% ethylene oxide, Aldrich Chem. Co., Inc., Milwaukee, WI
Solvent (A): **water** **H_2O** **7732-18-5**
Polymer (C): **maltodextrin**
Characterization: M_n/g.mol^{-1} = 990, M_w/g.mol^{-1} = 8800, Corn Products Brasil S/A, Mogi Guacu, SP, Brazil

Type of data: vapor-liquid equilibrium

T/K = 298.15

w_B	0.026	0.025	0.051	0.078	0.101
w_C	0.242	0.283	0.188	0.193	0.138
a_A	0.991	0.985	0.988	0.991	0.989

Polymer (B):	**polyglycerol (hyperbranched)**		**2004SEI**
Characterization:	M_n/g.mol^{-1} = 1400, M_w/g.mol^{-1} = 2100, 20 hydroxyl groups, synthesized in the laboratory		
Solvent (A):	**water**	**H_2O**	**7732-18-5**
Solvent (C):	**ethanol**	**C_2H_6O**	**64-17-5**

Type of data: vapor-liquid equilibrium

T/K = 363.15

w_B	0.20	0.20	0.20	0.20	0.20	0.20	0.40	0.40	0.40
x_C	0.300	0.500	0.750	0.851	0.899	0.951	0.202	0.300	0.404
y_C	0.582	0.667	0.804	0.876	0.916	0.953	0.563	0.618	0.674
w_B	0.40	0.40	0.40	0.40	0.40	0.60	0.60	0.60	0.60
x_C	0.650	0.853	0.900	0.946	0.980	0.301	0.500	0.749	0.840
y_C	0.786	0.910	0.932	0.965	0.986	0.635	0.745	0.873	0.924
w_B	0.60	0.60	0.70	0.70	0.70	0.70	0.70	0.70	
x_C	0.900	0.948	0.299	0.494	0.699	0.851	0.897	0.942	
y_C	0.951	0.970	0.709	0.770	0.872	0.940	0.959	0.977	

Polymer (B):	**polyglycerol (hyperbranched)**		**2003SE2**
Characterization:	M_n/g.mol^{-1} = 2000, M_w/g.mol^{-1} = 3000, 28 hydroxyl groups, synthesized in the laboratory		
Solvent (A):	**water**	**H_2O**	**7732-18-5**
Solvent (C):	**2-propanol**	**C_3H_8O**	**67-63-0**

Type of data: vapor-liquid equilibrium

T/K = 353.15

w_B	0.40	0.40	0.40	0.40	0.40	0.40	0.60	0.60	0.60
x_C	0.202	0.508	0.659	0.744	0.852	0.951	0.205	0.508	0.653
y_C	0.558	0.675	0.771	0.807	0.869	0.946	0.626	0.767	0.839
w_B	0.60	0.60	0.60	0.60	0.70	0.70	0.70	0.70	0.70
x_C	0.776	0.850	0.954	0.981	0.197	0.477	0.652	0.694	0.749
y_C	0.885	0.924	0.972	0.988	0.637	0.783	0.858	0.875	0.896
w_B	0.70	0.70							
x_C	0.893	0.958							
y_C	0.953	0.974							

Polymer (B):	**poly(propylene glycol)**		**2004SAL**
Characterization:	M_n/g.mol^{-1} = 405, Aldrich Chem. Co., Inc., Milwaukee, WI		
Solvent (A):	**water**	**H_2O**	**7732-18-5**
Salt (C):	**ammonium sulfate**	**$(NH_4)_2SO_4$**	**7783-20-2**

Type of data: vapor-liquid equilibrium

continued

continued

T/K = 298.15

w_A	0.9086	0.8684	0.7635	0.8883	0.8393	0.7025	0.8898	0.8137	0.6666
w_B	0.0551	0.1087	0.2365	0.0550	0.1230	0.2975	0.0450	0.1505	0.3334
w_C	0.0363	0.0229	0.0000	0.0567	0.0377	0.0000	0.0652	0.0358	0.0000
a_A	0.9855	0.9855	0.9855	0.9804	0.9804	0.9804	0.9773	0.9773	0.9773
w_A	0.8653	0.8239	0.6104	0.8199	0.7748	0.5492	0.8275	0.7148	0.4799
w_B	0.0603	0.1205	0.3896	0.1103	0.1705	0.4508	0.0815	0.2376	0.5201
w_C	0.0744	0.0556	0.0000	0.0698	0.0547	0.0000	0.0910	0.0476	0.0000
a_A	0.9729	0.9729	0.9729	0.9674	0.9674	0.9674	0.9633	0.9633	0.9633
w_A	0.8050	0.6792	0.4309	0.7676	0.6910	0.3495	0.7497	0.6127	0.3062
w_B	0.1003	0.2701	0.5691	0.1244	0.2352	0.6505	0.1274	0.3197	0.6938
w_C	0.0947	0.0507	0.0000	0.1080	0.0738	0.0000	0.1229	0.0676	0.0000
a_A	0.9564	0.9564	0.9564	0.9454	0.9454	0.9454	0.9423	0.9423	0.9423

Polymer (B): **poly(propylene glycol)** **2010CAR**
Characterization: M_n/g.mol^{-1} = 400, Aldrich Chem. Co., Inc., Milwaukee, WI
Solvent (A): **water** **H_2O** **7732-18-5**
Polymer (C): **maltodextrin**
Characterization: M_n/g.mol^{-1} = 1500, M_w/g.mol^{-1} = 20500,
Corn Products Brasil S/A, Mogi Guacu, SP, Brazil

Type of data: vapor-liquid equilibrium

T/K = 298.15

w_B	0.104	0.025	0.051	0.150	0.150
w_C	0.185	0.381	0.285	0.134	0.188
a_A	0.989	0.982	0.984	0.982	0.979

Polymer (B): **poly(propylene glycol)** **1998ZAF**
Characterization: M_n/g.mol^{-1} = 404, Aldrich Chem. Co., Inc., Milwaukee, WI
Solvent (A): **water** **H_2O** **7732-18-5**
Salt (C): **magnesium sulfate** **$MgSO_4$** **7487-88-9**

Type of data: vapor-liquid equilibrium

T/K = 298.15

c_B/(mol/kg A)	0.1292	0.2597	0.5538	0.1587	0.2764	0.7366	0.1851	0.2882
c_C/(mol/kg A)	0.3277	0.1907	0.0000	0.4253	0.2909	0.0000	0.5068	0.3869
a_A	0.9897	0.9897	0.9897	0.9860	0.9860	0.9860	0.9833	0.9833
c_B/(mol/kg A)	0.9127	0.2024	0.3882	1.0453	0.2021	0.4753	1.1931	0.2129
c_C/(mol/kg A)	0.0000	0.5777	0.3568	0.0000	0.6575	0.3523	0.0000	0.7118
a_A	0.9833	0.9820	0.9820	0.9820	0.9799	0.9799	0.9799	0.9776
c_B/(mol/kg A)	0.5613	1.3148	0.2169	0.7158	1.5521	0.2372	0.7728	1.8376
c_C/(mol/kg A)	0.3464	0.0000	0.8467	0.3521	0.0000	0.8683	0.3447	0.0000
a_A	0.9776	0.9776	0.9743	0.9743	0.9743	0.9728	0.9728	0.9728

Polymer (B):	**poly(propylene glycol)**		**2004SAL**
Characterization:	M_n/g.mol^{-1} = 405, Aldrich Chem. Co., Inc., Milwaukee, WI		
Solvent (A):	**water**	**H_2O**	**7732-18-5**
Salt (C):	**sodium sulfate**	**Na_2SO_4**	**7757-82-6**

Type of data: vapor-liquid equilibrium

T/K = 298.15

w_A	0.9579	0.9355	0.8045	0.9449	0.8948	0.7820	0.9272	0.8587	0.7339
w_B	0.0183	0.0509	0.1955	0.0250	0.0902	0.2180	0.0386	0.1216	0.2661
w_C	0.0238	0.0136	0.0000	0.0301	0.0150	0.0000	0.0342	0.0197	0.0000
a_A	0.9895	0.9895	0.9895	0.9872	0.9872	0.9872	0.9844	0.9844	0.9844
w_A	0.9077	0.7755	0.6894	0.8722	0.7356	0.6273	0.8470	0.7310	0.5573
w_B	0.0489	0.2122	0.3106	0.0736	0.2491	0.3727	0.0920	0.2410	0.4427
w_C	0.0434	0.0123	0.0000	0.0542	0.0153	0.0000	0.0610	0.0280	0.0000
a_A	0.9810	0.9810	0.9810	0.9766	0.9766	0.9766	0.9706	0.9706	0.9706
w_A	0.8286	0.6718	0.5035	0.8348	0.7636	0.4785	0.7941	0.7423	0.4247
w_B	0.1004	0.3002	0.4965	0.0820	0.1815	0.5215	0.1117	0.1897	0.5753
w_C	0.0710	0.0280	0.0000	0.0832	0.0549	0.0000	0.0942	0.0680	0.0000
a_A	0.9659	0.9659	0.9659	0.9655	0.9655	0.9655	0.9561	0.9561	0.9561
w_A	0.7761	0.7104	0.3960						
w_B	0.1205	0.2202	0.6040						
w_C	0.1034	0.0694	0.0000						
a_A	0.9542	0.9542	0.9542						

Polymer (B):	**poly(propylene oxide)**		**2007SAD**
Characterization:	M_n/g.mol^{-1} = 400, PPO400, Fluka AG, Buchs, Switzerland		
Solvent (A):	**water**	**H_2O**	**7732-18-5**
Salt (C):	**potassium citrate**	**$C_6H_5K_3O_7$**	**866-84-2**

Type of data: vapor-liquid equilibrium

T/K = 293.15

c_B/(mol/kg A)	0.3928	0.2513	0.1589	0.0000	0.5645	0.2880	0.1913	0.0000
c_C/(mol/kg A)	0.0000	0.0339	0.0624	0.1341	0.0000	0.0389	0.0749	0.1635
a_A	0.9922	0.9922	0.9922	0.9922	0.9917	0.9917	0.9917	0.9917
c_B/(mol/kg A)	0.6706	0.4327	0.2189	0.0000	0.7675	0.4692	0.2365	0.0000
c_C/(mol/kg A)	0.0000	0.0539	0.1282	0.2499	0.0000	0.0629	0.1384	0.2951
a_A	0.9872	0.9872	0.9872	0.9872	0.9846	0.9846	0.9846	0.9846
c_B/(mol/kg A)	0.9626	0.6066	0.2809	0.0000	1.0653	0.5327	0.2498	0.0000
c_C/(mol/kg A)	0.0000	0.0773	0.1860	0.3547	0.0000	0.1198	0.2333	0.4019
a_A	0.9799	0.9799	0.9799	0.9799	0.9785	0.9785	0.9785	0.9785
c_B/(mol/kg A)	1.0853	0.6518	0.3289	0.0000	1.3622	0.5606	0.3144	0.0000
c_C/(mol/kg A)	0.0000	0.0873	0.1927	0.4078	0.0000	0.1889	0.2910	0.5220
a_A	0.9791	0.9791	0.9791	0.9791	0.9721	0.9721	0.9721	0.9721

continued

continued

c_B/(mol/kg A)	1.4014	0.4780	0.2344	0.0000	1.9956	0.8696	0.4055	0.0000
c_C/(mol/kg A)	0.0000	0.2069	0.3180	0.5416	0.0000	0.1953	0.3782	0.6689
a_A	0.9692	0.9692	0.9692	0.9692	0.9648	0.9648	0.9648	0.9648
T/K = 298.15								
c_B/(mol/kg A)	0.4509	0.2727	0.1742	0.0000	0.5343	0.2621	0.1481	0.0000
c_C/(mol/kg A)	0.0000	0.0387	0.0682	0.1506	0.0000	0.0539	0.0960	0.1686
a_A	0.9921	0.9921	0.9921	0.9921	0.9903	0.9903	0.9903	0.9903
c_B/(mol/kg A)	0.6510	0.4168	0.2026	0.0000	0.7417	0.4666	0.2265	0.0000
c_C/(mol/kg A)	0.0000	0.0378	0.1103	0.2100	0.0000	0.0422	0.1231	0.2388
a_A	0.9880	0.9880	0.9880	0.9880	0.9868	0.9868	0.9868	0.9868
c_B/(mol/kg A)	0.7968	0.5116	0.2487	0.0000	0.8697	0.5557	0.2671	0.0000
c_C/(mol/kg A)	0.0000	0.0462	0.1354	0.2649	0.0000	0.0504	0.1451	0.2867
a_A	0.9855	0.9855	0.9855	0.9855	0.9845	0.9845	0.9845	0.9845
c_B/(mol/kg A)	1.0246	0.5960	0.2651	0.0000	1.2668	0.7174	0.3211	0.0000
c_C/(mol/kg A)	0.0000	0.0600	0.1606	0.3109	0.0000	0.0722	0.1948	0.3784
a_A	0.9832	0.9832	0.9832	0.9832	0.9805	0.9805	0.9805	0.9805
c_B/(mol/kg A)	1.4321	0.7929	0.3522	0.0000	1.5245	0.6866	0.4034	0.0000
c_C/(mol/kg A)	0.0000	0.0798	0.2130	0.4215	0.0000	0.1267	0.2149	0.4460
a_A	0.9787	0.9787	0.9787	0.9787	0.9775	0.9775	0.9775	0.9775
c_B/(mol/kg A)	2.0224	0.7752	0.4794	0.0000	2.1808	1.0793	0.4924	0.0000
c_C/(mol/kg A)	0.0000	0.1609	0.2498	0.5274	0.0000	0.1232	0.2842	0.5780
a_A	0.9734	0.9734	0.9734	0.9734	0.9699	0.9699	0.9699	0.9699
c_B/(mol/kg A)	2.3506	0.6018	0.5322	0.0000				
c_C/(mol/kg A)	0.0000	0.2553	0.2807	0.5891				
a_A	0.9691	0.9691	0.9691	0.9691				
T/K = 303.15								
c_B/(mol/kg A)	0.5197	0.2425	0.1398	0.0000	0.7425	0.3809	0.1563	0.0000
c_C/(mol/kg A)	0.0000	0.0501	0.0905	0.1628	0.0000	0.0643	0.1372	0.2212
a_A	0.9903	0.9903	0.9903	0.9903	0.9881	0.9881	0.9881	0.9881
c_B/(mol/kg A)	0.9775	0.4138	0.2321	0.0000	1.1321	0.5237	0.2105	0.0000
c_C/(mol/kg A)	0.0000	0.0852	0.1506	0.2660	0.0000	0.0884	0.1846	0.3058
a_A	0.9845	0.9845	0.9845	0.9845	0.9833	0.9833	0.9833	0.9833
c_B/(mol/kg A)	1.4111	0.6092	0.3310	0.0000	2.1151	0.7446	0.4216	0.0000
c_C/(mol/kg A)	0.0000	0.1026	0.1870	0.3616	0.0000	0.1371	0.2243	0.4516
a_A	0.9818	0.9818	0.9818	0.9818	0.9773	0.9773	0.9773	0.9773
c_B/(mol/kg A)	2.4087	0.7575	0.4536	0.0000	2.8961	0.8443	0.4993	0.0000
c_C/(mol/kg A)	0.0000	0.1561	0.2365	0.4826	0.0000	0.1754	0.2603	0.5237
a_A	0.9742	0.9742	0.9742	0.9742	0.9724	0.9724	0.9724	0.9724

continued

continued

T/K = 308.15

c_B/(mol/kg A)	0.5458	0.2941	0.1671	0.0000	0.6033	0.3049	0.1228	0.0000
c_C/(mol/kg A)	0.0000	0.0471	0.0825	0.1580	0.0000	0.0516	0.1077	0.1713
a_A	0.9909	0.9909	0.9909	0.9909	0.9906	0.9906	0.9906	0.9906
c_B/(mol/kg A)	0.6546	0.3268	0.1488	0.0000	0.9814	0.4940	0.2334	0.0000
c_C/(mol/kg A)	0.0000	0.0552	0.1130	0.1850	0.0000	0.0776	0.1558	0.2805
a_A	0.9898	0.9898	0.9898	0.9898	0.9849	0.9849	0.9849	0.9849
c_B/(mol/kg A)	1.4099	0.5274	0.3038	0.0000	1.5650	0.5841	0.3360	0.0000
c_C/(mol/kg A)	0.0000	0.0975	0.1606	0.3139	0.0000	0.0998	0.1628	0.3295
a_A	0.9839	0.9839	0.9839	0.9839	0.9829	0.9829	0.9829	0.9829
c_B/(mol/kg A)	1.8991	0.7778	0.3545	0.0000	3.2884	0.8558	0.3940	0.0000
c_C/(mol/kg A)	0.0000	0.0976	0.1987	0.3803	0.0000	0.1680	0.2847	0.4887
a_A	0.9794	0.9794	0.9794	0.9794	0.9742	0.9742	0.9742	0.9742

T/K = 313.15

c_B/(mol/kg A)	0.5800	0.3218	0.1607	0.0000	0.6880	0.3537	0.1742	0.0000
c_C/(mol/kg A)	0.0000	0.0373	0.0823	0.1536	0.0000	0.0409	0.0893	0.1621
a_A	0.9909	0.9909	0.9909	0.9909	0.9900	0.9900	0.9900	0.9900
c_B/(mol/kg A)	1.0431	0.4590	0.2165	0.0000	1.3081	0.5367	0.2497	0.0000
c_C/(mol/kg A)	0.0000	0.0533	0.1105	0.2100	0.0000	0.0624	0.1278	0.2460
a_A	0.9884	0.9884	0.9884	0.9884	0.9867	0.9867	0.9867	0.9867
c_B/(mol/kg A)	2.1460	0.6606	0.2687	0.0000	2.9861	0.7734	0.3374	0.0000
c_C/(mol/kg A)	0.0000	0.0766	0.1375	0.2893	0.0000	0.0898	0.1728	0.3314
a_A	0.9847	0.9847	0.9847	0.9847	0.9836	0.9836	0.9836	0.9836

Polymer (B): **poly(sodium acrylate)** **2004LAM, 2008LA2**
Characterization: M_n/g.mol^{-1} = 2600, M_w/g.mol^{-1} = 4300, M_z/g.mol^{-1} = 6100,
Rohagit SL 137, Roehm GmbH, Darmstadt, Germany
Solvent (A): **water** **H_2O** **7732-18-5**
Salt (C): **sodium chloride** **NaCl** **7647-14-5**

Type of data: vapor-liquid equilibrium

T/K = 298.15

w_B	0.2097	0.1667	0.1271	0.0872	0.0000	0.2756	0.2255	0.1740	0.1177
w_C	0.0000	0.0089	0.0168	0.0241	0.0380	0.0000	0.0148	0.0282	0.0413
a_A	0.9777	0.9777	0.9777	0.9777	0.9777	0.9617	0.9617	0.9617	0.9617
w_B	0.0000	0.3521	0.2895	0.2289	0.1580	0.0000	0.3939	0.3230	0.2614
w_C	0.0630	0.0000	0.0224	0.0426	0.0652	0.1011	0.0000	0.0289	0.0526
a_A	0.9617	0.9343	0.9343	0.9343	0.9343	0.9343	0.9140	0.9140	0.9140
w_B	0.1858	0.0000							
w_C	0.0797	0.1260							
a_A	0.9140	0.9140							

Polymer (B): **poly(sodium acrylate)** **2004LAM, 2008LA2**
Characterization: M_n/g.mol^{-1} = 6900, M_w/g.mol^{-1} = 17300, M_z/g.mol^{-1} = 39700, Sokalan SA 40, BASF AG, Ludwigshafen, Germany
Solvent (A): **water** **H_2O** **7732-18-5**
Salt (C): **sodium chloride** **NaCl** **7647-14-5**

Type of data: vapor-liquid equilibrium

T/K = 298.15

w_B	0.2228	0.2009	0.1793	0.1586	0.1382	0.1194	0.0999	0.0805	0.0000
w_C	0.0000	0.0040	0.0077	0.0112	0.0146	0.0179	0.0210	0.0243	0.0379
a_A	0.9778	0.9778	0.9778	0.9778	0.9778	0.9778	0.9778	0.9778	0.9778
w_B	0.2783	0.2491	0.2219	0.1938	0.1663	0.1395	0.1095	0.0789	0.0000
w_C	0.0000	0.0066	0.0125	0.0183	0.0238	0.0297	0.0352	0.0417	0.0570
a_A	0.9657	0.9657	0.9657	0.9657	0.9657	0.9657	0.9657	0.9657	0.9657
w_B	0.3273	0.2949	0.2587	0.2330	0.2000	0.1685	0.1333	0.0933	0.0000
w_C	0.0000	0.0087	0.0162	0.0239	0.0315	0.0388	0.0467	0.0556	0.0769
a_A	0.9522	0.9522	0.9522	0.9522	0.9522	0.9522	0.9522	0.9522	0.9522
w_B	0.3696	0.3426	0.3118	0.2783	0.2431	0.2063	0.1680	0.1288	0.0000
w_C	0.0000	0.0094	0.0192	0.0295	0.0399	0.0505	0.0612	0.0723	0.1077
a_A	0.9291	0.9291	0.9291	0.9291	0.9291	0.9291	0.9291	0.9291	0.9291

Polymer (B): **poly(sodium ethylenesulfonate)** **2004LAM, 2008LA2**
Characterization: M_n/g.mol^{-1} = 1600, M_w/g.mol^{-1} = 2800, M_z/g.mol^{-1} = 5100, Polysciences Europe GmbH, Eppelheim, Germany
Solvent (A): **water** **H_2O** **7732-18-5**
Salt (C): **sodium chloride** **NaCl** **7647-14-5**

Type of data: vapor-liquid equilibrium

T/K = 298.15

w_B	0.2551	0.2182	0.1810	0.1457	0.0000	0.3560	0.3033	0.2533	0.1899
w_C	0.0000	0.0080	0.0159	0.0233	0.0539	0.0000	0.0138	0.0267	0.0425
a_A	0.9677	0.9677	0.9677	0.9677	0.9677	0.9431	0.9431	0.9431	0.9431
w_B	0.0000	0.4497	0.3890	0.3262	0.2449	0.0000			
w_C	0.0895	0.0000	0.0192	0.0385	0.0632	0.1359			
a_A	0.9431	0.9054	0.9054	0.9054	0.9054	0.9054			

Polymer (B): **poly(sodium ethylenesulfonate)** **2004LAM, 2008LA2**
Characterization: M_n/g.mol^{-1} = 6900, M_w/g.mol^{-1} = 11800, M_z/g.mol^{-1} = 16200, Natriumpolat PIPU 005, Hoechst GmbH, Frankfurt, Germany
Solvent (A): **water** **H_2O** **7732-18-5**
Salt (C): **sodium chloride** **NaCl** **7647-14-5**

Type of data: vapor-liquid equilibrium

continued

continued

T/K = 298.15

w_B	0.3179	0.2686	0.2340	0.1998	0.1717	0.1465	0.1238	0.1028	0.0000
w_C	0.0000	0.0047	0.0083	0.0118	0.0147	0.0172	0.0194	0.0215	0.0337
a_A	0.9803	0.9803	0.9803	0.9803	0.9803	0.9803	0.9803	0.9803	0.9803
w_B	0.3911	0.3379	0.2862	0.2408	0.2028	0.1670	0.1335	0.1027	0.0000
w_C	0.0000	0.0073	0.0137	0.0194	0.0243	0.0288	0.0329	0.0368	0.0523
a_A	0.9687	0.9687	0.9687	0.9687	0.9687	0.9687	0.9687	0.9687	0.9687
w_B	0.4057	0.3590	0.3162	0.2706	0.2243	0.1817	0.1369	0.0892	0.0000
w_C	0.0000	0.0070	0.0134	0.0194	0.0253	0.0310	0.0370	0.0430	0.0554
a_A	0.9667	0.9667	0.9667	0.9667	0.9667	0.9667	0.9667	0.9667	0.9667
w_B	0.4462	0.4015	0.3553	0.3101	0.2632	0.2142	0.1628	0.1053	0.0000
w_C	0.0000	0.0082	0.0160	0.0234	0.0307	0.0383	0.0463	0.0551	0.0715
a_A	0.9559	0.9559	0.9559	0.9559	0.9559	0.9559	0.9559	0.9559	0.9559
w_B	0.4979	0.4560	0.4122	0.3675	0.3200	0.2686	0.2095	0.1371	0.0000
w_C	0.0000	0.0101	0.0197	0.0291	0.0387	0.0489	0.0607	0.0744	0.1008
a_A	0.9345	0.9345	0.9345	0.9345	0.9345	0.9345	0.9345	0.9345	0.9345

Polymer (B): **poly(sodium methacrylate)** **2004LAM, 2008LA2**
Characterization: M_n/g.mol^{-1} = 6100, M_w/g.mol^{-1} = 10000, M_z/g.mol^{-1} = 15400, Sigma-Aldrich Chemie GmbH, Steinheim, Germany
Solvent (A): **water** **H_2O** **7732-18-5**
Salt (C): **sodium chloride** **NaCl** **7647-14-5**

Type of data: vapor-liquid equilibrium

T/K = 298.15

w_B	0.2072	0.1913	0.1760	0.1618	0.1470	0.1308	0.1198	0.0000	0.2602
w_C	0.0000	0.0029	0.0056	0.0079	0.0108	0.0132	0.0161	0.0345	0.0000
a_A	0.9799	0.9799	0.9799	0.9799	0.9799	0.9799	0.9799	0.9799	0.9674
w_B	0.2350	0.2097	0.1839	0.1564	0.1298	0.1009	0.0695	0.0000	0.3015
w_C	0.0062	0.0120	0.0176	0.0233	0.0288	0.0346	0.0408	0.0544	0.0000
a_A	0.9674	0.9674	0.9674	0.9674	0.9674	0.9674	0.9674	0.9674	0.9528
w_B	0.2741	0.2490	0.2216	0.1931	0.1627	0.1266	0.0862	0.0000	0.3262
w_C	0.0085	0.0157	0.0231	0.0306	0.0380	0.0466	0.0562	0.0760	0.0000
a_A	0.9528	0.9528	0.9528	0.9528	0.9528	0.9528	0.9528	0.9528	0.9408
w_B	0.2974	0.2682	0.2392	0.2153	0.1816	0.1435	0.0971	0.0000	0.2439
w_C	0.0103	0.0197	0.0285	0.0355	0.0449	0.0552	0.0674	0.0926	0.0473
a_A	0.9408	0.9408	0.9408	0.9408	0.9408	0.9408	0.9408	0.9408	0.9142
w_B	0.2064	0.1640	0.1145	0.0000	0.2211	0.1757	0.1221	0.0000	
w_C	0.0597	0.0736	0.0899	0.1257	0.0646	0.0797	0.0979	0.1384	
a_A	0.9142	0.9142	0.9142	0.9142	0.9031	0.9031	0.9031	0.9031	

Polymer (B):	**poly(sodium methacrylate)**		**2004LAM, 2008LA2**
Characterization:	M_n/g.mol^{-1} = 14200, M_w/g.mol^{-1} = 20500, M_z/g.mol^{-1} = 27600, Polysciences Europe GmbH, Eppelheim, Germany		
Solvent (A):	**water**	**H_2O**	**7732-18-5**
Salt (C):	**sodium chloride**	**NaCl**	**7647-14-5**

Type of data: vapor-liquid equilibrium

T/K = 298.15

w_B	0.3017	0.2785	0.2567	0.2293	0.1983	0.1672	0.1317	0.0890	0.0000
w_C	0.0000	0.0078	0.0155	0.0228	0.0301	0.0383	0.0467	0.0564	0.0772
a_A	0.9519	0.9519	0.9519	0.9519	0.9519	0.9519	0.9519	0.9519	0.9519
w_B	0.3257	0.3052	0.2771	0.2491	0.2247	0.1904	0.1500	0.1032	0.0000
w_C	0.0000	0.0101	0.0197	0.0286	0.0362	0.0455	0.0556	0.0685	0.0952
a_A	0.9388	0.9388	0.9388	0.9388	0.9388	0.9388	0.9388	0.9388	0.9388

Polymer (B):	**poly(sodium styrenesulfonate)**		**2004LAM, 2008LA2**
Characterization:	M_n/g.mol^{-1} = 127000, M_w/g.mol^{-1} = 148000, M_z/g.mol^{-1} = 155000, Sigma-Aldrich Chemie GmbH, Steinheim, Germany		
Solvent (A):	**water**	**H_2O**	**7732-18-5**
Salt (C):	**sodium chloride**	**NaCl**	**7647-14-5**

Type of data: vapor-liquid equilibrium

T/K = 298.15

w_B	0.2475	0.2187	0.1932	0.1683	0.1439	0.1198	0.0947	0.0683	0.0000
w_C	0.0000	0.0038	0.0075	0.0111	0.0146	0.0182	0.0219	0.0257	0.0352
a_A	0.9794	0.9794	0.9794	0.9794	0.9794	0.9794	0.9794	0.9794	0.9794
w_B	0.3288	0.2910	0.2545	0.2196	0.1841	0.1477	0.1079	0.0632	0.0000
w_C	0.0000	0.0063	0.0123	0.0181	0.0238	0.0300	0.0366	0.0441	0.0548
a_A	0.9674	0.9674	0.9674	0.9674	0.9674	0.9674	0.9674	0.9674	0.9674
w_B	0.3909	0.3467	0.3056	0.2662	0.2267	0.1848	0.1375	0.0816	0.0000
w_C	0.0000	0.0083	0.0158	0.0230	0.0303	0.0382	0.0469	0.0574	0.0727
a_A	0.9551	0.9551	0.9551	0.9551	0.9551	0.9551	0.9551	0.9551	0.9551
w_B	0.4460	0.3956	0.3513	0.3075	0.2644	0.2194	0.1673	0.1012	0.0000
w_C	0.0000	0.0104	0.0196	0.0280	0.0365	0.0459	0.0568	0.0704	0.0913
a_A	0.9417	0.9417	0.9417	0.9417	0.9417	0.9417	0.9417	0.9417	0.9417
w_B	0.4867	0.4280	0.3800	0.3368	0.2935	0.2465	0.1908	0.1181	0.0000
w_C	0.0000	0.0127	0.0231	0.0326	0.0421	0.0524	0.0647	0.0810	0.1076
a_A	0.9292	0.9292	0.9292	0.9292	0.9292	0.9292	0.9292	0.9292	0.9292
w_B	0.2157	0.1667	0.1036	0.0126	0.0000	0.2205	0.1738	0.1072	0.0000
w_C	0.0895	0.1009	0.1173	0.1401	0.1432	0.1066	0.1200	0.1380	0.1677
a_A	0.8987	0.8987	0.8987	0.8987	0.8987	0.8748	0.8748	0.8748	0.8748

Polymer (B):	**poly(vinyl alcohol-*co*-sodium acrylate)**		**2004MAT**
Characterization:	40 mol% sodium acrylate, Sumitomo Chemical Co., Ltd., Japan		
Solvent (A):	**water**	H_2O	**7732-18-5**
Solvent (C):	**ethanol**	C_2H_6O	**64-17-5**

Type of data: vapor-liquid equilibrium

Comments: The mole fraction of ethanol in the liquid mixture, x_C, is given for the polymer-free solvent mixture. The copolymer was considered to be a physical gel, i.e., cross-linked by hydrogen bonds.

T/K = 298.15

w_B	0.05	0.05	0.05	0.05	0.05	0.10	0.10	0.10	0.10
x_C	0.400	0.500	0.600	0.700	0.800	0.400	0.500	0.600	0.700
y_C	0.661	0.699	0.735	0.785	0.839	0.676	0.721	0.760	0.796
P_A/kPa	2.39	2.24	2.07	1.71	1.30	2.39	2.04	1.86	1.60
P_C/kPa	4.65	5.21	5.75	6.26	6.77	4.99	5.26	5.89	6.23

w_B	0.10	0.15	0.15	0.15	0.15	0.15
x_C	0.800	0.400	0.500	0.600	0.700	0.800
y_C	0.858	0.697	0.733	0.782	0.822	0.860
P_A/kPa	1.15	2.20	2.03	1.68	1.41	1.08
P_C/kPa	6.93	5.07	5.57	6.01	6.51	6.64

Polymer (B):	**poly(vinyl alcohol-*co*-sodium acrylate)**		**1996MIS**
Characterization:	40 mol% sodium acrylate, Sumitomo Chemical Co., Ltd., Japan		
Solvent (A):	**water**	H_2O	**7732-18-5**
Solvent (C):	**1-propanol**	C_3H_8O	**71-23-8**

Type of data: vapor-liquid equilibrium

Comments: The copolymer was considered to be a physical gel, i.e., cross-linked by hydrogen bonds.

T/K = 298.15

w_A	0.5182	0.3912	0.2948	0.2192	0.1589	0.1082	0.0667	0.4909	0.3706
w_B	0.0500	0.0500	0.0500	0.0500	0.0500	0.0500	0.0500	0.1000	0.1000
w_C	0.4318	0.5588	0.6552	0.7308	0.7911	0.8418	0.8833	0.4091	0.5294
y_C	0.368	0.384	0.404	0.420	0.451	0.504	0.608	0.375	0.393
P_A/kPa	2.87	2.81	2.74	2.69	2.43	2.09	1.54	2.87	2.79
P_C/kPa	1.67	1.75	1.86	1.95	2.00	2.12	2.39	1.72	1.81

w_A	0.2833	0.2090	0.1526	0.1025	0.0628	0.4623	0.3510	0.2638	0.1962
w_B	0.1000	0.1000	0.1000	0.1000	0.1000	0.1500	0.1500	0.1500	0.1500
w_C	0.6167	0.6910	0.7474	0.7975	0.8372	0.3877	0.4990	0.5862	0.6538
y_C	0.400	0.438	0.489	0.547	0.629	0.379	0.403	0.421	0.481
P_A/kPa	2.85	2.30	2.13	1.92	1.37	2.85	2.75	2.63	2.19
P_C/kPa	1.90	1.79	2.04	2.32	2.32	1.76	1.86	1.91	2.03

continued

continued

w_A	0.1417	0.0977	0.0590
w_B	0.1500	0.1500	0.1500
w_C	0.7083	0.7523	0.7910
y_C	0.526	0.592	0.653
P_A/kPa	1.97	1.59	1.31
P_C/kPa	2.19	2.31	2.46

Polymer (B): **poly(vinyl alcohol-*co*-sodium acrylate)** **2004MAT**
Characterization: 40 mol% sodium acrylate, Sumitomo Chemical Co., Ltd., Japan
Solvent (A): **water** **H_2O** **7732-18-5**
Solvent (C): **2-propanol** **C_3H_8O** **67-63-0**

Type of data: vapor-liquid equilibrium

Comments: The mole fraction of 2-propanol in the liquid mixture, x_C, is given for the polymer-free solvent mixture. The copolymer was considered to be a physical gel, i.e., cross-linked by hydrogen bonds.

T/K = 298.15

w_B	0.05	0.05	0.05	0.05	0.05	0.05	0.05	0.10	0.10
x_C	0.200	0.299	0.399	0.500	0.599	0.700	0.797	0.200	0.300
y_C	0.558	0.599	0.610	0.632	0.669	0.727	0.784	0.560	0.591
P_A/kPa	2.72	2.63	2.44	2.40	2.23	1.81	1.47	2.76	2.57
P_C/kPa	3.44	3.68	3.82	4.13	4.50	4.81	5.35	3.51	3.72

w_B	0.10	0.10	0.10	0.10	0.10	0.15	0.15	0.15	0.15
x_C	0.400	0.500	0.599	0.700	0.799	0.200	0.299	0.400	0.500
y_C	0.618	0.650	0.691	0.740	0.804	0.579	0.603	0.630	0.689
P_A/kPa	2.50	2.34	1.99	1.73	1.28	2.69	2.62	2.42	2.14
P_C/kPa	4.04	4.34	4.46	4.93	5.24	3.70	3.98	4.12	4.73

w_B	0.15	0.15	0.15
x_C	0.600	0.700	0.800
y_C	0.718	0.761	0.821
P_A/kPa	1.94	1.57	1.15
P_C/kPa	4.94	4.99	5.28

Polymer (B): **poly(vinyl alcohol-*co*-vinyl propional)** **2005CSA**
Characterization: M_n/g.mol^{-1} = 76430, 7.0 mol% vinyl propional, prepared and fractionated in the laboratory from Poval 420, Kuraray Co., Japan
Solvent (A): **water** **H_2O** **7732-18-5**
Polymer (C): **poly(1-vinyl-2-pyrrolidinone)**
Characterization: M_n/g.mol^{-1} = 117000, fractionated in the laboratory from Fluka K90 PVP, Fluka AG, Buchs, Switzerland

Type of data: vapor-liquid equilibrium

continued

continued

Comments: w_B/w_C = 1/1 was kept constant.

T/K = 298.15

φ_{B+C}	0.01067	0.01413	0.01863	0.02328	0.02824	0.03641	0.04602
$\ln(a_A)$	$-5.74\ 10^{-6}$	$-6.32\ 10^{-6}$	$-1.09\ 10^{-5}$	$-1.74\ 10^{-5}$	$-2.23\ 10^{-5}$	$-3.77\ 10^{-5}$	$-5.45\ 10^{-5}$

Polymer (B): **poly(1-vinyl-2-pyrrolidinone)** **2005SA2**
Characterization: M_w/g.mol^{-1} = 10000, PVP K15,
Aldrich Chem. Co., Inc., Milwaukee, WI
Solvent (A): **water** H_2O **7732-18-5**
Salt (C): **ammonium sulfate** $(NH_4)_2SO_4$ **7783-20-2**

Type of data: vapor-liquid equilibrium

T/K = 298.15

w_B	0.2970	0.1578	0.0815	0.0000	0.3742	0.2897	0.1324	0.0000	0.4180
w_C	0.0000	0.0203	0.0316	0.0463	0.0000	0.0133	0.0403	0.0703	0.0000
a_A	0.9863	0.9863	0.9863	0.9863	0.9797	0.9797	0.9797	0.9797	0.9733
w_B	0.2407	0.0940	0.0000	0.4440	0.2978	0.1711	0.0000	0.4821	0.3446
w_C	0.0320	0.0650	0.0898	0.0000	0.0280	0.0590	0.1100	0.0000	0.0308
a_A	0.9733	0.9733	0.9733	0.9670	0.9670	0.9670	0.9670	0.9590	0.9590
w_B	0.1788	0.0000	0.4951	0.4025	0.1944	0.0000	0.5000	0.4048	0.1967
w_C	0.0753	0.1380	0.0000	0.0217	0.0779	0.1477	0.0000	0.0218	0.0788
a_A	0.9590	0.9590	0.9554	0.9554	0.9554	0.9554	0.9540	0.9540	0.9540
w_B	0.0000	0.5260	0.4528	0.2589	0.0000				
w_C	0.1510	0.0000	0.0208	0.0789	0.1802				
a_A	0.9540	0.9451	0.9451	0.9451	0.9451				

T/K = 308.15

w_B	0.3970	0.2590	0.0590	0.0000	0.4445	0.2828	0.1023	0.0000	0.4560
w_C	0.0000	0.0205	0.0578	0.0711	0.0000	0.0273	0.0670	0.0938	0.0000
a_A	0.9787	0.9787	0.9787	0.9787	0.9714	0.9714	0.9714	0.9714	0.9691
w_B	0.3250	0.0772	0.0000	0.4835	0.3760	0.1336	0.0000	0.4952	0.3496
w_C	0.0245	0.0807	0.1038	0.0000	0.0218	0.0818	0.1257	0.0000	0.0312
a_A	0.9691	0.9691	0.9691	0.9629	0.9629	0.9629	0.9629	0.9601	0.9601
w_B	0.1777	0.0000	0.5081	0.4094	0.1953	0.0000	0.5289	0.3751	0.1950
w_C	0.0749	0.1350	0.0000	0.0220	0.0783	0.1463	0.0000	0.0347	0.0861
a_A	0.9601	0.9601	0.9560	0.9560	0.9560	0.9560	0.9504	0.9504	0.9504
w_B	0.0000	0.5635	0.4226	0.2697	0.0000				
w_C	0.1559	0.0000	0.0397	0.0931	0.2017				
a_A	0.9504	0.9371	0.9371	0.9371	0.9371				

continued

continued

T/K = 318.15

w_B	0.3330	0.1606	0.0826	0.0000	0.4025	0.2657	0.1078	0.0000	0.4520
w_C	0.0000	0.0207	0.0321	0.0465	0.0000	0.0210	0.0502	0.0742	0.0000
a_A	0.9864	0.9864	0.9864	0.9864	0.9782	0.9782	0.9782	0.9782	0.9724
w_B	0.2513	0.0960	0.0000	0.4600	0.2593	0.1354	0.0000	0.4850	0.3164
w_C	0.0334	0.0664	0.0942	0.0000	0.0359	0.0606	0.0997	0.0000	0.0306
a_A	0.9724	0.9724	0.9724	0.9715	0.9715	0.9715	0.9715	0.9669	0.9669
w_B	0.1191	0.0000	0.5481	0.3726	0.1930	0.0000	0.5700	0.3996	0.2131
w_C	0.0779	0.1156	0.0000	0.0345	0.0852	0.1550	0.0000	0.0369	0.0941
a_A	0.9669	0.9669	0.9507	0.9507	0.9507	0.9507	0.9439	0.9439	0.9439
w_B	0.0000	0.5940	0.4868	0.2140	0.0000				
w_C	0.1748	0.0000	0.0282	0.1310	0.2164				
a_A	0.9439	0.9325	0.9325	0.9325	0.9325				

Polymer (B): **poly(1-vinyl-2-pyrrolidinone)** **2005SA1**
Characterization: M_w/g.mol^{-1} = 10000, PVP K15,
Aldrich Chem. Co., Inc., Milwaukee, WI
Solvent (A): **water** **H_2O** **7732-18-5**
Salt (C): **sodium chloride** **NaCl** **7647-14-5**

Type of data: vapor-liquid equilibrium

T/K = 298.15

w_A	0.7179	0.8097	0.8879	0.9513	0.9810	0.7090	0.7998	0.8835	0.9486
w_B	0.2821	0.1833	0.1000	0.0318	0.0000	0.2910	0.1928	0.1033	0.0336
w_C	0.0000	0.0070	0.0121	0.0169	0.0190	0.0000	0.0074	0.0132	0.0178
a_A	0.9890	0.9890	0.9890	0.9890	0.9890	0.9884	0.9884	0.9884	0.9884
w_A	0.9799	0.6142	0.8418	0.8429	0.9352	0.9637	0.5638	0.7542	0.8834
w_B	0.0000	0.3858	0.1338	0.1321	0.0302	0.0000	0.4362	0.2192	0.0733
w_C	0.0201	0.0000	0.0244	0.0250	0.0346	0.0363	0.0000	0.0266	0.0433
a_A	0.9884	0.9794	0.9794	0.9794	0.9794	0.9794	0.9685	0.9685	0.9685
w_A	0.9172	0.9473	0.5470	0.6335	0.7388	0.8249	0.9452	0.5150	0.6274
w_B	0.0346	0.0000	0.4530	0.3535	0.2333	0.1360	0.0000	0.4850	0.3534
w_C	0.0482	0.0527	0.0000	0.0130	0.0279	0.0391	0.0548	0.0000	0.0192
a_A	0.9685	0.9685	0.9669	0.9669	0.9669	0.9669	0.9669	0.9588	0.9588
w_A	0.7542	0.8395	0.9331	0.5056	0.5843	0.6982	0.7794	0.9266	0.4650
w_B	0.2060	0.1083	0.0000	0.4944	0.4010	0.2670	0.1713	0.0000	0.5350
w_C	0.0398	0.0522	0.0669	0.0000	0.0147	0.0348	0.0493	0.0734	0.0000
a_A	0.9588	0.9588	0.9588	0.9543	0.9543	0.9543	0.9543	0.9543	0.9417
w_A	0.5628	0.6710	0.8021	0.9092	0.4059	0.5060	0.6110	0.7316	0.8890
w_B	0.4172	0.2869	0.1290	0.0000	0.5941	0.4716	0.3412	0.1933	0.0000
w_C	0.0200	0.0421	0.0689	0.0908	0.0000	0.0224	0.0478	0.0751	0.1110
a_A	0.9417	0.9417	0.9417	0.9417	0.9259	0.9259	0.9259	0.9259	0.9259

continued

continued

T/K = 308.15

w_A	0.6755	0.8059	0.8737	0.9359	0.9766	0.6015	0.7333	0.8217	0.8990
w_B	0.3245	0.1839	0.1109	0.0442	0.0000	0.3985	0.2526	0.1550	0.0697
w_C	0.0000	0.0102	0.0154	0.0199	0.0234	0.0000	0.0141	0.0233	0.0313
a_A	0.9865	0.9865	0.9865	0.9865	0.9865	0.9774	0.9774	0.9774	0.9774
w_A	0.9615	0.5162	0.6593	0.7149	0.8515	0.9406	0.5023	0.6165	0.6928
w_B	0.0000	0.4838	0.3205	0.2565	0.1016	0.0000	0.4977	0.3653	0.2782
w_C	0.0385	0.0000	0.0202	0.0286	0.0469	0.0594	0.0000	0.0182	0.0290
a_A	0.9774	0.9640	0.9640	0.9640	0.9640	0.9640	0.9603	0.9603	0.9603
w_A	0.8456	0.9351	0.4703	0.5743	0.7070	0.8155	0.9197	0.4298	0.5437
w_B	0.1024	0.0000	0.5297	0.4055	0.2505	0.1224	0.0000	0.5702	0.4313
w_C	0.0520	0.0649	0.0000	0.0202	0.0425	0.0621	0.0803	0.0000	0.0250
a_A	0.9603	0.9603	0.9496	0.9496	0.9496	0.9496	0.9496	0.9329	0.9329
w_A	0.6061	0.7543	0.8974	0.4163	0.5075	0.6301	0.7601	0.8904	0.3727
w_B	0.3544	0.1745	0.0000	0.5837	0.4692	0.3191	0.1591	0.0000	0.6273
w_C	0.0395	0.0712	0.1026	0.0000	0.0233	0.0508	0.0808	0.1096	0.0000
a_A	0.9329	0.9329	0.9329	0.9273	0.9273	0.9273	0.9273	0.9273	0.9051
w_A	0.5350	0.6248	0.7060	0.8643					
w_B	0.4183	0.3040	0.2011	0.0000					
w_C	0.0467	0.0712	0.0929	0.1357					
a_A	0.9051	0.9051	0.9051	0.9051					

T/K = 318.15

w_A	0.6295	0.7703	0.8625	0.9369	0.9707	0.5525	0.7249	0.8143	0.8940
w_B	0.3705	0.2191	0.1189	0.0365	0.0000	0.4475	0.2584	0.1579	0.0678
w_C	0.0000	0.0106	0.0186	0.0266	0.0293	0.0000	0.0167	0.0278	0.0382
a_A	0.9830	0.9830	0.9830	0.9830	0.9830	0.9733	0.9733	0.9733	0.9733
w_A	0.9548	0.5396	0.7279	0.8146	0.9154	0.9517	0.4772	0.6476	0.7555
w_B	0.0000	0.4604	0.2492	0.1527	0.0401	0.0000	0.5228	0.3272	0.2036
w_C	0.0452	0.0000	0.0229	0.0327	0.0445	0.0483	0.0000	0.0252	0.0409
a_A	0.9733	0.9713	0.9713	0.9713	0.9713	0.9713	0.9589	0.9589	0.9589
w_A	0.8609	0.9327	0.4690	0.5695	0.7448	0.8527	0.9290	0.4092	0.5573
w_B	0.0831	0.0000	0.5310	0.4150	0.2134	0.0894	0.0000	0.5908	0.4110
w_C	0.0560	0.0673	0.0000	0.0155	0.0418	0.0579	0.0710	0.0000	0.0317
a_A	0.9589	0.9589	0.9563	0.9563	0.9563	0.9563	0.9563	0.9336	0.9336
w_A	0.6725	0.8000	0.8980	0.3905	0.4608	0.6237	0.7440	0.8863	0.3835
w_B	0.2727	0.1195	0.0000	0.6095	0.5225	0.3217	0.1735	0.0000	0.6165
w_C	0.0548	0.0805	0.1020	0.0000	0.0167	0.0546	0.0825	0.1137	0.0000
a_A	0.9336	0.9336	0.9336	0.9243	0.9243	0.9243	0.9243	0.9243	0.9214
w_A	0.5298	0.6139	0.7677	0.8828					
w_B	0.4356	0.3313	0.1427	0.0000					
w_C	0.0346	0.0548	0.0896	0.1172					
a_A	0.9214	0.9214	0.9214	0.9214					

Polymer (B): **poly(1-vinyl-2-pyrrolidinone)** **2006SA3**
Characterization: M_w/g.mol^{-1} = 10000, PVP K15,
Aldrich Chem. Co., Inc., Milwaukee, WI
Solvent (A): **water** **H_2O** **7732-18-5**
Salt (C): **sodium citrate** **$C_6H_5Na_3O_7$** **68-04-2**

Type of data: vapor-liquid equilibrium

T/K = 298.15

w_A	0.6800	0.7846	0.8391	0.9320	0.6480	0.7618	0.8476	0.9214	0.6220
w_B	0.3200	0.1936	0.1247	0.0000	0.3520	0.2158	0.1041	0.0000	0.3780
w_C	0.0000	0.0218	0.0362	0.0680	0.0000	0.0224	0.0483	0.0786	0.0000
a_A	0.9852	0.9852	0.9852	0.9852	0.9828	0.9828	0.9828	0.9828	0.9804
w_A	0.7539	0.8339	0.9110	0.5924	0.7025	0.7838	0.8827	0.5810	0.6724
w_B	0.2186	0.1124	0.0000	0.4076	0.2632	0.1486	0.0000	0.4190	0.2989
w_C	0.0275	0.0537	0.0890	0.0000	0.0343	0.0676	0.1173	0.0000	0.0287
a_A	0.9804	0.9804	0.9804	0.9740	0.9740	0.9740	0.9740	0.9729	0.9729
w_A	0.7621	0.8776							
w_B	0.1737	0.0000							
w_C	0.0642	0.1224							
a_A	0.9729	0.9729							

T/K = 308.15

w_A	0.6499	0.7402	0.8353	0.9294	0.6215	0.7155	0.8087	0.9123	0.5950
w_B	0.3501	0.2469	0.1284	0.0000	0.3785	0.2650	0.1450	0.0000	0.4050
w_C	0.0000	0.0129	0.0363	0.0706	0.0000	0.0195	0.0463	0.0877	0.0000
a_A	0.9838	0.9838	0.9838	0.9838	0.9807	0.9807	0.9807	0.9807	0.9776
w_A	0.7256	0.7768	0.8882	0.5850	0.6966	0.7844	0.8912	0.5490	0.6607
w_B	0.2407	0.1703	0.0000	0.4150	0.2785	0.1617	0.0000	0.4510	0.3081
w_C	0.0337	0.0529	0.1118	0.0000	0.0249	0.0539	0.1088	0.0000	0.0312
a_A	0.9776	0.9776	0.9776	0.9763	0.9763	0.9763	0.9763	0.9707	0.9707
w_A	0.7467	0.8657	0.5220	0.6454	0.7322	0.8378			
w_B	0.1809	0.0000	0.4780	0.3120	0.1814	0.0000			
w_C	0.0724	0.1343	0.0000	0.0426	0.0864	0.1622			
a_A	0.9707	0.9707	0.9640	0.9640	0.9640	0.9640			

T/K = 318.15

w_A	0.6650	0.8186	0.8857	0.9382	0.6270	0.7829	0.8593	0.9201	0.5890
w_B	0.3350	0.1570	0.0720	0.0000	0.3730	0.1879	0.0887	0.0000	0.4110
w_C	0.0000	0.0244	0.0423	0.0618	0.0000	0.0292	0.0520	0.0799	0.0000
a_A	0.9863	0.9863	0.9863	0.9863	0.9823	0.9823	0.9823	0.9823	0.9785
w_A	0.7235	0.7857	0.8952	0.5630	0.6647	0.7608	0.8781	0.5480	0.6667
w_B	0.2459	0.1621	0.0000	0.4370	0.3113	0.1786	0.0000	0.4520	0.3001
w_C	0.0306	0.0522	0.1048	0.0000	0.0240	0.0606	0.1219	0.0000	0.0332
a_A	0.9785	0.9785	0.9785	0.9747	0.9747	0.9747	0.9747	0.9721	0.9721

continued

continued

w_A	0.7645	0.8689	0.5060	0.6585	0.7294	0.8486
w_B	0.1645	0.0000	0.4940	0.2933	0.1954	0.0000
w_C	0.0710	0.1311	0.0000	0.0482	0.0752	0.1514
a_A	0.9721	0.9721	0.9654	0.9654	0.9654	0.9654

Polymer (B): **poly(1-vinyl-2-pyrrolidinone)** **2006SA2**
Characterization: M_w/g.mol^{-1} = 10000, PVP K15,
Aldrich Chem. Co., Inc., Milwaukee, WI
Solvent (A): **water** H_2O **7732-18-5**
Salt (C): **monosodium phosphate** NaH_2PO_4 **7558-80-7**

Type of data: vapor-liquid equilibrium

T/K = 298.15

w_A	0.6915	0.8246	0.8864	0.9520	0.6808	0.7942	0.8908	0.9475	0.6807
w_B	0.3085	0.1544	0.0815	0.0000	0.3192	0.1864	0.0714	0.0000	0.3193
w_C	0.0000	0.0210	0.0321	0.0480	0.0000	0.0194	0.0378	0.0525	0.0000
a_A	0.9872	0.9872	0.9872	0.9872	0.9859	0.9859	0.9859	0.9859	0.9857
w_A	0.8091	0.8733	0.9468	0.6648	0.7894	0.8596	0.9415	0.6142	0.7303
w_B	0.1680	0.0909	0.0000	0.3352	0.1853	0.1007	0.0000	0.3858	0.2443
w_C	0.0229	0.0358	0.0532	0.0000	0.0253	0.0397	0.0585	0.0000	0.0254
a_A	0.9857	0.9857	0.9857	0.9848	0.9848	0.9848	0.9848	0.9787	0.9787
w_A	0.8427	0.9191	0.6025	0.7127	0.8315	0.9107	0.5277	0.6292	0.7535
w_B	0.1028	0.0000	0.3975	0.2612	0.1081	0.0000	0.4723	0.3370	0.1581
w_C	0.0545	0.0809	0.0000	0.0261	0.0604	0.0893	0.0000	0.0338	0.0884
a_A	0.9787	0.9787	0.9766	0.9766	0.9766	0.9766	0.9614	0.9614	0.9614
w_A	0.8514	0.5053	0.6001	0.7149	0.8304				
w_B	0.0000	0.4947	0.3635	0.1829	0.0000				
w_C	0.1486	0.0000	0.0364	0.1022	0.1696				
a_A	0.9614	0.9561	0.9561	0.9561	0.9561				

T/K = 308.15

w_A	0.6632	0.7854	0.8691	0.9485	0.6000	0.7289	0.8163	0.9212	0.5650
w_B	0.3368	0.1941	0.0967	0.0000	0.4000	0.2452	0.1358	0.0000	0.4350
w_C	0.0000	0.0205	0.0342	0.0515	0.0000	0.0259	0.0479	0.0788	0.0000
a_A	0.9855	0.9855	0.9855	0.9855	0.9783	0.9783	0.9783	0.9783	0.9730
w_A	0.7471	0.8280	0.8973	0.5328	0.6510	0.7475	0.8728	0.5275	0.6645
w_B	0.2071	0.0974	0.0000	0.4672	0.3175	0.1868	0.0000	0.4725	0.2964
w_C	0.0458	0.0746	0.1027	0.0000	0.0315	0.0657	0.1272	0.0000	0.0391
a_A	0.9730	0.9730	0.9730	0.9671	0.9671	0.9671	0.9671	0.9658	0.9658
w_A	0.7293	0.8704	0.4915	0.6132	0.7148	0.8372	0.4560	0.5629	0.6914
w_B	0.2075	0.0000	0.5085	0.3466	0.2018	0.0000	0.5440	0.3971	0.1942
w_C	0.0632	0.1296	0.0000	0.0402	0.0834	0.1628	0.0000	0.0400	0.1144
a_A	0.9658	0.9658	0.9557	0.9557	0.9557	0.9557	0.9453	0.9453	0.9453

continued

continued

w_A	0.8028	0.4492	0.5604	0.6560	0.7929	0.4163	0.5004	0.6184	0.7379
w_B	0.0000	0.5508	0.3939	0.2433	0.0000	0.5837	0.4538	0.2401	0.0000
w_C	0.1972	0.0000	0.0457	0.1007	0.2071	0.0000	0.0458	0.1415	0.2621
a_A	0.9453	0.9431	0.9431	0.9431	0.9431	0.9273	0.9273	0.9273	0.9273

T/K = 318.15

w_A	0.7037	0.7866	0.8833	0.9548	0.6390	0.7554	0.8478	0.9321	0.6060
w_B	0.2963	0.1990	0.0854	0.0000	0.3610	0.2239	0.1095	0.0000	0.3940
w_C	0.0000	0.0144	0.0313	0.0452	0.0000	0.0207	0.0427	0.0679	0.0000
a_A	0.9891	0.9891	0.9891	0.9891	0.9834	0.9834	0.9834	0.9834	0.9794
w_A	0.7284	0.8260	0.9200	0.5565	0.6739	0.7558	0.9040	0.5560	0.6888
w_B	0.2463	0.1261	0.0000	0.4435	0.3029	0.2001	0.0000	0.4440	0.2822
w_C	0.0253	0.0479	0.0800	0.0000	0.0232	0.0441	0.0960	0.0000	0.0290
a_A	0.9794	0.9794	0.9794	0.9735	0.9735	0.9735	0.9735	0.9741	0.9741
w_A	0.7921	0.9009	0.5168	0.6063	0.7147	0.8815	0.4670	0.5627	0.6776
w_B	0.1507	0.0000	0.4832	0.3733	0.2299	0.0000	0.5330	0.4080	0.2457
w_C	0.0572	0.0991	0.0000	0.0204	0.0554	0.1185	0.0000	0.0293	0.0767
a_A	0.9741	0.9741	0.9679	0.9679	0.9679	0.9679	0.9558	0.9558	0.9558
w_A	0.8275	0.4570	0.5685	0.6683	0.8293				
w_B	0.0000	0.5430	0.3991	0.2559	0.0000				
w_C	0.1725	0.0000	0.0324	0.0758	0.1707				
a_A	0.9558	0.9530	0.9530	0.9530	0.9530				

Polymer (B): **poly(1-vinyl-2-pyrrolidinone)** **2010SA2**
Characterization: M_w/g.mol^{-1} = 10000, PVP K15,
Aldrich Chem. Co., Inc., Milwaukee, WI
Solvent (A): **water** H_2O **7732-18-5**
Salt (C): **sodium 1-pentanesulfonate** $C_5H_{11}NaO_3S$ **22767-49-3**

Type of data: vapor-liquid equilibrium

T/K = 298.15

w_B	0.3837	0.1894	0.1078	0.0358	0.0000	0.4209	0.2299	0.1376	0.0473
w_C	0.0000	0.0383	0.0499	0.0584	0.0616	0.0000	0.0464	0.0637	0.0773
a_A	0.9871	0.9871	0.9871	0.9871	0.9871	0.9820	0.9820	0.9820	0.9820
w_B	0.0000	0.4513	0.2626	0.1634	0.0581	0.0000	0.4858	0.3320	0.1920
w_C	0.0830	0.0000	0.0531	0.0757	0.0946	0.1034	0.0000	0.0521	0.0914
a_A	0.9820	0.9770	0.9770	0.9770	0.9770	0.9770	0.9697	0.9697	0.9697
w_B	0.0727	0.0000	0.4990	0.3149	0.2075	0.0786	0.0000	0.5418	0.3792
w_C	0.1195	0.1323	0.0000	0.0639	0.0961	0.1284	0.1439	0.0000	0.0664
a_A	0.9697	0.9697	0.9666	0.9666	0.9666	0.9666	0.9666	0.9542	0.9542
w_B	0.2514	0.1069	0.0000	0.5829	0.4226	0.2930	0.1394	0.0000	0.5856
w_C	0.1156	0.1660	0.2006	0.0000	0.0740	0.1347	0.2163	0.3157	0.0000
a_A	0.9542	0.9542	0.9542	0.9389	0.9389	0.9389	0.9389	0.9389	0.9373

continued

continued

w_B	0.4405	0.3221	0.1853	0.0000	0.6237	0.4740	0.3455	0.1798	0.0000
w_C	0.0678	0.1239	0.1951	0.3265	0.0000	0.0830	0.1588	0.2792	0.4294
a_A	0.9373	0.9373	0.9373	0.9373	0.9125	0.9125	0.9125	0.9125	0.9125
w_B	0.3855	0.1922	0.1115	0.0374	0.0000	0.4325	0.2384	0.1440	0.0501
w_C	0.0000	0.0388	0.0516	0.0611	0.0626	0.0000	0.0482	0.0667	0.0818
a_A	0.9870	0.9870	0.9870	0.9870	0.9870	0.9808	0.9808	0.9808	0.9808
w_B	0.0000	0.4623	0.3055	0.1700	0.0621	0.0000	0.4942	0.3376	0.2225
w_C	0.0883	0.0000	0.0479	0.0809	0.1025	0.1123	0.0000	0.0519	0.0856
a_A	0.9808	0.9749	0.9749	0.9749	0.9749	0.9749	0.9691	0.9691	0.9691
w_B	0.1079	0.0000	0.5161	0.3667	0.2226	0.0879	0.0000	0.5470	0.3960
w_C	0.1137	0.1346	0.0000	0.0575	0.1059	0.1452	0.1655	0.0000	0.0609
a_A	0.9691	0.9691	0.9614	0.9614	0.9614	0.9614	0.9614	0.9537	0.9537
w_B	0.2780	0.1467	0.0000	0.5835	0.4262	0.2979	0.1433	0.0000	0.5905
w_C	0.1069	0.1545	0.2032	0.0000	0.0746	0.1369	0.2224	0.3301	0.0000
a_A	0.9537	0.9537	0.9537	0.9365	0.9365	0.9365	0.9365	0.9365	0.9316
w_B	0.4420	0.2916	0.1764	0.0000	0.6299	0.4867	0.3374	0.2124	0.0000
w_C	0.0723	0.1480	0.2179	0.3614	0.0000	0.0797	0.1733	0.2625	0.4413
a_A	0.9316	0.9316	0.9316	0.9316	0.9083	0.9083	0.9083	0.9083	0.9083

2.3. Classical mass-fraction Henry's constants of water vapor in molten polymers

Polymer (B): **polyester (hyperbranched)** **2010DO1**
Characterization: M_n/g.mol^{-1} = 3200, fatty acid modified,
Boltorn H2004, density (298.15) = 1.0765 cm^3/g, T_g/K = 215.0,
Perstorp Specialty Chemicals AB, Perstorp, Sweden

Solvent (A)	T/ K	$H_{A,B}$/ MPa
water	308.15	0.1720
water	318.15	0.2652
water	328.15	0.4233
water	338.15	0.6126
water	348.15	0.7904

Polymer (B): **polyester (hyperbranched)** **2009DOM**
Characterization: M_n/g.mol^{-1} = 6500, M_w/g.mol^{-1} = 9750,
fatty acid modified, Boltorn U3000,
Perstorp Specialty Chemicals AB, Perstorp, Sweden

Solvent (A)	T/ K	$H_{A,B}$/ MPa
water	308.15	0.4151
water	318.15	0.6126
water	328.15	0.9089
water	338.15	1.188
water	348.15	1.618

Polymer (B): **polyester (hyperbranched)** **2010DO2**
Characterization: M_n/g.mol^{-1} = 9000, fatty acid modified,
Boltorn W3000, density (308.15) = 1.0408 cm^3/g, T_g/K = 205.2,
Perstorp Specialty Chemicals AB, Perstorp, Sweden

Solvent (A)	T/ K	$H_{A,B}$/ MPa
water	308.15	0.1457
water	318.15	0.2509
water	328.15	0.3991
water	338.15	0.5919
water	348.15	0.9623

Polymer (B): **poly(L-lactic acid-*co*-glycolic acid)** **2006ESE**
Characterization: M_n/g.mol^{-1} = 57500, 35 wt% glycolide, T_g/K = 315,
ρ = 1.24 g/cm^3, Aldrich Chem. Co., Inc., Milwaukee, WI

Solvent (A)	T/ K	$H_{A,B}$/ MPa
water	453.15	4.928
water	463.15	6.877
water	473.15	9.359
water	493.15	16.33

2.4. References

1996LIN Lin, D.-Q., Zhu, Z.-Q., Mei, L.-H., and Yang, L.-R., Isopiestic determination of the water activities of poly(ethylene glycol) + salt + water systems at 25°C, *J. Chem. Eng. Data*, 41, 1040, 1996.

1996MIS Mishima, K., Matsuyama, K., Kutsumi, M., Komorita, N., Tokuyasu, T., Miyake, Y., and Taylor, F., Effect of vinyl alcohol + sodium acrylate copolymer gel on the vapor-liquid equilibrium of 1-propanol + water, *J. Chem. Eng. Data*, 41, 953, 1996.

1996TSU Tsuji, T., Hasegawa, K., Hiaki, T., and Hongo, M., Isothermal vapor-liquid equilibria for the 2-propanol + water system containing poly(ethylene glycol) at 298.15 K, *J. Chem. Eng. Data*, 41, 956, 1996.

1998ZAF Zafarani-Moattar, M.T. and Salabat, A., Measurement and correlation of viscosities, densities, and water activities for the system poly(propylene glycol) + $MgSO_4$ + H_2O at 25°C, *J. Solution Chem.*, 27, 663, 1998.

2003MET Metz, S.J., van der Vegt, N.F.A., Mulder, M.H.V., and Wessling, M., Thermodynamics of water vapor sorption in poly(ethylene oxide) poly(butylene terephthalate) block copolymers (experimental data by S.J. Metz and M. Wessling), *J. Phys. Chem. B*, 107, 13629, 2003.

2003SAL Salabat, A. and Nasirzadeh, K., Measurement and prediction of water activity in PEG + $(NH_4)_2SO_4$ + H_2O systems using polymer scaling laws, *J. Mol. Liq.*, 103-104, 349, 2003.

2003SE2 Seiler, M., Buggert, M., Kavarnou, A., and Arlt, W., From alcohols to hyperbranched polymers: The influence of differently branched additives on the vapor-liquid equilibria of selected azeotropic systems, *J. Chem. Eng. Data*, 48, 933, 2003.

2003SE3 Seiler, M., Köhler, D., and Arlt, W., Hyperbranched polymers: New selective solvents for extractive distillation and solvent extraction, *Separation Purification Technol.*, 30, 179, 2003.

2004ARC Arce, A., Fornasiero, F., Rodriguez, O., Radke, C.J., and Prausnitz, J.M., Sorption and transport of water vapor in thin polymer films at 35°C, *Phys. Chem. Chem. Phys.*, 6, 103, 2004.

2004LAM Lammertz, S., Thermodynamische Eigenschaften wässriger Polyelektrolytlösungen, *Dissertation*, TU Kaiserslautern, 2004.

2004MAT Matsuyama, K. and Mishima, K., Effect of vinyl alcohol + sodium acrylate copolymer gel on the vapor-liquid equilibrium compositions of ethanol + water and 2-propanol + water systems, *J. Chem. Eng. Data*, 49, 1688, 2004.

2004OLI Oliveira, N.S., Oliveira, J., Gomes, T., Ferreira, A., Dorgan, J., and Marrucho, I.M., Gas sorption in poly(lactic acid) and packaging materials, *Fluid Phase Equil.*, 222-223, 317, 2004.

2004PAL Palamara, J.E., Zielinski, J.M., Hamedi, M., Duda, J.L., and Danner, R.P., Vapor-liquid equilibria of water, methanol, and methyl acetate in poly(vinyl acetate) and partially and fully hydrolyzed poly(vinyl alcohol), *Macromolecules*, 37, 6189, 2004.

2004SAD Sadeghi, R. and Zafarani-Moattar, M.T., Thermodynamics of aqueous solutions of polyvinylpyrrolidone, *J. Chem. Thermodyn.*, 36, 665, 2004.

2004SAL Salabat, A., Dashti, H., and Nasirzadeh, K., Measurement and correlation of water activities and refractive indices for the systems PPG 425 + $(NH_4)_2SO_4$ + H_2O and PPG 425 + Na_2SO_4 + H_2O at 298.15 K, *J. Chem. Eng. Data*, 49, 980, 2004.

2004SEI Seiler, M., Rolker, J., Mokrushina, L.V., Kautz, H., Frey, H., and Arlt, W., Vapor–liquid equilibria in dendrimer and hyperbranched polymer solutions: experimental data and modeling using UNIFAC-FV, *Fluid Phase Equil.*, 221, 83, 2004.6

2005CSA Csaki, K.F., Nagy, M., and Csempesz, F., Influence of the chain composition on the thermodynamic properties of binary and ternary polymer solutions (experimental data by F. Csempesz), *Langmuir*, 21, 761, 2005.

2005NIN Ninni, L., Meirelles, A.J.A., and Maurer, G., Thermodynamic properties of aqueous solutions of maltodextrins from laser-light scattering, calorimetry and isopiestic investigations, *Carbohydrate Polym.*, 59, 289, 2005.

2005SA1 Sadeghi, R. and Motamedi, M., Isopiestic determination of water activity in the poly(vinylpyrrolidone) + NaCl + H_2O system at different temperatures, *J. Chem. Eng. Data*, 50, 508, 2005.

2005SA2 Sadeghi, R., Vapor–liquid equilibria of the polyvinylpyrrolidone + $(NH_4)_2SO_4$ + H_2O system at different temperatures, *Fluid Phase Equil.*, 233, 176, 2005.

2005SUG Sugaya, R., Thermodynamics of aqueous biopolymer solutions and fractionation of dextran, Dissertation, Johannes Gutenberg Universität Mainz, 2005.

2006ESE Eser, H. and Tihminlioglu, F., Determination of thermodynamic and transport properties of solvents and non solvents in poly(L-lactide-*co*-glycolide), *J. Appl. Polym. Sci.*, 102, 2426, 2006.

2006OLI Oliveira, N.S., Goncalves, C.M., Coutinho, J.A.P., Ferreira, A., Dorgan, J., and Marrucho, I.M., Carbon dioxide, ethylene and water vapor sorption in poly(lactic acid), *Fluid Phase Equil.*, 250, 116, 2006.

2006SA1 Sadeghi, R. and Ziamajidi, F., Water activities of ternary mixtures of poly(ethylene glycol), NaCl and water over the temperature range of 293.15 K to 313.15 K, *J. Chem. Thermodyn.*, 38, 1335, 2006.

2006SA2 Sadeghi, R., Measurement and correlation of vapor-liquid equilibria of the poly(vinyl pyrrolidone) + NaH_2PO_4 + H_2O system at different temperatures, *CALPHAD*, 30, 53, 2006.

2006SA3 Sadeghi, R., Vapor-liquid equilibrium in aqueous systems containing poly(vinyl pyrrolidone) and sodium citrate at different temperatures. Experiment and modeling, *Fluid Phase Equil.*, 249, 33, 2006.

2006SA4 Sadeghi, R. and Ziamajidi, F., Thermodynamic properties of aqueous polypropylene oxide 400 solutions from isopiestic measurements over a range of temperatures, *Fluid Phase Equil.*, 249, 165, 2006.

2007CHA Cha, D.-H., Lee, J., Im, J., and Kim, H., (Vapor + liquid) equilibria for the {water + poly(ethylene glycol diacetyl ether) (PEGDAE) and methanol + PEGDAE} systems, *J. Chem. Thermodyn.*, 39, 483, 2007.

2007KAZ Kazemia, S., Zafarani-Moattar, M.T., Taghikhani, V., and Ghotbi, C., Measurement and correlation of vapor-liquid equilibria of the aqueous poly(ethylene glycol) + sodium citrate and poly(ethylene glycol) + potassium citrate systems, *Fluid Phase Equil.*, 262, 137, 2007

2007KUL Kulagina, G.S., Chalykh, A.E., Gerasimov, V.K., Chalykh, K.A., and Puryaeva, T.P., Sorption of water by poly(vinyl alcohol), *Polym. Sci., Ser. A*, 49, 425, 2007.

2007SAD Sadeghi, R. and Ziamajidi, F., Vapor-liquid equilibria of binary tripotassium citrate + water and ternary polypropylene oxide 400 + tripotassium citrate + water systems from isopiestic measurements over a range of temperatures, *Fluid Phase Equil.*, 255, 46, 2007.

2007SER Se, R.A.G. and Aznar, M., Vapor-liquid equilibrium of polymer + solvent systems: Experimental data and thermodynamic modeling, *Polymer*, 48, 5646, 2007.

2008LA1 Lammertz, S., Pessoa Filho, P.A., and Maurer, G., Thermodynamics of aqueous solutions of polyelectrolytes: Experimental results for the activity of water in aqueous solutions of some single synthetic polyelectrolytes, *J. Chem. Eng. Data*, 53, 1564, 2008.

2008LA2 Lammertz, S., Pessoa Filho, P.A., and Maurer, G., Thermodynamics of aqueous solutions of polyelectrolytes: Experimental results for the activity of water in aqueous solutions of a single synthetic polyelectrolyte and sodium chloride, *J. Chem. Eng. Data*, 53, 1796, 2008.

2008SAD Sadeghi, R., Hosseini, R., and Jamehbozorg, B., Effect of sodium phosphate salts on the thermodynamic properties of aqueous solutions of poly(ethylene oxide) 6000 at different temperatures, *J. Chem. Thermodyn.*, 40, 1364, 2008.

2008ZAF Zafarani-Moattar, M.T., Hamzehzadeh, S., and Hosseinzadeh, S., Phase diagrams for liquid-liquid equilibrium of ternary poly(ethylene glycol) + disodium tartrate aqueous system and vapor-liquid equilibrium of constituting binary aqueous systems at T = (298.15, 308.15, and 318.15) K. Experiment and correlation, *Fluid Phase Equil.*, 268, 142, 2008.

2009DOM Domanska, U. and Zolek-Tryznowska, Z., Thermodynamic properties of hyperbranched polymer, Boltorn U3000, using inverse gas chromatography, *J. Phys. Chem. B*, 113, 15312, 2009.

2009FO1 Foroutan, M. and Khomami, M.H., Influence of copolymer molar mass on the thermodynamic properties of aqueous solution of an amphiphilic copolymer, *J. Chem. Eng. Data*, 54, 861, 2009.

2009FO2 Foruotan, M. and Zarrabi, M., Influence of molar mass of polymer on the solvent activity for binary system of poly(*N*-vinylcaprolactam) and water, *J. Chem. Thermodyn.*, 41, 448, 2009.

2010CAR Carareto, N.D.D., Monteiro Filho, E.S., Pessoa Filho, P.A., and Meirelles, A.J.A., Water activity of aqueous solutions of ethylene oxide-propylene oxide block copolymers and maltodextrins, *Brazil. J. Chem. Eng.*, 27, 173, 2010.

2010DO1 Domanska, U. and Zolek-Tryznowska, Z., Measurements of mass-fraction activity coefficient at infinite dilution of aliphatic and aromatic hydrocarbons, thiophene, alcohols, water, ethers, and ketones in hyperbranched polymer, Boltorn H2004, using inverse gas chromatography, *J. Chem. Thermodyn.*, 42, 363, 2010.

2010DO2 Domanska, U. and Zolek-Tryznowska, Z., Mass-fraction activity coefficients at infinite dilution measurements for organic solutes and water in the hyperbranched polymer Boltorn W3000 using inverse gas chromatography, *J. Chem. Eng. Data*, 55, 1258, 2010.

2010SA1 Sadeghi, R., Golabiazar, R., and Ziaii, M., Vapor-liquid equilibria, density, speed of sound, and refractive index of sodium tungstate in water and in aqueous solutions of poly(ethyleneglycol) 6000, *J. Chem. Eng. Data*, 55, 125, 2010.

2010SA2 Sadeghi, R. and Ziaii, M., Thermodynamic investigation of the systems poly(ethylene glycol) + sodium pentane-1-sulfonate + water and poly(vinyl pyrrolidone) + sodium pentane-1-sulfonate + water, *J. Colloid Interface Sci.*, 346, 107, 2010.

2010SA3 Sadeghi, R. and Hosseini, R., Vapor-liquid equilibrium properties of sodium n-heptyl sulfonate in water and in aqueous solutions of poly(ethylene glycol) at different temperatures, *J. Iran. Chem. Soc.*, 7, 621, 2010.

2011JEC Jeck, S., Scharfer, P., Schabel, W., and Kind, M., Water sorption in poly(vinyl alcohol) membranes: An experimental and numerical study of solvent diffusion in a crosslinked polymer, *Chem. Eng. Process.*, 50, 543, 2011.

2011MO2 Morariu, S., Nichifor, M., Eckelt, J., and Wolf, B.A., Dextran-based polycations: Thermodynamic interaction with water as compared with unsubstituted dextran, 2 – Flory/Huggins interaction parameter, *Macromol. Chem. Phys.*, 212, 1932, 2011.

2011SA1 Sadeghi, R. and Shahebrahimi, Y., Vapor-liquid equilibria of aqueous polymer solutions from vapor-pressure osmometry and isopiestic measurements, *J. Chem. Eng. Data*, 56, 789, 2011.

2011SA2 Sadeghi, R. and Shahebrahimi, Y., Vapor pressure osmometry determination of solvent activities of different aqueous and nonaqueous polymer solutions at 318.15 K, *J. Chem. Eng. Data*, 56, 2946, 2011.

2011ZAF Zafarani-Moattar, M.T. and Tohidifar, N., Water activities in binary and ternary aqueous systems of poly(ethylene glycol) dimethyl ether 2000 and poly(propylene glycol) 400 at T = (298.15, 308.15, and 318.15) K, *J. Chem. Eng. Data*, 56, 3567, 2011.

3. LIQUID-LIQUID EQUILIBRIUM (LLE) DATA OF AQUEOUS POLYMER SOLUTIONS

3.1. Cloud-point and/or coexistence curves of quasibinary solutions

Polymer (B): **poly(*N*-acryloylasparaginamide)** **2011GLA**
Characterization: see table, synthesized in the laboratory
Solvent (A): **water** **H_2O** **7732-18-5**

Type of data: cloud points (UCST-behavior)

M_n/ g.mol^{-1}	M_w/M_n	w_B	T/ K
17300	1.26	0.010	277.45
27800	1.26	0.010	283.75
60800	1.27	0.010	291.05
100400	1.22	0.010	299.35
118700	1.26	0.010	297.45

Polymer (B): **poly(*N*,*N*-dimethylacrylamide-*co*-allyl methacrylate)** **2005YI2**
Characterization: M_n/g.mol^{-1} = 9200, M_w/M_n = 1.9,
14.0 mol% allyl methacrylate, synthesized in the laboratory
Solvent (A): **water** **H_2O** **7732-18-5**

Type of data: cloud points (LCST-behavior)

w_B 0.005 T/K 345.15

Polymer (B): **poly(*N*,*N*-dimethylacrylamide-*co*-allyl methacrylate)** **2005YI2**
Characterization: M_n/g.mol^{-1} = 10000, M_w/M_n = 2.2,
19.0 mol% allyl methacrylate, synthesized in the laboratory
Solvent (A): **water** **H_2O** **7732-18-5**

Type of data: cloud points (LCST-behavior)

w_B 0.005 T/K 327.15

Polymer (B): **poly(*N*,*N*-dimethylacrylamide-*co*-allyl methacrylate)** **2005YI2**

Characterization: M_n/g.mol^{-1} = 12000, M_w/M_n = 2.3, 21.0 mol% allyl methacrylate, synthesized in the laboratory

Solvent (A): **water** **H_2O** **7732-18-5**

Type of data: cloud points (LCST-behavior)

w_B	0.005	T/K	313.75

Polymer (B): **poly(*N*,*N*-dimethylacrylamide-*co*-allyl methacrylate)** **2005YI2**

Characterization: M_n/g.mol^{-1} = 13000, M_w/M_n = 2.2, 23.0 mol% allyl methacrylate, synthesized in the laboratory

Solvent (A): **water** **H_2O** **7732-18-5**

Type of data: cloud points (LCST-behavior)

w_B	0.005	T/K	302.55

Polymer (B): **poly(*N*,*N*-dimethylacrylamide-*co*-allyl methacrylate)** **2005YI2**

Characterization: M_n/g.mol^{-1} = 13000, M_w/M_n = 2.3, 28.0 mol% allyl methacrylate, synthesized in the laboratory

Solvent (A): **water** **H_2O** **7732-18-5**

Type of data: cloud points (LCST-behavior)

w_B	0.005	T/K	296.25

Polymer (B): **poly(*N*,*N*-dimethylacrylamide-*co*-allyl methacrylate)** **2005YI2**

Characterization: M_n/g.mol^{-1} = 12000, M_w/M_n = 2.1, 30.0 mol% allyl methacrylate, synthesized in the laboratory

Solvent (A): **water** **H_2O** **7732-18-5**

Type of data: cloud points (LCST-behavior)

w_B	0.005	T/K	288.85

Polymer (B): **poly(*N*,*N*-dimethylacrylamide-*co*-*N*-phenyl-acrylamide)** **2005YI1**

Characterization: M_n/g.mol^{-1} = 9700, M_w/M_n = 1.09, 12.0 mol% *N*-phenylacrylamide, synthesized in the laboratory

Solvent (A): **water** **H_2O** **7732-18-5**

Type of data: cloud points (LCST-behavior)

w_B	0.01	T/K	343.25

Polymer (B): **poly(*N,N*-dimethylacrylamide-*co*-*N*-phenyl-acrylamide)** **2005YI1**
Characterization: M_n/g.mol^{-1} = 2000, M_w/M_n = 1.08, 15.9 mol% *N*-phenylacrylamide, synthesized in the laboratory
Solvent (A): **water** H_2O **7732-18-5**

Type of data: cloud points (LCST-behavior)

w_B	0.01	T/K	312.45

Polymer (B): **poly(*N,N*-dimethylacrylamide-*co*-*N*-phenyl-acrylamide)** **2005YI1**
Characterization: M_n/g.mol^{-1} = 10200, M_w/M_n = 1.07, 15.9 mol% *N*-phenylacrylamide, synthesized in the laboratory
Solvent (A): **water** H_2O **7732-18-5**

Type of data: cloud points (LCST-behavior)

w_B	0.01	T/K	316.65

Polymer (B): **poly(*N,N*-dimethylacrylamide-*co*-*N*-phenyl-acrylamide)** **2005YI1**
Characterization: M_n/g.mol^{-1} = 3500, M_w/M_n = 1.06, 16.2 mol% *N*-phenylacrylamide, synthesized in the laboratory
Solvent (A): **water** H_2O **7732-18-5**

Type of data: cloud points (LCST-behavior)

w_B	0.01	T/K	313.35

Polymer (B): **poly(*N,N*-dimethylacrylamide-*co*-*N*-phenyl-acrylamide)** **2005YI1**
Characterization: M_n/g.mol^{-1} = 4700, M_w/M_n = 1.05, 16.3 mol% *N*-phenylacrylamide, synthesized in the laboratory
Solvent (A): **water** H_2O **7732-18-5**

Type of data: cloud points (LCST-behavior)

w_B	0.01	T/K	312.25

Polymer (B): **poly(*N,N*-dimethylacrylamide-*co*-*N*-phenyl-acrylamide)** **2005YI1**
Characterization: M_n/g.mol^{-1} = 4600, M_w/M_n = 1.07, 21.7 mol% *N*-phenylacrylamide, synthesized in the laboratory
Solvent (A): **water** H_2O **7732-18-5**

Type of data: cloud points (LCST-behavior)

w_B	0.01	T/K	290.35

Polymer (B): **poly(*N,N*-dimethylacrylamide-*co*-*N*-phenyl-acrylamide)** **2005YI1**

Characterization: M_n/g.mol^{-1} = 8600, M_w/M_n = 1.07, 21.7 mol% *N*-phenylacrylamide, synthesized in the laboratory

Solvent (A): **water** **H_2O** **7732-18-5**

Type of data: cloud points (LCST-behavior)

w_B 0.01 T/K 293.25

Polymer (B): **poly(*N,N*-dimethylacrylamide-*co*-*N*-phenyl-acrylamide)** **2005YI1**

Characterization: M_n/g.mol^{-1} = 3200, M_w/M_n = 1.07, 21.8 mol% *N*-phenylacrylamide, synthesized in the laboratory

Solvent (A): **water** **H_2O** **7732-18-5**

Type of data: cloud points (LCST-behavior)

w_B 0.01 T/K 289.45

Polymer (B): **poly(*N,N*-dimethylacrylamide-*co*-*N*-phenyl-acrylamide)** **2005YI1**

Characterization: M_n/g.mol^{-1} = 10600, M_w/M_n = 1.07, 22.0 mol% *N*-phenylacrylamide, synthesized in the laboratory

Solvent (A): **water** **H_2O** **7732-18-5**

Type of data: cloud points (LCST-behavior)

w_B 0.01 T/K 293.25

Polymer (B): **poly[2-(*N,N*-dimethylamino)ethyl methacrylate]** **2010DO1**

Characterization: M_n/g.mol^{-1} = 7700, M_w/M_n = 1.36, synthesized in the laboratory

Solvent (A): **water** **H_2O** **7732-18-5**

Type of data: cloud points (LCST-behavior)

pH = 9.27

w_B	0.005	0.010	0.020
T/K	308.05	303.25	301.85

Polymer (B): **poly[2-(*N,N*-dimethylamino)ethyl methacrylate-*b*-ethylene glycol-*b*-(2-(*N,N*-dimethylamino) ethyl methacrylate]** **2010DO1**

Characterization: M_n/g.mol^{-1} = 16500, M_w/M_n = 1.22, 13.2 wt% PEG, PDMAEMA42-PEG45-PDMAEMA42

Solvent (A): **water** **H_2O** **7732-18-5**

Type of data: cloud points (LCST-behavior)

pH = 9.27

w_B	0.005	0.010	0.020
T/K	313.65	310.75	308.35

Polymer (B):	**poly[2-(*N,N*-dimethylamino)ethyl methacrylate-*b*-ethylene glycol-*b*-(2-(*N,N*-dimethylamino) ethyl methacrylate]**	**2010DO1**
Characterization:	M_n/g.mol^{-1} = 22200, M_w/M_n = 1.28, 9.6 wt% PEG, PDMAEMA60-PEG45-PDMAEMA60	
Solvent (A):	**water** **H_2O**	**7732-18-5**

Type of data: cloud points (LCST-behavior)

pH = 9.27

w_B	0.005	0.010	0.020
T/K	311.05	310.55	306.95

Polymer (B):	**polyester (hyperbranched)**	**2011ZEI1, 2011ZEI2**
Characterization:	M_w/g.mol^{-1} = 2100, M_w/M_n = 1.3, Boltorn H20, Perstorp Specialty Chemicals AB, Perstorp, Sweden	
Solvent (A):	**water** **H_2O**	**7732-18-5**

Type of data: cloud points (UCST-behavior)

w_B	0.0086	0.0390	0.0558	0.106	0.150	0.200	0.250	0.300	0.350
T/K	350.65	354.40	359.65	366.75	371.65	371.65	371.65	371.65	368.65
w_B	0.400	0.450	0.500						
T/K	365.95	364.65	360.65						

Polymer (B):	**poly[2-(2-ethoxyethoxy)ethyl vinyl ether]**	**2005MAT**
Characterization:	M_n/g.mol^{-1} = 21900, M_w/g.mol^{-1} = 29100, M_z/g.mol^{-1} = 37900, synthesized in the laboratory	
Solvent (A):	**water** **H_2O**	**7732-18-5**

Type of data: cloud points (LCST-behavior)

c_B/(g/cm^3)	0.00509	0.01046	0.02051	0.03068	0.03925	0.05177	0.07312	0.10907
T/K	314.59	314.31	314.08	313.95	313.87	313.84	313.76	313.74
c_B/(g/cm^3)	0.15243	0.18909	0.23467	0.27401	0.30624	0.33915	0.37663	
T/K	313.69	313.69	313.62	313.59	313.56	313.52	313.47	

Type of data: coexistence data (tie lines, LCST-behavior)

Comments: The total feed concentration of the polymer is c_B/(g/cm^3) = 0.155. The corresponding cloud point is at 313.69 K.

T/K	314.15	313.95
c_B/(g/cm^3) (sol phase)	0.06489	0.09485
c_B/(g/cm^3) (gel phase)	0.31356	0.29823

Polymer (B): **poly[2-(2-ethoxyethoxy)ethyl vinyl ether]** **2005MAT**
Characterization: M_n/g.mol^{-1} = 30400, M_w/g.mol^{-1} = 38300, M_z/g.mol^{-1} = 49800
Solvent (A): **water** **H_2O** **7732-18-5**

Type of data: cloud points (LCST-behavior)

c_B/(g/cm^3)	0.01056	0.01927	0.02942	0.04314	0.05927	0.10567	0.11490	0.13087
T/K	313.87	313.77	313.69	313.66	313.56	313.52	313.59	313.62
c_B/(g/cm^3)	0.14401	0.18764	0.25401	0.31642	0.36205	0.40431	0.44235	0.48308
T/K	313.63	313.58	313.60	313.69	313.70	313.69	313.70	313.65
c_B/(g/cm^3)	0.53165	0.53421						
T/K	313.58	313.55						

Type of data: coexistence data (tie lines, LCST-behavior)

Comments: The total feed concentration of the polymer is c_B/(g/cm^3) = 0.182. The corresponding cloud point is at 313.6 K.

T/K	314.05	313.85	313.65
c_B/(g/cm^3) (sol phase)	0.05280	0.06173	0.08848
c_B/(g/cm^3) (gel phase)	0.32450	0.30427	0.24758

Polymer (B): **poly(ethoxytriethylene glycol methacrylate)** **2008BEC**
Characterization: M_n/g.mol^{-1} = 8500, M_w/M_n = 1.18
Solvent (A): **water** **H_2O** **7732-18-5**

Type of data: cloud points (LCST-behavior)

c_B/(mg/ml)	5.0	5.0	5.0
pH	4	7	10
T/K	301.45	293.15	294.45

Polymer (B): **poly(ethoxytriethylene glycol methacrylate)** **2008BEC**
Characterization: M_n/g.mol^{-1} = 13500, M_w/M_n = 1.21
Solvent (A): **water** **H_2O** **7732-18-5**

Type of data: cloud points (LCST-behavior)

c_B/(mg/ml)	5.0	5.0	5.0
pH	4	7	10
T/K	300.75	294.75	295.95

Polymer (B): **poly[ethylene glycol-*b*-(2-(*N*,*N*-dimethylamino) ethyl methacrylate]** **2010DO1**
Characterization: M_n/g.mol^{-1} = 12300, M_w/M_n = 1.35, 17.5 wt% PEG, PEG45-PDMAEMA60, synthesized in the laboratory
Solvent (A): **water** **H_2O** **7732-18-5**

Type of data: cloud points (LCST-behavior)

pH = 9.27

w_B	0.005	0.010	0.020
T/K	326.35	312.85	309.15

Polymer (B): **poly[ethylene glycol-*b*-(2-(*N,N*-dimethylamino) ethyl methacrylate]** **2010DO1**
Characterization: M_n/g.mol^{-1} = 17000, M_w/M_n = 1.36, 13.2 wt% PEG, PEG45-PDMAEMA84, synthesized in the laboratory
Solvent (A): **water** **H_2O** **7732-18-5**

Type of data: cloud points (LCST-behavior)

pH = 9.27

w_B	0.005	0.010	0.020
T/K	323.25	311.25	308.75

Polymer (B): **poly[ethylene glycol-*b*-(2-(*N,N*-dimethylamino) ethyl methacrylate]** **2010DO1**
Characterization: M_n/g.mol^{-1} = 18800, M_w/M_n = 1.31, 27.5 wt% PEG, PEG113-PDMAEMA84, synthesized in the laboratory
Solvent (A): **water** **H_2O** **7732-18-5**

Type of data: cloud points (LCST-behavior)

pH = 9.27

w_B	0.010	0.020
T/K	313.95	309.85

Polymer (B): **poly(ethylene oxide-*co*-allyl glycidyl ether)** **2011MAN**
Characterization: see table, synthesized in the laboratory
Solvent (A): **water** **H_2O** **7732-18-5**

Type of data: cloud points (LCST-behavior)

copolymer	M_n/ g.mol^{-1}	M_w/M_n	comonomer content/mol%	c_B/ mg/mL	T/ K
Bn2NP(EO52-*co*-AGE5)	3900	1.13	9	5	346.05
Bn2NP(EO52-*co*-AGE5)	3900	1.13	9	10	347.65
Bn2NP(EO52-*co*-AGE5)	3900	1.13	9	20	346.95
Bn2NP(EO102-*co*-AGE16)	4600	1.11	14	5	335.55
MeOP(EO114-*co*-AGE29)	7000	1.04	20	5	342.95

Polymer (B): **poly(ethylene oxide-*co*-*N,N*-dibenzyl aminoglycidyl)** **2011MAN**
Characterization: see table, synthesized in the laboratory
Solvent (A): **water** **H_2O** **7732-18-5**

Type of data: cloud points (LCST-behavior)

continued

continued

copolymer	M_n/ g.mol^{-1}	M_w/M_n	comonomer content/mol%	c_B/ mg/mL	T/ K
MeOP(EO110-*co*-DBAG2)	4900	1.15	2	5	344.45
MeOP(EO112-*co*-DBAG5)	5400	1.14	4	5	330.05
MeOP(EO112-*co*-DBAG5)	5400	1.14	4	10	330.95
MeOP(EO112-*co*-DBAG5)	5400	1.14	4	20	331.25
MeOP(EO90-*co*-DBAG15)	5800	1.17	9	5	295.05

Polymer (B): **poly(ethylene oxide-*co*-ethoxyl vinyl glycidyl ether)** **2011MAN**
Characterization: see table, synthesized in the laboratory
Solvent (A): **water** **H_2O** **7732-18-5**

Type of data: cloud points (LCST-behavior)

copolymer	M_n/ g.mol^{-1}	M_w/M_n	comonomer content/mol%	c_B/ mg/mL	T/ K
MeOBn2NP(EO104-*co*-EVGE6)	2400	1.06	5	5	355.65
MeOBn2NP(EO115-*co*-EVGE11)	2500	1.08	9	5	341.85
MeOBn2NP(EO115-*co*-EVGE11)	2500	1.08	9	10	342.75
MeOBn2NP(EO120-*co*-EVGE30)	5130	1.08	20	5	304.25
MeOBn2NP(EO120-*co*-EVGE30)	5130	1.08	20	10	308.45
MeOBn2NP(EO120-*co*-EVGE30)	5130	1.08	20	20	307.45
MeOBn2NP(EO89-*co*-EVGE30)	5100	1.11	25	5	282.15

Polymer (B): **poly(ethylene oxide-*co*-isopropylidene glyceryl glycidyl ether)** **2011MAN**
Characterization: see table, synthesized in the laboratory
Solvent (A): **water** **H_2O** **7732-18-5**

Type of data: cloud points (LCST-behavior)

copolymer	M_n/ g.mol^{-1}	M_w/M_n	comonomer content/mol%	c_B/ mg/mL	T/ K
MeOBn2NP(EO265-*co*-IGG26)	8200	1.11	9	5	353.95
MeOBn2NP(EO180-*co*-IGG32)	9750	1.15	15	5	339.85
MeOBn2NP(EO47-*co*-IGG17)	2800	1.08	27	5	319.95
MeOBn2NP(EO47-*co*-IGG17)	2800	1.08	27	10	321.15
MeOBn2NP(EO47-*co*-IGG17)	2800	1.08	27	20	320.05

Polymer (B): **poly(ethylene oxide-*b*-propylene oxide-*b*-ethylene oxide)** **2005ZH2**
Characterization: M_n/g.mol^{-1} = 4950, 30 wt% ethylene oxide, EO17PO60EO17, BASF Chemical Company, Florham Park, NJ
Solvent (A): **water** **H_2O** **7732-18-5**

Type of data: cloud points (LCST-behavior)

w_B	0.10	T/K	325.15

Polymer (B): **poly(ethylene oxide-*b*-propylene oxide-*b*-ethylene oxide)** **2005ZH2**
Characterization: M_n/g.mol^{-1} = 2200, 40 wt% ethylene oxide, EO10PO23EO10, BASF Chemical Company, Florham Park, NJ
Solvent (A): **water** **H_2O** **7732-18-5**

Type of data: cloud points (LCST-behavior)

w_B	0.10	T/K	346.15

Polymer (B): **poly(ethylene oxide-*b*-propylene oxide-*b*-ethylene oxide)** **2005ZH2**
Characterization: M_n/g.mol^{-1} = 2900, 40 wt% ethylene oxide, EO13PO30EO13, BASF Chemical Company, Florham Park, NJ
Solvent (A): **water** **H_2O** **7732-18-5**

Type of data: cloud points (LCST-behavior)

w_B	0.10	T/K	333.15

Polymer (B): **poly(ethylene oxide-*b*-propylene oxide-*b*-ethylene oxide)** **2005ZH2**
Characterization: M_n/g.mol^{-1} = 5900, 40 wt% ethylene oxide, EO27PO61EO27, BASF Chemical Company, Florham Park, NJ
Solvent (A): **water** **H_2O** **7732-18-5**

Type of data: cloud points (LCST-behavior)

w_B	0.10	T/K	351.15

Polymer (B): **poly(ethylene oxide-*b*-propylene oxide-*b*-ethylene oxide)** **2005ZH2**
Characterization: M_n/g.mol^{-1} = 4600, 50 wt% ethylene oxide, EO26PO40EO26, BASF Chemical Company, Florham Park, NJ
Solvent (A): **water** **H_2O** **7732-18-5**

Type of data: cloud points (LCST-behavior)

w_B	0.10	T/K	359.15

Polymer (B): **poly(ethylene oxide-*b*-propylene oxide-*b*-ethylene oxide)** **2005ZH2**
Characterization: M_n/g.mol^{-1} = 6500, 50 wt% ethylene oxide, EO37PO56EO37, BASF Chemical Company, Florham Park, NJ
Solvent (A): **water** **H_2O** **7732-18-5**

Type of data: cloud points (LCST-behavior)

w_B	0.10	T/K	367.15

Polymer (B): **poly(2-hydroxypropyl acrylate)** **2008EGG**
Characterization: M_n/g.mol^{-1} = 8200, M_w/g.mol^{-1} = 9500, synthesized in the laboratory
Solvent (A): **water** **H_2O** **7732-18-5**

Type of data: cloud points (LCST-behavior)

w_B	0.005	0.010
T/K	299.85	294.55

Polymer (B): **poly(2-hydroxypropyl acrylate)** **2008EGG**
Characterization: M_n/g.mol^{-1} = 11100, M_w/g.mol^{-1} = 13400, synthesized in the laboratory
Solvent (A): **water** **H_2O** **7732-18-5**

Type of data: cloud points (LCST-behavior)

w_B	0.005	0.010
T/K	299.85	294.55

Polymer (B): **poly(2-hydroxypropyl acrylate-*co*-*N*-acryloylmorpholine)** **2008EGG**
Characterization: see table, synthesized in the laboratory
Solvent (A): **water** **H_2O** **7732-18-5**

Type of data: cloud points (LCST-behavior)

M_n/ g.mol^{-1}	M_w/M_n	composition ratio wt% (NAM/HPA)	w_B	T/ K
8500	1.23	49/51	0.005	352.65
8500	1.23	49/51	0.010	339.05
8300	1.20	40/60	0.005	335.85
8300	1.20	40/60	0.010	326.15
8800	1.20	30/70	0.005	322.35
8800	1.20	30/70	0.010	319.45
8400	1.20	19/82	0.005	314.65
8400	1.20	19/82	0.010	304.05
8100	1.16	08/92	0.005	337.05
8100	1.16	08/92	0.010	298.45

continued

continued

8200	1.16	0/100	0.005	299.85
8200	1.16	0/100	0.010	294.55

Polymer (B): **poly(2-hydroxypropyl acrylate-*co*-*N,N*-dimethylacrylamide)** **2008EGG**
Characterization: see table, synthesized in the laboratory
Solvent (A): **water** H_2O **7732-18-5**

Type of data: cloud points (LCST-behavior)

M_n/ g.mol^{-1}	M_w/M_n	composition ratio wt% (DMA/HPA)	w_B	T/ K
9800	1.27	43/57	0.010	356.05
9600	1.26	34/66	0.005	344.75
9600	1.26	34/66	0.010	335.45
10900	1.24	26/74	0.005	328.95
10900	1.24	26/74	0.010	321.85
10700	1.22	18/82	0.005	319.85
10700	1.22	18/82	0.010	311.75
10500	1.20	10/90	0.005	308.45
10500	1.20	10/90	0.010	301.65
11100	1.21	0/100	0.005	299.85
11100	1.21	0/100	0.010	294.55

Polymer (B): **poly(*N*-isopropylacrylamide)** **2009KOB**
Characterization: M_n/g.mol^{-1} = 4880, M_w/g.mol^{-1} = 5470, synthesized in the laboratory
Solvent (A): **water** H_2O **7732-18-5**

Type of data: cloud points (LCST-behavior)

w_B	0.0050	0.0100	0.0200	0.0304	0.0385	0.0503	0.0745	0.0907
T/K	297.70	295.00	292.50	291.50	291.40	291.40	290.90	290.80

Polymer (B): **poly(*N*-isopropylacrylamide)** **2007HOU**
Characterization: M_n/g.mol^{-1} = 5680, M_w/g.mol^{-1} = 7210, synthesized in the laboratory
Solvent (A): **water** H_2O **7732-18-5**

Type of data: cloud points (LCST-behavior)

c_B/(mg/ml)	1	2	4	5	10
T/K	310.65	310.15	308.45	307.35	306.35

Polymer (B): **poly(*N*-isopropylacrylamide)** **2009KOB**
Characterization: M_n/g.mol^{-1} = 7740, M_w/g.mol^{-1} = 8980, synthesized in the laboratory
Solvent (A): **water** **H_2O** **7732-18-5**

Type of data: cloud points (LCST-behavior)

w_B	0.0051	0.0100	0.0201	0.0306	0.0407	0.0497	0.0693	0.0848
T/K	297.95	296.46	295.49	295.11	294.97	294.93	294.92	294.63

Polymer (B): **poly(*N*-isopropylacrylamide)** **2007HOU**
Characterization: M_n/g.mol^{-1} = 11820, M_w/g.mol^{-1} = 15640, synthesized in the laboratory
Solvent (A): **water** **H_2O** **7732-18-5**

Type of data: cloud points (LCST-behavior)

c_B/(mg/ml)	1	2	4	5	10
T/K	308.65	308.45	308.35	307.35	306.35

Polymer (B): **poly(*N*-isopropylacrylamide)** **2009KOB**
Characterization: M_n/g.mol^{-1} = 19700, M_w/g.mol^{-1} = 22100, synthesized in the laboratory
Solvent (A): **water** **H_2O** **7732-18-5**

Type of data: cloud points (LCST-behavior)

w_B	0.0049	0.0099	0.0200	0.0301	0.0403	0.0494	0.0716	0.0838
T/K	301.81	301.20	300.20	299.45	299.42	299.33	298.27	297.75

Polymer (B): **poly(*N*-isopropylacrylamide)** **2007HOU**
Characterization: M_n/g.mol^{-1} = 22330, M_w/g.mol^{-1} = 46030, synthesized in the laboratory
Solvent (A): **water** **H_2O** **7732-18-5**

Type of data: cloud points (LCST-behavior)

c_B/(mg/ml)	1	2	4	5	10
T/K	308.35	308.25	307.35	307.25	306.35

Polymer (B): **poly(*N*-isopropylacrylamide)** **2009KOB**
Characterization: M_n/g.mol^{-1} = 28800, M_w/g.mol^{-1} = 31100, synthesized in the laboratory
Solvent (A): **water** **H_2O** **7732-18-5**

Type of data: cloud points (LCST-behavior)

w_B	0.0050	0.0098	0.0199	0.0299	0.0394	0.0493	0.0715	0.0825
T/K	303.88	303.73	303.44	303.39	303.18	302.88	302.75	302.40

Polymer (B): **poly(*N*-isopropylacrylamide)** **2008KAW**
Characterization: M_n/g.mol^{-1} = 39700, M_w/g.mol^{-1} = 46500, synthesized and fractionated in the laboratory
Solvent (A): **water** **H_2O** **7732-18-5**

Type of data: cloud points (LCST-behavior)

w_B	0.0050	0.0101	0.0202	0.0293	0.0398	0.0496	0.0752	0.0971
T/K	304.77	304.47	304.24	304.16	303.96	303.84	303.63	303.50

Polymer (B): **poly(*N*-isopropylacrylamide)** **2008KAW**
Characterization: M_n/g.mol^{-1} = 40500, M_w/g.mol^{-1} = 51800, synthesized and fractionated in the laboratory
Solvent (A): **water** **H_2O** **7732-18-5**

Type of data: cloud points (LCST-behavior)

w_B	0.0049	0.0099	0.0199	0.0297	0.0397	0.0494	0.0747	0.0932
T/K	304.89	304.45	304.03	303.83	303.55	303.43	303.41	303.24

Polymer (B): **poly(*N*-isopropylacrylamide)** **2008KAW**
Characterization: M_n/g.mol^{-1} = 45400, M_w/g.mol^{-1} = 51700, synthesized and fractionated in the laboratory
Solvent (A): **water** **H_2O** **7732-18-5**

Type of data: cloud points (LCST-behavior)

w_B	0.0050	0.0099	0.0200	0.0306	0.0401	0.0504	0.0750	0.0946
T/K	305.86	305.53	305.24	305.11	305.00	304.89	304.70	304.59

Polymer (B): **poly(*N*-isopropylacrylamide)** **2008KAW**
Characterization: M_n/g.mol^{-1} = 54200, M_w/g.mol^{-1} = 64500, synthesized and fractionated in the laboratory
Solvent (A): **water** **H_2O** **7732-18-5**

Type of data: cloud points (LCST-behavior)

w_B	0.0049	0.0098	0.0191	0.0300	0.0393	0.0504	0.0702	0.0903
T/K	306.06	305.89	305.72	305.59	305.49	305.38	305.24	305.10

Polymer (B): **poly(*N*-isopropylacrylamide)** **2009KOB**
Characterization: M_n/g.mol^{-1} = 64600, M_w/g.mol^{-1} = 72300, synthesized in the laboratory
Solvent (A): **water** **H_2O** **7732-18-5**

Type of data: cloud points (LCST-behavior)

w_B	0.0050	0.0100	0.0200	0.0298	0.0401	0.0494	0.0695	0.0911
T/K	306.11	305.92	305.64	305.19	305.01	304.84	304.81	304.75

Polymer (B): **poly(*N*-isopropylacrylamide)** **2011ISE**
Characterization: M_n/g.mol^{-1} = 64600, M_w/g.mol^{-1} = 104000, sulfonate end groups, synthesized in the laboratory by redox polymerization
Solvent (A): **water** **H_2O** **7732-18-5**

Type of data: cloud points (LCST-behavior)

w_B	0.0050	0.0100	0.0199	0.0301	0.0399	0.0505	0.0705	0.0896
T/K	306.94	306.82	306.70	306.63	06.56	306.52	306.44	306.35

Polymer (B): **poly(*N*-isopropylacrylamide)** **2011POS**
Characterization: M_n/g.mol^{-1} = 73800, M_w/g.mol^{-1} = 104200, synthesized in the laboratory
Solvent (A): **water** **H_2O** **7732-18-5**

Type of data: cloud points (LCST-behavior)

w_B	0.05	0.04	0.01	0.006	0.003	0.0015	0.0010	0.00075	0.0003
T/K	306.75	307.05	307.15	307.15	307.45	307.55	307.85	307.35	308.25

Comments: RI detection

w_B	0.03354	0.02499	0.02326	0.01705	0.01480	0.01266	0.01005	0.00638	0.00353
T/K	306.95	307.05	306.95	307.05	306.95	306.85	306.95	307.05	307.25
w_B	0.00056	0.00032	0.00004	0.00002					
T/K	307.85	307.75	308.55	309.05					

Comments: Method visual detection

Polymer (B): **poly(*N*-isopropylacrylamide)** **2008KAW**
Characterization: M_n/g.mol^{-1} = 80700, M_w/g.mol^{-1} = 94400, synthesized and fractionated in the laboratory
Solvent (A): **water** **H_2O** **7732-18-5**

Type of data: cloud points (LCST-behavior)

w_B	0.0051	0.0102	0.0201	0.0298	0.0403	0.0493	0.0739	0.0857
T/K	305.95	305.84	305.67	305.58	305.51	305.44	305.32	305.25

Polymer (B): **poly(*N*-isopropylacrylamide)** **2008KAW**
Characterization: M_n/g.mol^{-1} = 106500, M_w/g.mol^{-1} = 131000, synthesized and fractionated in the laboratory
Solvent (A): **water** **H_2O** **7732-18-5**

Type of data: cloud points (LCST-behavior)

w_B	0.0052	0.0102	0.0201	0.0304	0.0405	0.0507	0.0743	0.0847
T/K	305.69	305.55	305.40	305.32	305.25	305.15	305.02	304.98

Polymer (B): **poly(*N*-isopropylacrylamide)** **2008KAW**
Characterization: M_n/g.mol^{-1} = 108000, M_w/g.mol^{-1} = 127000, synthesized and fractionated in the laboratory

continued

continued

Solvent (A): **water** H_2O **7732-18-5**

Type of data: cloud points (LCST-behavior)

w_B	0.0058	0.0114	0.0233	0.0304	0.0445	0.0502	0.0703	0.0917
T/K	304.72	304.51	304.31	304.22	304.06	304.04	303.88	303.74

Polymer (B): **poly(*N*-isopropylacrylamide)** **2008KAW**
Characterization: M_n/g.mol^{-1} = 126000, M_w/g.mol^{-1} = 144000, synthesized and fractionated in the laboratory
Solvent (A): **water** H_2O **7732-18-5**

Type of data: cloud points (LCST-behavior)

w_B	0.0047	0.0091	0.0193	0.0305	0.0405	0.0509	0.0759	0.0951
T/K	304.67	304.53	304.34	304.23	304.14	304.06	303.93	303.82

Polymer (B): **poly(*N*-isopropylacrylamide)** **2011ISE**
Characterization: M_n/g.mol^{-1} = 140000, M_w/g.mol^{-1} = 186000, sulfonate end groups, synthesized in the laboratory by redox polymerization
Solvent (A): **water** H_2O **7732-18-5**

Type of data: cloud points (LCST-behavior)

w_B	0.0050	0.0099	0.0200	0.0300	0.0399	0.0505	0.0695	0.0892
T/K	306.75	306.64	306.54	306.46	306.37	306.31	306.18	306.09

Polymer (B): **poly(*N*-isopropylacrylamide)** **2005RAY**
Characterization: see table, synthesized in the laboratory
Solvent (A): **water** H_2O **7732-18-5**

Type of data: cloud points (LCST-behavior)

M_n/ g.mol^{-1}	M_w/M_n	tacticity % meso diads	w_B	T/ K
34000	1.20	45	0.01	304.25
37500	1.20	47	0.01	303.65
37200	1.20	49	0.01	303.25
39900	1.21	51	0.01	302.65
36900	1.23	53	0.01	301.65
35200	1.25	57	0.01	300.85
39800	1.21	62	0.01	297.85
39300	1.24	66	0.01	290.15

Polymer (B): **poly(*N*-isopropylacrylamide)** **2005XIA**
Characterization: see table, synthesized in the laboratory
Solvent (A): **water** H_2O **7732-18-5**

continued

continued

Type of data: cloud points (LCST-behavior)

M_n/ g.mol^{-1}	M_w/M_n	w_B	T/ K
2800	1.07	0.01	317.15
5000	1.15	0.01	312.05
6500	1.09	0.01	309.45
6700	1.16	0.01	309.55
10900	1.11	0.01	308.65
15700	1.13	0.01	307.75
26500	1.16	0.01	306.45
28900	2.00	0.01	304.35

Polymer (B): **poly(*N*-isopropylacrylamide-*co*-acrylamide)** **2006SHE**
Characterization: 2.2 mol% acrylamide, synthesized in the laboratory
Solvent (A): **water** **H_2O** **7732-18-5**

Type of data: cloud points (LCST-behavior)

c_B/(g/cm^3) 0.005 T/K 307.55

Polymer (B): **poly(*N*-isopropylacrylamide-*co*-acrylamide)** **2006SHE**
Characterization: 4.5 mol% acrylamide, synthesized in the laboratory
Solvent (A): **water** **H_2O** **7732-18-5**

Type of data: cloud points (LCST-behavior)

c_B/(g/cm^3) 0.005 T/K 309.45

Polymer (B): **poly(*N*-isopropylacrylamide-*co*-acrylamide)** **2006SHE**
Characterization: 6.6 mol% acrylamide, synthesized in the laboratory
Solvent (A): **water** **H_2O** **7732-18-5**

Type of data: cloud points (LCST-behavior)

c_B/(g/cm^3) 0.005 T/K 311.15

Polymer (B): **poly(*N*-isopropylacrylamide-*co*-acrylamide)** **2006SHE**
Characterization: M_η/g.mol^{-1} = 15600, 7.4 mol% acrylamide, synthesized in the laboratory
Solvent (A): **water** **H_2O** **7732-18-5**

Type of data: cloud points (LCST-behavior)

c_B/(g/cm^3) 0.005 T/K 311.35

Polymer (B): **poly(*N*-isopropylacrylamide-*co*-acrylamide)** **2006SHE**
Characterization: M_η/g.mol^{-1} = 16900, 8.1 mol% acrylamide, synthesized in the laboratory
Solvent (A): **water** H_2O **7732-18-5**

Type of data: cloud points (LCST-behavior)

c_B/(g/cm^3) 0.005 *T*/K 312.15

Polymer (B): **poly(*N*-isopropylacrylamide-*co*-acrylamide)** **2006SHE**
Characterization: M_η/g.mol^{-1} = 9700, 14.4 mol% acrylamide, synthesized in the laboratory
Solvent (A): **water** H_2O **7732-18-5**

Type of data: cloud points (LCST-behavior)

c_B/(g/cm^3) 0.005 *T*/K 316.85

Polymer (B): **poly(*N*-isopropylacrylamide-*co*-acrylamide-*co*-2-hydroxyethyl methacrylate)** **2010ZH1**
Characterization: see table, synthesized in the laboratory
Solvent (A): **water** H_2O **7732-18-5**

Type of data: cloud points (LCST-behavior)

w_B = 0.008 was kept constant

NIPAM mol%	AM mol%	HEMA mol%	*T*/ K
75	24	1	321.15
70	29	1	326.95
65	34	1	333.95
60	39	1	341.35
60	37	3	334.25
60	35	5	323.15
60	33	7	317.85
58	39	3	335.65
56	39	5	327.45
53	39	8	321.35

Polymer (B): **poly(*N*-isopropylacrylamide-*co*-acrylic acid)** **1999JON**
Characterization: 32 mol% acrylic acid, synthesized in the laboratory
Solvent (A): **water** H_2O **7732-18-5**

Type of data: cloud points (LCST-behavior)

w_B	0.005	0.005	0.005	0.005	0.005	0.005	0.005
pH	1.0	1.5	2.0	2.5	3.0	3.5	4.0
T/K	<273.15	284.15	287.15	291.15	295.15	301.15	316.15

Polymer (B): **poly(*N*-isopropylacrylamide-*co*-acrylic acid)** **1999JON**
Characterization: 54 mol% acrylic acid, synthesized in the laboratory
Solvent (A): **water** H_2O **7732-18-5**

Type of data: cloud points (LCST-behavior)

w_B	0.005	0.005	0.005	0.005	0.005
pH	1.0	1.5	2.0	2.5	3.0
T/K	<273.15	283.15	288.15	299.15	318.15

Polymer (B): **poly(*N*-isopropylacrylamide-*co*-acrylic acid)** **1999JON**
Characterization: 80 mol% acrylic acid, synthesized in the laboratory
Solvent (A): **water** H_2O **7732-18-5**

Type of data: cloud points (LCST-behavior)

w_B	0.005	0.005	0.005	0.005	0.005
pH	1.0	1.5	2.0	2.5	3.0
T/K	<273.15	276.15	289.15	299.15	324.15

Polymer (B): **poly(*N*-isopropylacrylamide-*b*-ε-caprolactone-*b*-*N*-isopropylacrylamide)** **2008LOH**
Characterization: M_n/g.mol^{-1} = 17420, M_w/M_n = 1.54, 14.0 wt% ε-caprolactone, synthesized in the laboratory
Solvent (A): **water** H_2O **7732-18-5**

Type of data: cloud points (LCST-behavior)

c_B/(g/dm^3) 0.250 T/K 310.05

Polymer (B): **poly(*N*-isopropylacrylamide-*b*-ε-caprolactone-*b*-*N*-isopropylacrylamide)** **2008LOH**
Characterization: M_n/g.mol^{-1} = 6830, M_w/M_n = 1.36, 24.4 wt% ε-caprolactone, synthesized in the laboratory
Solvent (A): **water** H_2O **7732-18-5**

Type of data: cloud points (LCST-behavior)

c_B/(g/dm^3) 0.250 T/K 304.95

Polymer (B): **poly(*N*-isopropylacrylamide-*b*-ε-caprolactone-*b*-*N*-isopropylacrylamide)** **2008LOH**
Characterization: M_n/g.mol^{-1} = 5210, M_w/M_n = 1.68, 34.4 wt% ε-caprolactone, synthesized in the laboratory
Solvent (A): **water** H_2O **7732-18-5**

Type of data: cloud points (LCST-behavior)

c_B/(g/dm^3) 0.250 T/K 303.75

Polymer (B): **poly(*N*-isopropylacrylamide-*co*-*N*,*N*-dimethylacrylamide)** **2006SHE**
Characterization: 10.7 mol% *N*,*N*-dimethylacrylamide, synthesized in the laboratory
Solvent (A): **water** **H_2O** **7732-18-5**

Type of data: cloud points (LCST-behavior)

c_B/(g/cm^3)	0.005	T/K	310.05

Polymer (B): **poly(*N*-isopropylacrylamide-*co*-*N*,*N*-dimethylacrylamide)** **2003BAR**
Characterization: 13.0 mol% *N*,*N*-dimethylacrylamide, synthesized in the laboratory
Solvent (A): **water** **H_2O** **7732-18-5**

Type of data: cloud points (LCST-behavior)

w_B	0.01	T/K	309.15

Polymer (B): **poly(*N*-isopropylacrylamide-*co*-*N*,*N*-dimethylacrylamide)** **2006SHE**
Characterization: 14.7 mol% *N*,*N*-dimethylacrylamide, synthesized in the laboratory
Solvent (A): **water** **H_2O** **7732-18-5**

Type of data: cloud points (LCST-behavior)

c_B/(g/cm^3)	0.005	T/K	311.25

Polymer (B): **poly(*N*-isopropylacrylamide-*co*-*N*,*N*-dimethylacrylamide)** **2006SHE**
Characterization: 17.3 mol% *N*,*N*-dimethylacrylamide, synthesized in the laboratory
Solvent (A): **water** **H_2O** **7732-18-5**

Type of data: cloud points (LCST-behavior)

c_B/(g/cm^3)	0.005	T/K	312.95

Polymer (B): **poly(*N*-isopropylacrylamide-*co*-*N*,*N*-dimethylacrylamide)** **2006SHE**
Characterization: 21.2 mol% *N*,*N*-dimethylacrylamide, synthesized in the laboratory
Solvent (A): **water** **H_2O** **7732-18-5**

Type of data: cloud points (LCST-behavior)

c_B/(g/cm^3)	0.005	T/K	314.25

Polymer (B):	**poly(*N*-isopropylacrylamide-*co*-*N,N*-dimethylacrylamide)**	**2003BAR**
Characterization:	27.0 mol% *N,N*-dimethylacrylamide, synthesized in the laboratory	
Solvent (A):	**water** **H_2O**	**7732-18-5**

Type of data: cloud points (LCST-behavior)

w_B	0.01	T/K	312.15

Polymer (B):	**poly(*N*-isopropylacrylamide-*co*-*N,N*-dimethylacrylamide)**	**2003BAR**
Characterization:	30.0 mol% *N,N*-dimethylacrylamide, synthesized in the laboratory	
Solvent (A):	**water** **H_2O**	**7732-18-5**

Type of data: cloud points (LCST-behavior)

w_B	0.01	T/K	315.15

Polymer (B):	**poly(*N*-isopropylacrylamide-*co*-*N,N*-dimethylacrylamide)**	**2006SHE**
Characterization:	31.4 mol% *N,N*-dimethylacrylamide, synthesized in the laboratory	
Solvent (A):	**water** **H_2O**	**7732-18-5**

Type of data: cloud points (LCST-behavior)

c_B/(g/cm^3)	0.005	T/K	319.15

Polymer (B):	**poly(*N*-isopropylacrylamide-*co*-*N,N*-dimethylacrylamide)**	**2003BAR**
Characterization:	50.0 mol% *N,N*-dimethylacrylamide, synthesized in the laboratory	
Solvent (A):	**water** **H_2O**	**7732-18-5**

Type of data: cloud points (LCST-behavior)

w_B	0.01	T/K	323.15

Polymer (B):	**poly(*N*-isopropylacrylamide-*co*-*N,N*-dimethylacrylamide)**	**2003BAR**
Characterization:	60.0 mol% *N,N*-dimethylacrylamide, synthesized in the laboratory	
Solvent (A):	**water** **H_2O**	**7732-18-5**

Type of data: cloud points (LCST-behavior)

w_B	0.01	T/K	336.15

Polymer (B):	**poly(*N*-isopropylacrylamide-*co*-*N,N*-dimethylacrylamide)**	**2003BAR**
Characterization:	66.0 mol% *N,N*-dimethylacrylamide, synthesized in the laboratory	
Solvent (A):	**water** **H_2O**	**7732-18-5**

Type of data: cloud points (LCST-behavior)

w_B	0.01	T/K	345.15

Polymer (B):	**poly(*N*-isopropylacrylamide-*co*-*N*-glycineacrylamide)**	**2000PRI**
Characterization:	M_n/g.mol^{-1} = 30000, M_w/g.mol^{-1} = 77000, 20 mol% *N*-glycine-acrylamide, synthesized in the laboratory	
Solvent (A):	**water** **H_2O**	**7732-18-5**

Type of data: cloud points (LCST-behavior)

c_B/(g/l)	1.0	1.0	1.0	1.0
pH	2.77	3.83	4.25	5.10
T/K	303.55	305.85	307.55	311.45

Comments: pH values were prepared by buffer solutions from 0.1 M citric acid and 0.1 M sodium hydroxide at constant ionic strength of 0.1 M NaCl.

Polymer (B):	**poly(*N*-isopropylacrylamide-*co*-2-hydroxyethyl methacrylate)**	**2006SHE**
Characterization:	9.4 mol% 2-hydroxyethyl methacrylate, synthesized in the laboratory	
Solvent (A):	**water** **H_2O**	**7732-18-5**

Type of data: cloud points (LCST-behavior)

c_B/(g/cm^3)	0.005	T/K	303.45

Polymer (B):	**poly(*N*-isopropylacrylamide-*co*-2-hydroxyethyl methacrylate)**	**2006SHE**
Characterization:	17.0 mol% 2-hydroxyethyl methacrylate, synthesized in the laboratory	
Solvent (A):	**water** **H_2O**	**7732-18-5**

Type of data: cloud points (LCST-behavior)

c_B/(g/cm^3)	0.005	T/K	299.75

Polymer (B):	**poly(*N*-isopropylacrylamide-*co*-2-hydroxyethyl methacrylate)**	**2006SHE**
Characterization:	25.8 mol% 2-hydroxyethyl methacrylate, synthesized in the laboratory	
Solvent (A):	**water** **H_2O**	**7732-18-5**

continued

continued

Type of data: cloud points (LCST-behavior)

c_B/(g/cm^3)	0.005	*T*/K	294.55

Polymer (B): **poly(*N*-isopropylacrylamide-*co*-2-hydroxyethyl methacrylate)** **2006SHE**

Characterization: 34.9 mol% 2-hydroxyethyl methacrylate, synthesized in the laboratory

Solvent (A): **water** **H_2O** **7732-18-5**

Type of data: cloud points (LCST-behavior)

c_B/(g/cm^3)	0.005	*T*/K	290.15

Polymer (B): **poly(*N*-isopropylacrylamide-*co*-2-methacryloamidohistidine)** **2003KAL**

Characterization: see table, synthesized in the laboratory

Solvent (A): **water** **H_2O** **7732-18-5**

Type of data: cloud points (LCST-behavior)

w_B 0.005 was kept constant

Mol% MAH	M_η/ g.mol^{-1}	*T*/K pH = 4.0	pH = 7.4	pH = 9.0
1.26	62376	303.05	304.35	303.75
2.48	62790	303.95	305.85	304.45
4.85	62082	304.45	306.55	304.85
9.24	62541	304.75	307.25	305.15
4.85	68142	304.25	305.85	304.65
4.85	77590	304.05	305.65	304.55

Polymer (B): **poly(*N*-isopropylacrylamide-*co*-8-methacryloyloxyoctanoic acid)** **2004SA1**

Characterization: M_w/g.mol^{-1} = 354000, 7.1 mol% 8-methacryloyloxy-octanoic acid, synthesized in the laboratory

Solvent (A): **water** **H_2O** **7732-18-5**

Type of data: cloud points (LCST-behavior)

c_B/(mg/ml)	5.0	5.0	5.0	5.0	5.0
pH	*	7	8	9	10
T/K	290.95	299.65	307.65	310.75	315.05

Comments: pH values were prepared by buffer solutions at constant ionic strength of 0.1 M. * is deionized water. Cloud points were detected by using modulated DSC.

Polymer (B): **poly(*N*-isopropylacrylamide-*co*-8-methacryloyloxyoctanoic acid)** **2004SA1**
Characterization: M_w/g.mol^{-1} = 318000, 12.4 mol% 8-methacryloyloxy-octanoic acid, synthesized in the laboratory
Solvent (A): **water** **H_2O** **7732-18-5**

Type of data: cloud points (LCST-behavior)

c_B/(mg/ml)	5.0	5.0
pH	7	8
T/K	294.25	313.35

Comments: pH values were prepared by buffer solutions at constant ionic strength of 0.1 M. Cloud points were detected by using modulated DSC.

Polymer (B): **poly(*N*-isopropylacrylamide-*co*-8-methacryloyloxyoctanoic acid)** **2004SA1**
Characterization: M_w/g.mol^{-1} = 275000, 17.4 mol% 8-methacryloyloxy-octanoic acid, synthesized in the laboratory
Solvent (A): **water** **H_2O** **7732-18-5**

Type of data: cloud points (LCST-behavior)

c_B/(mg/ml)	5.0
pH	8
T/K	314.75

Comments: pH values were prepared by buffer solutions at constant ionic strength of 0.1 M. Cloud points were detected by using modulated DSC.

Polymer (B): **poly(*N*-isopropylacrylamide-*co*-5-methacryloyloxypentanoic acid)** **2004SA1**
Characterization: M_w/g.mol^{-1} = 291000, 6.7 mol% 5-methacryloyloxy-pentanoic acid, synthesized in the laboratory
Solvent (A): **water** **H_2O** **7732-18-5**

Type of data: cloud points (LCST-behavior)

c_B/(mg/ml)	5.0	5.0	5.0	5.0	5.0
pH	*	7	8	9	10
T/K	299.45	305.45	307.55	308.15	306.65

Comments: pH values were prepared by buffer solutions at constant ionic strength of 0.1 M. * is deionized water. Cloud points were detected by using modulated DSC.

Polymer (B): **poly(*N*-isopropylacrylamide-*co*-5-methacryloyloxypentanoic acid)** **2004SA1**
Characterization: M_w/g.mol^{-1} = 340000, 10.8 mol% 5-methacryloyloxy-pentanoic acid, synthesized in the laboratory
Solvent (A): **water** **H_2O** **7732-18-5**

continued

continued

Type of data: cloud points (LCST-behavior)

c_B/(mg/ml)	5.0	5.0	5.0	5.0
pH	*	7	8	9
T/K	283.25	307.65	310.25	315.15

Comments: pH values were prepared by buffer solutions at constant ionic strength of 0.1 M. * is deionized water. Cloud points were detected by using modulated DSC.

Polymer (B): **poly(*N*-isopropylacrylamide-*co*-5-methacryloyloxypentanoic acid)** **2004SA1**

Characterization: M_w/g.mol^{-1} = 241000, 17.9 mol% 5-methacryloyloxypentanoic acid, synthesized in the laboratory

Solvent (A): **water** **H_2O** **7732-18-5**

Type of data: cloud points (LCST-behavior)

c_B/(mg/ml)	5.0	5.0
pH	7	8
T/K	307.55	316.95

Comments: pH values were prepared by buffer solutions at constant ionic strength of 0.1 M. Cloud points were detected by using modulated DSC.

Polymer (B): **poly(*N*-isopropylacrylamide-*co*-5-methacryloyloxypentanoic acid)** **2004SA1**

Characterization: M_w/g.mol^{-1} = 310000, 23.7 mol% 5-methacryloyloxypentanoic acid, synthesized in the laboratory

Solvent (A): **water** **H_2O** **7732-18-5**

Type of data: cloud points (LCST-behavior)

c_B/(mg/ml)	5.0	5.0
pH	7	8
T/K	304.55	325.95

Comments: pH values were prepared by buffer solutions at constant ionic strength of 0.1 M. Cloud points were detected by using modulated DSC.

Polymer (B): **poly(*N*-isopropylacrylamide-*co*-11-methacryloyloxyundecanoic acid)** **2004SA1**

Characterization: M_w/g.mol^{-1} = 287000, 6.5 mol% 11-methacryloyloxyundecanoic acid, synthesized in the laboratory

Solvent (A): **water** **H_2O** **7732-18-5**

Type of data: cloud points (LCST-behavior)

c_B/(mg/ml)	5.0	5.0	5.0	5.0	5.0
pH	*	7	8	9	10
T/K	279.15	290.25	306.65	301.65	305.85

Comments: pH values were prepared by buffer solutions at constant ionic strength of 0.1 M. * is deionized water. Cloud points were detected by using modulated DSC.

Polymer (B): **poly(*N*-isopropylacrylamide-*co*-11-methacryloyloxyundecanoic acid)** **2004SA1**
Characterization: M_w/g.mol^{-1} = 289000, 11.2 mol% 11-methacryloyloxyundecanoic acid, synthesized in the laboratory
Solvent (A): **water** **H_2O** **7732-18-5**

Type of data: cloud points (LCST-behavior)

c_B/(mg/ml)	5.0	5.0
pH	8	9
T/K	306.35	308.85

Comments: pH values were prepared by buffer solutions at constant ionic strength of 0.1 M. Cloud points were detected by using modulated DSC.

Polymer (B): **poly(*N*-isopropylacrylamide-*co*-11-methacryloyloxyundecanoic acid)** **2004SA1**
Characterization: M_w/g.mol^{-1} = 231000, 17.2 mol% 11-methacryloyloxyundecanoic acid, synthesized in the laboratory
Solvent (A): **water** **H_2O** **7732-18-5**

Type of data: cloud points (LCST-behavior)

c_B/(mg/ml)	5.0
pH	8
T/K	306.85

Comments: pH values were prepared by buffer solutions at constant ionic strength of 0.1 M. Cloud points were detected by using modulated DSC.

Polymer (B): **poly(*N*-isopropylacrylamide-*co*-4-pentenoic acid)** **1995CHE**
Characterization: M_w/g.mol^{-1} = 240000, 0.7 mol% 4-pentenoic acid
Solvent (A): **water** **H_2O** **7732-18-5**

Type of data: cloud points (LCST-behavior)

c_B/(mg/ml)	2.0	2.0
pH	4.0	7.5
T/K	303.05	306.15

Comments: pH values were prepared by phosphate buffer solutions.

Polymer (B): **poly(*N*-isopropylacrylamide-*co*-4-pentenoic acid)** **1995CHE**
Characterization: M_w/g.mol^{-1} = 280000, 1.3 mol% 4-pentenoic acid
Solvent (A): **water** **H_2O** **7732-18-5**

Type of data: cloud points (LCST-behavior)

c_B/(mg/ml)	2.0	2.0
pH	4.0	7.5
T/K	302.85	307.45

Comments: pH values were prepared by phosphate buffer solutions.

Polymer (B): **poly(*N*-isopropylacrylamide-*co*-4-pentenoic acid)** **1995CHE**
Characterization: M_w/g.mol^{-1} = 370000, 2.3 mol% 4-pentenoic acid
Solvent (A): **water** **H_2O** **7732-18-5**

Type of data: cloud points (LCST-behavior)

c_B/(mg/ml)	2.0	2.0
pH	4.0	7.5
T/K	302.05	309.65

Comments: pH values were prepared by phosphate buffer solutions.

Polymer (B): **poly(*N*-isopropylacrylamide-*co*-4-pentenoic acid)** **1995CHE**
Characterization: M_w/g.mol^{-1} = 400000, 6.5 mol% 4-pentenoic acid
Solvent (A): **water** **H_2O** **7732-18-5**

Type of data: cloud points (LCST-behavior)

c_B/(mg/ml)	2.0
pH	4.0
T/K	300.45

Comments: pH values were prepared by phosphate buffer solutions.

Polymer (B): **poly(*N*-isopropylacrylamide-*co*-4-pentenoic acid)** **1995CHE**
Characterization: M_w/g.mol^{-1} = 300000, 10.8 mol% 4-pentenoic acid
Solvent (A): **water** **H_2O** **7732-18-5**

Type of data: cloud points (LCST-behavior)

c_B/(mg/ml)	2.0
pH	4.0
T/K	297.45

Comments: pH values were prepared by phosphate buffer solutions.

Polymer (B): **poly(*N*-isopropylacrylamide-*co*-4-pentenoic acid)** **1995CHE**
Characterization: M_w/g.mol^{-1} = 220000, 21.7 mol% 4-pentenoic acid
Solvent (A): **water** **H_2O** **7732-18-5**

Type of data: cloud points (LCST-behavior)

c_B/(mg/ml)	2.0
pH	4.0
T/K	292.35

Comments: pH values were prepared by phosphate buffer solutions.

Polymer (B): **poly(*N*-isopropylacrylamide-*co*-4-pentenoic acid)** **1995CHE**
Characterization: M_w/g.mol^{-1} = 190000, 28.6 mol% 4-pentenoic acid
Solvent (A): **water** **H_2O** **7732-18-5**

Type of data: cloud points (LCST-behavior)

c_B/(mg/ml)	2.0
pH	4.0
T/K	277.15

Comments: pH values were prepared by phosphate buffer solutions.

Polymer (B): **poly(*N*-isopropylacrylamide-*co*-sodium 2-acrylamido-2-methyl-1-propanesulfonate-*co*-*N*-*tert*-butylacrylamide)** **2007SH1**
Characterization: synthesized in the laboratory
Solvent (A): **water** **H_2O** **7732-18-5**

Type of data: cloud points (LCST-behavior)

Comments: molar ratio of NIPAm/AMPS/NTBAm = 100/4.3/0

w_B	0.002	0.004	0.006	0.008	0.010
T/K	323.45	320.05	318.15	316.25	315.65

Comments: molar ratio of NIPAm/AMPS/NTBAm = 100/4.1/3.1

w_B	0.002	0.004	0.006	0.008	0.010
T/K	321.35	317.35	315.25	313.45	312.85

Comments: molar ratio of NIPAm/AMPS/NTBAm = 100/4.0/9.5

w_B	0.002	0.004	0.006	0.008	0.010
T/K	318.35	314.35	312.05	310.55	309.95

Comments: molar ratio of NIPAm/AMPS/NTBAm = 100/3.5/18.4

w_B	0.002	0.004	0.006	0.008	0.010
T/K	314.45	310.65	308.15	307.95	306.55

Polymer (B): **poly(*N*-isopropylacrylamide-*co*-p-vinylphenylboronic acid)** **2005CIM**
Characterization: 3.2 mol% p-vinylphenylboronic acid, synthesized in the laboratory
Solvent (A): **water** **H_2O** **7732-18-5**

Type of data: cloud points (LCST-behavior)

pH	4.0	5.0	7.4
T/K	299.75	300.05	300.95

Polymer (B): **poly(*N*-isopropylacrylamide-*co*-p-vinylphenylboronic acid)** **2005CIM**
Characterization: 7.3 mol% p-vinylphenylboronic acid, synthesized in the laboratory
Solvent (A): **water** H_2O **7732-18-5**

Type of data: cloud points (LCST-behavior)

pH	4.0	5.0	7.4
T/K	299.05	299.35	300.65

Polymer (B): **poly(*N*-isopropylacrylamide-*co*-p-vinylphenylboronic acid)** **2005CIM**
Characterization: 31.1 mol% p-vinylphenylboronic acid, synthesized in the laboratory
Solvent (A): **water** H_2O **7732-18-5**

Type of data: cloud points (LCST-behavior)

pH	4.0	5.0	7.4
T/K	298.75	298.95	299.45

Polymer (B): **poly(methoxydiethylene glycol methacrylate)** **2008BEC**
Characterization: M_n/g.mol^{-1} = 6300, M_w/M_n = 1.31, synthesized in the laboratory
Solvent (A): **water** H_2O **7732-18-5**

Type of data: cloud points (LCST-behavior)

c_B/(mg/ml)	5.0	5.0
pH	7	10
T/K	293.75	294.75

Polymer (B): **poly(methoxydiethylene glycol methacrylate)** **2008BEC**
Characterization: M_n/g.mol^{-1} = 20400, M_w/M_n = 1.21, synthesized in the laboratory
Solvent (A): **water** H_2O **7732-18-5**

Type of data: cloud points (LCST-behavior)

c_B/(mg/ml)	5.0	5.0
pH	7	10
T/K	294.95	296.25

Polymer (B): **poly(methoxydiethylene glycol methacrylate)** **2004KIT**
Characterization: M_n/g.mol^{-1} = 21600, M_w/g.mol^{-1} = 31900, synthesized in the laboratory
Solvent (A): **water** H_2O **7732-18-5**

Type of data: cloud points (LCST-behavior)

c_B/(g/l) 1.0 *T*/K 298.45

Polymer (B):	**poly(methoxydiethylene glycol methacrylate-*co*-dodecyl methacrylate)**		**2004KIT**
Characterization:	M_n/g.mol^{-1} = 32100, M_w/g.mol^{-1} = 52600, 9.43 mol% methoxydiethylene glycol methacrylate, molar ratio = 1/9.61, synthesized in the laboratory		
Solvent (A):	**water**	**H_2O**	**7732-18-5**

Type of data: cloud points (LCST-behavior)

c_B/(g/l)	1.0	T/K	292.65

Polymer (B):	**poly(methoxydiethylene glycol methacrylate-*co*-methoxyoligoethylene glycol methacrylate)**		**2004KIT**
Characterization:	M_n/g.mol^{-1} = 46600, M_w/g.mol^{-1} = 116000, 2.75 mol% methoxyoligoethylene glycol(9 EO units) methacrylate, molar ratio = 1/35.4, synthesized in the laboratory		
Solvent (A):	**water**	**H_2O**	**7732-18-5**

Type of data: cloud points (LCST-behavior)

c_B/(g/l)	1.0	T/K	300.85

Polymer (B):	**poly(methoxydiethylene glycol methacrylate-*co*-methoxyoligoethylene glycol methacrylate)**		**2004KIT**
Characterization:	M_n/g.mol^{-1} = 35200, M_w/g.mol^{-1} = 67200, 4.12 mol% methoxyoligoethylene glycol(9 EO units) methacrylate, molar ratio = 1/23.3, synthesized in the laboratory		
Solvent (A):	**water**	**H_2O**	**7732-18-5**

Type of data: cloud points (LCST-behavior)

c_B/(g/l)	1.0	T/K	302.85

Polymer (B):	**poly(methoxydiethylene glycol methacrylate-*co*-methoxyoligoethylene glycol methacrylate)**		**2004KIT**
Characterization:	M_n/g.mol^{-1} = 13100, M_w/g.mol^{-1} = 20100, 9.69 mol% methoxyoligoethylene glycol(9 EO units) methacrylate, molar ratio = 1/9.32, synthesized in the laboratory		
Solvent (A):	**water**	**H_2O**	**7732-18-5**

Type of data: cloud points (LCST-behavior)

c_B/(g/l)	1.0	T/K	312.45

Polymer (B): **poly(methoxydiethylene glycol methacrylate-*co*-methoxyoligoethylene glycol methacrylate)** **2004KIT**

Characterization: M_n/g.mol^{-1} = 39700, M_w/g.mol^{-1} = 76300, 1.83 mol% methoxyoligoethylene glycol(23 EO units) methacrylate, molar ratio = 1/53.5, synthesized in the laboratory

Solvent (A): **water** **H_2O** **7732-18-5**

Type of data: cloud points (LCST-behavior)

c_B/(g/l)	1.0	*T*/K	302.75

Polymer (B): **poly(methoxydiethylene glycol methacrylate-*co*-methoxyoligoethylene glycol methacrylate)** **2004KIT**

Characterization: M_n/g.mol^{-1} = 18000, M_w/g.mol^{-1} = 34200, 4.40 mol% methoxyoligoethylene glycol(23 EO units) methacrylate, molar ratio = 1/21.7, synthesized in the laboratory

Solvent (A): **water** **H_2O** **7732-18-5**

Type of data: cloud points (LCST-behavior)

c_B/(g/l)	1.0	*T*/K	309.15

Polymer (B): **poly(methoxydiethylene glycol methacrylate-*co*-methoxyoligoethylene glycol methacrylate)** **2004KIT**

Characterization: M_n/g.mol^{-1} = 33200, M_w/g.mol^{-1} = 56800, 9.43 mol% methoxyoligoethylene glycol(23 EO units) methacrylate, molar ratio = 1/9.61, synthesized in the laboratory

Solvent (A): **water** **H_2O** **7732-18-5**

Type of data: cloud points (LCST-behavior)

c_B/(g/l)	1.0	*T*/K	334.65

Polymer (B): **poly(methoxyoligoethylene glycol methacrylate)** **2008BEC**

Characterization: M_n/g.mol^{-1} = 9300, M_w/M_n = 1.19, synthesized in the laboratory

Solvent (A): **water** **H_2O** **7732-18-5**

Type of data: cloud points (LCST-behavior)

c_B/(mg/ml)	5.0	5.0	5.0
pH	4	7	10
T/K	380.15	366.85	369.75

Polymer (B): **poly(methoxyoligoethylene glycol methacrylate)** **2008BEC**
Characterization: M_n/g.mol^{-1} = 18300, M_w/M_n = 1.26, synthesized in the laboratory
Solvent (A): **water** H_2O **7732-18-5**

Type of data: cloud points (LCST-behavior)

c_B/(mg/ml)	5.0	5.0	5.0
pH	4	7	10
T/K	366.35	362.95	365.95

Polymer (B): **poly[oligo(ethylene glycol) methylacrylate-*co*-oligo(propylene glycol) methacrylate]** **2009WA2**
Characterization: see table, synthesized in the laboratory, oligomers have M_n(OEGM)/g.mol^{-1} = 475, M_n(OPGM)/g.mol^{-1} = 430
Solvent (A): **water** H_2O **7732-18-5**

Type of data: cloud points (LCST-behavior)

c_B/(mg/ml) 3.0 was kept constant, phosphate buffer, pH = 7.4

M_n/ g.mol^{-1}	M_w/M_n	composition ratio mol% (OEGM/OPGM)	*T*/ K
33500	1.43	79:21	343.15
32700	1.37	56:44	332.15
31700	1.33	22:78	309.15
16700	1.39	22:78	308.15
30300	1.69	21:79	307.15
15500	1.61	22:78	309.15
160000	1.72	22:78	309.15

Polymer (B): **poly(propylene oxide-*b*-ethylene oxide-*b*-propylene oxide)** **2011HUF**
Characterization: M_n/g.mol^{-1} = 2670, M_w/M_n = 1.06, ρ = 1.048 g/cm^3, PPO_{14}-PEO_{24}-PPO_{14}, reverse Pluronic 17R4, BASF Chemical Company, Florham Park, NJ
Solvent (A): **deuterium oxide** D_2O **7789-20-0**

Type of data: cloud points (LCST-behavior)

w_B	0.14000	0.22708	0.23300	0.33477	0.38782	0.36890	0.43211	0.47778	0.02819
φ_B	0.147	0.236	0.243	0.346	0.399	0.382	0.444	0.490	0.029
T/K	314.16	316.34	316.78	318.48	319.71	319.09	321.18	323.08	322.61

w_B	0.04777	0.11970	0.15814	0.17286	0.22708	0.26511	0.29935
φ_B	0.050	0.125	0.165	0.180	0.236	0.275	0.310
T/K	310.10	306.70	315.04	315.54	316.41	317.34	317.84

continued

continued

Type of data: coexistence data (tie lines, LCST-behavior)

φ_B (feed)	0.382	0.382	0.382	0.382	0.243	0.243	0.243	0.243
T/K	321.9906	320.0166	319.2438	320.6097	318.690	320.669	322.643	317.700
φ_B (top phase)	0.473	0.415	0.392	0.437	0.408	0.460	0.499	0.366
φ_B (bottom phase)	0.053	0.075	0.099	0.070	0.079	0.054	0.036	0.103
φ_B (feed)	0.243	0.243	0.243	0.243	0.243	0.243	0.147	0.147
T/K	317.204	318.125	317.5307	317.1346	317.0369	316.9377	315.0315	316.0236
φ_B (top phase)	0.333	0.388	0.358	0.327	0.315	0.301	0.350	0.384
φ_B (bottom phase)	0.125	0.097	0.113	0.135	0.146	0.155	0.119	0.094
φ_B (feed)	0.147	0.147	0.147	0.147	0.147	0.147	0.147	0.147
T/K	317.0162	318.0083	318.0096	318.9969	319.9840	320.9719	321.9610	322.9438
φ_B (top phase)	0.409	0.433	0.433	0.453	0.472	0.488	0.505	0.520
φ_B (bottom phase)	0.081	0.069	0.066	0.058	0.050	0.042	0.035	0.030
φ_B (feed)	0.147	0.147	0.147	0.147				
T/K	317.5075	316.5149	315.7251	315.3766				
φ_B (top phase)	0.423	0.400	0.377	0.364				
φ_B (bottom phase)	0.073	0.087	0.098	0.106				

Polymer (B): **poly(propylene oxide-*b*-ethylene oxide-*b*-propylene oxide)** **2011HUF**

Characterization: M_n/g.mol^{-1} = 2670, M_w/M_n = 1.06, ρ = 1.048 g/cm^3, PPO_{14}-PEO_{24}-PPO_{14}, reverse Pluronic 17R4, BASF Chemical Company, Florham Park, NJ

Solvent (A): **water** **H_2O** **7732-18-5**

Type of data: cloud points (LCST-behavior)

w_B	0.084	0.169	0.234	0.255	0.105	0.130	0.332
φ_B	0.080	0.163	0.226	0.246	0.101	0.125	0.322
T/K	314.05	316.40	317.99	317.85	313.72	314.81	319.68

Type of data: coexistence data (tie lines, LCST-behavior)

φ_B (feed)	0.322	0.322	0.322	0.322	0.322	0.246	0.246	0.246
T/K	319.696	320.691	322.666	325.615	330.466	321.662	320.662	319.682
φ_B (top phase)	0.113	0.086	0.060	0.035	0.014	0.064	0.076	0.091
φ_B (bottom phase)	0.349	0.396	0.450	0.508	0.581	0.449	0.426	0.398
φ_B (feed)	0.246	0.246	0.246	0.246	0.246	0.246	0.246	0.246
T/K	319.1849	318.6892	318.1918	317.9978	317.8925	324.6070	328.0328	318.0885
φ_B (top phase)	0.099	0.114	0.137	0.157	0.178	0.041	0.026	0.153
φ_B (bottom phase)	0.377	0.354	0.319	0.294	0.273	0.504	0.559	0.308

continued

continued

φ_B (feed)	0.246	0.246	0.246	0.226	0.226	0.226	0.226	0.226
T/K	317.9408	317.8667	318.3867	318.4773	318.5944	318.8951	319.8841	318.2973
φ_B (top phase)	0.168	0.194	0.128	0.126	0.124	0.111	0.089	0.139
φ_B (bottom phase)	0.286	0.261	0.337	0.329	0.335	0.349	0.391	0.311
φ_B (feed)	0.226	0.226	0.226	0.226	0.226	0.226	0.226	0.226
T/K	318.1984	318.0998	318.0215	318.0006	321.1685	319.3867	318.0632	321.8507
φ_B (top phase)	0.151	0.162	0.182	0.192	0.069	0.096	0.168	0.053
φ_B (bottom phase)	0.298	0.282	0.261	0.249	0.428	0.373	0.274	0.458
φ_B (feed)	0.163	0.163	0.163	0.163	0.163	0.163	0.163	0.163
T/K	317.0974	317.3956	317.8921	318.8826	320.3691	321.8507	317.0018	316.8037
φ_B (top phase)	0.123	0.113	0.100	0.084	0.066	0.053	0.125	0.135
φ_B (bottom phase)	0.329	0.343	0.363	0.394	0.430	0.458	0.326	0.314
φ_B (feed)	0.163	0.163	0.163	0.163	0.125	0.125	0.125	0.125
T/K	316.5057	323.4270	325.3930	327.3451	315.7073	315.4598	316.2041	317.6920
φ_B (top phase)	0.155	0.042	0.026	0.023	0.111	0.115	0.102	0.084
φ_B (bottom phase)	0.289	0.485	0.518	0.546	0.356	0.351	0.368	0.398
φ_B (feed)	0.125	0.125	0.125	0.125	0.125	0.125	0.125	0.125
T/K	318.6826	319.6742	320.6631	321.6498	322.6359	324.6017	325.5854	326.5656
φ_B (top phase)	0.073	0.065	0.056	0.051	0.044	0.034	0.030	0.026
φ_B (bottom phase)	0.417	0.436	0.453	0.468	0.483	0.511	0.527	0.542
φ_B (feed)	0.125	0.125	0.125	0.125	0.101	0.101	0.101	0.101
T/K	327.5435	323.6225	316.7057	317.6905	315.7059	325.5798	315.7078	317.6909
φ_B (top phase)	0.023	0.038	0.097	0.084	0.089	0.029	0.088	0.072
φ_B (bottom phase)	0.553	0.499	0.385	0.401	0.396	0.534	0.393	0.422
φ_B (feed)	0.101	0.101	0.101	0.080	0.080	0.080	0.080	0.080
T/K	318.6835	316.7002	319.6726	316.704	317.200	318.190	319.179	316.710
φ_B (top phase)	0.065	0.080	0.058	0.069	0.066	0.062	0.054	0.069
φ_B (bottom phase)	0.436	0.403	0.452	0.436	0.447	0.456	0.466	0.437
φ_B (feed)	0.080	0.080	0.080					
T/K	316.108	315.651	315.255					
φ_B (top phase)	0.072	0.073	0.075					
φ_B (bottom phase)	0.439	0.442	0.442					

Polymer (B): **poly(*N*-vinylcaprolactam)** **2003OKH**
Characterization: M_w/g.mol^{-1} = 4500, synthesized in the laboratory
Solvent (A): **water** **H_2O** **7732-18-5**

Type of data: cloud points (LCST-behavior)

c_B/(g/l) 1.0 T/K 303.15 (in 0.05 M NaCl aqueous solution)

Polymer (B): **poly(*N*-vinylcaprolactam)** **2005LAU**
Characterization: M_w/g.mol^{-1} = 330000, synthesized in the laboratory
Solvent (A): **water** **H_2O** **7732-18-5**

Type of data: cloud points (LCST-behavior)

c_B/(g/l)	1.0	*T*/K	306.25

Polymer (B): **poly(*N*-vinylcaprolactam)** **2005DUB**
Characterization: synthesized in the laboratory
Solvent (A): **water** **H_2O** **7732-18-5**

Type of data: cloud points (LCST-behavior)

c_B/(g/l)	1.6	2.2	2.5	3.0	3.7	5.0	7.5
T/K	306.05	305.35	305.45	305.25	305.35	305.95	306.05

Polymer (B): **poly(*N*-vinylcaprolactam-*co*-methacrylic acid)** **2003OKH**
Characterization: M_w/g.mol^{-1} = 4100, 9 mol% methacrylic acid, synthesized in the laboratory
Solvent (A): **water** **H_2O** **7732-18-5**

Type of data: cloud points (LCST-behavior)

c_B/(g/l)	1.0	*T*/K	303.15	(in 0.05 M NaCl aqueous solution)

Polymer (B): **poly(*N*-vinylcaprolactam-*co*-methacrylic acid)** **2003OKH**
Characterization: 12 mol% methacrylic acid, synthesized in the laboratory
Solvent (A): **water** **H_2O** **7732-18-5**

Type of data: cloud points (LCST-behavior)

c_B/(g/l)	1.0	*T*/K	310.65	(in 0.05 M NaCl aqueous solution)

Polymer (B): **poly(*N*-vinylcaprolactam-*co*-methacrylic acid)** **2003OKH**
Characterization: 18 mol% methacrylic acid, synthesized in the laboratory
Solvent (A): **water** **H_2O** **7732-18-5**

Type of data: cloud points (LCST-behavior)

c_B/(g/l)	1.0	*T*/K	313.85	(in 0.05 M NaCl aqueous solution)

Polymer (B): **poly(*N*-vinylcaprolactam-*co*-methacrylic acid)** **2003OKH**
Characterization: M_w/g.mol^{-1} = 4100, 37 mol% methacrylic acid
Solvent (A): **water** **H_2O** **7732-18-5**

Type of data: cloud points (LCST-behavior)

c_B/(g/l)	1.0	*T*/K	304.15	(in 0.05 M NaCl aqueous solution)

Polymer (B): **poly[*N*-vinylcaprolactam-*g*-poly(ethyleneoxidoxyalkyl methacrylate)]** **2005LAU**
Characterization: M_w/g.mol^{-1} = 71000, 6.3 wt% poly(ethyleneoxidoxyalkyl methacrylate), MAC11EO42, synthesized in the laboratory
Solvent (A): **water** **H_2O** **7732-18-5**

Type of data: cloud points (LCST-behavior)

c_B/(g/l) 1.0 *T*/K 306.25

Polymer (B): **poly[*N*-vinylcaprolactam-*g*-poly(ethyleneoxidoxyalkyl methacrylate)]** **2005LAU**
Characterization: M_w/g.mol^{-1} = 310000, 13.0 wt% poly(ethyleneoxidoxyalkyl methacrylate), MAC11EO42, synthesized in the laboratory
Solvent (A): **water** **H_2O** **7732-18-5**

Type of data: cloud points (LCST-behavior)

c_B/(g/l) 1.0 *T*/K 306.25

Polymer (B): **poly[*N*-vinylcaprolactam-*g*-poly(ethyleneoxidoxyalkyl methacrylate)]** **2005LAU**
Characterization: M_w/g.mol^{-1} = 250000, 15.8 wt% poly(ethyleneoxidoxyalkyl methacrylate), MAC11EO42, synthesized in the laboratory
Solvent (A): **water** **H_2O** **7732-18-5**

Type of data: cloud points (LCST-behavior)

c_B/(g/l) 1.0 *T*/K 306.55

Polymer (B): **poly[*N*-vinylcaprolactam-*g*-poly(ethyleneoxidoxyalkyl methacrylate)]** **2005LAU**
Characterization: M_w/g.mol^{-1} = 300000, 18.3 wt% poly(ethyleneoxidoxyalkyl methacrylate), MAC11EO42, synthesized in the laboratory
Solvent (A): **water** **H_2O** **7732-18-5**

Type of data: cloud points (LCST-behavior)

c_B/(g/l) 1.0 *T*/K 306.65

Polymer (B): **poly[*N*-vinylcaprolactam-*g*-poly(ethyleneoxidoxyalkyl methacrylate)]** **2005LAU**
Characterization: M_w/g.mol^{-1} = 260000, 34.0 wt% poly(ethyleneoxidoxyalkyl methacrylate), MAC11EO42, synthesized in the laboratory
Solvent (A): **water** **H_2O** **7732-18-5**

Type of data: cloud points (LCST-behavior)

c_B/(g/l) 1.0 *T*/K 306.95

Polymer (B): **poly(vinyl methyl ether)** **2005NIE**
Characterization: M_w/g.mol^{-1} = 21000
Solvent (A): **deuterium oxide** **D_2O** **7789-20-0**

Type of data: spinodal points (LCST-behavior)

w_B	0.1	0.2	0.3	0.4	0.5	0.6	0.7	0.8
T/K	306.25	307.35	308.32	304.37	305.15	302.80	298.47	305.20

Polymer (B): **poly(vinyl methyl ether)** **2006NIE**
Characterization: –
Solvent (A): **deuterium oxide** **D_2O** **7789-20-0**

Type of data: spinodal points (UCST-behavior)

w_B	0.6897	0.7620	0.7954	0.8004	0.8492	0.8843	0.9000
T/K	184.95	236.45	248.05	240.45	233.95	242.15	219.95

Type of data: spinodal points (LCST-behavior)

w_B	0.1	0.2	0.3	0.4	0.5	0.6	0.6897	0.7	0.762
T/K	306.85	309.45	309.65	306.95	305.55	303.65	293.95	298.45	301.95
w_B	0.7954	0.8004							
T/K	297.65	302.75							

Polymer (B): **poly(vinyl methyl ether)** **2006LOO**
Characterization: M_n/g.mol^{-1} = 8000, M_w/g.mol^{-1} = 20000, synthesized in the laboratory
Solvent (A): **water** **H_2O** **7732-18-5**

Type of data: cloud points (LCST-behavior)

w_B	0.02	0.1	0.2	0.3	0.35	0.4	0.5	0.6	0.7
T/K	310.75	307.35	308.65	309.65	310.45	307.15	302.65	301.55	302.15

Polymer (B): **poly(vinyl methyl ether)** **2005VER**
Characterization: M_n/g.mol^{-1} = 12000, M_w/g.mol^{-1} = 15000, synthesized in the laboratory
Solvent (A): **water** **H_2O** **7732-18-5**

Type of data: cloud points (LCST-behavior)

c_B/(g/l) 1.0 T/K 308.15

Polymer (B): **poly(vinyl methyl ether-*b*-vinyl isobutyl ether)** **2005VER**
Characterization: M_n/g.mol^{-1} = 9800, M_n/g.mol^{-1}(PVME-block) = 9300, M_n/g.mol^{-1}(PVIBE-block) = 500, M_w/g.mol^{-1} = 11270, 3 mol% vinyl isobutyl ether, PVME160-*b*-PVIBE5, synthesized in the laboratory
Solvent (A): **water** **H_2O** **7732-18-5**

continued

continued

Type of data: cloud points (LCST-behavior)

c_B/(g/l)	1.0	T/K	305.15

Polymer (B): **poly(vinyl methyl ether-*b*-vinyl isobutyl ether)** **2005VER**
Characterization: M_n/g.mol^{-1} = 5900, M_n/g.mol^{-1}(PVME-block) = 5000, M_n/g.mol^{-1}(PVIBE-block) = 900, M_w/g.mol^{-1} = 6785, 10 mol% vinyl isobutyl ether, PVME85-*b*-PVIBE9, synthesized in the laboratory
Solvent (A): **water** **H_2O** **7732-18-5**

Type of data: cloud points (LCST-behavior)

c_B/(g/l)	1.0	T/K	312.15

Polymer (B): **poly(vinyl methyl ether-*b*-vinyl isobutyl ether-*b*-vinyl methyl ether)** **2005VER**
Characterization: M_n/g.mol^{-1} = 9500, M_n/g.mol^{-1}(PVME-block) = 4500, M_n/g.mol^{-1}(PVIBE-block) = 1000, M_w/g.mol^{-1} = 11210, 7 mol% vinyl isobutyl ether, PVME65-*b*-PVIBE10-*b*-PVME65, synthesized in the laboratory
Solvent (A): **water** **H_2O** **7732-18-5**

Type of data: cloud points (LCST-behavior)

c_B/(g/l)	1.0	T/K	314.15

3.2. Table of systems where binary LLE data were published only in graphical form as phase diagrams or related figures

Polymer (B)	Solvent (A)	Ref.
N-Carboxyethylchitosan	water	2010PEN
Chitosan-*g*-poly(*N*-isopropylacrylamide)	water	2009LIX
	water	2009REC
	water	2010BAO
Deca(ethylene glycol) mono(nonylphenyl) ether	water	2010KIM
Deca(ethylene glycol) monotridecyl ether	water	2010KIM
Dextran-*g*-poly(*N*-isopropylacrylamide)	water	2009PAT
Ethylcellulose-*g*-[poly(ethylene glycol) methyl ether methacrylate]	water	2008LIY
Ethyl(hydroxyethyl)cellulose	water	2003LUT
	water	2004OLS
Gelatine	water	2005GUP
Hepta(ethylene glycol) monododecyl ether	water	2005SHI
	water	2010KIM

Polymer (B)	Solvent (A)	Ref.
Hepta(ethylene glycol) monohexadecyl ether		
	water	2005EIN
Hepta(ethylene glycol) monooctadecyl ether		
	water	2005EIN
Hepta(ethylene glycol) monotetradecyl ether		
	water	2005SHI
Hexa(ethylene glycol) monodecyl ether		
	water	2005IMA
	water	2011BIA
Hexa(ethylene glycol) monododecyl ether		
	water	2004YOS
	water	2011BIA
Hexa(ethylene glycol) monohexadecyl ether		
	water	2006EIN
Hexa(ethylene glycol) monotetradecyl ether		
	water	2004YOS
Hydroxyethylcellulose-*g*-poly(*N*-isopropyl-acrylamide)		
	water	2011PHA
Hydroxypropylcellulose		
	water	2007VS1
	water	2007VS2
	water	2008VSH
	water	2010KUL
	water	2011FET
Hydroxypropylcellulose-*g*-poly(2-(*N,N*-dimethyl-amino)ethyl methacrylate)		
	water	2010MAL

Polymer (B)	Solvent (A)	Ref.
Hydroxypropylcellulose-*g*-poly(*N*-isopropyl-acrylamide)-*g*-poly(acrylic acid)	water	2009LIX
Methylcellulose	water	2005SP1
	water	2006SCH
	water	2011FET
Methyl(hydroxypropyl)cellulose	water	2006SCH
	water	2008SI2
	water	2009VIR
	water	2011FET
Nona(ethylene glycol) monododecyl ether	water	2010KIM
Octa(ethylene glycol) monododecyl ether	water	2011BIA
Octa(ethylene glycol) monohexadecyl ether	water	2005HAM
Octa(ethylene glycol) monooctadecyl ether	water	2005HAM
Octa(ethylene glycol) monotetradecyl ether	water	2005HAM
Penta(ethylene glycol) monodecyl ether	water	2005IMA
Penta(ethylene glycol) monododecyl ether	water	2005SHI
Penta(ethylene glycol) monooctyl ether	water	2011BIA

Polymer (B)	**Solvent (A)**	**Ref.**
Penta(ethylene glycol) monotetradecyl ether		
	water	2006EIN
Poly(acrylamide-*co*-hydroxypropyl acrylate)		
	water	1975TAY
Poly(acrylamide-*co*-*N*-isopropylacrylamide)		
	water	2004EEC
	water	2007ERB
Poly(acrylamide-*co*-*N*-isopropylmethacrylamide)		
	deuterium oxide	2010KOU
Poly(3-acrylamido-3-deoxy-1,2:5,6-di-O-isopropylidene-α-D-glucofuranose-*co*-*N*-isopropylacrylamide)		
	water	2009SHI
Poly(acrylic acid-*co*-*N*-isopropylacrylamide)		
	water	1993OTA
	water	1997YOO
	water	2001BUL
	water	2004MA1
	water	2006WEN
	water	2010SIB
Poly[acrylic acid-*g*-poly(*N*-isopropylacrylamide)]		
	water	2008LI1
Poly[*N*-(3-acryloyloxypropyl)pyrrolidone]		
	deuterium oxide	2010SU1
	water	2010SU1
Poly(*N*-acryloyl-L-proline methyl ester-*b*-*N*-acryloyl-4-*trans*-hydroxy-L-proline)		
	water	2010MO1
Poly(*N*-acryloylpyrrolidine)		
	water	2007SKR

Polymer (B)	**Solvent (A)**	**Ref.**
Poly(*N*-acryloylpyrrolidine-*b*-*N*-isopropyl-acrylamide)	water	2007SKR
Poly(*N*-acryloyl-L-valine *N'*-methylamide)	water	2010LI2
Poly[allylamine-*g*-poly(*N*-isopropylacrylamide)]	water	2005GAO
Poly(amidoamine) dendrimer	water water	2006HAB 2006TON
Polyaspartamide copolymers	water	2009MOO
Poly(m-benzamide)	water	2006SUG
Poly(5,6-benzo-2-methylene-1,3-dioxepane-*co*-*N*-isopropylacrylamide)	water	2007REN
Poly[1,3-bis(dimethylamino)propan-2-yl methacrylate]	water	2009YUB
Poly[1-(bis(3-(dimethylamino)propyl)amino)-propan-2-yl methacrylate]	water	2009YUB
Poly{2-*tert*-butoxy-*N*-[2-(methacryloyloxy)ethyl]-*N,N*-dimethyl-2-oxoethanammonium chloride-*co*-*N*-isopropylacrylamide}	water	2010GON
Poly(*N*-*tert*-butylacrylamide-*co*-acrylamide)	water	2009MAH

Polymer (B)	**Solvent (A)**	**Ref.**
Poly(*N*-*tert*-butylacrylamide-*co*-*N,N*-dimethylacrylamide)	water	1999LIU
Poly(*N*-*tert*-butylacrylamide-*co*-*N*-ethylacrylamide)	water	1999LIU
Poly(butyl acrylate-*co*-*N*-isopropylacrylamide)	water	2010ZHE
Poly(*tert*-butyl acrylate)-*g*-poly[oligo(ethylene glycol) methyl ether methacrylate)]	water	2011SON
Poly(butylene oxide-*b*-ethylene oxide)	water	2004KEL
	water	2005CHA
Poly(butylene oxide-*b*-ethylene oxide-*b*-butylene oxide)	water	2004KEL
Poly(*tert*-butyl glycidylether)-*b*-polyglycidol star polymer	water	2011LIB
Poly(4-*tert*-butyl styrene-*b*-*N*-isopropylacrylamide-*b*-4-*tert*-butyl styrene)	water	2010BIV
Poly(ε-caprolactone-*b*-ethylene glycol-*b*-ε-caprolactone)	water	2005BAE
	water	2006LUC
Poly(ε-caprolactone-*b*-*N*-isopropylacrylamide-*b*-ε-caprolactone)	water	2008CHA

Polymer (B)	Solvent (A)	Ref.
Poly[*N*-cyclopropylacrylamide-*co*-4-(2-phenyl-diazenyl)benzamido-*N*-(2-aminoethyl)acrylamide]	water	2009JOC
Poly(diacetone acrylamide-*co*-acrylamide)	water	1975TAY
Poly(diacetone acrylamide-*co*-hydroxyethyl acrylate)	water	1975TAY
Poly(3,5-dibromobenzyl acrylate-*b*-*N*-isopropyl-acrylamide-*b*-3,5-dibromobenzyl acrylate)	water	2010BIV
Poly(*N,N*-diethylacrylamide)	water	2010KA2
Poly(*N,N*-diethylacrylamide-*co*-acrylamide)	water	1999LIU
Poly(*N,N*-diethylacrylamide-*co*-acrylic acid)	water	2004MA1
	water	2008FAN
Poly(*N,N*-diethylacrylamide-*co*-*N*-acryloxy-succinimide)	water	2010ZH2
Poly(*N,N*-diethylacrylamide-*co*-*N,N*-dimethyl-acrylamide)	water	1999LIU
Poly(*N,N*-diethylacrylamide-*co*-*N*-ethylacrylamide)	water	1999LIU
Poly(*N,N*-diethylacrylamide-*co*-methacrylic acid)	water	2003LI1

Polymer (B)	Solvent (A)	Ref.
Poly[*N,N*-diethylacrylamide-*co*-4-(2-phenyl-diazenyl)benzamido-*N*-(2-aminoethyl)acrylamide]	water	2009JOC
Poly[2-(*N,N*-diethylamino)ethyl methacrylate]	water	2010SCH
Poly[2-(*N,N*-diethylamino)ethyl methacrylate] star polymer	water	2010SCH
Poly[2-(*N,N*-diethylamino)ethyl methacrylate-*b*-*N*-isopropylacrylamide]	water	2008LI3
Poly[di(ethylene glycol) methacrylate]	water	2010LUZ
Poly[di(ethylene glycol) methacrylate-*co*-oligo(ethylene glycol) methacrylate]	water	2010LUZ
Poly[di(ethylene glycol) methyl ether methacrylate-*co*-oligo(ethylene glycol) methyl ether methacrylate]	water	2010DO2
Poly[di(ethylene glycol) methyl ether methacrylate]-*b*-tri(ethylene glycol) methyl ether methacrylate]	water	2008YAM
Poly[di(ethylene glycol) methyl ether methacrylate]-*co*-tri(ethylene glycol) methyl ether methacrylate]	water	2008YAM
Poly(*N,N*-dimethylacrylamide)	water	2007SKR
	water	2011FIS

Polymer (B)	Solvent (A)	Ref.
Poly(*N,N*-dimethylacrylamide-*co*-glycidyl methacrylate)	water	2003YIN
Poly(*N,N*-dimethylacrylamide-*b*-*N*-isopropyl-acrylamide)	water	2007SKR
Poly(*N,N*-dimethylacrylamide-*b*-*N*-isopropyl-acrylamide-*b*-*N*-acryloylpyrrolidine)	water	2007SKR
Poly[2-(2-(2-(*N,N*-dimethylamino)ethoxy)ethyl)-methylamino)ethyl acrylate]	water	2009YUB
Poly[2-(*N,N*-dimethylamino)ethyl methacrylate]	water	2006VER
	water	2007FOU
	water	2007PL2
	water	2007PL3
(azo end-groups)	water	2010TAN
	water	2010ZH3
Poly[2-(*N,N*-dimethylamino)ethyl methacrylate-*b*-acrylic acid]	water	2011XIO
Poly[2-(*N,N*-dimethylamino)ethyl methacrylate-*b*-(2,2,3,4,4,4-hexafluorobutyl methacrylate)]	water	2009LI1
Poly[2-(*N,N*-dimethylamino)ethyl methacrylate-*co*-2-hydroxyethyl methacrylate]	water	2011GAL
Poly[2-(*N,N*-dimethylamino)ethyl methacrylate-*b*-*N*-isopropylacrylamide]	water	2008LI3

Polymer (B)	Solvent (A)	Ref.
Poly[2-(*N,N*-dimethylamino)ethyl methacrylate-*g*-poly(*N*-isopropylacrylamide]	water	2008LI2
	water	2009LI2
Poly{2-(*N,N*-dimethylamino)ethyl methacrylate-*co*-4-methyl-[7-(methacryloyl)oxyethyloxy]coumarin}	water	2011ZH3
Poly[2-(*N,N*-dimethylamino)ethyl methacrylate-*b*-(2,2,3,3,4,4,5,5-octafluoropentyl methacrylate)]	water	2009LI1
Poly[2-(*N,N*-dimethylamino)ethyl methacrylate-*b*-oligo(ethylene glycol) methacrylate-*b*-butyl methacrylate]	water	2010WAR
Poly[2-(*N,N*-dimethylamino)ethyl methacrylate-*stat*-oligo(ethylene glycol) methyl ether methacrylate]	water	2007FOU
Poly[2-(*N,N*-dimethylamino)ethyl methacrylate-*b*-(2,2,2-trifluoroethyl methacrylate)]	water	2009LI1
Poly[2-(*N,N*-dimethylamino)ethyl methacrylate-*co*-*N*-vinylcaprolactam]	water	2006VER
Poly[2-(5,5-dimethyl-1,3-dioxan-2-yloxy)ethyl acrylate-*co*-oligo(ethylene glycol) acrylate]	water	2010QIA
Poly{*N*-[(2,2-dimethyl-1,3-dioxolane)methyl]-acrylamide}	water	2008ZOU

Polymer (B)	Solvent (A)	Ref.
Poly(3,3-dimethyl-1-vinyl-2-pyrrolidinone)		
	water	2010CH4
Poly[2-(1,3-dioxan-2-yloxy)ethyl acrylate-*co*-oligo(ethylene glycol) acrylate]		
	water	2010QIA
Poly[3-(1,3-dioxolan-2-yl)ethyl-1-vinyl-2-pyrrolidinone]		
	water	2010CH4
Poly(divinyl ether-*alt*-maleic anhydride)		
	water	2004VOL
	water	2005IZU
Polyester dendrimer		
	water	2006PAR
Polyesteramides, hyperbranched		
	water	2011KEL
Poly(ether amine)		
	water	2010YUB
Poly[2-(2-ethoxyethoxy)ethyl acrylate]		
	water	2007SKR
Poly[2-(2-ethoxyethoxy)ethyl vinyl ether]		
	water	2005MAT
Poly[2-(2-ethoxyethoxy)ethyl vinyl ether-*b*-(2-methoxyethyl vinyl ether)]		
	deuterium oxide	2006OSA
	deuterium oxide	2010SHI
	water	2006OSA
Poly[*N*-(2-ethoxyethyl)acrylamide]		
	water	2009MA3

Polymer (B)	**Solvent (A)**	**Ref.**
Poly[*N*-(2-ethoxyethyl)acrylamide-*co*-*N*-isopropylacrylamide]		
	water	2009MA3
Poly[*N*-(3-ethoxypropyl)acrylamide]		
	water	2005UGU
Poly(*N*-ethylacrylamide)		
	water	1975TAY
	water	2003XUE
Poly(*N*-ethylacrylamide-*co*-*N*-isopropylacrylamide)		
	water	1975TAY
Poly(ethylaminophosphazene hydrochloride)		
	water	2010GRI
Poly(ethylene glycol)		
	water	2004DOR
Poly(ethylene glycol-*b*-*N*-acryloyl-2,2-dimethyl-1,3-oxazolidine)		
	water	2011CUI
Poly(ethylene glycol-*b*-ε-caprolactone)		
	water	2006LUC
Poly(ethylene glycol-*b*-*N*-isopropylacrylamide)		
	water	2005MOT
	water	2005ZH3
Poly(ethylene glycol-*b*-*N*-isopropylacrylamide-*b*-ε-caprolactone)		
	water	2011CHE2
Poly(ethylene glycol-*stat*-propylene glycol) monobutyl ether		
	water	2005AUB

Polymer (B)	Solvent (A)	Ref.
Poly(ethylene glycol-*alt*-trehalose)	water	2011EIS
Poly(ethylene glycol-*b*-4-vinylpyridine-*b*-*N*-isopropylacrylamide)	water	2005ZH4
Poly(ethylene glycol) dimethyl ether	water	2004DOR
Poly[(ethylene glycol) methyl ether methacrylate]	water	2007FOU
Poly(ethylene glycol) monododecyl ether	water	2005YAM
Poly(ethylene glycol) monomethyl ether	water	2004DOR
Poly(ethylene oxide-*b*-n-alkyl glycidyl carbamate-*b*-ethylene oxide)	water	2007DIM
Poly(ethylene oxide-*b*-butylene oxide)	water	2004KEL
	water	2005CHA
Poly[ethylene oxide-*b*-(ethylene oxide-*co*-propylene oxide)-*b*-DL-lactide]	water	2005AUB
Poly(ethylene oxide-*b*-*N*-isopropylacrylamide)	water	2004NED
	water	2006QIN
Poly[ethylene oxide-*b*-(DL-lactide-*co*-glycolide)-*b*-ethylene oxide]	water	2004PA1

Polymer (B)	**Solvent (A)**	**Ref.**
Poly(ethylene oxide-*b*-propylene oxide)		
	water	2007MAN
Poly(ethylene oxide-*co*-propylene oxide)		
	water	2003CAM
	water	2005AUB
Poly(ethylene oxide-*b*-propylene oxide-*b*-ethylene oxide)		
	deuterium oxide	2010KLO
	water	2004VAR
	water	2008BER
	water	2008SHA
	water	2009ALV
	water	2009KOS
	water	2009NAN
	water	2010TSU
	water	2010WAN
Poly[(ethylene oxide-*co*-propylene oxide)-*b*-DL-lactide]		
	water	2005AUB
Poly(ethylene oxide-*b*-propylene oxide) monododecyl ether		
	water	2011ZH2
Poly(ethylene oxide-*b*-styrene oxide)		
	water	2003YAN
Poly(ethylenimine)		
	water	2007LIU
Poly(ethylenimine) (branched, *N*-acylated)		
	water	2011KIM
Poly(ethylenimine) (hyperbranched, isobutyramide residues)		
	water	2011WA2

Polymer (B)	**Solvent (A)**	**Ref.**
Poly(ethyl ethylene phosphate-*b*-propylene oxide-*b*-ethyl ethylene phosphate)	water	2009WA3
Poly[3-ethyl-3-(hydroxymethyl)-oxetane]	water	2010CH2
Poly{3-ethyl-3-(hydroxymethyl)-oxetane-*g*-poly[2-(*N,N*-diethylamino)ethyl methacrylate]}	water	2010CH2
Poly[3-ethyl-3-(hydroxymethyl)-oxetane-*g*-poly(ethylene oxide)]	water	2010CH2
Poly(2-ethylhexyl acrylate-*b*-*N*-isopropyl-acrylamide-*b*-2-ethylhexyl acrylate)	water	2010BIV
Poly(*N*-ethyl-*N*-methylacrylamide)	water	2008XUJ
	water	2009CAO
Poly[*N*-ethyl-*N*-methylacrylamide-*co*-4-(2-phenyl-diazenyl)benzamido-*N*-(2-aminoethyl)acrylamide]	water	2009JOC
Poly(*N,N'*-ethyl methylene malonamide)	water	2010MO2
Poly(2-ethyl-2-oxazoline)	water	2003CHR
	water	2007PAR
	water	2008HO1
Poly(2-ethyl-2-oxazoline) star polymer	water	2011KOW

Polymer (B)	Solvent (A)	Ref.
Poly[2-ethyl-2-oxazoline-*co*-2-(dec-9-enyl)-2-oxazoline] sugar-substituted		
	water	2011KEM
Poly(2-ethyl-2-oxazoline-*b*-2-nonyl-2-oxazoline)		
	water	2009LAM
Poly(2-ethyl-2-oxazoline-*co*-2-nonyl-2-oxazoline)		
	water	2009LAM
Poly(2-ethyl-2-oxazoline-*co*-2-propyl-2-oxazoline)		
	water	2007PAR
	water	2008HO1
Poly(*N*-ethyl-2-propionamidoacrylamide)		
	water	2010BER
	water	2010MO2
Poly(3-ethyl-1-vinyl-2-pyrrolidinone)		
	water	2009TRE
	water	2010CH4
Poly(3-ethyl-1-vinyl-2-pyrrolidinone-*co*-1-vinyl-2-pyrrolidinone)		
	water	2009TRE
Poly[L-glutamic acid-*g*-oligo(ethylene glycol monomethyl ether)]		
	water	2011CHE1
Poly[L-glutamic acid-*g*-poly(*N*-isopropylacrylamide)]		
	water	2008HE2
Poly(γ-glutamine), α-*N*-substituted		
	water	2003TAC
Poly[glycerol-*co*-3-methyl-3-(hydroxymethyl)-oxetane]		
	water	2011XIA

Polymer (B)	**Solvent (A)**	**Ref.**
Poly(glycidol-*b*-*N*-isopropylacrylamide)	water	2010MEN
Poly[glycidol-*co*-oligo(ethylene glycol) monoethyl ether], hyperbranched	water	2010KOJ
Poly(glycidol-*b*-propylene oxide-*b*-glycidol)	water	2006HAL
Poly(hexylamine methacrylamide)	water	2010CH3
Poly(hexylamine methacrylamide-*co*-*N*-isoproylacrylamide)	water	2010CH3
Poly(2-hydroxyethyl acrylate-*co*-butyl acrylate)	water	2007MUN
Poly(2-hydroxyethyl acrylate-*co*-2-hydroxyethyl methacrylate)	water	2008KHU
Poly(2-hydroxyethyl acrylate-*co*-2-hydroxypropyl acrylate)	water	1975TAY
Poly(2-hydroxyethyl acrylate-*co*-vinyl butyl ether)	water	2006MUN
Poly(2-hydroxyethyl methacrylate) (amino group-functionalized)	water	2011JUN
Poly(2-hydroxyethyl methacrylate), poly(amidoamine) dendronized	water	2010GAO

Polymer (B)	**Solvent (A)**	**Ref.**
Poly[2-hydroxyethyl methacrylate-*co*-2-(*N,N*-dimethylamino)ethyl methacrylate]	water	2011GAL
	water	2011LON
Poly(2-hydroxyethyl methacrylate-*b*-*N*-isopropylacrylamide)	water	2006CAO
Poly(2-hydroxyethyl methacrylate-*co*-methacrylic acid)	water	2011LON
Poly{2-hydroxyethyl methacrylate-*co*-[3-(methacryloylamino)propyl]trimethylammonium chloride}	water	2011LON
Poly{2-hydroxyethyl methacrylate-*co*-[2-(methacryloyloxy)ethyl]trimethylammonium chloride}	water	2011LON
Poly[2-hydroxyethyl methacrylate-*g*-poly(ethylene glycol]	water	2005ZH1
Poly(2-hydroxyisopropyl acrylate-*co*-aminoethyl methacrylate)	water	2009DEN
Poly(*N*-hydroxymethylacrylamide-*co*-*N*-isopropylacrylamide-*co*-butyl acrylate)	water	2010ZHO
Poly(*N*-(1-hydroxymethyl)propylmethacrylamide)	water	2005SET
Poly(2-hydroxypropyl acrylate-*co*-aminoethyl methacrylate)	water	2009DEN

Polymer (B)	Solvent (A)	Ref.
Poly(isobutyl vinyl ether-*co*-2-hydroxyethyl vinyl ether)		
	water	2004SUG
Poly(*N*-isopropylacrylamide)		
	deuterium oxide	2004MAO
	deuterium oxide	2010SHI
(end-group modified)	deuterium oxide	2011RUG
	water	2002DIN
	water	2003STI
	water	2004DU2
	water	2004MAO
	water	2004SAN
	water	2004SHI
	water	2005DUR
	water	2005STA
	water	2005TIE
	water	2006DUA
	water	2006KUJ
	water	2006OKA
	water	2006PLU
	water	2006XIA
	water	2007DE2
	water	2007SHI
	water	2007SKR
	water	2007VA2
	water	2008BER
	water	2008BUR
	water	2008KAT
	water	2008ZHO
	water	2009CAO
(urea end-functionalized)	water	2009FUC
	water	2009HIR
	water	2009KOB
	water	2009MA3
	water	2009OSA
	water	2009PAM
	water	2009TAU
	water	2010DUB
	water	2010SHE
	water	2010SHI
	water	2010YEX
(methanolyzed)	water	2011MIZ
(isotactic-rich)	water	2011NAK

Polymer (B)	**Solvent (A)**	**Ref.**
Poly(*N*-isopropylacrylamide), cyclic		
	water	2007QIU
	water	2007XUJ
Poly(*N*-isopropylacrylamide), star		
	water	2009XUJ
Poly[*N*-isopropylacrylamide-*b*-*N*-(acetylimino)ethylene]		
	water	2003DAV
Poly(*N*-isopropylacrylamide-*co*-acrylamide)		
	water	2004EEC
	water	2007ERB
Poly(*N*-isopropylacrylamide-*co*-6-acrylaminohexanoic acid)		
	water	2000KUC
Poly(*N*-isopropylacrylamide-*co*-3-acrylaminopropanoic acid)		
	water	2000KUC
Poly(*N*-isopropylacrylamide-*co*-11-acrylaminoundecanoic acid)		
	water	2000KUC
Poly(*N*-isopropylacrylamide-*co*-acrylic acid)		
	water	1993OTA
	water	1997YOO
	water	2001BUL
	water	2004MA1
	water	2006WEN
	water	2010SIB
Poly(*N*-isopropylacrylamide-*co*-acrylic acid-*co*-ethyl methacrylate)		
	water	2005TIE

Polymer (B)	Solvent (A)	Ref.
Poly(*N*-isopropylacrylamide-*co*-acryloyloxypropylphosphinic acid)	water	2004NON
Poly(*N*-isopropylacrylamide-*b*-*N*-acryloyl-pyrrolidine-*b*-*N,N*-dimethylacrylamide)	water	2007SKR
Poly(*N*-isopropylacrylamide-*co*-*N*-adamantyl-acrylamide)	water	2008WIN
Poly(*N*-isopropylacrylamide-*co*-benzo-15-crown-5-acrylamide)	water	2008MIP
Poly(*N*-isopropylacrylamide-*b*-*N*-benzyloxycarbonyl-4-hydroxy-L-proline)	water	2010LE2
Poly[*N*-isopropylacrylamide-*co*-(*N*-(R,S)-*sec*-butylacrylamide)]	water	2009LIP
Poly[*N*-isopropylacrylamide-*co*-(*N*-(S)-*sec*-butylacrylamide)]	water	2009LIP
Poly(*N*-isopropylacrylamide-*co*-butyl acrylate)	water	2010ZHE
Poly(*N*-isopropylacrylamide-*b*-ε-caprolactone)	water	2006CHO
Poly(*N*-isopropylacrylamide-*co*-*N,N*-diethylacrylamide)	water	2009MA2

Polymer (B)	**Solvent (A)**	**Ref.**
Poly(*N*-isopropylacrylamide-*co*-*N,N*-dimethylacrylamide)	water	2005LIU
Poly(*N*-isopropylacrylamide-*b*-*N,N*-dimethyl-acrylamide-*b*-*N*-acryloylpyrrolidine)	water	2007SKR
Poly[(*N*-isopropylacrylamide-*co*-*N,N*-dimethyl-acrylamide)-*b*-(DL-lactide-*co*-glycolide)]	water	2005LIU
Poly(*N*-isopropylacrylamide-*co*-*N,N*-dimethyl-aminopropylmethacrylamide)	water	2010SIB
Poly[*N*-isopropylacrylamide-*b*-3'-(1',2':5',6'-di-O-isopropylidene-α-D-glucofuranosyl)-6-methacrylamido hexanoate]	water	2007OEZ
Poly[*N*-isopropylacrylamide-*co*-3'-(1',2':5',6'-di-O-isopropylidene-α-D-glucofuranosyl)-6-methacrylamido hexanoate]	water	2007OEZ
Poly[*N*-isopropylacrylamide-*b*-3'-(1',2':5',6'-di-O-isopropylidene-α-D-glucofuranosyl)-6-methacrylamido undecanoate]	water	2007OEZ
Poly[*N*-isopropylacrylamide-*co*-3'-(1',2':5',6'-di-O-isopropylidene-α-D-glucofuranosyl)-6-methacrylamido undecanoate]	water	2007OEZ
Poly[*N*-isopropylacrylamide-*b*-2-(*N,N*-dimethyl-amino)ethyl methacrylate]	water	2008LI3

Polymer (B)	**Solvent (A)**	**Ref.**
Poly[*N*-isopropylacrylamide-*co*-2-(*N,N*-dimethyl-amino)ethyl methacrylate-*co*-butyl methacylate]	water	2004TAK
Poly(*N*-isopropylacrylamide-*co*-*N*-dodecyl-acrylamide)	water	2008WIN
Poly(*N*-isopropylacrylamide-*b*-ethylene glycol)	water	2007VA2
	water	2009TAU
Poly(*N*-isopropylacrylamide-*b*-ethylene glycol-*b*-*N*-isopropylacrylamide)	water	2011MAL
Poly(*N*-isopropylacrylamide-*b*-ethylene oxide)	water	2004NED
Poly(*N*-isopropylacrylamide-*b*-ethylenimine)	water	2002DIN
Poly{*N*-isopropylacrylamide-*b*-[(L-glutamic acid)-*co*-(γ-benzyl-L-glutamate)]}	water	2008HE1
Poly(*N*-isopropylacrylamide-*alt*-2-hydroxyethyl methacrylate)	water	2008FAR
Poly(*N*-isopropylacrylamide-*co*-2-hydroxyethyl methacrylate)	water	2009ZHA
Poly[*N*-isopropylacrylamide-*co*-(2-hydroxyisopropyl)acrylamide]	water	2006MA1
	water	2006MA2

Polymer (B)	**Solvent (A)**	**Ref.**
Poly(*N*-isopropylacrylamide-*co*- *N*-hydroxymethylacrylamide)		
	water	2006DI2
	water	2008KOT
Poly(*N*-isopropylacrylamide-*b*- 4-hydroxy-L-proline)		
	water	2010LE2
Poly(*N*-isopropylacrylamide-*co*-3H-imidazole- 4-carbodithioic acid 4-vinylbenzyl ester)		
	water	2005CAR
Poly{*N*-isopropylacrylamide-*b*-[*N*-isopropyl- acrylamide-*co*-*N*-(hydroxymethyl)acrylamide)]}		
	water	2009KOT
Poly[*N*-isopropylacrylamide-*b*-*N*-isopropylacryl- amide(isotactic)-*b*-*N*-isopropylacrylamide]		
	water	2008NU1
	water	2008NU2
Poly[*N*-isopropylacrylamide(isotactic)-*b*- *N*-isopropylacrylamide-*b*-*N*-isopropyl- acrylamide(isotactic)]		
	water	2008NU1
	water	2008NU2
Poly(*N*-isopropylacrylamide-*co*-*N*-isopropyl- methacrylamide)		
	water	2005STA
Poly(*N*-isopropylacrylamide-*b*-DL-lactide)		
	water	2003LI2
Poly(*N*-isopropylacrylamide-*b*-L-lysine)		
	water	2008ZHA

Polymer (B)	Solvent (A)	Ref.
Poly(*N*-isopropylacrylamide-*co*-maleic acid)	water	2004WEI
Poly(*N*-isopropylacrylamide-*co*-maleic anhydride)	water	2007FRA
Poly(*N*-isopropylacrylamide-*co*-maleimide)	water	2007FRA
Poly[*N*-isopropylacrylamide-*co*-(p-methacryl-amido)-acetophenone thiosemicarbazone]	water	2005LIC
Poly{*N*-isopropylacrylamide-*b*-3-[*N*-(3-meth-acrylamidopropyl)-*N,N*-dimethylammonio]-propane sulfate}	water	2002VIR
Poly{*N*-isopropylacrylamide-*co*-3-[*N*-(3-meth-acrylamidopropyl)-*N,N*-dimethylammonio]-propane sulfate}	water	2005NED
Poly(*N*-isopropylacrylamide-*co*-methacrylic acid)	water	2006YIN
Poly(*N*-isopropylacrylamide-*co*-methacrylic acid-*co*-octadecyl acrylate)	water	2011ZHO
Poly(*N*-isopropylacrylamide-*co*-*N*-methacryloyl-L-leucine)	water	2000BIG
Poly{*N*-isopropylacrylamide-*b*-[2-(methacryloyl-oxy)ethyl phosphorylcholine]-*b*-*N*-isopropyl-acrylamide}	water	2009CRI

Polymer (B)	Solvent (A)	Ref.
Poly(*N*-isopropylacrylamide-*co*-8-methacryloyl-oxyoctanoic acid methyl ester)		
	water	2004SA1
Poly(*N*-isopropylacrylamide-*co*-5-methacryloyl-oxypentanoic acid methyl ester)		
	water	2004SA1
Poly(*N*-isopropylacrylamide-*co*-11-methacryloyl-oxyundecanoic acid methyl ester)		
	water	2004SA1
Poly[*N*-isopropylacrylamide-*co*-methoxy-poly(ethylene glycol) monomethacrylate]		
	water	2006KI1
	water	2006KI2
Poly(*N*-isopropylacrylamide-*b*-4-methyl-ε-caprolactone)		
	water	2010LE1
Poly(*N*-isopropylacrylamide-*b*-methyl methacrylate)		
	water	2006WEI
Poly(*N*-isopropylacrylamide-*co*-*N*-methyl-*N*-vinylacetamide)		
	water	2004EEC
Poly[*N*-isopropylacrylamide-*co*-oligo(ethylene glycol) monomethacrylate]		
	deuterium oxide	2003KOH
	deuterium oxide	2005KOH
	water	2004ALA
	water	2006ALA
Poly[*N*-isopropylacrylamide-*co*-oligo(ethylene glycol) monomethacrylate-*co*-dodecyl methacrylate]		
	water	2005VIE

Polymer (B)	Solvent (A)	Ref.
Poly(*N*-isopropylacrylamide-*co*-octadecyl acrylate)	water	2000SHI
Poly(*N*-isopropylacrylamide-*co*-4-pentenoic acid)	water	1999KUN
Poly(*N*-isopropylacrylamide-*b*-4-phenyl-ε-caprolactone)	water	2010LE1
Poly[*N*-isopropylacrylamide-*co*-4-(2-phenyl-diazenyl)benzamido-*N*-(2-aminoethyl)acrylamide]	water	2009JOC
Poly[*N*-isopropylacrylamide-*g*-poly(*N*-(acetylimino)ethylene)]	water	2003DAV
Poly[*N*-isopropylacrylamide-*g*-poly(ethylene glycol)]	water	2005BIS
	water	2007VA2
Poly[*N*-isopropylacrylamide-*g*-poly(2-ethyl-2-oxazoline)]	water	2010RUE
Poly[*N*-isopropylacrylamide-*g*-poly(2-(2-methoxy-carbonylethyl)ethyl-2-oxazoline)]	water	2010RUE
Poly(*N*-isopropylacrylamide-*b*-*N*-propylacrylamide)	water	2009CAO
Poly(*N*-isopropylacrylamide-*co*-*N*-propylacryl-amide)	water	2004MAO
Poly(*N*-isopropylacrylamide-*co*-propylacrylic acid)	water	2006YIN

Polymer (B)	**Solvent (A)**	**Ref.**
Poly(*N*-isopropylacrylamide-*b*-propylene oxide-*b*-*N*-isopropylacrylamide)		
	water	2004HAS
Poly(*N*-isopropylacrylamide-*co*-sodium 2-acrylamido-2-methyl-1-propanesulfonate)		
	water	2003NOW
	water	2008MAS
Poly(*N*-isopropylacrylamide-*co*-sodium acrylate)		
	water	2006MYL
Poly(*N*-isopropylacrylamide-*co*-sodium styrenesulfonate)		
	water	2004NOW
Poly(*N*-isopropylacrylamide-*co*-styrenesulfonate)		
	water	2011MCF
Poly{[*N*-isopropylacrylamide-*co*-3-(trimethoxysilyl)propyl methacrylate]-*b*-(2-(*N,N*-diethylamino)ethyl methacrylate)}		
	water	2009CHA1
Poly(*N*-isopropylacrylamide-*co*-*N*-vinylacetamide)		
	water	2004EEC
Poly(*N*-isopropylacrylamide-*co*-1-vinylimidazole)		
	water	2005BIS
Poly(*N*-isopropylacrylamide-*co*-vinyl laurate)		
	water	2005CAO
Poly(*N*-isopropylacrylamide-*co*-p-vinylphenylboronic acid)		
	water	2005CIM
Poly(*N*-isopropylacrylamide-*b*-4-vinylpyridine)		
	water	2007XUY

Polymer (B)	Solvent (A)	Ref.
Poly(*N*-isopropylacrylamide-*co*-1-vinyl-2-pyrrolidinone)	water	2004EEC
	water	2006DI2
	water	2006GEE
Poly(*N*-isopropylmethacrylamide)	water	2005SP2
	water	2005STA
	water	2006KIR
Poly(*N*-isopropylmethacrylamide-*co*-acrylamide)	deuterium oxide	2010KOU
Poly(*N*-isopropylmethacrylamide-*co*-sodium methacrylate)	deuterium oxide	2005SP2
Poly(2-isopropyl-2-oxazoline)	water	2007PAR
	water	2008HU1
	water	2008HU2
	water	2010ZH5
Poly(2-isopropyl-2-oxazoline) (endfunctionalized)	water	2004PA2
	water	2008HU2
	water	2009OBE
Poly(2-isopropyl-2-oxazoline-*co*-2-butyl-2-oxazoline)	water	2008HU1
Poly(2-isopropyl-2-oxazoline-*co*-2-nonyl-2-oxazoline)	water	2008HU1

Polymer (B)	**Solvent (A)**	**Ref.**
Poly(2-isopropyl-2-oxazoline-*co*-2-propyl-2-oxazoline)		
	water	2007PAR
	water	2008HU1
Poly(L-lysine isophthalamide)		
	water	2005YUE
Poly(maleic anhydride-*alt*-*tert*-butylstyrene)-*g*-poly(ethylene glycol) monomethyl ether		
	water	2002YIN
Poly(maleic anhydride-*alt*-styrene)-*g*-poly(ethylene glycol) monomethyl ether		
	water	2002YIN
Poly[methacrylic acid-*co*-butyl methacrylate-*co*-poly(ethylene glycol) monomethyl ether methacrylate]		
	water	2005JON
Poly(methacrylic acid-*co*-glycidyl methacrylate-*co*-poly(ethylene glycol) monomethyl ether methacrylate)		
	water	2005JON
Poly(methacrylic acid-*co*-lauryl methacrylate-*co*-poly(ethylene glycol) monomethyl ether methacrylate)		
	water	2005JON
Poly[methacrylic acid-*g*-oligo(2-ethyl-2-oxazoline)]		
	water	2010WEB
Poly(*N*-methacryloyl-L-β-isopropylasparagine)		
	water	2010LUO
Poly[*N*-(2-methacryloyloxyethyl)pyrrolidone]		
	deuterium oxide	2010SU1
	water	2010SU1

Polymer (B)	Solvent (A)	Ref.
Poly[*N*-(3-methacryloyloxypropyl)pyrrolidone]	deuterium oxide	2010SU1
	water	2010SU1
Poly[2-(2-methoxyethoxy)ethyl acrylate]	water	2006HU1
	water	2009MA1
Poly[2-(2-methoxyethoxy)ethyl methacrylate]	water	2007SKR
	water	2009MA1
	water	2011HIR
Poly[2-(2-methoxyethoxy)ethyl methacrylate] star polymers	water	2011HIR
Poly[2-(2-methoxyethoxy)ethyl methacrylate-*b*-(2,2-dimethyl-1,3-dioxolan-4-yl)methyl methacrylate] star polymers	water	2011HIR
Poly[2-(2-methoxyethoxy)ethyl methacrylate-*co*-oligo(ethylene glycol) methacrylate]	water	2006LUT
	water	2007LUT
Poly[2-(2-methoxyethoxy)ethyl methacrylate-*co*-oligo(ethylene glycol) methacrylate-*b*-*N*-(3-(*N,N*-dimethylamino)propyl)methacrylamide]	water	2011TIA
Poly[2-(2-methoxyethoxy)ethyl methacrylate-*co*-oligo(ethylene glycol) methyl ether methacrylate]	water	2007SKR
	water	2011PAR
Poly[2-(2-methoxyethoxy)ethyl methacrylate-*co*-oligo(ethylene glycol) methyl ether methacrylate-*b*-glycidyl methacrylate]	water	2011PAR

Polymer (B)	Solvent (A)	Ref.
Poly(2-methoxyethyl vinyl ether)	deuterium oxide	2007MAE
	water	2007MAE
Poly[methoxypoly(ethylene glycol)-*b*-ε-caprolactone]	water	2006KI2
Poly[methoxytri(ethylene glycol) acrylate-*b*-4-vinylbenzyl methoxytris(oxyethylene) ether]	water	2006HU2
Poly(methylacrylamide-*alt*-hydroxyethyl methacrylate)	water	2010FAR
Poly(3-methyl-1-vinyl-2-pyrrolidinone)	water	2010CH4
Poly(octadecyl acrylate-*b*-*N*-isopropylacrylamide-*b*-octadecyl acrylate)	water	2010BIV
Poly[oligo(ethylene glycol)-*alt*-succinic anhydride]	water	2011FEN
Poly[oligo(ethylene glycol) diglycidyl ether-*co*-piperazine-*co*-oligo(propylene glycol) diglycidyl ether)]	water	2009REN
Poly[oligo(ethylene glycol) ethyl ether methacrylate]	water	2008ISH
Poly[oligo(ethylene glycol) methacrylate] (end-group modified)	water	2010ROT

Polymer (B)	Solvent (A)	Ref.
Poly[oligo(ethylene glycol) methacrylate-*b*-butyl methacrylate-*b*-2-(*N,N*-dimethylamino)ethyl methacrylate	water	2010WAR
Poly[oligo(ethylene glycol) methacrylate-*co*-butyl methacrylate-*co*-2-(*N,N*-dimethylamino)ethyl methacrylate]	water	2010WAR
Poly[oligo(ethylene glycol) methacrylate-*b*-2-(*N,N*-dimethylamino)ethyl methacrylate]	water	2010WAR
Poly[oligo(ethylene glycol) methacrylate-*b*-2-(*N,N*-dimethylamino)ethyl methacrylate-*b*-butyl methacrylate]	water	2010WAR
Poly[oligo(ethylene glycol) methacrylate-*co*-oligo(propylene glycol) methacrylate]	water	2009WA2
Poly[oligo(ethylene glycol) methyl ether acrylate]	water	2007SKR
Poly[oligo(ethylene glycol) methyl ether methacrylate]	water	2007SKR
	water	2008ISH
	water	2008YAM
Poly[oligo(ethylene glycol) methyl ether methylacrylate-*co*-ethylene glycol dimethacrylate-*co*-oligo(propylene glycol) methacrylate]	water	2009TAI
Poly[oligo(ethylene glycol) methyl ether methacrylate-*b*-*N*-isopropyl methacrylamide]	water	2010JOC

Polymer (B)	Solvent (A)	Ref.
Poly[oligo(2-ethyl-2-oxazoline)methacrylate]		
	water	2009WEB
Poly[oligo(2-ethyl-2-oxazoline)methacrylate-*stat*-methacrylic acid]		
	water	2011WEB
Poly[oligo(2-ethyl-2-oxazoline)methacrylate-*stat*-methyl methacrylate]		
	water	2011WEB
	water	2009WEB
Poly[4-(oligooxyethylene)styrene]		
	water	2006HU1
Polypeptide		
	water	2003YAM
	water	2004MEY
Poly[perfluoroalkylacrylate-*co*-poly(ethylene oxide) methacrylate]		
	water	2006SHA
Poly[phosphoester-*b*-2-(*N,N*-dimethylamino)ethyl methacrylate]		
	water	2010LI1
Poly(*N*-propylacrylamide)		
	water	2004MAO
	water	2005VA1
	water	2008HUH
	water	2009CAO
Poly(*N*-propylacrylamide-*b*-*N*-isopropylacrylamide)		
	water	2009CAO
Poly(*N*-propylacrylamide-*b*-*N*-isopropylacrylamide-*b*-*N,N*-ethylmethylacrylamide)		
	water	2009CAO

Polymer (B)	Solvent (A)	Ref.
Poly(propylene glycol)	water	2010ELI
Poly(propylene glycol) bis(2-aminopropyl ether)	water (1M HCl)	2006INO
Poly(propylene oxide-*b*-ethylene oxide-*b*-propylene oxide)	water	2004DER
Poly(2-propyl-2-oxazoline)	water	2007PAR
	water	2008HO1
	water	2008HU1
Poly(styrene-*b*-*N*-isopropylacrylamide)	water	2008TRO
	water	2011SHI
Poly(styrene-*b*-*N*-isopropylacrylamide-*b*-styrene)	water	2010BIV
Poly(styrene-*alt*-maleic anhydride)	water	2006QIU
Poly[(styrene-*alt*-maleic anhydride)-*g*-oligo(oxypropylene)amine]	water	2009LIN
Poly[(styrene-*alt*-maleic anhydride)-*g*-poly(amidoamine) dendrons]	water	2009GAO
Poly(styrenesulfonate-*b*-*N*-isopropylacrylamide)	water	2009TAU
Poly(sulfobetaine methacrylate-*co*-*N*-isopropylacrylamide)	water	2009CHA2

Polymer (B)	**Solvent (A)**	**Ref.**
Poly(*N*-tetrahydrofurfurylacrylamide)	water	2010MAE
Poly(*N*-tetrahydrofurfurylacrylamide-*co*-*N*-isopropylacrylamide)	water	2010MAE
Poly(*N*-tetrahydrofurfurylmethacrylamide)	water	2010MAE
Poly(*N*-tetrahydrofurfurylmethacrylamide-*co*-*N*-isopropylacrylamide)	water	2010MAE
Poly(urethane amine)	water	2005IHA
Poly(*N*-vinylacetamide-*co*-acrylic acid)	water	2006MOR
Poly(*N*-vinylacetamide-*co*-methyl acrylate)	water	2004MOR
Poly(*N*-vinylacetamide-*co*-vinyl acetate)	water	2003SET
Poly(vinyl alcohol) acetals	water	1975TAY
Poly(vinyl alcohol)-*g*-poly(p-dioxanone)	water	2011WUG
Poly(vinyl alcohol-*co*-sodium acrylate)	water	2008PAN
Poly(vinylamine)	water	2009CHE1

Polymer (B)	Solvent (A)	Ref.
Poly(vinylamine-*co*-vinylamine boronate)		
	water	2009CHE1
Poly(*N*-vinylcaprolactam)		
	water	2004DU1
	water	2004LAU
	water	2006VER
(carboxylated)	water	2011CAO
Poly(*N*-vinylcaprolactam-*co*-acrylic acid)		
	water	2003SHT
Poly[*N*-vinylcaprolactam-*g*-poly(ethylene oxide)]		
	water	2003VE1
	water	2004DU1
	water	2005KJ1
Poly[*N*-vinylcaprolactam-*g*-poly(tetrahydrofuran)]		
	water	2003VE1
Poly(*N*-vinylcaprolactam-*co*-1-vinylimidazole)		
	water	2006LOZ
	water	2007SH2
Poly(*N*-vinylcaprolactam-*co*-1-vinyl-2-methylimidazole)		
	water	2007SH2
Poly(*N*-vinylformamide-*co*-vinyl acetate)		
	water	2003SET
Poly(vinyl methyl ether)		
	deuterium oxide	2005VA2
	water	2005VA1
	water	2006VAN
	water	2007VA1
	water	2009DUR
	water	2011ASS

Polymer (B)	Solvent (A)	Ref.
Poly(*N*-vinylpiperidone)	water	2011IEO
Poly(*N*-vinylpiperidone-*b*-vinyl acetate)	water	2011IEO
Poly[4-vinylpyridine-*g*-poly(ethylene oxide)]	water	2008REN
Poly(5-vinyltetrazole-*co*-2-methyl-5-vinyltetrazole)	water	2009KIZ
Poly(5-vinyltetrazole-*co*-2-nonyl-5-vinyltetrazole)	water	2009KIZ
Poly(5-vinyltetrazole-*co*-2-pentyl-5-vinyltetrazole)	water	2009KIZ
Poly(5-vinyltetrazole-*co*-1,1,7-trihydrododeca-fluoroheptyl methacrylate)	water	2009KIZ
Poly(5-vinyltetrazole-*co*-*N*-vinylcaprolactam)	water	2010KIZ
Poly(5-vinyltetrazole-*co*-1-vinylimidazole)	water	2009KIZ
Poly(1-vinyl-1,2,3-triazole)	water	2003TSY
Poly(1-vinyl-1,2,4-triazole)	water	2003TSY
Tetra(ethylene glycol) monodecyl ether	water	2004RAN

Polymer (B)	Solvent (A)	Ref.
Tetra(ethylene glycol) monooctyl ether	water	2011BIA
N,N,N-Trimethylchitosan chloride-*g*-poly(*N*-isopropylacrylamide)	water	2007MAO

3.3. Cloud-point and/or coexistence curves of quasiternary solutions containing water and at least one polymer

3.3.1. Nonelectrolyte solutions

Polymer (B): **dextran** **2005SUG**
Characterization: M_n/g.mol^{-1} = 27800, M_w/g.mol^{-1} = 59500,
ρ = 1.376 g/cm^3 (298 K),
Polymer Standard Service GmbH, Mainz, Germany
Solvent (A): **water** H_2O **7732-18-5**
Solvent (C): **acetic acid** $C_2H_4O_2$ **64-19-7**

Type of data: cloud points

T/K = 298.15

w_A	0.414	0.411	0.409
w_B	0.029	0.017	0.010
w_C	0.556	0.572	0.581

Polymer (B): **dextran** **2005SUG**
Characterization: M_n/g.mol^{-1} = 300000, M_w/g.mol^{-1} = 2100000,
ρ = 2.009 g/cm^3 (298 K),
Amersham Pharmacia Biotech AB, Uppsala, Sweden
Solvent (A): **water** H_2O **7732-18-5**
Polymer (C): **bovine serum albumin**
Characterization: M_w/g.mol^{-1} = 66000, Acros Organics, New Jersey

Type of data: cloud points

T/K = 298.15

w_A	0.656	0.706	0.751	0.780	0.808	0.843	0.862	0.882	0.912
w_B	0.001	0.001	0.010	0.019	0.035	0.044	0.057	0.060	0.075
w_C	0.343	0.293	0.239	0.201	0.157	0.112	0.081	0.059	0.013

w_A	0.899	0.875	0.851	0.685
w_B	0.099	0.124	0.148	0.314
w_C	0.002	0.002	0.001	0.001

Polymer (B): **dextran** **2005SUG**
Characterization: M_n/g.mol^{-1} = 27800, M_w/g.mol^{-1} = 59500,
ρ = 1.376 g/cm^3 (298 K),
Polymer Standard Service GmbH, Mainz, Germany

continued

continued

Solvent (A): **water** H_2O **7732-18-5**
Solvent (C): ***N,N*-dimethylacetamide** C_4H_9NO **127-19-5**

Type of data: cloud points

T/K = 298.15

w_A	0.241	0.223	0.204
w_B	0.018	0.010	0.007
w_C	0.741	0.767	0.789

Polymer (B): **dextran** **2005SUG**
Characterization: M_n/g.mol^{-1} = 27800, M_w/g.mol^{-1} = 59500, ρ = 1.376 g/cm^3 (298 K), Polymer Standard Service GmbH, Mainz, Germany
Solvent (A): **water** H_2O **7732-18-5**
Solvent (C): **ethanol** C_2H_6O **64-17-5**

Type of data: cloud points

T/K = 298.15

w_A	0.617	0.627	0.631
w_B	0.043	0.023	0.017
w_C	0.340	0.350	0.352

Polymer (B): **dextran** **2005SUG**
Characterization: M_n/g.mol^{-1} = 5500, M_w/g.mol^{-1} = 11100, ρ = 1.306 g/cm^3 (298 K), Polymer Standard Service GmbH, Mainz, Germany
Solvent (A): **water** H_2O **7732-18-5**
Solvent (C): **methanol** CH_4O **67-56-1**

Type of data: cloud points

T/K = 298.15

w_A	0.438	0.474	0.481	0.471	0.474	0.471	0.466	0.491	0.481
w_B	0.186	0.099	0.081	0.064	0.076	0.070	0.070	0.090	0.085
w_C	0.376	0.427	0.438	0.465	0.450	0.459	0.463	0.419	0.434

w_A	0.471	0.472	0.480	0.468	0.481
w_B	0.078	0.072	0.065	0.070	0.014
w_C	0.451	0.456	0.454	0.463	0.504

Polymer (B): **dextran** **2005SUG**
Characterization: M_n/g.mol^{-1} = 27800, M_w/g.mol^{-1} = 59500, ρ = 1.376 g/cm^3 (298 K), Polymer Standard Service GmbH, Mainz, Germany

continued

continued

Solvent (A):	**water**	H_2O	**7732-18-5**
Solvent (C):	**methanol**	CH_4O	**67-56-1**

Type of data: cloud points

T/K = 298.15

w_A	0.674	0.687	0.688	0.542	0.576	0.607	0.627	0.664	0.696
w_B	0.091	0.072	0.059	0.291	0.244	0.208	0.165	0.125	0.073
w_C	0.236	0.241	0.254	0.167	0.180	0.186	0.208	0.211	0.231
w_A	0.709	0.707	0.725	0.000	0.510				
w_B	0.048	0.030	0.013	0.946	0.157				
w_C	0.243	0.263	0.262	0.054	0.333				

critical concentrations: $w_{A,\,crit}$ = 0.510, $w_{B,\,crit}$ = 0.157, $w_{C,\,crit}$ = 0.333

Polymer (B):	**dextran**		**2005SUG**
Characterization:	M_n/g.mol^{-1} = 359000, M_w/g.mol^{-1} = 3420000, ρ = 2.050 g/cm^3 (298 K), Polymer Standard Service GmbH, Mainz, Germany		
Solvent (A):	**water**	H_2O	**7732-18-5**
Solvent (C):	**methanol**	CH_4O	**67-56-1**

Type of data: cloud points

T/K = 298.15

w_A	0.706	0.679	0.733
w_B	0.002	0.001	0.003
w_C	0.291	0.320	0.264

Polymer (B):	**dextran**		**2005SUG**
Characterization:	M_n/g.mol^{-1} = 27800, M_w/g.mol^{-1} = 59500, ρ = 1.376 g/cm^3 (298 K), Polymer Standard Service GmbH, Mainz, Germany		
Solvent (A):	**water**	H_2O	**7732-18-5**
Solvent (C):	**2-propanol**	C_3H_8O	**67-63-0**

Type of data: cloud points

T/K = 298.15

w_A	0.659	0.669	0.672
w_B	0.066	0.043	0.027
w_C	0.274	0.288	0.300

Polymer (B): **dextran** **2005SUG**
Characterization: M_n/g.mol^{-1} = 5500, M_w/g.mol^{-1} = 11100,
ρ = 1.306 g/cm^3 (298 K),
Polymer Standard Service GmbH, Mainz, Germany
Solvent (A): **water** **H_2O** **7732-18-5**
Solvent (C): **2-propanone** **C_3H_6O** **67-64-1**

Type of data: cloud points

T/K = 298.15

w_A	0.635	0.654	0.661	0.621	0.516	0.569	0.552	0.584	0.609
w_B	0.079	0.049	0.033	0.025	0.302	0.220	0.176	0.165	0.147
w_C	0.286	0.297	0.307	0.354	0.182	0.211	0.272	0.251	0.244

w_A	0.634
w_B	0.101
w_C	0.266

Polymer (B): **dextran** **2005SUG**
Characterization: M_n/g.mol^{-1} = 27800, M_w/g.mol^{-1} = 59500,
ρ = 1.376 g/cm^3 (298 K),
Polymer Standard Service GmbH, Mainz, Germany
Solvent (A): **water** **H_2O** **7732-18-5**
Solvent (C): **2-propanone** **C_3H_6O** **67-64-1**

Type of data: cloud points

T/K = 298.15

w_A	0.674	0.687	0.688	0.542	0.576	0.607	0.627	0.664	0.696
w_B	0.091	0.072	0.059	0.291	0.244	0.208	0.165	0.125	0.073
w_C	0.236	0.241	0.254	0.167	0.180	0.186	0.208	0.211	0.231

w_A	0.709	0.707	0.725
w_B	0.048	0.030	0.013
w_C	0.243	0.263	0.262

critical concentrations: $w_{A,\,crit}$ = 0.640, $w_{B,\,crit}$ = 0.153, $w_{C,\,crit}$ = 0.206

Polymer (B): **dextran** **2005SUG**
Characterization: M_n/g.mol^{-1} = 359000, M_w/g.mol^{-1} = 3420000,
ρ = 2.050 g/cm^3 (298 K),
Polymer Standard Service GmbH, Mainz, Germany
Solvent (A): **water** **H_2O** **7732-18-5**
Solvent (C): **2-propanone** **C_3H_6O** **67-64-1**

Type of data: cloud points

T/K = 298.15

w_A	0.800	0.788	0.792
w_B	0.010	0.006	0.003
w_C	0.189	0.206	0.205

Polymer (B): **dextran** **2005SUG**
Characterization: M_n/g.mol^{-1} = 27800, M_w/g.mol^{-1} = 59500, ρ = 1.376 g/cm^3 (298 K), Polymer Standard Service GmbH, Mainz, Germany
Solvent (A): **water** H_2O **7732-18-5**
Solvent (C): **tetrahydrofuran** C_4H_8O **109-99-9**

Type of data: cloud points

T/K = 298.15

w_A	0.541	0.571
w_B	0.071	0.018
w_C	0.389	0.411

Polymer (B): **polyester (hyperbranched)** **2011ZEI1, 2011ZEI3**
Characterization: M_w/g.mol^{-1} = 2100, M_w/M_n = 1.3, Boltorn H20, Perstorp Specialty Chemicals AB, Perstorp, Sweden
Solvent (A): **water** H_2O **7732-18-5**
Solvent (C): **1-propanol** C_3H_8O **71-23-8**

Type of data: cloud points (water-rich region of the system)

T/K = 353.15

w_A	0.835	0.768	0.805	0.877	0.788	0.921	0.990
w_B	0.055	0.122	0.070	0.036	0.094	0.018	0.010
w_C	0.109	0.110	0.125	0.086	0.118	0.061	0.000

Type of data: cloud points (water-lean region of the system)

T/K = 353.15

w_A	0.125	0.120	0.126	0.084	0.074	0.000
w_B	0.123	0.116	0.110	0.089	0.074	0.033
w_C	0.752	0.764	0.764	0.828	0.851	0.967

Polymer (B): **polyester (hyperbranched, aliphatic)** **2003SEI**
Characterization: M_n/g.mol^{-1} = 5650, M_w/g.mol^{-1} = 10500, hydroxyl functionalities are esterified with saturated fatty acids, Boltorn H3200, Perstorp Specialty Chemicals AB, Sweden
Solvent (A): **water** H_2O **7732-18-5**
Solvent (C): **tetrahydrofuran** C_4H_8O **109-99-9**

Type of data: cloud points

T/K = 321.15

w_A	0.0638	0.0737	0.0789	0.0796	0.0831	0.0868	0.1046
w_B	0.1883	0.1461	0.1179	0.1004	0.0879	0.0746	0.0445
w_C	0.7479	0.7802	0.8032	0.8200	0.8290	0.8386	0.8509

continued

continued

T/K = 334.15

w_A	0.0569	0.0637	0.0736	0.0805	0.0856	0.0887	0.0892	0.0992	0.1031
w_B	0.2401	0.2144	0.1646	0.1414	0.1214	0.1121	0.1018	0.0732	0.0664
w_C	0.7030	0.7219	0.7618	0.7781	0.7930	0.7992	0.8090	0.8276	0.8305

Type of data: coexistence data (tie lines)

Phase I			Phase II		
w_A	w_B	w_C	w_A	w_B	w_C
T/K = 321.15					
0.0121	0.7589	0.2290	0.8553	0.0014	0.1434
0.0136	0.6757	0.3107	0.8041	0.0011	0.1948
0.0180	0.5947	0.3873	0.7317	0.0031	0.2652
0.0216	0.5173	0.4611	0.6763	0.0019	0.3218
0.0232	0.4722	0.5046	0.6325	0.0014	0.3660
0.0255	0.4584	0.5162	0.5876	0.0019	0.4105
0.0274	0.4327	0.5399	0.4916	0.0049	0.5035
0.0259	0.4212	0.5529	0.4195	0.0002	0.5803
0.0272	0.4132	0.5596	0.3519	0.0073	0.6408
0.0304	0.3920	0.5776	0.2758	0.0002	0.7240
0.0361	0.3468	0.6171	0.2283	0.0065	0.7652
0.0370	0.3199	0.6431	0.1953	0.0108	0.7939
0.0372	0.3031	0.6597	0.1321	0.0196	0.8483
T/K = 334.15					
0.0052	0.8940	0.1008	0.9226	0.0010	0.0763
0.0131	0.7204	0.2665	0.8669	0.0012	0.1319
0.0131	0.6850	0.3019	0.8441	0.0013	0.1546
0.0159	0.6100	0.3740	0.8093	0.0005	0:1901
0.0247	0.5013	0.4740	0.7129	0.0024	0.2847
0.0320	0.4382	0.5298	0.5968	0.0010	0.4022
0.0306	0.3801	0.5893	0.4323	0.0040	0.5637
0.0338	0.3425	0.6237	0.2820	0.0119	0.7061
0.0335	0.3195	0.6470	0.1398	0.0514	0.8088

Polymer (B): **poly(ethylene glycol)** **2007WAN**
Characterization: M_w/g.mol^{-1} = 1450, Acros Organics, New Jersey
Solvent (A): **water** **H_2O** **7732-18-5**
Polymer (C): **bovine serum albumin**
Characterization: M_w/g.mol^{-1} = 66400, Sigma Chemical Co., Inc., St. Louis, MO

continued

continued

Comments: 0.1 mol/l sodium acetate buffer solution gives a constant pH = 5.2.

Type of data: cloud points

c_B/(mg/ml)	52	58	64	68	72	75	79	54
c_C/(mg/ml)	150	150	150	150	150	150	150	180
T/K	262.0	265.6	270.4	274.0	277.1	280.8	284.8	265.7
c_B/(mg/ml)	61	65	70	74	79	47	52	57
c_C/(mg/ml)	180	180	180	180	180	220	220	220
T/K	270.8	275.0	280.1	284.7	290.1	266.2	270.4	276.2
c_B/(mg/ml)	63	69	74	79	32	35	40	44
c_C/(mg/ml)	220	220	220	220	330	330	330	330
T/K	283.8	291.2	299.7	307.7	268.2	274.2	282.3	290.1
c_B/(mg/ml)	47	22	23	27	31	37		
c_C/(mg/ml)	330	400	400	400	400	400		
T/K	294.7	262.7	266.3	271.7	283.3	295.6		

Type of data: coexistence data (tie lines)

Phase I		Phase II	
c_B/(mg/ml)	c_C/(mg/ml)	c_B/(mg/ml)	c_C/(mg/ml)
T/K = 268			
76	97	32	370
91	40	28	410
100	35	28	440
T/K = 271			
73	120	38	350
87	80	32	430
98	48	29	500
T/K = 278			
82	90	34	380
91	76	35	410
110	35	31	460

Polymer (B): **poly(ethylene glycol)** **2004HOP**
Characterization: M_w/g.mol^{-1} = 8000, Sigma Chemical Co., Inc., St. Louis, MO
Solvent (A): **water** H_2O **7732-18-5**
Polymer (C): **dextran**
Characterization: M_n/g.mol^{-1} = 53000, M_w/g.mol^{-1} = 148000, Sigma Chemical Co., Inc., St. Louis, MO

continued

continued

Type of data: cloud points

T/K = 295 phosphate buffered solutions, pH = 7.65

w_A	0.9340	0.9303	0.9259	0.9206	0.9128	0.9049	0.8930	0.8816	0.8536
w_B	0.0591	0.0555	0.0518	0.0485	0.0441	0.0389	0.0322	0.0250	0.0152
w_C	0.0068	0.0142	0.0223	0.0309	0.0431	0.0562	0.0748	0.0933	0.1312

Type of data: coexistence data (tie lines)

T/K = 295 phosphate buffered solutions, pH = 7.65

Phase I			Phase II		
w_A	w_B	w_C	w_A	w_B	w_C
0.8015	0.0108	0.1876	0.8958	0.0940	0.0102
0.8144	0.0129	0.1727	0.9019	0.0853	0.0128
0.8255	0.0141	0.1605	0.9056	0.0784	0.0160
0.8486	0.0152	0.1361	0.9093	0.0693	0.0214
0.8646	0.0202	0.1151	0.9116	0.0605	0.0280
0.8864	0.0318	0.0817	0.9129	0.0423	0.0448

Polymer (B): **poly(ethylene glycol)** **2006SAR**
Characterization: M_n/g.mol^{-1} = 6000, Merck, Darmstadt, Germany
Solvent (A): **water** **H_2O** **7732-18-5**
Polymer (C): **poly(acrylic acid)**
Characterization: M_n/g.mol^{-1} = 2100, Aldrich Chem. Co., Inc., Milwaukee, WI

Type of data: cloud points (LCST behavior)

T/K = 293.15

w_A	0.6318	0.6516	0.6618	0.6727	0.6809	0.6880	0.6920	0.6947	0.7032
w_B	0.3183	0.2820	0.2644	0.2435	0.2300	0.2193	0.2097	0.2018	0.1834
w_C	0.0499	0.0664	0.0738	0.0838	0.0891	0.0927	0.0983	0.1035	0.1134
w_A	0.7166	0.7226	0.7289	0.7295	0.7304	0.7329	0.7313	0.7336	0.7350
w_B	0.1501	0.1296	0.1180	0.1052	0.0960	0.0843	0.0804	0.0732	0.0677
w_C	0.1333	0.1478	0.1531	0.1653	0.1736	0.1828	0.1883	0.1932	0.1973

T/K = 303.15

w_A	0.6404	0.6608	0.6734	0.7005	0.7063	0.7116	0.7205	0.7266	0.7323
w_B	0.3162	0.2840	0.2631	0.2110	0.1970	0.1835	0.1580	0.1457	0.1337
w_C	0.0434	0.0552	0.0635	0.0885	0.0967	0.1049	0.1215	0.1277	0.1340
w_A	0.7357	0.7401	0.7438	0.7451	0.7445	0.7444	0.7441	0.7440	0.7464
w_B	0.1246	0.1143	0.1009	0.0958	0.0921	0.0902	0.0866	0.0819	0.0802
w_C	0.1397	0.1456	0.1553	0.1591	0.1634	0.1654	0.1693	0.1741	0.1734

continued

continued

T/K = 313.15

w_A	0.6430	0.6705	0.6913	0.7077	0.7204	0.7366	0.7389	0.7420	0.7489
w_B	0.3195	0.2755	0.2453	0.2161	0.1911	0.1651	0.1585	0.1503	0.1324
w_C	0.0375	0.0540	0.0634	0.0762	0.0885	0.0983	0.1026	0.1077	0.1187
w_A	0.7563	0.7582	0.7558	0.7582	0.7557	0.7527			
w_B	0.1031	0.0939	0.0904	0.0833	0.0818	0.0739			
w_C	0.1406	0.1479	0.1538	0.1585	0.1625	0.1734			

Polymer (B): **poly(ethylene glycol)** **2006YAN**
Characterization: M_n/g.mol^{-1} = 4000, M_w/g.mol^{-1} = 4200,
Fluka AG, Buchs, Switzerland
Solvent (A): **water** **H_2O** **7732-18-5**
Polymer (C): **poly(ethylenimine)**
Characterization: M_n/g.mol^{-1} = 10000, M_w/g.mol^{-1} = 25000,
Aldrich Chem. Co., Inc., Milwaukee, WI

Type of data: cloud points (LCST behavior)

T/K = 298.15

pH = 5.3

w_A	0.9191	0.9296	0.9535	0.9570	0.9565	0.9535	0.9498	0.9378	0.9273
w_B	0.0789	0.0677	0.0400	0.0295	0.0223	0.0180	0.0152	0.0106	0.0068
w_C	0.0020	0.0027	0.0065	0.0135	0.0212	0.0285	0.0350	0.0516	0.0659
w_A	0.9070	0.8981							
w_B	0.0037	0.0036							
w_C	0.0893	0.0983							

pH = 7.5

w_A	0.8968	0.9011	0.9035	0.9082	0.9178	0.9216	0.9251	0.9248	0.9209
w_B	0.0118	0.0121	0.0137	0.0197	0.0297	0.0334	0.0406	0.0541	0.0653
w_C	0.0914	0.0868	0.0828	0.0721	0.0525	0.0450	0.0343	0.0211	0.0138
w_A	0.9163	0.9143							
w_B	0.0738	0.0771							
w_C	0.0099	0.0086							

pH = 9.2

w_A	0.8665	0.8719	0.8810	0.8834	0.8875	0.8910	0.8911	0.8911	0.8903
w_B	0.1204	0.1141	0.0994	0.0917	0.0851	0.0658	0.0620	0.0619	0.0518
w_C	0.0131	0.0140	0.0196	0.0249	0.0274	0.0432	0.0469	0.0470	0.0579
w_A	0.8872	0.8847							
w_B	0.0445	0.0253							
w_C	0.0683	0.0900							

continued

continued

Type of data: coexistence data (tie lines)

Total system			Top phase			Bottom phase		
w_A	w_B	w_C	w_A	w_B	w_C	w_A	w_B	w_C
T/K = 298.15	pH = 5.3							
0.9437	0.0300	0.0263	0.9533	0.0355	0.0112	0.8893	0.0036	0.1071
0.9314	0.0345	0.0341	0.9485	0.0449	0.0066	0.8666	0.0031	0.1303
0.9384	0.0204	0.0412	0.9562	0.0311	0.0127	0.9111	0.0038	0.0851
0.9162	0.0416	0.0422	0.9413	0.0535	0.0052	0.8319	0.0017	0.1664
0.9015	0.0498	0.0487	0.9325	0.0642	0.0033	0.7988	0.0019	0.1993
T/K = 298.15	pH = 7.5							
0.9203	0.0400	0.0397	0.9261	0.0449	0.0290	0.8849	0.0093	0.1058
0.9098	0.0458	0.0444	0.9230	0.0600	0.0170	0.8714	0.0063	0.1223
0.9009	0.0546	0.0445	0.9156	0.0755	0.0089	0.8646	0.0058	0.1296
0.8836	0.0619	0.0545	0.8981	0.0934	0.0085	0.8615	0.0060	0.1325
0.8731	0.0679	0.0590	0.8879	0.1048	0.0073	0.8501	0.0061	0.1438
T/K = 298.15	pH = 9.2							
0.8853	0.0576	0.0571	0.8827	0.0887	0.0286	0.8871	0.0280	0.0849
0.8781	0.0621	0.0598	0.8750	0.1081	0.0169	0.8811	0.0220	0.0969
0.8621	0.0699	0.0680	0.8570	0.1317	0.0113	0.8695	0.0125	0.1180
0.8564	0.0749	0.0687	0.8494	0.1408	0.0098	0.8619	0.0113	0.1268
0.8427	0.0799	0.0774	0.8336	0.1569	0.0095	0.8507	0.0085	0.1408

Polymer (B): **poly(ethylene glycol)** **2004PER**
Characterization: M_n/g.mol^{-1} = 8000
Solvent (A): **water** H_2O **7732-18-5**
Polymer (C): **poly(vinyl acetate-*co*-vinyl alcohol)**
Characterization: M_n/g.mol^{-1} = 10000, 12.0 mol% vinyl acetate, PVA 10000, 88% hydrolyzed, Scientific Polymer Products, Inc., Ontario, NY

Type of data: coexistence data (tie lines)

T/K = 298.15

Total system			Top phase			Bottom phase		
w_A	w_B	w_C	w_A	w_B	w_C	w_A	w_B	w_C
0.8640	0.0430	0.0930	0.8780	0.0620	0.0600	0.8540	0.0290	0.1170
0.8530	0.0480	0.0990	0.8810	0.0810	0.0380	0.8300	0.0200	0.1500
0.8420	0.0530	0.1050	0.8770	0.0930	0.0300	0.8100	0.0150	0.1750
0.8210	0.0600	0.1190	0.8650	0.1110	0.0240	0.7790	0.0130	0.2080

continued

continued

Plait-point composition: $w_A = 0.8790 + w_B = 0.0800 + w_C = 0.0410$

Polymer (B):	**poly(ethylene glycol)**		**2006CHE**
Characterization:	M_n/g.mol^{-1} = 6000		
Solvent (A):	**water**	**H_2O**	**7732-18-5**
Polymer (C):	**xanthan**		
Characterization:	Monsanto		

Type of data: coexistence data (tie lines)

T/K = 298.15

Total system			Top phase			Bottom phase		
w_A	w_B	w_C	w_A	w_B	w_C	w_A	w_B	w_C
0.9780	0.0200	0.0020	0.9595	0.0400	0.0005	0.9806	0.0170	0.0024
0.9775	0.0200	0.0025	0.9525	0.0450	0.0025	0.9814	0.0160	0.0026
0.9770	0.0200	0.0030	0.9462	0.0480	0.0058	0.9812	0.0162	0.0028
0.9765	0.0200	0.0035	0.9440	0.0490	0.0070	0.9791	0.0180	0.0029
0.9760	0.0200	0.0040	0.9363	0.0500	0.0137	0.9805	0.0167	0.0028

Polymer (B):	**poly(ethylene oxide-*co*-propylene oxide)**		**2004PER**
Characterization:	M_n/g.mol^{-1} = 3900, 50.0 mol% ethylene oxide, Ucon 50-HB-5100, Union Carbide Corp., NY		
Solvent (A):	**water**	**H_2O**	**7732-18-5**
Polymer (C):	**hydroxypropylstarch**		
Characterization:	M_n/g.mol^{-1} = 100000, Reppe Glykos AB, Vaxjo, Sweden		

Type of data: coexistence data (tie lines)

T/K = 295.15

Total system			Top phase			Bottom phase		
w_A	w_B	w_C	w_A	w_B	w_C	w_A	w_B	w_C
0.8319	0.0473	0.1208	0.8517	0.0599	0.0884	0.8091	0.0236	0.1673
0.7902	0.0804	0.1294	0.8414	0.1236	0.0350	0.7222	0.0175	0.2603
0.8032	0.0627	0.1341	0.8467	0.0946	0.0587	0.7504	0.0193	0.2303
0.7899	0.0709	0.1392	0.8431	0.1142	0.0427	0.7313	0.0180	0.2507
0.8048	0.0450	0.1502	0.8502	0.0822	0.0676	0.7801	0.0207	0.1992

Plait-point composition: $w_A = 0.8401 + w_B = 0.0369 + w_C = 0.1230$

Polymer (B): **poly(ethylene oxide-*co*-propylene oxide)** **2005BOL**
Characterization: M_n/g.mol^{-1} = 1059, M_η/g.mol^{-1} = 1228, 50 mol% ethylene oxide, Dow Chemical Co., San Lorenzo, Argentina
Solvent (A): **water** **H_2O** **7732-18-5**
Polymer (C): **maltodextrin**
Characterization: M_n/g.mol^{-1} = 838, M_η/g.mol^{-1} = 922, Polimerosa, Kasdorf SA, Buenos Aires, Argentina

Type of data: coexistence data (tie lines)

T/K = 297.15

Total system			Top phase			Bottom phase		
w_A	w_B	w_C	w_A	w_B	w_C	w_A	w_B	w_C
0.6610	0.0950	0.2440	0.7360	0.1640	0.1000	0.5919	0.0306	0.3775
0.6730	0.0930	0.2340	0.7480	0.1550	0.0970	0.6070	0.0312	0.3618
0.6860	0.0890	0.2250	0.7470	0.1376	0.1154	0.6330	0.0366	0.3304
0.7020	0.0860	0.2120	0.7552	0.1400	0.1048	0.6604	0.0460	0.2936
0.7240	0.0790	0.1970	0.7660	0.1145	0.1195	0.6835	0.0386	0.2779
0.7370	0.0750	0.1880	0.7696	0.0967	0.1337	0.6974	0.0404	0.2622

Polymer (B): **poly(ethylene oxide-*co*-propylene oxide)** **2004PER**
Characterization: M_n/g.mol^{-1} = 3900, 50.0 mol% ethylene oxide, Ucon 50-HB-5100, Union Carbide Corp., NY
Solvent (A): **water** **H_2O** **7732-18-5**
Polymer (C): **poly(vinyl acetate-*co*-vinyl alcohol)**
Characterization: M_n/g.mol^{-1} = 10000, 12.0 mol% vinyl acetate, PVA 10000, 88% hydrolyzed, Scientific Polymer Products, Inc., Ontario, NY

Type of data: coexistence data (tie lines)

T/K = 295.15

Total system			Top phase			Bottom phase		
w_A	w_B	w_C	w_A	w_B	w_C	w_A	w_B	w_C
0.7958	0.1047	0.0995	0.8077	0.1273	0.0650	0.7827	0.0214	0.1959
0.7612	0.1323	0.1065	0.7790	0.1931	0.0279	0.7375	0.0174	0.2451
0.7799	0.1104	0.1097	0.7970	0.1617	0.0413	0.7642	0.0180	0.2178
0.7493	0.1327	0.1180	0.7710	0.2067	0.0223	0.7129	0.0168	0.2703

Plait-point composition: w_A = 0.8100 + w_B = 0.0600 + w_C = 0.1300

Polymer (B): **poly(ethylene oxide-*b*-propylene oxide-*b*-ethylene oxide)** **2010MAH**
Characterization: M_w/g.mol^{-1} = 2036, (EO)2.5(PO)31(EO)2.5, Aldrich Chem. Co., Inc., Milwaukee, WI
Solvent (A): **water** **H_2O** **7732-18-5**
Component (C): **L-alanine** **$C_3H_7NO_2$** **56-41-7**

Type of data: cloud points (LCST-behavior)

w_B = 0.01 in the binary aqueous solution was kept constant

c_C/(mol/l)	0.00	0.05	0.10	0.25	0.50
T/K	300.45	299.35	298.35	297.25	296.05

Polymer (B): **poly(ethylene oxide-*b*-propylene oxide-*b*-ethylene oxide)** **2010MAH**
Characterization: M_w/g.mol^{-1} = 2900, 40.0 wt% ethylene oxide, L64, (EO)13(PO)30(EO)13, Aldrich Chem. Co., Inc., Milwaukee, WI
Solvent (A): **water** **H_2O** **7732-18-5**
Component (C): **L-alanine** **$C_3H_7NO_2$** **56-41-7**

Type of data: cloud points (LCST-behavior)

w_B = 0.01 in the binary aqueous solution was kept constant

c_C/(mol/l)	0.00	0.05	0.10	0.25	0.50
T/K	330.25	323.05	321.65	320.35	316.35

Polymer (B): **poly(ethylene oxide-*b*-propylene oxide-*b*-ethylene oxide)** **2004TAD, 2005TAD**
Characterization: M_n/g.mol^{-1} = 4750, 37.0 mol% ethylene oxide, (EO)17-(PO)58-(EO)17, P103, Aldrich Chem. Co., Inc., Milwaukee, WI
Solvent (A): **water** **H_2O** **7732-18-5**
Polymer (C): **dextran**
Characterization: M_n/g.mol^{-1} = 8200, M_w/g.mol^{-1} = 11600, Dextran 19, Sigma Chemical Co., Inc., St. Louis, MO

Type of data: coexistence data (tie lines)

T/K = 298.15

Total system			Top phase			Bottom phase		
w_A	w_B	w_C	w_A	w_B	w_C	w_A	w_B	w_C
0.8183	0.1015	0.0802	0.8450	0.0460	0.1090	0.7829	0.1634	0.0537
0.8210	0.0904	0.0886	0.8503	0.0276	0.1221	0.7657	0.1916	0.0427
0.8073	0.1129	0.0798	0.8518	0.0230	0.1252	0.7569	0.2036	0.0395

Polymer (B): **poly(ethylene oxide-*b*-propylene oxide-*b*-ethylene oxide)** **2004TAD, 2005TAD**
Characterization: M_n/g.mol^{-1} = 4750, 37.0 mol% ethylene oxide, (EO)17-(PO)58-(EO)17, P103, Aldrich Chem. Co., Inc., Milwaukee, WI
Solvent (A): **water** **H_2O** **7732-18-5**
Polymer (C): **dextran**
Characterization: M_n/g.mol^{-1} = 236000, M_w/g.mol^{-1} = 410000, Dextran 400, Sigma Chemical Co., Inc., St. Louis, MO

Type of data: cloud points

T/K = 298.15

w_A	0.9001	0.9103
w_B	0.0705	0.0343
w_C	0.0294	0.0554

Type of data: coexistence data (tie lines)

T/K = 298.15

Total system			Top phase			Bottom phase		
w_A	w_B	w_C	w_A	w_B	w_C	w_A	w_B	w_C
0.9081	0.0524	0.0395	0.9219	0.0203	0.0578	0.8687	0.1207	0.0106
0.8893	0.0602	0.0505	0.9079	0.0230	0.0691	0.8488	0.1401	0.0111
0.8570	0.0729	0.0701	0.8735	0.0236	0.1029	0.8152	0.1805	0.0043

Polymer (B): **poly(ethylene oxide-*b*-propylene oxide-*b*-ethylene oxide)** **2004TAD, 2005TAD**
Characterization: M_n/g.mol^{-1} = 6500, 54.4 mol% ethylene oxide, (EO)37-(PO)62-(EO)37, F105, ICI Surfactants, Cleveland, UK
Solvent (A): **water** **H_2O** **7732-18-5**
Polymer (C): **dextran**
Characterization: M_n/g.mol^{-1} = 8200, M_w/g.mol^{-1} = 11600, Dextran 19, Sigma Chemical Co., Inc., St. Louis, MO

Type of data: coexistence data (tie lines)

T/K = 298.15

Total system			Top phase			Bottom phase		
w_A	w_B	w_C	w_A	w_B	w_C	w_A	w_B	w_C
0.7710	0.1169	0.1121	0.7821	0.0022	0.2157	0.7601	0.2239	0.0160
0.8200	0.0800	0.1000	0.8470	0.0083	0.1447	0.7894	0.1835	0.0271
0.7998	0.0901	0.1101	0.8238	0.0025	0.1737	0.7706	0.2085	0.0209

Polymer (B): **poly(ethylene oxide-*b*-propylene oxide-*b*-ethylene oxide)** **2004TAD, 2005TAD**
Characterization: M_n/g.mol^{-1} = 6500, 54.4 mol% ethylene oxide, (EO)37-(PO)62-(EO)37, F105, ICI Surfactants, Cleveland, UK
Solvent (A): **water** **H_2O** **7732-18-5**
Polymer (C): **dextran**
Characterization: M_n/g.mol^{-1} = 236000, M_w/g.mol^{-1} = 410000, Dextran 400, Sigma Chemical Co., Inc., St. Louis, MO

Type of data: cloud points

T/K = 298.15

w_A	0.8991	0.9095
w_B	0.0720	0.0350
w_C	0.0289	0.0555

Type of data: coexistence data (tie lines)

T/K = 298.15

Total system			Top phase			Bottom phase		
w_A	w_B	w_C	w_A	w_B	w_C	w_A	w_B	w_C
0.8957	0.0442	0.0601	0.9025	0.0142	0.0833	0.8685	0.1236	0.0079
0.8712	0.0601	0.0687	0.8833	0.0124	0.1043	0.8464	0.1485	0.0051
0.8994	0.0511	0.0495	0.9115	0.0173	0.0712	0.8845	0.1027	0.0128
0.8250	0.0875	0.0875	0.8364	0.0076	0.1560	0.8203	0.1767	0.0030

Polymer (B): **poly(ethylene oxide-*b*-propylene oxide-*b*-ethylene oxide)** **2004TAD, 2005TAD**
Characterization: M_w/g.mol^{-1} = 8530, 83.5 mol% ethylene oxide, (EO)76-(PO)30-(EO)76, F68, Aldrich Chem. Co., Inc., Milwaukee, WI
Solvent (A): **water** **H_2O** **7732-18-5**
Polymer (C): **dextran**
Characterization: M_n/g.mol^{-1} = 8200, M_w/g.mol^{-1} = 11600, Dextran 19, Sigma Chemical Co., Inc., St. Louis, MO

Type of data: cloud points

T/K = 298.15

w_A	0.8750	0.8649	0.8629	0.8525	0.8401	0.8174	0.8021
w_B	0.1006	0.0826	0.0637	0.0537	0.0405	0.0235	0.0155
w_C	0.0244	0.0525	0.0734	0.0938	0.1194	0.1591	0.1824

continued

continued

Type of data: coexistence data (tie lines)

T/K = 298.15

Total system			Top phase			Bottom phase		
w_A	w_B	w_C	w_A	w_B	w_C	w_A	w_B	w_C
0.8405	0.0700	0.0895	0.8039	0.0236	0.1725	0.8591	0.1035	0.0374
0.8196	0.0799	0.1005	0.7692	0.0101	0.2207	0.8512	0.1307	0.0181
0.8031	0.0899	0.1070	0.7556	0.0070	0.2374	0.8343	0.1481	0.0176

Polymer (B): **poly(ethylene oxide-*b*-propylene oxide-*b*-ethylene oxide)** **2004TAD, 2005TAD**
Characterization: M_w/g.mol^{-1} = 8530, 83.5 mol% ethylene oxide, (EO)76-(PO)30-(EO)76, F68, Aldrich Chem. Co., Inc., Milwaukee, WI
Solvent (A): **water** **H_2O** **7732-18-5**
Polymer (C): **dextran**
Characterization: M_n/g.mol^{-1} = 236000, M_w/g.mol^{-1} = 410000, Dextran 400, Sigma Chemical Co., Inc., St. Louis, MO

Type of data: cloud points

T/K = 298.15

w_A	0.9201	0.9099	0.8892
w_B	0.0404	0.0351	0.0229
w_C	0.0395	0.0550	0.0879

Type of data: coexistence data (tie lines)

T/K = 298.15

Total system			Top phase			Bottom phase		
w_A	w_B	w_C	w_A	w_B	w_C	w_A	w_B	w_C
0.8913	0.0396	0.0691	0.8713	0.0146	0.1141	0.9219	0.0657	0.0124
0.8699	0.0602	0.0699	0.8198	0.0002	0.1800	0.9008	0.0949	0.0043
0.8414	0.0892	0.0694	0.7608	0.0087	0.2305	0.8750	0.1222	0.0028

Polymer (B): **poly(ethylene oxide-*b*-propylene oxide-*b*-ethylene oxide)** **2004TAD, 2005TAD**
Characterization: M_n/g.mol^{-1} = 14000, 83.6 mol% ethylene oxide, (EO)127-(PO)50-(EO)127, P108, ICI Surfactants, Cleveland, UK

continued

continued

Solvent (A): **water** **H_2O** **7732-18-5**
Polymer (C): **dextran**
Characterization: M_n/g.mol^{-1} = 8200, M_w/g.mol^{-1} = 11600, Dextran 19, Sigma Chemical Co., Inc., St. Louis, MO

Type of data: coexistence data (tie lines)

T/K = 298.15

Total system			Top phase			Bottom phase		
w_A	w_B	w_C	w_A	w_B	w_C	w_A	w_B	w_C
0.8499	0.0600	0.0901	0.8433	0.0432	0.1135	0.8595	0.1038	0.0367
0.8389	0.0700	0.0911	0.8283	0.0323	0.1394	0.8463	0.1288	0.0249
0.8210	0.0800	0.0990	0.8017	0.0235	0.1748	0.8330	0.1485	0.0185

Polymer (B): **poly(ethylene oxide-*b*-propylene oxide-*b*-ethylene oxide)** **2004TAD, 2005TAD**
Characterization: M_n/g.mol^{-1} = 14000, 83.6 mol% ethylene oxide, (EO)127-(PO)50-(EO)127, P108, ICI Surfactants, Cleveland, UK
Solvent (A): **water** **H_2O** **7732-18-5**
Polymer (C): **dextran**
Characterization: M_n/g.mol^{-1} = 236000, M_w/g.mol^{-1} = 410000, Dextran 400, Sigma Chemical Co., Inc., St. Louis, MO

Type of data: cloud points

T/K = 298.15

w_A	0.9256	0.9152
w_B	0.0368	0.0157
w_C	0.0376	0.0691

Type of data: coexistence data (tie lines)

T/K = 298.15

Total system			Top phase			Bottom phase		
w_A	w_B	w_C	w_A	w_B	w_C	w_A	w_B	w_C
0.9191	0.0313	0.0496	0.9059	0.0102	0.0839	0.9324	0.0512	0.0164
0.8912	0.0494	0.0594	0.8700	0.0083	0.1217	0.9118	0.0832	0.0050
0.8715	0.0692	0.0593	0.8357	0.0037	0.1606	0.8951	0.1019	0.0030

Polymer (B): **poly(ethylene oxide-*b*-propylene oxide-*b*-ethylene oxide)** **2004TAD, 2005TAD**

Characterization: M_n/g.mol^{-1} = 4800, 83.8 mol% ethylene oxide, (EO)44-(PO)17-(EO)44, F38, ICI Surfactants, Cleveland, UK

Solvent (A): **water** **H_2O** **7732-18-5**

Polymer (C): **dextran**

Characterization: M_n/g.mol^{-1} = 8200, M_w/g.mol^{-1} = 11600, Dextran 19, Sigma Chemical Co., Inc., St. Louis, MO

Type of data: cloud points

T/K = 298.15

w_A	0.8377	0.8508	0.8628	0.8578	0.8631	0.8620	0.8602	0.8506	0.8459
w_B	0.1534	0.1399	0.1187	0.1199	0.1109	0.1101	0.1021	0.0932	0.0863
w_C	0.0089	0.0093	0.0185	0.0223	0.0260	0.0279	0.0377	0.0562	0.0678

w_A	0.8434	0.8365
w_B	0.0817	0.0838
w_C	0.0749	0.0797

Type of data: coexistence data (tie lines)

T/K = 298.15

Total system			Top phase			Bottom phase		
w_A	w_B	w_C	w_A	w_B	w_C	w_A	w_B	w_C
0.8239	0.0853	0.0908	0.7786	0.0000	0.2214	0.8441	0.1087	0.0472
0.7903	0.0998	0.1099	0.7343	0.0000	0.2657	0.8235	0.1548	0.0217
0.7497	0.1303	0.1200	0.6673	0.0000	0.3327	0.7922	0.1950	0.0128
0.7596	0.1202	0.1202	0.6759	0.0000	0.3241	0.8006	0.1849	0.0145

Polymer (B): **poly(ethylene oxide-*b*-propylene oxide-*b*-ethylene oxide)** **2004TAD, 2005TAD**

Characterization: M_n/g.mol^{-1} = 4800, 83.8 mol% ethylene oxide, (EO)44-(PO)17-(EO)44, F38, ICI Surfactants, Cleveland, UK

Solvent (A): **water** **H_2O** **7732-18-5**

Polymer (C): **dextran**

Characterization: M_n/g.mol^{-1} = 236000, M_w/g.mol^{-1} = 410000, Dextran 400, Sigma Chemical Co., Inc., St. Louis, MO

Type of data: cloud points

T/K = 298.15

w_A	0.8973	0.8944	0.8830
w_B	0.0656	0.0456	0.0380
w_C	0.0371	0.0600	0.0790

continued

continued

Type of data: coexistence data (tie lines)

T/K = 298.15

Total system			Top phase			Bottom phase		
w_A	w_B	w_C	w_A	w_B	w_C	w_A	w_B	w_C
0.8805	0.0499	0.0696	0.8527	0.0264	0.1209	0.9092	0.0757	0.0151
0.8600	0.0700	0.0700	0.8016	0.0149	0.1835	0.8997	0.0950	0.0053
0.8407	0.0896	0.0697	0.7693	0.0028	0.2279	0.8712	0.1247	0.0041

Polymer (B): **poly(ethylene oxide-*b*-propylene oxide-*b*-ethylene oxide)** **2010MAH**
Characterization: M_w/g.mol^{-1} = 2036, (EO)2.5(PO)31(EO)2.5, Aldrich Chem. Co., Inc., Milwaukee, WI
Solvent (A): **water** **H_2O** **7732-18-5**
Component (C): **glycine** **$C_2H_5NO_2$** **56-40-6**

Type of data: cloud points (LCST-behavior)

w_B = 0.01 in the binary aqueous solution was kept constant

c_C/(mol/l)	0.00	0.05	0.10	0.25	0.50
T/K	300.45	299.75	298.55	297.25	295.25

Polymer (B): **poly(ethylene oxide-*b*-propylene oxide-*b*-ethylene oxide)** **2010MAH**
Characterization: M_w/g.mol^{-1} = 2900, 40.0 wt% ethylene oxide, L64, (EO)13(PO)30(EO)13, Aldrich Chem. Co., Inc., Milwaukee, WI
Solvent (A): **water** **H_2O** **7732-18-5**
Component (C): **glycine** **$C_2H_5NO_2$** **56-40-6**

Type of data: cloud points (LCST-behavior)

w_B = 0.01 in the binary aqueous solution was kept constant

c_C/(mol/l)	0.00	0.05	0.10	0.25	0.50
T/K	330.25	322.25	319.75	317.45	315.35

Polymer (B): **poly(ethylene oxide-*b*-propylene oxide-*b*-ethylene oxide)** **2010MAH**
Characterization: M_w/g.mol^{-1} = 2036, (EO)2.5(PO)31(EO)2.5, Aldrich Chem. Co., Inc., Milwaukee, WI
Solvent (A): **water** **H_2O** **7732-18-5**
Component (C): **glycylglycine** **$C_4H_8N_2O_3$** **556-50-3**

continued

continued

Type of data: cloud points (LCST-behavior)

$w_B = 0.01$ in the binary aqueous solution was kept constant

c_C/(mol/l)	0.00	0.05	0.10	0.25	0.50
T/K	300.45	299.05	297.15	295.65	293.25

Polymer (B): **poly(ethylene oxide-*b*-propylene oxide-*b*-ethylene oxide)** **2010MAH**
Characterization: M_w/g.mol^{-1} = 2900, 40.0 wt% ethylene oxide, L64, (EO)13(PO)30(EO)13, Aldrich Chem. Co., Inc., Milwaukee, WI
Solvent (A): **water** **H_2O** **7732-18-5**
Component (C): **glycylglycine** **$C_4H_8N_2O_3$** **556-50-3**

Type of data: cloud points (LCST-behavior)

$w_B = 0.01$ in the binary aqueous solution was kept constant

c_C/(mol/l)	0.00	0.05	0.10	0.25	0.50
T/K	330.25	320.95	317.95	314.85	310.65

Polymer (B): **poly(ethylene oxide-*b*-propylene oxide-*b*-ethylene oxide)** **2010MAH**
Characterization: M_w/g.mol^{-1} = 2036, (EO)2.5(PO)31(EO)2.5, Aldrich Chem. Co., Inc., Milwaukee, WI
Solvent (A): **water** **H_2O** **7732-18-5**
Component (C): **glycyl-DL-valine** **$C_7H_{14}N_2O_3$** **2325-17-9**

Type of data: cloud points (LCST-behavior)

$w_B = 0.01$ in the binary aqueous solution was kept constant

c_C/(mol/l)	0.00	0.05	0.10	0.25	0.50
T/K	300.45	298.85	297.75	295.95	293.05

Polymer (B): **poly(ethylene oxide-*b*-propylene oxide-*b*-ethylene oxide)** **2010MAH**
Characterization: M_w/g.mol^{-1} = 2900, 40.0 wt% ethylene oxide, L64, (EO)13(PO)30(EO)13, Aldrich Chem. Co., Inc., Milwaukee, WI
Solvent (A): **water** **H_2O** **7732-18-5**
Component (C): **glycyl-DL-valine** **$C_7H_{14}N_2O_3$** **2325-17-9**

Type of data: cloud points (LCST-behavior)

$w_B = 0.01$ in the binary aqueous solution was kept constant

c_C/(mol/l)	0.00	0.05	0.10	0.25	0.50
T/K	330.25	319.65	317.45	313.45	310.05

Polymer (B):	**poly(ethylene oxide-*b*-propylene oxide-*b*-ethylene oxide)**		**2010MAH**
Characterization:	M_w/g.mol^{-1} = 2036, (EO)2.5(PO)31(EO)2.5, Aldrich Chem. Co., Inc., Milwaukee, WI		
Solvent (A):	**water**	H_2O	**7732-18-5**
Component (C):	**L-proline**	$C_5H_9NO_2$	**147-85-3**

Type of data: cloud points (LCST-behavior)

w_B = 0.01 in the binary aqueous solution was kept constant

c_C/(mol/l)	0.00	0.05	0.10	0.25	0.50
T/K	300.45	299.55	298.05	296.55	294.35

Polymer (B):	**poly(ethylene oxide-*b*-propylene oxide-*b*-ethylene oxide)**		**2010MAH**
Characterization:	M_w/g.mol^{-1} = 2900, 40.0 wt% ethylene oxide, L64, (EO)13(PO)30(EO)13, Aldrich Chem. Co., Inc., Milwaukee, WI		
Solvent (A):	**water**	H_2O	**7732-18-5**
Component (C):	**L-proline**	$C_5H_9NO_2$	**147-85-3**

Type of data: cloud points (LCST-behavior)

w_B = 0.01 in the binary aqueous solution was kept constant

c_C/(mol/l)	0.00	0.05	0.10	0.25	0.50
T/K	330.25	320.35	318.25	314.35	311.85

Polymer (B):	**poly(ethylene oxide-*b*-propylene oxide-*b*-ethylene oxide)**		**2010MAH**
Characterization:	M_w/g.mol^{-1} = 2036, (EO)2.5(PO)31(EO)2.5, Aldrich Chem. Co., Inc., Milwaukee, WI		
Solvent (A):	**water**	H_2O	**7732-18-5**
Component (C):	**L-threonine**	$C_4H_9NO_3$	**72-19-5**

Type of data: cloud points (LCST-behavior)

w_B = 0.01 in the binary aqueous solution was kept constant

c_C/(mol/l)	0.00	0.05	0.10	0.25	0.50
T/K	300.45	299.45	298.15	297.05	294.55

Polymer (B):	**poly(ethylene oxide-*b*-propylene oxide-*b*-ethylene oxide)**		**2010MAH**
Characterization:	M_w/g.mol^{-1} = 2900, 40.0 wt% ethylene oxide, L64, (EO)13(PO)30(EO)13, Aldrich Chem. Co., Inc., Milwaukee, WI		
Solvent (A):	**water**	H_2O	**7732-18-5**
Component (C):	**L-threonine**	$C_4H_9NO_3$	**72-19-5**

continued

continued

Type of data: cloud points (LCST-behavior)

$w_B = 0.01$ in the binary aqueous solution was kept constant

c_C/(mol/l)	0.00	0.05	0.10	0.25	0.50
T/K	330.25	320.55	317.45	315.35	312.35

Polymer (B): **poly(ethylene oxide-*b*-propylene oxide-*b*-ethylene oxide)** **2010MAH**

Characterization: M_w/g.mol^{-1} = 2036, (EO)2.5(PO)31(EO)2.5, Aldrich Chem. Co., Inc., Milwaukee, WI

Solvent (A): **water** **H_2O** **7732-18-5**

Component (C): **L-valine** **$C_5H_{11}NO_2$** **72-18-4**

Type of data: cloud points (LCST-behavior)

$w_B = 0.01$ in the binary aqueous solution was kept constant

c_C/(mol/l)	0.00	0.05	0.10	0.25	0.50
T/K	300.45	299.25	298.35	297.15	294.95

Polymer (B): **poly(ethylene oxide-*b*-propylene oxide-*b*-ethylene oxide)** **2010MAH**

Characterization: M_w/g.mol^{-1} = 2900, 40.0 wt% ethylene oxide, L64, (EO)13(PO)30(EO)13, Aldrich Chem. Co., Inc., Milwaukee, WI

Solvent (A): **water** **H_2O** **7732-18-5**

Component (C): **L-valine** **$C_5H_{11}NO_2$** **72-18-4**

Type of data: cloud points (LCST-behavior)

$w_B = 0.01$ in the binary aqueous solution was kept constant

c_C/(mol/l)	0.00	0.05	0.10	0.25	0.50
T/K	330.25	321.75	318.45	316.75	314.25

Polymer (B): **poly(*N*-ethyl-*N*-methylacrylamide)** **2004PAN**

Characterization: M_w/g.mol^{-1} = <5000, M_w/M_n < 1.5, synthesized in the laboratory by different methods

Solvent (A): **water** **H_2O** **7732-18-5**

Solvent (C): **methanol** **CH_4O** **67-56-1**

Type of data: cloud points (LCST-behavior)

Comments: The sample was synthesized by chain transfer polymerization.

$w_B = 0.01$ in the binary aqueous solution was kept constant

c_C/(mol/l)	0.0	0.5	1.0	1.5
T/K	345.2	338.2	336.2	334.2

continued

continued

Comments: The sample was synthesized by anionic polymerization (butyl end group).

w_B = 0.01 in the binary aqueous solution was kept constant

c_C/(mol/l)	0.0	0.5	1.0	1.5
T/K	348.2	337.2	336.2	335.2

Polymer (B): **poly(*N*-ethyl-*N*-methylacrylamide)** **2004PAN**
Characterization: M_w/g.mol^{-1} = <5000, M_w/M_n < 1.5,
synthesized in the laboratory by different methods
Solvent (A): **water** H_2O **7732-18-5**
Solvent (C): **2-methyl-1-propanol** $C_4H_{10}O$ **78-83-1**

Type of data: cloud points (LCST-behavior)

Comments: The sample was synthesized by chain transfer polymerization.

w_B = 0.01 in the binary aqueous solution was kept constant

c_C/(mol/l)	0.000	0.125	0.250	0.375	0.500
T/K	345.2	340.2	336.2	332.2	330.2

Comments: The sample was synthesized by anionic polymerization (butyl end group).

w_B = 0.01 in the binary aqueous solution was kept constant

c_C/(mol/l)	0.00	0.25	0.50
T/K	348.2	334.7	321.2

Polymer (B): **poly(*N*-ethyl-*N*-methylacrylamide)** **2004PAN**
Characterization: M_w/g.mol^{-1} = <5000, M_w/M_n < 1.5,
synthesized in the laboratory by different methods
Solvent (A): **water** H_2O **7732-18-5**
Solvent (C): **2-propanol** C_3H_8O **67-63-0**

Type of data: cloud points (LCST-behavior)

w_B = 0.01 in the binary aqueous solution was kept constant

c_C/(mol/l)	0.0	0.5	1.0	1.5	2.0
T/K	345.2	340.2	340.2	336.7	335.7

Polymer (B): **poly(*N*-isopropylacrylamide)** **2007HOU**
Characterization: M_n/g.mol^{-1} = 5680, M_w/g.mol^{-1} = 7210,
synthesized in the laboratory
Solvent (A): **water** H_2O **7732-18-5**
Polymer (C): **bovine serum albumin**
Characterization: M_n/g.mol^{-1} = 63000

Type of data: cloud points (LCST-behavior)

c_B/(mg/ml)	5	5	5
c_C/(mg/ml)	1	5	10
T/K	308.95	309.75	310.25

Polymer (B):	**poly(*N*-isopropylacrylamide)**	**2007HOU**
Characterization:	M_n/g.mol^{-1} = 11820, M_w/g.mol^{-1} = 15640, synthesized in the laboratory	
Solvent (A):	**water** H_2O	**7732-18-5**
Polymer (C):	**bovine serum albumin**	
Characterization:	M_n/g.mol^{-1} = 63000	

Type of data: cloud points (LCST-behavior)

c_B/(mg/ml)	5	5	5
c_C/(mg/ml)	1	5	10
T/K	307.65	308.15	309.15

Polymer (B):	**poly(*N*-isopropylacrylamide)**	**2007HOU**
Characterization:	M_n/g.mol^{-1} = 22330, M_w/g.mol^{-1} = 46030, synthesized in the laboratory	
Solvent (A):	**water** H_2O	**7732-18-5**
Polymer (C):	**bovine serum albumin**	
Characterization:	M_n/g.mol^{-1} = 63000	

Type of data: cloud points (LCST-behavior)

c_B/(mg/ml)	5	5	5
c_C/(mg/ml)	1	5	10
T/K	307.95	307.95	309.15

Polymer (B):	**poly(*N*-isopropylacrylamide)**	**2007HOU**
Characterization:	M_n/g.mol^{-1} = 5680, M_w/g.mol^{-1} = 7210, synthesized in the laboratory	
Solvent (A):	**water** H_2O	**7732-18-5**
Polymer (C):	**poly(ethylene glycol)**	
Characterization:	M_n/g.mol^{-1} = 3400	

Type of data: cloud points (LCST-behavior)

c_B/(mg/ml)	5	5	5	5	5
c_C/(mg/ml)	1	2	4	5	10
T/K	307.45	307.65	308.45	308.55	309.45

Polymer (B):	**poly(*N*-isopropylacrylamide)**	**2007HOU**
Characterization:	M_n/g.mol^{-1} = 11820, M_w/g.mol^{-1} = 15640, synthesized in the laboratory	
Solvent (A):	**water** H_2O	**7732-18-5**
Polymer (C):	**poly(ethylene glycol)**	
Characterization:	M_n/g.mol^{-1} = 3400	

Type of data: cloud points (LCST-behavior)

c_B/(mg/ml)	5	5	5	5	5
c_C/(mg/ml)	1	2	4	5	10
T/K	307.45	307.45	308.25	308.45	309.25

Polymer (B): **poly(*N*-isopropylacrylamide)** **2007HOU**
Characterization: M_n/g.mol^{-1} = 22330, M_w/g.mol^{-1} = 46030, synthesized in the laboratory
Solvent (A): **water** **H_2O** **7732-18-5**
Polymer (C): **poly(ethylene glycol)**
Characterization: M_n/g.mol^{-1} = 3400

Type of data: cloud points (LCST-behavior)

c_B/(mg/ml)	5	5	5	5	5
c_C/(mg/ml)	1	2	4	5	10
T/K	307.25	307.35	308.55	308.55	309.35

Polymer (B): **poly(*N*-isopropylacrylamide-*co*-butyl acrylate)** **2010LIY**
Characterization: M_n/g.mol^{-1} = 2200, synthesized in the laboratory
Solvent (A): **water** **H_2O** **7732-18-5**
Polymer (C): **poly[2-(*N,N*-dimethylamino)ethyl methacrylate-*co*-acrylic acid-*co*-butyl methacrylate]**
Characterization: M_n/g.mol^{-1} = 37000, synthesized in the laboratory

Type of data: coexistence data (tie lines)

Top phase			Bottom phase		
w_A	w_B	w_C	w_A	w_B	w_C
T/K = 283.15					
0.9416	0.0433	0.0151	0.9500	0.0129	0.0371
0.9394	0.0495	0.0111	0.9496	0.0116	0.0388
0.9339	0.0600	0.0061	0.9476	0.0101	0.0423
0.9267	0.0690	0.0043	0.9457	0.0081	0.0462
0.9189	0.0788	0.0023	0.9422	0.0075	0.0503
T/K = 293.15					
0.9449	0.0430	0.0121	0.9535	0.0113	0.0352
0.9415	0.0495	0.0090	0.9523	0.0102	0.0375
0.9341	0.0601	0.0058	0.9490	0.0090	0.0420
0.9268	0.0689	0.0043	0.9457	0.0081	0.0462
0.9150	0.0827	0.0023	0.9400	0.0075	0.0525

Polymer (B): **poly[oligo(ethylene glycol) methyl ether methacrylate]** **2011ROT**
Characterization: M_n/g.mol^{-1} = 23200, synthesized in the laboratory
Solvent (A): **water** **H_2O** **7732-18-5**
Solvent (C): **2-propanol** **C_3H_8O** **67-63-0**

Type of data: cloud points

continued

continued

c_B/(g/dm^3) = 5	was kept constant					
φ_A	0.000	0.005	0.01	0.015	0.02	0.03
T/K	308.75	301.65	296.25	288.65	283.65	266.45

Comments: φ_A is the volume fraction of water in the solvent mixture.

Polymer (B): **poly(propylene glycol)** **2007SIL**
Characterization: M_n/g.mol^{-1} = 400, Sigma Chemical Co., Inc., St. Louis, MO
Solvent (A): **water** H_2O **7732-18-5**
Component (C): **fructose** $C_6H_{12}O_6$ **57-48-7**

Type of data: coexistence data (tie lines)

Total system			Top phase			Bottom phase		
w_A	w_B	w_C	w_A	w_B	w_C	w_A	w_B	w_C
T/K = 288.15								
0.4400	0.3500	0.2100	0.2659	0.6897	0.0444	0.4806	0.2203	0.2991
0.3417	0.3979	0.2604	0.1829	0.7863	0.0308	0.4549	0.0415	0.5036
0.3406	0.4792	0.1802	0.1428	0.8379	0.0193	0.4672	0.1243	0.4085
0.3654	0.4199	0.2147	0.1793	0.7872	0.0335	0.4807	0.1233	0.3960
T/K = 298.15								
0.4307	0.3493	0.2200	0.2782	0.6714	0.0504	0.4948	0.1872	0.3180
0.3423	0.3983	0.2594	0.1775	0.8004	0.0221	0.4762	0.0114	0.5124
0.3396	0.4800	0.1804	0.1441	0.8302	0.0257	0.4998	0.0320	0.4682
0.3649	0.4201	0.2150	0.1823	0.7941	0.0236	0.5155	0.0309	0.4536

Polymer (B): **poly(propylene glycol)** **2007SIL**
Characterization: M_n/g.mol^{-1} = 400, Sigma Chemical Co., Inc., St. Louis, MO
Solvent (A): **water** H_2O **7732-18-5**
Component (C): **sucrose** $C_{12}H_{22}O_{11}$ **57-50-1**

Type of data: coexistence data (tie lines)

Total system			Top phase			Bottom phase		
w_A	w_B	w_C	w_A	w_B	w_C	w_A	w_B	w_C
T/K = 288.15								
0.3404	0.4770	0.1826	0.2056	0.7694	0.0250	0.4514	0.1774	0.3712
0.3692	0.4509	0.1799	0.2455	0.7221	0.0324	0.4433	0.2035	0.3532
0.3210	0.4788	0.2002	0.1823	0.7921	0.0256	0.4427	0.1201	0.4372
0.2900	0.4790	0.2310	0.1615	0.8237	0.0148	0.4128	0.1170	0.4702

continued

continued

T/K = 298.15

0.3403	0.4799	0.1798	0.2409	0.7296	0.0295	0.4072	0.2662	0.3266
0.3696	0.4494	0.1810	0.2338	0.7399	0.0263	0.4374	0.3116	0.2510
0.4019	0.4300	0.1681	0.2653	0.7075	0.0272	0.4579	0.3088	0.2333
0.3172	0.4768	0.2060	0.1863	0.7955	0.0182	0.4205	0.2016	0.3779

T/K = 318.15

0.5342	0.3071	0.1587	0.1914	0.7745	0.0341	0.6195	0.1740	0.2065
0.4491	0.3515	0.1994	0.1774	0.7990	0.0236	0.5638	0.1237	0.3125
0.3704	0.4504	0.1792	0.1545	0.8287	0.0168	0.4896	0.1354	0.3750
0.4201	0.3603	0.2196	0.2087	0.7708	0.0205	0.5008	0.1494	0.3498

Polymer (B): **poly(pyrrolidinoacrylamide)** **2004PAN**
Characterization: M_w/g.mol^{-1} = <5000, M_w/M_n < 1.5,
synthesized in the laboratory
Solvent (A): **water** **H_2O** **7732-18-5**
Solvent (C): **methanol** **CH_4O** **67-56-1**

Type of data: cloud points (LCST-behavior)

w_B = 0.01 in the binary aqueous solution was kept constant

c_C/(mol/l)	0.0	0.5	1.0	1.5	2.0
T/K	331.2	337.2	335.7	335.7	340.2

Polymer (B): **poly(pyrrolidinoacrylamide)** **2004PAN**
Characterization: M_w/g.mol^{-1} = <5000, M_w/M_n < 1.5,
synthesized in the laboratory
Solvent (A): **water** **H_2O** **7732-18-5**
Solvent (C): **2-methyl-1-propanol** **$C_4H_{10}O$** **78-83-1**

Type of data: cloud points (LCST-behavior)

w_B = 0.01 in the binary aqueous solution was kept constant

Comments: The sample was synthesized by chain transfer polymerization.

c_C/(mol/l)	0.000	0.125	0.250	0.375	0.500
T/K	345.2	340.2	337.2	333.2	331.2

Comments: The sample was synthesized by anionic polymerization (butyl end group).

c_C/(mol/l)	0.000	0.125	0.250	0.375	0.500
T/K	331.2	327.7	326.2	324.2	321.2

Polymer (B): **poly(pyrrolidinoacrylamide)** **2004PAN**
Characterization: M_w/g.mol^{-1} = <5000, M_w/M_n < 1.5, synthesized in the laboratory
Solvent (A): **water** **H_2O** **7732-18-5**
Solvent (C): **2-propanol** **C_3H_8O** **67-63-0**

Type of data: cloud points (LCST-behavior)

w_B = 0.01 in the binary aqueous solution was kept constant

c_C/(mol/l)	0.0	0.5	1.0	1.5	2.0
T/K	331.2	331.2	331.2	330.2	327.2

Polymer (B): **pullulan** **2004ECK**
Characterization: M_n/g.mol^{-1} = 150000, M_w = 430000, M_z = 1550000, linear polysaccharid mainly composed of α-(1-6) D-maltotriose units, Polymer Standard Service, Mainz, Germany
Solvent (A): **water** **H_2O** **7732-18-5**
Solvent (C): **2-propanol** **C_3H_8O** **67-63-0**

Type of data: cloud points

T/K = 298.15

w_A	0.6927	0.7117
w_B	0.0766	0.0488
w_C	0.2306	0.2395

Polymer (B): **pullulan** **2004ECK**
Characterization: M_n/g.mol^{-1} = 150000, M_w = 430000, M_z = 1550000, linear polysaccharid mainly composed of α-(1-6) D-maltotriose units, Polymer Standard Service, Mainz, Germany
Solvent (A): **water** **H_2O** **7732-18-5**
Solvent (C): **2-propanone** **C_3H_6O** **67-64-1**

Type of data: cloud points

T/K = 298.15

w_A	0.6194	0.6471	0.6752	0.6849	0.6989	0.7109	0.7157	0.7273	0.7339
w_B	0.2377	0.2149	0.1411	0.1211	0.1002	0.0790	0.0835	0.0605	0.0550
w_C	0.1429	0.1379	0.1836	0.1940	0.2009	0.2101	0.2008	0.2122	0.2112
w_A	0.7447	0.7506	0.7517	0.7564	0.7595	0.7620	0.7629	0.7663	0.7710
w_B	0.0428	0.0404	0.0350	0.0300	0.0262	0.0232	0.0204	0.0155	0.0123
w_C	0.2125	0.2090	0.2134	0.2136	0.2143	0.2148	0.2167	0.2182	0.2167
w_A	0.7717	0.7744	0.7746	0.7752					
w_B	0.0095	0.0078	0.0067	0.0058					
w_C	0.2189	0.2178	0.2187	0.2190					

continued

continued

Type of data: coexistence data (tie lines)

T/K = 298.15

Total system			Sol phase			Gel phase		
w_A	w_B	w_C	w_A	w_B	w_C	w_A	w_B	w_C
0.6182	0.0059	0.3759	0.6209	0.0013	0.3778	0.3641	0.4325	0.2033
0.6871	0.0764	0.2364	0.7321	0.0197	0.2482	0.5403	0.2628	0.1969
0.6944	0.0065	0.2992	0.6974	0.0017	0.3009	0.4484	0.3929	0.1587
0.6987	0.0776	0.2236	0.7405	0.0353	0.2242	0.6405	0.1730	0.1865
0.7188	0.0547	0.2265	0.7357	0.0270	0.2373	0.6342	0.1904	0.1754
0.7285	0.0266	0.2449	0.7365	0.0115	0.2520	0.5560	0.2651	0.1789
0.7633	0.0089	0.2278	0.7640	0.0077	0.2283	0.6786	0.1525	0.1689

critical concentrations: $w_{A,\,crit} = 0.692$, $w_{B,\,crit} = 0.111$, $w_{C,\,crit} = 0.197$

Polymer (B): **pullulan** **2004ECK**
Characterization: M_n/g.mol^{-1} = 150000, M_w = 430000, M_z = 1550000, linear polysaccharid mainly composed of α-(1-6) D-maltotriose units, Polymer Standard Service, Mainz, Germany
Solvent (A): **water** H_2O **7732-18-5**
Solvent (C): **tetrahydrofuran** C_4H_8O **109-99-9**

Type of data: cloud points

T/K = 298.15

w_A	0.6545	0.6644	0.6703
w_B	0.0724	0.0573	0.0483
w_C	0.2731	0.2783	0.2814

3.3.2. Electrolyte solutions

Polymer (B):	**methyl(hydroxypropyl)cellulose**		**2011SAR**
Characterization:	M_w/g.mol^{-1} = 10000, 9 wt% hydroxypropyl content, Fluka AG, Buchs, Switzerland		
Solvent (A):	**water**	**H_2O**	**7732-18-5**
Salt (C):	**potassium bromide**	**KBr**	**7758-02-3**

Type of data: cloud points (LCST behavior)

c_B/(mol/kg H_2O)	0.00005	0.00010	0.00020	0.00050	0.00100
c_C/(mol/kg H_2O)	0.531	0.531	0.531	0.531	0.531
T/K	335.15	332.15	331.15	331.15	329.15

Polymer (B):	**methyl(hydroxypropyl)cellulose**		**2011SAR**
Characterization:	M_w/g.mol^{-1} = 10000, 9 wt% hydroxypropyl content, Fluka AG, Buchs, Switzerland		
Solvent (A):	**water**	**H_2O**	**7732-18-5**
Salt (C):	**potassium chloride**	**KCl**	**7447-40-7**

Type of data: cloud points (LCST behavior)

c_B/(mol/kg H_2O)	0.00005	0.00010	0.00020	0.00050	0.00100
c_C/(mol/kg H_2O)	0.519	0.519	0.519	0.519	0.519
T/K	327.15	325.15	325.15	325.65	322.15

Polymer (B):	**methyl(hydroxypropyl)cellulose**		**2011SAR**
Characterization:	M_w/g.mol^{-1} = 10000, 9 wt% hydroxypropyl content, Fluka AG, Buchs, Switzerland		
Solvent (A):	**water**	**H_2O**	**7732-18-5**
Salt (C):	**sodium bromide**	**NaBr**	**7647-15-6**

Type of data: cloud points (LCST behavior)

c_B/(mol/kg H_2O)	0.00005	0.00010	0.00020	0.00050	0.00100
c_C/(mol/kg H_2O)	0.527	0.527	0.527	0.527	0.527
T/K	335.15	333.15	332.15	331.65	329.15

Polymer (B):	**methyl(hydroxypropyl)cellulose**		**2011SAR**
Characterization:	M_w/g.mol^{-1} = 10000, 9 wt% hydroxypropyl content, Fluka AG, Buchs, Switzerland		
Solvent (A):	**water**	**H_2O**	**7732-18-5**
Salt (C):	**sodium chloride**	**NaCl**	**7647-14-5**

Type of data: cloud points (LCST behavior)

c_B/(mol/kg H_2O)	0.00005	0.00010	0.00020	0.00050	0.00100
c_C/(mol/kg H_2O)	0.515	0.515	0.515	0.515	0.515
T/K	328.15	326.15	322.15	324.15	324.15

Polymer (B):	**methyl(hydroxypropyl)cellulose**		**2011SAR**
Characterization:	M_w/g.mol^{-1} = 10000, 9 wt% hydroxypropyl content, Fluka AG, Buchs, Switzerland		
Solvent (A):	**water**	**H_2O**	**7732-18-5**
Salt (C):	**sodium nitrate**	**$NaNO_3$**	**7631-99-4**

Type of data: cloud points (LCST behavior)

c_B/(mol/kg H_2O)	0.00005	0.00010	0.00020	0.00050	0.00100
c_C/(mol/kg H_2O)	0.522	0.522	0.522	0.522	0.522
T/K	337.15	334.15	333.15	333.15	331.15

Polymer (B):	**methyl(hydroxypropyl)cellulose**		**2011SAR**
Characterization:	M_w/g.mol^{-1} = 10000, 9 wt% hydroxypropyl content, Fluka AG, Buchs, Switzerland		
Solvent (A):	**water**	**H_2O**	**7732-18-5**
Salt (C):	**sodium phosphate**	**Na_3PO_4**	**7601-54-9**

Type of data: cloud points (LCST behavior)

c_B/(mol/kg H_2O)	0.00005	0.00010	0.00020	0.00050	0.00100
c_C/(mol/kg H_2O)	0.0509	0.0509	0.0509	0.0509	0.0509
T/K	328.15	326.15	322.15	323.15	324.15

Polymer (B):	**methyl(hydroxypropyl)cellulose**		**2011SAR**
Characterization:	M_w/g.mol^{-1} = 10000, 9 wt% hydroxypropyl content, Fluka AG, Buchs, Switzerland		
Solvent (A):	**water**	**H_2O**	**7732-18-5**
Salt (C):	**sodium sulfate**	**Na_2SO_4**	**7757-82-6**

Type of data: cloud points (LCST behavior)

c_B/(mol/kg H_2O)	0.00005	0.00010	0.00020	0.00050	0.00100
c_C/(mol/kg H_2O)	0.0503	0.0503	0.0503	0.0503	0.0503
T/K	332.15	329.15	325.65	328.15	326.15

Polymer (B):	**poly(*N*-acryloyl-L-aspartic acid *N'*-propylamide)**		**2004KUR**
Characterization:	M_n/g.mol^{-1} = 1000000, M_w/g.mol^{-1} = 1600000, synthesized in the laboratory		
Solvent (A):	**water**	**H_2O**	**7732-18-5**
Salt (C):	**ammonium acetate**	**$(NH_4)C_2H_3O_2$**	**631-61-8**

Type of data: cloud points (LCST behavior)

Comments: The concentration of the acetate is given by a 0.1 M buffer solution.

w_B	0.01	0.01	0.01	0.01
pH	3.7	3.85	4.0	4.3
T/K	283.15	291.15	297.15	329.15

Polymer (B): **poly(*N*-acryloyl-L-glutamic acid *N'*-propylamide)** **2004KUR**
Characterization: M_n/g.mol^{-1} = 947000, M_w/g.mol^{-1} = 1800000, synthesized in the laboratory
Solvent (A): **water** **H_2O** **7732-18-5**
Salt (C): **ammonium acetate** **$(NH_4)C_2H_3O_2$** **631-61-8**

Type of data: cloud points (LCST behavior)

Comments: The concentration of the acetate is given by a 0.1 M buffer solution.

w_B	0.01	0.01	0.01	0.01
pH	4.3	4.45	4.6	4.75
T/K	275.15	283.15	291.15	303.15

Polymer (B): **poly(*N*,*N*-dimethylacrylamide-*co*-*tert*-butylacrylamide)** **2009FO2**
Characterization: M_n/g.mol^{-1} = 1700, unspecified comonomer content, synthesized in the laboratory
Solvent (A): **water** **H_2O** **7732-18-5**
Salt (C): **monoammonium phosphate** **$(NH_4)H_2PO_4$** **7722-76-1**

Type of data: cloud points (LCST behavior)

T/K = 338.15

w_A	0.8770	0.8750	0.8814	0.8687	0.8513	0.8441	0.8387
w_B	0.0874	0.0930	0.0927	0.1124	0.1297	0.1373	0.1430
w_C	0.0356	0.0320	0.0259	0.0189	0.0190	0.0186	0.0183

Polymer (B): **poly(*N*,*N*-dimethylacrylamide-*co*-*tert*-butylacrylamide)** **2009FO2**
Characterization: M_n/g.mol^{-1} = 1700, unspecified comonomer content, synthesized in the laboratory
Solvent (A): **water** **H_2O** **7732-18-5**
Salt (C): **monopotassium phosphate** **KH_2PO_4** **7778-77-0**

Type of data: cloud points (LCST behavior)

T/K = 338.15

w_A	0.8943	0.8877	0.8848	0.8732	0.8662	0.8611	0.8561
w_B	0.0547	0.0840	0.0917	0.1044	0.1119	0.1172	0.1224
w_C	0.0510	0.0283	0.0235	0.0224	0.0219	0.0217	0.0215

Polymer (B): **poly(*N*,*N*-dimethylacrylamide-*co*-*tert*-butylacrylamide)** **2009FO2**
Characterization: M_n/g.mol^{-1} = 1700, unspecified comonomer content, synthesized in the laboratory
Solvent (A): **water** **H_2O** **7732-18-5**
Salt (C): **monosodium carbonate** **$NaHCO_3$** **144-55-8**

continued

continued

Type of data: cloud points (LCST behavior)

T/K = 338.15

w_A	0.9242	0.9370	0.9482	0.9400	0.9345
w_B	0.0198	0.0230	0.0268	0.0354	0.0417
w_C	0.0560	0.0400	0.0250	0.0246	0.0238

Polymer (B): **poly(*N,N*-dimethylacrylamide-*co*-*tert*-butylacrylamide)** **2009FO2**
Characterization: M_n/g.mol^{-1} = 1700, unspecified comonomer content, synthesized in the laboratory
Solvent (A): **water** H_2O **7732-18-5**
Salt (C): **sodium chloride** **NaCl** **7647-14-5**

Type of data: cloud points (LCST behavior)

T/K = 338.15

w_A	0.9330	0.9268	0.9214	0.9083	0.8956	0.8866	0.8787	0.8731	0.8696
w_B	0.0390	0.0488	0.0588	0.0726	0.0854	0.0953	0.1036	0.1094	0.1129
w_C	0.0280	0.0244	0.0198	0.0191	0.0190	0.0181	0.0177	0.0175	0.0175

Polymer (B): **poly(*N*-ethylacrylamide)** **2003XUE**
Characterization: M_n/g.mol^{-1} = 3300, M_w/g.mol^{-1} = 5400, synthesized in the laboratory
Solvent (A): **water** H_2O **7732-18-5**
Salt (C): **potassium chloride** **KCl** **7447-40-7**

Type of data: cloud points (LCST behavior)

w_B	0.01	0.01	0.01	0.01	0.01
w_C	0.000	0.001	0.005	0.010	0.150
T/K	359.15	357.15	355.15	353.15	313.15

Polymer (B): **poly(*N*-ethylacrylamide)** **2003XUE**
Characterization: M_n/g.mol^{-1} = 3300, M_w/g.mol^{-1} = 5400, synthesized in the laboratory
Solvent (A): **water** H_2O **7732-18-5**
Component (C): **sodium dodecyl sulfate** $C_{12}H_{25}NaO_4S$ **151-21-3**

Type of data: cloud points (LCST behavior)

w_B	0.01	0.01	0.01	0.01
w_C	0.000	0.001	0.005	0.010
T/K	354.15	366.15	377.15	389.15

Polymer (B):	**poly(ethylene glycol)**		**2009REG**
Characterization:	M_n/g.mol^{-1} = 5400, Merck-Schuchardt, Hohenbrunn, Germany		
Solvent (A):	**water**	H_2O	**7732-18-5**
Salt (C):	**ammonium citrate**	$(NH_4)_3C_6H_5O_7$	**3458-72-8**

Type of data: cloud points

T/K = 298.15

w_A	0.4850	0.5193	0.5445	0.5655	0.5905	0.6163	0.6333	0.6511	0.6637
w_B	0.4702	0.4307	0.4009	0.3759	0.3452	0.3134	0.2920	0.2691	0.2519
w_C	0.0448	0.0500	0.0546	0.0586	0.0643	0.0703	0.0747	0.0798	0.0844
w_A	0.6790	0.6885	0.6987	0.7065	0.7669	0.7868	0.7830	0.7782	
w_B	0.2322	0.2187	0.2049	0.1933	0.0940	0.0344	0.0111	0.0050	
w_C	0.0888	0.0928	0.0964	0.1002	0.1391	0.1788	0.2059	0.2168	

T/K = 303.15

w_A	0.5271	0.5688	0.6121	0.6418	0.6675	0.6885	0.7059	0.7204	0.7324
w_B	0.4294	0.3800	0.3270	0.2889	0.2546	0.2255	0.2005	0.1788	0.1600
w_C	0.0435	0.0512	0.0609	0.0693	0.0779	0.0860	0.0936	0.1008	0.1076
w_A	0.7409	0.7481	0.7547	0.7550	0.7592	0.7634	0.7667	0.7691	0.7712
w_B	0.1461	0.1338	0.1219	0.1214	0.1135	0.1052	0.0987	0.0935	0.0888
w_C	0.1130	0.1181	0.1234	0.1236	0.1273	0.1314	0.1346	0.1374	0.1400
w_A	0.7730	0.7747	0.7758	0.7812	0.7865	0.7840			
w_B	0.0848	0.0807	0.0781	0.0631	0.0310	0.0140			
w_C	0.1422	0.1446	0.1461	0.1557	0.1825	0.2020			

T/K = 308.15

w_A	0.5191	0.5720	0.6125	0.6451	0.6711	0.6905	0.7089	0.7240	0.7349
w_B	0.4498	0.3865	0.3364	0.2960	0.2628	0.2363	0.2124	0.1910	0.1742
w_C	0.0311	0.0415	0.0511	0.0589	0.0661	0.0732	0.0787	0.0850	0.0909
w_A	0.7438	0.7512	0.7576	0.7628	0.7668	0.7701	0.7730	0.7753	0.7774
w_B	0.1600	0.1477	0.1365	0.1270	0.1193	0.1127	0.1067	0.1016	0.0967
w_C	0.0962	0.1011	0.1059	0.1102	0.1139	0.1172	0.1203	0.1231	0.1259
w_A	0.7784	0.7838	0.7774						
w_B	0.0943	0.0580	0.0026						
w_C	0.1273	0.1582	0.2200						

T/K = 313.15

w_A	0.5178	0.5763	0.6134	0.6457	0.6715	0.6924	0.7072	0.7236	0.7354
w_B	0.4554	0.3875	0.3434	0.3039	0.2715	0.2445	0.2247	0.2025	0.1858
w_C	0.0268	0.0362	0.0432	0.0504	0.0570	0.0631	0.0681	0.0739	0.0788
w_A	0.7452	0.7533	0.7603	0.7680	0.7735	0.7780	0.7817	0.7827	0.8004
w_B	0.1709	0.1579	0.1465	0.1349	0.1255	0.1177	0.1110	0.1091	0.0716
w_C	0.0839	0.0888	0.0932	0.0971	0.1010	0.1043	0.1073	0.1082	0.1280

continued

continued

w_A	0.8087	0.8090	0.8036	0.7988
w_B	0.0456	0.0307	0.0153	0.0100
w_C	0.1457	0.1603	0.1811	0.1912

T/K = 318.15

w_A	0.5413	0.5951	0.6367	0.6672	0.6894	0.7077	0.7228	0.7355	0.7493
w_B	0.4330	0.3732	0.3257	0.2900	0.2630	0.2399	0.2205	0.2037	0.1853
w_C	0.0257	0.0317	0.0376	0.0428	0.0476	0.0524	0.0567	0.0608	0.0654
w_A	0.7585	0.7693	0.7770	0.7839	0.7910	0.7980	0.8016	0.8163	0.8244
w_B	0.1718	0.1570	0.1456	0.1348	0.1239	0.1128	0.1069	0.0761	0.0531
w_C	0.0697	0.0737	0.0774	0.0813	0.0851	0.0892	0.0915	0.1076	0.1225
w_A	0.8309	0.8091	0.7874						
w_B	0.0151	0.0046	0.0022						
w_C	0.1540	0.1863	0.2104						

Type of data: coexistence data (tie lines)

Total system			Top phase			Bottom phase		
w_A	w_B	w_C	w_A	w_B	w_C	w_A	w_B	w_C
T/K = 298.15								
0.74	0.12	0.14	0.6985	0.2041	0.0973	0.7862	0.0090	0.2049
0.71	0.15	0.14	0.6605	0.2547	0.0848	0.7726	0.0038	0.2236
0.68	0.18	0.14	0.6263	0.3001	0.0736	0.7548	0.0032	0.2421
0.65	0.21	0.14	0.5888	0.3489	0.0623	0.7399	0.0035	0.2566
0.62	0.24	0.14	0.5518	0.3959	0.0523	0.7164	0.0069	0.2768
T/K = 303.15								
0.74	0.12	0.14	0.6840	0.2299	0.0861	0.7916	0.0031	0.2053
0.71	0.15	0.14	0.6460	0.2804	0.0736	0.7752	0.0042	0.2207
0.68	0.18	0.14	0.6100	0.3239	0.0661	0.7613	0.0033	0.2354
0.65	0.21	0.14	0.5795	0.3632	0.0573	0.7429	0.0029	0.2542
0.62	0.24	0.14	0.5447	0.4042	0.0511	0.7210	0.0057	0.2733
T/K = 308.15								
0.74	0.12	0.14	0.6657	0.2607	0.0736	0.7970	0.0049	0.1980
0.71	0.15	0.14	0.6282	0.3095	0.0623	0.7815	0.0069	0.2116
0.68	0.18	0.14	0.5933	0.3494	0.0573	0.7643	0.0058	0.2299
0.65	0.21	0.14	0.5601	0.3913	0.0486	0.7482	0.0055	0.2463
0.62	0.24	0.14	0.5275	0.4315	0.0411	0.7343	0.0025	0.2632
T/K = 313.15								
0.74	0.12	0.14	0.6488	0.2970	0.0542	0.7998	0.0026	0.1977
0.71	0.15	0.14	0.6173	0.3229	0.0598	0.7860	0.0021	0.2119
0.68	0.18	0.14	0.5748	0.3729	0.0523	0.7635	0.0070	0.2295
0.65	0.21	0.14	0.5432	0.4095	0.0473	0.7500	0.0048	0.2452
0.62	0.24	0.14	0.5153	0.4411	0.0436	0.7348	0.0027	0.2625

Polymer (B):	**poly(ethylene glycol)**		**2003HUD**
Characterization:	M_w/g.mol^{-1} = 2000, Aldrich Chem. Co., Inc., Milwaukee, WI		
Solvent (A):	**water**	**H_2O**	**7732-18-5**
Salt (C):	**ammonium sulfate**	**$(NH_4)_2SO_4$**	**7783-20-2**

Type of data: cloud points

T/K = 298.15

w_A	0.6000	0.6309	0.6590	0.6847	0.7083	0.7249	0.7402	0.7527	0.7647
w_B	0.3621	0.3156	0.2839	0.2508	0.2183	0.1935	0.1709	0.1511	0.1333
w_C	0.0379	0.0535	0.0571	0.0645	0.0734	0.0816	0.0889	0.0962	0.1020
w_A	0.7593	0.7627	0.7714	0.7789	0.7857	0.7923	0.7984	0.8010	0.8047
w_B	0.1329	0.1328	0.1184	0.1050	0.0933	0.0816	0.0712	0.0622	0.0549
w_C	0.1078	0.1045	0.1102	0.1161	0.1210	0.1261	0.1304	0.1368	0.1404
w_A	0.8079	0.8113	0.8134	0.8157	0.8189	0.8186	0.8156	0.8133	0.8102
w_B	0.0481	0.0415	0.0357	0.0299	0.0223	0.0167	0.0131	0.0093	0.0059
w_C	0.1440	0.1472	0.1509	0.1544	0.1588	0.1647	0.1713	0.1774	0.1839
w_A	0.8090	0.8088	0.8082	0.7807					
w_B	0.0035	0.0014	0.0005	0.0013					
w_C	0.1875	0.1898	0.1913	0.2180					

Type of data: coexistence data (tie lines)

T/K = 298.15

Total system			Top phase			Bottom phase		
w_A	w_B	w_C	w_A	w_B	w_C	w_A	w_B	w_C
0.7022	0.1955	0.1023	0.6287	0.3241	0.0472	0.8081	0.0104	0.1815
0.6777	0.1929	0.1294	0.5615	0.4090	0.0295	0.7806	0.0013	0.2181
0.6521	0.1917	0.1562	0.5186	0.4600	0.0214	0.7473	0.95210^{-5}	0.2526
0.6265	0.1880	0.1855	0.4928	0.4897	0.0175	0.7098	0.22710^{-5}	0.2902
0.6000	0.1857	0.2143	0.4563	0.5307	0.0130	0.6774	0.39110^{-6}	0.3226

Polymer (B):	**poly(ethylene glycol)**		**2007GRA**
Characterization:	M_n/g.mol^{-1} = 2000, M_w/g.mol^{-1} = 2080, Merck, Darmstadt, Germany		
Solvent (A):	**water**	**H_2O**	**7732-18-5**
Salt (C):	**ammonium sulfate**	**$(NH_4)_2SO_4$**	**7783-20-2**

Type of data: cloud points

T/K = 298.15

w_A	0.35628	0.37232	0.39482	0.41214	0.42644	0.43969	0.45609	0.46143	0.48257
w_B	0.63610	0.61990	0.59598	0.57737	0.56165	0.54728	0.52940	0.52259	0.49959
w_C	0.00762	0.00778	0.00920	0.01049	0.01191	0.01303	0.01451	0.01598	0.01784

continued

continued

w_A	0.51805	0.53081	0.54227	0.55372	0.56413	0.57368	0.58359	0.59244	0.59551
w_B	0.45925	0.44633	0.43378	0.42053	0.40803	0.39701	0.38562	0.37534	0.36792
w_C	0.02270	0.02286	0.02395	0.02575	0.02784	0.02931	0.03079	0.03222	0.03657
w_A	0.60947	0.61868	0.62647	0.63625	0.64203	0.64952	0.66617	0.68617	0.69744
w_B	0.35266	0.34029	0.33022	0.31889	0.30838	0.29765	0.27494	0.25025	0.23315
w_C	0.03787	0.04103	0.04331	0.04486	0.04959	0.05283	0.05889	0.06358	0.06941
w_A	0.70922	0.71945	0.73067	0.73853	0.75282	0.75967	0.76603	0.77182	0.77713
w_B	0.21643	0.20069	0.18772	0.17661	0.15480	0.14439	0.13398	0.12405	0.11459
w_C	0.07435	0.07986	0.08161	0.08486	0.09238	0.09594	0.09999	0.10413	0.10828
w_A	0.77970	0.78662	0.79242	0.79951	0.80438	0.80890	0.81147	0.81342	0.81264
w_B	0.10688	0.09385	0.07827	0.06645	0.05636	0.04673	0.03531	0.02860	0.01948
w_C	0.11342	0.11953	0.12931	0.13404	0.13926	0.14437	0.15322	0.15798	0.16788
w_A	0.80941								
w_B	0.01320								
w_C	0.17739								

Type of data: coexistence data (tie lines)

T/K = 298.15

Total system			Top phase			Bottom phase		
w_A	w_B	w_C	w_A	w_B	w_C	w_A	w_B	w_C
0.72053	0.15300	0.12647	0.62700	0.33104	0.04196	0.79944	0.00559	0.19497
0.66195	0.19805	0.14000	0.56165	0.40909	0.02926	0.76084	0.00033	0.23883
0.60700	0.23525	0.15775	0.48091	0.50325	0.01584	0.71357	0.00001	0.28642
0.55300	0.26825	0.17875	0.43149	0.55579	0.01272	0.66732	0.000004	0.33268
0.51102	0.29524	0.19374	0.38707	0.60294	0.00999	0.63010	0.00000	0.36990
0.48575	0.31000	0.20425	0.35912	0.63156	0.00932	0.60892	0.00000	0.39108
0.47000	0.32000	0.21000	0.34255	0.64942	0.00803	0.59251	0.00000	0.40749

Polymer (B): **poly(ethylene glycol)** **2007GRA**
Characterization: M_n/g.mol^{-1} = 3740, M_w/g.mol^{-1} = 3820,
Merck, Darmstadt, Germany
Solvent (A): **water** H_2O **7732-18-5**
Salt (C): **ammonium sulfate** $(NH_4)_2SO_4$ **7783-20-2**

Type of data: cloud points

T/K = 298.15

w_A	0.35457	0.37932	0.40273	0.42521	0.43593	0.46364	0.48285	0.48770	0.51115
w_B	0.63913	0.61294	0.58758	0.56350	0.55062	0.52212	0.50147	0.49414	0.46830
w_C	0.00630	0.00774	0.00969	0.01129	0.01345	0.01424	0.01568	0.01816	0.02055

continued

continued

w_A	0.53085	0.54610	0.56175	0.58106	0.59469	0.60769	0.61978	0.63013	0.63950
w_B	0.44578	0.42810	0.40982	0.38781	0.37190	0.35684	0.34287	0.33028	0.31932
w_C	0.02337	0.02580	0.02843	0.03113	0.03341	0.03547	0.03735	0.03959	0.04118
w_A	0.64954	0.66102	0.66682	0.67716	0.68896	0.69658	0.70444	0.71119	0.71685
w_B	0.30558	0.29152	0.28259	0.27034	0.25624	0.24668	0.23688	0.22828	0.22099
w_C	0.04488	0.04746	0.05059	0.05250	0.05480	0.05674	0.05868	0.06053	0.06216
w_A	0.72234	0.73161	0.73704	0.74245	0.74808	0.75432	0.75989	0.76561	0.77136
w_B	0.21343	0.20080	0.19335	0.18577	0.17777	0.16892	0.16001	0.15147	0.14254
w_C	0.06423	0.06759	0.06961	0.07178	0.07415	0.07676	0.08010	0.08292	0.08610
w_A	0.77786	0.78575	0.79308	0.80046	0.80803	0.81517	0.82219	0.82975	0.83512
w_B	0.13265	0.12141	0.11012	0.09819	0.08597	0.07456	0.06265	0.04948	0.03890
w_C	0.08949	0.09284	0.09680	0.10135	0.10600	0.11027	0.11516	0.12077	0.12598
w_A	0.84394	0.84466	0.84399	0.83119	0.82522	0.82006			
w_B	0.02261	0.01445	0.00855	0.00259	0.00193	0.00120			
w_C	0.13345	0.14089	0.14746	0.16622	0.17285	0.17874			

Type of data: coexistence data (tie lines)

T/K = 298.15

Total system			Top phase			Bottom phase		
w_A	w_B	w_C	w_A	w_B	w_C	w_A	w_B	w_C
0.71322	0.16832	0.11846	0.60489	0.35873	0.03638	0.80909	0.00022	0.19069
0.65678	0.20888	0.13434	0.54813	0.42504	0.02683	0.76530	0.00000	0.23470
0.60254	0.24649	0.15097	0.49523	0.48602	0.01875	0.71930	0.00000	0.28070
0.54943	0.27788	0.17269	0.43307	0.55494	0.01199	0.66643	0.00000	0.33357

Polymer (B): **poly(ethylene glycol)** **2011KHA**
Characterization: M_n/g.mol^{-1} = 4000
Solvent (A): **water** H_2O **7732-18-5**
Salt (C): **ammonium sulfate** $(NH_4)_2SO_4$ **7783-20-2**

Type of data: coexistence data (tie lines)

T/K = 298.15

Top phase			Bottom phase		
w_A	w_B	w_C	w_A	w_B	w_C
0.6060	0.3541	0.0399	0.6977	0.0771	0.2252
0.5840	0.3782	0.0378	0.6866	0.0573	0.2561
0.5667	0.3984	0.0349	0.6766	0.0549	0.2685
0.4340	0.5418	0.0242	0.6029	0.0299	0.3672

Polymer (B): **poly(ethylene glycol)** **2005HEY**
Characterization: M_n/g.mol^{-1} = 7264, DP = 165 EO-units, Aldrich Chem. Co., Inc., Milwaukee, WI
Solvent (A): **water** H_2O **7732-18-5**
Salt (C): **ammonium sulfate** $(NH_4)_2SO_4$ **7783-20-2**

Type of data: cloud points

T/K = 298.15

z_B/(base mol/kg H_2O)	7.18	5.79	4.76	3.74	3.29	2.67	2.10
c_C/(mol/kg H_2O)	0.562	0.619	0.668	0.729	0.763	0.810	0.858
z_B/(base mol/kg H_2O)	1.57	1.26	1.01	0.689	0.558	0.158	0.117
c_C/(mol/kg H_2O)	0.908	0.933	0.965	0.996	1.02	1.14	1.17

Type of data: coexistence data (tie lines)

T/K = 298.15

Top phase		Bottom phase	
z_B/ base mol/kg H_2O	c_C/ mol/kg H_2O	z_B/ base mol/kg H_2O	c_C/ mol/kg H_2O
4.597	0.675	0.670	0.990
5.395	0.652	0.628	0.996
5.505	0.636	0.363	1.050
6.118	0.606	0.222	1.104
6.515	0.590	0.218	1.106
6.833	0.578	0.187	1.124
7.196	0.566	0.152	1.148

Comments: Concentrations are molal, i.e., per kg water; base moles are calculated with the molar mass of 44 g/mol per EO-unit.

Polymer (B): **poly(ethylene glycol)** **2007GRA**
Characterization: M_n/g.mol^{-1} = 9040, M_w/g.mol^{-1} = 9890, Merck, Darmstadt, Germany
Solvent (A): **water** H_2O **7732-18-5**
Salt (C): **ammonium sulfate** $(NH_4)_2SO_4$ **7783-20-2**

Type of data: cloud points

T/K = 298.15

w_A	0.49105	0.52324	0.54371	0.56066	0.57912	0.59202	0.60536	0.61692	0.62850
w_B	0.48879	0.45533	0.43272	0.41444	0.39296	0.37821	0.36318	0.35030	0.33712
w_C	0.02016	0.02143	0.02357	0.02490	0.02792	0.02977	0.03146	0.03278	0.03438

continued

continued

w_A	0.63697	0.64637	0.65542	0.66439	0.67193	0.67828	0.69091	0.70060	0.70964
w_B	0.32678	0.31582	0.30502	0.29471	0.28617	0.27785	0.26404	0.25254	0.24115
w_C	0.03625	0.03781	0.03956	0.04090	0.04190	0.04387	0.04505	0.04686	0.04921
w_A	0.71066	0.71806	0.72702	0.73695	0.74823	0.76045	0.77470	0.78509	0.79624
w_B	0.24006	0.23034	0.21911	0.20664	0.19253	0.17726	0.15883	0.14470	0.12928
w_C	0.04928	0.05160	0.05387	0.05641	0.05924	0.06229	0.06647	0.07021	0.07448
w_A	0.80959	0.81615	0.82338	0.82953	0.83643	0.84448	0.85294	0.86019	0.86598
w_B	0.11210	0.10263	0.09223	0.08263	0.07215	0.06029	0.04767	0.03613	0.02586
w_C	0.07831	0.08122	0.08439	0.08784	0.09142	0.09523	0.09939	0.10368	0.10816
w_A	0.86989	0.87145	0.86993	0.86498					
w_B	0.01738	0.01121	0.00643	0.00530					
w_C	0.11273	0.11734	0.12364	0.12972					

Type of data: coexistence data (tie lines)

T/K = 298.15

Total system			Top phase			Bottom phase		
w_A	w_B	w_C	w_A	w_B	w_C	w_A	w_B	w_C
0.78500	0.11000	0.10500	0.67183	0.28535	0.04282	0.85577	0.00057	0.14366
0.73000	0.15000	0.12000	0.61358	0.35376	0.03266	0.81788	0.00000	0.18212
0.68800	0.18000	0.13200	0.55423	0.42067	0.02510	0.78927	0.00000	0.21073
0.64100	0.21300	0.14600	0.50303	0.47662	0.02035	0.75264	0.00000	0.24736
0.60400	0.24000	0.15600	0.47567	0.50544	0.01889	0.72104	0.00000	0.27896

Polymer (B): **poly(ethylene glycol)** **2005MAB**
Characterization: M_n/g.mol^{-1} = 400, synthesis grade
Solvent (A): **water** H_2O **7732-18-5**
Salt (C): **cesium carbonate** Cs_2CO_3 **534-17-8**

Type of data: cloud points

T/K = 298.15

w_A	0.5186	0.5530	0.5512	0.5517	0.5496	0.5580	0.5587	0.5601	0.5480
w_B	0.01127	0.02706	0.03201	0.03910	0.04950	0.06324	0.09091	0.1082	0.1121
w_C	0.4701	0.4199	0.4168	0.4092	0.4009	0.3788	0.3504	0.3317	0.3399
w_A	0.5518	0.5421	0.5314	0.5330	0.5211	0.5103	0.4991	0.4499	0.4300
w_B	0.1420	0.1633	0.1846	0.2267	0.2466	0.2766	0.3205	0.4154	0.4249
w_C	0.3062	0.2946	0.2840	0.2403	0.2323	0.2131	0.1804	0.1347	0.1451

continued

continued

Type of data: coexistence data (tie lines)

T/K = 298.15

Top phase			Bottom phase		
w_A	w_B	w_C	w_A	w_B	w_C
0.4694	0.4311	0.09955	0.5602	0.05435	0.3855
0.4748	0.3717	0.1535	0.5539	0.1230	0.3231
0.4801	0.3522	0.1677	0.5445	0.1291	0.3264
0.5191	0.2898	0.1911	0.5524	0.1544	0.2932

Polymer (B): **poly(ethylene glycol)** **2005MAB**
Characterization: M_n/g.mol^{-1} = 1000, synthesis grade
Solvent (A): **water** **H_2O** **7732-18-5**
Salt (C): **cesium carbonate** **Cs_2CO_3** **534-17-8**

Type of data: cloud points

T/K = 298.15

w_A	0.5356	0.5803	0.6006	0.6178	0.6305	0.6578	0.6433	0.6586	0.6466
w_B	0.00083	0.00224	0.00428	0.00811	0.01827	0.03054	0.04240	0.05837	0.07182
w_C	0.4636	0.4175	0.3951	0.3741	0.3512	0.3117	0.3143	0.2830	0.2816
w_A	0.6382	0.6356	0.6347	0.6226	0.6262	0.6150	0.6107	0.6054	0.5918
w_B	0.08856	0.1081	0.1289	0.1449	0.1592	0.1654	0.1940	0.2121	0.2381
w_C	0.2732	0.2563	0.2364	0.2325	0.2146	0.2196	0.1953	0.1825	0.1701
w_A	0.5786	0.5629	0.5285	0.5095	0.4970	0.4567	0.4030		
w_B	0.2769	0.3033	0.3638	0.3767	0.4094	0.4567	0.5504		
w_C	0.1445	0.1338	0.1077	0.1138	0.09362	0.08660	0.04662		

Type of data: coexistence data (tie lines)

T/K = 298.15

Top phase			Bottom phase		
w_A	w_B	w_C	w_A	w_B	w_C
0.4721	0.4483	0.07957	0.5883	0.01494	0.3968
0.4808	0.4334	0.08583	0.5897	0.01457	0.3957
0.5245	0.3704	0.1051	0.6167	0.01586	0.3674
0.5283	0.3657	0.1060	0.6288	0.02576	0.3454

Polymer (B):	**poly(ethylene glycol)**		**2005MAB**
Characterization:	M_n/g.mol^{-1} = 4000, synthesis grade		
Solvent (A):	**water**	**H_2O**	**7732-18-5**
Salt (C):	**cesium carbonate**	**Cs_2CO_3**	**534-17-8**

Type of data: cloud points

T/K = 298.15

w_A	0.6656	0.6960	0.5269	0.7078	0.7167	0.5824	0.7288	0.7276	0.7316
w_B	0.00116	0.00227	0.00294	0.00381	0.00720	0.00800	0.01535	0.02714	0.04136
w_C	0.3332	0.3017	0.4702	0.2884	0.2761	0.4096	0.2559	0.2453	0.2270
w_A	0.7337	0.7248	0.7158	0.7100	0.6991	0.6862	0.6514	0.6029	0.5636
w_B	0.05664	0.07597	0.09957	0.1180	0.1427	0.1668	0.2279	0.3024	0.3526
w_C	0.2097	0.1992	0.1846	0.1720	0.1582	0.1470	0.1207	0.09466	0.08384
w_A	0.5401	0.5200	0.5017	0.4206					
w_B	0.3825	0.4117	0.4338	0.5333					
w_C	0.07740	0.06832	0.06454	0.04606					

T/K = 308.15

w_A	0.6160	0.6972	0.7002	0.7161	0.7294	0.7494	0.7580	0.7503	0.7450
w_B	0.00030	0.00136	0.00222	0.00387	0.00516	0.01426	0.03044	0.04540	0.06588
w_C	0.3837	0.3014	0.2976	0.2800	0.2654	0.2363	0.2116	0.2043	0.1891
w_A	0.7363	0.7326	0.7212	0.7096	0.6960	0.6843	0.6706	0.6130	0.5774
w_B	0.08265	0.1043	0.1208	0.1470	0.1581	0.1929	0.2132	0.2869	0.3408
w_C	0.1811	0.1631	0.1580	0.1434	0.1459	0.1228	0.1162	0.1001	0.08177
w_A	0.5578	0.5360	0.4225	0.3001					
w_B	0.3677	0.3870	0.5288	0.6694					
w_C	0.07450	0.07699	0.04875	0.03047					

T/K = 318.15

w_A	0.6722	0.7256	0.7216	0.7515	0.7552	0.7612	0.7549	0.7395	0.7323
w_B	0.00039	0.00148	0.00227	0.00446	0.00876	0.03061	0.04791	0.08597	0.1060
w_C	0.3274	0.2729	0.2761	0.2440	0.2360	0.2082	0.1972	0.1745	0.1617
w_A	0.7305	0.7139	0.6977	0.6825	0.6808	0.5740	0.5535	0.4788	0.4355
w_B	0.1313	0.1516	0.1807	0.1991	0.2105	0.3526	0.3836	0.4743	0.5255
w_C	0.1382	0.1345	0.1216	0.1184	0.1087	0.07340	0.06286	0.04688	0.03904
w_A	0.3127	0.2168							
w_B	0.6647	0.7704							
w_C	0.02259	0.01285							

continued

continued

Type of data: coexistence data (tie lines)

Top phase			Bottom phase		
w_A	w_B	w_C	w_A	w_B	w_C
T/K = 298.15					
0.5216	0.4058	0.07257	0.5824	0.00800	0.4096
0.5719	0.3431	0.08503	0.6537	0.01021	0.3361
0.5861	0.3251	0.08876	0.6881	0.01246	0.2994
0.5921	0.3190	0.08895	0.6751	0.01301	0.3119
T/K = 308.15					
0.3166	0.6509	0.03251	0.7467	0.03880	0.2145
0.4788	0.4618	0.05942	0.7201	0.11990	0.1600
0.5231	0.4072	0.06970	0.7323	0.09281	0.1749
T/K = 318.15					
0.3903	0.5781	0.03156	0.5946	0.00342	0.4020
0.4319	0.5258	0.04235	0.6530	0.00532	0.3417
0.5184	0.4236	0.05804	0.7211	0.00763	0.2713

Polymer (B): **poly(ethylene glycol)** **2004HUM**
Characterization: M_n/g.mol^{-1} = 1000, synthesis grade
Solvent (A): **water** **H_2O** **7732-18-5**
Salt (C): **cesium sulfate** **Cs_2SO_4** **10294-54-9**

Type of data: cloud points

T/K = 298.15

w_A	0.51650	0.52220	0.54638	0.58319	0.60535	0.61097	0.62178	0.61815	0.62480
w_B	0.00120	0.00140	0.00382	0.00861	0.01895	0.02853	0.05222	0.08175	0.1133
w_C	0.4823	0.4764	0.4498	0.4082	0.3757	0.3605	0.3260	0.3001	0.2619
w_A	0.62020	0.60680	0.61120	0.60490	0.59420	0.58760	0.57800	0.55570	0.54740
w_B	0.1219	0.1487	0.1664	0.2022	0.2214	0.2540	0.2769	0.3154	0.3318
w_C	0.2579	0.2445	0.2224	0.1929	0.1844	0.1584	0.1451	0.1289	0.1208
w_A	0.52330	0.47522	0.45865	0.44309					
w_B	0.3726	0.4481	0.4552	0.4804					
w_C	0.1041	0.07668	0.08615	0.07651					

continued

continued

Type of data: coexistence data (tie lines)

T/K = 298.15

Top phase			Bottom phase		
w_A	w_B	w_C	w_A	w_B	w_C
0.6284	0.1888	0.1828	0.6348	0.04795	0.3173
0.6191	0.2088	0.1721	0.6360	0.03899	0.3250
0.5763	0.3198	0.1039	0.5779	0.001232	0.4209
0.5809	0.3210	0.09807	0.5837	0.005622	0.4107

Polymer (B): **poly(ethylene glycol)** **2004HUM**
Characterization: M_n/g.mol^{-1} = 4000, synthesis grade
Solvent (A): **water** H_2O **7732-18-5**
Salt (C): **cesium sulfate** Cs_2SO_4 **10294-54-9**

Type of data: cloud points

T/K = 298.15

w_A	0.62943	0.65968	0.66466	0.67133	0.67353	0.67856	0.69990	0.69936	0.70849
w_B	0.00077	0.00182	0.00264	0.00357	0.00377	0.00384	0.01520	0.01524	0.02491
w_C	0.3698	0.3385	0.3327	0.3251	0.3227	0.3176	0.2849	0.2854	0.2666
w_A	0.70883	0.70326	0.7001	0.6931	0.6923	0.6909	0.6808	0.6610	0.6597
w_B	0.05977	0.07414	0.1009	0.1173	0.1331	0.1507	0.1699	0.1983	0.2130
w_C	0.2314	0.2226	0.1990	0.1896	0.1746	0.1584	0.1493	0.1407	0.1273
w_A	0.6502	0.6366	0.6216	0.6062	0.58875	0.58077	0.57115	0.56371	
w_B	0.2311	0.2571	0.2833	0.3077	0.33764	0.34526	0.3647	0.3766	
w_C	0.1187	0.1063	0.09510	0.08610	0.07361	0.07397	0.06415	0.05969	

T/K = 308.15

w_A	0.64720	0.68340	0.68817	0.69945	0.72231	0.73885	0.73786	0.72992	0.72582
w_B	0.00080	0.00280	0.00383	0.00515	0.01669	0.03025	0.03034	0.04318	0.05278
w_C	0.3520	0.3138	0.3080	0.2954	0.2610	0.2309	0.2318	0.2269	0.2214
w_A	0.71917	0.71320	0.70780	0.69460	0.68410	0.66350	0.66420	0.65180	0.63758
w_B	0.06323	0.1016	0.1203	0.1458	0.1748	0.2132	0.2230	0.2436	0.2715
w_C	0.2176	0.1852	0.1719	0.1596	0.1411	0.1233	0.1128	0.1046	0.09092
w_A	0.62254	0.60612	0.58914	0.57256	0.53608				
w_B	0.2890	0.3170	0.3392	0.3646	0.4055				
w_C	0.08846	0.07688	0.07166	0.06284	0.05842				

continued

continued

T/K = 318.15

w_A	0.68455	0.69936	0.71185	0.71966	0.72210	0.73742	0.75320	0.74580	0.74163
w_B	0.00095	0.00195	0.00316	0.00425	0.00500	0.01138	0.02480	0.03630	0.03767
w_C	0.3145	0.2987	0.2850	0.2761	0.2729	0.2512	0.2220	0.2179	0.2207
w_A	0.75580	0.74192	0.73637	0.71321	0.7147	0.7079	0.6868	0.6634	0.6540
w_B	0.05140	0.06778	0.08903	0.11819	0.1266	0.1522	0.1997	0.2374	0.2456
w_C	0.1928	0.1903	0.1746	0.1686	0.1587	0.1399	0.1135	0.0992	0.1004
w_A	0.62986	0.61907	0.56408	0.51309	0.49592				
w_B	0.2850	0.3093	0.3840	0.4454	0.4623				
w_C	0.08514	0.07163	0.05192	0.04151	0.04178				

Type of data: coexistence data (tie lines)

Top phase			Bottom phase		
w_A	w_B	w_C	w_A	w_B	w_C
T/K = 298.15					
0.5319	0.4077	0.0604	0.5624	0.00391	0.4337
0.5827	0.3253	0.0920	0.6147	0.01812	0.3672
0.5967	0.2906	0.1127	0.6270	0.01071	0.3623
0.6149	0.2594	0.1257	0.6561	0.02040	0.3235
T/K = 308.15					
0.5367	0.4180	0.0453	0.5675	0.01054	0.4220
0.5808	0.3557	0.0635	0.6169	0.005945	0.3772
0.6097	0.3047	0.0856	0.6606	0.007271	0.3321
0.6495	0.2581	0.0924	0.7010	0.006239	0.2928
T/K = 318.15					
0.5256	0.4430	0.0314	0.5815	0.003283	0.4152
0.5248	0.4418	0.0334	0.5810	0.002519	0.4165
0.5840	0.3625	0.0535	0.6481	0.002356	0.3495
0.5938	0.3387	0.0678	0.6906	0.005739	0.3037
0.6368	0.2806	0.0826	0.7633	0.001302	0.2345

Polymer (B): **poly(ethylene glycol)** **2004HUM**
Characterization: M_n/g.mol^{-1} = 10000, synthesis grade
Solvent (A): **water** H_2O **7732-18-5**
Salt (C): **cesium sulfate** Cs_2SO_4 **10294-54-9**

Type of data: cloud points

continued

continued

T/K = 298.15

w_A	0.57597	0.58547	0.65020	0.66893	0.6550	0.6753	0.6836	0.7087	0.7281
w_B	0.4011	0.3592	0.2718	0.2514	0.2433	0.2163	0.1955	0.1712	0.1318
w_C	0.02293	0.05533	0.07800	0.07967	0.1017	0.1084	0.1209	0.1201	0.1401
w_A	0.7373	0.7299	0.73968	0.75396	0.76720	0.76459	0.75928	0.75829	0.63516
w_B	0.1179	0.1058	0.08082	0.05214	0.03080	0.01801	0.01172	0.00341	0.00314
w_C	0.1448	0.1643	0.1795	0.1939	0.2020	0.2174	0.2290	0.2383	0.3617
w_A	0.74333	0.73346	0.71430						
w_B	0.00167	0.00104	0.00050						
w_C	0.2550	0.2655	0.2852						

Type of data: coexistence data (tie lines)

T/K = 298.15

Top phase			Bottom phase		
w_A	w_B	w_C	w_A	w_B	w_C
0.7087	0.1712	0.1201	0.7811	0.01773	0.2012
0.6689	0.2514	0.07967	0.7063	0.02502	0.2687
0.6016	0.3688	0.02961	0.6352	0.003138	0.3617
0.5760	0.4011	0.02293	0.5875	0.004145	0.4083

Polymer (B): **poly(ethylene glycol)** **2008MOH**
Characterization: M_n/g.mol^{-1} = 20000, Merck, Darmstadt, Germany
Solvent (A): **water** H_2O **7732-18-5**
Salt (C): **copper sulfate** $CuSO_4$ **7758-98-7**

Type of data: cloud points

T/K = 290.15

w_A	0.6318	0.6376	0.6591	0.6699	0.7034	0.7165	0.7189	0.7410	0.7589
w_B	0.3412	0.3350	0.3105	0.2981	0.2596	0.2448	0.2418	0.2168	0.1957
w_C	0.0270	0.0274	0.0304	0.0320	0.0370	0.0387	0.0393	0.0422	0.0454
w_A	0.7607	0.7843	0.8033	0.8169	0.8250	0.8338	0.8410	0.8439	0.8467
w_B	0.1934	0.1643	0.1370	0.1167	0.1038	0.0895	0.0768	0.0712	0.0663
w_C	0.0459	0.0514	0.0597	0.0664	0.0712	0.0767	0.0822	0.0849	0.0870
w_A	0.8478	0.8482	0.8426	0.8305	0.8189	0.8119	0.8047	0.7979	0.7916
w_B	0.0625	0.0582	0.0497	0.0460	0.0410	0.0383	0.0360	0.0320	0.0286
w_C	0.0897	0.0936	0.1077	0.1235	0.1401	0.1498	0.1593	0.1701	0.1798

T/K = 299.15

w_A	0.6247	0.6651	0.6745	0.6838	0.6953	0.7107	0.7171	0.7196	0.7454
w_B	0.3531	0.3074	0.2974	0.2863	0.2725	0.2550	0.2476	0.2446	0.2157
w_C	0.0222	0.0275	0.0281	0.0299	0.0322	0.0343	0.0353	0.0358	0.0389

continued

continued

w_A	0.7630	0.7865	0.8058	0.8182	0.8274	0.8343	0.8403	0.8463	0.8498
w_B	0.1945	0.1652	0.1394	0.1214	0.1078	0.0971	0.0875	0.0795	0.0718
w_C	0.0425	0.0483	0.0548	0.0604	0.0648	0.0686	0.0722	0.0742	0.0784
w_A	0.8524	0.8506	0.8402	0.8365	0.8354	0.8338	0.8204	0.8130	
w_B	0.0638	0.0492	0.0398	0.0375	0.0366	0.0350	0.0300	0.0260	
w_C	0.0838	0.1002	0.1200	0.1260	0.1280	0.1312	0.1496	0.1610	

T/K = 308.15

w_A	0.6178	0.6579	0.6613	0.6622	0.6926	0.6943	0.7183	0.7206	0.7510
w_B	0.3627	0.3178	0.3138	0.3127	0.2794	0.2769	0.2505	0.2477	0.2128
w_C	0.0195	0.0243	0.0249	0.0251	0.0280	0.0288	0.0312	0.0317	0.0362
w_A	0.7687	0.7854	0.8018	0.8178	0.8281	0.8383	0.8465	0.8507	0.8550
w_B	0.1925	0.1711	0.1500	0.1282	0.1139	0.0977	0.0864	0.0798	0.0732
w_C	0.0388	0.0435	0.0482	0.0540	0.0580	0.0640	0.0671	0.0695	0.0718
w_A	0.8576	0.8647	0.8568	0.8549	0.8491	0.8487	0.8473	0.8347	
w_B	0.0670	0.0401	0.0258	0.0250	0.0201	0.0195	0.0190	0.0100	
w_C	0.0754	0.0952	0.1174	0.1201	0.1308	0.1318	0.1337	0.1553	

T/K = 317.15

w_A	0.6025	0.6374	0.6589	0.6672	0.7067	0.7191	0.7183	0.7210	0.7291
w_B	0.3825	0.3436	0.3197	0.3107	0.2672	0.2537	0.2537	0.2510	0.2421
w_C	0.0150	0.0190	0.0214	0.0221	0.0261	0.0272	0.0280	0.0280	0.0288
w_A	0.7513	0.7622	0.7720	0.7798	0.7907	0.7970	0.8049	0.8100	0.8173
w_B	0.2167	0.2042	0.1919	0.1828	0.1690	0.1612	0.1512	0.1448	0.1350
w_C	0.0320	0.0336	0.0361	0.0374	0.0403	0.0418	0.0439	0.0452	0.0477
w_A	0.8232	0.8621	0.8717	0.8624	0.8581	0.8543	0.8539	0.8505	
w_B	0.1273	0.0655	0.0258	0.0139	0.0099	0.0059	0.0047	0.0047	
w_C	0.0495	0.0724	0.1025	0.1237	0.1320	0.1398	0.1414	0.1448	

Type of data: coexistence data (tie lines)

Total system			Top phase			Bottom phase		
w_A	w_B	w_C	w_A	w_B	w_C	w_A	w_B	w_C
T/K = 290.15								
0.7950	0.1520	0.0530	0.7610	0.1930	0.0460	0.8420	0.0500	0.1080
0.8280	0.0910	0.0810	0.7410	0.2170	0.0420	0.8310	0.0460	0.1230
0.7800	0.1540	0.0660	0.7030	0.2600	0.0370	0.8190	0.0410	0.1400
0.7560	0.1800	0.0640	0.6700	0.2980	0.0320	0.8120	0.0380	0.1500
0.7350	0.2140	0.0510	0.6600	0.3100	0.0300	0.8050	0.0360	0.1590
0.7510	0.2020	0.0470	0.6380	0.3350	0.0270	0.7980	0.0320	0.1700
0.7620	0.1460	0.0920	0.6320	0.3410	0.0270	0.7910	0.0290	0.1800

continued

continued

T/K = 299.15								
0.7850	0.1540	0.0610	0.7430	0.2210	0.0360	0.8560	0.0400	0.1040
0.7420	0.2030	0.0550	0.7150	0.2510	0.0340	0.8390	0.0400	0.1210
0.7890	0.1410	0.0700	0.6760	0.2920	0.0320	0.8370	0.0370	0.1260
0.7660	0.1750	0.0590	0.6510	0.3210	0.0280	0.8260	0.0360	0.1380
0.7510	0.1950	0.0540	0.6320	0.3410	0.0270	0.8120	0.0300	0.1580
0.7720	0.1360	0.0920	0.6090	0.3690	0.0220	0.8020	0.0260	0.1720
T/K = 308.15								
0.7875	0.1445	0.0680	0.7212	0.2500	0.0288	0.8690	0.0260	0.1050
0.7880	0.1540	0.0580	0.6930	0.2790	0.0280	0.8630	0.0250	0.1120
0.7530	0.1920	0.0550	0.6819	0.2930	0.0251	0.8650	0.0200	0.1150
0.7610	0.1820	0.0570	0.6541	0.3210	0.0249	0.8490	0.0190	0.1320
0.7500	0.1940	0.0560	0.6366	0.3390	0.0244	0.8300	0.0190	0.1510
0.7730	0.1320	0.0950	0.6055	0.3750	0.0195	0.8170	0.0100	0.1730
T/K = 317.15								
0.7720	0.1870	0.0410	0.7210	0.2510	0.0280	0.8720	0.0260	0.1020
0.8110	0.1070	0.0820	0.7070	0.2670	0.0260	0.8620	0.0140	0.1240
0.7500	0.1940	0.0560	0.6770	0.3010	0.0220	0.8580	0.0100	0.1320
0.7880	0.1450	0.0670	0.6590	0.3200	0.0210	0.8470	0.0060	0.1470
0.7760	0.1480	0.0760	0.6370	0.3440	0.0190	0.8320	0.0050	0.1630
0.7730	0.1320	0.0950	0.6030	0.3820	0.0150	0.8100	0.0050	0.1850

Polymer (B):	**poly(ethylene glycol)**		**2008OL3**
Characterization:	M_w/g.mol^{-1} = 4000, Isofar, Rio de Janeiro, Brazil		
Solvent (A):	**water**	**H_2O**	**7732-18-5**
Salt (C):	**copper sulfate**	**$CuSO_4$**	**7758-98-7**

Type of data: coexistence data (tie lines)

Total system			Top phase			Bottom phase		
w_A	w_B	w_C	w_A	w_B	w_C	w_A	w_B	w_C
T/K = 278.15								
0.7833	0.1151	0.1017	0.7653	0.1736	0.0611	0.7980	0.0667	0.1353
0.7760	0.1201	0.1039	0.7539	0.1956	0.0505	0.7933	0.0574	0.1492
0.7678	0.1250	0.1072	0.7447	0.2084	0.0469	0.7867	0.0552	0.1582
0.7607	0.1300	0.1093	0.7353	0.2263	0.0384	0.7800	0.0565	0.1635
0.7537	0.1341	0.1122	0.7267	0.2368	0.0365	0.7733	0.0592	0.1675

continued

continued

T/K = 283.15

0.7978	0.1140	0.0882	0.7902	0.1507	0.0592	0.8224	0.0432	0.1344
0.7889	0.1201	0.0910	0.7785	0.1701	0.0514	0.8192	0.0335	0.1473
0.7823	0.1248	0.0929	0.7635	0.1903	0.0463	0.8095	0.0342	0.1564
0.7720	0.1300	0.0980	0.7468	0.2163	0.0369	0.8000	0.0317	0.1683
0.7690	0.1350	0.0960	0.7445	0.2177	0.0378	0.7986	0.0312	0.1702

T/K = 308.15

0.7817	0.1151	0.1032	0.7453	0.2108	0.0439	0.8067	0.0538	0.1395
0.7769	0.1200	0.1031	0.7380	0.2195	0.0425	0.8047	0.0512	0.1441
0.7688	0.1251	0.1062	0.7247	0.2382	0.0371	0.7947	0.0480	0.1574
0.7617	0.1300	0.1083	0.7158	0.2495	0.0348	0.7893	0.0489	0.1617
0.7555	0.1351	0.1094	0.7084	0.2596	0.0319	0.7813	0.0506	0.1681

T/K = 318.15

0.7841	0.1200	0.0959	0.7511	0.2026	0.0463	0.8130	0.0381	0.1489
0.7791	0.1249	0.0960	0.7306	0.2338	0.0357	0.8081	0.0333	0.1587
0.7620	0.1371	0.1010	0.7141	0.2594	0.0265	0.7968	0.0347	0.1685
0.7489	0.1479	0.1030	0.6963	0.2821	0.0217	0.7858	0.0404	0.1738
0.7372	0.1570	0.1059	0.6833	0.2968	0.0198	0.7691	0.0463	0.1846

Polymer (B): **poly(ethylene glycol)** **2011REG**
Characterization: M_n/g.mol^{-1} = 1800
Solvent (A): **water** H_2O **7732-18-5**
Salt (C): **diammonium hydrogen citrate** $(NH_4)_2HC_6H_5O_7$ **3012-65-5**

Type of data: cloud points

T/K = 298.15

w_A	0.5051	0.5443	0.5509	0.5631	0.5695	0.5760	0.5873	0.5852	0.5922
w_B	0.1460	0.1520	0.1550	0.1550	0.1600	0.1630	0.1760	0.1840	0.1910
w_C	0.3489	0.3037	0.2941	0.2819	0.2705	0.2610	0.2367	0.2308	0.2168
w_A	0.5901	0.5941	0.5924	0.5885	0.5822	0.5704	0.5615	0.5532	0.5353
w_B	0.2010	0.2080	0.2351	0.2460	0.2593	0.2819	0.3034	0.3193	0.3530
w_C	0.2089	0.1979	0.1725	0.1655	0.1585	0.1477	0.1351	0.1275	0.1117
w_A	0.5258	0.5125	0.4946	0.4772	0.4505	0.4270	0.4031	0.3931	
w_B	0.3770	0.4050	0.4400	0.4707	0.5103	0.5412	0.5746	0.5890	
w_C	0.0972	0.0825	0.0654	0.0521	0.0392	0.0318	0.0223	0.0179	

T/K = 303.15

w_A	0.5202	0.5823	0.5842	0.5902	0.5899	0.5978	0.5999	0.6042	0.6015
w_B	0.1240	0.1461	0.1502	0.1520	0.1570	0.1610	0.1701	0.1770	0.2000
w_C	0.3558	0.2716	0.2656	0.2578	0.2531	0.2412	0.2300	0.2188	0.1985

continued

continued

w_A	0.6054	0.6043	0.6056	0.6039	0.5993	0.5957	0.5869	0.5828	0.5745
w_B	0.2030	0.2100	0.2140	0.2230	0.2340	0.2523	0.2720	0.2821	0.2980
w_C	0.1916	0.1857	0.1804	0.1731	0.1667	0.1520	0.1411	0.1351	0.1275
w_A	0.5638	0.5568	0.5513	0.5397	0.5306	.4908	.4649	0.4349	0.4127
w_B	0.3170	0.3310	0.3450	0.3670	0.3870	0.4586	0.4969	0.5351	0.5640
w_C	0.1192	0.1122	0.1037	0.0933	0.0824	0.0506	0.0382	0.0300	0.0233

T/K = 308.15

w_A	0.5505	0.5851	0.5975	0.6100	0.6047	0.6093	0.6103	0.6141	0.6135
w_B	0.1150	0.1290	0.1300	0.1400	0.1460	0.1500	0.1610	0.1690	0.1790
w_C	0.3345	0.2859	0.2725	0.2500	0.2493	0.2407	0.2287	0.2169	0.2075
w_A	0.6175	0.6196	0.6163	0.6148	0.6130	0.6122	0.6061	0.5976	0.5920
w_B	0.1950	0.2030	0.2120	0.2160	0.2257	0.2330	0.2450	0.2620	0.2750
w_C	0.1875	0.1774	0.1717	0.1692	0.1613	0.1548	0.1489	0.1404	0.1330
w_A	0.5819	0.5759	0.5631	0.5496	0.5194	0.4978	0.4745	0.4532	0.4302
w_B	0.2950	0.3100	0.3380	0.3680	0.4210	0.4560	0.4908	0.5200	0.5490
w_C	0.1231	0.1141	0.0989	0.0824	0.0596	0.0462	0.0347	0.0268	0.0208

T/K = 313.15

w_A	0.5417	0.6085	0.6132	0.6174	0.6253	0.6306	0.6338	0.6351	0.6360
w_B	0.1020	0.1181	0.1190	0.1240	0.1310	0.1350	0.1390	0.1540	0.1640
w_C	0.3563	0.2734	0.2678	0.2586	0.2437	0.2344	0.2272	0.2109	0.2000
w_A	0.6359	0.6289	0.6279	0.6262	0.6171	0.6152	0.6087	0.6052	0.5941
w_B	0.1830	0.2050	0.2080	0.2160	0.2380	0.2450	0.2581	0.2682	0.2889
w_C	0.1811	0.1661	0.1641	0.1578	0.1449	0.1398	0.1332	0.1266	0.1170
w_A	0.5880	0.5829	0.5622	0.5450	0.5164	0.5043	0.4862	0.4690	0.4394
w_B	0.3020	0.3154	0.3574	0.3875	0.4373	0.4560	0.4810	0.5047	0.5412
w_C	0.1100	0.1017	0.0804	0.0675	0.0463	0.0397	0.0328	0.0263	0.0194

T/K = 318.15

w_A	0.5630	0.5857	0.6010	0.6066	0.6200	0.6215	0.6370	0.6430	0.6474
w_B	0.0970	0.0940	0.0990	0.0976	0.1000	0.1060	0.1130	0.1170	0.1390
w_C	0.3400	0.3203	0.3000	0.2958	0.2800	0.2725	0.2500	0.2400	0.2136
w_A	0.6480	0.6478	0.6467	0.6425	0.6435	0.6400	0.6392	0.6331	0.6278
w_B	0.1720	0.1760	0.1900	0.2080	0.2141	0.2240	0.2308	0.2420	0.2530
w_C	0.1800	0.1762	0.1633	0.1495	0.1424	0.1360	0.1300	0.1249	0.1192
w_A	0.6196	0.6138	0.6037	0.5877	0.5602	0.5419	0.5057	0.4862	0.4527
w_B	0.2697	0.2810	0.3000	0.3314	0.3812	0.4081	0.4620	0.4864	0.5289
w_C	0.1107	0.1052	0.0963	0.0809	0.0586	0.0500	0.0323	0.0274	0.0184

continued

continued

Type of data: coexistence data (tie lines)

Total system			Top phase			Bottom phase		
w_A	w_B	w_C	w_A	w_B	w_C	w_A	w_B	w_C
T/K = 298.15								
0.58	0.20	0.22	0.5444	0.3330	0.1226	0.5823	0.1740	0.2437
0.56	0.22	0.22	0.5283	0.3630	0.1087	0.5715	0.1560	0.2725
0.53	0.25	0.22	0.4975	0.4350	0.0675	0.5472	0.1560	0.2968
0.50	0.28	0.22	0.4644	0.4870	0.0486	0.5240	0.1560	0.3200
T/K = 303.15								
0.58	0.20	0.22	0.5547	0.3550	0.0903	0.6069	0.1620	0.2311
0.56	0.22	0.22	0.5371	0.3860	0.0769	0.5974	0.1320	0.2706
0.53	0.25	0.22	0.5119	0.4290	0.0591	0.5823	0.1280	0.2897
T/K = 308.15								
0.60	0.18	0.22	0.5547	0.3550	0.0903	0.6064	0.1380	0.2556
0.58	0.20	0.22	0.5371	0.3860	0.0769	0.5926	0.1340	0.2734
0.56	0.22	0.22	0.5119	0.4290	0.0591	0.5782	0.1280	0.2938
T/K = 313.15								
0.62	0.16	0.22	0.5682	0.3380	0.0938	0.6175	0.1412	0.2413
0.60	0.18	0.22	0.5370	0.3940	0.0690	0.5997	0.1402	0.2601
0.58	0.20	0.22	0.4978	0.4600	0.0422	0.5802	0.1377	0.2821
0.56	0.22	0.22	0.4807	0.4860	0.0333	0.5723	0.1368	0.2909
T/K = 318.15								
0.64	0.14	0.22	0.5791	0.3400	0.0809	0.6443	0.1170	0.2387
0.62	0.16	0.22	0.5620	0.3740	0.0640	0.6291	0.1150	0.2559
0.60	0.18	0.22	0.5389	0.4130	0.0481	0.6206	0.0990	0.2804
0.58	0.20	0.22	0.5138	0.4510	0.0352	0.6082	0.1030	0.2888
0.56	0.22	0.22	0.4842	0.4870	0.0288	0.5968	0.1000	0.3032

Polymer (B): **poly(ethylene glycol)** **2008AMA**
Characterization: M_n/g.mol^{-1} = 4000, Merck-Schuchardt, Germany
Solvent (A): **water** H_2O **7732-18-5**
Salt (C): **diammonium phosphate** $(NH_4)_2HPO_4$ **7783-28-0**

Type of data: cloud points

T/K = 298.15

w_A	0.6061	0.6240	0.6549	0.6682	0.6802	0.6912	0.7012	0.7103	0.7186
w_B	0.3713	0.3515	0.3162	0.3004	0.2862	0.2732	0.2603	0.2486	0.2388
w_C	0.0226	0.0245	0.0289	0.0314	0.0336	0.0356	0.0385	0.0411	0.0426

continued

continued

w_A	0.7263	0.7333	0.7397	0.7512	0.7610	0.7655	0.7695	0.7801	0.7833
w_B	0.2289	0.2191	0.2108	0.1953	0.1820	0.1757	0.1699	0.1542	0.1497
w_C	0.0448	0.0476	0.0495	0.0535	0.0570	0.0588	0.0606	0.0657	0.0670
w_A	0.7888	0.7937	0.7984	0.8002	0.8040	0.8057	0.8116	0.8168	0.8224
w_B	0.1412	0.1333	0.1267	0.1232	0.1171	0.1143	0.1057	0.0970	0.0875
w_C	0.0700	0.0730	0.0749	0.0766	0.0789	0.0800	0.0827	0.0862	0.0901
w_A	0.8292	0.8361	0.8415	0.8458	0.8504	0.8539	0.8567	0.8579	0.8573
w_B	0.0763	0.0648	0.0550	0.0463	0.0369	0.0288	0.0212	0.0155	0.0120
w_C	0.0945	0.0991	0.1035	0.1079	0.1127	0.1173	0.1221	0.1266	0.1307

T/K = 303.15

w_A	0.6051	0.6396	0.6679	0.6911	0.7105	0.7267	0.7404	0.7520	0.7619
w_B	0.3753	0.3369	0.3043	0.2764	0.2522	0.2319	0.2140	0.1980	0.1843
w_C	0.0196	0.0235	0.0278	0.0325	0.0373	0.0414	0.0456	0.0500	0.0538
w_A	0.7706	0.7782	0.7847	0.7906	0.7956	0.8001	0.8078	0.8142	0.8217
w_B	0.1724	0.1619	0.1523	0.1437	0.1358	0.1287	0.1166	0.1065	0.0942
w_C	0.0570	0.0599	0.0630	0.0657	0.0686	0.0712	0.0756	0.0793	0.0841
w_A	0.8259	0.8296	0.8311	0.8379	0.8400	0.8436	0.8452	0.8534	0.8595
w_B	0.0874	0.0816	0.0789	0.0680	0.0644	0.0582	0.0554	0.0394	0.0248
w_C	0.0867	0.0888	0.0900	0.0941	0.0956	0.0982	0.0994	0.1072	0.1157

T/K = 308.15

w_A	0.5842	0.6226	0.6542	0.6800	0.7014	0.7192	0.7343	0.7472	0.7579
w_B	0.4001	0.3586	0.3219	0.2920	0.2661	0.2436	0.2250	0.2083	0.1932
w_C	0.0157	0.0188	0.0239	0.0280	0.0325	0.0372	0.0407	0.0445	0.0489
w_A	0.7674	0.7830	0.7892	0.8000	0.8087	0.8184	0.8237	0.8303	0.8330
w_B	0.1806	0.1593	0.1502	0.1348	0.1221	0.1067	0.0983	0.0882	0.0818
w_C	0.0520	0.0577	0.0606	0.0652	0.0692	0.0749	0.0780	0.0815	0.0852
w_A	0.8377	0.8439	0.8568	0.8637					
w_B	0.0745	0.0649	0.0407	0.0259					
w_C	0.0878	0.0912	0.1025	0.1104					

T/K = 318.15

w_A	0.6033	0.6388	0.6675	0.6913	0.7280	0.7546	0.7650	0.7824	0.7963
w_B	0.3833	0.3432	0.3107	0.2827	0.2386	0.2058	0.1921	0.1694	0.1519
w_C	0.0134	0.0180	0.0218	0.0260	0.0334	0.0396	0.0429	0.0482	0.0518
w_A	0.8069	0.8155	0.8223	0.8335	0.8398	0.8448	0.8489	0.8546	0.8626
w_B	0.1368	0.1242	0.1136	0.0972	0.0877	0.0798	0.0731	0.0642	0.0498
w_C	0.0563	0.0603	0.0641	0.0693	0.0725	0.0754	0.0780	0.0812	0.0876
w_A	0.8722	0.8773	0.8781	0.8781					
w_B	0.0335	0.0224	0.0161	0.0102					
w_C	0.0943	0.1003	0.1058	0.1117					

continued

continued

Type of data: coexistence data (tie lines)

Top phase			Bottom phase		
w_A	w_B	w_C	w_A	w_B	w_C
T/K = 303.15					
0.6848	0.2858	0.0294	0.8188	0.0180	0.1632
0.6362	0.3466	0.0172	0.8077	0.0005	0.1918
0.6140	0.3718	0.0142	0.7881	0.0025	0.2094
T/K = 308.15					
0.6761	0.2976	0.0263	0.8262	0.0122	0.1616
0.6571	0.3177	0.0252	0.8179	0.0107	0.1714
0.6128	0.3652	0.0220	0.7905	0.0109	0.1986
T/K = 313.15					
0.7043	0.2672	0.0285	0.8370	0.0114	0.1516
0.6466	0.3310	0.0224	0.8204	0.0096	0.1700
0.6247	0.3555	0.0198	0.8114	0.0088	0.1798

Polymer (B): **poly(ethylene glycol)** **2011KHA**
Characterization: M_n/g.mol^{-1} = 4000
Solvent (A): **water** H_2O **7732-18-5**
Salt (C): **diammonium phosphate** $(NH_4)_2HPO_4$ **7783-28-0**

Type of data: coexistence data (tie lines)

T/K = 298.15

Top phase			Bottom phase		
w_A	w_B	w_C	w_A	w_B	w_C
0.7062	0.2489	0.0449	0.7320	0.0578	0.2102
0.6303	0.3452	0.0245	0.6698	0.0479	0.2823
0.6082	0.3699	0.0219	0.6658	0.0441	0.2901
0.5717	0.4098	0.0185	0.6210	0.0269	0.3521

Polymer (B): **poly(ethylene glycol)** **2008ZA4**
Characterization: M_n/g.mol^{-1} = 4230, Merck, Darmstadt, Germany
Solvent (A): **water** H_2O **7732-18-5**
Salt (C): **dipotassium oxalate** $C_2K_2O_4$ **583-52-8**

continued

continued

Type of data: cloud points

T/K = 298.15

w_A	0.7097	0.7154	0.7225	0.7319	0.7387	0.7455	0.7554	0.7645	0.7756
w_B	0.2138	0.2058	0.1960	0.1840	0.1734	0.1618	0.1476	0.1322	0.1139
w_C	0.0765	0.0788	0.0815	0.0841	0.0879	0.0927	0.0970	0.1033	0.1105
w_A	0.7870	0.7989	0.8088	0.8154	0.8223	0.8268			
w_B	0.0951	0.0752	0.0580	0.0446	0.0321	0.0226			
w_C	0.1179	0.1259	0.1332	0.1400	0.1456	0.1506			

Type of data: coexistence data (tie lines)

T/K = 298.15

Top phase			Bottom phase		
w_A	w_B	w_C	w_A	w_B	w_C
0.6930	0.2356	0.0714	0.8258	0.0144	0.1598
0.6394	0.3128	0.0478	0.8040	0.0127	0.1833
0.6043	0.3584	0.0373	0.7853	0.0109	0.2038
0.5700	0.3999	0.0301	0.7636	0.0096	0.2268
0.5394	0.4388	0.0218	0.7422	0.0076	0.2502
0.5287	0.4524	0.0189	0.7304	0.0064	0.2632

Polymer (B): **poly(ethylene glycol)** **2004HAG**
Characterization: M_n/g.mol^{-1} = 1500, Merck, Darmstadt, Germany
Solvent (A): **water** **H_2O** **7732-18-5**
Salt (C): **dipotassium phosphate** **K_2HPO_4** **7758-11-4**

Type of data: coexistence data (tie lines)

T/K = 298.15 pH = 9.1

Total system			Top phase			Bottom phase		
w_A	w_B	w_C	w_A	w_B	w_C	w_A	w_B	w_C
0.7486	0.1399	0.1115	0.6587	0.3038	0.0375	0.8014	0.0401	0.1585
0.7574	0.1399	0.1027	0.6835	0.2723	0.0442	0.8063	0.0489	0.1448
0.7382	0.1600	0.1018	0.6657	0.2958	0.0385	0.8067	0.0493	0.1440
0.7664	0.1400	0.0936	0.7318	0.2111	0.0571	0.8084	0.0754	0.1162
0.7475	0.1600	0.0925	0.6899	0.2646	0.0455	0.8119	0.0519	0.1362
0.7374	0.1799	0.0827	0.6907	0.2625	0.0468	0.8122	0.0513	0.1365
0.7383	0.1801	0.0816	0.6657	0.2964	0.0379	0.8114	0.0400	0.1486
0.7186	0.1998	0.0816	0.6624	0.2969	0.0407	0.8167	0.0288	0.1545

Polymer (B): **poly(ethylene glycol)** **2004HAG**
Characterization: M_n/g.mol^{-1} = 4000, Merck, Darmstadt, Germany
Solvent (A): **water** H_2O **7732-18-5**
Salt (C): **dipotassium phosphate** K_2HPO_4 **7758-11-4**

Type of data: coexistence data (tie lines)

Total system			Top phase			Bottom phase		
w_A	w_B	w_C	w_A	w_B	w_C	w_A	w_B	w_C
T/K = 298.15	pH = 7.2 (prepared by a mixture of dibasic to monobasic potassium phosphate)							
0.8028	0.0887	0.1085	0.7175	0.2364	0.0461	0.8381	0.0272	0.1347
0.8014	0.0954	0.1032	0.7267	0.2236	0.0497	0.8438	0.0260	0.1302
0.8096	0.0957	0.0947	0.7507	0.1842	0.0651	0.8433	0.0385	0.1182
0.7968	0.1090	0.0942	0.7410	0.2051	0.0539	0.8460	0.0230	0.1310
0.8013	0.1090	0.0897	0.7659	0.1727	0.0614	0.8404	0.0435	0.1161
0.7924	0.1227	0.0849	0.7530	0.1892	0.0578	0.8440	0.0385	0.1175
T/K = 298.15	pH = 9.1							
0.7674	0.1249	0.1077	0.6733	0.2955	0.0312	0.8349	0.0077	0.1574
0.7809	0.1117	0.1074	0.6797	0.2878	0.0325	0.8385	0.0133	0.1482
0.7760	0.1266	0.0974	0.6914	0.2724	0.0362	0.8448	0.0121	0.1431
0.7894	0.1134	0.0972	0.7037	0.2571	0.0392	0.8454	0.0148	0.1398
0.7961	0.1136	0.0903	0.7215	0.2283	0.0502	0.8524	0.0154	0.1322
0.7848	0.1282	0.0870	0.7187	0.2384	0.0429	0.8472	0.0223	0.1305
0.7986	0.1144	0.0870	0.7314	0.2194	0.0492	0.8317	0.0403	0.1280
0.7708	0.1424	0.0868	0.7060	0.2553	0.0387	0.8456	0.0150	0.1394
T/K = 298.15	pH = 10.8 (prepared by addition of sodium hydroxide)							
0.7600	0.1200	0.1200	0.6678	0.3110	0.0212	0.8035	0.0262	0.1703
0.7550	0.1350	0.1100	0.6568	0.3186	0.0246	0.8176	0.0151	0.1673
0.7647	0.1353	0.1000	0.6325	0.3416	0.0259	0.8277	0.0150	0.1573
0.7805	0.1197	0.0998	0.6228	0.3282	0.0490	0.8532	0.0210	0.1258
0.7901	0.1200	0.0899	0.6800	0.2684	0.0516	0.8759	0.0109	0.1132
0.7752	0.1350	0.0898	0.6953	0.2593	0.0454	0.8427	0.0099	0.1474

Polymer (B): **poly(ethylene glycol)** **2008YAN**
Characterization: M_w/g.mol^{-1} = 10000, M_w/M_n = 1.05,
Fluka AG, Buchs, Switzerland
Solvent (A): **water** H_2O **7732-18-5**
Salt (C): **dipotassium phosphate** K_2HPO_4 **7758-11-4**

continued

continued

Type of data: cloud points

T/K = 298.15

w_A	0.59693	0.63215	0.64683	0.67248	0.68392	0.70191	0.71562	0.72662	0.74114
w_B	0.38480	0.34430	0.32710	0.29940	0.28650	0.26530	0.24810	0.23480	0.21770
w_C	0.01827	0.02355	0.02607	0.02812	0.02958	0.03279	0.03628	0.03858	0.04116
w_A	0.75763	0.77718	0.78413	0.79327	0.80073	0.81438	0.82680	0.83841	0.85288
w_B	0.19670	0.17300	0.16310	0.15130	0.14080	0.12270	0.10580	0.08919	0.06859
w_C	0.04567	0.04982	0.05277	0.05543	0.05847	0.06292	0.06740	0.07240	0.07853
w_A	0.86001	0.87183	0.88366	0.88597	0.88821	0.88779	0.88592	0.85155	0.81331
w_B	0.05615	0.03959	0.02075	0.01293	0.00769	0.00711	0.00148	0.00095	0.00069
w_C	0.08384	0.08858	0.09559	0.10110	0.10410	0.10510	0.11260	0.14750	0.18600
w_A	0.78029	0.75651	0.71865						
w_B	0.00051	0.00039	0.00025						
w_C	0.21920	0.24310	0.28110						

Polymer (B): **poly(ethylene glycol)** **2006MOH**
Characterization: M_n/g.mol^{-1} = 10000, M_w = 11000, Merck, Darmstadt, Germany
Solvent (A): **water** **H_2O** **7732-18-5**
Salt (C): **dipotassium phosphate** **K_2HPO_4** **7758-11-4**

Type of data: cloud points

w_B/w_A = 0.1222 was kept constant

w_C	0.0520	0.0577	0.0638	0.0696	0.0731	0.0780
T/K	322.8	315.2	306.0	298.9	293.7	278.0

w_B/w_A = 0.1117 was kept constant

w_C	0.0551	0.0602	0.0641	0.0665	0.0708	0.0750
T/K	325.7	318.5	313.6	309.7	303.9	298.3

Polymer (B): **poly(ethylene glycol)** **2008YAN**
Characterization: M_w/g.mol^{-1} = 20000, M_w/M_n = 1.05, Fluka AG, Buchs, Switzerland
Solvent (A): **water** **H_2O** **7732-18-5**
Salt (C): **dipotassium phosphate** **K_2HPO_4** **7758-11-4**

Type of data: cloud points

T/K = 298.15

w_A	0.60370	0.62681	0.64598	0.66614	0.68845	0.71240	0.71812	0.72728	0.74227
w_B	0.37690	0.35180	0.32960	0.30660	0.27960	0.25200	0.24590	0.23450	0.21640
w_C	0.01940	0.02139	0.02442	0.02726	0.03195	0.03560	0.03598	0.03822	0.04133
w_A	0.76174	0.77516	0.79033	0.79930	0.81389	0.82545	0.83783	0.85077	0.86071
w_B	0.19420	0.17760	0.15880	0.14790	0.12840	0.11350	0.09690	0.07898	0.06491
w_C	0.04406	0.04724	0.05087	0.05280	0.05771	0.06105	0.06527	0.07025	0.07438

continued

continued

w_A	0.87120	0.87592	0.88205	0.88552	0.89726	0.89693	0.89224	0.88557	0.84297
w_B	0.05070	0.04327	0.03513	0.02887	0.00834	0.00392	0.00112	0.00087	0.00069
w_C	0.07810	0.08081	0.08282	0.08561	0.09440	0.09915	0.10664	0.11356	0.15634
w_A	0.81142	0.79544	0.74940						
w_B	0.00060	0.00054	0.00039						
w_C	0.18798	0.20402	0.25021						

Polymer (B): **poly(ethylene glycol)** **2004SHA**
Characterization: M_n/g.mol^{-1} = 4900-5100,
Guangzhou Chemical Reagent Corporation, China
Solvent (A): **water** H_2O **7732-18-5**
Salt (C): **dipotassium phosphate/monopotassium phosphate**
K_2HPO_4/H_2KPO_4 **7758-11-4/7778-77-0**

Type of data: coexistence data (tie lines)

T/K = 298.15

Total system			Top phase				Bottom phase			
w_A	w_B	w_C	w_A	w_B	w_C	pH	w_A	w_B	w_C	pH
0.7876	0.1080	0.1044	0.6934	0.2324	0.0742	6.90	0.8394	0.0354	0.1252	6.70
0.7489	0.1458	0.1053	0.6896	0.2272	0.0832	6.93	0.8211	0.0458	0.1331	6.92
0.7533	0.1428	0.1039	0.6317	0.3038	0.0645	7.54	0.8222	0.0523	0.1255	7.39
0.7118	0.1705	0.1177	0.5345	0.4125	0.0530	8.10	0.7968	0.0456	0.1576	7.85
0.7041	0.1902	0.1057	0.5478	0.3941	0.0581	8.28	0.8210	0.0366	0.1424	7.98

Polymer (B): **poly(ethylene glycol)** **2004SHA**
Characterization: M_n/g.mol^{-1} = 5500-6500,
Guangzhou Chemical Reagent Corporation, China
Solvent (A): **water** H_2O **7732-18-5**
Salt (C): **dipotassium phosphate/monopotassium phosphate**
K_2HPO_4/H_2KPO_4 **7758-11-4/7778-77-0**

Type of data: coexistence data (tie lines)

T/K = 298.15

Total system			Top phase				Bottom phase			
w_A	w_B	w_C	w_A	w_B	w_C	pH	w_A	w_B	w_C	pH
0.7756	0.1131	0.1113	0.6901	0.2323	0.0776	6.70	0.8359	0.0296	0.1345	6.70
0.7751	0.1131	0.1118	0.6849	0.2387	0.0764	6.95	0.8354	0.0322	0.1324	6.90
0.7701	0.1345	0.0954	0.6630	0.2757	0.0613	7.48	0.8343	0.0471	0.1186	7.34
0.7266	0.1537	0.1197	0.5379	0.4136	0.0485	7.78	0.8067	0.0370	0.1563	7.72
0.7331	0.1698	0.0971	0.5792	0.3676	0.0532	7.90	0.8342	0.0366	0.1292	7.91

Polymer (B): **poly(ethylene glycol)** **2004SHA**
Characterization: M_n/g.mol^{-1} = 9000-11000,
Guangzhou Chemical Reagent Corporation, China
Solvent (A): **water** **H_2O** **7732-18-5**
Salt (C): **dipotassium phosphate/monopotassium phosphate**
K_2HPO_4/H_2KPO_4 **7758-11-4/7778-77-0**

Type of data: coexistence data (tie lines)

T/K = 298.15

Total system			Top phase				Bottom phase			
w_A	w_B	w_C	w_A	w_B	w_C	pH	w_A	w_B	w_C	pH
0.7648	0.1154	0.1198	0.6731	0.2317	0.0952	6.76	0.8219	0.0403	0.1378	6.79
0.7709	0.1362	0.0929	0.7114	0.2185	0.0701	7.02	0.8395	0.0444	0.1161	6.94
0.7958	0.1077	0.0965	0.6931	0.2506	0.0563	7.56	0.8443	0.0412	0.1145	7.41
0.7424	0.1633	0.0943	0.6118	0.3407	0.0475	7.99	0.8342	0.0358	0.1300	7.79
0.7379	0.1710	0.0911	0.5865	0.3653	0.0482	8.10	0.8429	0.0331	0.1240	7.96

Polymer (B): **poly(ethylene glycol)** **2008YAN**
Characterization: M_w/g.mol^{-1} = 10000, M_w/M_n = 1.05,
Fluka AG, Buchs, Switzerland
Solvent (A): **water** **H_2O** **7732-18-5**
Salt (C): **dipotassium phosphate/monopotassium phosphate**
K_2HPO_4/H_2KPO_4 **7758-11-4/7778-77-0**

Comments: The salt mixture is composed of each 50% dipotassium phosphate or monopotassium phosphate.

Type of data: cloud points

T/K = 298.15

w_A	0.59347	0.61415	0.63111	0.64721	0.66379	0.67604	0.68744	0.69980	0.71448
w_B	0.38380	0.36080	0.34100	0.32230	0.30300	0.28700	0.27240	0.25840	0.24090
w_C	0.02273	0.02505	0.02789	0.03049	0.03321	0.03696	0.04016	0.04180	0.04462
w_A	0.72481	0.74519	0.75349	0.76150	0.77558	0.79631	0.81094	0.81801	0.82470
w_B	0.22760	0.20200	0.19010	0.17670	0.15760	0.12850	0.10720	0.09572	0.08449
w_C	0.04759	0.05281	0.05641	0.06180	0.06682	0.07519	0.08186	0.08627	0.09081
w_A	0.83103	0.84116	0.84894	0.85081	0.85713	0.86056	0.86343	0.85472	0.83964
w_B	0.07115	0.05524	0.04036	0.03469	0.02117	0.01564	0.00807	0.00228	0.00096
w_C	0.09782	0.10360	0.11070	0.11450	0.12170	0.12380	0.12850	0.14300	0.15940
w_A	0.77130	0.74674	0.73133						
w_B	0.00060	0.00046	0.00027						
w_C	0.22810	0.25280	0.26840						

Polymer (B): **poly(ethylene glycol)** **2008YAN**
Characterization: M_w/g.mol^{-1} = 20000, M_w/M_n = 1.05,
Fluka AG, Buchs, Switzerland
Solvent (A): **water** **H_2O** **7732-18-5**
Salt (C): **dipotassium phosphate/monopotassium phosphate**
K_2HPO_4/H_2KPO_4 **7758-11-4/7778-77-0**

Comments: The salt mixture is composed of each 50% dipotassium phosphate or monopotassium phosphate.

Type of data: cloud points

T/K = 298.15

w_A	0.61777	0.63430	0.66527	0.68250	0.70715	0.72026	0.73518	0.74177	0.75302
w_B	0.35460	0.33620	0.29980	0.27990	0.24920	0.23380	0.21450	0.20640	0.19240
w_C	0.02763	0.02950	0.03493	0.03760	0.04365	0.04594	0.05032	0.05183	0.05458
w_A	0.76358	0.77800	0.78706	0.79562	0.80221	0.81286	0.82022	0.82498	0.83132
w_B	0.17840	0.16010	0.14780	0.13650	0.12690	0.11190	0.10100	0.09246	0.08261
w_C	0.05802	0.06190	0.06514	0.06788	0.07089	0.07524	0.07878	0.08256	0.08607
w_A	0.83711	0.84310	0.85392	0.86336	0.87008	0.87345	0.87255	0.86787	0.84222
w_B	0.07327	0.06421	0.04853	0.03314	0.02182	0.00965	0.00495	0.00143	0.00098
w_C	0.08962	0.09269	0.09755	0.10350	0.10810	0.11690	0.12250	0.13070	0.15680
w_A	0.82712	0.78855	0.76721						
w_B	0.00088	0.00065	0.00049						
w_C	0.17200	0.21080	0.23230						

Polymer (B): **poly(ethylene glycol)** **2008ZA4**
Characterization: M_n/g.mol^{-1} = 4230, Merck, Darmstadt, Germany
Solvent (A): **water** **H_2O** **7732-18-5**
Salt (C): **dipotassium tartrate** **$C_4H_4K_2O_6$** **6100-19-2**

Type of data: cloud points

T/K = 298.15

w_A	0.6536	0.6621	0.6686	0.6787	0.6926	0.7074	0.7201	0.7331	0.7406
w_B	0.2677	0.2562	0.2468	0.2330	0.2124	0.1909	0.1703	0.1481	0.1291
w_C	0.0787	0.0817	0.0846	0.0883	0.0950	0.1017	0.1096	0.1188	0.1303
w_A	0.7499	0.7599	0.7600	0.7643	0.7694	0.7736	0.7759	0.7777	0.7801
w_B	0.1104	0.0934	0.0859	0.0750	0.0646	0.0560	0.0480	0.0403	0.0332
w_C	0.1397	0.1467	0.1541	0.1607	0.1660	0.1704	0.1761	0.1820	0.1867
w_A	0.7812	0.7814	0.7818	0.7800	0.7783				
w_B	0.0310	0.0267	0.0226	0.0180	0.0120				
w_C	0.1878	0.1919	0.1956	0.2020	0.2097				

continued

continued

Type of data: coexistence data (tie lines)

T/K = 298.15

Top phase			Bottom phase		
w_A	w_B	w_C	w_A	w_B	w_C
0.6273	0.3028	0.0699	0.7616	0.0090	0.2294
0.5893	0.3524	0.0583	0.7202	0.0073	0.2725
0.5550	0.3907	0.0543	0.6821	0.0050	0.3129
0.5133	0.4375	0.0492	0.6468	0.0042	0.3490
0.4670	0.4897	0.0433	0.6118	0.0022	0.3860
0.3759	0.5881	0.0360	0.5539	0.0018	0.4443

Polymer (B): **poly(ethylene glycol)** **2010ZA2**
Characterization: M_n/g.mol^{-1} = 2000, Merck, Darmstadt, Germany
Solvent (A): **water** **H_2O** **7732-18-5**
Salt (C): **disodium phosphate** **Na_2HPO_4** **7558-79-4**

Type of data: cloud points

T/K = 298.15

w_A	0.6100	0.6456	0.6722	0.7110	0.7317	0.7405	0.7475	0.7528	0.7576
w_B	0.3797	0.3391	0.3089	0.2626	0.2372	0.2261	0.2172	0.2104	0.2041
w_C	0.0103	0.0153	0.0189	0.0264	0.0311	0.0334	0.0353	0.0368	0.0383
w_A	0.7608	0.7648	0.7680	0.7708	0.7761				
w_B	0.1997	0.1945	0.1902	0.1864	0.1794				
w_C	0.0395	0.0407	0.0418	0.0428	0.0445				

T/K = 318.15

w_A	0.7835	0.7653	0.7448	0.7179	0.6996	0.6790	0.6603	0.6405	0.6138
w_B	0.2030	0.2196	0.2376	0.2626	0.2788	0.2969	0.3129	0.3286	0.3520
w_C	0.0135	0.0151	0.0176	0.0195	0.0216	0.0241	0.0268	0.0309	0.0342
w_A	0.5942								
w_B	0.3689								
w_C	0.0369								

Polymer (B): **poly(ethylene glycol)** **2005HEY**
Characterization: M_n/g.mol^{-1} = 7264, DP = 165 EO-units, Aldrich Chem. Co., Inc., Milwaukee, WI
Solvent (A): **water** **H_2O** **7732-18-5**
Salt (C): **disodium phosphate** **Na_2HPO_4** **7558-79-4**

continued

continued

Type of data: cloud points

T/K = 298.15

z_B/(base mol/kg H_2O)	10.2	7.67	6.16	4.76	3.19	3.15	2.45
c_C/(mol/kg H_2O)	0.170	0.213	0.266	0.313	0.350	0.390	0.430
z_B/(base mol/kg H_2O)	1.84	1.14	0.482	0.313	0.774	0.116	0.170
c_C/(mol/kg H_2O)	0.468	0.519	0.590	0.614	0.490	0.670	0.650

Type of data: coexistence data (tie lines)

T/K = 298.15

Top phase		Bottom phase	
z_B/ base mol/kg H_2O	c_C/ mol/kg H_2O	z_B/ base mol/kg H_2O	c_C/ mol/kg H_2O
3.743	0.352	0.910	0.539
4.061	0.337	0.613	0.560
4.534	0.317	0.448	0.582
4.686	0.311	0.275	0.616
5.203	0.292	0.200	0.638
5.561	0.280	0.188	0.643
5.788	0.273	0.152	0.659

Comments: Concentrations are molal, i.e., per kg water; base moles are calculated with the molar mass of 44 g/mol per EO-unit.

Polymer (B): **poly(ethylene glycol)** **2006MOH**
Characterization: M_n/g.mol^{-1} = 10000, M_w = 11000, Merck, Darmstadt, Germany
Solvent (A): **water** **H_2O** **7732-18-5**
Salt (C): **disodium phosphate** **Na_2HPO_4** **7558-79-4**

Type of data: cloud points

w_B/w_A = 0.1114 was kept constant

w_C	0.0444	0.0459	0.0490	0.0532	0.0562	0.0600
T/K	322.6	320.0	314.7	306.5	301.5	291.2

Polymer (B): **poly(ethylene glycol)** **2011PAT**
Characterization: M_w/g.mol^{-1} = 1500, Sigma Chemical Co., Inc., St. Louis, MO
Solvent (A): **water** **H_2O** **7732-18-5**
Salt (C): **disodium succinate** **$C_4H_4Na_2O_4$** **150-90-3**

Type of data: coexistence data (tie lines)

continued

continued

Total system			Top phase			Bottom phase		
w_A	w_B	w_C	w_A	w_B	w_C	w_A	w_B	w_C
T/K = 298.15								
0.6392	0.2318	0.1290	0.5746	0.3374	0.0879	0.7545	0.0370	0.2084
0.6241	0.2425	0.1334	0.5479	0.3699	0.0822	0.7511	0.0360	0.2129
0.6189	0.2470	0.1341	0.5393	0.3816	0.0791	0.7494	0.0344	0.2161
0.6088	0.2531	0.1381	0.5218	0.4053	0.0728	0.7431	0.0206	0.2364
T/K = 313.15								
0.6608	0.2154	0.1237	0.5113	0.4099	0.0789	0.7712	0.0735	0.1553
0.6447	0.2291	0.1262	0.4840	0.4407	0.0753	0.7692	0.0632	0.1676
0.6192	0.2507	0.1301	0.4524	0.4788	0.0688	0.7642	0.0510	0.1849
0.5978	0.2694	0.1328	0.3967	0.5438	0.0595	0.7610	0.0459	0.1931

Polymer (B): **poly(ethylene glycol)** **2005ZA3**
Characterization: M_n/g.mol^{-1} = 5890, Merck, Darmstadt, Germany
Solvent (A): **water** H_2O **7732-18-5**
Salt (C): **disodium succinate** $C_4H_4Na_2O_4$ **150-90-3**

Type of data: cloud points

T/K = 298.15

w_A	0.4889	0.5251	0.5560	0.5751	0.5784	0.5953	0.6246	0.6542	0.6833
w_B	0.4750	0.4348	0.3978	0.3743	0.3701	0.3497	0.3142	0.2764	0.2383
w_C	0.0361	0.0401	0.0462	0.0506	0.0515	0.0550	0.0612	0.0694	0.0784
w_A	0.7123	0.7293	0.7422	0.7475	0.7699	0.7778	0.7851	0.7975	0.8063
w_B	0.1992	0.1746	0.1552	0.1325	0.1151	0.1018	0.0899	0.0697	0.0551
w_C	0.0885	0.0961	0.1026	0.1200	0.1150	0.1204	0.1250	0.1328	0.1386
w_A	0.8117	0.8168	0.8174	0.8194	0.8213	0.8230			
w_B	0.0456	0.0371	0.0359	0.0313	0.0285	0.0240			
w_C	0.1427	0.1461	0.1467	0.1493	0.1502	0.1530			

Type of data: coexistence data (tie lines)

T/K = 298.15

Top phase			Bottom phase		
w_A	w_B	w_C	w_A	w_B	w_C
0.4945	0.4649	0.0406	0.7550	0.0038	0.2412
0.5117	0.4464	0.0419	0.7633	0.0034	0.2333
0.5658	0.3850	0.0492	0.7826	0.0064	0.2110
0.6042	0.3390	0.0568	0.7999	0.0026	0.1975
0.6436	0.2901	0.0663	0.8143	0.0038	0.1819

Polymer (B): **poly(ethylene glycol)** **2011PAT**
Characterization: M_w/g.mol^{-1} = 1500, Sigma Chemical Co., Inc., St. Louis, MO
Solvent (A): **water** **H_2O** **7732-18-5**
Salt (C): **disodium tartrate** **$C_4H_4Na_2O_6$** **868-18-8**

Type of data: coexistence data (tie lines)

Total system			Top phase			Bottom phase		
w_A	w_B	w_C	w_A	w_B	w_C	w_A	w_B	w_C
T/K = 283.15								
0.6797	0.2007	0.1196	0.6391	0.3129	0.0480	0.7565	0.0289	0.2146
0.6674	0.2109	0.1217	0.6211	0.3353	0.0436	0.7419	0.0178	0.2403
0.6513	0.2224	0.1264	0.6075	0.3537	0.0387	0.7381	0.0116	0.2503
0.6295	0.2389	0.1316	0.5818	0.3870	0.0312	0.7071	0.0086	0.2843
0.6035	0.2589	0.1376	0.5469	0.4248	0.0282	0.7023	0.0055	0.2922
T/K = 298.15								
0.7028	0.1832	0.1140	0.6478	0.2862	0.0660	0.7818	0.0273	0.1909
0.6724	0.2130	0.1146	0.6184	0.3265	0.0551	0.7703	0.0154	0.2143
0.6519	0.2286	0.1195	0.5805	0.3765	0.0430	0.7500	0.0090	0.2410
0.6179	0.2458	0.1363	0.5370	0.4292	0.0338	0.7195	0.0085	0.2720
0.6040	0.2550	0.1410	0.5259	0.4425	0.0315	0.7113	0.0071	0.2815
T/K = 313.15								
0.6637	0.1932	0.1431	0.6201	0.3180	0.0619	0.7366	0.0173	0.2461
0.6407	0.2114	0.1479	0.5753	0.3732	0.0515	0.7207	0.0089	0.2704
0.6145	0.2310	0.1545	0.5467	0.4093	0.0440	0.7022	0.0044	0.2934
0.5931	0.2470	0.1598	0.5112	0.4494	0.0394	0.6856	0.0026	0.3118
0.5746	0.2588	0.1666	0.4971	0.4686	0.0343	0.6689	0.0006	0.3305

Polymer (B): **poly(ethylene glycol)** **2008ZA2**
Characterization: M_n/g.mol^{-1} = 4230, Merck, Darmstadt, Germany
Solvent (A): **water** **H_2O** **7732-18-5**
Salt (C): **disodium tartrate** **$C_4H_4Na_2O_6$** **868-18-8**

Type of data: cloud points

T/K = 298.15

w_A	0.5172	0.5501	0.5910	0.6300	0.6610	0.6953	0.7203	0.7199	0.7372
w_B	0.4597	0.4212	0.3725	0.3245	0.2857	0.2413	0.2082	0.2082	0.1826
w_C	0.0231	0.0287	0.0365	0.0455	0.0533	0.0634	0.0715	0.0719	0.0802
w_A	0.7388	0.7535	0.7663	0.7769	0.7853	0.7937	0.8030	0.8128	0.8195
w_B	0.1805	0.1584	0.1385	0.1204	0.1056	0.0915	0.0750	0.0573	0.0440
w_C	0.0807	0.0881	0.0952	0.1027	0.1091	0.1148	0.1220	0.1299	0.1365

continued

continued

w_A	0.8250	0.8288	0.8300	0.8294
w_B	0.0303	0.0197	0.0110	0.0068
w_C	0.1447	0.1515	0.1590	0.1638

T/K = 308.15

w_A	0.4485	0.4948	0.5518	0.5988	0.6411	0.6760	0.7057	0.7274	0.7426
w_B	0.5320	0.4819	0.4188	0.3641	0.3137	0.2701	0.2321	0.2025	0.1812
w_C	0.0195	0.0233	0.0294	0.0371	0.0452	0.0539	0.0622	0.0701	0.0762
w_A	0.7608	0.7626	0.7788	0.7944	0.8096	0.8244			
w_B	0.1537	0.1516	0.1271	0.1029	0.0788	0.0551			
w_C	0.0855	0.0858	0.0941	0.1027	0.1116	0.1205			

T/K = 318.15

w_A	0.4906	0.5222	0.5517	0.5983	0.6346	0.6653	0.6931	0.7144	0.7356
w_B	0.4837	0.4504	0.4185	0.3663	0.3237	0.2864	0.2519	0.2242	0.1959
w_C	0.0257	0.0274	0.0298	0.0354	0.0417	0.0483	0.0550	0.0614	0.0685
w_A	0.7511	0.7675	0.7845	0.8021	0.8152	0.8292	0.8394	0.8469	0.8498
w_B	0.1746	0.1517	0.1269	0.1011	0.0812	0.0589	0.0418	0.0282	0.0164
w_C	0.0743	0.0808	0.0886	0.0968	0.1036	0.1119	0.1188	0.1249	0.1338

Type of data: coexistence data (tie lines)

Top phase			Bottom phase		
w_A	w_B	w_C	w_A	w_B	w_C
T/K = 298.15					
0.6911	0.2487	0.0602	0.8170	0.0092	0.1738
0.6680	0.2769	0.0551	0.8059	0.0088	0.1853
0.6445	0.3060	0.0495	0.7972	0.0087	0.1941
0.6100	0.3487	0.0413	0.7816	0.0079	0.2105
0.5701	0.3968	0.0331	0.7678	0.0070	0.2252
0.5214	0.4545	0.0241	0.7464	0.0063	0.2473
T/K = 308.15					
0.6367	0.3193	0.0440	0.8263	0.0093	0.1644
0.6010	0.3589	0.0401	0.8078	0.0076	0.1846
0.5856	0.3792	0.0352	0.8036	0.0067	0.1897
0.5347	0.4379	0.0274	0.7836	0.0059	0.2105
0.4979	0.4781	0.0240	0.7653	0.0056	0.2291
0.4683	0.5102	0.0215	0.7468	0.0051	0.2481

continued

continued

T/K = 318.15

0.6662	0.2842	0.0496	0.8359	0.0110	0.1531
0.5936	0.3707	0.0357	0.8111	0.0077	0.1812
0.5623	0.4064	0.0313	0.7984	0.0063	0.1953
0.5171	0.4559	0.0270	0.7855	0.0057	0.2088
0.4639	0.5104	0.0257	0.7657	0.0052	0.2291
0.4237	0.5512	0.0251	0.7448	0.0048	0.2504

Polymer (B): **poly(ethylene glycol)** **2008MA1**
Characterization: M_w/g.mol^{-1} = 1500, Sigma Chemical Co., Inc., St. Louis, MO
Solvent (A): **water** H_2O **7732-18-5**
Salt (C): **lithium sulfate** Li_2SO_4 **10377-48-7**

Type of data: coexistence data (tie lines)

Total system			Top phase			Bottom phase		
w_A	w_B	w_C	w_A	w_B	w_C	w_A	w_B	w_C
T/K = 283.15								
0.6586	0.2343	0.1071	0.5769	0.3763	0.0468	0.7535	0.0633	0.1832
0.6429	0.2463	0.1108	0.5534	0.4055	0.0411	0.7469	0.0592	0.1939
0.6260	0.2595	0.1145	0.5321	0.4325	0.0354	0.7378	0.0579	0.2043
0.6086	0.2728	0.1186	0.4936	0.4778	0.0286	0.7336	0.0527	0.2137
0.5818	0.2945	0.1237	0.4668	0.5094	0.0238	0.7082	0.0546	0.2372
T/K = 298.15								
0.6800	0.2241	0.0959	0.6026	0.3444	0.0530	0.7761	0.0685	0.1554
0.6560	0.2434	0.1006	0.5667	0.3918	0.0415	0.7655	0.0660	0.1685
0.6288	0.2646	0.1066	0.5235	0.4440	0.0325	0.7477	0.0591	0.1932
0.6065	0.2823	0.1112	0.4904	0.4838	0.0258	0.7369	0.0568	0.2063
0.5806	0.3031	0.1163	0.4459	0.5326	0.0215	0.7228	0.0560	0.2212
T/K = 313.15								
0.6800	0.2241	0.0959	0.5611	0.4024	0.0365	0.7869	0.0541	0.1590
0.6560	0.2434	0.1006	0.5182	0.4528	0.0290	0.7691	0.0523	0.1786
0.6288	0.2646	0.1066	0.4786	0.5012	0.0202	0.7583	0.0499	0.1918
0.6065	0.2823	0.1112	0.4415	0.5408	0.0177	0.7442	0.0464	0.2094
0.5806	0.3031	0.1163	0.4168	0.5679	0.0153	0.7370	0.0426	0.2204

Polymer (B): **poly(ethylene glycol)** **2003HUD**
Characterization: M_w/g.mol^{-1} = 2000, Aldrich Chem. Co., Inc., Milwaukee, WI
Solvent (A): **water** H_2O **7732-18-5**
Salt (C): **lithium sulfate** Li_2SO_4 **10377-48-7**

continued

continued

Type of data: cloud points

T/K = 298.15

w_A	0.6146	0.6607	0.7436	0.7850	0.8103	0.8155	0.8192	0.8207	0.8223
w_B	0.3420	0.2800	0.1700	0.1053	0.0656	0.0554	0.0512	0.0474	0.0445
w_C	0.0434	0.0593	0.0864	0.1097	0.1241	0.1291	0.1296	0.1319	0.1332
w_A	0.8240	0.8250	0.8259	0.8268	0.8275	0.8284	0.8289	0.8292	0.8313
w_B	0.0411	0.0387	0.0357	0.0340	0.0321	0.0304	0.0292	0.0272	0.0250
w_C	0.1349	0.1363	0.1384	0.1392	0.1404	0.1412	0.1419	0.1436	0.1437
w_A	0.8317	0.8295	0.8290	0.8186	0.7977				
w_B	0.0210	0.0164	0.0100	0.0084	0.0048				
w_C	0.1473	0.1541	0.1610	0.1730	0.1975				

Type of data: coexistence data (tie lines)

T/K = 298.15

Total system			Top phase			Bottom phase		
w_A	w_B	w_C	w_A	w_B	w_C	w_A	w_B	w_C
0.7216	0.1931	0.0853	0.6771	0.2584	0.0645	0.8207	0.0482	0.1311
0.7180	0.1922	0.0898	0.6391	0.3094	0.0515	0.8278	0.0290	0.1432
0.7094	0.1913	0.0993	0.6123	0.3450	0.0427	0.8247	0.0085	0.1668
0.7048	0.1900	0.1052	0.5806	0.3864	0.0330	0.8212	0.0059	0.1729

Polymer (B): **poly(ethylene glycol)** **2007CAR**
Characterization: M_w/g.mol^{-1} = 4000, Isofar, Rio de Janeiro, Brazil
Solvent (A): **water** **H_2O** **7732-18-5**
Salt (C): **lithium sulfate** **Li_2SO_4** **10377-48-7**

Type of data: coexistence data (tie lines)

Total system			Top phase			Bottom phase		
w_A	w_B	w_C	w_A	w_B	w_C	w_A	w_B	w_C
T/K = 278.15								
0.6882	0.2373	0.0745	0.6334	0.3282	0.0384	0.8199	0.0354	0.1447
0.6690	0.2485	0.0825	0.5944	0.3705	0.0351	0.8097	0.0381	0.1522
0.6350	0.2729	0.0921	0.5368	0.4427	0.0205	0.7726	0.0345	0.1929
0.6041	0.2957	0.1002	0.4904	0.4954	0.0142	0.7685	0.0324	0.1991
0.5758	0.3130	0.1112	0.4157	0.5754	0.0089	0.7363	0.0359	0.2278

continued

continued

T/K = 298.15

0.6838	0.2426	0.0736	0.6052	0.3596	0.0352	0.8268	0.0263	0.1469
0.6587	0.2598	0.0815	0.5620	0.4102	0.0278	0.8063	0.0270	0.1667
0.6254	0.2846	0.0900	0.5227	0.4548	0.0225	0.7854	0.0324	0.1822
0.5960	0.3048	0.0992	0.4779	0.5042	0.0179	0.7643	0.0435	0.1922
0.5496	0.3433	0.1071	0.4277	0.5603	0.0120	0.7343	0.0434	0.2223

T/K = 308.15

0.7034	0.2279	0.0687	0.6369	0.3290	0.0341	0.8399	0.0529	0.1072
0.6673	0.2553	0.0774	0.5744	0.4016	0.0240	0.8195	0.0268	0.1537
0.6379	0.2753	0.0868	0.5263	0.4547	0.0190	0.8090	0.0252	0.1658
0.6104	0.2950	0.0946	0.4868	0.4996	0.0136	0.7766	0.0384	0.1850
0.5765	0.3190	0.1045	0.4416	0.5470	0.0114	0.7515	0.0411	0.2074

T/K = 318.15

0.7021	0.2281	0.0698	0.5970	0.3801	0.0229	0.8626	0.0017	0.1357
0.6850	0.2420	0.0730	0.5719	0.4088	0.0193	0.8379	0.0084	0.1537
0.6586	0.2590	0.0824	0.5308	0.4517	0.0175	0.8170	0.0021	0.1809
0.6349	0.2745	0.0906	0.4903	0.4970	0.0127	0.7955	0.0049	0.1996
0.6109	0.2927	0.0964	0.4577	0.5317	0.0106	0.7773	0.0139	0.2088

Polymer (B): **poly(ethylene glycol)** **2004GRA**
Characterization: M_n/g.mol^{-1} = 4000, Merck, Darmstadt, Germany
Solvent (A): **water** H_2O **7732-18-5**
Salt (C): **lithium sulfate** Li_2SO_4 **10377-48-7**

Type of data: cloud points

T/K = 278.15

w_A	0.7358	0.7465	0.8082	0.8099	0.7908	0.7789	0.7661	0.7443	0.7191
w_B	0.0030	0.0046	0.0066	0.0242	0.0696	0.0892	0.1120	0.1482	0.1849
w_C	0.2612	0.2489	0.1852	0.1659	0.1396	0.1319	0.1219	0.1075	0.0960
w_A	0.6813	0.6590	0.5508	0.4864	0.4415				
w_B	0.2376	0.2678	0.4066	0.4865	0.5397				
w_C	0.0811	0.0732	0.0426	0.0271	0.0188				

T/K = 298.15

w_A	0.7529	0.7673	0.7990	0.8021	0.8143	0.8202	0.8270	0.8305	0.8352
w_B	0.0007	0.0008	0.0014	0.0020	0.0023	0.0033	0.0052	0.0084	0.0232
w_C	0.2464	0.2319	0.1996	0.1959	0.1834	0.1765	0.1678	0.1611	0.1416
w_A	0.8334	0.8277	0.8185	0.8108	0.7941	0.7704	0.7425	0.7054	0.6614
w_B	0.0302	0.0436	0.0657	0.0791	0.1031	0.1360	0.1744	0.2254	0.2835
w_C	0.1364	0.1287	0.1158	0.1101	0.1028	0.0936	0.0831	0.0692	0.0551

continued

continued

w_A	0.6197	0.5772	0.5340	0.4887	0.4425	0.3954
w_B	0.3337	0.3848	0.4369	0.4887	0.5408	0.5932
w_C	0.0466	0.0380	0.0291	0.0226	0.0167	0.0114

T/K = 318.15

w_A	0.7488	0.8549	0.8578	0.8523	0.8476	0.8186	0.7819	0.7614	0.7317
w_B	0.0027	0.0045	0.0120	0.0315	0.0476	0.0908	0.1430	0.1704	0.2107
w_C	0.2485	0.1406	0.1302	0.1162	0.1048	0.0906	0.0751	0.0682	0.0576

w_A	0.6801	0.6293	0.5857	0.5403	0.4934	0.4446	0.3975
w_B	0.2755	0.3389	0.3904	0.4420	0.4933	0.5458	0.5963
w_C	0.0444	0.0318	0.0239	0.0177	0.0133	0.0096	0.0062

Type of data: coexistence data (tie lines)

Total system			Top phase			Bottom phase		
w_A	w_B	w_C	w_A	w_B	w_C	w_A	w_B	w_C
T/K = 278.15								
0.5650	0.3000	0.1350	0.4465	0.5338	0.0197	0.7171	0.0000	0.2829
0.6040	0.2630	0.1330	0.4884	0.4836	0.0280	0.7418	0.0000	0.2582
0.6603	0.2130	0.1267	0.5470	0.4119	0.0411	0.7820	0.0000	0.2180
0.7080	0.1690	0.1230	0.6186	0.3213	0.0601	0.8072	0.0000	0.1928
T/K = 298.15								
0.5040	0.3690	0.1270	0.3915	0.5973	0.0112	0.6861	0.0000	0.3139
0.5530	0.3190	0.1280	0.4208	0.5653	0.0139	0.7243	0.0000	0.2757
0.6070	0.2750	0.1180	0.4830	0.4955	0.0215	0.7616	0.0000	0.2384
0.6710	0.2190	0.1100	0.5734	0.3900	0.0366	0.7961	0.0000	0.2039
T/K = 318.15								
0.4862	0.3898	0.1240	0.3648	0.6299	0.0053	0.6836	0.0000	0.3164
0.5520	0.3250	0.1230	0.4244	0.5676	0.0080	0.7230	0.0000	0.2770
0.5971	0.2999	0.1030	0.4594	0.5303	0.0103	0.7762	0.0000	0.2238
0.6770	0.2350	0.0880	0.5541	0.4263	0.0196	0.8280	0.0000	0.1720

Polymer (B): **poly(ethylene glycol)** **2008MA2**
Characterization: M_w/g.mol^{-1} = 6000, Sigma Chemical Co., Inc., St. Louis, MO
Solvent (A): **water** H_2O **7732-18-5**
Salt (C): **lithium sulfate** Li_2SO_4 **10377-48-7**

Type of data: coexistence data (tie lines)

continued

continued

Total system			Top phase			Bottom phase		
w_A	w_B	w_C	w_A	w_B	w_C	w_A	w_B	w_C
T/K = 283.15								
0.7002	0.2136	0.0862	0.6126	0.3552	0.0322	0.8085	0.0364	0.1551
0.6804	0.2291	0.0905	0.5768	0.3963	0.0269	0.7973	0.0396	0.1631
0.6537	0.2510	0.0953	0.5431	0.4340	0.0229	0.7802	0.0401	0.1797
0.6320	0.2682	0.0998	0.5139	0.4665	0.0196	0.7661	0.0447	0.1892
0.5997	0.2945	0.1058	0.4745	0.5093	0.0162	0.7438	0.0455	0.2107
T/K = 298.15								
0.6910	0.2254	0.0836	0.5930	0.3781	0.0289	0.8159	0.0310	0.1531
0.6715	0.2414	0.0871	0.5609	0.4141	0.0250	0.8003	0.0402	0.1595
0.6475	0.2604	0.0921	0.5297	0.4506	0.0197	0.7833	0.0414	0.1753
0.6233	0.2794	0.0973	0.5009	0.4796	0.0195	0.7700	0.0386	0.1914
0.5933	0.3033	0.1034	0.4567	0.5291	0.0142	0.7499	0.0442	0.2059
T/K = 313.15								
0.6809	0.2365	0.0826	0.5645	0.4129	0.0226	0.8150	0.0317	0.1533
0.6599	0.2530	0.0871	0.5256	0.4570	0.0174	0.8046	0.0340	0.1614
0.6363	0.2725	0.0912	0.5179	0.4651	0.0170	0.7852	0.0338	0.1810
0.6047	0.2983	0.0970	0.4615	0.5240	0.0145	0.7722	0.0348	0.1930
0.5802	0.3181	0.1017	0.4311	0.5576	0.0113	0.7518	0.0436	0.2046

Polymer (B): **poly(ethylene glycol)** **2008SI2**
Characterization: M_n/g.mol^{-1} = 8000, Sigma Chemical Co., Inc., St. Louis, MO
Solvent (A): **water** **H_2O** **7732-18-5**
Salt (C): **lithium sulfate** **Li_2SO_4** **10377-48-7**

Type of data: coexistence data (tie lines)

T/K = 296.15

Total system			Top phase			Bottom phase		
w_A	w_B	w_C	w_A	w_B	w_C	w_A	w_B	w_C
0.7750	0.1300	0.0950	0.7113	0.2296	0.0591	0.8517	0.0102	0.1381
0.7600	0.1400	0.1000	0.6858	0.2628	0.0514	0.8428	0.0031	0.1541
0.7450	0.1500	0.1050	0.6643	0.2907	0.0450	0.8306	0.0008	0.1686

Polymer (B): **poly(ethylene glycol)** **2010MA1**
Characterization: M_n/g.mol^{-1} = 400, Sigma-Aldrich, Inc., St. Louis, MO
Solvent (A): **water** H_2O **7732-18-5**
Salt (C): **magnesium sulfate** $MgSO_4$ **7487-88-9**

Type of data: coexistence data (tie lines)

Total system			Top phase			Bottom phase		
w_A	w_B	w_C	w_A	w_B	w_C	w_A	w_B	w_C
T/K = 298.15								
0.7512	0.1411	0.1077	0.7071	0.2422	0.0507	0.7939	0.0408	0.1653
0.7350	0.1555	0.1095	0.6850	0.2859	0.0291	0.7832	0.0261	0.1907
0.7126	0.1745	0.1129	0.6493	0.3315	0.0192	0.7737	0.0185	0.2078
0.6952	0.1779	0.1269	0.6293	0.3561	0.0146	0.7601	0.0010	0.2399
T/K = 308.15								
0.7551	0.1384	0.1065	0.7150	0.2389	0.0461	0.7944	0.0382	0.1674
0.7303	0.1554	0.1143	0.6779	0.2947	0.0274	0.7827	0.0165	0.2008
0.7148	0.1664	0.1188	0.6564	0.3252	0.0184	0.7721	0.0080	0.2199
0.6984	0.1777	0.1239	0.6300	0.3575	0.0125	0.7642	0.0003	0.2355
T/K = 318.15								
0.7527	0.1392	0.1081	0.7014	0.2639	0.0347	0.8028	0.0155	0.1817
0.7302	0.1533	0.1165	0.6729	0.3037	0.0234	0.7902	0.0020	0.2098
0.7205	0.1618	0.1177	0.6621	0.3220	0.0159	0.7801	0.0010	0.2199
0.6986	0.1794	0.1220	0.6307	0.3579	0.0114	0.7673	0.0000	0.2327

Polymer (B): **poly(ethylene glycol)** **2008MA1**
Characterization: M_w/g.mol^{-1} = 1500, Sigma Chemical Co., Inc., St. Louis, MO
Solvent (A): **water** H_2O **7732-18-5**
Salt (C): **magnesium sulfate** $MgSO_4$ **7487-88-9**

Type of data: coexistence data (tie lines)

Total system			Top phase			Bottom phase		
w_A	w_B	w_C	w_A	w_B	w_C	w_A	w_B	w_C
T/K = 283.15								
0.6920	0.2569	0.0511	0.6588	0.3134	0.0278	0.7770	0.0798	0.1432
0.6670	0.2769	0.0561	0.6181	0.3627	0.0192	0.7679	0.0710	0.1611
0.6405	0.2984	0.0611	0.5879	0.3983	0.0138	0.7573	0.0631	0.1796
0.6139	0.3200	0.0661	0.5468	0.4435	0.0097	0.7481	0.0569	0.1950
0.5873	0.3416	0.0711	0.5163	0.4770	0.0067	0.7260	0.0541	0.2199

continued

continued

T/K = 298.15

0.6670	0.2769	0.0561	0.6110	0.3727	0.0163	0.7707	0.0675	0.1618
0.6405	0.2984	0.0611	0.5604	0.4280	0.0116	0.7592	0.0629	0.1779
0.6139	0.3200	0.0661	0.5347	0.4572	0.0081	0.7534	0.0584	0.1882
0.5873	0.3416	0.0711	0.5149	0.4795	0.0056	0.7362	0.0534	0.2104
0.5607	0.3632	0.0761	0.4738	0.5224	0.0038	0.7247	0.0483	0.2270

T/K = 313.15

0.6670	0.2769	0.0561	0.6027	0.3843	0.0130	0.7822	0.0552	0.1626
0.6405	0.2984	0.0611	0.5913	0.4001	0.0086	0.7678	0.0528	0.1794
0.6139	0.3200	0.0661	0.5319	0.4618	0.0063	0.7621	0.0499	0.1880
0.5873	0.3416	0.0711	0.4889	0.5068	0.0043	0.7420	0.0466	0.2114
0.5607	0.3632	0.0761	0.4617	0.5350	0.0033	0.7333	0.0448	0.2219

Polymer (B):	**poly(ethylene glycol)**		**2010AZI**
Characterization:	M_w/g.mol^{-1} = 1500, Merck, Darmstadt, Germany		
Solvent (A):	**water**	**H_2O**	**7732-18-5**
Salt (C):	**magnesium sulfate**	**$MgSO_4$**	**7487-88-9**

Type of data: cloud points

T/K = 308.15

w_A	0.5818	0.6164	0.6461	0.6544	0.6876	0.7014	0.7057	0.7137	0.7188
w_B	0.4120	0.3771	0.3437	0.3331	0.2949	0.2771	0.2712	0.2613	0.2560
w_C	0.0062	0.0065	0.0102	0.0125	0.0175	0.0215	0.0231	0.0250	0.0252
w_A	0.7306	0.8039	0.8380	0.7981	0.7979	0.7950	0.7912		
w_B	0.2409	0.1312	0.0605	0.0456	0.0409	0.0414	0.0416		
w_C	0.0285	0.0649	0.1015	0.1563	0.1612	0.1636	0.1672		

T/K = 313.15

w_A	0.6188	0.6701	0.6842	0.7782	0.7875	0.7952	0.8027	0.8082	0.8139
w_B	0.3625	0.3055	0.2881	0.1775	0.1643	0.1537	0.1431	0.1345	0.1259
w_C	0.0187	0.0244	0.0277	0.0443	0.0482	0.0511	0.0542	0.0573	0.0602
w_A	0.8218	0.8255	0.8267	0.8349	0.8403	0.8493	0.8535		
w_B	0.1151	0.1104	0.1074	0.0950	0.0869	0.0719	0.0650		
w_C	0.0631	0.0641	0.0659	0.0701	0.0728	0.0788	0.0815		

T/K = 318.15

w_A	0.6061	0.6287	0.6934	0.7492	0.7858	0.8052	0.8197	0.8323	0.8386
w_B	0.3803	0.3560	0.2902	0.2239	0.1747	0.1461	0.1247	0.1070	0.0976
w_C	0.0136	0.0153	0.0164	0.0269	0.0395	0.0487	0.0556	0.0607	0.0638
w_A	0.8472	0.8490	0.8480	0.8166	0.7591	0.7226	0.6655		
w_B	0.0848	0.0744	0.0700	0.0381	0.0463	0.0543	0.0643		
w_C	0.0680	0.0766	0.0820	0.1453	0.1946	0.2231	0.2702		

continued

continued

Type of data: coexistence data (tie lines)

Total system			Top phase			Bottom phase		
w_A	w_B	w_C	w_A	w_B	w_C	w_A	w_B	w_C
T/K = 308.15								
0.7975	0.1325	0.0700	0.8380	0.0605	0.1015	0.7912	0.0416	0.1672
0.8271	0.0799	0.0930	0.7218	0.2268	0.0514	0.7979	0.0409	0.1612
0.8518	0.0478	0.1004	0.7270	0.2321	0.0409	0.7950	0.0414	0.1636
0.6956	0.2680	0.0364	0.7931	0.0876	0.1193	0.7981	0.0456	0.1563
T/K = 313.15								
0.7783	0.0443	0.1774	0.7500	0.2070	0.0430	0.7404	0.0495	0.2101
0.7703	0.0719	0.1578	0.6842	0.2881	0.0277	0.6552	0.0639	0.2809
0.8237	0.0952	0.0811	0.6701	0.3055	0.0244	0.6635	0.0623	0.2742
0.7705	0.0910	0.1385	0.6188	0.3625	0.0187	0.7319	0.0525	0.2156
T/K = 318.15								
0.8324	0.0898	0.0778	0.7411	0.2238	0.0351	0.7591	0.0463	0.1946
0.7706	0.1463	0.0831	0.6287	0.3560	0.0153	0.8166	0.0381	0.1453
0.6959	0.2270	0.0771	0.5843	0.3911	0.0246	0.6655	0.0643	0.2702
0.7400	0.1772	0.0828	0.6061	0.3803	0.0136	0.7226	0.0543	0.2231

Polymer (B): **poly(ethylene glycol)** **2008MA2**
Characterization: M_w/g.mol^{-1} = 6000, Sigma Chemical Co., Inc., St. Louis, MO
Solvent (A): **water** H_2O **7732-18-5**
Salt (C): **magnesium sulfate** $MgSO_4$ **7487-88-9**

Type of data: coexistence data (tie lines)

Total system			Top phase			Bottom phase		
w_A	w_B	w_C	w_A	w_B	w_C	w_A	w_B	w_C
T/K = 283.15								
0.7407	0.1815	0.0778	0.6685	0.3157	0.0158	0.8224	0.0291	0.1485
0.7156	0.2015	0.0829	0.6401	0.3484	0.0115	0.8071	0.0257	0.1672
0.6935	0.2188	0.0877	0.6135	0.3782	0.0083	0.7832	0.0397	0.1771
0.6682	0.2388	0.0930	0.5800	0.4138	0.0062	0.7637	0.0481	0.1882
0.6397	0.2613	0.0990	0.5525	0.4435	0.0040	0.7424	0.0455	0.2121

continued

continued

T/K = 298.15

0.7385	0.1852	0.0763	0.6563	0.3294	0.0143	0.8298	0.0250	0.1452
0.7176	0.2009	0.0815	0.6304	0.3590	0.0106	0.8132	0.0279	0.1589
0.6926	0.2199	0.0875	0.6018	0.3905	0.0077	0.7917	0.0345	0.1738
0.6650	0.2405	0.0945	0.5697	0.4246	0.0057	0.7700	0.0319	0.1981
0.6365	0.2620	0.1015	0.5363	0.4600	0.0037	0.7488	0.0390	0.2122

T/K = 313.15

0.7324	0.1926	0.0750	0.6437	0.3458	0.0105	0.8346	0.0153	0.1501
0.7051	0.2143	0.0806	0.6110	0.3813	0.0077	0.8146	0.0181	0.1673
0.6834	0.2319	0.0847	0.5843	0.4094	0.0063	0.7964	0.0280	0.1756
0.6599	0.2510	0.0891	0.5581	0.4377	0.0042	0.7746	0.0433	0.1821
0.6341	0.2710	0.0949	0.5323	0.4652	0.0025	0.7540	0.0423	0.2037

Polymer (B): **poly(ethylene glycol)** **2010SA1**
Characterization: M_w/g.mol^{-1} = 6000, Aldrich Chem. Co., Inc., Milwaukee, WI
Solvent (A): **water** **H_2O** **7732-18-5**
Salt (C): **magnesium sulfate** **$MgSO_4$** **7487-88-9**

Type of data: coexistence data (tie lines)

T/K = 298.15 pH = 5.5

Top phase			Bottom phase		
w_A	w_B	w_C	w_A	w_B	w_C
0.701	0.279	0.0200	0.843	0.045	0.112
0.673	0.313	0.0140	0.841	0.039	0.120
0.617	0.375	0.0080	0.831	0.021	0.148
0.560	0.435	0.0050	0.810	0.012	0.178

Polymer (B): **poly(ethylene glycol)** **2005HEY**
Characterization: M_n/g.mol^{-1} = 7264, DP = 165 EO-units, Aldrich Chem. Co., Inc., Milwaukee, WI
Solvent (A): **water** **H_2O** **7732-18-5**
Salt (C): **magnesium sulfate** **$MgSO_4$** **7487-88-9**

Type of data: cloud points

T/K = 298.15

z_B/(base mol/kg H_2O)	13.1	10.9	9.71	7.65	6.19	4.32	3.56
c_C/(mol/kg H_2O)	0.141	0.209	0.265	0.316	0.426	0.585	0.665
z_B/(base mol/kg H_2O)	3.12	2.45	1.76	1.14	0.860	0.565	0.393
c_C/(mol/kg H_2O)	0.711	0.800	0.884	0.946	1.03	1.09	1.14

continued

continued

Type of data: coexistence data (tie lines)

T/K = 298.15

Top phase		Bottom phase	
z_B/ base mol/kg H_2O	c_C/ mol/kg H_2O	z_B/ base mol/kg H_2O	c_C/ mol/kg H_2O
7.875	0.343	0.514	1.086
8.321	0.321	0.329	1.132
8.426	0.317	0.346	1.145
8.773	0.300	0.305	1.177
8.853	0.297	0.293	1.184
9.335	0.276	0.246	1.216
9.439	0.272	0.222	1.235

Comments: Concentrations are molal, i.e., per kg water; base moles are calculated with the molar mass of 44 g/mol per EO-unit.

Polymer (B): **poly(ethylene glycol)** **2009CUN**
Characterization: M_n/g.mol^{-1} = 8000
Solvent (A): **water** **H_2O** **7732-18-5**
Salt (C): **magnesium sulfate** **$MgSO_4$** **7487-88-9**

Type of data: coexistence data (tie lines)

Total system			Top phase			Bottom phase		
w_A	w_B	w_C	w_A	w_B	w_C	w_A	w_B	w_C
T/K = 298.15								
0.7850	0.1399	0.0751	0.69468	0.28521	0.02011	0.84248	0.02805	0.12947
0.7699	0.1500	0.0800	0.67241	0.31233	0.01526	0.83085	0.03057	0.13858
0.7553	0.1598	0.0849	0.64923	0.33767	0.01310	0.81714	0.03387	0.14899
0.7394	0.1694	0.0912	0.62684	0.36274	0.01042	0.80369	0.03489	0.16142
0.7246	0.1798	0.0956	0.60869	0.38137	0.00994	0.79040	0.02588	0.18372
0.7096	0.1899	0.1005	0.58833	0.40463	0.00704	0.77658	0.03651	0.18691
0.6951	0.2001	0.1049	0.57070	0.42463	0.00467	0.76048	0.04400	0.19552
0.6803	0.2098	0.1099	0.54829	0.45026	0.00145	0.75861	0.03561	0.20578
0.6647	0.2202	0.1151	0.52917	0.47019	0.00064	0.74305	0.03736	0.21959

continued

continued

T/K = 323.15

0.7850	0.1400	0.0750	0.64888	0.34293	0.00819	0.85013	0.02592	0.12395
0.7701	0.1499	0.0800	0.63302	0.36172	0.00526	0.84442	0.02282	0.13276
0.7548	0.1601	0.0850	0.60660	0.38834	0.00506	0.83273	0.02405	0.14322
0.7400	0.1699	0.0901	0.58948	0.40738	0.00314	0.81946	0.02613	0.15441
0.7251	0.1801	0.0948	0.56567	0.43223	0.00210	0.80510	0.02889	0.16601
0.7101	0.1897	0.1001	0.54426	0.45347	0.00227	0.79230	0.02911	0.17859
0.6947	0.2003	0.1050	0.52611	0.47216	0.00173	0.78179	0.03035	0.18786
0.6795	0.2103	0.1102	0.51147	0.48610	0.00243	0.76887	0.03506	0.19607
0.6654	0.2200	0.1147	0.49424	0.50260	0.00316	0.75673	0.03910	0.20417

Polymer (B): **poly(ethylene glycol)** **2005CAS**
Characterization: M_w/g.mol^{-1} = 8000
Solvent (A): **water** **H_2O** **7732-18-5**
Salt (C): **magnesium sulfate** **$MgSO_4$** **7487-88-9**

Type of data: coexistence data (tie lines)

T/K = 308.15

Total system			Top phase			Bottom phase		
w_A	w_B	w_C	w_A	w_B	w_C	w_A	w_B	w_C
0.8099	0.1151	0.0750	0.8633	0.0250	0.1117	0.7034	0.2783	0.0183
0.7196	0.1603	0.1201	0.7741	0.0407	0.1852	0.5543	0.4415	0.0042
0.6998	0.1702	0.1300	0.7545	0.0427	0.2028	0.5034	0.4932	0.0034
0.6786	0.1802	0.1412	0.7234	0.0902	0.1864	0.5052	0.4918	0.0030

Polymer (B): **poly(ethylene glycol)** **2008RAS**
Characterization: M_n/g.mol^{-1} = 10000, Merck, Darmstadt, Germany
Solvent (A): **water** **H_2O** **7732-18-5**
Salt (C): **magnesium sulfate** **$MgSO_4$** **7487-88-9**

Type of data: cloud points

T/K = 295.15

w_A	0.43016	0.48647	0.56770	0.58078	0.61434	0.62099	0.66747	0.68457	0.70004
w_B	0.56538	0.50803	0.42435	0.41104	0.37571	0.36802	0.31744	0.29702	0.27893
w_C	0.00446	0.00550	0.00795	0.00818	0.00995	0.01099	0.01509	0.01841	0.02103
w_A	0.72235	0.73363	0.75798	0.76054	0.77732	0.78418	0.78896	0.79735	0.79952
w_B	0.25363	0.23916	0.21004	0.20677	0.18473	0.17585	0.16903	0.15807	0.15509
w_C	0.02402	0.02721	0.03198	0.03269	0.03795	0.03997	0.04201	0.04458	0.04539

continued

continued

w_A	0.81003	0.81867	0.82690	0.83403	0.83850	0.84091	0.84343	0.84667	0.84649
w_B	0.13962	0.12748	0.11571	0.10462	0.09689	0.09186	0.08679	0.08321	0.08165
w_C	0.05035	0.05385	0.05739	0.06135	0.06461	0.06723	0.06978	0.07012	0.07186
w_A	0.84765	0.85025	0.85230	0.85364	0.85387	0.85552	0.85680	0.85923	0.86341
w_B	0.07995	0.07563	0.07209	0.07051	0.06885	0.06638	0.06359	0.06014	0.05177
w_C	0.07240	0.07412	0.07561	0.07585	0.07728	0.07810	0.07961	0.08063	0.08482
w_A	0.86427	0.86506	0.86583	0.86649	0.86769	0.86720	0.86804	0.86856	0.86902
w_B	0.05028	0.04874	0.04731	0.04586	0.04426	0.04439	0.04336	0.04221	0.04102
w_C	0.08545	0.08620	0.08686	0.08765	0.08805	0.08841	0.08860	0.08923	0.08996
w_A	0.86966	0.87027	0.87046	0.87131	0.87154	0.87152	0.87195	0.87237	0.87277
w_B	0.04002	0.03909	0.03853	0.03726	0.03695	0.03624	0.03540	0.03455	0.03385
w_C	0.09032	0.09064	0.09101	0.09143	0.09151	0.09224	0.09265	0.09308	0.09338
w_A	0.87401	0.87358	0.87352	0.87440	0.87478	0.87514	0.87883	0.87842	0.87834
w_B	0.03245	0.03222	0.03186	0.03091	0.03037	0.02981	0.01986	0.01759	0.01498
w_C	0.09354	0.09420	0.09462	0.09469	0.09485	0.09505	0.10131	0.10399	0.10668
w_A	0.87842	0.87764	0.87687	0.87569	0.87623	0.87680	0.86526	0.83839	0.81669
w_B	0.01188	0.00951	0.00689	0.00492	0.00421	0.00339	0.00271	0.00250	0.00230
w_C	0.10970	0.11285	0.11624	0.11939	0.11956	0.11981	0.13203	0.15911	0.18101
w_A	0.80766	0.73532	0.73336						
w_B	0.00210	0.00190	0.00170						
w_C	0.19024	0.26278	0.26494						

T/K = 301.15

w_A	0.39970	0.42755	0.46191	0.57272	0.59847	0.60595	0.63144	0.67888	0.69379
w_B	0.59809	0.57014	0.53564	0.42207	0.39502	0.38704	0.36007	0.30710	0.29010
w_C	0.00221	0.00231	0.00245	0.00521	0.00651	0.00701	0.00849	0.01402	0.01611
w_A	0.76904	0.76892	0.79976	0.83740	0.85834	0.86765	0.87708	0.89089	
w_B	0.19845	0.19811	0.15895	0.10599	0.06635	0.05001	0.03375	0.00901	
w_C	0.03251	0.03297	0.04129	0.05661	0.07531	0.08234	0.08917	0.10010	

T/K = 305.15

w_A	0.55711	0.64274	0.72686	0.80480	0.81623	0.83712	0.84984	0.85474	0.85868
w_B	0.43990	0.35001	0.25469	0.15598	0.13841	0.10994	0.08779	0.07911	0.07135
w_C	0.00299	0.00725	0.01845	0.03922	0.04536	0.05294	0.06237	0.06615	0.06997
w_A	0.86097	0.86427	0.86777	0.87033					
w_B	0.06701	0.06062	0.05501	0.05101					
w_C	0.07202	0.07511	0.07722	0.07866					

T/K = 311.15

w_A	0.51899	0.56341	0.66789	0.72093	0.76029	0.78168	0.80894	0.82564	0.83710
w_B	0.47945	0.43428	0.32504	0.26406	0.21859	0.19001	0.15541	0.12998	0.11401
w_C	0.00156	0.00231	0.00707	0.01501	0.02112	0.02831	0.03565	0.04438	0.04889

continued

continued

w_A	0.84696	0.85438	0.85941	0.86095	0.86539	0.86997	0.87360	0.87895	0.88369
w_B	0.10003	0.08803	0.07811	0.07394	0.06752	0.06010	0.05456	0.04446	0.03619
w_C	0.05301	0.05759	0.06248	0.06511	0.06709	0.06993	0.07184	0.07659	0.08012

Type of data: coexistence data (tie lines)

Total system			Top phase			Bottom phase		
w_A	w_B	w_C	w_A	w_B	w_C	w_A	w_B	w_C
T/K = 295.15								
0.6101	0.2477	0.1422	0.4586	0.5400	0.0014	0.7348	0.0054	0.2598
0.6027	0.2562	0.1411	0.4070	0.5902	0.0028	0.7118	0.0079	0.2803
0.6655	0.2837	0.0508	0.5687	0.4303	0.0010	0.8143	0.0045	0.1812
0.7214	0.2408	0.0378	0.6907	0.2972	0.0121	0.8628	0.0051	0.1321
0.6670	0.2549	0.0781	0.5197	0.4712	0.0091	0.7749	0.0047	0.2204
0.6767	0.2761	0.0472	0.6192	0.3678	0.0130	0.8361	0.0046	0.1593
T/K = 301.15								
0.5986	0.2502	0.1512	0.3769	0.6203	0.0028	0.7127	0.0171	0.2702
0.6085	0.2507	0.1408	0.4273	0.5701	0.0026	0.7460	0.0140	0.2400
0.6904	0.2501	0.0595	0.6527	0.3402	0.0071	0.8367	0.0132	0.1501
0.7337	0.2339	0.0324	0.6938	0.2940	0.0122	0.8760	0.0150	0.1090
0.6747	0.2463	0.0790	0.5762	0.4220	0.0018	0.7988	0.0111	0.1901
0.6725	0.2780	0.0495	0.6268	0.3701	0.0031	0.8240	0.0120	0.1640
T/K = 305.15								
0.6057	0.2438	0.1505	0.4555	0.5400	0.0045	0.7098	0.0152	0.2750
0.6240	0.2210	0.1550	0.4249	0.5711	0.0040	0.6835	0.0161	0.3004
0.6497	0.1889	0.1614	0.5840	0.4109	0.0051	0.7537	0.0162	0.2301
0.6651	0.1696	0.1653	0.6948	0.2900	0.0152	0.8053	0.0144	0.1803
0.6584	0.1780	0.1636	0.5456	0.4503	0.0041	0.7393	0.0105	0.2502
0.6893	0.1395	0.1712	0.6312	0.3678	0.0010	0.7858	0.0141	0.2001
T/K = 311.15								
0.6378	0.2417	0.1205	0.4726	0.5251	0.0023	0.7625	0.0271	0.2104
0.6599	0.2156	0.1245	0.5151	0.4830	0.0019	0.7867	0.0230	0.1903
0.6812	0.1902	0.1286	0.5404	0.4578	0.0018	0.7995	0.0303	0.1702
0.6898	0.1800	0.1302	0.5989	0.4001	0.0010	0.8341	0.0251	0.1408
0.6991	0.1689	0.1320	0.6183	0.3804	0.0013	0.8448	0.0270	0.1282
0.7105	0.1553	0.1342	0.6664	0.3295	0.0041	0.8727	0.0262	0.1011

Polymer (B):	**poly(ethylene glycol)**		**2003HUD**
Characterization:	M_w/g.mol^{-1} = 2000, Aldrich Chem. Co., Inc., Milwaukee, WI		
Solvent (A):	**water**	**H_2O**	**7732-18-5**
Salt (C):	**manganese sulfate**	**$MnSO_4$**	**7785-87-7**

continued

continued

Type of data: cloud points

T/K = 298.15

w_A	0.6830	0.7140	0.7450	0.7615	0.7733	0.7777	0.7829	0.7888	0.7914
w_B	0.2774	0.2335	0.1820	0.1482	0.1218	0.1116	0.0995	0.0855	0.0786
w_C	0.0396	0.0525	0.0730	0.0903	0.1049	0.1107	0.1176	0.1257	0.1300
w_A	0.7945	0.7961	0.7975	0.7988	0.8000	0.8012	0.8010	0.8013	0.7998
w_B	0.0704	0.0654	0.0606	0.0553	0.0511	0.0451	0.0406	0.0295	0.0208
w_C	0.1351	0.1385	0.1419	0.1459	0.1489	0.1537	0.1584	0.1692	0.1794
w_A	0.7972	0.7926	0.7903	0.7786	0.7738				
w_B	0.0210	0.0138	0.0124	0.0087	0.0048				
w_C	0.1818	0.1936	0.1973	0.2127	0.2214				

Type of data: coexistence data (tie lines)

T/K = 298.15

Total system			Top phase			Bottom phase		
w_A	w_B	w_C	w_A	w_B	w_C	w_A	w_B	w_C
0.7169	0.1893	0.0938	0.6742	0.2909	0.0349	0.7910	0.0132	0.1958
0.7137	0.1878	0.0985	0.6623	0.3070	0.0307	0.7895	0.0121	0.1984
0.7041	0.1867	0.1092	0.6421	0.3333	0.0246	0.7801	0.0069	0.2130
0.7011	0.1853	0.1136	0.6314	0.3467	0.0219	0.7784	0.0062	0.2154
0.6954	0.1845	0.1201	0.6170	0.3643	0.0187	0.7738	0.0048	0.2214

Polymer (B): **poly(ethylene glycol)** **2008YAN**
Characterization: M_w/g.mol^{-1} = 10000, M_w/M_n = 1.05, Fluka AG, Buchs, Switzerland
Solvent (A): **water** **H_2O** **7732-18-5**
Salt (C): **monopotassium phosphate** **KH_2PO_4** **7778-77-0**

Type of data: cloud points

T/K = 298.15

w_A	0.58195	0.59928	0.60949	0.62269	0.64821	0.63196	0.65890	0.67329	0.68665
w_B	0.37970	0.35790	0.34650	0.32830	0.29530	0.31760	0.28010	0.26220	0.24360
w_C	0.03835	0.04282	0.04401	0.04901	0.05649	0.05044	0.06100	0.06451	0.06975
w_A	0.69550	0.70256	0.71133	0.72008	0.72776	0.73432	0.74184	0.75197	0.76220
w_B	0.23110	0.22040	0.20770	0.19660	0.18570	0.17630	0.16400	0.14920	0.13340
w_C	0.07340	0.07704	0.08097	0.08332	0.08654	0.08938	0.09416	0.09883	0.10440
w_A	0.76800	0.77310	0.78130	0.78921	0.79718	0.80357	0.81153	0.81565	0.81718
w_B	0.12430	0.11610	0.10070	0.08789	0.07342	0.06173	0.04527	0.03035	0.02222
w_C	0.10770	0.11080	0.11800	0.12290	0.12940	0.13470	0.14320	0.15400	0.16060

continued

continued

w_A	0.81927	0.81994	0.81716
w_B	0.01383	0.00686	0.00174
w_C	0.16690	0.17320	0.18110

Polymer (B): **poly(ethylene glycol)** **2003NOZ**
Characterization: M_n/g.mol^{-1} = 10000, Merck, Darmstadt, Germany
Solvent (A): **water** H_2O **7732-18-5**
Salt (C): **monopotassium phosphate** KH_2PO_4 **7778-77-0**

Type of data: cloud points

w_B/w_A = 0.1111 was kept constant

w_C	0.126365	0.123766	0.121355	0.119020	0.115040
T/K	304.05	305.95	307.85	309.75	313.15

Polymer (B): **poly(ethylene glycol)** **2003NOZ**
Characterization: M_n/g.mol^{-1} = 15000, Merck, Darmstadt, Germany
Solvent (A): **water** H_2O **7732-18-5**
Salt (C): **monopotassium phosphate** KH_2PO_4 **7778-77-0**

Type of data: cloud points

w_B/w_A = 0.0712 was kept constant

w_C	0.134032	0.124564	0.111819	0.104877	0.094627	0.084267	0.073669
T/K	298.95	305.35	311.15	317.85	324.05	331.35	337.15

Polymer (B): **poly(ethylene glycol)** **2008YAN**
Characterization: M_w/g.mol^{-1} = 20000, M_w/M_n = 1.05
Fluka AG, Buchs, Switzerland
Solvent (A): **water** H_2O **7732-18-5**
Salt (C): **monopotassium phosphate** KH_2PO_4 **7778-77-0**

Type of data: cloud points

T/K = 298.15

w_A	0.58056	0.60011	0.60721	0.61762	0.63502	0.64884	0.66593	0.67664	0.69196
w_B	0.38600	0.36270	0.35380	0.34070	0.31790	0.29960	0.27690	0.26220	0.24080
w_C	0.03344	0.03719	0.03899	0.04168	0.04708	0.05156	0.05717	0.06116	0.06724
w_A	0.70116	0.71798	0.72771	0.73582	0.74431	0.75449	0.76486	0.77260	0.78240
w_B	0.22710	0.20550	0.19090	0.18130	0.16660	0.15190	0.13710	0.12500	0.11010
w_C	0.07174	0.07652	0.08139	0.08288	0.08909	0.09361	0.09804	0.10240	0.10750
w_A	0.79136	0.79690	0.80657	0.80933	0.81841	0.82338	0.82997	0.83190	0.83367
w_B	0.09594	0.08570	0.07173	0.06457	0.05139	0.04062	0.02803	0.01940	0.00983
w_C	0.11270	0.11740	0.12170	0.12610	0.13020	0.13600	0.14200	0.14870	0.15650
w_A	0.83181	0.82323	0.80879						
w_B	0.00349	0.00037	0.00021						
w_C	0.16470	0.17640	0.19100						

Polymer (B): **poly(ethylene glycol)** **2003NOZ**
Characterization: M_n/g.mol^{-1} = 15000, Merck, Darmstadt, Germany
Solvent (A): **water** **H_2O** **7732-18-5**
Salt (C): **monosodium phosphate** **NaH_2PO_4** **7558-80-7**

Type of data: cloud points

w_B/w_A = 0.0712 was kept constant

w_C	0.108183	0.102417	0.094370	0.087807	0.080801	0.073519	0.068224
T/K	303.75	307.75	314.55	320.05	325.35	330.25	334.95

Polymer (B): **poly(ethylene glycol)** **2006SAR**
Characterization: M_n/g.mol^{-1} = 6000, Merck, Darmstadt, Germany
Solvent (A): **water** **H_2O** **7732-18-5**
Polymer (C): **poly(acrylic acid)**
Characterization: M_n/g.mol^{-1} = 2100, Aldrich Chem. Co., Inc., Milwaukee, WI

Type of data: cloud points (LCST behavior)

T/K = 293.15

w_A	0.6318	0.6516	0.6618	0.6727	0.6809	0.6880	0.6920	0.6947	0.7032
w_B	0.3183	0.2820	0.2644	0.2435	0.2300	0.2193	0.2097	0.2018	0.1834
w_C	0.0499	0.0664	0.0738	0.0838	0.0891	0.0927	0.0983	0.1035	0.1134
w_A	0.7166	0.7226	0.7289	0.7295	0.7304	0.7329	0.7313	0.7336	0.7350
w_B	0.1501	0.1296	0.1180	0.1052	0.0960	0.0843	0.0804	0.0732	0.0677
w_C	0.1333	0.1478	0.1531	0.1653	0.1736	0.1828	0.1883	0.1932	0.1973

T/K = 303.15

w_A	0.6404	0.6608	0.6734	0.7005	0.7063	0.7116	0.7205	0.7266	0.7323
w_B	0.3162	0.2840	0.2631	0.2110	0.1970	0.1835	0.1580	0.1457	0.1337
w_C	0.0434	0.0552	0.0635	0.0885	0.0967	0.1049	0.1215	0.1277	0.1340
w_A	0.7357	0.7401	0.7438	0.7451	0.7445	0.7444	0.7441	0.7440	0.7464
w_B	0.1246	0.1143	0.1009	0.0958	0.0921	0.0902	0.0866	0.0819	0.0802
w_C	0.1397	0.1456	0.1553	0.1591	0.1634	0.1654	0.1693	0.1741	0.1734

T/K = 313.15

w_A	0.6430	0.6705	0.6913	0.7077	0.7204	0.7366	0.7389	0.7420	0.7489
w_B	0.3195	0.2755	0.2453	0.2161	0.1911	0.1651	0.1585	0.1503	0.1324
w_C	0.0375	0.0540	0.0634	0.0762	0.0885	0.0983	0.1026	0.1077	0.1187
w_A	0.7563	0.7582	0.7558	0.7582	0.7557	0.7527			
w_B	0.1031	0.0939	0.0904	0.0833	0.0818	0.0739			
w_C	0.1406	0.1479	0.1538	0.1585	0.1625	0.1734			

Polymer (B): **poly(ethylene glycol)** **2009GRU**
Characterization: M_n/g.mol^{-1} = 6700, M_w/g.mol^{-1} = 6950,
Polyglycol 6000S, Hoechst GmbH, Frankfurt, Germany
Solvent (A): **water** **H_2O** **7732-18-5**
Polymer (C): **poly(sodium acrylate)**
Characterization: M_n/g.mol^{-1} = 2600, M_w/g.mol^{-1} = 4300, M_z/g.mol^{-1} = 6100,
Rohagit SL 137, Roehm GmbH, Darmstadt, Germany

Type of data: cloud points

T/K = 298.15

w_A	0.592	0.685	0.780	0.817	0.829	0.827	0.810	0.770	0.692
w_B	0.014	0.022	0.025	0.037	0.075	0.081	0.126	0.173	0.276
w_C	0.394	0.293	0.195	0.146	0.096	0.092	0.064	0.057	0.032

w_A	0.597	0.503	0.453	0.422
w_B	0.373	0.460	0.530	0.578
w_C	0.030	0.037	0.017	0.000

T/K = 323.15

w_A	0.596	0.694	0.794	0.825	0.863	0.839	0.837	0.779	0.696
w_B	0.007	0.008	0.007	0.006	0.008	0.068	0.080	0.179	0.285
w_C	0.397	0.298	0.199	0.169	0.129	0.093	0.083	0.042	0.019

w_A	0.617	0.600	0.502
w_B	0.367	0.388	0.485
w_C	0.016	0.012	0.013

Type of data: coexistence data (tie lines)

Total system			Top phase			Bottom phase		
w_A	w_B	w_C	w_A	w_B	w_C	w_A	w_B	w_C
T/K = 298.15								
0.743	0.129	0.128	0.684	0.286	0.030	0.781	0.028	0.191
0.700	0.150	0.150	0.633	0.331	0.036	0.745	0.023	0.232
0.637	0.196	0.167	0.560	0.406	0.034	0.690	0.018	0.292
T/K = 323.15								
0.820	0.060	0.120	0.753	0.212	0.035	0.841	0.007	0.152
0.770	0.100	0.130	0.682	0.297	0.021	0.805	0.008	0.187
0.680	0.170	0.150	0.592	0.396	0.012	0.731	0.008	0.261
0.600	0.220	0.180	0.502	0.485	0.013	0.673	0.008	0.319

Polymer (B): **poly(ethylene glycol)** **2009GRU**
Characterization: M_n/g.mol^{-1} = 34400, M_w/g.mol^{-1} = 35700,
Polyglycol 35000S, Hoechst GmbH, Frankfurt, Germany
Solvent (A): **water** **H_2O** **7732-18-5**
Polymer (C): **poly(sodium acrylate)**
Characterization: M_n/g.mol^{-1} = 2600, M_w/g.mol^{-1} = 4300, M_z/g.mol^{-1} = 6100,
Rohagit SL 137, Roehm GmbH, Darmstadt, Germany

Type of data: cloud points

T/K = 298.15

w_A	0.631	0.719	0.810	0.865	0.873	0.864	0.839	0.831	0.775
w_B	0.005	0.004	0.003	0.005	0.025	0.056	0.103	0.108	0.184
w_C	0.364	0.277	0.187	0.130	0.102	0.080	0.058	0.061	0.041

w_A	0.709
w_B	0.264
w_C	0.027

T/K = 323.15

w_A	0.601	0.700	0.799	0.898	0.901	0.896	0.875	0.861	0.786
w_B	0.002	0.002	0.003	0.003	0.013	0.029	0.066	0.088	0.188
w_C	0.397	0.298	0.198	0.099	0.086	0.075	0.059	0.051	0.026

w_A	0.698	0.600	0.501
w_B	0.291	0.395	0.496
w_C	0.011	0.005	0.003

Type of data: coexistence data (tie lines)

Total system			Top phase			Bottom phase		
w_A	w_B	w_C	w_A	w_B	w_C	w_A	w_B	w_C
T/K = 298.15								
0.847	0.080	0.073	0.827	0.125	0.048	0.873	0.011	0.116
0.800	0.100	0.100	0.748	0.221	0.031	0.833	0.003	0.164
0.736	0.151	0.113	0.687	0.284	0.029	0.783	0.003	0.214
T/K = 323.15								
0.860	0.050	0.090	0.785	0.191	0.024	0.881	0.004	0.115
0.790	0.100	0.110	0.699	0.289	0.012	0.830	0.004	0.166
0.700	0.150	0.150	0.592	0.401	0.007	0.751	0.003	0.246
0.600	0.200	0.200	0.466	0.530	0.004	0.665	0.003	0.332

Polymer (B): **poly(ethylene glycol)** **2009GRU**
Characterization: M_n/g.mol^{-1} = 52900, Polymer Standard Service, Germany
Solvent (A): **water** **H_2O** **7732-18-5**
Polymer (C): **poly(sodium acrylate)**
Characterization: M_n/g.mol^{-1} = 2600, M_w/g.mol^{-1} = 4300, M_z/g.mol^{-1} = 6100, Rohagit SL 137, Roehm GmbH, Darmstadt, Germany

Type of data: cloud points

T/K = 298.15

w_A	0.603	0.702	0.799	0.868	0.875	0.878	0.843	0.779
w_B	0.003	0.002	0.003	0.003	0.012	0.031	0.081	0.180
w_C	0.394	0.296	0.198	0.129	0.113	0.091	0.076	0.041

T/K = 323.15

w_A	0.604	0.611	0.701	0.707	0.800	0.800	0.900	0.899	0.898
w_B	0.002	0.011	0.001	0.007	0.001	0.003	0.001	0.003	0.019
w_C	0.394	0.378	0.298	0.286	0.199	0.197	0.099	0.098	0.083
w_A	0.904	0.890	0.858	0.858	0.825	0.824	0.783		
w_B	0.031	0.042	0.086	0.088	0.133	0.135	0.189		
w_C	0.065	0.068	0.056	0.054	0.042	0.041	0.028		

Type of data: coexistence data (tie lines)

Total system			Top phase			Bottom phase		
w_A	w_B	w_C	w_A	w_B	w_C	w_A	w_B	w_C
T/K = 298.15								
0.849	0.051	0.100	0.800	0.154	0.046	0.861	0.002	0.137
0.810	0.080	0.110	0.750	0.219	0.031	0.831	0.002	0.167
0.749	0.101	0.150	0.671	0.313	0.016	0.774	0.002	0.224
0.680	0.150	0.170	0.596	0.400	0.004	0.716	0.002	0.282
T/K = 323.15								
0.859	0.051	0.090	0.780	0.192	0.028	0.872	0.003	0.125
0.790	0.100	0.110	0.708	0.270	0.022	0.825	0.002	0.173
0.699	0.150	0.151	0.596	0.400	0.004	0.745	0.002	0.253
0.599	0.202	0.199	0.474	0.522	0.004	0.653	0.002	0.345

Polymer (B): **poly(ethylene glycol)** **2009GRU**
Characterization: M_n/g.mol^{-1} = 103000, Polymer Standard Service, Germany
Solvent (A): **water** **H_2O** **7732-18-5**
Polymer (C): **poly(sodium acrylate)**
Characterization: M_n/g.mol^{-1} = 2600, M_w/g.mol^{-1} = 4300, M_z/g.mol^{-1} = 6100, Rohagit SL 137, Roehm GmbH, Darmstadt, Germany

continued

continued

Type of data: cloud points

T/K = 298.15

w_A	0.602	0.703	0.800	0.850	0.849	0.878	0.895	0.868	0.899
w_B	0.003	0.003	0.003	0.086	0.002	0.003	0.019	0.057	0.005
w_C	0.395	0.294	0.197	0.064	0.149	0.119	0.086	0.075	0.096

T/K = 323.15

w_A	0.608	0.608	0.704	0.706	0.800	0.899	0.899	0.912	0.914
w_B	0.003	0.007	0.003	0.007	0.004	0.001	0.002	0.010	0.022
w_C	0.389	0.385	0.293	0.287	0.196	0.100	0.099	0.078	0.064
w_A	0.896	0.894	0.862	0.863	0.836				
w_B	0.042	0.054	0.086	0.091	0.130				
w_C	0.062	0.052	0.052	0.046	0.034				

Type of data: coexistence data (tie lines)

Total system			Top phase			Bottom phase		
w_A	w_B	w_C	w_A	w_B	w_C	w_A	w_B	w_C
T/K = 298.15								
0.850	0.050	0.100	0.825	0.120	0.055	0.866	0.002	0.132
0.750	0.100	0.150	0.677	0.312	0.011	0.779	0.002	0.219
0.680	0.150	0.170	0.597	0.398	0.005	0.719	0.002	0.279
T/K = 323.15								
0.860	0.050	0.090	0.788	0.192	0.020	0.879	0.002	0.119
0.790	0.100	0.110	0.703	0.282	0.015	0.829	0.003	0.168
0.699	0.151	0.150	0.575	0.419	0.006	0.754	0.004	0.242
0.739	0.101	0.160	0.627	0.365	0.008	0.778	0.004	0.218

Polymer (B): **poly(ethylene glycol)** **2009GRU**
Characterization: M_n/g.mol^{-1} = 6700, M_w/g.mol^{-1} = 6950,
Polyglycol 6000S, Hoechst GmbH, Frankfurt, Germany
Solvent (A): **water** **H_2O** **7732-18-5**
Polymer (C: **poly(sodium acrylate)**
Characterization: M_n/g.mol^{-1} = 6900, M_w/g.mol^{-1} = 17300, M_z/g.mol^{-1} = 39700,
Sokalan SA 40, BASF AG, Ludwigshafen, Germany

Type of data: cloud points

T/K = 298.15

w_A	0.591	0.696	0.741	0.840	0.849	0.842	0.834	0.775	0.586
w_B	0.017	0.007	0.012	0.011	0.024	0.067	0.094	0.190	0.391
w_C	0.392	0.297	0.247	0.149	0.127	0.091	0.072	0.035	0.023

continued

continued

w_A	0.491	0.421
w_B	0.496	0.579
w_C	0.013	0.000

T/K = 323.15

w_A	0.595	0.693	0.795	0.799	0.876	0.884	0.886	0.862	0.791
w_B	0.005	0.011	0.005	0.002	0.006	0.017	0.032	0.080	0.182
w_C	0.400	0.296	0.200	0.199	0.118	0.099	0.082	0.058	0.027

w_A	0.700	0.693	0.594	0.496
w_B	0.291	0.300	0.399	0.496
w_C	0.009	0.007	0.007	0.008

Type of data: coexistence data (tie lines)

Total system			Top phase			Bottom phase		
w_A	w_B	w_C	w_A	w_B	w_C	w_A	w_B	w_C
T/K = 298.15								
0.750	0.150	0.100	0.788	0.174	0.038	0.838	0.016	0.146
0.752	0.150	0.098	0.720	0.248	0.032	0.789	0.011	0.200
0.709	0.150	0.141	0.664	0.312	0.024	0.740	0.011	0.249
0.649	0.181	0.170	0.595	0.386	0.019	0.690	0.011	0.299
T/K = 323.15								
0.840	0.050	0.110	0.779	0.200	0.021	0.854	0.004	0.142
0.770	0.120	0.110	0.713	0.281	0.006	0.810	0.004	0.186
0.649	0.201	0.150	0.579	0.412	0.009	0.701	0.004	0.295
0.530	0.270	0.200	0.438	0.556	0.006	0.609	0.003	0.388

Polymer (B): **poly(ethylene glycol)** **2009GRU**
Characterization: M_n/g.mol^{-1} = 34400, M_w/g.mol^{-1} = 35700,
Polyglycol 35000S, Hoechst GmbH, Frankfurt, Germany
Solvent (A): **water** **H_2O** **7732-18-5**
Polymer (C: **poly(sodium acrylate)**
Characterization: M_n/g.mol^{-1} = 6900, M_w/g.mol^{-1} = 17300, M_z/g.mol^{-1} = 39700,
Sokalan SA 40, BASF AG, Ludwigshafen, Germany

Type of data: cloud points

T/K = 298.15

w_A	0.684	0.498	0.599	0.697	0.794	0.887	0.777	0.850	0.877
w_B	0.298	0.005	0.007	0.006	0.007	0.015	0.194	0.095	0.045
w_C	0.018	0.497	0.394	0.297	0.199	0.098	0.029	0.055	0.078

continued

continued

T/K = 323.15

w_A	0.504	0.594	0.693	0.699	0.784	0.868	0.898	0.918	0.897
w_B	0.489	0.399	0.297	0.293	0.196	0.096	0.046	0.004	0.002
w_C	0.007	0.007	0.010	0.008	0.020	0.036	0.056	0.078	0.101

w_A	0.890	0.796	0.693	0.594	0.598
w_B	0.010	0.004	0.010	0.008	0.003
w_C	0.100	0.200	0.297	0.398	0.399

Type of data: coexistence data (tie lines)

Total system			Top phase			Bottom phase		
w_A	w_B	w_C	w_A	w_B	w_C	w_A	w_B	w_C
T/K = 298.15								
0.850	0.050	0.100	0.805	0.152	0.043	0.863	0.011	0.126
0.799	0.101	0.100	0.756	0.219	0.025	0.829	0.010	0.161
0.747	0.153	0.100	0.708	0.275	0.017	0.791	0.010	0.199
0.700	0.150	0.150	0.648	0.336	0.016	0.737	0.008	0.255
T/K = 323.15								
0.870	0.050	0.080	0.806	0.175	0.019	0.888	0.004	0.108
0.700	0.150	0.150	0.622	0.370	0.008	0.744	0.004	0.252
0.589	0.231	0.180	0.506	0.487	0.007	0.652	0.004	0.344
0.739	0.151	0.110	0.670	0.320	0.010	0.787	0.004	0.209
0.610	0.190	0.200	0.508	0.485	0.007	0.658	0.004	0.338

Polymer (B): **poly(ethylene glycol)** **2009GRU**
Characterization: M_n/g.mol^{-1} = 52900, Polymer Standard Service, Germany
Solvent (A): **water** **H_2O** **7732-18-5**
Polymer (C: **poly(sodium acrylate)**
Characterization: M_n/g.mol^{-1} = 6900, M_w/g.mol^{-1} = 17300, M_z/g.mol^{-1} = 39700, Sokalan SA 40, BASF AG, Ludwigshafen, Germany

Type of data: cloud points

T/K = 298.15

w_A	0.608	0.700	0.800	0.889	0.789	0.867	0.889	0.897	0.836
w_B	0.008	0.006	0.008	0.016	0.188	0.083	0.045	0.028	0.128
w_C	0.384	0.294	0.192	0.095	0.023	0.050	0.066	0.075	0.036

T/K = 323.15

w_A	0.899	0.801	0.702	0.599	0.906	0.879	0.842	0.914	0.914
w_B	0.001	0.001	0.001	0.001	0.035	0.078	0.127	0.018	0.002
w_C	0.100	0.198	0.297	0.400	0.059	0.043	0.031	0.068	0.084

continued

continued

Type of data: coexistence data (tie lines)

Total system			Top phase			Bottom phase		
w_A	w_B	w_C	w_A	w_B	w_C	w_A	w_B	w_C
T/K = 298.15								
0.850	0.050	0.100	0.800	0.176	0.024	0.860	0.009	0.131
0.800	0.080	0.120	0.746	0.239	0.015	0.818	0.008	0.174
0.730	0.120	0.150	0.671	0.323	0.006	0.755	0.006	0.239
0.650	0.150	0.200	0.573	0.420	0.007	0.680	0.006	0.314
T/K = 323.15								
0.870	0.050	0.080	0.802	0.167	0.031	0.888	0.002	0.110
0.800	0.100	0.100	0.723	0.264	0.013	0.838	0.002	0.160
0.700	0.150	0.150	0.616	0.378	0.006	0.742	0.002	0.256
0.590	0.230	0.180	0.496	0.501	0.003	0.649	0.002	0.349

Polymer (B): **poly(ethylene glycol)** **2009GRU**
Characterization: M_n/g.mol^{-1} = 103000, Polymer Standard Service, Germany
Solvent (A): **water** **H_2O** **7732-18-5**
Polymer (C: **poly(sodium acrylate)**
Characterization: M_n/g.mol^{-1} = 6900, M_w/g.mol^{-1} = 17300, M_z/g.mol^{-1} = 39700, Sokalan SA 40, BASF AG, Ludwigshafen, Germany

Type of data: cloud points

T/K = 298.15

w_A	0.607	0.704	0.800	0.862	0.896	0.898	0.905	0.883
w_B	0.009	0.009	0.003	0.101	0.041	0.003	0.021	0.080
w_C	0.384	0.287	0.197	0.037	0.063	0.099	0.074	0.037

T/K = 323.15

w_A	0.593	0.612	0.693	0.703	0.801	0.891	0.900	0.925
w_B	0.004	0.002	0.006	0.001	0.002	0.003	0.001	0.019
w_C	0.403	0.386	0.301	0.296	0.197	0.106	0.099	0.056
w_A	0.912	0.881	0.873	0.834				
w_B	0.035	0.084	0.092	0.139				
w_C	0.053	0.035	0.035	0.027				

Type of data: coexistence data (tie lines)

continued

continued

Total system			Top phase			Bottom phase		
w_A	w_B	w_C	w_A	w_B	w_C	w_A	w_B	w_C
T/K = 298.15								
0.849	0.050	0.101	0.800	0.179	0.021	0.860	0.003	0.137
0.800	0.080	0.120	0.746	0.244	0.010	0.820	0.003	0.177
T/K = 323.15								
0.870	0.050	0.080	0.807	0.174	0.019	0.889	0.003	0.108
0.800	0.100	0.100	0.729	0.263	0.008	0.835	0.003	0.162

Polymer (B): **poly(ethylene glycol)** **2009GRU**
Characterization: M_n/g.mol^{-1} = 6700, M_w/g.mol^{-1} = 6950, Polyglycol 6000S, Hoechst GmbH, Frankfurt, Germany

Solvent (A): **water** H_2O **7732-18-5**
Polymer (C): **poly(sodium ethylenesulfonate)**
Characterization: M_n/g.mol^{-1} = 1600, M_w/g.mol^{-1} = 2800, M_z/g.mol^{-1} = 5100, Polysciences Europe GmbH, Eppelheim, Germany

Type of data: cloud points

T/K = 298.15

w_A	0.489	0.480	0.469	0.471	0.471	0.457	0.441	0.431	0.421
w_B	0.349	0.364	0.414	0.414	0.445	0.488	0.533	0.559	0.579
w_C	0.162	0.156	0.117	0.115	0.084	0.055	0.026	0.010	0.000

T/K = 323.15

w_A	0.750	0.800	0.847	0.867	0.853	0.859	0.851	0.794	0.703
w_B	0.002	0.002	0.006	0.023	0.068	0.062	0.073	0.176	0.284
w_C	0.248	0.198	0.147	0.110	0.079	0.079	0.076	0.030	0.013
w_A	0.606	0.509							
w_B	0.385	0.480							
w_C	0.009	0.011							

Type of data: coexistence data (tie lines)

T/K = 323.15

Total system			Top phase			Bottom phase		
w_A	w_B	w_C	w_A	w_B	w_C	w_A	w_B	w_C
0.786	0.060	0.154	0.765	0.185	0.050	0.795	0.002	0.203
0.718	0.100	0.182	0.683	0.306	0.011	0.738	0.001	0.261
0.668	0.150	0.182	0.632	0.334	0.034	0.698	0.001	0.301
0.575	0.201	0.224	0.534	0.456	0.010	0.614	0.001	0.385

Polymer (B): **poly(ethylene glycol)** **2009GRU**
Characterization: M_n/g.mol^{-1} = 34400, M_w/g.mol^{-1} = 35700, Polyglycol 35000S, Hoechst GmbH, Frankfurt, Germany
Solvent (A): **water** **H_2O** **7732-18-5**
Polymer (C): **poly(sodium ethylenesulfonate)**
Characterization: M_n/g.mol^{-1} = 1600, M_w/g.mol^{-1} = 2800, M_z/g.mol^{-1} = 5100, Polysciences Europe GmbH, Eppelheim, Germany

Type of data: cloud points

T/K = 298.15

w_A	0.746	0.793	0.845	0.870	0.876	0.879	0.859	0.791	0.702
w_B	0.005	0.009	0.006	0.006	0.027	0.033	0.073	0.162	0.283
w_C	0.249	0.198	0.149	0.124	0.097	0.088	0.068	0.047	0.015

w_A	0.603
w_B	0.392
w_C	0.005

T/K = 323.15

w_A	0.749	0.799	0.899	0.906	0.895	0.872	0.795	0.702	0.603
w_B	0.003	0.002	0.002	0.019	0.045	0.081	0.184	0.290	0.392
w_C	0.248	0.199	0.099	0.075	0.060	0.047	0.021	0.008	0.005

w_A	0.507
w_B	0.486
w_C	0.007

Type of data: coexistence data (tie lines)

Total system			Top phase			Bottom phase		
w_A	w_B	w_C	w_A	w_B	w_C	w_A	w_B	w_C
T/K = 298.15								
0.810	0.050	0.140	0.798	0.130	0.072	0.814	0.008	0.178
0.755	0.091	0.154	0.742	0.206	0.052	0.764	0.006	0.230
0.682	0.150	0.168	0.670	0.315	0.015	0.690	0.006	0.304
0.590	0.200	0.210	0.573	0.419	0.008	0.602	0.004	0.394
T/K = 323.15								
0.838	0.050	0.112	0.809	0.151	0.040	0.852	0.004	0.144
0.770	0.090	0.140	0.727	0.241	0.032	0.788	0.004	0.208
0.716	0.130	0.154	0.671	0.322	0.007	0.750	0.004	0.246
0.638	0.180	0.182	0.604	0.389	0.007	0.668	0.003	0.329
0.710	0.140	0.150	0.663	0.330	0.007	0.754	0.003	0.243
0.789	0.101	0.110	0.752	0.236	0.012	0.819	0.004	0.177

Polymer (B): **poly(ethylene glycol)** **2009GRU**
Characterization: M_n/g.mol^{-1} = 52900, Polymer Standard Service, Germany
Solvent (A): **water** **H_2O** **7732-18-5**
Polymer (C): **poly(sodium ethylenesulfonate)**
Characterization: M_n/g.mol^{-1} = 1600, M_w/g.mol^{-1} = 2800, M_z/g.mol^{-1} = 5100, Polysciences Europe GmbH, Eppelheim, Germany

Type of data: cloud points

T/K = 298.15

w_A	0.751	0.800	0.848	0.885	0.873	0.861	0.835	0.794
w_B	0.005	0.006	0.006	0.018	0.053	0.076	0.112	0.179
w_C	0.244	0.194	0.146	0.097	0.074	0.063	0.053	0.027

T/K = 323.15

w_A	0.750	0.800	0.900	0.909	0.899	0.873	0.838
w_B	0.001	0.001	0.001	0.019	0.038	0.079	0.130
w_C	0.249	0.199	0.099	0.072	0.063	0.048	0.032

Type of data: coexistence data (tie lines)

Total system			Top phase			Bottom phase		
w_A	w_B	w_C	w_A	w_B	w_C	w_A	w_B	w_C
T/K = 298.15								
0.810	0.050	0.140	0.797	0.149	0.054	0.809	0.006	0.185
0.756	0.090	0.154	0.744	0.211	0.045	0.757	0.005	0.238
T/K = 323.15								
0.838	0.050	0.112	0.799	0.163	0.038	0.854	0.003	0.143
0.770	0.090	0.140	0.723	0.239	0.038	0.794	0.003	0.203
0.716	0.130	0.154	0.671	0.321	0.008	0.746	0.004	0.250
0.637	0.181	0.182	0.588	0.411	0.001	0.679	0.005	0.316

Polymer (B): **poly(ethylene glycol)** **2009GRU**
Characterization: M_n/g.mol^{-1} = 103000, Polymer Standard Service, Germany
Solvent (A): **water** **H_2O** **7732-18-5**
Polymer (C): **poly(sodium ethylenesulfonate)**
Characterization: M_n/g.mol^{-1} = 1600, M_w/g.mol^{-1} = 2800, M_z/g.mol^{-1} = 5100, Polysciences Europe GmbH, Eppelheim, Germany

Type of data: cloud points

T/K = 298.15

w_A	0.752	0.800	0.849	0.890	0.878	0.900	0.877
w_B	0.006	0.005	0.005	0.020	0.091	0.040	0.005
w_C	0.242	0.195	0.146	0.090	0.031	0.060	0.118

continued

continued

T/K = 323.15

w_A	0.752	0.802	0.900	0.917	0.902	0.876	0.845
w_B	0.001	0.001	0.001	0.009	0.043	0.081	0.131
w_C	0.247	0.197	0.099	0.074	0.055	0.043	0.024

Type of data: coexistence data (tie lines)

Total system			Top phase			Bottom phase		
w_A	w_B	w_C	w_A	w_B	w_C	w_A	w_B	w_C
T/K = 298.15								
0.810	0.050	0.140	0.793	0.165	0.042	0.809	0.004	0.187
0.757	0.090	0.153	0.741	0.258	0.001	0.762	0.004	0.234
T/K = 323.15								
0.838	0.050	0.112	0.811	0.130	0.059	0.850	0.001	0.149
0.800	0.060	0.140	0.753	0.242	0.005	0.818	0.001	0.181

Polymer (B): **poly(ethylene glycol)** **2009GRU**
Characterization: M_n/g.mol^{-1} = 6700, M_w/g.mol^{-1} = 6950,
Polyglycol 6000S, Hoechst GmbH, Frankfurt, Germany
Solvent (A): **water** **H_2O** **7732-18-5**
Polymer (C): **poly(sodium ethylenesulfonate)**
Characterization: M_n/g.mol^{-1} = 6900, M_w/g.mol^{-1} = 11800, M_z/g.mol^{-1} = 16200,
Natriumpolat PIPU 005, Hoechst GmbH, Frankfurt, Germany

Type of data: cloud points

T/K = 298.15

w_A	0.598	0.691	0.787	0.803	0.808	0.799	0.766	0.692	0.597
w_B	0.009	0.013	0.017	0.057	0.095	0.111	0.189	0.293	0.395
w_C	0.393	0.296	0.196	0.140	0.097	0.090	0.045	0.015	0.008

w_A	0.512	0.421
w_B	0.481	0.579
w_C	0.007	0.000

T/K = 323.15

w_A	0.595	0.692	0.796	0.809	0.814	0.798	0.796	0.790	0.762
w_B	0.007	0.011	0.005	0.014	0.042	0.090	0.116	0.148	0.190
w_C	0.398	0.297	0.199	0.177	0.144	0.112	0.088	0.062	0.048

w_A	0.682	0.589	0.595
w_B	0.286	0.394	0.399
w_C	0.032	0.017	0.006

continued

continued

Type of data: coexistence data (tie lines)

Total system			Top phase			Bottom phase		
w_A	w_B	w_C	w_A	w_B	w_C	w_A	w_B	w_C
T/K = 298.15								
0.744	0.111	0.145	0.755	0.211	0.034	0.725	0.005	0.270
0.703	0.145	0.152	0.720	0.260	0.020	0.680	0.005	0.315
0.602	0.200	0.198	0.620	0.370	0.010	0.575	0.005	0.420
T/K = 323.15								
0.780	0.100	0.120	0.768	0.182	0.050	0.784	0.004	0.212
0.716	0.145	0.139	0.700	0.274	0.026	0.718	0.004	0.278
0.669	0.180	0.151	0.661	0.316	0.023	0.674	0.004	0.322
0.609	0.201	0.190	0.598	0.388	0.014	0.607	0.004	0.389

Polymer (B): **poly(ethylene glycol)** **2009GRU**
Characterization: M_n/g.mol^{-1} = 34400, M_w/g.mol^{-1} = 35700, Polyglycol 35000S, Hoechst GmbH, Frankfurt, Germany
Solvent (A): **water** **H_2O** **7732-18-5**
Polymer (C): **poly(sodium ethylenesulfonate)**
Characterization: M_n/g.mol^{-1} = 6900, M_w/g.mol^{-1} = 11800, M_z/g.mol^{-1} = 16200, Natriumpolat PIPU 005, Hoechst GmbH, Frankfurt, Germany

Type of data: cloud points

T/K = 298.15

w_A	0.600	0.595	0.696	0.794	0.848	0.846	0.866	0.845	0.846
w_B	0.002	0.008	0.007	0.006	0.002	0.005	0.016	0.060	0.068
w_C	0.398	0.397	0.297	0.200	0.150	0.149	0.118	0.095	0.086
w_A	0.841	0.837	0.818	0.782	0.690				
w_B	0.093	0.101	0.133	0.196	0.294				
w_C	0.066	0.062	0.049	0.022	0.016				

T/K = 323.15

w_A	0.496	0.595	0.699	0.796	0.827	0.817	0.816	0.790	0.770
w_B	0.006	0.006	0.005	0.005	0.025	0.090	0.094	0.145	0.193
w_C	0.498	0.399	0.296	0.199	0.148	0.093	0.090	0.065	0.037
w_A	0.694	0.598							
w_B	0.295	0.399							
w_C	0.011	0.003							

continued

continued

Type of data: coexistence data (tie lines)

Total system			Top phase			Bottom phase		
w_A	w_B	w_C	w_A	w_B	w_C	w_A	w_B	w_C
T/K = 298.15								
0.829	0.070	0.101	0.791	0.184	0.025	0.830	0.003	0.167
0.750	0.110	0.140	0.725	0.256	0.019	0.756	0.003	0.241
0.680	0.180	0.140	0.665	0.321	0.014	0.686	0.003	0.311
0.620	0.200	0.180	0.611	0.374	0.015	0.622	0.003	0.375
0.709	0.151	0.140	0.694	0.293	0.013	0.715	0.003	0.282
T/K = 323.15								
0.800	0.050	0.150	0.800	0.130	0.070	0.795	0.005	0.200
0.750	0.100	0.150	0.760	0.210	0.030	0.740	0.005	0.255
0.700	0.150	0.150	0.705	0.285	0.010	0.680	0.005	0.315
0.700	0.200	0.100	0.705	0.285	0.010	0.683	0.005	0.312
0.724	0.135	0.141	0.735	0.243	0.022	0.712	0.003	0.285
0.600	0.201	0.199	0.577	0.003	0.420	0.615	0.380	0.005

Polymer (B): **poly(ethylene glycol)** **2009GRU**
Characterization: M_n/g.mol^{-1} = 52900, Polymer Standard Service, Germany
Solvent (A): **water** **H_2O** **7732-18-5**
Polymer (C): **poly(sodium ethylenesulfonate)**
Characterization: M_n/g.mol^{-1} = 6900, M_w/g.mol^{-1} = 11800, M_z/g.mol^{-1} = 16200, Natriumpolat PIPU 005, Hoechst GmbH, Frankfurt, Germany

Type of data: cloud points

T/K = 298.15

w_A	0.595	0.700	0.799	0.803	0.825	0.825	0.831	0.823	0.807
w_B	0.002	0.001	0.003	0.011	0.026	0.034	0.065	0.081	0.111
w_C	0.403	0.299	0.198	0.186	0.149	0.141	0.104	0.096	0.082

w_A	0.811	0.801
w_B	0.132	0.155
w_C	0.057	0.044

T/K = 323.15

w_A	0.601	0.700	0.800	0.869	0.877	0.870	0.848	0.817
w_B	0.001	0.001	0.001	0.002	0.009	0.038	0.083	0.130
w_C	0.398	0.299	0.199	0.129	0.114	0.092	0.069	0.053

continued

continued

Type of data: coexistence data (tie lines)

Total system			Top phase			Bottom phase		
w_A	w_B	w_C	w_A	w_B	w_C	w_A	w_B	w_C
T/K = 298.15								
0.800	0.050	0.150	0.794	0.161	0.045	0.793	0.002	0.205
0.750	0.100	0.150	0.754	0.214	0.032	0.739	0.002	0.259
0.680	0.150	0.170	0.691	0.298	0.011	0.659	0.002	0.339
0.600	0.200	0.200	0.613	0.384	0.003	0.569	0.001	0.430
T/K = 323.15								
0.820	0.070	0.110	0.825	0.124	0.051	0.845	0.002	0.153
0.666	0.147	0.187	0.675	0.308	0.017	0.674	0.001	0.325

Polymer (B): **poly(ethylene glycol)** **2009GRU**
Characterization: M_n/g.mol^{-1} = 103000, Polymer Standard Service, Germany
Solvent (A): **water** H_2O **7732-18-5**
Polymer (C): **poly(sodium ethylenesulfonate)**
Characterization: M_n/g.mol^{-1} = 6900, M_w/g.mol^{-1} = 11800, M_z/g.mol^{-1} = 16200, Natriumpolat PIPU 005, Hoechst GmbH, Frankfurt, Germany

Type of data: cloud points

T/K = 298.15

w_A	0.599	0.702	0.793	0.866	0.848	0.860	0.870	0.852	0.867
w_B	0.001	0.001	0.003	0.034	0.007	0.021	0.044	0.089	0.065
w_C	0.400	0.297	0.204	0.100	0.145	0.119	0.086	0.059	0.068

T/K = 323.15

w_A	0.610	0.709	0.798	0.822	0.852	0.875	0.887	0.870
w_B	0.003	0.005	0.002	0.134	0.083	0.045	0.019	0.001
w_C	0.387	0.286	0.200	0.044	0.065	0.080	0.094	0.129

Type of data: coexistence data (tie lines)

Total system			Top phase			Bottom phase		
w_A	w_B	w_C	w_A	w_B	w_C	w_A	w_B	w_C
T/K = 298.15								
0.800	0.050	0.150	0.794	0.162	0.044	0.792	0.008	0.200
0.770	0.080	0.150	0.775	0.191	0.034	0.762	0.008	0.230
0.829	0.051	0.120	0.826	0.128	0.046	0.826	0.008	0.166

continued

continued

T/K = 323.15

0.819	0.071	0.110	0.794	0.170	0.036	0.830	0.002	0.168
0.680	0.180	0.140	0.652	0.343	0.005	0.696	0.002	0.302

Polymer (B): **poly(ethylene glycol)** **2009GRU**
Characterization: M_n/g.mol^{-1} = 6700, M_w/g.mol^{-1} = 6950, Polyglycol 6000S, Hoechst GmbH, Frankfurt, Germany
Solvent (A): **water** **H_2O** **7732-18-5**
Polymer (C): **poly(sodium methacrylate)**
Characterization: M_n/g.mol^{-1} = 6100, M_w/g.mol^{-1} = 10000, M_z/g.mol^{-1} = 15400, Sigma-Aldrich Chemie GmbH, Steinheim, Germany

Type of data: cloud points

T/K = 298.15

w_A	0.695	0.746	0.796	0.843	0.858	0.862	0.836	0.820	0.788
w_B	0.007	0.006	0.004	0.008	0.025	0.042	0.091	0.122	0.174
w_C	0.298	0.248	0.200	0.149	0.117	0.096	0.073	0.058	0.038

w_A	0.747	0.700	0.603	0.454	0.422
w_B	0.226	0.285	0.381	0.540	0.578
w_C	0.027	0.015	0.016	0.006	0.000

T/K = 323.15

w_A	0.694	0.795	0.845	0.875	0.876	0.873	0.852	0.787	0.746
w_B	0.009	0.006	0.007	0.008	0.026	0.043	0.076	0.179	0.232
w_C	0.297	0.199	0.148	0.117	0.098	0.084	0.072	0.034	0.022

w_A	0.700	0.603
w_B	0.284	0.380
w_C	0.016	0.017

Type of data: coexistence data (tie lines)

Total system			Top phase			Bottom phase		
w_A	w_B	w_C	w_A	w_B	w_C	w_A	w_B	w_C
T/K = 298.15								
0.817	0.070	0.113	0.784	0.188	0.028	0.826	0.006	0.168
0.767	0.120	0.113	0.729	0.252	0.019	0.790	0.006	0.204
0.706	0.171	0.123	0.665	0.318	0.017	0.737	0.008	0.255
0.616	0.220	0.164	0.558	0.436	0.006	0.659	0.006	0.335

continued

continued

T/K = 323.15

0.847	0.050	0.103	0.793	0.172	0.035	0.861	0.007	0.132
0.787	0.100	0.113	0.732	0.243	0.025	0.821	0.006	0.173
0.696	0.150	0.154	0.616	0.373	0.011	0.743	0.006	0.251
0.646	0.200	0.154	0.571	0.413	0.016	0.706	0.007	0.287

Polymer (B): **poly(ethylene glycol)** **2009GRU**
Characterization: M_n/g.mol^{-1} = 34400, M_w/g.mol^{-1} = 35700,
Polyglycol 35000S, Hoechst GmbH, Frankfurt, Germany
Solvent (A): **water** H_2O **7732-18-5**
Polymer (C): **poly(sodium methacrylate)**
Characterization: M_n/g.mol^{-1} = 6100, M_w/g.mol^{-1} = 10000, M_z/g.mol^{-1} = 15400,
Sigma-Aldrich Chemie GmbH, Steinheim, Germany

Type of data: cloud points

T/K = 298.15

w_A	0.696	0.797	0.822	0.865	0.882	0.882	0.858	0.844	0.781
w_B	0.005	0.003	0.013	0.006	0.020	0.045	0.080	0.096	0.176
w_C	0.299	0.200	0.165	0.129	0.098	0.073	0.062	0.060	0.043
w_A	0.704	0.599							
w_B	0.271	0.396							
w_C	0.025	0.005							

T/K = 323.15

w_A	0.696	0.797	0.847	0.896	0.910	0.898	0.868	0.833	0.789
w_B	0.006	0.004	0.004	0.004	0.011	0.040	0.083	0.133	0.184
w_C	0.298	0.199	0.149	0.100	0.079	0.062	0.049	0.034	0.027
w_A	0.748	0.699	0.606	0.516					
w_B	0.236	0.288	0.383	0.474					
w_C	0.016	0.013	0.011	0.010					

Type of data: coexistence data (tie lines)

Total system			Top phase			Bottom phase		
w_A	w_B	w_C	w_A	w_B	w_C	w_A	w_B	w_C
T/K = 298.15								
0.849	0.051	0.100	0.807	0.147	0.046	0.871	0.009	0.120
0.802	0.140	0.058	0.781	0.192	0.027	0.846	0.004	0.150
0.780	0.100	0.120	0.733	0.239	0.028	0.811	0.003	0.186
0.710	0.140	0.150	0.655	0.327	0.018	0.752	0.002	0.246

continued

continued

T/K = 323.15

0.868	0.050	0.082	0.809	0.167	0.024	0.892	0.007	0.101
0.816	0.081	0.103	0.741	0.243	0.016	0.849	0.004	0.147
0.747	0.130	0.123	0.664	0.326	0.010	0.792	0.005	0.203
0.646	0.200	0.154	0.561	0.435	0.004	0.666	0.050	0.284

Polymer (B): **poly(ethylene glycol)** **2009GRU**
Characterization: M_n/g.mol^{-1} = 6700, M_w/g.mol^{-1} = 6950,
Polyglycol 6000S, Hoechst GmbH, Frankfurt, Germany
Solvent (A): **water** **H_2O** **7732-18-5**
Polymer (C): **poly(sodium methacrylate)**
Characterization: M_n/g.mol^{-1} = 14200, M_w/g.mol^{-1} = 20500, M_z/g.mol^{-1} = 27600,
Polysciences Europe GmbH, Eppelheim, Germany

Type of data: cloud points

T/K = 298.15

w_A	0.686	0.787	0.866	0.818	0.783	0.653	0.602	0.511	0.453
w_B	0.020	0.017	0.038	0.124	0.181	0.338	0.388	0.478	0.543
w_C	0.294	0.196	0.096	0.058	0.036	0.009	0.010	0.011	0.004

w_A	0.422
w_B	0.578
w_C	0.000

T/K = 323.15

w_A	0.800	0.792	0.876	0.885	0.856	0.789	0.701	0.605
w_B	0.002	0.010	0.005	0.016	0.079	0.178	0.286	0.378
w_C	0.198	0.198	0.119	0.099	0.065	0.033	0.013	0.017

Type of data: coexistence data (tie lines)

Total system			Top phase			Bottom phase		
w_A	w_B	w_C	w_A	w_B	w_C	w_A	w_B	w_C
T/K = 298.15								
0.820	0.080	0.100	0.780	0.179	0.041	0.829	0.019	0.152
0.755	0.130	0.115	0.716	0.269	0.015	0.783	0.017	0.200
0.714	0.151	0.135	0.667	0.318	0.015	0.742	0.018	0.240
0.644	0.200	0.156	0.587	0.409	0.004	0.684	0.016	0.300
T/K = 323.15								
0.846	0.050	0.104	0.782	0.197	0.021	0.858	0.005	0.137
0.774	0.101	0.125	0.699	0.289	0.012	0.806	0.005	0.189
0.694	0.150	0.156	0.609	0.381	0.010	0.738	0.005	0.257
0.603	0.220	0.177	0.512	0.479	0.009	0.671	0.005	0.324

Polymer (B): **poly(ethylene glycol)** **2009GRU**
Characterization: M_n/g.mol^{-1} = 34400, M_w/g.mol^{-1} = 35700, Polyglycol 35000S, Hoechst GmbH, Frankfurt, Germany
Solvent (A): **water** **H_2O** **7732-18-5**
Polymer (C): **poly(sodium methacrylate)**
Characterization: M_n/g.mol^{-1} = 14200, M_w/g.mol^{-1} = 20500, M_z/g.mol^{-1} = 27600, Polysciences Europe GmbH, Eppelheim, Germany

Type of data: cloud points

T/K = 298.15

w_A	0.694	0.795	0.844	0.605	0.694	0.779	0.833	0.857	0.844
w_B	0.008	0.008	0.063	0.390	0.277	0.167	0.068	0.033	0.007
w_C	0.298	0.197	0.093	0.005	0.029	0.054	0.099	0.110	0.149

T/K = 323.15

w_A	0.691	0.791	0.800	0.894	0.917	0.908	0.869	0.790	0.700
w_B	0.008	0.009	0.002	0.006	0.004	0.026	0.085	0.182	0.288
w_C	0.301	0.200	0.198	0.100	0.079	0.066	0.046	0.028	0.012

w_A	0.605
w_B	0.383
w_C	0.012

Type of data: coexistence data (tie lines)

Total system			Top phase			Bottom phase		
w_A	w_B	w_C	w_A	w_B	w_C	w_A	w_B	w_C
T/K = 298.15								
0.718	0.200	0.082	0.692	0.281	0.027	0.764	0.008	0.228
0.750	0.110	0.140	0.698	0.277	0.025	0.780	0.007	0.213
0.800	0.060	0.140	0.745	0.204	0.051	0.814	0.008	0.178
0.690	0.140	0.170	0.627	0.366	0.007	0.728	0.007	0.265
T/K = 323.15								
0.749	0.101	0.150	0.635	0.357	0.008	0.792	0.008	0.200
0.680	0.170	0.150	0.570	0.416	0.014	0.746	0.008	0.246
0.850	0.075	0.075	0.789	0.193	0.017	0.872	0.008	0.120

Polymer (B): **poly(ethylene glycol)** **2003HUD**
Characterization: M_w/g.mol^{-1} = 2000, Aldrich Chem. Co., Inc., Milwaukee, WI
Solvent (A): **water** **H_2O** **7732-18-5**
Salt (C): **potassium carbonate** **K_2CO_3** **584-08-7**

Type of data: cloud points

continued

continued

T/K = 298.15

w_A	0.6504	0.6949	0.7322	0.7453	0.7586	0.7724	0.7821	0.7864	0.7916
w_B	0.2952	0.2423	0.2003	0.1782	0.1608	0.1439	0.1322	0.1252	0.1170
w_C	0.0544	0.0628	0.0675	0.0765	0.0806	0.0837	0.0857	0.0884	0.0914
w_A	0.7945	0.8006	0.8050	0.8079	0.8092	0.8117	0.8146	0.8156	0.8169
w_B	0.1111	0.1041	0.0996	0.0964	0.0932	0.0906	0.0880	0.0863	0.0841
w_C	0.0944	0.0953	0.0954	0.0957	0.0976	0.0977	0.0974	0.0981	0.0990
w_A	0.8185	0.8199	0.8219	0.8394	0.8475	0.8475	0.8476	0.8475	0.8460
w_B	0.0821	0.0770	0.0740	0.0559	0.0424	0.0210	0.0181	0.0148	0.0115
w_C	0.0994	0.1031	0.1041	0.1047	0.1101	0.1315	0.1343	0.1377	0.1425
w_A	0.8437	0.8399	0.8316						
w_B	0.0080	0.0059	0.0042						
w_C	0.1483	0.1542	0.1642						

Type of data: coexistence data (tie lines)

T/K = 298.15

Total system			Top phase			Bottom phase		
w_A	w_B	w_C	w_A	w_B	w_C	w_A	w_B	w_C
0.7159	0.1907	0.0934	0.6342	0.3157	0.0501	0.8385	0.0033	0.1582
0.6909	0.1870	0.1221	0.5474	0.4173	0.0353	0.8074	0.0001	0.1925
0.6668	0.1835	0.1497	0.5081	0.4622	0.0297	0.7712	0.0001	0.2287
0.6213	0.1768	0.2019	0.4039	0.5776	0.0185	0.7172	0.0000	0.2828
0.5999	0.1734	0.2267	0.3871	0.5959	0.0170	0.6872	0.0000	0.3128
0.5790	0.1705	0.2505	0.3405	0.6459	0.0136	0.6646	0.0000	0.3354

Polymer (B): **poly(ethylene glycol)** **2005TAB**
Characterization: M_w/g.mol^{-1} = 4000, M_w/M_n = 1.1,
Merck, Darmstadt, Germany
Solvent (A): **water** H_2O **7732-18-5**
Salt (C): **potassium chloride** **KCl** **7447-40-7**

Type of data: cloud points

T/K = 333.15

w_B	0.0076	0.0076	0.0132	0.0255	0.0359	0.0400	0.0533	0.0630	0.0705
w_C	0.2749	0.2949	0.2502	0.2238	0.2169	0.2195	0.2070	0.2018	0.1964
w_B	0.0782	0.0788	0.0805	0.0888	0.0979	0.0987	0.1074	0.1185	0.1261
w_C	0.1935	0.1967	0.1942	0.1905	0.1871	0.1917	0.1879	0.1836	0.1813
w_B	0.1338	0.1480	0.2348	0.3280	0.4373	0.4473			
w_C	0.1783	0.1731	0.1488	0.1339	0.1291	0.1191			

Comments: The first and the last line indicate saturated solutions in liquid-liquid-solid equilibrium.

continued

continued

Type of data: coexistence data (tie lines)

T/K = 333.15

Total system			Top phase			Bottom phase		
w_A	w_B	w_C	w_A	w_B	w_C	w_A	w_B	w_C
0.7055	0.0930	0.2015	0.6164	0.2348	0.1488	0.7405	0.0400	0.2195
0.6791	0.1008	0.2201	0.5381	0.3280	0.1339	0.7366	0.0132	0.2502
0.6101	0.1498	0.2401	0.4336	0.4473	0.1191	0.7175	0.0076	0.2749

Comments: The last line indicates saturated solutions in liquid-liquid-solid equilibrium.

Polymer (B): poly(ethylene glycol) **2007JAY**
Characterization: M_n/g.mol^{-1} = 1900-2200, Merck, Darmstadt, Germany
Solvent (A): water **H_2O** **7732-18-5**
Salt (C): potassium citrate **$K_3C_6H_5O_7$** **866-84-2**

Type of data: cloud points

T/K = 298.15

w_A	0.5301	0.5718	0.6078	0.6353	0.6566	0.6717	0.6726	0.6853	0.6953
w_B	0.0438	0.0561	0.0759	0.0931	0.1097	0.1187	0.1223	0.1346	0.1456
w_C	0.4261	0.3721	0.3163	0.2716	0.2337	0.2096	0.2051	0.1801	0.1591
w_A	0.7016	0.7030	0.7093	0.7143	0.7243	0.7266	0.7247	0.7132	0.6916
w_B	0.1543	0.1551	0.1631	0.1697	0.1918	0.2191	0.2498	0.2787	0.3040
w_C	0.1441	0.1419	0.1276	0.1160	0.0839	0.0543	0.0255	0.0081	0.0044

T/K = 308.15

w_A	0.5489	0.5746	0.5914	0.6311	0.6427	0.6679	0.6616	0.6766	0.6777
w_B	0.0420	0.0496	0.0551	0.0679	0.0715	0.0868	0.0824	0.0914	0.0934
w_C	0.4091	0.3758	0.3535	0.3010	0.2858	0.2453	0.2560	0.2320	0.2289
w_A	0.6937	0.7064	0.7050	0.7198	0.7141	0.7213	0.7275	0.7346	0.7475
w_B	0.1022	0.1151	0.1131	0.1297	0.1215	0.1310	0.1368	0.1467	0.1639
w_C	0.2041	0.1785	0.1819	0.1505	0.1644	0.1477	0.1357	0.1187	0.0886
w_A	0.7536	0.7544	0.7524	0.7499	0.7417	0.7338	0.7085	0.6810	
w_B	0.1902	0.2010	0.2176	0.2246	0.2403	0.2559	0.2847	0.3146	
w_C	0.0562	0.0446	0.0300	0.0255	0.0180	0.0103	0.0068	0.0044	

T/K = 318.15

w_A	0.5356	0.5525	0.5773	0.6044	0.6185	0.6508	0.6822	0.7074	0.7155
w_B	0.0353	0.0380	0.0391	0.0452	0.0493	0.0592	0.0708	0.0807	0.0851
w_C	0.4291	0.4095	0.3836	0.3504	0.3322	0.2900	0.2470	0.2119	0.1994

continued

continued

w_A	0.7362	0.7454	0.7561	0.7724	0.7761	0.7670	0.7691	0.7413	0.7315
w_B	0.0961	0.1016	0.1096	0.1327	0.1722	0.1926	0.2057	0.2392	0.2586
w_C	0.1677	0.1530	0.1343	0.0949	0.0517	0.0404	0.0252	0.0195	0.0099
w_A	0.6903								
w_B	0.3004								
w_C	0.0093								

Type of data: coexistence data (tie lines)

Top phase w_A	w_B	w_C	Bottom phase w_A	w_B	w_C
T/K = 298.15					
0.6148	0.3034	0.0818	0.7138	0.0593	0.2269
0.5959	0.3338	0.0703	0.7148	0.0291	0.2561
0.5769	0.3631	0.0600	0.7025	0.0049	0.2926
T/K = 308.15					
0.6507	0.2760	0.0733	0.7267	0.0125	0.2608
0.6323	0.3002	0.0675	0.7188	0.0100	0.2712
0.6021	0.3385	0.0594	0.7031	0.0100	0.2869
T/K = 318.15					
0.6673	0.2705	0.0622	0.7621	0.0157	0.2222
0.6185	0.3339	0.0476	0.7277	0.0106	0.2617
0.5899	0.3673	0.0428	0.7122	0.0005	0.2873

Polymer (B): **poly(ethylene glycol)** **2003HUD**
Characterization: M_w/g.mol^{-1} = 1000, Aldrich Chem. Co., Inc., Milwaukee, WI
Solvent (A): **water** H_2O **7732-18-5**
Salt (C): **potassium phosphate** K_3PO_4 **7778-53-2**

Type of data: cloud points

T/K = 298.15

w_A	0.6376	0.6751	0.6886	0.7051	0.7199	0.7317	0.7437	0.7560	0.7638
w_B	0.3212	0.2655	0.2488	0.2234	0.1989	0.1805	0.1603	0.1396	0.1269
w_C	0.0412	0.0594	0.0626	0.0715	0.0812	0.0878	0.0960	0.1044	0.1093
w_A	0.7719	0.7799	0.7828	0.7866	0.7902	0.7929	0.7959	0.7984	0.8006
w_B	0.1140	0.0996	0.0955	0.0887	0.0819	0.0760	0.0710	0.0662	0.0611
w_C	0.1141	0.1205	0.1217	0.1247	0.1279	0.1311	0.1331	0.1354	0.1383
w_A	0.8029	0.8045	0.8059	0.8089	0.8106	0.8116	0.8116	0.8123	0.8094
w_B	0.0570	0.0529	0.0491	0.0422	0.0367	0.0319	0.0267	0.0190	0.0142
w_C	0.1401	0.1426	0.1450	0.1489	0.1527	0.1565	0.1617	0.1687	0.1764

continued

continued

w_A	0.7861
w_B	0.00701
w_C	0.2069

Type of data: coexistence data (tie lines)

T/K = 298.15

Total system			Top phase			Bottom phase		
w_A	w_B	w_C	w_A	w_B	w_C	w_A	w_B	w_C
0.7161	0.1896	0.0943	0.6528	0.3002	0.0470	0.8102	0.0254	0.1644
0.6802	0.1832	0.1366	0.5548	0.4312	0.0140	0.7723	0.0009	0.2268
0.6465	0.1773	0.1762	0.4957	0.4988	0.0055	0.7289	0.28410^{-4}	0.2711
0.6149	0.1717	0.2134	0.4382	0.5604	0.0014	0.6930	0.43310^{-6}	0.3070

Polymer (B): **poly(ethylene glycol)** **2008CAR**
Characterization: M_w/g.mol^{-1} = 1500, Isofar, Rio de Janeiro, Brazil
Solvent (A): **water** **H_2O** **7732-18-5**
Salt (C): **potassium phosphate** **K_3PO_4** **7778-53-2**

Type of data: coexistence data (tie lines)

Total system			Top phase			Bottom phase		
w_A	w_B	w_C	w_A	w_B	w_C	w_A	w_B	w_C
T/K = 278.15								
0.7355	0.1585	0.1060	0.6964	0.2413	0.0623	0.7574	0.0467	0.1959
0.7082	0.1711	0.1207	0.6597	0.2946	0.0457	0.6882	0.0796	0.2321
0.6895	0.1786	0.1319	0.6375	0.3237	0.0388	0.6746	0.0878	0.2376
0.6555	0.1960	0.1485	0.5984	0.3717	0.0299	0.6539	0.0699	0.2761
0.6277	0.2084	0.1639	0.5692	0.4109	0.0199	0.6522	0.0210	0.3268
T/K = 298.15								
0.7503	0.1502	0.0995	0.7181	0.2137	0.0682	0.7652	0.0705	0.1643
0.7276	0.1597	0.1127	0.6723	0.2778	0.0499	0.7395	0.0534	0.2071
0.7107	0.1696	0.1197	0.6438	0.3113	0.0450	0.7272	0.0563	0.2165
0.6903	0.1796	0.1301	0.6181	0.3451	0.0368	0.7157	0.0378	0.2465
0.6668	0.1932	0.1400	0.5864	0.3813	0.0323	0.5991	0.1276	0.2733
T/K = 308.15								
0.7437	0.1449	0.1114	0.7038	0.2341	0.0622	0.7769	0.0396	0.1836
0.7353	0.1582	0.1065	0.6908	0.2509	0.0583	0.7743	0.0471	0.1786
0.7169	0.1665	0.1166	0.6442	0.3083	0.0475	0.7602	0.0288	0.2110
0.7016	0.1733	0.1251	0.6071	0.3530	0.0400	0.7198	0.0483	0.2319
0.6845	0.1868	0.1287	0.5982	0.3676	0.0342	0.6754	0.0599	0.2647

continued

continued

T/K = 318.15

0.7386	0.1544	0.1070	0.6502	0.3077	0.0421	0.7597	0.0161	0.2242
0.7216	0.1632	0.1152	0.7180	0.2460	0.0359	0.6400	0.1176	0.2424
0.6994	0.1755	0.1250	0.4766	0.4917	0.0317	0.6882	0.0596	0.2522
0.6760	0.1869	0.1371	0.4872	0.4856	0.0272	0.7132	0.0022	0.2845
0.6421	0.1967	0.1612	0.4999	0.4702	0.0299	0.7269	0.0068	0.2663

Polymer (B): **poly(ethylene glycol)** **2003HUD**
Characterization: M_w/g.mol^{-1} = 2000, Aldrich Chem. Co., Inc., Milwaukee, WI
Solvent (A): **water** H_2O **7732-18-5**
Salt (C): **potassium phosphate** K_3PO_4 **7778-53-2**

Type of data: cloud points

T/K = 298.15

w_A	0.6000	0.6329	0.6671	0.6975	0.7176	0.7330	0.7490	0.7627	0.7740
w_B	0.3715	0.3315	0.2926	0.2593	0.2289	0.2116	0.1902	0.1709	0.1532
w_C	0.0285	0.0356	0.0403	0.0432	0.0535	0.0554	0.0608	0.0664	0.0728
w_A	0.7833	0.7939	0.8032	0.8109	0.8171	0.8242	0.8278	0.8305	0.8343
w_B	0.1363	0.1220	0.1072	0.0937	0.0814	0.0704	0.0592	0.0540	0.0487
w_C	0.0804	0.0841	0.0896	0.0954	0.1015	0.1054	0.1130	0.1155	0.1170
w_A	0.8375	0.8387	0.8392	0.8423	0.8441	0.8458	0.8461	0.8462	0.8415
w_B	0.0423	0.0375	0.0331	0.0291	0.0253	0.0218	0.0185	0.0158	0.0159
w_C	0.1202	0.1238	0.1277	0.1286	0.1306	0.1324	0.1354	0.1380	0.1426
w_A	0.8408	0.8411	0.8409	0.8376	0.8389	0.8371	0.8356	0.8318	0.8248
w_B	0.0138	0.0122	0.0103	0.0085	0.0072	0.0061	0.0046	0.0032	0.0018
w_C	0.1454	0.1467	0.1488	0.1539	0.1539	0.1568	0.1598	0.1650	0.1734

Type of data: coexistence data (tie lines)

T/K = 298.15

Total system			Top phase			Bottom phase		
w_A	w_B	w_C	w_A	w_B	w_C	w_A	w_B	w_C
0.7166	0.1888	0.0946	0.6129	0.3569	0.0302	0.8294	0.0059	0.1647
0.6804	0.1824	0.1372	0.5340	0.4458	0.0202	0.7818	0.00013	0.2181
0.6471	0.1760	0.1769	0.4857	0.4985	0.0158	0.7348	0.000008	0.2652
0.6150	0.1709	0.2141	0.4414	0.5460	0.0126	0.6940	0.0000	0.3060
0.5853	0.1656	0.2491	0.3881	0.6023	0.0096	0.6601	0.0000	0.3399
0.5544	0.1636	0.2820	0.3611	0.6306	0.0083	0.6243	0.0000	0.3757

Polymer (B): **poly(ethylene glycol)** **2003HUD**
Characterization: M_w/g.mol^{-1} = 3400, Aldrich Chem. Co., Inc., Milwaukee, WI
Solvent (A): **water** **H_2O** **7732-18-5**
Salt (C): **potassium phosphate** **K_3PO_4** **7778-53-2**

Type of data: cloud points

T/K = 298.15

w_A	0.6750	0.7170	0.7549	0.7874	0.8099	0.8267	0.8392	0.8483	0.8550
w_B	0.2923	0.2409	0.1918	0.1476	0.1135	0.0873	0.0686	0.0544	0.0432
w_C	0.0327	0.0421	0.0533	0.0650	0.0766	0.0860	0.0922	0.0973	0.1018
w_A	0.8592	0.8601	0.8626	0.8675	0.8629	0.8598	0.8571		
w_B	0.0352	0.0339	0.0290	0.0163	0.0105	0.0096	0.0091		
w_C	0.1056	0.1060	0.1084	0.1162	0.1266	0.1306	0.1338		

Type of data: coexistence data (tie lines)

T/K = 298.15

Total system			Top phase			Bottom phase		
w_A	w_B	w_C	w_A	w_B	w_C	w_A	w_B	w_C
0.7161	0.1897	0.0942	0.6136	0.3664	0.0200	0.8261	0.0002	0.1737
0.6801	0.1833	0.1366	0.5361	0.4535	0.0104	0.7778	0.0000	0.2222
0.6464	0.1774	0.1762	0.4889	0.5044	0.0067	0.7319	0.0000	0.2681
0.6149	0.1718	0.2133	0.4414	0.5545	0.0041	0.6928	0.0000	0.3072

Polymer (B): **poly(ethylene glycol)** **2007CAR**
Characterization: M_w/g.mol^{-1} = 4000, Isofar, Rio de Janeiro, Brazil
Solvent (A): **water** **H_2O** **7732-18-5**
Salt (C): **potassium phosphate** **K_3PO_4** **7778-53-2**

Type of data: coexistence data (tie lines)

Total system			Top phase			Bottom phase		
w_A	w_B	w_C	w_A	w_B	w_C	w_A	w_B	w_C
T/K = 278.15								
0.7319	0.1731	0.0950	0.6799	0.2792	0.0409	0.7857	0.0120	0.2023
0.7034	0.1930	0.1036	0.6513	0.3114	0.0373	0.7520	0.0188	0.2292
0.6693	0.2147	0.1160	0.6147	0.3555	0.0298	0.6855	0.0393	0.2752
0.6367	0.2352	0.1281	0.5776	0.3963	0.0261	0.6772	0.0318	0.2910
0.6102	0.2521	0.1377	0.5466	0.4302	0.0232	0.6489	0.0042	0.3469

continued

continued

T/K = 298.15

0.7295	0.1756	0.0949	0.6697	0.2926	0.0377	0.8123	0.0018	0.1859
0.6916	0.2017	0.1067	0.6199	0.3583	0.0218	0.7664	0.0116	0.2220
0.6657	0.2189	0.1154	0.5924	0.3931	0.0145	0.7413	0.0121	0.2466
0.6417	0.2334	0.1249	0.5692	0.4181	0.0127	0.7186	0.0076	0.2738
0.6083	0.2552	0.1365	0.5290	0.4655	0.0055	0.6736	0.0232	0.3032

T/K = 308.15

0.7356	0.1727	0.0917	0.6667	0.3078	0.0255	0.8197	0.0018	0.1785
0.7083	0.1910	0.1007	0.6309	0.3496	0.0195	0.7910	0.0162	0.1928
0.6777	0.2110	0.1113	0.5940	0.3877	0.0183	0.7590	0.0206	0.2204
0.6419	0.2360	0.1221	0.5570	0.4287	0.0143	0.7245	0.0350	0.2405
0.6105	0.2525	0.1370	0.5240	0.4604	0.0156	0.6816	0.0426	0.2758

T/K = 318.15

0.7469	0.1666	0.0865	0.6651	0.3056	0.0293	0.8105	0.0331	0.1564
0.7156	0.1860	0.0984	0.6188	0.3571	0.0241	0.7695	0.0484	0.1821
0.6883	0.2054	0.1063	0.5984	0.3812	0.0204	0.7435	0.0396	0.2169
0.6488	0.2311	0.1201	0.5508	0.4332	0.0160	0.6955	0.0653	0.2392
0.6298	0.2450	0.1252	0.5390	0.4469	0.0141	0.6975	0.0354	0.2671

Polymer (B): **poly(ethylene glycol)** **2006MOH**
Characterization: M_n/g.mol^{-1} = 10000, M_w = 11000,
Merck, Darmstadt, Germany
Solvent (A): **water** H_2O **7732-18-5**
Salt (C): **potassium phosphate** K_3PO_4 **7778-53-2**

Type of data: cloud points

w_C/w_A = 0.0488 was kept constant

w_B	0.1503	0.1617	0.1816	0.1900	0.2005	0.2250
T/K	329.0	327.6	325.5	324.2	323.2	320.3

w_B/w_A = 0.0711 was kept constant

w_C	0.0589	0.0600	0.0639	0.0683	0.0728	0.0800
T/K	324.2	322.8	318.3	312.9	308.2	299.7

Polymer (B): **poly(ethylene glycol)** **2003NOZ**
Characterization: M_n/g.mol^{-1} = 15000, Merck, Darmstadt, Germany
Solvent (A): **water** H_2O **7732-18-5**
Salt (C): **potassium phosphate** K_3PO_4 **7778-53-2**

Type of data: cloud points

w_B/w_A = 0.0712 was kept constant

w_C	0.073357	0.068231	0.064211	0.059009	0.053599	0.047861	0.043912
T/K	301.15	309.35	312.65	316.15	323.75	329.15	335.05

Polymer (B):	**poly(ethylene glycol)**		**2008ZA4**
Characterization:	M_n/g.mol^{-1} = 4230, Merck, Darmstadt, Germany		
Solvent (A):	**water**	H_2O	**7732-18-5**
Salt (C):	**potassium sodium tartrate**	$C_4H_4KNaO_6$	**304-59-6**

Type of data: cloud points

T/K = 298.15

w_A	0.6654	0.6689	0.6722	0.6755	0.6815	0.6874	0.6965	0.7089	0.7220
w_B	0.2708	0.2663	0.2616	0.2567	0.2484	0.2405	0.2271	0.2094	0.1901
w_C	0.0638	0.0648	0.0662	0.0678	0.0701	0.0721	0.0764	0.0817	0.0879
w_A	0.7424	0.7545	0.7676	0.7798	0.7909	0.7999	0.8040	0.8088	
w_B	0.1585	0.1359	0.1118	0.0887	0.0681	0.0506	0.0369	0.0269	
w_C	0.0991	0.1096	0.1206	0.1315	0.1410	0.1495	0.1591	0.1643	

Type of data: coexistence data (tie lines)

T/K = 298.15

Top phase			Bottom phase		
w_A	w_B	w_C	w_A	w_B	w_C
0.6960	0.2280	0.0760	0.8034	0.0133	0.1833
0.6751	0.2548	0.0701	0.7916	0.0090	0.1994
0.6180	0.3270	0.0550	0.7700	0.0049	0.2251
0.5849	0.3670	0.0481	0.7389	0.0024	0.2587
0.5485	0.4111	0.0404	0.7103	0.0013	0.2884
0.5115	0.4501	0.0384	0.6805	0.0008	0.3187

Polymer (B):	**poly(ethylene glycol)**		**2005HEY**
Characterization:	M_n/g.mol^{-1} = 7264, DP = 165 EO-units, Aldrich Chem. Co., Inc., Milwaukee, WI		
Solvent (A):	**water**	H_2O	**7732-18-5**
Salt (C):	**sodium carbonate**	Na_2CO_3	**497-19-8**

Type of data: cloud points

T/K = 298.15

z_B/(base mol/kg H_2O)	7.27	5.88	4.53	3.56	2.48	1.80	1.39
c_C/(mol/kg H_2O)	0.294	0.326	0.391	0.434	0.499	0.541	0.572
z_B/(base mol/kg H_2O)	0.991	0.793	0.644	0.459	0.259	0.106	0.055
c_C/(mol/kg H_2O)	0.603	0.621	0.635	0.657	0.692	0.734	0.786

Type of data: coexistence data (tie lines)

continued

continued

T/K = 298.15

Top phase		Bottom phase	
z_B/ base mol/kg H_2O	c_C/ mol/kg H_2O	z_B/ base mol/kg H_2O	c_C/ mol/kg H_2O
4.657	0.380	0.302	0.674
5.058	0.362	0.234	0.690
5.293	0.358	0.232	0.690
5.515	0.347	0.133	0.726
5.685	0.342	0.107	0.740
6.721	0.313	0.053	0.787

Comments: Concentrations are molal, i.e., per kg water; base moles are calculated with the molar mass of 44 g/mol per EO-unit.

Polymer (B): **poly(ethylene glycol)** **2006MOH**
Characterization: M_n/g.mol^{-1} = 10000, M_w = 11000, Merck, Darmstadt, Germany
Solvent (A): **water** **H_2O** **7732-18-5**
Salt (C): **sodium carbonate** **Na_2CO_3** **497-19-8**

Type of data: cloud points

w_B/w_A = 0.1000 was kept constant

w_C	0.0353	0.0385	0.0435	0.0465	0.0490	0.0518
T/K	320.6	313.8	303.3	295.5	289.8	282.9

Polymer (B): **poly(ethylene glycol)** **2005TAB**
Characterization: M_w/g.mol^{-1} = 4000, M_w/M_n = 1.1, Merck, Darmstadt, Germany
Solvent (A): **water** **H_2O** **7732-18-5**
Salt (C): **sodium chloride** **NaCl** **7647-14-5**

Type of data: cloud points

T/K = 333.15

w_B	0.0073	0.0270	0.0306	0.0341	0.0490	0.0614	0.0590	0.0675	0.0726
w_C	0.2705	0.2207	0.2128	0.2068	0.2005	0.1940	0.1950	0.1919	0.1877
w_B	0.0795	0.0854	0.0914	0.0980	0.1062	0.1215	0.1278	0.1378	0.3106
w_C	0.1876	0.1844	0.1817	0.1795	0.1725	0.1728	0.1700	0.1664	0.1305
w_B	0.4340	0.4445							
w_C	0.1180	0.1207							

Comments: The first and the last line indicate saturated solutions in liquid-liquid-solid equilibrium.

continued

continued

Type of data: coexistence data (tie lines)

T/K = 333.15

Total system			Top phase			Bottom phase		
w_A	w_B	w_C	w_A	w_B	w_C	w_A	w_B	w_C
0.7096	0.0899	0.2005	0.5589	0.3106	0.1305	0.7523	0.0270	0.2207
0.6799	0.1000	0.2201	0.4799	0.3973	0.1228	0.7378	0.0093	0.2529
0.5874	0.1541	0.2585	0.4348	0.4445	0.1207	0.7222	0.0073	0.2705

Comments: The last line indicates saturated solutions in liquid-liquid-solid equilibrium.

Polymer (B): **poly(ethylene glycol)** **1994HOG**
Characterization: M_w/g.mol^{-1} = 8000, Aldrich Chem. Co., Inc., Milwaukee, WI
Solvent (A): **water** H_2O **7732-18-5**
Salt (C): **sodium chloride** **NaCl** **7647-14-5**

Type of data: coexistence data (tie lines)

T/K = 301

Top phase			Bottom phase		
w_A	w_B	w_C	w_A	w_B	w_C
0.672	0.194	0.134	0.799	0.011	0.190
0.669	0.200	0.131	0.797	0.019	0.184
0.612	0.263	0.125	0.790	0.005	0.205
0.521	0.362	0.117	0.750	0.001	0.249
0.487	0.399	0.114	0.7415	0.0005	0.2580
0.448	0.434	0.118	0.7309	0.0001	0.2690

Plait-point composition: w_A = 0.760 + w_B = 0.090 + w_C = 0.150

Polymer (B): **poly(ethylene glycol)** **2006TUB**
Characterization: M_n/g.mol^{-1} = 600, Sigma Chemical Co., Inc., St. Louis, MO
Solvent (A): **water** H_2O **7732-18-5**
Salt (C): **sodium citrate** $Na_3C_6H_5O_7$ **68-04-2**

Type of data: coexistence data (tie lines)

T/K = 295.15

continued

continued

Total system			Top phase			Bottom phase		
w_A	w_B	w_C	w_A	w_B	w_C	w_A	w_B	w_C
pH = 5.20								
0.6875	0.1714	0.1411	0.6567	0.2431	0.1002	0.7179	0.1008	0.1813
0.6750	0.1783	0.1467	0.6245	0.2952	0.0803	0.7205	0.0731	0.2064
0.6619	0.1855	0.1526	0.6002	0.3317	0.0681	0.7182	0.0521	0.2297
pH = 8.20								
0.7303	0.1613	0.1084	0.6724	0.2620	0.0656	0.7786	0.0773	0.1441
0.7161	0.1698	0.1141	0.6357	0.3116	0.0527	0.7848	0.0486	0.1666
0.6999	0.1795	0.1206	0.5981	0.3604	0.0415	0.7848	0.0287	0.1865
pH = 9.20								
0.7143	0.1742	0.1115	0.6826	0.2387	0.0787	0.7609	0.0793	0.1598
0.7071	0.1789	0.1140	0.6544	0.2804	0.0652	0.7621	0.0731	0.1648
0.7016	0.1820	0.1164	0.6399	0.3008	0.0593	0.7633	0.0632	0.1735
0.6932	0.1871	0.1197	0.6200	0.3282	0.0518	0.7635	0.0515	0.1850

Polymer (B): **poly(ethylene glycol)** **2008ALV**
Characterization: M_w/g.mol^{-1} = 600, Hoechst GmbH, Frankfurt, Germany
Solvent (A): **water** H_2O **7732-18-5**
Salt (C): **sodium citrate** $Na_3C_6H_5O_7$ **68-04-2**

Type of data: coexistence data (tie lines)

T/K = 298.15

Total system			Top phase			Bottom phase		
w_A	w_B	w_C	w_A	w_B	w_C	w_A	w_B	w_C
0.4995	0.3005	0.2000	0.4320	0.5520	0.0160	0.5870	0.0320	0.3810
0.5402	0.2800	0.1798	0.4586	0.5220	0.0194	0.6170	0.0160	0.3670
0.5801	0.2610	0.1589	0.5140	0.4570	0.0290	0.6570	0.0190	0.3240
0.6090	0.2400	0.1510	0.5446	0.4170	0.0384	0.6750	0.0560	0.2690

Polymer (B): **poly(ethylene glycol)** **2006TUB**
Characterization: M_n/g.mol^{-1} = 1000, Sigma Chemical Co., Inc., St. Louis, MO
Solvent (A): **water** H_2O **7732-18-5**
Salt (C): **sodium citrate** $Na_3C_6H_5O_7$ **68-04-2**

Type of data: coexistence data (tie lines)

continued

continued

T/K = 295.15

Total system			Top phase			Bottom phase		
w_A	w_B	w_C	w_A	w_B	w_C	w_A	w_B	w_C
pH = 5.20								
0.7129	0.1529	0.1342	0.6654	0.2420	0.0926	0.7494	0.0844	0.1662
0.7084	0.1553	0.1363	0.6489	0.2663	0.0848	0.7531	0.0719	0.1750
0.7011	0.1592	0.1397	0.6309	0.2921	0.0770	0.7569	0.0537	0.1894
pH = 8.20								
0.7776	0.0988	0.1236	0.6804	0.2585	0.0611	0.7920	0.0751	0.1329
0.7709	0.1018	0.1273	0.6697	0.2729	0.0574	0.7984	0.0552	0.1464
0.7637	0.1050	0.1313	0.6477	0.3020	0.0503	0.8015	0.0408	0.1577
0.7560	0.1084	0.1356	0.6259	0.3301	0.0440	0.8027	0.0289	0.1684
0.7479	0.1120	0.1401	0.5884	0.3772	0.0344	0.8025	0.0212	0.1763
pH = 9.20								
0.7491	0.1486	0.1023	0.6893	0.2461	0.0646	0.7973	0.0700	0.1327
0.7407	0.1536	0.1057	0.6647	0.2788	0.0565	0.8040	0.0493	0.1467
0.7327	0.1583	0.1090	0.6396	0.3112	0.0492	0.8067	0.0369	0.1564
0.7260	0.1623	0.1117	0.6230	0.3321	0.0449	0.8077	0.0276	0.1647

Polymer (B):	**poly(ethylene glycol)**		**2006TUB**
Characterization:	M_n/g.mol^{-1} = 1450, Sigma Chemical Co., Inc., St. Louis, MO		
Solvent (A):	**water**	**H_2O**	**7732-18-5**
Salt (C):	**sodium citrate**	**$Na_3C_6H_5O_7$**	**68-04-2**

Type of data: coexistence data (tie lines)

T/K = 295.15

Total system			Top phase			Bottom phase		
w_A	w_B	w_C	w_A	w_B	w_C	w_A	w_B	w_C
pH = 5.20								
0.7384	0.1501	0.1115	0.7145	0.1973	0.0882	0.7731	0.0816	0.1453
0.7302	0.1548	0.1150	0.6900	0.2345	0.0755	0.7786	0.0588	0.1626
0.7255	0.1575	0.1170	0.6773	0.2530	0.0697	0.7797	0.0502	0.1701
0.7164	0.1627	0.1209	0.6555	0.2836	0.0609	0.7798	0.0368	0.1834

continued

continued

pH = 8.20

0.8026	0.0934	0.1040	0.7405	0.2026	0.0569	0.8247	0.0545	0.1208
0.7938	0.0975	0.1087	0.7102	0.2429	0.0469	0.8271	0.0397	0.1332
0.7807	0.1037	0.1156	0.6697	0.2941	0.0362	0.8262	0.0257	0.1481
0.7736	0.1071	0.1193	0.6501	0.3181	0.0318	0.8246	0.0199	0.1555

pH = 9.20

0.7386	0.1897	0.0717	0.7164	0.2282	0.0554	0.8208	0.0471	0.1321
0.7221	0.2017	0.0762	0.6779	0.2784	0.0437	0.8225	0.0275	0.1500
0.7130	0.2083	0.0787	0.6563	0.3055	0.0382	0.8220	0.0214	0.1566

Polymer (B): **poly(ethylene glycol)** **2011PAT**
Characterization: M_w/g.mol^{-1} = 1500, Sigma Chemical Co., Inc., St. Louis, MO
Solvent (A): **water** H_2O **7732-18-5**
Salt (C): **sodium citrate** $Na_3C_6H_5O_7$ **68-04-2**

Type of data: coexistence data (tie lines)

Total system			Top phase			Bottom phase		
w_A	w_B	w_C	w_A	w_B	w_C	w_A	w_B	w_C
T/K = 283.15								
0.7318	0.1466	0.1216	0.6788	0.2517	0.0694	0.7992	0.0310	0.1698
0.6996	0.1740	0.1265	0.6318	0.3231	0.0451	0.7611	0.0113	0.2277
0.6651	0.1987	0.1363	0.5957	0.3706	0.0337	0.7342	0.0031	0.2627
0.6263	0.2261	0.1475	0.5545	0.4224	0.0230	0.6916	0.0010	0.3074
0.5997	0.2442	0.1560	0.5247	0.4563	0.0189	0.6830	0.0005	0.3165
T/K = 298.15								
0.7320	0.1463	0.1217	0.6848	0.2517	0.0635	0.7937	0.0184	0.1878
0.6980	0.1749	0.1271	0.6330	0.3231	0.0439	0.7725	0.0046	0.2229
0.6644	0.1990	0.1366	0.5992	0.3706	0.0302	0.7494	0.0018	0.2488
0.6416	0.2168	0.1416	0.5563	0.4224	0.0213	0.7171	0.0001	0.2828
0.6005	0.2430	0.1565	0.5267	0.4563	0.0170	0.6970	0.0001	0.3029
T/K = 313.15								
0.7317	0.1466	0.1216	0.6443	0.3062	0.0494	0.8027	0.0082	0.1891
0.6980	0.1749	0.1272	0.6012	0.3635	0.0354	0.7818	0.0035	0.2147
0.6643	0.1992	0.1365	0.5618	0.4126	0.0256	0.7591	0.0004	0.2405
0.6267	0.2257	0.1476	0.5215	0.4587	0.0197	0.7189	0.0001	0.2810
0.5997	0.2443	0.1560	0.4950	0.4899	0.0151	0.6943	0.0001	0.3056

Polymer (B): **poly(ethylene glycol)** **2008ALV**
Characterization: M_w/g.mol^{-1} = 1500, Clariant, Germany
Solvent (A): **water** H_2O **7732-18-5**
Salt (C): **sodium citrate** $Na_3C_6H_5O_7$ **68-04-2**

Type of data: coexistence data (tie lines)

T/K = 298.15

Total system			Top phase			Bottom phase		
w_A	w_B	w_C	w_A	w_B	w_C	w_A	w_B	w_C
0.6500	0.2606	0.0894	0.4644	0.5210	0.0146	0.6787	0.0003	0.3210
0.6569	0.2398	0.1033	0.4920	0.4890	0.0190	0.7069	0.0001	0.2930
0.6589	0.2207	0.1204	0.5280	0.4460	0.0260	0.7258	0.0052	0.2690
0.6803	0.2000	0.1197	0.5760	0.3900	0.0340	0.7530	0.0080	0.2390
0.6737	0.1800	0.1463	0.6220	0.3340	0.0440	0.7740	0.0160	0.2100
0.7305	0.1597	0.1098	0.6820	0.2470	0.0710	0.7820	0.0460	0.1720

Polymer (B): **poly(ethylene glycol)** **2008CAR**
Characterization: M_w/g.mol^{-1} = 1500, Isofar, Rio de Janeiro, Brazil
Solvent (A): **water** H_2O **7732-18-5**
Salt (C): **sodium citrate** $Na_3C_6H_5O_7$ **68-04-2**

Type of data: coexistence data (tie lines)

Total system			Top phase			Bottom phase		
w_A	w_B	w_C	w_A	w_B	w_C	w_A	w_B	w_C
T/K = 278.15								
0.7098	0.1834	0.1068	0.6396	0.3056	0.0548	0.6908	0.1319	0.1773
0.6803	0.2040	0.1157	0.6043	0.3504	0.0453	0.6556	0.1216	0.2228
0.6500	0.2245	0.1255	0.5650	0.4059	0.0291	0.6158	0.1270	0.2572
0.6203	0.2452	0.1345	0.5577	0.4231	0.0192	0.5863	0.1411	0.2726
0.5957	0.2626	0.1417	0.5318	0.4492	0.0190	0.5704	0.1369	0.2927
T/K = 298.15								
0.6967	0.1832	0.1201	0.7525	0.1902	0.0573	0.7544	0.0380	0.2076
0.6677	0.2024	0.1299	0.6845	0.2769	0.0386	0.6597	0.0892	0.2511
0.6373	0.2226	0.1401	0.6855	0.2813	0.0332	0.6329	0.1030	0.2641
0.6078	0.2425	0.1497	0.6417	0.3341	0.0242	0.6127	0.1062	0.2811
0.5764	0.2634	0.1602	0.5859	0.3923	0.0218	0.6077	0.0994	0.2929

Polymer (B): **poly(ethylene glycol)** **2008OL1**
Characterization: M_w/g.mol^{-1} = 1500, Isofar, Rio de Janeiro, Brazil
Solvent (A): **water** H_2O **7732-18-5**
Salt (C): **sodium citrate** $Na_3C_6H_5O_7$ **68-04-2**

pH = 8.24

Type of data: coexistence data (tie lines)

Total system			Top phase			Bottom phase		
w_A	w_B	w_C	w_A	w_B	w_C	w_A	w_B	w_C
T/K = 283.15								
0.7127	0.1843	0.1030	0.6650	0.3046	0.0303	0.7588	0.0176	0.2235
0.6804	0.2039	0.1157	0.6291	0.3483	0.0226	0.7269	0.0169	0.2562
0.6503	0.2244	0.1254	0.5968	0.3787	0.0245	0.7062	0.0123	0.2816
0.6203	0.2452	0.1345	0.5652	0.4171	0.0176	0.6836	0.0150	0.3014
0.5959	0.2625	0.1416	0.5291	0.4505	0.0204	0.6573	0.0215	0.3212
T/K = 308.15								
0.6969	0.1831	0.1200	0.6634	0.3029	0.0337	0.7525	0.0304	0.2171
0.6676	0.2025	0.1299	0.6257	0.3489	0.0254	0.7303	0.0294	0.2403
0.6374	0.2226	0.1400	0.5849	0.3935	0.0216	0.7042	0.0206	0.2752
0.6078	0.2425	0.1497	0.5522	0.4320	0.0158	0.6850	0.0141	0.3010
0.5762	0.2635	0.1603	0.5234	0.4645	0.0121	0.6590	0.0143	0.3267
T/K = 318.15								
0.6966	0.1834	0.1200	0.6425	0.3220	0.0354	0.7481	0.0363	0.2156
0.6680	0.2022	0.1298	0.6197	0.3552	0.0251	0.7239	0.0431	0.2330
0.6374	0.2226	0.1400	0.5885	0.3909	0.0206	0.6966	0.0343	0.2690
0.6079	0.2425	0.1496	0.5405	0.4450	0.0145	0.6872	0.0317	0.2811
0.5770	0.2631	0.1599	0.5245	0.4641	0.0114	0.6583	0.0351	0.3066

Polymer (B): **poly(ethylene glycol)** **2005MUR**
Characterization: M_n/g.mol^{-1} = 2000, Merck, Darmstadt, Germany
Solvent (A): **water** H_2O **7732-18-5**
Salt (C): **sodium citrate** $Na_3C_6H_5O_7$ **68-04-2**

Type of data: cloud points

T/K = 298.15

w_A	0.5245	0.5509	0.5779	0.5894	0.6204	0.6239	0.6424	0.6677	0.6602
w_B	0.4470	0.4155	0.3824	0.3681	0.3311	0.3235	0.2987	0.2645	0.2700
w_C	0.0285	0.0336	0.0397	0.0425	0.0485	0.0526	0.0589	0.0678	0.0698

continued

continued

w_A	0.6750	0.6893	0.7008	0.7060	0.7172	0.7241	0.7286	0.7379	0.7424
w_B	0.2525	0.2314	0.2150	0.2039	0.1844	0.1734	0.1627	0.1465	0.1363
w_C	0.0725	0.0793	0.0842	0.0901	0.0984	0.1025	0.1087	0.1156	0.1213
w_A	0.7479	0.7497	0.7494	0.7521	0.7530	0.7540	0.7548	0.7553	0.7555
w_B	0.1165	0.1105	0.1090	0.1012	0.0972	0.0921	0.0863	0.0813	0.0756
w_C	0.1356	0.1398	0.1416	0.1467	0.1498	0.1539	0.1589	0.1634	0.1689
w_A	0.7554	0.7550	0.7472	0.7410	0.7403				
w_B	0.0691	0.0648	0.0473	0.0401	0.0395				
w_C	0.1755	0.1802	0.2055	0.2189	0.2202				

T/K = 303.15

w_A	0.5350	0.5690	0.6271	0.6398	0.6573	0.6670	0.6820	0.6880	0.7005
w_B	0.4380	0.3974	0.3244	0.3076	0.2838	0.2702	0.2485	0.2395	0.2202
w_C	0.0270	0.0336	0.0485	0.0526	0.0589	0.0628	0.0695	0.0725	0.0793
w_A	0.7109	0.7175	0.7261	0.7320	0.7375	0.7427	0.7468	0.7549	0.7563
w_B	0.2033	0.1920	0.1766	0.1653	0.1541	0.1427	0.1329	0.1093	0.1042
w_C	0.0858	0.0905	0.0973	0.1027	0.1084	0.1146	0.1203	0.1358	0.1395
w_A	0.7576	0.7584	0.7589	0.7597	0.7603	0.7605	0.7605	0.7601	0.7573
w_B	0.0988	0.0947	0.0919	0.0865	0.0810	0.0760	0.0723	0.0654	0.0530
w_C	0.1436	0.1469	0.1492	0.1538	0.1587	0.1635	0.1672	0.1745	0.1897
w_A	0.7557	0.7536	0.7410						
w_B	0.0487	0.0447	0.0305						
w_C	0.1956	0.2017	0.2285						

T/K = 308.15

w_A	0.5581	0.5957	0.6248	0.6444	0.6634	0.6833	0.6956	0.7060	0.7214
w_B	0.4151	0.3685	0.3337	0.3067	0.2810	0.2546	0.2346	0.2186	0.1941
w_C	0.0268	0.0358	0.0415	0.0489	0.0556	0.0621	0.0698	0.0754	0.0845
w_A	0.7331	0.7409	0.7428	0.7566	0.7574	0.7607	0.7634	0.7616	0.7650
w_B	0.1682	0.1577	0.1483	0.1269	0.1183	0.1106	0.1064	0.1028	0.0952
w_C	0.0987	0.1014	0.1089	0.1165	0.1243	0.1287	0.1302	0.1356	0.1398
w_A	0.7675	0.7632	0.7689						
w_B	0.0873	0.0883	0.0788						
w_C	0.1452	0.1485	0.1523						

T/K = 313.15

w_A	0.5842	0.6159	0.6403	0.6629	0.6802	0.6969	0.7115	0.7207	0.7309
w_B	0.3892	0.3503	0.3192	0.2892	0.2652	0.2410	0.2187	0.2039	0.1866
w_C	0.0266	0.0338	0.0405	0.0479	0.0546	0.0621	0.0698	0.0754	0.0825
w_A	0.7491	0.7516	0.7575	0.7620	0.7662	0.7690	0.7699	0.7710	0.7722
w_B	0.1522	0.1470	0.1336	0.1225	0.1105	0.1013	0.0979	0.0934	0.0880
w_C	0.0987	0.1014	0.1089	0.1155	0.1233	0.1297	0.1322	0.1356	0.1398

continued

continued

w_A	0.7731	0.7738	0.7743	0.7747	0.7747
w_B	0.0827	0.0777	0.0734	0.0663	0.0641
w_C	0.1442	0.1485	0.1523	0.1590	0.1612

T/K = 318.15

w_A	0.6054	0.6166	0.6330	0.6636	0.6767	0.6903	0.6956	0.7064	0.7124
w_B	0.3701	0.3569	0.3358	0.2966	0.2775	0.2601	0.2521	0.2368	0.2278
w_C	0.0245	0.0265	0.0312	0.0398	0.0458	0.0496	0.0523	0.0568	0.0598
w_A	0.7163	0.7227	0.7307	0.7353	0.7462	0.7467	0.7527	0.7567	0.7660
w_B	0.2212	0.2088	0.1991	0.1891	0.1745	0.1677	0.1579	0.1487	0.1356
w_C	0.0625	0.0685	0.0702	0.0756	0.0793	0.0856	0.0894	0.0946	0.0984
w_A	0.7631	0.7705	0.7700	0.7753	0.7811	0.7770	0.7748	0.7771	0.7805
w_B	0.1346	0.1209	0.1172	0.1056	0.0958	0.1045	0.1049	0.0973	0.0897
w_C	0.1023	0.1086	0.1128	0.1191	0.1231	0.1185	0.1203	0.1256	0.1298
w_A	0.7797	0.7819	0.7798	0.7815					
w_B	0.0867	0.0796	0.0768	0.0691					
w_C	0.1336	0.1385	0.1434	0.1494					

Polymer (B): **poly(ethylene glycol)** **2005MUR, 2006PER**
Characterization: M_n/g.mol^{-1} = 2000, Merck, Darmstadt, Germany
Solvent (A): **water** H_2O **7732-18-5**
Salt (C): **sodium citrate** $Na_3C_6H_5O_7$ **68-04-2**

Type of data: coexistence data (tie lines)

Top phase			Bottom phase		
w_A	w_B	w_C	w_A	w_B	w_C
T/K = 298.15					
0.6644	0.2679	0.0677	0.7533	0.0417	0.2050
0.6356	0.3078	0.0566	0.7352	0.0323	0.2325
0.6068	0.3443	0.0489	0.7202	0.0258	0.2540
0.5779	0.3832	0.0389	0.7019	0.0170	0.2811
T/K = 308.15					
0.6856	0.2486	0.0658	0.7816	0.0599	0.1585
0.6495	0.2989	0.0516	0.7670	0.0430	0.1900
0.6137	0.3422	0.0441	0.7519	0.0330	0.2151
0.5919	0.3744	0.0337	0.7428	0.0209	0.2363
0.5632	0.4104	0.0264	0.7277	0.0115	0.2608

continued

continued

T/K = 318.15

0.6559	0.3053	0.0388	0.7968	0.0467	0.1565
0.6272	0.3405	0.0323	0.7856	0.0376	0.1768
0.5949	0.3831	0.0220	0.7788	0.0236	0.1976
0.5734	0.4078	0.0188	0.7691	0.0181	0.2128
0.5520	0.4343	0.0137	0.7591	0.0104	0.2305

Polymer (B): **poly(ethylene glycol)** **2010PER**
Characterization: M_n/g.mol^{-1} = 2000, Merck, Darmstadt, Germany
Solvent (A): **water** H_2O **7732-18-5**
Salt (C): **sodium citrate** $Na_3C_6H_5O_7$ **68-04-2**

Type of data: coexistence data (tie lines)

Total system			Top phase			Bottom phase		
w_A	w_B	w_C	w_A	w_B	w_C	w_A	w_B	w_C
T/K = 298.15								
0.725	0.125	0.150	0.6644	0.2679	0.0677	0.7533	0.0417	0.2050
0.700	0.125	0.175	0.6356	0.3078	0.0566	0.7352	0.0323	0.2325
0.675	0.125	0.200	0.6068	0.3443	0.0489	0.7202	0.0258	0.2540
0.650	0.125	0.225	0.5779	0.3832	0.0389	0.7019	0.0170	0.2811
T/K = 308.15								
0.750	0.125	0.125	0.6856	0.2486	0.0658	0.7816	0.0599	0.1585
0.725	0.125	0.150	0.6495	0.2989	0.0516	0.7670	0.0430	0.1900
0.700	0.125	0.175	0.6137	0.3422	0.0441	0.7519	0.0330	0.2151
0.675	0.125	0.200	0.5919	0.3744	0.0337	0.7428	0.0209	0.2363
0.650	0.125	0.225	0.5632	0.4104	0.0264	0.7277	0.0115	0.2608
T/K = 318.15								
0.750	0.125	0.125	0.6559	0.3053	0.0388	0.7968	0.0467	0.1565
0.725	0.125	0.150	0.6272	0.3405	0.0323	0.7856	0.0376	0.1768
0.700	0.125	0.175	0.5949	0.3831	0.0220	0.7788	0.0236	0.1976
0.675	0.125	0.200	0.5734	0.4078	0.0188	0.7691	0.0181	0.2128
0.650	0.125	0.225	0.5520	0.4343	0.0137	0.7591	0.0104	0.2305

Polymer (B): **poly(ethylene glycol)** **2008ALV**
Characterization: M_w/g.mol^{-1} = 3000, Fluka AG, Buchs, Switzerland
Solvent (A): **water** H_2O **7732-18-5**
Salt (C): **sodium citrate** $Na_3C_6H_5O_7$ **68-04-2**

Type of data: coexistence data (tie lines)

continued

continued

T/K = 298.15

Total system			Top phase			Bottom phase		
w_A	w_B	w_C	w_A	w_B	w_C	w_A	w_B	w_C
0.6403	0.2197	0.1400	0.5119	0.4690	0.0191	0.7367	0.0013	0.2620
0.6691	0.1999	0.1310	0.5426	0.4340	0.0234	0.7594	0.0016	0.2390
0.6993	0.1797	0.1210	0.5864	0.3840	0.0296	0.7831	0.0039	0.2130
0.7280	0.1590	0.1130	0.6290	0.3330	0.0380	0.8033	0.0067	0.1900
0.7404	0.1550	0.1046	0.6718	0.2840	0.0442	0.8160	0.0100	0.1740
0.7748	0.1251	0.1001	0.7213	0.2100	0.0687	0.8218	0.0372	0.1410

Polymer (B): **poly(ethylene glycol)** **2006TUB**
Characterization: M_n/g.mol^{-1} = 3350, Sigma Chemical Co., Inc., St. Louis, MO
Solvent (A): **water** H_2O **7732-18-5**
Salt (C): **sodium citrate** $Na_3C_6H_5O_7$ **68-04-2**

Type of data: coexistence data (tie lines)

T/K = 295.15

Total system			Top phase			Bottom phase		
w_A	w_B	w_C	w_A	w_B	w_C	w_A	w_B	w_C
pH = 5.20								
0.7935	0.1012	0.1053	0.7626	0.1596	0.0778	0.8178	0.0552	0.1270
0.7868	0.1045	0.1087	0.7432	0.1876	0.0692	0.8208	0.0396	0.1396
0.7800	0.1078	0.1122	0.7228	0.2156	0.0616	0.8211	0.0303	0.1486
0.7727	0.1114	0.1159	0.7030	0.2419	0.0551	0.8202	0.0226	0.1572
pH = 8.20								
0.8302	0.0889	0.0809	0.7812	0.1664	0.0524	0.8554	0.0491	0.0955
0.8228	0.0928	0.0844	0.7535	0.2021	0.0444	0.8601	0.0340	0.1059
0.8148	0.0970	0.0882	0.7270	0.2351	0.0379	0.8616	0.0235	0.1149
0.8060	0.1016	0.0924	0.6986	0.2695	0.0319	0.8611	0.0155	0.1234
pH = 9.20								
0.8286	0.0818	0.0896	0.7910	0.1441	0.0649	0.8456	0.0536	0.1008
0.8233	0.0844	0.0923	0.7660	0.1778	0.0562	0.8498	0.0412	0.1090
0.8174	0.0872	0.0954	0.7412	0.2100	0.0488	0.8516	0.0321	0.1163
0.8115	0.0900	0.0985	0.7285	0.2261	0.0454	0.8518	0.0239	0.1243

Polymer (B):	**poly(ethylene glycol)**		**2009PER**
Characterization:	M_n/g.mol^{-1} = 4000, Merck-Schuchardt, Hohenbrunn, Germany		
Solvent (A):	**water**	**H_2O**	**7732-18-5**
Salt (C):	**sodium citrate**	**$Na_3C_6H_5O_7$**	**68-04-2**

Type of data: cloud points

T/K = 298.15

w_A	0.5385	0.5659	0.5929	0.6054	0.6354	0.6399	0.6584	0.6837	0.6829
w_B	0.4400	0.4085	0.3754	0.3611	0.3241	0.3165	0.2917	0.2595	0.2593
w_C	0.0215	0.0256	0.0317	0.0335	0.0405	0.0436	0.0499	0.0568	0.0578
w_A	0.6940	0.7093	0.7208	0.7270	0.7392	0.7471	0.7526	0.7629	0.7684
w_B	0.2455	0.2244	0.2080	0.1969	0.1774	0.1664	0.1557	0.1395	0.1293
w_C	0.0605	0.0663	0.0712	0.0761	0.0834	0.0865	0.0917	0.0976	0.1023
w_A	0.7799	0.7837	0.7844	0.7871	0.7880	0.7900	0.7908	0.7913	0.7935
w_B	0.1095	0.1035	0.1020	0.0942	0.0902	0.0851	0.0793	0.0743	0.0686
w_C	0.1106	0.1128	0.1136	0.1187	0.1218	0.1249	0.1299	0.1344	0.1379
w_A	0.7974	0.7980	0.7992	0.7950	0.7812	0.7689	0.7500	0.7310	
w_B	0.0621	0.0578	0.0403	0.0331	0.0347	0.0270	0.0235	0.0147	
w_C	0.1405	0.1442	0.1605	0.1719	0.1841	0.2041	0.2265	0.2543	

T/K = 308.15

w_A	0.5664	0.5872	0.6051	0.6362	0.6751	0.6964	0.7149	0.7356	0.7487
w_B	0.4120	0.3874	0.3656	0.3295	0.2815	0.2521	0.2291	0.2024	0.1831
w_C	0.0216	0.0254	0.0293	0.0343	0.0434	0.0515	0.0560	0.0620	0.0682
w_A	0.7607	0.7726	0.7828	0.7889	0.7915	0.7934	0.8010	0.8055	0.8093
w_B	0.1628	0.1441	0.1278	0.1189	0.1111	0.1071	0.0917	0.0829	0.0782
w_C	0.0765	0.0833	0.0894	0.0922	0.0974	0.0995	0.1073	0.1116	0.1125
w_A	0.8151	0.8189	0.8212	0.8244	0.8011	0.7815	0.7815	0.7704	
w_B	0.0646	0.0568	0.0512	0.0455	0.0189	0.0135	0.0135	0.0096	
w_C	0.1203	0.1243	0.1276	0.1301	0.1800	0.2050	0.2050	0.2200	

T/K = 318.15

w_A	0.5491	0.5849	0.6085	0.6354	0.6537	0.6769	0.6954	0.7095	0.7211
w_B	0.4345	0.3925	0.3662	0.3358	0.3134	0.2855	0.2626	0.2437	0.2283
w_C	0.0164	0.0226	0.0253	0.0288	0.0329	0.0376	0.0420	0.0468	0.0506
w_A	0.7342	0.7450	0.7522	0.7580	0.7670	0.7724	0.7783	0.7815	0.7868
w_B	0.2118	0.1957	0.1845	0.1762	0.1658	0.1583	0.1502	0.1448	0.1380
w_C	0.0540	0.0593	0.0633	0.0658	0.0672	0.0693	0.0715	0.0737	0.0752
w_A	0.7906	0.7943	0.7982	0.8047	0.8110	0.8169	0.8221	0.8239	0.8265
w_B	0.1324	0.1279	0.1222	0.1134	0.1040	0.0956	0.0881	0.0824	0.0757
w_C	0.0770	0.0778	0.0796	0.0819	0.0850	0.0875	0.0898	0.0937	0.0978
w_A	0.8279	0.8344	0.8398	0.8445	0.8494	0.8518	0.8518	0.8458	0.8224
w_B	0.0704	0.0590	0.0485	0.0388	0.0310	0.0238	0.0185	0.0178	0.0186
w_C	0.1017	0.1066	0.1117	0.1167	0.1196	0.1244	0.1297	0.1364	0.1590

continued

continued

w_A	0.8105	0.7952	0.7763
w_B	0.0175	0.0148	0.0057
w_C	0.1720	0.1900	0.2180

Type of data: coexistence data (tie lines)

Top phase			Bottom phase		
w_A	w_B	w_C	w_A	w_B	w_C
T/K = 298.15					
0.6775	0.2615	0.0610	0.7822	0.0337	0.1841
0.6440	0.3060	0.0500	0.7689	0.0270	0.2041
0.6103	0.3502	0.0395	0.7500	0.0235	0.2265
0.5827	0.3863	0.0310	0.7310	0.0147	0.2543
T/K = 308.15					
0.6853	0.2647	0.0500	0.8225	0.0295	0.1480
0.6450	0.3150	0.0400	0.8011	0.0189	0.1800
0.6135	0.3515	0.0350	0.7815	0.0135	0.2050
0.5824	0.3901	0.0275	0.7704	0.0096	0.2200
T/K = 318.15					
0.6734	0.2876	0.0390	0.8224	0.0186	0.1590
0.6420	0.3250	0.0330	0.8105	0.0175	0.1720
0.6024	0.3706	0.0270	0.7952	0.0148	0.1900
0.5614	0.4176	0.0210	0.7763	0.0057	0.2180

Polymer (B):	**poly(ethylene glycol)**		**2008OL1**
Characterization:	M_w/g.mol^{-1} = 4000, Isofar, Rio de Janeiro, Brazil		
Solvent (A):	**water**	**H_2O**	**7732-18-5**
Salt (C):	**sodium citrate**	**$Na_3C_6H_5O_7$**	**68-04-2**

Type of data: coexistence data (tie lines)

Total system			Top phase			Bottom phase		
w_A	w_B	w_C	w_A	w_B	w_C	w_A	w_B	w_C
T/K = 278.15	pH = 8.24							
0.7101	0.1832	0.1067	0.6593	0.3037	0.0370	0.7648	0.0266	0.2086
0.6806	0.2037	0.1157	0.6295	0.3435	0.0269	0.7624	0.0050	0.2326
0.6502	0.2243	0.1256	0.5835	0.3939	0.0226	0.7283	0.0093	0.2625
0.6205	0.2450	0.1345	0.5501	0.4307	0.0192	0.7046	0.0065	0.2888
0.5959	0.2624	0.1417	0.5321	0.4534	0.0146	0.6764	0.0084	0.3152

continued

continued

T/K = 283.15 pH = 8.24

0.7101	0.1832	0.1067	0.6593	0.3037	0.0370	0.7648	0.0266	0.2086
0.6806	0.2037	0.1157	0.6295	0.3435	0.0269	0.7624	0.0050	0.2326
0.6502	0.2243	0.1256	0.5835	0.3939	0.0226	0.7283	0.0093	0.2625
0.6205	0.2450	0.1345	0.5501	0.4307	0.0192	0.7046	0.0065	0.2888
0.5959	0.2624	0.1417	0.5321	0.4534	0.0146	0.6764	0.0084	0.3152

T/K = 298.15 pH = 8.24

0.6969	0.1829	0.1202	0.6130	0.3628	0.0242	0.7202	0.0019	0.2286
0.6682	0.2020	0.1299	0.5933	0.3875	0.0192	0.6885	0.0000	0.2562
0.6378	0.2221	0.1401	0.5497	0.4377	0.0126	0.6771	0.0000	0.2731
0.6084	0.2416	0.1500	0.5228	0.4669	0.0103	0.6317	0.0000	0.3089
0.5770	0.2629	0.1602	0.4850	0.5059	0.0091	0.5977	0.0000	0.3448

T/K = 308.15 pH = 8.24

0.6968	0.1831	0.1201	0.5922	0.3853	0.0225	0.7723	0.0000	0.2277
0.6673	0.2027	0.1299	0.5577	0.4249	0.0175	0.7432	0.0000	0.2568
0.6367	0.2220	0.1414	0.5429	0.4465	0.0106	0.7169	0.0000	0.2831
0.6077	0.2422	0.1501	0.5059	0.4833	0.0108	0.6881	0.0000	0.3119
0.5762	0.2632	0.1605	0.4905	0.5006	0.0090	0.6434	0.0000	0.3566

T/K = 318.15 pH = 8.24

0.6968	0.1831	0.1200	0.5870	0.3985	0.0145	0.7741	0.0000	0.2259
0.6677	0.2024	0.1300	0.5513	0.4336	0.0151	0.7605	0.0000	0.2395
0.6370	0.2225	0.1405	0.5365	0.4520	0.0116	0.7240	0.0000	0.2760
0.6079	0.2424	0.1496	0.4912	0.4998	0.0090	0.6947	0.0000	0.3053
0.5764	0.2633	0.1603	0.4797	0.5133	0.0070	0.6654	0.0000	0.3346

Polymer (B): **poly(ethylene glycol)** **2004ZA2**
Characterization: M_n/g.mol^{-1} = 5890, Merck, Darmstadt, Germany
Solvent (A): **water** **H_2O** **7732-18-5**
Salt (C): **sodium citrate** **$Na_3C_6H_5O_7$** **68-04-2**

Type of data: cloud points

T/K = 298.15

w_A	0.5352	0.5660	0.6091	0.6142	0.6317	0.6444	0.6608	0.6757	0.6919
w_B	0.4503	0.4159	0.3663	0.3598	0.3398	0.3246	0.3051	0.2865	0.2666
w_C	0.0145	0.0181	0.0246	0.0260	0.0285	0.0310	0.0341	0.0378	0.0415
w_A	0.7042	0.7182	0.7288	0.7405	0.7511	0.7874	0.7999	0.8120	0.8217
w_B	0.2508	0.2328	0.2189	0.2034	0.1892	0.1370	0.1160	0.0946	0.0777
w_C	0.0450	0.0490	0.0523	0.0561	0.0597	0.0756	0.0841	0.0934	0.1006
w_A	0.8291	0.8348	0.8409	0.8424	0.8468	0.8510	0.8534	0.8527	
w_B	0.0644	0.0547	0.0461	0.0406	0.0343	0.0240	0.0165	0.0110	
w_C	0.1065	0.1105	0.1130	0.1170	0.1189	0.1250	0.1301	0.1363	

continued

continued

T/K = 308.15

w_A	0.5684	0.5902	0.6091	0.6402	0.6811	0.7044	0.7229	0.7436	0.7587
w_B	0.4140	0.3894	0.3676	0.3315	0.2835	0.2541	0.2311	0.2044	0.1851
w_C	0.0176	0.0204	0.0233	0.0283	0.0354	0.0415	0.0460	0.0520	0.0562
w_A	0.7727	0.7856	0.7968	0.8029	0.8075	0.8104	0.8142	0.8173	0.8200
w_B	0.1648	0.1461	0.1298	0.1209	0.1141	0.1091	0.1036	0.0990	0.0947
w_C	0.0625	0.0683	0.0734	0.0762	0.0784	0.0805	0.0822	0.0837	0.0853
w_A	0.8228	0.8265	0.8293	0.8310	0.8330	0.8352	0.8371	0.8393	0.8419
w_B	0.0899	0.0849	0.0802	0.0774	0.0740	0.0708	0.0676	0.0640	0.0598
w_C	0.0873	0.0886	0.0905	0.0916	0.0930	0.0940	0.0953	0.0967	0.0983
w_A	0.8441	0.8462	0.8478	0.8494	0.8541	0.8579	0.8623	0.8640	0.8645
w_B	0.0565	0.0532	0.0503	0.0475	0.0400	0.0337	0.0232	0.0195	0.0154
w_C	0.0994	0.1006	0.1019	0.1031	0.1059	0.1084	0.1145	0.1165	0.1201
w_A	0.8649	0.8642	0.8620						
w_B	0.0121	0.0086	0.0058						
w_C	0.1230	0.1272	0.1322						

T/K = 318.15

w_A	0.5571	0.5929	0.6155	0.6424	0.6607	0.6849	0.7044	0.7195	0.7321
w_B	0.4305	0.3885	0.3622	0.3318	0.3094	0.2815	0.2586	0.2397	0.2243
w_C	0.0124	0.0186	0.0223	0.0258	0.0299	0.0336	0.0370	0.0408	0.0436
w_A	0.7452	0.7580	0.7652	0.7730	0.7810	0.7864	0.7923	0.7975	0.8028
w_B	0.2078	0.1917	0.1805	0.1702	0.1598	0.1523	0.1442	0.1368	0.1300
w_C	0.0470	0.0503	0.0543	0.0568	0.0592	0.0613	0.0635	0.0657	0.0672
w_A	0.8066	0.8103	0.8142	0.8207	0.8270	0.8329	0.8381	0.8419	0.8465
w_B	0.1244	0.1199	0.1142	0.1054	0.0960	0.0876	0.0801	0.0744	0.0677
w_C	0.0690	0.0698	0.0716	0.0739	0.0770	0.0795	0.0818	0.0837	0.0858
w_A	0.8499	0.8574	0.8638	0.8695	0.8744	0.8768	0.8768	0.8708	
w_B	0.0624	0.0510	0.0405	0.0308	0.0230	0.0158	0.0105	0.0098	
w_C	0.0877	0.0916	0.0957	0.0997	0.1026	0.1074	0.1127	0.1194	

Type of data: coexistence data (tie lines)

Top phase			Bottom phase		
w_A	w_B	w_C	w_A	w_B	w_C
T/K = 298.15					
0.5865	0.3885	0.0250	0.7590	0.0052	0.2358
0.6063	0.3676	0.0261	0.7830	0.0032	0.2138
0.6324	0.3380	0.0296	0.8053	0.0031	0.1916
0.6503	0.3193	0.0304	0.8171	0.0030	0.1799
0.6692	0.2949	0.0359	0.8245	0.0040	0.1715
0.6923	0.2686	0.0391	0.8378	0.0068	0.1554
0.7326	0.2134	0.0540	0.8521	0.0117	0.1362

continued

continued

T/K = 308.15

0.5493	0.4344	0.0163	0.7756	0.0001	0.2243
0.5670	0.4145	0.0185	0.7841	0.0009	0.2150
0.6067	0.3716	0.0217	0.7961	0.0020	0.2019
0.6327	0.3417	0.0256	0.8173	0.0033	0.1794
0.6806	0.2833	0.0361	0.8453	0.0042	0.1505
0.7181	0.2374	0.0445	0.8633	0.0125	0.1242

T/K = 318.15

0.5423	0.4474	0.0103	0.7827	0.0027	0.2146
0.5512	0.4373	0.0115	0.7920	0.0003	0.2077
0.5674	0.4183	0.0143	0.8040	0.0007	0.1953
0.5932	0.3881	0.0187	0.8138	0.0002	0.1860
0.6196	0.3585	0.0219	0.8231	0.0020	0.1749
0.6470	0.3267	0.0263	0.8324	0.0020	0.1656
0.6820	0.2854	0.0326	0.8404	0.0040	0.1556
0.7217	0.2373	0.0410	0.8583	0.0076	0.1341

Polymer (B): **poly(ethylene glycol)** **2007PER**
Characterization: M_n/g.mol^{-1} = 6000, Merck, Darmstadt, Germany
Solvent (A): **water** H_2O **7732-18-5**
Salt (C): **sodium citrate** $Na_3C_6H_5O_7$ **68-04-2**

Type of data: cloud points

T/K = 293.15

w_A	0.5071	0.5230	0.5515	0.5758	0.5971	0.6165	0.6333	0.6488	0.6625
w_B	0.4823	0.4637	0.4305	0.4029	0.3787	0.3553	0.3363	0.3170	0.2998
w_C	0.0106	0.0133	0.0180	0.0213	0.0242	0.0282	0.0304	0.0342	0.0377
w_A	0.6748	0.6961	0.7137	0.7284	0.7464	0.7605	0.7749	0.7855	0.7958
w_B	0.2850	0.2574	0.2350	0.2159	0.1922	0.1720	0.1505	0.1325	0.1157
w_C	0.0402	0.0465	0.0513	0.0557	0.0614	0.0675	0.0746	0.0820	0.0885
w_A	0.8032	0.8090	0.8135	0.8182	0.8245	0.8293	0.8327	0.8378	0.8399
w_B	0.1021	0.0912	0.0824	0.0757	0.0645	0.0563	0.0498	0.0406	0.0339
w_C	0.0947	0.0998	0.1041	0.1061	0.1110	0.1144	0.1175	0.1216	0.1262
w_A	0.8427	0.8420	0.8361	0.8280	0.8130	0.7875	0.7753	0.7371	0.6830
w_B	0.0273	0.0167	0.0105	0.0058	0.0027	0.0026	0.0023	0.0022	0.0020
w_C	0.1300	0.1413	0.1534	0.1662	0.1843	0.2099	0.2224	0.2607	0.3150
w_A	0.6203	0.6006							
w_B	0.0016	0.0016							
w_C	0.3781	0.3978							

continued

continued

T/K = 303.15

w_A	0.5057	0.5368	0.5631	0.5862	0.6065	0.6248	0.6410	0.6554	0.6687
w_B	0.4857	0.4493	0.4193	0.3930	0.3698	0.3483	0.3292	0.3128	0.2966
w_C	0.0086	0.0139	0.0176	0.0208	0.0237	0.0269	0.0298	0.0318	0.0347
w_A	0.6805	0.6914	0.7011	0.7102	0.7184	0.7259	0.7328	0.7392	0.7451
w_B	0.2821	0.2688	0.2559	0.2449	0.2349	0.2249	0.2161	0.2076	0.2001
w_C	0.0374	0.0398	0.0430	0.0449	0.0467	0.0492	0.0511	0.0532	0.0548
w_A	0.7506	0.7557	0.7604	0.7647	0.7685	0.7759	0.7823	0.7881	0.7932
w_B	0.1931	0.1863	0.1797	0.1736	0.1675	0.1571	0.1478	0.1399	0.1323
w_C	0.0563	0.0580	0.0599	0.0617	0.0640	0.0670	0.0699	0.0720	0.0745
w_A	0.7972	0.8011	0.8065	0.8108	0.8149	0.8183	0.8211	0.8257	0.8292
w_B	0.1251	0.1188	0.1108	0.1035	0.0972	0.0915	0.0864	0.0792	0.0730
w_C	0.0777	0.0801	0.0827	0.0857	0.0879	0.0902	0.0925	0.0951	0.0978
w_A	0.8331	0.8367	0.8399	0.8436	0.8465	0.8500	0.8525	0.8543	0.8554
w_B	0.0669	0.0608	0.0550	0.0493	0.0447	0.0392	0.0348	0.0313	0.0283
w_C	0.1000	0.1025	0.1051	0.1071	0.1088	0.1108	0.1127	0.1144	0.1163
w_A	0.8566	0.8570	0.8580	0.8585	0.8588	0.8589	0.8589	0.8599	0.8549
w_B	0.0252	0.0226	0.0205	0.0188	0.0173	0.0161	0.0144	0.0109	0.0067
w_C	0.1182	0.1204	0.1215	0.1227	0.1239	0.1250	0.1267	0.1292	0.1384
w_A	0.8441	0.8191	0.7918	0.7747	0.7161	0.7003			
w_B	0.0040	0.0030	0.0025	0.0022	0.0020	0.0014			
w_C	0.1519	0.1779	0.2057	0.2231	0.2819	0.2983			

T/K = 313.15

w_A	0.5120	0.5313	0.5480	0.5775	0.6022	0.6131	0.6233	0.6328	0.6436
w_B	0.4799	0.4572	0.4379	0.4039	0.3757	0.3635	0.3521	0.3414	0.3292
w_C	0.0081	0.0115	0.0141	0.0186	0.0221	0.0234	0.0246	0.0258	0.0272
w_A	0.6530	0.6592	0.6763	0.6829	0.6911	0.6973	0.7032	0.7146	0.7195
w_B	0.3177	0.3108	0.2904	0.2831	0.2730	0.2656	0.2585	0.2441	0.2385
w_C	0.0293	0.0300	0.0333	0.0340	0.0359	0.0371	0.0383	0.0413	0.0420
w_A	0.7246	0.7364	0.7420	0.7445	0.7532	0.7566	0.7660	0.7718	0.7770
w_B	0.2317	0.2166	0.2096	0.2062	0.1952	0.1906	0.1784	0.1700	0.1627
w_C	0.0437	0.0470	0.0484	0.0493	0.0516	0.0528	0.0556	0.0582	0.0603
w_A	0.7818	0.7861	0.7902	0.7975	0.8052	0.8105	0.8140	0.8183	0.8221
w_B	0.1554	0.1498	0.1438	0.1333	0.1229	0.1150	0.1099	0.1035	0.0979
w_C	0.0628	0.0641	0.0660	0.0692	0.0719	0.0745	0.0761	0.0782	0.0800
w_A	0.8255	0.8286	0.8299	0.8340	0.8392	0.8427	0.8456	0.8499	0.8559
w_B	0.0926	0.0882	0.0858	0.0803	0.0720	0.0664	0.0616	0.0547	0.0454
w_C	0.0819	0.0832	0.0843	0.0857	0.0888	0.0909	0.0928	0.0954	0.0987

continued

continued

w_A	0.8609	0.8633	0.8657	0.8698	0.8700	0.8698	0.8690	0.8680	0.8629
w_B	0.0369	0.0326	0.0287	0.0213	0.0193	0.0169	0.0145	0.0096	0.0063
w_C	0.1022	0.1041	0.1056	0.1089	0.1107	0.1133	0.1165	0.1224	0.1308
w_A	0.8566	0.8494	0.8198	0.7923	0.7751	0.7162	0.7006		
w_B	0.0045	0.0023	0.0017	0.0016	0.0013	0.0011	0.0011		
w_C	0.1389	0.1483	0.1785	0.2061	0.2236	0.2827	0.2983		

Type of data: coexistence data (tie lines)

Top phase			Bottom phase		
w_A	w_B	w_C	w_A	w_B	w_C
T/K = 293.15					
0.6611	0.3024	0.0365	0.7738	0.0186	0.2076
0.6305	0.3391	0.0304	0.7491	0.0093	0.2416
0.5866	0.3903	0.0231	0.7383	0.0081	0.2536
0.5647	0.4149	0.0204	0.7277	0.0061	0.2662
0.5499	0.4321	0.0180	0.7175	0.0020	0.2805
0.5250	0.4614	0.0136	0.6929	0.0012	0.3059
T/K = 303.15					
0.6943	0.2649	0.0408	0.8007	0.0100	0.1893
0.6785	0.2853	0.0362	0.7964	0.0091	0.1945
0.6487	0.3212	0.0301	0.7891	0.0079	0.2030
0.6360	0.3349	0.0291	0.7758	0.0057	0.2185
0.6204	0.3531	0.0265	0.7668	0.0018	0.2314
0.6035	0.3723	0.0242	0.7599	0.0012	0.2389
T/K = 313.15					
0.7212	0.2345	0.0443	0.8266	0.0099	0.1635
0.6755	0.2915	0.0330	0.8163	0.0023	0.1814
0.6287	0.3451	0.0262	0.7931	0.0022	0.2047
0.6033	0.3741	0.0226	0.7759	0.0021	0.2220
0.5849	0.3947	0.0204	0.7663	0.0019	0.2318
0.5347	0.4532	0.0121	0.7503	0.0013	0.2484

Polymer (B): **poly(ethylene glycol)** **2006TUB**
Characterization: M_n/g.mol^{-1} = 8000, Sigma Chemical Co., Inc., St. Louis, MO
Solvent (A): **water** H_2O **7732-18-5**
Salt (C): **sodium citrate** $Na_3C_6H_5O_7$ **68-04-2**

Type of data: coexistence data (tie lines)

T/K = 295.15

continued

continued

Total system			Top phase			Bottom phase		
w_A	w_B	w_C	w_A	w_B	w_C	w_A	w_B	w_C
pH = 5.20								
0.8391	0.0426	0.1183	0.7919	0.1392	0.0689	0.8503	0.0268	0.1229
0.8373	0.0439	0.1188	0.7677	0.1716	0.0607	0.8498	0.0210	0.1292
0.8278	0.0465	0.1257	0.7290	0.2208	0.0502	0.8468	0.0130	0.1402
0.8226	0.0479	0.1295	0.6841	0.2756	0.0403	0.8452	0.0108	0.1440
0.8168	0.0494	0.1338	0.6674	0.2956	0.0370	0.8422	0.0075	0.1503
pH = 8.20								
0.8617	0.0586	0.0797	0.8080	0.1305	0.0615	0.8806	0.0333	0.0861
0.8575	0.0604	0.0821	0.7744	0.1698	0.0558	0.8845	0.0248	0.0907
0.8528	0.0624	0.0848	0.7473	0.2009	0.0518	0.8867	0.0179	0.0954
0.8476	0.0646	0.0878	0.7124	0.2403	0.0473	0.8874	0.0129	0.0997
pH = 9.20								
0.8527	0.0586	0.0887	0.7679	0.1870	0.0451	0.8816	0.0149	0.1035
0.8431	0.0624	0.0945	0.7175	0.2471	0.0354	0.8802	0.0078	0.1120
0.8354	0.0655	0.0991	0.6872	0.2827	0.0301	0.8778	0.0034	0.1188

Polymer (B): **poly(ethylene glycol)** **2005ZA3**
Characterization: M_n/g.mol^{-1} = 5890, Merck, Darmstadt, Germany
Solvent (A): **water** **H_2O** **7732-18-5**
Salt (C): **sodium formate** **$CHNaO_2$** **141-53-7**

Type of data: cloud points

T/K = 298.15

w_A	0.4799	0.5121	0.5383	0.5549	0.5690	0.5838	0.5965	0.6108	0.6353
w_B	0.4192	0.3797	0.3475	0.3265	0.3076	0.2875	0.2704	0.2505	0.2164
w_C	0.1009	0.1082	0.1142	0.1186	0.1234	0.1287	0.1331	0.1387	0.1483
w_A	0.6469	0.6567	0.6642	0.6717	0.6795	0.6861	0.6942	0.7023	0.7109
w_B	0.2002	0.1863	0.1745	0.1633	0.1520	0.1406	0.1277	0.1147	0.1018
w_C	0.1529	0.1570	0.1613	0.1650	0.1685	0.1733	0.1781	0.1830	0.1873
w_A	0.7176	0.7243	0.7312	0.7392	0.7486				
w_B	0.0904	0.0793	0.0682	0.0548	0.0388				
w_C	0.1920	0.1964	0.2006	0.2060	0.2126				

Type of data: coexistence data (tie lines)

T/K = 298.15

continued

continued

Top phase			Bottom phase		
w_A	w_B	w_C	w_A	w_B	w_C
0.4554	0.4466	0.0980	0.7013	0.0009	0.2978
0.4993	0.3936	0.1071	0.7194	0.0120	0.2686
0.5210	0.3662	0.1128	0.7283	0.0185	0.2532
0.5453	0.3397	0.1150	0.7364	0.0213	0.2423
0.5788	0.2926	0.1286	0.7452	0.0313	0.2235

Polymer (B): **poly(ethylene glycol)** **2003HUD**
Characterization: M_w/g.mol^{-1} = 2000, Aldrich Chem. Co., Inc., Milwaukee, WI
Solvent (A): **water** H_2O **7732-18-5**
Salt (C): **sodium hydroxide** $NaOH$ **1310-73-2**

Type of data: cloud points

T/K = 298.15

w_A	0.5369	0.6240	0.6957	0.7542	0.7923	0.8034	0.8124	0.8219	0.8312
w_B	0.4392	0.3427	0.2602	0.1899	0.1432	0.1296	0.1186	0.1067	0.0951
w_C	0.0239	0.0333	0.0441	0.0559	0.0645	0.0670	0.0690	0.0714	0.0737
w_A	0.8400	0.8508	0.8566	0.8631	0.8682	0.8825	0.8865	0.8889	0.8852
w_B	0.0841	0.0704	0.0630	0.0546	0.0480	0.0279	0.0200	0.0104	0.0071
w_C	0.0759	0.0788	0.0804	0.0823	0.0838	0.0896	0.0935	0.1007	0.1077

Type of data: coexistence data (tie lines)

T/K = 298.15

Total system			Top phase			Bottom phase		
w_A	w_B	w_C	w_A	w_B	w_C	w_A	w_B	w_C
0.7348	0.1925	0.0727	0.5643	0.4092	0.0265	0.8817	0.0057	0.1126
0.7209	0.1896	0.0895	0.4879	0.4941	0.0180	0.8656	0.0005	0.1339
0.7074	0.1868	0.1058	0.4416	0.5445	0.0139	0.8462	0.00023	0.1538

Polymer (B): **poly(ethylene glycol)** **2004ZA1**
Characterization: M_n/g.mol^{-1} = 5890, Merck, Darmstadt, Germany
Solvent (A): **water** H_2O **7732-18-5**
Salt (C): **sodium hydroxide** $NaOH$ **1310-73-2**

Type of data: cloud points

continued

continued

T/K = 298.15

w_A	0.5587	0.5935	0.6181	0.6419	0.6621	0.6515	0.7080	0.7243	0.7418
w_B	0.4239	0.3873	0.3607	0.3352	0.3133	0.3249	0.2625	0.2441	0.2246
w_C	0.0174	0.0192	0.0212	0.0229	0.0246	0.0236	0.0295	0.0316	0.0336
w_A	0.7539	0.7647	0.7741	0.7837	0.7925	0.8011	0.8106	0.8195	0.8278
w_B	0.2111	0.1990	0.1885	0.1775	0.1675	0.1577	0.1468	0.1365	0.1269
w_C	0.0350	0.0363	0.0374	0.0388	0.0400	0.0412	0.0426	0.0440	0.0453
w_A	0.8358	0.8437	0.8507	0.8567	0.8622	0.8860	0.8946	0.9029	0.9111
w_B	0.1177	0.1086	0.1005	0.0933	0.0869	0.0582	0.0479	0.0375	0.0270
w_C	0.0465	0.0477	0.0488	0.0500	0.0509	0.0558	0.0575	0.0596	0.0619
w_A	0.9175								
w_B	0.0179								
w_C	0.0646								

T/K = 308.15

w_A	0.5905	0.6079	0.6210	0.6325	0.6947	0.7244	0.7498	0.7734	0.7945
w_B	0.3920	0.3738	0.3599	0.3477	0.2808	0.2485	0.2204	0.1942	0.1706
w_C	0.0175	0.0183	0.0191	0.0198	0.0245	0.0271	0.0298	0.0324	0.0349
w_A	0.8110	0.8254	0.8375	0.8476	0.8568	0.8646	0.8713	0.8776	0.8826
w_B	0.1517	0.1352	0.1213	0.1096	0.0989	0.0900	0.0821	0.0749	0.0690
w_C	0.0373	0.0394	0.0412	0.0428	0.0443	0.0454	0.0466	0.0475	0.0484
w_A	0.8870	0.8907	0.8931	0.8958	0.9000	0.9046	0.9082	0.9129	0.9198
w_B	0.0639	0.0595	0.0564	0.0533	0.0480	0.0425	0.0382	0.0325	0.0238
w_C	0.0491	0.0498	0.0505	0.0509	0.0520	0.0529	0.0536	0.0546	0.0564
w_A	0.9249								
w_B	0.0173								
w_C	0.0578								

T/K = 318.15

w_A	0.5458	0.5587	0.5688	0.5785	0.5893	0.5999	0.6188	0.6268	0.6430
w_B	0.4409	0.4275	0.4170	0.4068	0.3953	0.3843	0.3639	0.3556	0.3380
w_C	0.0133	0.0138	0.0142	0.0147	0.0154	0.0158	0.0173	0.0176	0.0190
w_A	0.6711	0.6999	0.7297	0.7520	0.7730	0.7922	0.8094	0.8229	0.8359
w_B	0.3080	0.2772	0.2449	0.2208	0.1976	0.1765	0.1573	0.1420	0.1275
w_C	0.0209	0.0229	0.0254	0.0272	0.0294	0.0313	0.0333	0.0351	0.0366
w_A	0.8465	0.8560	0.8639	0.8707	0.8861	0.8925	0.9114	0.9281	0.9330
w_B	0.1154	0.1047	0.0958	0.0881	0.0708	0.0635	0.0419	0.0219	0.0156
w_C	0.0381	0.0393	0.0403	0.0412	0.0431	0.0440	0.0467	0.0500	0.0514
w_A	0.9340								
w_B	0.0125								
w_C	0.0535								

continued

continued

Type of data: coexistence data (tie lines)

Top phase			Bottom phase		
w_A	w_B	w_C	w_A	w_B	w_C
T/K = 298.15					
0.4784	0.5088	0.0128	0.8824	0.0000	0.1176
0.5316	0.4523	0.0161	0.8942	0.0025	0.1033
0.5670	0.4149	0.0181	0.9003	0.0047	0.0950
0.5997	0.3795	0.0208	0.9124	0.0000	0.0876
0.6163	0.3622	0.0215	0.9127	0.0025	0.0848
0.6767	0.2963	0.0270	0.9151	0.0092	0.0757
0.7311	0.2359	0.0330	0.9114	0.0218	0.0668
T/K = 308.15					
0.4752	0.5140	0.0108	0.8918	0.0000	0.1082
0.5056	0.4820	0.0124	0.8985	0.0000	0.1015
0.5395	0.4454	0.0151	0.9027	0.0054	0.0919
0.5653	0.4180	0.0167	0.9117	0.0051	0.0832
0.6160	0.3651	0.0189	0.9190	0.0050	0.0760
0.9211	0.0087	0.0702	0.6570	0.3206	0.0224
0.9228	0.0109	0.0663	0.6866	0.2892	0.0242
0.9176	0.0220	0.0604	0.7267	0.2451	0.0282
T/K = 318.15					
0.4786	0.5106	0.0108	0.9061	0.0000	0.0939
0.4891	0.4998	0.0111	0.9094	0.0010	0.0896
0.5047	0.4839	0.0114	0.9145	0.0000	0.0855
0.9220	0.0000	0.0780	0.5468	0.4398	0.0134
0.9300	0.0000	0.0700	0.5795	0.4059	0.0146
0.9352	0.0000	0.0648	0.6225	0.3606	0.0169
0.9351	0.0096	0.0553	0.7110	0.2655	0.0235

Polymer (B): **poly(ethylene glycol)** **2005HEY**
Characterization: M_n/g.mol^{-1} = 7264, DP = 165 EO-units,
Aldrich Chem. Co., Inc., Milwaukee, WI
Solvent (A): **water** **H_2O** **7732-18-5**
Salt (C): **sodium hydroxide** **NaOH** **1310-73-2**

Type of data: cloud points

T/K = 298.15

z_B/(base mol/kg H_2O)	8.63	9.89	8.01	6.43	5.08	4.14	3.61
c_C/(mol/kg H_2O)	1.04	0.961	1.07	1.15	1.23	1.30	1.35
z_B/(base mol/kg H_2O)	3.12	2.62	2.14	1.86	1.34	0.544	0.283
c_C/(mol/kg H_2O)	1.40	1.45	1.50	1.53	1.59	1.71	1.80

continued

continued

Type of data: coexistence data (tie lines)

T/K = 298.15

Top phase		Bottom phase	
z_B/ base mol/kg H_2O	c_C/ mol/kg H_2O	z_B/ base mol/kg H_2O	c_C/ mol/kg H_2O
5.200	1.215	0.567	1.692
6.027	1.162	0.415	1.749
6.445	1.137	0.362	1.774
6.787	1.119	0.345	1.778
6.849	1.115	0.321	1.796
7.235	1.096	0.263	1.834
7.580	1.079	0.164	1.927

Comments: Concentrations are molal, i.e., per kg water; base moles are calculated with the molar mass of 44 g/mol per EO-unit.

Polymer (B): **poly(ethylene glycol)** **2011JI1**
Characterization: M_n/g.mol^{-1} = 3500-4500, Sigma-Aldrich, Inc., St. Louis, MO
Solvent (A): **water** **H_2O** **7732-18-5**
Salt (C): **sodium nitrate** **$NaNO_3$** **7631-99-4**

Type of data: cloud points

w_A	0.5500	0.6000	0.6500	0.7000	0.5680	0.6160	0.6640	0.7120	0.3700
w_B	0.2000	0.1500	0.1000	0.0500	0.1920	0.1440	0.0960	0.0480	0.4000
w_C	0.2500	0.2500	0.2500	0.2500	0.2400	0.2400	0.2400	0.2400	0.2300
T/K	317.805	331.353	342.917	354.977	327.808	339.118	348.979	360.593	307.718
w_A	0.4700	0.5200	0.5860	0.6320	0.6780	0.7240	0.3974	0.4931	0.5409
w_B	0.3000	0.2500	0.1840	0.1380	0.0920	0.0460	0.3826	0.2869	0.2391
w_C	0.2300	0.2300	0.2300	0.2300	0.2300	0.2300	0.2200	0.2200	0.2200
T/K	317.359	326.892	337.043	347.215	354.854	365.987	319.900	326.037	335.147
w_A	0.6040	0.6480	0.6920	0.7360	0.4247	0.5160	0.5617	0.6219	0.6640
w_B	0.1760	0.1320	0.0880	0.0440	0.3653	0.2740	0.2283	0.1680	0.1260
w_C	0.2200	0.2200	0.2200	0.2200	0.2100	0.2100	0.2100	0.2100	0.2100
T/K	345.534	353.497	361.001	370.960	330.402	334.009	342.635	353.389	359.802
w_A	0.7060	0.7480	0.4522	0.5391	0.5826	0.6399	0.6800	0.7199	0.7599
w_B	0.0840	0.0420	0.3478	0.2609	0.2174	0.1600	0.1200	0.0800	0.0400
w_C	0.2100	0.2100	0.2000	0.2000	0.2000	0.2000	0.2000	0.2000	0.2000
T/K	366.636	376.321	339.048	343.882	351.700	361.331	366.265	372.455	382.133

continued

continued

w_A	0.4796	0.5622	0.6034	0.6580	0.6960	0.7340	0.5070	0.5852	0.6243
w_B	0.3304	0.2478	0.2065	0.1520	0.1140	0.0760	0.3130	0.2348	0.1957
w_C	0.1900	0.1900	0.1900	0.1900	0.1900	0.1900	0.1800	0.1800	0.1800
T/K	347.388	350.967	359.970	367.174	371.183	378.338	355.967	359.549	366.173
w_A	0.6761	0.7120	0.7481	0.5343	0.6082	0.6452	0.6941	0.7280	0.5618
w_B	0.1440	0.1080	0.0720	0.2957	0.2218	0.1848	0.1360	0.1020	0.2782
w_C	0.1800	0.1800	0.1800	0.1700	0.1700	0.1700	0.1700	0.1700	0.1600
T/K	372.577	377.170	383.132	363.733	367.574	373.310	379.646	383.241	370.699
w_A	0.6312	0.6661	0.5891	0.6543					
w_B	0.2087	0.1739	0.2609	0.1957					
w_C	0.1600	0.1600	0.1500	0.1500					
T/K	375.408	379.441	378.644	382.105					

Polymer (B):	**poly(ethylene glycol)**		**2011JI2**
Characterization:	M_n/g.mol^{-1} = 3500-4500, Sigma-Aldrich, Inc., St. Louis, MO		
Solvent (A):	**water**	**H_2O**	**7732-18-5**
Salt (C):	**sodium nitrate**	**$NaNO_3$**	**7631-99-4**

Type of data: cloud points

T/K = 288.15

w_A	0.3682	0.3743	0.3860	0.4042	0.4249	0.4462	0.4652	0.4854	0.5016
w_B	0.3948	0.3879	0.3755	0.3562	0.3339	0.3084	0.2851	0.2581	0.2355
w_C	0.2370	0.2378	0.2385	0.2396	0.2412	0.2454	0.2497	0.2565	0.2629
w_A	0.5160	0.5292	0.5437	0.5573	0.5691	0.5793	0.5880	0.5923	0.5957
w_B	0.2144	0.1948	0.1709	0.1491	0.1300	0.1122	0.0966	0.0815	0.0689
w_C	0.2696	0.2760	0.2854	0.2936	0.3009	0.3085	0.3154	0.3262	0.3354
w_A	0.6000	0.6038	0.6016	0.6029	0.6011	0.6004	0.5956	0.5932	0.5612
w_B	0.0575	0.0476	0.0363	0.0289	0.0235	0.0166	0.0124	0.0093	0.0033
w_C	0.3425	0.3486	0.3621	0.3682	0.3754	0.3830	0.3920	0.3975	0.4355
w_A	0.5448								
w_B	0.0002								
w_C	0.4550								

T/K = 308.15

w_A	0.3711	0.3777	0.3901	0.4061	0.4240	0.4432	0.4609	0.4776	0.4930
w_B	0.4014	0.3939	0.3814	0.3639	0.3445	0.3238	0.3030	0.2827	0.2629
w_C	0.2275	0.2284	0.2285	0.2300	0.2315	0.2330	0.2361	0.2397	0.2441
w_A	0.5072	0.5207	0.5336	0.5425	0.5548	0.5719	0.5879	0.6003	0.6089
w_B	0.2443	0.2254	0.2075	0.1954	0.1741	0.1475	0.1219	0.0967	0.0768
w_C	0.2485	0.2539	0.2589	0.2621	0.2711	0.2806	0.2902	0.3030	0.3143
w_A	0.6148	0.6180	0.6199	0.6208	0.6218	0.6198	0.6174	0.6148	0.6118
w_B	0.0599	0.0469	0.0353	0.0247	0.0128	0.0113	0.0104	0.0098	0.0092
w_C	0.3253	0.3351	0.3448	0.3545	0.3654	0.3689	0.3722	0.3754	0.3790

continued

continued

w_A	0.6089	0.6059	0.5998	0.5971
w_B	0.0086	0.0077	0.0070	0.0064
w_C	0.3825	0.3864	0.3932	0.3965

Type of data: coexistence data (tie lines)

Total system			Top phase			Bottom phase		
w_A	w_B	w_C	w_A	w_B	w_C	w_A	w_B	w_C
T/K = 288.15								
0.4691	0.1974	0.3335	0.4138	0.3460	0.2402	0.5356	0.0015	0.4629
0.4946	0.1781	0.3273	0.4438	0.3133	0.2429	0.5570	0.0043	0.4387
0.5090	0.1716	0.3194	0.4671	0.2841	0.2488	0.5669	0.0080	0.4251
0.5328	0.1508	0.3164	0.4928	0.2480	0.2592	0.5786	0.0150	0.4064
0.5589	0.1253	0.3158	0.5376	0.1821	0.2803	0.5959	0.0318	0.3723
T/K = 308.15								
0.4505	0.2007	0.3488	0.3644	0.4017	0.2339	0.5509	0.0001	0.4490
0.4805	0.1820	0.3375	0.3998	0.3642	0.2360	0.5760	0.0009	0.4231
0.5116	0.1619	0.3265	0.4323	0.3276	0.2401	0.5960	0.0024	0.4016
0.5374	0.1445	0.3181	0.4710	0.2877	0.2413	0.6110	0.0058	0.3832
0.5677	0.1176	0.3147	0.5074	0.2446	0.2480	0.6248	0.0136	0.3616

Polymer (B): **poly(ethylene glycol)** **2010JIM**
Characterization: M_n/g.mol^{-1} = 4000, Merck, Darmstadt, Germany
Solvent (A): **water** H_2O **7732-18-5**
Salt (C): **sodium perchlorate** $NaClO_4$ **7601-89-0**

Type of data: cloud points

T/K = 288.15

w_A	0.1654	0.1893	0.2191	0.2578	0.2999	0.3307	0.3560	0.3603	0.3656
w_B	0.4859	0.4733	0.4543	0.4262	0.3923	0.3650	0.3400	0.3391	0.3324
w_C	0.3487	0.3374	0.3266	0.3160	0.3078	0.3043	0.3040	0.3006	0.3020
w_A	0.3828	0.3938	0.3999	0.4125	0.4245	0.4300	0.4425	0.4473	0.4524
w_B	0.3161	0.3041	0.2995	0.2862	0.2738	0.2680	0.2542	0.2483	0.2440
w_C	0.3011	0.3021	0.3006	0.3013	0.3017	0.3020	0.3033	0.3044	0.3036
w_A	0.4624	0.4700	0.4741	0.4857	0.4906	0.5006	0.5100	0.5281	0.5489
w_B	0.2324	0.2229	0.2192	0.2052	0.1978	0.1855	0.1732	0.1491	0.1230
w_C	0.3052	0.3071	0.3067	0.3091	0.3116	0.3139	0.3168	0.3228	0.3281
w_A	0.5662	0.5830	0.5969	0.6055	0.6084	0.6044	0.5877	0.5611	
w_B	0.0990	0.0745	0.0549	0.0402	0.0249	0.0102	0.0047	0.0019	
w_C	0.3348	0.3425	0.3482	0.3543	0.3667	0.3854	0.4076	0.4370	

continued

continued

T/K = 298.15

w_A	0.1350	0.1766	0.2294	0.2650	0.3124	0.3549	0.3628	0.3647	0.3767
w_B	0.4902	0.4760	0.4468	0.4216	0.3856	0.3483	0.3369	0.3350	0.3231
w_C	0.3748	0.3474	0.3238	0.3134	0.3020	0.2968	0.3003	0.3003	0.3002
w_A	0.3792	0.3947	0.4084	0.4245	0.4431	0.4486	0.4710	0.4749	0.4886
w_B	0.3208	0.3056	0.2916	0.2752	0.2555	0.2482	0.2250	0.2193	0.2053
w_C	0.3000	0.2997	0.3000	0.3003	0.3014	0.3032	0.3040	0.3058	0.3061
w_A	0.4925	0.4985	0.5148	0.5222	0.5381	0.5521	0.5612	0.5658	0.5707
w_B	0.1996	0.1935	0.1733	0.1652	0.1448	0.1247	0.1146	0.1081	0.0986
w_C	0.3079	0.3080	0.3119	0.3126	0.3171	0.3232	0.3242	0.3261	0.3307
w_A	0.5820	0.5881	0.6000	0.6013	0.6096	0.6116	0.6009	0.5789	0.5657
w_B	0.0851	0.0744	0.0562	0.0554	0.0405	0.0251	0.0098	0.0050	0.0026
w_C	0.3329	0.3375	0.3438	0.3433	0.3499	0.3633	0.3893	0.4161	0.4317

T/K = 308.15

w_A	0.1240	0.1656	0.2184	0.2540	0.3014	0.3020	0.3319	0.3439	0.3554
w_B	0.4952	0.4810	0.4518	0.4266	0.3906	0.3859	0.3606	0.3533	0.3402
w_C	0.3808	0.3534	0.3298	0.3194	0.3080	0.3121	0.3075	0.3028	0.3044
w_A	0.3768	0.3833	0.4093	0.4369	0.4583	0.4820	0.5009	0.5233	0.5433
w_B	0.3217	0.3152	0.2888	0.2613	0.2394	0.2095	0.1901	0.1628	0.1343
w_C	0.3015	0.3015	0.3019	0.3018	0.3023	0.3085	0.3090	0.3139	0.3224
w_A	0.5412	0.5620	0.5616	0.5664	0.5692	0.5754	0.5786	0.5772	0.5837
w_B	0.1317	0.1120	0.1119	0.1085	0.1031	0.0952	0.0901	0.0894	0.0826
w_C	0.3271	0.3260	0.3265	0.3251	0.3277	0.3294	0.3313	0.3334	0.3337
w_A	0.6114	0.6118	0.6140	0.6138	0.6154	0.6034	0.5791	0.5611	0.5450
w_B	0.0388	0.0307	0.0300	0.0297	0.0288	0.0080	0.0041	0.0028	0.0020
w_C	0.3498	0.3575	0.3560	0.3565	0.3558	0.3886	0.4168	0.4361	0.4530

Type of data: coexistence data (tie lines)

Total system			Top phase			Bottom phase		
w_A	w_B	w_C	w_A	w_B	w_C	w_A	w_B	w_C
T/K = 288.15								
0.4249	0.1498	0.4253	0.1853	0.4671	0.3476	0.5066	0.0016	0.4918
0.4662	0.1271	0.4067	0.2227	0.4416	0.3357	0.5470	0.0017	0.4513
0.5103	0.0989	0.3908	0.2677	0.4094	0.3229	0.5799	0.0020	0.4181
0.5446	0.0755	0.3799	0.3175	0.3679	0.3146	0.5970	0.0043	0.3987
0.5769	0.0519	0.3712	0.3716	0.3291	0.2993	0.6102	0.0093	0.3805

continued

continued

T/K = 298.15

0.3784	0.1569	0.4647	0.1701	0.4721	0.3578	0.4720	0.0015	0.5265
0.4250	0.1346	0.4404	0.2304	0.4370	0.3326	0.5126	0.0015	0.4859
0.4655	0.1125	0.4220	0.2675	0.4152	0.3173	0.5356	0.0015	0.4629
0.5174	0.0867	0.3959	0.3139	0.3863	0.2998	0.5709	0.0019	0.4272
0.5657	0.0554	0.3789	0.3916	0.3234	0.2850	0.5920	0.0088	0.3992

T/K = 308.15

0.4253	0.1133	0.4614	0.2024	0.4498	0.3478	0.4907	0.0015	0.5078
0.4763	0.0879	0.4358	0.2512	0.4152	0.3336	0.5311	0.0016	0.4673
0.5264	0.0624	0.4112	0.3034	0.3724	0.3242	0.5692	0.0017	0.4291
0.5752	0.0383	0.3865	0.3524	0.3360	0.3116	0.6011	0.0030	0.3959

Polymer (B): **poly(ethylene glycol)** **2010ZA1**
Characterization: M_n/g.mol^{-1} = 2000, Merck, Darmstadt, Germany
Solvent (A): **water** **H_2O** **7732-18-5**
Salt (C): **sodium phosphate** **Na_3PO_4** **7601-54-9**

Type of data: cloud points

T/K = 298.15

w_A	0.7186	0.7483	0.7672	0.7823	0.7977	0.8184	0.8374	0.8564	0.8764
w_B	0.2628	0.2277	0.2047	0.1856	0.1656	0.1375	0.1106	0.0833	0.0531
w_C	0.0186	0.0240	0.0281	0.0321	0.0367	0.0441	0.0520	0.0603	0.0705

w_A	0.8880
w_B	0.0333
w_C	0.0787

Polymer (B): **poly(ethylene glycol)** **2005HEY**
Characterization: M_n/g.mol^{-1} = 7264, DP = 165 EO-units,
Aldrich Chem. Co., Inc., Milwaukee, WI
Solvent (A): **water** **H_2O** **7732-18-5**
Salt (C): **sodium phosphate** **Na_3PO_4** **7601-54-9**

Type of data: cloud points

T/K = 298.15

z_B/(base mol/kg H_2O)	11.0	9.34	6.82	5.51	3.83	2.89	2.25
c_C/(mol/kg H_2O)	0.074	0.092	0.127	0.151	0.190	0.220	0.249

z_B/(base mol/kg H_2O)	1.90	1.32	0.89	0.562	0.418	0.318	0.143
c_C/(mol/kg H_2O)	0.261	0.285	0.309	0.323	0.338	0.348	0.371

Type of data: coexistence data (tie lines)

continued

continued

T/K = 298.15

Top phase		Bottom phase	
z_B/ base mol/kg H_2O	c_C/ mol/kg H_2O	z_B/ base mol/kg H_2O	c_C/ mol/kg H_2O
3.096	0.212	0.556	0.324
3.687	0.193	0.445	0.334
4.261	0.177	0.303	0.352
4.620	0.169	0.221	0.368
4.984	0.161	0.187	0.376
5.302	0.154	0.179	0.379

Comments: Concentrations are molal, i.e., per kg water; base moles are calculated with the molar mass of 44 g/mol per EO-unit.

Polymer (B): **poly(ethylene glycol)** **2010MA1**
Characterization: M_n/g.mol^{-1} = 400, Sigma-Aldrich, Inc., St. Louis, MO
Solvent (A): **water** **H_2O** **7732-18-5**
Salt (C): **sodium sulfate** **Na_2SO_4** **7757-82-6**

Type of data: coexistence data (tie lines)

Total system			Top phase			Bottom phase		
w_A	w_B	w_C	w_A	w_B	w_C	w_A	w_B	w_C
T/K = 298.15								
0.6694	0.1978	0.1328	0.6062	0.3588	0.0350	0.7297	0.0387	0.2316
0.6582	0.2009	0.1409	0.5979	0.3728	0.0293	0.7163	0.0296	0.2541
0.6301	0.2249	0.1450	0.5558	0.4197	0.0245	0.7038	0.0312	0.2650
0.6103	0.2352	0.1545	0.5308	0.4490	0.0202	0.6858	0.0228	0.2876
T/K = 308.15								
0.6662	0.2017	0.1321	0.6006	0.3619	0.0375	0.7299	0.0424	0.2277
0.6524	0.2142	0.1334	0.5803	0.3893	0.0304	0.7227	0.0405	0.2368
0.6286	0.2279	0.1435	0.5493	0.4264	0.0243	0.7047	0.0298	0.2655
0.6014	0.2445	0.1541	0.5167	0.4628	0.0205	0.6863	0.0252	0.2885
T/K = 318.15								
0.6794	0.1941	0.1265	0.6166	0.3473	0.0361	0.7421	0.0401	0.2178
0.6482	0.2143	0.1375	0.5759	0.3963	0.0278	0.7202	0.0314	0.2484
0.6280	0.2246	0.1474	0.5538	0.4213	0.0249	0.7041	0.0270	0.2689
0.6116	0.2336	0.1548	0.5328	0.4466	0.0206	0.6921	0.0193	0.2886

Polymer (B): **poly(ethylene glycol)** **1994HOG**
Characterization: M_w/g.mol^{-1} = 1000, Aldrich Chem. Co., Inc., Milwaukee, WI
Solvent (A): **water** **H_2O** **7732-18-5**
Salt (C): **sodium sulfate** **Na_2SO_4** **7757-82-6**

Type of data: coexistence data (tie lines)

T/K = 301

Top phase			Bottom phase		
w_A	w_B	w_C	w_A	w_B	w_C
0.7011	0.2572	0.0417	0.8178	0.0328	0.1494
0.6051	0.3635	0.0314	0.8049	0.0152	0.1799
0.5386	0.4316	0.0298	0.7857	0.0079	0.2064

Plait-point composition: w_A = 0.809 + w_B = 0.106 + w_C = 0.082

Polymer (B): **poly(ethylene glycol)** **2008MA1**
Characterization: M_w/g.mol^{-1} = 1500, Sigma Chemical Co., Inc., St. Louis, MO
Solvent (A): **water** **H_2O** **7732-18-5**
Salt (C): **sodium sulfate** **Na_2SO_4** **7757-82-6**

Type of data: coexistence data (tie lines)

Total system			Top phase			Bottom phase		
w_A	w_B	w_C	w_A	w_B	w_C	w_A	w_B	w_C
T/K = 278.15								
0.7994	0.1107	0.0899	0.7441	0.1998	0.0519	0.9058	0.0444	0.1301
0.7911	0.1165	0.0925	0.7173	0.2343	0.0395	0.8353	0.0396	0.1233
0.7824	0.1226	0.0950	0.7040	0.2424	0.0407	0.8471	0.0111	0.1539
0.7746	0.1277	0.0977	0.6919	0.2728	0.0301	0.8333	0.0074	0.1592
0.7652	0.1349	0.0999	0.6829	0.2806	0.0296	0.8288	0.0061	0.1658
T/K = 298.15								
0.7241	0.1937	0.0822	0.6480	0.3236	0.0284	0.8190	0.0247	0.1563
0.7034	0.2094	0.0872	0.6333	0.3451	0.0216	0.8011	0.0212	0.1777
0.6818	0.2250	0.0932	0.6011	0.3794	0.0195	0.7869	0.0186	0.1945
0.6476	0.2504	0.1020	0.5513	0.4341	0.0146	0.7726	0.0103	0.2171
0.6263	0.2661	0.1076	0.5251	0.4629	0.0120	0.7583	0.0059	0.2358

continued

continued

T/K = 305.65								
0.7241	0.1937	0.0822	0.6460	0.3258	0.0282	0.8368	0.0155	0.1477
0.7034	0.2094	0.0872	0.6029	0.3738	0.0233	0.8154	0.0159	0.1687
0.6818	0.2250	0.0932	0.5800	0.3987	0.0213	0.7976	0.0130	0.1894
0.6476	0.2504	0.1020	0.5415	0.4433	0.0152	0.7762	0.0104	0.2134
0.6263	0.2661	0.1076	0.5084	0.4794	0.0122	0.7549	0.0102	0.2349
T/K = 313.15								
0.7241	0.1937	0.0822	0.6335	0.3358	0.0307	0.8368	0.0160	0.1472
0.7034	0.2094	0.0872	0.5864	0.3907	0.0229	0.8154	0.0143	0.1703
0.6818	0.2250	0.0932	0.5466	0.4341	0.0193	0.8047	0.0140	0.1813
0.6476	0.2504	0.1022	0.5192	0.4654	0.0154	0.7797	0.0131	0.2072
0.6263	0.2661	0.1076	0.5014	0.4879	0.0107	0.7655	0.0091	0.2254

Polymer (B): **poly(ethylene glycol)** **2004HAG**
Characterization: M_n/g.mol^{-1} = 1500, Merck, Darmstadt, Germany
Solvent (A): **water** H_2O **7732-18-5**
Salt (C): **sodium sulfate** Na_2SO_4 **7757-82-6**

Type of data: coexistence data (tie lines)

T/K = 298.15 pH = 4.7

Total system			Top phase			Bottom phase		
w_A	w_B	w_C	w_A	w_B	w_C	w_A	w_B	w_C
0.7749	0.1202	0.1049	0.6712	0.2978	0.0310	0.8447	0.0129	0.1424
0.8006	0.1004	0.0990	0.7190	0.2353	0.0457	0.8396	0.0392	0.1212
0.7623	0.1403	0.0974	0.6642	0.3026	0.0332	0.8350	0.0203	0.1447
0.7887	0.1200	0.0913	0.7168	0.2354	0.0478	0.8368	0.0433	0.1199
0.7695	0.1398	0.0907	0.6887	0.2751	0.0362	0.8376	0.0258	0.1366
0.7499	0.1602	0.0899	0.6606	0.3087	0.0307	0.8356	0.0176	0.1468
0.7759	0.1403	0.0838	0.7154	0.2385	0.0461	0.8389	0.0378	0.1233
0.7634	0.1602	0.0764	0.7119	0.2426	0.0455	0.8354	0.0419	0.1227

Polymer (B): **poly(ethylene glycol)** **1994HOG**
Characterization: M_w/g.mol^{-1} = 3350, American Chemicals Ltd., Montreal, Quebec, Canada
Solvent (A): **water** H_2O **7732-18-5**
Salt (C): **sodium sulfate** Na_2SO_4 **7757-82-6**

Type of data: coexistence data (tie lines)

T/K = 301

continued

continued

Top phase			Bottom phase		
w_A	w_B	w_C	w_A	w_B	w_C
0.5088	0.4814	0.0098	0.7415	0.00850	0.2500
0.6598	0.3151	0.0251	0.83635	0.00985	0.1538
0.6678	0.3087	0.0235	0.8450	0.01190	0.1431
0.6855	0.2875	0.0270	0.8505	0.01210	0.1374
0.6982	0.2731	0.0287	0.8556	0.01150	0.1329
0.7042	0.2687	0.0271	0.85755	0.01425	0.1282
0.7270	0.2374	0.0356	0.8624	0.01360	0.1240
0.7342	0.2285	0.0373	0.8634	0.02060	0.1160
0.7754	0.1788	0.0458	0.8655	0.03570	0.0988

Plait-point composition: $w_A = 0.839 + w_B = 0.093 + w_C = 0.068$

T/K = 308

Top phase			Bottom phase		
w_A	w_B	w_C	w_A	w_B	w_C
0.79113	0.17407	0.03480	0.79020	0.00270	0.20710
0.73185	0.24190	0.02625	0.83860	0.00420	0.15720
0.6610	0.3182	0.0208	0.87100	0.00800	0.12100
0.5760	0.4048	0.0192	0.87820	0.02400	0.09780

Plait-point composition: $w_A = 0.870 + w_B = 0.077 + w_C = 0.053$

Polymer (B): **poly(ethylene glycol)** **2007CAR**
Characterization: M_w/g.mol^{-1} = 4000, Isofar, Rio de Janeiro, Brazil
Solvent (A): **water** **H_2O** **7732-18-5**
Salt (C): **sodium sulfate** **Na_2SO_4** **7757-82-6**

Type of data: coexistence data (tie lines)

Total system			Top phase			Bottom phase		
w_A	w_B	w_C	w_A	w_B	w_C	w_A	w_B	w_C
T/K = 278.15								
0.8338	0.0860	0.0802	0.7698	0.1806	0.0496	0.8591	0.0176	0.1233
0.7971	0.1127	0.0902	0.7047	0.2633	0.0320	0.8079	0.0407	0.1514
0.7763	0.1238	0.0999	0.6724	0.3050	0.0226	0.7883	0.0375	0.1742
0.7419	0.1486	0.1095	0.6167	0.3657	0.0176	0.7900	0.0165	0.1935
0.7102	0.1717	0.1181	0.6045	0.3805	0.0150	0.7565	0.0205	0.2230

continued

continued

T/K = 298.15								
0.8362	0.0873	0.0765	0.7854	0.1650	0.0496	0.8806	0.0161	0.1033
0.7975	0.1113	0.0912	0.6831	0.2873	0.0296	0.8520	0.0064	0.1416
0.7716	0.1279	0.1005	0.6465	0.3331	0.0204	0.8472	0.0013	0.1515
0.7462	0.1498	0.1040	0.6163	0.3682	0.0155	0.8167	0.0085	0.1748
0.7169	0.1723	0.1108	0.5741	0.4159	0.0100	0.8040	0.0056	0.1904
T/K = 308.15								
0.8341	0.0864	0.0795	0.7435	0.2170	0.0395	0.8809	0.0172	0.1019
0.7980	0.1119	0.0901	0.6711	0.3032	0.0257	0.8674	0.0010	0.1316
0.7722	0.1281	0.0997	0.6310	0.3500	0.0190	0.8485	0.0052	0.1463
0.7412	0.1520	0.1068	0.5913	0.3933	0.0154	0.8293	0.0032	0.1675
0.7230	0.1690	0.1080	0.5809	0.4046	0.0145	0.8194	0.0036	0.1770
T/K = 318.15								
0.7412	0.1928	0.0660	0.6786	0.2925	0.0289	0.8615	0.0083	0.1302
0.7025	0.2242	0.0733	0.6291	0.3500	0.0209	0.8554	0.0024	0.1422
0.6723	0.2490	0.0787	0.5874	0.3934	0.0192	0.8298	0.0054	0.1648
0.6328	0.2814	0.0858	0.5556	0.4309	0.0135	0.8141	0.0002	0.1857
0.5680	0.3331	0.0989	0.4897	0.5008	0.0095	0.7722	0.0013	0.2265

Polymer (B): **poly(ethylene glycol)** **2004HAG**
Characterization: M_n/g.mol^{-1} = 4000, Merck, Darmstadt, Germany
Solvent (A): **water** H_2O **7732-18-5**
Salt (C): **sodium sulfate** Na_2SO_4 **7757-82-6**

Type of data: coexistence data (tie lines)

Total system			Top phase			Bottom phase		
w_A	w_B	w_C	w_A	w_B	w_C	w_A	w_B	w_C
T/K = 298.15 pH = 5.3								
0.7500	0.1501	0.0999	0.6099	0.3733	0.0168	0.8371	0.0062	0.1567
0.7650	0.1352	0.0998	0.6122	0.3721	0.0157	0.8393	0.0091	0.1516
0.7604	0.1496	0.0900	0.6313	0.3494	0.0193	0.8440	0.0059	0.1501
0.7753	0.1350	0.0897	0.6464	0.3334	0.0202	0.8500	0.0169	0.1331
0.7851	0.1349	0.0800	0.6764	0.2990	0.0246	0.8607	0.0147	0.1246
0.8000	0.1200	0.0800	0.6914	0.2793	0.0293	0.8716	0.0092	0.1192
0.7401	0.1799	0.0800	0.6307	0.3491	0.0202	0.8378	0.0078	0.1544
0.7702	0.1499	0.0799	0.6600	0.3156	0.0244	0.8557	0.0047	0.1396
0.7801	0.1499	0.0700	0.6949	0.2761	0.0290	0.8696	0.0013	0.1291
0.7947	0.1353	0.0700	0.7235	0.2428	0.0337	0.8749	0.0130	0.1121

continued

continued

T/K = 298.15 pH = 9.2

0.7715	0.1365	0.0920	0.6330	0.3398	0.0272	0.8517	0.0017	0.1466
0.7766	0.1363	0.0871	0.6545	0.3158	0.0297	0.8529	0.0089	0.1382
0.7798	0.1365	0.0837	0.6616	0.3126	0.0258	0.8605	0.0097	0.1298
0.7848	0.1364	0.0788	0.6759	0.2964	0.0277	0.8518	0.0216	0.1266
0.7883	0.1363	0.0754	0.6919	0.2713	0.0368	0.8702	0.0097	0.1201
0.7939	0.1362	0.0699	0.7153	0.2510	0.0337	0.8671	0.0190	0.1139

Polymer (B): **poly(ethylene glycol)** **2008MA2**
Characterization: M_w/g.mol^{-1} = 6000, Sigma Chemical Co., Inc., St. Louis, MO
Solvent (A): **water** H_2O **7732-18-5**
Salt (C): **sodium sulfate** Na_2SO_4 **7757-82-6**

Type of data: coexistence data (tie lines)

Total system			Top phase			Bottom phase		
w_A	w_B	w_C	w_A	w_B	w_C	w_A	w_B	w_C
T/K = 283.15								
0.7584	0.1635	0.0781	0.6844	0.2928	0.0228	0.8429	0.0172	0.1399
0.7333	0.1831	0.0836	0.6577	0.3228	0.0195	0.8241	0.0137	0.1622
0.7154	0.1964	0.0882	0.6344	0.3500	0.0156	0.8131	0.0118	0.1751
0.6906	0.2153	0.0941	0.6122	0.3741	0.0137	0.7916	0.0091	0.1993
0.6721	0.2294	0.0985	0.5886	0.3993	0.0121	0.7800	0.0090	0.2110
T/K = 298.15								
0.7510	0.1682	0.0808	0.6610	0.3153	0.0237	0.8486	0.0072	0.1442
0.7338	0.1817	0.0845	0.6382	0.3422	0.0196	0.8379	0.0072	0.1549
0.7092	0.2014	0.0894	0.6116	0.3719	0.0165	0.8184	0.0067	0.1749
0.6902	0.2164	0.0934	0.5857	0.4007	0.0136	0.8108	0.0069	0.1823
0.6624	0.2394	0.0982	0.5663	0.4196	0.0141	0.7923	0.0088	0.1989
T/K = 313.15								
0.7485	0.1718	0.0797	0.6399	0.3409	0.0192	0.8542	0.0058	0.1400
0.7255	0.1900	0.0845	0.6164	0.3671	0.0165	0.8410	0.0050	0.1540
0.7056	0.2055	0.0889	0.5914	0.3942	0.0144	0.8263	0.0054	0.1683
0.6815	0.2246	0.0939	0.5599	0.4277	0.0124	0.8117	0.0044	0.1839
0.6598	0.2418	0.0984	0.5437	0.4454	0.0109	0.7950	0.0025	0.2025

Polymer (B):	**poly(ethylene glycol)**		**2005HEY**
Characterization:	M_n/g.mol^{-1} = 7264, DP = 165 EO-units, Aldrich Chem. Co., Inc., Milwaukee, WI		
Solvent (A):	**water**	**H_2O**	**7732-18-5**
Salt (C):	**sodium sulfate**	**Na_2SO_4**	**7757-82-6**

Type of data: cloud points

T/K = 298.15

z_B/(base mol/kg H_2O)	10.9	10.2	8.09	6.20	4.13	3.13	2.75
c_C/(mol/kg H_2O)	0.224	0.225	0.288	0.346	0.424	0.486	0.508
z_B/(base mol/kg H_2O)	2.04	1.24	0.735	0.478	0.225	0.035	0.170
c_C/(mol/kg H_2O)	0.555	0.618	0.659	0.691	0.739	0.809	0.650

Type of data: coexistence data (tie lines)

T/K = 298.15

Top phase		Bottom phase	
z_B/ base mol/kg H_2O	c_C/ mol/kg H_2O	z_B/ base mol/kg H_2O	c_C/ mol/kg H_2O
5.381	0.369	0.849	0.645
4.718	0.396	1.283	0.621
4.882	0.389	1.144	0.627
5.850	0.352	0.624	0.666
6.270	0.338	0.483	0.685
6.533	0.330	0.491	0.683
6.804	0.321	0.327	0.714

Comments: Concentrations are molal, i.e., per kg water; base moles are calculated with the molar mass of 44 g/mol per EO-unit.

Polymer (B):	**poly(ethylene glycol)**		**2009CUN**
Characterization:	M_n/g.mol^{-1} = 8000		
Solvent (A):	**water**	**H_2O**	**7732-18-5**
Salt (C):	**sodium sulfate**	**Na_2SO_4**	**7757-82-6**

Type of data: coexistence data (tie lines)

Total system			Top phase			Bottom phase		
w_A	w_B	w_C	w_A	w_B	w_C	w_A	w_B	w_C

T/K = 298.15

continued

continued

0.7848	0.1402	0.0751	0.68891	0.28623	0.02486	0.87643	0.00420	0.11937
0.7698	0.1500	0.0803	0.66357	0.31467	0.02176	0.86325	0.00422	0.13253
0.7549	0.1601	0.0851	0.64226	0.33826	0.01948	0.85380	0.00260	0.14360
0.7399	0.1700	0.0901	0.62353	0.35872	0.01775	0.84675	0.00156	0.15169
0.7249	0.1802	0.0949	0.60558	0.37882	0.01560	0.83510	0.00190	0.16300
0.7099	0.1900	0.1001	0.58860	0.39862	0.01278	0.82060	0.00130	0.17810
0.6945	0.2004	0.1051	0.57029	0.41818	0.01153	0.80995	0.00229	0.18776
0.6788	0.2107	0.1106	0.55362	0.43693	0.00945	0.79455	0.00343	0.20202
0.6645	0.2203	0.1152	0.54032	0.45075	0.00893	0.78345	0.00303	0.21352

T/K = 323.15

0.7849	0.1400	0.0751	0.62720	0.35490	0.01790	0.88470	0.00080	0.11450
0.7700	0.1500	0.0800	0.60630	0.37830	0.01540	0.87570	0.00090	0.12340
0.7549	0.1599	0.0851	0.58557	0.40068	0.01375	0.86450	0.00070	0.13480
0.7397	0.1701	0.0902	0.56420	0.42230	0.01350	0.85320	0.00164	0.14516
0.7249	0.1801	0.0950	0.54911	0.43921	0.01168	0.84297	0.00184	0.15519
0.7099	0.1900	0.1001	0.53198	0.45856	0.00946	0.83184	0.00159	0.16657
0.6950	0.2000	0.1051	0.51385	0.47826	0.00789	0.82182	0.00169	0.17649
0.6798	0.2101	0.1101	0.49758	0.49486	0.00756	0.81184	0.00247	0.18569
0.6648	0.2202	0.1150	0.48498	0.50954	0.00548	0.80209	0.00265	0.19526

Polymer (B): **poly(ethylene glycol)** **2007ROD**
Characterization: M_w/g.mol^{-1} = 8000, Sigma Chemical Co., Inc., St. Louis, MO
Solvent (A): **water** H_2O **7732-18-5**
Salt (C): **sodium sulfate** Na_2SO_4 **7757-82-6**

Type of data: coexistence data (tie lines)

T/K = 298.15

Total system			Top phase			Bottom phase		
w_A	w_B	w_C	w_A	w_B	w_C	w_A	w_B	w_C
0.7883	0.1279	0.0838	0.6894	0.2877	0.0229	0.8468	0.0333	0.1199
0.7977	0.1247	0.0776	0.7063	0.2689	0.0248	0.8530	0.0374	0.1096
0.8087	0.1186	0.0727	0.7294	0.2436	0.0270	0.8602	0.0373	0.1025
0.8144	0.1148	0.0708	0.7414	0.2291	0.0295	0.8619	0.0403	0.0978
0.8177	0.1129	0.0694	0.7521	0.2164	0.0315	0.8657	0.0373	0.0970

Polymer (B): **poly(ethylene glycol)** **2005CAS**
Characterization: M_w/g.mol^{-1} = 8000
Solvent (A): **water** H_2O **7732-18-5**
Salt (C): **sodium sulfate** Na_2SO_4 **7757-82-6**

continued

continued

Type of data: coexistence data (tie lines)

T/K = 308.15

Total system			Top phase			Bottom phase		
w_A	w_B	w_C	w_A	w_B	w_C	w_A	w_B	w_C
0.7643	0.1037	0.1320	0.8254	0.0013	0.1733	0.5952	0.3904	0.0144
0.6999	0.1300	0.0701	0.8882	0.0020	0.1098	0.6943	0.2804	0.0253
0.6898	0.1499	0.1603	0.7710	0.0015	0.2275	0.5560	0.4092	0.0348
0.6698	0.1601	0.1701	0.7566	0.0012	0.2422	0.4511	0.5369	0.0120
0.6500	0.1700	0.1800	0.7380	0.0014	0.2606	0.4538	0.5234	0.0228

Polymer (B): **poly(ethylene glycol)** **1994HOG**
Characterization: M_w/g.mol^{-1} = 8000, Aldrich Chem. Co., Inc., Milwaukee, WI
Solvent (A): **water** H_2O **7732-18-5**
Salt (C): **sodium sulfate** Na_2SO_4 **7757-82-6**

Type of data: coexistence data (tie lines)

T/K = 301

Top phase			Bottom phase		
w_A	w_B	w_C	w_A	w_B	w_C
0.77772	0.19018	0.03210	0.8772	0.0504	0.0724
0.7320	0.2411	0.0269	0.8765	0.0254	0.0981
0.6426	0.3345	0.0229	0.85595	0.00985	0.1342
0.5865	0.3951	0.0184	0.8414	0.0050	0.1536
0.5396	0.4451	0.0153	0.82239	0.00001	0.1776

Plait-point composition: $w_A = 0.864 + w_B = 0.085 + w_C = 0.051$

Polymer (B): **poly(ethylene glycol)** **2008MAL**
Characterization: M_n/g.mol^{-1} = 600, Sigma Chemical Co., Inc., St. Louis, MO
Solvent (A): **water** H_2O **7732-18-5**
Salt (C): **sodium tartrate** $C_4H_5NaO_6$ **14475-11-7**

Type of data: coexistence data (tie lines)

T/K = 298.15 pH = 4.90

Comments: Sodium tartrate stock solution was prepared from tartaric acid solution by the addition of the appropriate quantities of sodium hydroxide to adjust pH = 4.90.

continued

continued

Total system			Top phase			Bottom phase		
w_A	w_B	w_C	w_A	w_B	w_C	w_A	w_B	w_C
0.6833	0.1950	0.1217	0.6413	0.2863	0.0724	0.7270	0.0732	0.1998
0.6782	0.2000	0.1218	0.6364	0.2943	0.0693	0.7270	0.0742	0.1988
0.6653	0.2062	0.1285	0.6169	0.3253	0.0578	0.7259	0.0561	0.2180
0.6833	0.1950	0.1217	0.5167	0.4310	0.0523	0.7245	0.0487	0.2268

Polymer (B): **poly(ethylene glycol)** **2008MAL**
Characterization: M_n/g.mol^{-1} = 1000, Sigma Chemical Co., Inc., St. Louis, MO
Solvent (A): **water** **H_2O** **7732-18-5**
Salt (C): **sodium tartrate** **$C_4H_5NaO_6$** **14475-11-7**

Type of data: coexistence data (tie lines)

T/K = 298.15 pH = 4.90

Comments: Sodium tartrate stock solution was prepared from tartaric acid solution by the addition of the appropriate quantities of sodium hydroxide to adjust pH = 4.90.

Total system			Top phase			Bottom phase		
w_A	w_B	w_C	w_A	w_B	w_C	w_A	w_B	w_C
0.6782	0.2000	0.1218	0.6581	0.2587	0.0832	0.7688	0.0785	0.1527
0.6653	0.2062	0.1285	0.6173	0.3151	0.0676	0.7781	0.0525	0.1694
0.6587	0.2120	0.1293	0.5771	0.3693	0.0536	0.7828	0.0272	0.1900
0.6666	0.2067	0.1267	0.5409	0.4175	0.0416	0.7820	0.0133	0.2047

Polymer (B): **poly(ethylene glycol)** **2008MAL**
Characterization: M_n/g.mol^{-1} = 2000, Sigma Chemical Co., Inc., St. Louis, MO
Solvent (A): **water** **H_2O** **7732-18-5**
Salt (C): **sodium tartrate** **$C_4H_5NaO_6$** **14475-11-7**

Type of data: coexistence data (tie lines)

T/K = 298.15 pH = 4.90

Comments: Sodium tartrate stock solution was prepared from tartaric acid solution by the addition of the appropriate quantities of sodium hydroxide to adjust pH = 4.90.

continued

continued

Total system			Top phase			Bottom phase		
w_A	w_B	w_C	w_A	w_B	w_C	w_A	w_B	w_C
0.7663	0.1250	0.1087	0.7102	0.2293	0.0605	0.8128	0.0385	0.1487
0.7500	0.1300	0.1200	0.6632	0.2914	0.0454	0.8102	0.0182	0.1716
0.7200	0.1500	0.1300	0.6053	0.3639	0.0308	0.7995	0.0018	0.1987

Polymer (B): **poly(ethylene glycol)** **2008MAL**
Characterization: M_n/g.mol^{-1} = 4000, Sigma Chemical Co., Inc., St. Louis, MO
Solvent (A): **water** **H_2O** **7732-18-5**
Salt (C): **sodium tartrate** **$C_4H_5NaO_6$** **14475-11-7**

Type of data: coexistence data (tie lines)

T/K = 298.15 pH = 4.90

Comments: Sodium tartrate stock solution was prepared from tartaric acid solution by the addition of the appropriate quantities of sodium hydroxide to adjust pH = 4.90.

Total system			Top phase			Bottom phase		
w_A	w_B	w_C	w_A	w_B	w_C	w_A	w_B	w_C
0.7766	0.0937	0.1297	0.7046	0.2424	0.0530	0.8486	0.0170	0.1344
0.7686	0.0948	0.1366	0.6893	0.2621	0.0486	0.8479	0.0111	0.1410
0.7634	0.0975	0.1391	0.6810	0.2728	0.0462	0.8460	0.0052	0.1488
0.7548	0.0995	0.1457	0.6671	0.2907	0.0422	0.8425	0.0007	0.1568

Polymer (B): **poly(ethylene glycol)** **2008MAL**
Characterization: M_n/g.mol^{-1} = 6000, Sigma Chemical Co., Inc., St. Louis, MO
Solvent (A): **water** **H_2O** **7732-18-5**
Salt (C): **sodium tartrate** **$C_4H_5NaO_6$** **14475-11-7**

Type of data: coexistence data (tie lines)

T/K = 298.15 pH = 4.90

Comments: Sodium tartrate stock solution was prepared from tartaric acid solution by the addition of the appropriate quantities of sodium hydroxide to adjust pH = 4.90.

continued

continued

Total system			Top phase			Bottom phase		
w_A	w_B	w_C	w_A	w_B	w_C	w_A	w_B	w_C
0.8220	0.0830	0.0950	0.7646	0.1753	0.0601	0.8545	0.0308	0.1147
0.7998	0.1000	0.1002	0.7128	0.2402	0.0470	0.8542	0.0124	0.1334
0.7919	0.1050	0.1031	0.6907	0.2671	0.0422	0.8517	0.0091	0.1392
0.7790	0.1145	0.1065	0.6563	0.3086	0.0351	0.8484	0.0048	0.1468

Polymer (B): **poly(ethylene glycol)** **2008MAL**
Characterization: M_n/g.mol^{-1} = 8000, Sigma Chemical Co., Inc., St. Louis, MO
Solvent (A): **water** H_2O **7732-18-5**
Salt (C): **sodium tartrate** $C_4H_5NaO_6$ **14475-11-7**

Type of data: coexistence data (tie lines)

T/K = 298.15 pH = 4.90

Comments: Sodium tartrate stock solution was prepared from tartaric acid solution by the addition of the appropriate quantities of sodium hydroxide to adjust pH = 4.90.

Total system			Top phase			Bottom phase		
w_A	w_B	w_C	w_A	w_B	w_C	w_A	w_B	w_C
0.8050	0.1000	0.0950	0.7241	0.2262	0.0497	0.8606	0.0118	0.1276
0.7721	0.1400	0.0879	0.6922	0.2653	0.0425	0.8572	0.0066	0.1362
0.7800	0.1200	0.1000	0.6771	0.2832	0.0397	0.8535	0.0034	0.1431
0.7600	0.1400	0.1000	0.6506	0.3146	0.0348	0.8476	0.0003	0.1521

Polymer (B): **poly(ethylene glycol)** **2010SAD**
Characterization: M_n/g.mol^{-1} = 6000
Solvent (A): **water** H_2O **7732-18-5**
Salt (C): **sodium tungstate** Na_2WO_4 **13472-45-2**

Type of data: cloud points

T/K = 298.15

w_A	0.4950	0.5202	0.5513	0.5769	0.5834	0.5867	0.6061	0.6099	0.6254
w_B	0.4532	0.4208	0.3797	0.3444	0.3354	0.3306	0.3028	0.2972	0.2741
w_C	0.0518	0.0590	0.0690	0.0787	0.0812	0.0827	0.0911	0.0929	0.1005
w_A	0.6380	0.6450	0.6593	0.6626	0.6784	0.6922	0.7034	0.7121	0.7136
w_B	0.2549	0.2441	0.2214	0.2162	0.1899	0.1662	0.1456	0.1288	0.1258
w_C	0.1071	0.1109	0.1193	0.1212	0.1317	0.1416	0.1510	0.1591	0.1606

continued

continued

w_A	0.7205	0.7228	0.7266	0.7298	0.7308	0.7338	0.7357	0.7359	0.7376
w_B	0.1112	0.1063	0.0974	0.0893	0.0868	0.0782	0.0722	0.0712	0.0653
w_C	0.1683	0.1709	0.1760	0.1809	0.1824	0.1880	0.1921	0.1929	0.1971

w_A	0.7386	0.7394	0.7394	0.7400	0.7407	0.7405	0.7398	0.7384
w_B	0.0609	0.0566	0.0565	0.0527	0.0451	0.0376	0.0332	0.0282
w_C	0.2005	0.2040	0.2041	0.2073	0.2142	0.2219	0.2270	0.2334

T/K = 303.15

w_A	0.4928	0.5159	0.5355	0.5575	0.5795	0.5983	0.6150	0.6352	0.6513
w_B	0.4612	0.4327	0.4079	0.3795	0.3503	0.3246	0.3013	0.2722	0.2481
w_C	0.0460	0.0514	0.0566	0.0630	0.0702	0.0771	0.0837	0.0926	0.1006

w_A	0.6649	0.6763	0.6868	0.6960	0.7041	0.7099	0.7174	0.7183	0.7236
w_B	0.2273	0.2093	0.1923	0.1770	0.1630	0.1527	0.1388	0.1371	0.1268
w_C	0.1078	0.1144	0.1209	0.1270	0.1329	0.1374	0.1438	0.1446	0.1496

w_A	0.7265	0.7331	0.7353	0.7377	0.7432	0.7476	0.7508	0.7522	0.7520
w_B	0.1211	0.1073	0.1023	0.0969	0.0828	0.0694	0.0565	0.0444	0.0359
w_C	0.1524	0.1596	0.1624	0.1654	0.1740	0.1830	0.1927	0.2034	0.2121

w_A	0.7505	0.7479	0.7444	0.7407
w_B	0.0289	0.0226	0.0179	0.0142
w_C	0.2206	0.2295	0.2377	0.2451

T/K = 308.15

w_A	0.4736	0.4946	0.5195	0.5492	0.5758	0.6024	0.6226	0.6435	0.6584
w_B	0.4854	0.4602	0.4298	0.3925	0.3582	0.3227	0.2949	0.2653	0.2435
w_C	0.0410	0.0452	0.0507	0.0583	0.0660	0.0749	0.0825	0.0912	0.0981

w_A	0.6755	0.6895	0.7011	0.7120	0.7206	0.7281	0.7339	0.7385	0.7433
w_B	0.2179	0.1960	0.1774	0.1593	0.1444	0.1307	0.1197	0.1106	0.1006
w_C	0.1066	0.1145	0.1215	0.1287	0.1350	0.1412	0.1464	0.1509	0.1561

w_A	0.7468	0.7537	0.7579	0.7604	0.7614	0.7614	0.7603
w_B	0.0929	0.0756	0.0625	0.0519	0.0433	0.0343	0.0281
w_C	0.1603	0.1707	0.1796	0.1877	0.1953	0.2043	0.2116

T/K = 313.15

w_A	0.5101	0.5284	0.5405	0.5619	0.5853	0.6082	0.6306	0.6504	0.6682
w_B	0.4430	0.4204	0.4053	0.3781	0.3476	0.3172	0.2866	0.2588	0.2333
w_C	0.0469	0.0512	0.0542	0.0600	0.0671	0.0746	0.0828	0.0908	0.0985

w_A	0.6827	0.6975	0.7091	0.7185	0.7279	0.7356	0.7424	0.7462	0.7515
w_B	0.2119	0.1896	0.1714	0.1564	0.1409	0.1277	0.1155	0.1084	0.0981
w_C	0.1054	0.1129	0.1195	0.1251	0.1312	0.1367	0.1421	0.1454	0.1504

w_A	0.7535	0.7577	0.7639	0.7687	0.7712	0.7724	0.7723	0.7717	0.7707
w_B	0.0941	0.0853	0.0707	0.0566	0.0466	0.0376	0.0309	0.0267	0.0229
w_C	0.1524	0.1570	0.1654	0.1747	0.1822	0.1900	0.1968	0.2016	0.2064

continued

continued

T/K = 318.15

w_A	0.5202	0.5454	0.5652	0.5780	0.6001	0.6204	0.6460	0.6648	0.6820
w_B	0.4366	0.4057	0.3809	0.3646	0.3360	0.3091	0.2744	0.2481	0.2235
w_C	0.0432	0.0489	0.0539	0.0574	0.0639	0.0705	0.0796	0.0871	0.0945
w_A	0.6977	0.7106	0.7214	0.7316	0.7408	0.7481	0.7544	0.7595	0.7611
w_B	0.2004	0.1809	0.1641	0.1479	0.1326	0.1202	0.1089	0.0992	0.0961
w_C	0.1019	0.1085	0.1145	0.1205	0.1266	0.1317	0.1367	0.1413	0.1428
w_A	0.7658	0.7720	0.7762	0.7795	0.7813	0.7824	0.7816	0.7803	
w_B	0.0866	0.0728	0.0621	0.0518	0.0444	0.0364	0.0248	0.0203	
w_C	0.1476	0.1552	0.1617	0.1687	0.1743	0.1812	0.1936	0.1994	

Type of data: coexistence data (tie lines)

Total system			Top phase			Bottom phase		
w_A	w_B	w_C	w_A	w_B	w_C	w_A	w_B	w_C
T/K = 303.15								
0.6843	0.1528	0.1629	0.6129	0.3080	0.0791	0.7431	0.0117	0.2452
0.6577	0.1631	0.1792	0.5812	0.3493	0.0695	0.7235	0.0158	0.2607
0.6288	0.1734	0.1978	0.5269	0.4186	0.0545	0.6954	0.0173	0.2873
0.6121	0.1816	0.2063	0.5109	0.4377	0.0514	0.6798	0.0194	0.3008
0.5892	0.1917	0.2191	0.4852	0.4687	0.0461	0.6599	0.0182	0.3219
0.5695	0.2064	0.2241	0.4488	0.5124	0.0388	0.6429	0.0203	0.3368
0.5303	0.2318	0.2379	0.4051	0.5579	0.0370	0.6047	0.0184	0.3769
T/K = 308.15								
0.6871	0.1731	0.1398	0.6409	0.2719	0.0872	0.7467	0.0177	0.2356
0.6590	0.1882	0.1528	0.6002	0.3277	0.0721	0.7266	0.0170	0.2564
0.6423	0.1972	0.1605	0.5690	0.3613	0.0697	0.7106	0.0171	0.2723
0.6160	0.2099	0.1741	0.5321	0.4042	0.0637	0.6876	0.0176	0.2948
0.5942	0.2224	0.1834	0.4883	0.4715	0.0402	0.6707	0.0136	0.3157
0.5666	0.2373	0.1961	0.4658	0.4879	0.0463	0.6535	0.0146	0.3319
T/K = 313.15								
0.7206	0.1337	0.1457	0.6464	0.2669	0.0867	0.7726	0.0349	0.1925
0.7048	0.1420	0.1532	0.6272	0.2942	0.0786	0.7616	0.0215	0.2169
0.6820	0.1586	0.1594	0.5935	0.3377	0.0688	0.7489	0.0132	0.2379
0.6626	0.1682	0.1692	0.5598	0.3813	0.0589	0.7319	0.0146	0.2535
0.6421	0.1797	0.1782	0.5334	0.4132	0.0534	0.7160	0.0129	0.2711
0.6162	0.1987	0.1851	0.5129	0.4358	0.0513	0.6927	0.0159	0.2914
0.5914	0.2178	0.1908	0.4891	0.4654	0.0455	0.6740	0.0182	0.3078

continued

continued

T/K = 318.15

0.6975	0.1716	0.1309	0.6315	0.2988	0.0697	0.7859	0.0155	0.1986
0.6860	0.1780	0.1360	0.6051	0.3291	0.0658	0.7692	0.0200	0.2108
0.6690	0.1856	0.1454	0.5619	0.3853	0.0528	0.7539	0.0231	0.2230
0.6362	0.1983	0.1655	0.5277	0.4229	0.0494	0.7237	0.0206	0.2557
0.6201	0.1968	0.1831	0.5017	0.4501	0.0482	0.6984	0.0304	0.2712
0.5965	0.2160	0.1875	0.4702	0.4907	0.0391	0.6868	0.0191	0.2941
0.5674	0.2334	0.1992	0.4294	0.5369	0.0337	0.6673	0.0168	0.3159

Polymer (B): **poly(ethylene glycol)** **2003HUD**
Characterization: M_w/g.mol^{-1} = 2000, Aldrich Chem. Co., Inc., Milwaukee, WI
Solvent (A): **water** H_2O **7732-18-5**
Salt (C): **zinc sulfate** $ZnSO_4$ **7733-02-0**

Type of data: cloud points

T/K = 298.15

w_A	0.6062	0.6588	0.7139	0.7441	0.7626	0.7797	0.7914	0.7982	0.8049
w_B	0.3753	0.2952	0.2386	0.1895	0.1569	0.1123	0.0880	0.0677	0.0503
w_C	0.0185	0.0460	0.0475	0.0664	0.0805	0.1080	0.1206	0.1341	0.1448

w_A	0.8056	0.8028	0.8013	0.7814
w_B	0.0360	0.0252	0.0182	0.0030
w_C	0.1584	0.1720	0.1805	0.2156

Type of data: coexistence data (tie lines)

T/K = 298.15

Total system			Top phase			Bottom phase		
w_A	w_B	w_C	w_A	w_B	w_C	w_A	w_B	w_C
0.6982	0.1831	0.1187	0.6219	0.3554	0.0227	0.7776	0.0038	0.2186
0.6923	0.1825	0.1252	0.6035	0.3776	0.0189	0.7740	0.0031	0.2229
0.6888	0.1805	0.1307	0.5913	0.3919	0.0168	0.7708	0.0026	0.2266
0.6828	0.1800	0.1372	0.5845	0.3998	0.0157	0.7625	0.0016	0.2359

Polymer (B): **poly(ethylene glycol)** **2008OL2**
Characterization: M_w/g.mol^{-1} = 4000, Isofar, Rio de Janeiro, Brazil
Solvent (A): **water** H_2O **7732-18-5**
Salt (C): **zinc sulfate** $ZnSO_4$ **7733-02-0**

pH = 3.66

Type of data: coexistence data (tie lines)

continued

continued

Total system			Top phase			Bottom phase		
w_A	w_B	w_C	w_A	w_B	w_C	w_A	w_B	w_C
T/K = 278.15								
0.7034	0.1821	0.1145	0.6941	0.2854	0.0205	0.7450	0.0534	0.2016
0.6834	0.1980	0.1187	0.6607	0.3255	0.0138	0.7239	0.0486	0.2276
0.6501	0.2223	0.1276	0.6399	0.3521	0.0081	0.7022	0.0522	0.2456
0.6339	0.2343	0.1318	0.6190	0.3749	0.0061	0.6805	0.0559	0.2637
0.6082	0.2535	0.1383	0.5813	0.4144	0.0043	0.6576	0.0678	0.2746
T/K = 283.15								
0.6883	0.2168	0.0949	0.6640	0.3255	0.0105	0.7250	0.0565	0.2185
0.6723	0.2316	0.0961	0.6485	0.3431	0.0084	0.7179	0.0611	0.2210
0.6494	0.2481	0.1025	0.6160	0.3776	0.0064	0.6925	0.0587	0.2488
0.6363	0.2584	0.1053	0.6037	0.3912	0.0051	0.6843	0.0631	0.2526
0.6045	0.2825	0.1131	0.5742	0.4226	0.0032	0.6599	0.0650	0.2751
T/K = 298.15								
0.7204	0.1314	0.1482	0.6460	0.3335	0.0205	0.7613	0.0219	0.2167
0.6876	0.1376	0.1748	0.6190	0.3680	0.0131	0.7429	0.0191	0.2380
0.6539	0.1446	0.2015	0.5796	0.4112	0.0092	0.7175	0.0276	0.2548
0.6242	0.1495	0.2263	0.5512	0.4441	0.0047	0.6950	0.0376	0.2674
0.5994	0.1551	0.2455	0.5266	0.4692	0.0042	0.6857	0.0262	0.2881
T/K = 308.15								
0.7438	0.1816	0.0746	0.7153	0.2660	0.0187	0.8058	0.0380	0.1562
0.7127	0.2003	0.0870	0.6709	0.3180	0.0110	0.7773	0.0416	0.1811
0.6971	0.2098	0.0930	0.6495	0.3397	0.0108	0.7552	0.0462	0.1986
0.6627	0.2398	0.0975	0.6149	0.3786	0.0065	0.7244	0.0530	0.2225
0.6429	0.2510	0.1061	0.5994	0.3949	0.0056	0.7069	0.0498	0.2433
T/K = 318.15								
0.6947	0.1979	0.1074	0.6197	0.3219	0.0584	0.7769	0.0534	0.1697
0.6610	0.2241	0.1148	0.5865	0.3577	0.0558	0.7551	0.0519	0.1930
0.6438	0.2360	0.1202	0.5717	0.3701	0.0582	0.7242	0.0489	0.2270
0.6168	0.2569	0.1262	0.5352	0.4062	0.0585	0.7102	0.0541	0.2357
0.5880	0.2796	0.1324	0.4981	0.4315	0.0704	0.7195	0.0573	0.2232

Polymer (B):	**poly(ethylene glycol)**		**2005ZA3**
Characterization:	M_n/g.mol^{-1} = 5890, Merck, Darmstadt, Germany		
Solvent (A):	**water**	**H_2O**	**7732-18-5**
Salt (C):	**zinc sulfate**	**$ZnSO_4$**	**7733-02-0**

continued

continued

Type of data: cloud points

T/K = 298.15

w_A	0.6669	0.6825	0.7188	0.7447	0.7552	0.7676	0.7751	0.7842	0.7885
w_B	0.3143	0.2966	0.2508	0.2166	0.2025	0.1840	0.1737	0.1594	0.1533
w_C	0.0188	0.0209	0.0304	0.0387	0.0423	0.0484	0.0512	0.0564	0.0582
w_A	0.8055	0.8106	0.8220	0.8345	0.8430	0.8479	0.8479	0.8503	0.8523
w_B	0.1256	0.1161	0.0956	0.0713	0.0541	0.0441	0.0440	0.0384	0.0328
w_C	0.0689	0.0733	0.0824	0.0942	0.1029	0.1080	0.1081	0.1113	0.1149
w_A	0.8542	0.8543							
w_B	0.0282	0.0258							
w_C	0.1176	0.1199							

Polymer (B): **poly(ethylene glycol)** **2008MA2**
Characterization: M_w/g.mol^{-1} = 6000, Sigma Chemical Co., Inc., St. Louis, MO
Solvent (A): **water** **H_2O** **7732-18-5**
Salt (C): **zinc sulfate** **$ZnSO_4$** **7733-02-0**

Type of data: coexistence data (tie lines)

Total system			Top phase			Bottom phase		
w_A	w_B	w_C	w_A	w_B	w_C	w_A	w_B	w_C
T/K = 283.15								
0.7334	0.1802	0.0864	0.6929	0.2831	0.0240	0.8045	0.0031	0.1924
0.7115	0.1964	0.0921	0.6597	0.3215	0.0188	0.7918	0.0090	0.1992
0.6886	0.2137	0.0977	0.6386	0.3479	0.0135	0.7636	0.0070	0.2294
0.6619	0.2341	0.1040	0.6052	0.3833	0.0115	0.7438	0.0139	0.2423
0.6443	0.2474	0.1083	0.5824	0.4095	0.0081	0.7330	0.0194	0.2476
T/K = 298.15								
0.7408	0.1696	0.0896	0.6850	0.2931	0.0219	0.8151	0.0029	0.1820
0.7165	0.1887	0.0948	0.6615	0.3214	0.0171	0.7931	0.0076	0.1993
0.6900	0.2095	0.1005	0.6233	0.3643	0.0124	0.7701	0.0129	0.2170
0.6699	0.2251	0.1050	0.5941	0.3963	0.0096	0.7671	0.0035	0.2294
0.6373	0.2502	0.1125	0.5627	0.4299	0.0074	0.7322	0.0228	0.2450
T/K = 313.15								
0.7369	0.1721	0.0910	0.6667	0.3150	0.0183	0.8167	0.0062	0.1771
0.7125	0.1911	0.0964	0.6393	0.3466	0.0141	0.7960	0.0142	0.1898
0.6893	0.2094	0.1013	0.6225	0.3669	0.0106	0.7721	0.0145	0.2134
0.6653	0.2284	0.1063	0.5873	0.4049	0.0078	0.7573	0.0189	0.2238
0.6413	0.2475	0.1112	0.5510	0.4429	0.0061	0.7425	0.0265	0.2310

Polymer (B):	**poly(ethylene glycol)**		**2005HEY**
Characterization:	M_n/g.mol^{-1} = 7264, DP = 165 EO-units, Aldrich Chem. Co., Inc., Milwaukee, WI		
Solvent (A):	**water**	**H_2O**	**7732-18-5**
Salt (C):	**zinc sulfate**	**$ZnSO_4$**	**7733-02-0**

Type of data: cloud points

T/K = 298.15

z_B/(base mol/kg H_2O)	8.09	5.98	5.15	4.73	4.26	3.58	2.98
c_C/(mol/kg H_2O)	0.227	0.301	0.334	0.368	0.395	0.433	0.474
z_B/(base mol/kg H_2O)	2.56	2.14	1.72	1.33	0.931	0.581	0.287
c_C/(mol/kg H_2O)	0.522	0.555	0.598	0.642	0.689	0.783	0.804

Type of data: coexistence data (tie lines)

T/K = 298.15

Top phase		Bottom phase	
z_B/ base mol/kg H_2O	c_C/ mol/kg H_2O	z_B/ base mol/kg H_2O	c_C/ mol/kg H_2O
2.961	0.480	1.206	0.648
3.614	0.430	0.605	0.725
4.283	0.387	0.447	0.762
4.750	0.361	0.328	0.802
5.035	0.346	0.287	0.819
5.421	0.328	0.261	0.832

Comments: Concentrations are molal, i.e., per kg water; base moles are calculated with the molar mass of 44 g/mol per EO-unit.

Polymer (B):	**poly(ethylene glycol) dimethyl ether**		**2011SAD**
Characterization:	M_n/g.mol^{-1} = 698, Merck, Darmstadt, Germany		
Solvent (A):	**water**	**H_2O**	**7732-18-5**
Salt (C):	**diammonium phosphate**	**$(NH_4)_2HPO_4$**	**7783-28-0**

Type of data: cloud points

T/K = 298.15

w_A	0.7814	0.7832	0.7865	0.7876	0.7875	0.7858	0.7818	0.7752	0.7685
w_B	0.0224	0.0252	0.0326	0.0384	0.0465	0.0583	0.0737	0.0924	0.1117
w_C	0.1962	0.1916	0.1809	0.1740	0.1660	0.1559	0.1445	0.1324	0.1198
w_A	0.7606	0.7516	0.7420	0.7321	0.7201	0.7093	0.6980	0.6826	
w_B	0.1317	0.1534	0.1747	0.1950	0.2170	0.2352	0.2534	0.2762	
w_C	0.1077	0.0950	0.0833	0.0729	0.0629	0.0555	0.0486	0.0412	

continued

continued

T/K = 303.15

w_A	0.7705	0.7786	0.7826	0.7865	0.7884	0.7900	0.7906	0.7901	0.7853
w_B	0.0098	0.0149	0.0182	0.0210	0.0256	0.0298	0.0394	0.0518	0.0709
w_C	0.2197	0.2065	0.1992	0.1925	0.1860	0.1802	0.1700	0.1581	0.1438
w_A	0.7785	0.7701	0.7611	0.7510	0.7412	0.7308	0.7175	0.7034	0.6848
w_B	0.0892	0.1105	0.1325	0.1550	0.1763	0.1980	0.2210	0.2438	0.2725
w_C	0.1323	0.1194	0.1064	0.0940	0.0825	0.0712	0.0615	0.0528	0.0427

T/K = 308.15

w_A	0.7750	0.7821	0.7867	0.7905	0.7930	0.7936	0.7939	0.7913	0.7881
w_B	0.0089	0.0127	0.0163	0.0214	0.0281	0.0342	0.0430	0.0558	0.0710
w_C	0.2161	0.2052	0.1970	0.1881	0.1789	0.1722	0.1631	0.1529	0.1409
w_A	0.7809	0.7731	0.7636	0.7528	0.7423	0.7287	0.7156	0.7019	0.6867
w_B	0.0890	0.1074	0.1289	0.1521	0.1750	0.2001	0.2232	0.2455	0.2680
w_C	0.1301	0.1195	0.1075	0.0951	0.0827	0.0712	0.0612	0.0526	0.0453

T/K = 313.15

w_A	0.7704	0.7815	0.7863	0.7931	0.7965	0.7972	0.7948	0.7882	0.7777
w_B	0.0068	0.0105	0.0143	0.0191	0.0272	0.0389	0.0548	0.0762	0.1011
w_C	0.2228	0.2080	0.1994	0.1878	0.1763	0.1639	0.1504	0.1356	0.1212
w_A	0.7662	0.7544	0.7416	0.7286	0.7161	0.7032			
w_B	0.1261	0.1507	0.1758	0.2004	0.2219	0.2427			
w_C	0.1077	0.0949	0.0826	0.0710	0.0620	0.0541			

T/K = 318.15

w_A	0.7792	0.7876	0.7923	0.7980	0.8010	0.8012	0.7973	0.7900	0.7794
w_B	0.0089	0.0129	0.0164	0.0213	0.0292	0.0410	0.0582	0.0777	0.1018
w_C	0.2119	0.1995	0.1913	0.1807	0.1698	0.1578	0.1445	0.1323	0.1188
w_A	0.7677	0.7556	0.7425	0.7303	0.7180	0.7059	0.6931		
w_B	0.1261	0.1505	0.1751	0.1976	0.2185	0.2381	0.2576		
w_C	0.1062	0.0939	0.0824	0.0721	0.0635	0.0560	0.0493		

Type of data: coexistence data (tie lines)

Total system			Top phase			Bottom phase		
w_A	w_B	w_C	w_A	w_B	w_C	w_A	w_B	w_C
T/K = 298.15								
0.7363	0.1487	0.1150	0.6907	0.2632	0.0461	0.7856	0.0286	0.1858
0.7181	0.1587	0.1232	0.6687	0.2960	0.0353	0.7721	0.0094	0.2185
0.6993	0.1682	0.1325	0.6527	0.3191	0.0282	0.7402	0.0104	0.2494
0.6786	0.1803	0.1411	0.6364	0.3397	0.0239	0.7169	0.0087	0.2744
0.6589	0.1896	0.1515	0.6248	0.3559	0.0193	0.6926	0.0056	0.3018

continued

continued

T/K = 308.15

0.7154	0.1588	0.1258	0.6498	0.3188	0.0314	0.7811	0.0114	0.2075
0.6951	0.1684	0.1365	0.6363	0.3379	0.0258	0.7528	0.0038	0.2434
0.6728	0.1795	0.1477	0.6161	0.3630	0.0209	0.7258	0.0096	0.2646
0.6526	0.1899	0.1575	0.6072	0.3754	0.0174	0.6931	0.0070	0.2999

T/K = 318.15

0.7404	0.1464	0.1132	0.6752	0.2822	0.0426	0.7953	0.0299	0.1748
0.7141	0.1607	0.1252	0.6412	0.3302	0.0286	0.7774	0.0084	0.2142
0.6951	0.1689	0.1360	0.6161	0.3588	0.0251	0.7583	0.0035	0.2382
0.6727	0.1816	0.1457	0.5959	0.3844	0.0197	0.7395	0.0053	0.2552
0.6475	0.1954	0.1571	0.5813	0.4010	0.0177	0.7091	0.0019	0.2890

Polymer (B): **poly(ethylene glycol) dimethyl ether** **2011SAD**
Characterization: M_n/g.mol^{-1} = 2145, Merck, Darmstadt, Germany
Solvent (A): **water** H_2O **7732-18-5**
Salt (C): **diammonium phosphate** $(NH_4)_2HPO_4$ **7783-28-0**

Type of data: cloud points

T/K = 298.15

w_A	0.8185	0.8333	0.8368	0.8376	0.8337	0.8236	0.8064	0.8046	0.7945
w_B	0.0181	0.0217	0.0259	0.0301	0.0426	0.0659	0.0956	0.1007	0.1167
w_C	0.1634	0.1450	0.1373	0.1323	0.1237	0.1105	0.0980	0.0947	0.0888
w_A	0.7848	0.7817	0.7672	0.7630	0.7480	0.7400	0.7247	0.7124	0.7005
w_B	0.1343	0.1385	0.1631	0.1701	0.1923	0.2040	0.2248	0.2414	0.2566
w_C	0.0809	0.0798	0.0697	0.0669	0.0597	0.0560	0.0505	0.0462	0.0429
w_A	0.6712	0.6361	0.5999	0.5745	0.5473	0.5036			
w_B	0.2932	0.3331	0.3713	0.3976	0.4259	0.4702			
w_C	0.0356	0.0308	0.0288	0.0279	0.0268	0.0262			

T/K = 303.15

w_A	0.8042	0.8290	0.8376	0.8405	0.8406	0.8385	0.8332	0.8204	0.8193
w_B	0.0174	0.0201	0.0236	0.0271	0.0306	0.0380	0.0520	0.0768	0.0773
w_C	0.1784	0.1509	0.1388	0.1324	0.1288	0.1235	0.1148	0.1028	0.1034
w_A	0.8108	0.8056	0.8011	0.7908	0.7886	0.7765	0.7713	0.7589	0.7506
w_B	0.0926	0.1022	0.1093	0.1263	0.1305	0.1496	0.1585	0.1772	0.1892
w_C	0.0966	0.0922	0.0896	0.0829	0.0809	0.0739	0.0702	0.0639	0.0602
w_A	0.7354	0.7269	0.7099	0.6836	0.6589	0.6390	0.6265	0.6131	0.5983
w_B	0.2108	0.2230	0.2454	0.2785	0.3084	0.3308	0.3445	0.3591	0.3753
w_C	0.0538	0.0501	0.0447	0.0379	0.0327	0.0302	0.0290	0.0278	0.0264
w_A	0.5727	0.5438	0.5019						
w_B	0.4019	0.4320	0.4751						
w_C	0.0254	0.0242	0.0230						

continued

continued

T/K = 308.15

w_A	0.8062	0.8285	0.8444	0.8462	0.8458	0.8439	0.8437	0.8399	0.8316
w_B	0.0152	0.0190	0.0222	0.0261	0.0304	0.0346	0.0384	0.0464	0.0615
w_C	0.1786	0.1525	0.1334	0.1277	0.1238	0.1215	0.1179	0.1137	0.1069
w_A	0.8186	0.8170	0.8090	0.8032	0.7987	0.7862	0.7840	0.7707	0.7638
w_B	0.0843	0.0861	0.1000	0.1109	0.1170	0.1371	0.1411	0.1609	0.1722
w_C	0.0971	0.0969	0.0910	0.0859	0.0843	0.0767	0.0749	0.0684	0.0640
w_A	0.7529	0.7434	0.7308	0.7038	0.6795	0.6566	0.6396	0.6285	0.6162
w_B	0.1874	0.2014	0.2190	0.2544	0.2847	0.3119	0.3311	0.3433	0.3569
w_C	0.0597	0.0552	0.0502	0.0418	0.0358	0.0315	0.0293	0.0282	0.0269
w_A	0.5932	0.5648	0.5159						
w_B	0.3824	0.4116	0.4607						
w_C	0.0244	0.0236	0.0234						

T/K = 313.15

w_A	0.8215	0.8366	0.8463	0.8486	0.8461	0.8453	0.8432	0.8412	0.8363
w_B	0.0217	0.0258	0.0290	0.0334	0.0382	0.0425	0.0467	0.0508	0.0588
w_C	0.1568	0.1376	0.1247	0.1180	0.1157	0.1122	0.1101	0.1080	0.1049
w_A	0.8365	0.8321	0.8271	0.8286	0.8217	0.8144	0.8147	0.8064	0.7994
w_B	0.0596	0.0664	0.0747	0.0747	0.0839	0.0956	0.0962	0.1087	0.1205
w_C	0.1039	0.1015	0.0982	0.0967	0.0944	0.0900	0.0891	0.0849	0.0801
w_A	0.7978	0.7870	0.7830	0.7746	0.7663	0.7608	0.7495	0.7438	0.7289
w_B	0.1227	0.1391	0.1453	0.1576	0.1698	0.1784	0.1944	0.2024	0.2223
w_C	0.0795	0.0739	0.0717	0.0678	0.0639	0.0608	0.0561	0.0538	0.0488
w_A	0.7230	0.7027	0.6816	0.6675	0.6529	0.6402	0.6314	0.6189	0.6045
w_B	0.2301	0.2565	0.2824	0.2995	0.3162	0.3313	0.3414	0.3551	0.3705
w_C	0.0469	0.0408	0.0360	0.0330	0.0309	0.0285	0.0272	0.0260	0.0250
w_A	0.5846	0.5629	0.5388	0.5040					
w_B	0.3916	0.4144	0.4397	0.4755					
w_C	0.0238	0.0227	0.0215	0.0205					

T/K = 318.15

w_A	0.8103	0.8310	0.8446	0.8461	0.8436	0.8422	0.8405	0.8383	0.8371
w_B	0.0265	0.0292	0.0390	0.0490	0.0520	0.0563	0.0577	0.0634	0.0639
w_C	0.1632	0.1398	0.1164	0.1049	0.1044	0.1015	0.1018	0.0983	0.0990
w_A	0.8335	0.8337	0.8292	0.8290	0.8236	0.8230	0.8177	0.8145	0.8109
w_B	0.0699	0.0702	0.0771	0.0792	0.0859	0.0885	0.0952	0.1006	0.1058
w_C	0.0966	0.0961	0.0937	0.0918	0.0905	0.0885	0.0871	0.0849	0.0833
w_A	0.8069	0.8033	0.7966	0.7943	0.7866	0.7850	0.7771	0.7735	0.7651
w_B	0.1133	0.1177	0.1280	0.1313	0.1428	0.1456	0.1572	0.1622	0.1741
w_C	0.0798	0.0790	0.0754	0.0744	0.0706	0.0694	0.0657	0.0643	0.0608

continued

continued

w_A	0.7607	0.7480	0.7312	0.7140	0.7009	0.6845	0.6715	0.6584	0.6440
w_B	0.1803	0.1987	0.2209	0.2430	0.2603	0.2799	0.2957	0.3106	0.3271
w_C	0.0590	0.0533	0.0479	0.0430	0.0388	0.0356	0.0328	0.0310	0.0289
w_A	0.6283	0.5997	0.5803	0.5597	0.5300	0.5086			
w_B	0.3451	0.3755	0.3959	0.4175	0.4482	0.4712			
w_C	0.0266	0.0248	0.0238	0.0228	0.0218	0.0202			

Type of data: coexistence data (tie lines)

Total system			Top phase			Bottom phase		
w_A	w_B	w_C	w_A	w_B	w_C	w_A	w_B	w_C
T/K = 298.15								
0.7646	0.1335	0.1019	0.7064	0.2500	0.0436	0.8203	0.0147	0.1650
0.7423	0.1448	0.1129	0.6706	0.2921	0.0373	0.8047	0.0114	0.1839
0.7218	0.1545	0.1237	0.6404	0.3299	0.0297	0.7847	0.0065	0.2088
0.6963	0.1664	0.1373	0.6120	0.3599	0.0281	0.7586	0.0056	0.2358
0.6667	0.1826	0.1507	0.5935	0.3804	0.0261	0.7312	0.0088	0.2600
T/K = 303.15								
0.7647	0.1365	0.0988	0.7100	0.2467	0.0433	0.8256	0.0160	0.1584
0.7439	0.1454	0.1107	0.6782	0.2859	0.0359	0.8057	0.0128	0.1815
0.7215	0.1553	0.1232	0.6403	0.3291	0.0306	0.7882	0.0108	0.2010
0.6980	0.1662	0.1358	0.6104	0.3612	0.0284	0.7690	0.0037	0.2273
0.6665	0.1853	0.1482	0.5841	0.3904	0.0255	0.7442	0.0027	0.2531
T/K = 308.15								
0.7520	0.1461	0.1019	0.6901	0.2705	0.0394	0.8144	0.0151	0.1705
0.7275	0.1571	0.1154	0.6589	0.3071	0.0340	0.7995	0.0087	0.1918
0.6953	0.1731	0.1316	0.6134	0.3590	0.0276	0.7666	0.0081	0.2253
0.6710	0.1791	0.1499	0.5794	0.3959	0.0247	0.7385	0.0063	0.2552
0.6334	0.1987	0.1679	0.5407	0.4359	0.0234	0.7018	0.0074	0.2908
T/K = 313.15								
0.7644	0.1334	0.1022	0.6778	0.2863	0.0359	0.8219	0.0186	0.1595
0.7386	0.1442	0.1172	0.6418	0.3285	0.0297	0.8085	0.0080	0.1835
0.7086	0.1598	0.1316	0.6114	0.3632	0.0254	0.7819	0.0082	0.2099
0.6821	0.1739	0.1440	0.5885	0.3868	0.0247	0.7577	0.0098	0.2325
0.6424	0.1925	0.1651	0.5453	0.4322	0.0225	0.7178	0.0086	0.2736
T/K = 318.15								
0.7740	0.1258	0.1002	0.6885	0.2753	0.0362	0.8225	0.0282	0.1493
0.7484	0.1390	0.1126	0.6455	0.3253	0.0292	0.8126	0.0190	0.1684
0.7132	0.1601	0.1267	0.6116	0.3650	0.0234	0.7840	0.0144	0.2016
0.6859	0.1726	0.1415	0.5706	0.4063	0.0231	0.7642	0.0103	0.2255
0.6586	0.1859	0.1555	0.5537	0.4233	0.0230	0.7374	0.0086	0.2540

Polymer (B):	**poly(ethylene glycol) dimethyl ether**		**2011ZAF**
Characterization:	M_n/g.mol^{-1} = 2300, Merck, Darmstadt, Germany		
Solvent (A):	**water**	**H_2O**	**7732-18-5**
Salt (C):	**dipotassium phosphate**	**K_2HPO_4**	**7758-11-4**

Type of data: cloud points

T/K = 298.15

w_A	0.5625	0.6062	0.6343	0.6738	0.6968	0.7164	0.7362	0.7545	0.7695
w_B	0.4195	0.3704	0.3376	0.2905	0.2608	0.2357	0.2081	0.1816	0.1583
w_C	0.0180	0.0234	0.0281	0.0357	0.0424	0.0479	0.0557	0.0639	0.0722
w_A	0.7824	0.7948	0.8071	0.8165	0.8270	0.8335	0.8378	0.8409	
w_B	0.1372	0.1169	0.0963	0.0793	0.0603	0.0469	0.0367	0.0284	
w_C	0.0804	0.0883	0.0966	0.1042	0.1127	0.1196	0.1255	0.1307	

T/K = 303.15

w_A	0.6065	0.6166	0.6260	0.6418	0.6581	0.6753	0.6951	0.7159	0.7356
w_B	0.3699	0.3583	0.3474	0.3287	0.3094	0.2882	0.2635	0.2366	0.2099
w_C	0.0236	0.0251	0.0266	0.0295	0.0325	0.0365	0.0414	0.0475	0.0545
w_A	0.7545	0.7707	0.7854	0.7980	0.8079	0.8152	.8226	0.8276	0.8318
w_B	0.1834	0.1598	0.1377	0.1182	0.1020	0.0890	0.0761	0.0666	0.0585
w_C	0.0621	0.0695	0.0769	0.0838	0.0901	0.0958	0.1013	0.1058	0.1097
w_A	0.8361	0.8404							
w_B	0.0502	0.0411							
w_C	0.1137	0.1185							

T/K = 308.15

w_A	0.5994	0.6423	0.6715	0.6931	0.7078	0.7306	0.7475	0.7654	0.7745
w_B	0.3782	0.3287	0.2943	0.2676	0.2486	0.2184	0.1956	0.1703	0.1572
w_C	0.0224	0.0290	0.0342	0.0393	0.0436	0.0510	0.0569	0.0643	0.0683
w_A	0.7842	0.7929	0.7995	0.8087	0.8179	0.8275	0.8351	0.8416	0.8468
w_B	0.1419	0.1287	0.1180	0.1038	0.0887	0.0731	0.0594	0.0469	0.0355
w_C	0.0739	0.0784	0.0825	0.0875	0.0934	0.0994	0.1055	0.1115	0.1177

T/K = 318.15

w_A	0.5595	0.5834	0.6046	0.6338	0.6552	0.6792	0.6980	0.7121	0.7285
w_B	0.4233	0.3970	0.3737	0.3407	0.3158	0.2871	0.2643	0.2464	0.2256
w_C	0.0172	0.0196	0.0217	0.0255	0.0290	0.0337	0.0377	0.0415	0.0459
w_A	0.7433	0.7596	0.7738	0.7868	0.7996	0.8106	0.8208		
w_B	0.2058	0.1837	0.1638	0.1447	0.1260	0.1095	0.0940		
w_C	0.0509	0.0567	0.0624	0.0685	0.0744	0.0799	0.0852		

continued

continued

Type of data: coexistence data (tie lines)

Total system			Top phase			Bottom phase		
w_A	w_B	w_C	w_A	w_B	w_C	w_A	w_B	w_C
T/K = 298.15								
0.6852	0.1888	0.1260	0.5944	0.3840	0.0216	0.7694	0.0103	0.2203
0.6554	0.1901	0.1545	0.5547	0.4270	0.0183	0.7365	0.0090	0.2545
0.6280	0.1896	0.1824	0.5177	0.4673	0.0150	0.7041	0.0064	0.2895
0.5999	0.1897	0.2104	0.4819	0.5041	0.0140	0.6740	0.0048	0.3212
0.5783	0.1897	0.2320	0.4480	0.5395	0.0125	0.6427	0.0038	0.3535
0.5437	0.1897	0.2666	0.4067	0.5815	0.0118	0.6106	0.0024	0.3870
T/K = 303.15								
0.6844	0.1895	0.1261	0.5818	0.3962	0.0220	0.7704	0.0131	0.2165
0.6553	0.1901	0.1546	0.5404	0.4415	0.0181	0.7418	0.0101	0.2481
0.6288	0.1894	0.1818	0.5004	0.4842	0.0154	0.7111	0.0087	0.2802
0.5996	0.1898	0.2106	0.4627	0.5233	0.0140	0.6750	0.0077	0.3173
0.5747	0.1899	0.2354	0.4307	0.5561	0.0132	0.6421	0.0072	0.3507
0.5436	0.1898	0.2666	0.3916	0.5955	0.0129	0.6106	0.0053	0.3841
T/K = 308.15								
0.6837	0.1901	0.1262	0.5711	0.4071	0.0218	0.7723	0.0137	0.2140
0.6554	0.1900	0.1546	0.5284	0.4541	0.0175	0.7455	0.0098	0.2447
0.6296	0.1891	0.1813	0.4874	0.4971	0.0155	0.7154	0.0093	0.2753
0.5993	0.1898	0.2109	0.4493	0.5368	0.0139	0.6773	0.0088	0.3139
0.5711	0.1901	0.2388	0.4173	0.5695	0.0132	0.6433	0.0085	0.3482
0.5433	0.1900	0.2667	0.3799	0.6071	0.0130	0.6115	0.0067	0.3818
T/K = 318.15								
0.6836	0.1898	0.1266	0.5540	0.4262	0.0198	0.7800	0.0087	0.2113
0.6556	0.1901	0.1543	0.5103	0.4738	0.0159	0.7477	0.0059	0.2464
0.6278	0.1898	0.1824	0.4739	0.5112	0.0149	0.7161	0.0053	0.2786
0.6040	0.1878	0.2082	0.4408	0.5455	0.0137	0.6866	0.0053	0.3081
0.5721	0.1897	0.2382	0.4013	0.5866	0.0121	0.6507	0.0051	0.3442
0.5433	0.1897	0.2670	0.3667	0.6222	0.0111	0.6154	0.0044	0.3802

Polymer (B): **poly(ethylene glycol) dimethyl ether** **2010ZA1**
Characterization: M_n/g.mol^{-1} = 2300, Merck, Darmstadt, Germany
Solvent (A): **water** **H_2O** **7732-18-5**
Salt (C): **disodium phosphate** **Na_2HPO_4** **7558-79-4**

Type of data: cloud points

continued

continued

T/K = 298.15

w_A	0.7135	0.7165	0.7227	0.7270	0.7362	0.7379	0.7431	0.7479	0.7518
w_B	0.2585	0.2545	0.2471	0.2417	0.2308	0.2283	0.2221	0.2157	0.2107
w_C	0.0280	0.0290	0.0302	0.0313	0.0330	0.0338	0.0348	0.0364	0.0375
w_A	0.7573	0.7615	0.7656	0.7682	0.7714	0.7765	0.7771	0.7829	0.7830
w_B	0.2037	0.1984	0.1930	0.1891	0.1851	0.1782	0.1775	0.1695	0.1693
w_C	0.0390	0.0401	0.0414	0.0427	0.0435	0.0453	0.0454	0.0476	0.0477
w_A	0.7870	0.7875	0.7909	0.7933					
w_B	0.1637	0.1630	0.1585	0.1550					
w_C	0.0493	0.0495	0.0506	0.0517					

T/K = 308.15

w_A	0.7065	0.7166	0.7213	0.7261	0.7319	0.7345	0.7422	0.7426	0.7514
w_B	0.2662	0.2541	0.2485	0.2429	0.2358	0.2325	0.2231	0.2225	0.2116
w_C	0.0273	0.0293	0.0302	0.0310	0.0323	0.0330	0.0347	0.0349	0.0370
w_A	0.7522	0.7610	0.7610	0.7693	0.7761	0.7769	0.7821	0.7869	0.7901
w_B	0.2106	0.1995	0.1994	0.1888	0.1800	0.1791	0.1722	0.1658	0.1613
w_C	0.0372	0.0395	0.0396	0.0419	0.0439	0.0440	0.0457	0.0473	0.0486
w_A	0.7932								
w_B	0.1572								
w_C	0.0496								

T/K = 313.15

w_A	0.7015	0.7176	0.7359	0.7501	0.7632	0.7760	0.7861	0.7933
w_B	0.2536	0.2385	0.2217	0.2099	0.1991	0.1891	0.1819	0.1770
w_C	0.0449	0.0439	0.0424	0.0400	0.0377	0.0349	0.0320	0.0297

T/K = 318.15

w_A	0.7049	0.7180	0.7337	0.7492	0.7637	0.7777	0.7909
w_B	0.2670	0.2513	0.2327	0.2141	0.1961	0.1786	0.1619
w_C	0.0281	0.0307	0.0336	0.0367	0.0402	0.0437	0.0472

Type of data: coexistence data (tie lines)

Top phase			Bottom phase		
w_A	w_B	w_C	w_A	w_B	w_C
T/K = 298.15					
0.75290	0.20890	0.03820	0.86590	0.04200	0.09210
0.74254	0.22220	0.03526	0.87195	0.02973	0.09832
0.73570	0.23130	0.03300	0.87220	0.02680	0.10100
0.71818	0.25214	0.02968	0.87168	0.02342	0.10490
0.69540	0.28260	0.02200	0.86520	0.00860	0.12620
0.68590	0.29310	0.02100	0.86230	0.00670	0.13100

continued

continued

T/K = 308.15

0.74812	0.21638	0.03550	0.85762	0.05043	0.09195
0.73385	0.23430	0.03185	0.86000	0.04208	0.09792
0.72123	0.24870	0.03007	0.86421	0.03302	0.10277
0.70750	0.26526	0.02724	0.86784	0.02310	0.10906
0.70225	0.27208	0.02567	0.87117	0.01411	0.11472
0.66610	0.31620	0.01770	0.86770	0.00960	0.12270

T/K = 313.15

0.66034	0.32451	0.01515	0.85539	0.03057	0.11404
0.64065	0.34590	0.01345	0.85221	0.02211	0.12568
0.61966	0.36850	0.01184	0.84706	0.01149	0.14145
0.58660	0.40418	0.00922	0.83445	0.00552	0.16003
0.56834	0.42412	0.00754	0.81812	0.00317	0.17871

T/K = 318.15

0.69638	0.28905	0.01457	0.86554	0.02187	0.11259
0.67208	0.31347	0.01445	0.86657	0.01912	0.11431
0.65853	0.32908	0.01239	0.86563	0.01694	0.11743
0.62805	0.36358	0.00837	0.85005	0.01258	0.13737
0.61336	0.38115	0.00549	0.83276	0.01067	0.15657
0.57903	0.41662	0.00435	0.81142	0.00901	0.17957

Polymer (B): **poly(ethylene glycol) dimethyl ether** **2011SAD**
Characterization: M_n/g.mol^{-1} = 2145, Merck, Darmstadt, Germany
Solvent (A): **water** H_2O **7732-18-5**
Salt (C): **monoammonium phosphate** $(NH_4)H_2PO_4$ **7722-76-1**

Type of data: cloud points

T/K = 293.15

w_A	0.7050	0.7039	0.7024	0.7016	0.6991	0.6932	0.6902	0.6841	0.6759
w_B	0.0858	0.0893	0.0932	0.0963	0.1033	0.1170	0.1258	0.1403	0.1609
w_C	0.2092	0.2068	0.2044	0.2021	0.1976	0.1898	0.1840	0.1756	0.1632
w_A	0.6649	0.6523	0.6346	0.6130	0.5890	0.5636	0.5315		
w_B	0.1852	0.2131	0.2459	0.2822	0.3197	0.3569	0.4007		
w_C	0.1499	0.1346	0.1195	0.1048	0.0913	0.0795	0.0678		

T/K = 298.15

w_A	0.7182	0.7166	0.7162	0.7144	0.7111	0.7082	0.7045	0.6995	0.6933
w_B	0.0664	0.0706	0.0744	0.0789	0.0871	0.0945	0.1045	0.1173	0.1315
w_C	0.2154	0.2128	0.2094	0.2067	0.2018	0.1973	0.1910	0.1832	0.1752
w_A	0.6840	0.6738	0.6610	0.6450	0.6278	0.6057	0.5837	0.5599	0.5382
w_B	0.1540	0.1775	0.2041	0.2340	0.2635	0.2989	0.3317	0.3655	0.3949
w_C	0.1620	0.1487	0.1349	0.1210	0.1087	0.0954	0.0846	0.0746	0.0669

continued

continued

T/K = 303.15

w_A	0.7293	0.7276	0.7275	0.7241	0.7210	0.7168	0.7086	0.6993	0.6907
w_B	0.0541	0.0588	0.0630	0.0719	0.0802	0.0917	0.1105	0.1330	0.1506
w_C	0.2166	0.2136	0.2095	0.2040	0.1988	0.1915	0.1809	0.1677	0.1587
w_A	0.6810	0.6683	0.6549	0.6394	0.6230	0.6054	0.5847	0.5672	0.5403
w_B	0.1725	0.1979	0.2227	0.2505	0.2771	0.3038	0.3345	0.3591	0.3949
w_C	0.1465	0.1338	0.1224	0.1101	0.0999	0.0908	0.0808	0.0737	0.0648

T/K = 308.15

w_A	0.7394	0.7403	0.7392	0.7376	0.7362	0.7333	0.7259	0.7166	0.7071
w_B	0.0359	0.0383	0.0453	0.0513	0.0570	0.0678	0.0847	0.1066	0.1289
w_C	0.2247	0.2214	0.2155	0.2111	0.2068	0.1989	0.1894	0.1768	0.1640
w_A	0.6931	0.6789	0.6601	0.6407	0.6163	0.5903	0.5591	0.5266	
w_B	0.1573	0.1849	0.2205	0.2539	0.2918	0.3299	0.3725	0.4151	
w_C	0.1496	0.1362	0.1194	0.1054	0.0919	0.0798	0.0684	0.0583	

T/K = 313.15

w_A	0.7476	0.7482	0.7490	0.7483	0.7474	0.7454	0.7410	0.7343	0.7232
w_B	0.0257	0.0287	0.0314	0.0379	0.0446	0.0527	0.0655	0.0826	0.1052
w_C	0.2267	0.2231	0.2196	0.2138	0.2080	0.2019	0.1935	0.1831	0.1716
w_A	0.7087	0.6923	0.6741	0.6418	0.6150	0.5864	0.5553	0.5229	
w_B	0.1351	0.1672	0.2014	0.2553	0.2961	0.3370	0.3786	0.4206	
w_C	0.1562	0.1405	0.1245	0.1029	0.0889	0.0766	0.0661	0.0565	

T/K = 318.15

w_A	0.7527	0.7554	0.7559	0.7574	0.7575	0.7576	0.7561	0.7527	0.7470
w_B	0.0212	0.0243	0.0271	0.0299	0.0331	0.0393	0.0474	0.0579	0.0724
w_C	0.2261	0.2203	0.2170	0.2127	0.2094	0.2031	0.1965	0.1894	0.1806
w_A	0.7372	0.7212	0.7037	0.6834	0.6593	0.6335	0.6028	0.5689	0.5397
w_B	0.0945	0.1249	0.1577	0.1940	0.2341	0.2743	0.3183	0.3640	0.4032
w_C	0.1683	0.1539	0.1386	0.1226	0.1066	0.0922	0.0789	0.0671	0.0571

Polymer (B): **poly(ethylene glycol) dimethyl ether** **2010ZA2**
Characterization: M_n/g.mol^{-1} = 2300, Merck, Darmstadt, Germany
Solvent (A): **water** H_2O **7732-18-5**
Salt (C): **monosodium phosphate** NaH_2PO_4 **7558-80-7**

Type of data: cloud points

T/K = 298.15

w_A	0.7738	0.7679	0.7612	0.7533	0.7444	0.7343	0.7225	0.7086	0.6938
w_B	0.0767	0.0881	0.1012	0.1167	0.1335	0.1531	0.1740	0.1981	0.2218
w_C	0.1495	0.1440	0.1376	0.1300	0.1221	0.1126	0.1035	0.0933	0.0844
w_A	0.6776	0.6617	0.6266						
w_B	0.2468	0.2703	0.3182						
w_C	0.0756	0.0680	0.0552						

Polymer (B):	**poly(ethylene glycol) dimethyl ether**		**2010ZA3**
Characterization:	M_n/g.mol^{-1} = 2000, Merck, Darmstadt, Germany		
Solvent (A):	**water**	H_2O	**7732-18-5**
Salt (C):	**potassium phosphate**	K_3PO_4	**7778-53-2**

Type of data: cloud points

T/K = 298.15

w_A	0.5281	0.6028	0.6615	0.7015	0.7368	0.7652	0.7845	0.7997	0.8115
w_B	0.4597	0.3774	0.3098	0.2607	0.2142	0.1751	0.1456	0.1226	0.1040
w_C	0.0122	0.0198	0.0287	0.0378	0.0490	0.0597	0.0699	0.0777	0.0845
w_A	0.8212	0.8295	0.8362	0.8414	0.8457	0.8486	0.8508		
w_B	0.0881	0.0745	0.0633	0.0539	0.0458	0.0396	0.0343		
w_C	0.0907	0.0960	0.1005	0.1047	0.1085	0.1118	0.1149		

T/K = 303.15

w_A	0.5605	0.5852	0.6008	0.6365	0.6617	0.6863	0.7115	0.7353	0.7553
w_B	0.4248	0.3976	0.3803	0.3399	0.3102	0.2806	0.2494	0.2189	0.1921
w_C	0.0147	0.0172	0.0189	0.0236	0.0281	0.0331	0.0391	0.0458	0.0526
w_A	0.7750	0.7916	0.8069	0.8195	0.8292	0.8347	0.8403	0.8456	0.8504
w_B	0.1649	0.1408	0.1169	0.0973	0.0811	0.0699	0.0596	0.0503	0.0413
w_C	0.0601	0.0676	0.0762	0.0832	0.0897	0.0954	0.1001	0.1041	0.1083
w_A	0.8548								
w_B	0.0339								
w_C	0.1113								

T/K = 313.15

w_A	0.5710	0.6183	0.6750	0.7201	0.7521	0.7774	0.7974	0.8143	0.8283
w_B	0.4131	0.3604	0.2956	0.2396	0.1980	0.1630	0.1336	0.1083	0.0868
w_C	0.0159	0.0213	0.0294	0.0403	0.0499	0.0596	0.0690	0.0774	0.0849
w_A	0.8387	0.8463	0.8516	0.8554	0.8580	0.8596	0.8619		
w_B	0.0706	0.0578	0.0486	0.0415	0.0361	0.0315	0.0273		
w_C	0.0907	0.0959	0.0998	0.1031	0.1059	0.1089	0.1108		

T/K = 318.15

w_A	0.5924	0.6365	0.6699	0.6918	0.7158	0.7335	0.7519	0.7716	0.7857
w_B	0.3900	0.3409	0.3034	0.2777	0.2484	0.2265	0.2029	0.1767	0.1573
w_C	0.0176	0.0226	0.0267	0.0305	0.0358	0.0400	0.0452	0.0517	0.0570
w_A	0.8062	0.8205	0.8325	0.8430	0.8516	0.8579	0.8622	0.8659	0.8684
w_B	0.1293	0.1086	0.0908	0.0752	0.0621	0.0515	0.0438	0.0371	0.0319
w_C	0.0645	0.0709	0.0767	0.0818	0.0863	0.0906	0.0940	0.0970	0.0997

Type of data: coexistence data (tie lines)

continued

continued

Total system			Top phase			Bottom phase		
w_A	w_B	w_C	w_A	w_B	w_C	w_A	w_B	w_C
T/K = 298.15								
0.7417	0.1793	0.0790	0.6702	0.2995	0.0303	0.8392	0.0169	0.1439
0.7054	0.1883	0.1063	0.5910	0.3896	0.0194	0.8031	0.0130	0.1839
0.6775	0.1888	0.1337	0.5419	0.4434	0.0147	0.7727	0.0104	0.2169
0.6539	0.1873	0.1588	0.5064	0.4812	0.0124	0.7458	0.0095	0.2447
0.6212	0.1880	0.1908	0.4499	0.5405	0.0096	0.7106	0.0084	0.2810
0.6109	0.1897	0.1994	0.4348	0.5563	0.0089	0.6984	0.0076	0.2940
0.5594	0.1988	0.2418	0.3520	0.6411	0.0069	0.6518	0.0055	0.3427
T/K = 303.15								
0.7342	0.1857	0.0801	0.6569	0.3163	0.0268	0.8379	0.0172	0.1449
0.7068	0.1876	0.1056	0.5857	0.3967	0.0176	0.8043	0.0129	0.1828
0.6782	0.1884	0.1334	0.5063	0.4815	0.0122	0.7612	0.0092	0.2296
0.6482	0.1908	0.1610	0.4677	0.5220	0.0103	0.7337	0.0085	0.2578
0.6214	0.1902	0.1884	0.4240	0.5675	0.0085	0.7062	0.0076	0.2862
0.6005	0.1919	0.2076	0.3900	0.6022	0.0078	0.6821	0.0068	0.3111
0.5696	0.1915	0.2389	0.3020	0.6919	0.0061	0.6310	0.0049	0.3641
T/K = 308.15								
0.7299	0.1898	0.0803	0.6429	0.3340	0.0231	0.8359	0.0182	0.1459
0.7027	0.1900	0.1073	0.5792	0.4047	0.0161	0.8047	0.0133	0.1820
0.6763	0.1901	0.1336	0.4701	0.5200	0.0099	0.7494	0.0082	0.2424
0.6481	0.1917	0.1602	0.4268	0.5648	0.0084	0.7212	0.0077	0.2711
0.6281	0.1879	0.1840	0.3975	0.5949	0.0076	0.7016	0.0069	0.2915
0.5991	0.1895	0.2114	0.3449	0.6483	0.0068	0.6675	0.0063	0.3262
0.5719	0.1908	0.2373	0.2515	0.7429	0.0056	0.6098	0.0045	0.3857
T/K = 318.15								
0.7320	0.1903	0.0777	0.6311	0.3492	0.0197	0.8472	0.0161	0.1367
0.7221	0.1771	0.1008	0.5830	0.4017	0.0153	0.8256	0.0138	0.1606
0.6782	0.1887	0.1331	0.5093	0.4801	0.0106	0.7863	0.0119	0.2018
0.6500	0.1896	0.1604	0.4570	0.5354	0.0076	0.7541	0.0087	0.2372
0.6274	0.1885	0.1841	0.4208	0.5718	0.0074	0.7257	0.0097	0.2646
0.5976	0.1883	0.2141	0.3723	0.6213	0.0064	0.6941	0.0079	0.2980
0.5736	0.1881	0.2383	0.3338	0.6601	0.0061	0.6669	0.0075	0.3256

Polymer (B): **poly(ethylene glycol) dimethyl ether** **2009ZA2**
Characterization: M_n/g.mol^{-1} = 2000, Merck, Darmstadt, Germany
Solvent (A): **water** H_2O **7732-18-5**
Salt (C): **sodium carbonate** Na_2CO_3 **497-19-8**

continued

continued

Type of data: cloud points

T/K = 288.15

w_A	0.7067	0.7425	0.7819	0.8087	0.8340	0.8559	0.8643
w_B	0.2578	0.2155	0.1671	0.1319	0.0983	0.0683	0.0557
w_C	0.0355	0.0420	0.0510	0.0594	0.0677	0.0758	0.0800

T/K = 293.15

w_A	0.7111	0.7454	0.7792	0.8061	0.8360	0.8600
w_B	0.2534	0.2126	0.1720	0.1367	0.0974	0.0647
w_C	0.0355	0.0420	0.0488	0.0572	0.0666	0.0753

T/K = 298.15

w_A	0.6912	0.7378	0.7800	0.8184	0.8493	0.8684
w_B	0.2758	0.2221	0.1717	0.1226	0.0823	0.0559
w_C	0.0330	0.0401	0.0483	0.0590	0.0684	0.0757

T/K = 308.15

w_A	0.6911	0.7554	0.8013	0.8381	0.8586	0.8749	0.8839
w_B	0.2788	0.2044	0.1484	0.1013	0.0748	0.0526	0.0391
w_C	0.0301	0.0402	0.0503	0.0606	0.0666	0.0725	0.0770

T/K = 318.15

w_A	0.7376	0.7660	0.7996	0.8286	0.8536	0.8734	0.8866	0.8950	0.8992
w_B	0.2254	0.1930	0.1538	0.1191	0.0882	0.0629	0.0445	0.0313	0.0234
w_C	0.0370	0.0410	0.0466	0.0523	0.0582	0.0637	0.0689	0.0737	0.0774

Type of data: coexistence data (tie lines)

Top phase			Bottom phase		
w_A	w_B	w_C	w_A	w_B	w_C
T/K = 288.15					
0.6874	0.2810	0.0316	0.8632	0.0274	0.1094
0.6548	0.3188	0.0264	0.8527	0.0236	0.1237
0.6268	0.3515	0.0217	0.8420	0.0197	0.1383
0.5973	0.3839	0.0188	0.8313	0.0150	0.1537
0.5723	0.4117	0.0160	0.8157	0.0134	0.1709
0.5528	0.4329	0.0143	0.8006	0.0129	0.1865
T/K = 293.15					
0.6852	0.2847	0.0301	0.8625	0.0304	0.1071
0.6478	0.3265	0.0257	0.8482	0.0258	0.1260
0.6258	0.3511	0.0231	0.8410	0.0231	0.1359
0.5920	0.3879	0.0201	0.8313	0.0210	0.1477
0.5575	0.4253	0.0172	0.8225	0.0172	0.1603
0.5374	0.4465	0.0161	0.8083	0.0182	0.1735

continued

continued

T/K = 298.15

0.6654	0.3072	0.0274	0.8671	0.0241	0.1088
0.6392	0.3366	0.0242	0.8564	0.0186	0.1250
0.5937	0.3870	0.0193	0.8322	0.0150	0.1528
0.5652	0.4173	0.0175	0.8247	0.0118	0.1635
0.5341	0.4509	0.0150	0.8119	0.0115	0.1766

T/K = 308.15

0.6592	0.3148	0.0260	0.8808	0.0180	0.1012
0.6311	0.3459	0.0230	0.8716	0.0123	0.1161
0.5959	0.3831	0.0210	0.8579	0.0117	0.1304
0.5226	0.4600	0.0174	0.8371	0.0127	0.1502
0.5009	0.4823	0.0168	0.8236	0.0118	0.1646

T/K = 318.15

0.7014	0.2666	0.0320	0.8969	0.0133	0.0898
0.6504	0.3243	0.0253	0.8860	0.0097	0.1043
0.6159	0.3621	0.0220	0.8770	0.0082	0.1148
0.5802	0.4011	0.0187	0.8632	0.0080	0.1288
0.5436	0.4399	0.0165	0.8516	0.0071	0.1413

Polymer (B): **poly(ethylene glycol) dimethyl ether** **2010ZA2**
Characterization: M_n/g.mol^{-1} = 2300, Merck, Darmstadt, Germany
Solvent (A): **water** **H_2O** **7732-18-5**
Salt (C): **sodium phosphate** **Na_3PO_4** **7601-54-9**

Type of data: cloud points

T/K = 298.15

w_A	0.6812	0.6969	0.7131	0.7295	0.7464	0.7647	0.7799	0.7935	0.8054
w_B	0.2993	0.2814	0.2628	0.2437	0.2239	0.2022	0.1834	0.1662	0.1506
w_C	0.0195	0.0217	0.0241	0.0268	0.0297	0.0331	0.0367	0.0403	0.0440

w_A	0.8140	0.8218	0.8306	0.8451
w_B	0.1392	0.1286	0.1167	0.0966
w_C	0.0468	0.0496	0.0527	0.0583

Polymer (B): **poly(ethylene glycol) dimethyl ether** **2010ZA1**
Characterization: M_n/g.mol^{-1} = 2300, Merck, Darmstadt, Germany
Solvent (A): **water** **H_2O** **7732-18-5**
Salt (C): **sodium phosphate** **Na_3PO_4** **7601-54-9**

Type of data: cloud points

T/K = 298.15

w_A	0.6812	0.6969	0.7131	0.7295	0.7464	0.7647	0.7799	0.7935	0.8054
w_B	0.2993	0.2814	0.2628	0.2437	0.2239	0.2022	0.1834	0.1662	0.1506
w_C	0.0195	0.0217	0.0241	0.0268	0.0297	0.0331	0.0367	0.0403	0.0440

continued

continued

w_A	0.8140	0.8218	0.8306	0.8451
w_B	0.1392	0.1286	0.1167	0.0966
w_C	0.0468	0.0496	0.0527	0.0583

T/K = 308.15

w_A	0.6910	0.7095	0.7295	0.7438	0.7572	0.7705	0.7836	0.7977	0.8107
w_B	0.2948	0.2743	0.2518	0.2355	0.2200	0.2045	0.1889	0.1721	0.1557
w_C	0.0142	0.0162	0.0187	0.0207	0.0228	0.0250	0.0275	0.0302	0.0336
w_A	0.8169	0.8237	0.8311	0.8370	0.8422	0.8475	0.8526	0.8588	0.8636
w_B	0.1478	0.1389	0.1293	0.1215	0.1147	0.1076	0.1007	0.0923	0.0857
w_C	0.0353	0.0374	0.0396	0.0415	0.0431	0.0449	0.0467	0.0489	0.0507
w_A	0.8723	0.8754	0.8790	0.8813					
w_B	0.0738	0.0695	0.0646	0.0614					
w_C	0.0539	0.0551	0.0564	0.0573					

T/K = 313.15

w_A	0.7343	0.7514	0.7668	0.7810	0.7951	0.8056	0.8156	0.8281	0.8367
w_B	0.2453	0.2257	0.2076	0.1903	0.1727	0.1595	0.1467	0.1302	0.1190
w_C	0.0204	0.0229	0.0256	0.0287	0.0322	0.0349	0.0377	0.0417	0.0443
w_A	0.8564								
w_B	0.0925								
w_C	0.0511								

T/K = 318.15

w_A	0.7104	0.7308	0.7461	0.7545	0.7646	0.7724	0.7800	0.7898	0.7982
w_B	0.2697	0.2469	0.2292	0.2193	0.2075	0.1980	0.1886	0.1763	0.1656
w_C	0.0199	0.0223	0.0247	0.0262	0.0279	0.0296	0.0314	0.0339	0.0362
w_A	0.8054	0.8224	0.8280	0.8332	0.8382	0.8464	0.8514	0.8561	0.8632
w_B	0.1564	0.1342	0.1266	0.1198	0.1131	0.1020	0.0951	0.0886	0.0788
w_C	0.0382	0.0434	0.0454	0.0470	0.0487	0.0516	0.0535	0.0553	0.0580
w_A	0.8682								
w_B	0.0719								
w_C	0.0599								

Type of data: coexistence data (tie lines)

Top phase			Bottom phase		
w_A	w_B	w_C	w_A	w_B	w_C
T/K = 298.15					
0.75436	0.21474	0.03090	0.89460	0.00961	0.09579
0.72188	0.25339	0.02473	0.88597	0.00880	0.10523
0.69759	0.28149	0.02092	0.87800	0.00813	0.11387
0.68207	0.29902	0.01891	0.87100	0.00736	0.12164
0.66670	0.31620	0.01710	0.86366	0.00684	0.12950

continued

continued

T/K = 308.15

0.72624	0.25241	0.02135	0.88507	0.02864	0.08629
0.70655	0.27575	0.01770	0.88371	0.02244	0.09385
0.68880	0.29513	0.01607	0.88418	0.01537	0.10045
0.67815	0.30674	0.01511	0.88382	0.01290	0.10328
0.66697	0.31942	0.01361	0.87963	0.01066	0.10971

T/K = 313.15

0.68470	0.30210	0.01320	0.89380	0.00690	0.09930
0.65050	0.33940	0.01010	0.87380	0.00580	0.12040
0.62750	0.36400	0.00850	0.86320	0.00430	0.13250
0.59190	0.40110	0.00700	0.85440	0.00390	0.14170
0.56860	0.42560	0.00580	0.83680	0.00280	0.16040

T/K = 318.15

0.70207	0.28246	0.01547	0.89977	0.01412	0.08611
0.67646	0.31125	0.01229	0.89078	0.01426	0.09496
0.65632	0.33331	0.01037	0.88508	0.01393	0.10099
0.63811	0.35329	0.00860	0.87586	0.01430	0.10984
0.61671	0.37631	0.00698	0.86497	0.01390	0.12113

Polymer (B): **poly(ethylene glycol) dimethyl ether** **2008ZA3**
Characterization: M_n/g.mol^{-1} = 2000, Merck, Darmstadt, Germany
Solvent (A): **water** H_2O **7732-18-5**
Salt (C): **sodium sulfate** Na_2SO_4 **7757-82-6**

Type of data: cloud points

T/K = 288.15

w_A	0.9912	0.9879	0.9860	0.9817	0.9764	0.9658	0.9535	0.9380	0.9250
w_B	0.0026	0.0067	0.0092	0.0143	0.0201	0.0314	0.0443	0.0603	0.0737
w_C	0.0062	0.0054	0.0048	0.0040	0.0035	0.0028	0.0022	0.0017	0.0013
w_A	0.9147								
w_B	0.0843								
w_C	0.0010								

T/K = 298.15

w_A	0.4113	0.4700	0.5108	0.5546	0.5805	0.6346	0.6575	0.7107	0.7523
w_B	0.0035	0.0064	0.0089	0.0128	0.0154	0.0226	0.0263	0.0380	0.0514
w_C	0.5852	0.5236	0.4803	0.4326	0.4041	0.3428	0.3162	0.2513	0.1963
w_A	0.7870	0.8124	0.8318	0.8422	0.8501				
w_B	0.0663	0.0793	0.0906	0.0974	0.1034				
w_C	0.1467	0.1083	0.0776	0.0604	0.0465				

continued

continued

T/K = 308.15

w_A	0.4174	0.5181	0.5780	0.6362	0.6854	0.7250	0.7666	0.7968	0.8240
w_B	0.0049	0.0099	0.0156	0.0228	0.0318	0.0416	0.0554	0.0679	0.0808
w_C	0.5777	0.4720	0.4064	0.3410	0.2828	0.2334	0.1780	0.1353	0.0952

T/K = 318.15

w_A	0.4208	0.5349	0.6246	0.6946	0.7537	0.7993	0.8279	0.8462	0.8578
w_B	0.0049	0.0108	0.0195	0.0325	0.0477	0.0635	0.0751	0.0832	0.0900
w_C	0.5743	0.4543	0.3559	0.2729	0.1986	0.1372	0.0970	0.0706	0.0522

Type of data: coexistence data (tie lines)

Top phase			Bottom phase		
w_A	w_B	w_C	w_A	w_B	w_C
T/K = 288.15					
0.6804	0.2916	0.0280	0.8282	0.0274	0.1444
0.6500	0.3251	0.0249	0.8222	0.0211	0.1567
0.6351	0.3451	0.0198	0.8115	0.0180	0.1705
0.6060	0.3773	0.0167	0.7981	0.0182	0.1837
0.5801	0.4060	0.0139	0.7889	0.0170	0.1941
0.5575	0.4292	0.0133	0.7787	0.0155	0.2058
T/K = 298.15					
0.6471	0.3295	0.0234	0.8318	0.0210	0.1472
0.6183	0.3628	0.0189	0.8182	0.0181	0.1637
0.5767	0.4085	0.0148	0.7998	0.0159	0.1843
0.5607	0.4272	0.0121	0.7873	0.0133	0.1994
0.5354	0.4538	0.0108	0.7752	0.0108	0.2140
T/K = 308.15					
0.6466	0.3310	0.0224	0.8564	0.0175	0.1261
0.6262	0.3530	0.0208	0.8420	0.0104	0.1476
0.6029	0.3802	0.0169	0.8291	0.0100	0.1609
0.5790	0.4074	0.0136	0.8168	0.0097	0.1735
0.5564	0.4313	0.0123	0.8063	0.0074	0.1863
0.5373	0.4522	0.0105	0.7924	0.0066	0.2010
T/K = 318.15					
0.6503	0.3261	0.0236	0.8532	0.0235	0.1233
0.6056	0.3763	0.0181	0.8468	0.0124	0.1408
0.5829	0.4029	0.0142	0.8369	0.0083	0.1548
0.5537	0.4355	0.0108	0.8250	0.0054	0.1696
0.5391	0.4504	0.0105	0.8125	0.0039	0.1836
0.5226	0.4681	0.0093	0.7991	0.0040	0.1969

Polymer (B):	**poly(ethylene glycol) dimethyl ether**		**2011SAD**
Characterization:	M_n/g.mol^{-1} = 698, Merck, Darmstadt, Germany		
Solvent (A):	**water**	H_2O	**7732-18-5**
Salt (C):	**triammonium phosphate**	$(NH_4)_3PO_4$	**10361-65-6**

Type of data: cloud points

T/K = 298.15

w_A	0.8326	0.8327	0.8323	0.8309	0.8291	0.8256	0.8194	0.8139	0.8065
w_B	0.0267	0.0302	0.0334	0.0384	0.0458	0.0561	0.0699	0.0821	0.0972
w_C	0.1407	0.1371	0.1343	0.1307	0.1251	0.1183	0.1107	0.1040	0.0963
w_A	0.7968	0.7867	0.7751	0.7633	0.7502	0.7366	0.7208		
w_B	0.1149	0.1338	0.1542	0.1748	0.1958	0.2166	0.2390		
w_C	0.0883	0.0795	0.0707	0.0619	0.0540	0.0468	0.0402		

T/K = 303.15

w_A	0.8328	0.8337	0.8343	0.8341	0.8327	0.8297	0.8228	0.8143	0.8031
w_B	0.0189	0.0225	0.0258	0.0313	0.0394	0.0505	0.0659	0.0840	0.1053
w_C	0.1483	0.1438	0.1399	0.1346	0.1279	0.1198	0.1113	0.1017	0.0916
w_A	0.7904	0.7777	0.7641	0.7503	0.7363	0.7218	0.7069		
w_B	0.1278	0.1502	0.1733	0.1952	0.2163	0.2370	0.2574		
w_C	0.0818	0.0721	0.0626	0.0545	0.0474	0.0412	0.0357		

T/K = 308.15

w_A	0.8327	0.8354	0.8365	0.8385	0.8385	0.8370	0.8353	0.8278	0.8205
w_B	0.0160	0.0189	0.0216	.0243	0.0285	0.0353	0.0415	0.0617	0.0763
w_C	0.1513	0.1457	0.1419	0.1372	0.1330	0.1277	0.1232	0.1105	0.1032
w_A	0.8108	0.8005	0.7886	0.7763	0.7642	0.7510	0.7378	0.7229	0.7073
w_B	0.0943	0.1126	0.1331	0.1538	0.1734	0.1942	0.2140	0.2351	0.2565
w_C	0.0949	0.0869	0.0783	0.0699	0.0624	0.0548	0.0482	0.0420	0.0362

T/K = 313.15

w_A	0.8398	0.8402	0.8404	0.8411	0.8379	0.8333	0.8272	0.8182	0.8078
w_B	0.0132	0.0189	0.0221	0.0279	0.0421	0.0539	0.0673	0.0837	0.1020
w_C	0.1470	0.1409	0.1375	0.1310	0.1200	0.1128	0.1055	0.0981	0.0902
w_A	0.7960	0.7830	0.7692	0.7567	0.7432	0.7295	0.7166	0.7016	
w_B	0.1223	0.1437	0.1656	0.1856	0.2059	0.2256	0.2436	0.2635	
w_C	0.0817	0.0733	0.0652	0.0577	0.0509	0.0449	0.0398	0.0349	

T/K = 318.15

w_A	0.8364	0.8419	0.8441	0.8448	0.8437	0.8398	0.8342	0.8252	0.8126
w_B	0.0109	0.0160	0.0214	0.0266	0.0333	0.0431	0.0561	0.0741	0.0959
w_C	0.1527	0.1421	0.1345	0.1286	0.1230	0.1171	0.1097	0.1007	0.0915
w_A	0.7983	0.7837	0.7687	0.7524	0.7366	0.7215	0.7091	0.6949	
w_B	0.1200	0.1441	0.1675	0.1921	0.2153	0.2366	0.2537	0.2720	
w_C	0.0817	0.0722	0.0638	0.0555	0.0481	0.0419	0.0372	0.0331	

Polymer (B):	**poly(ethylene glycol) dimethyl ether**	**2011SAD**
Characterization:	M_n/g.mol^{-1} = 2145, Merck, Darmstadt, Germany	
Solvent (A):	**water** H_2O	**7732-18-5**
Salt (C):	**triammonium phosphate** $(NH_4)_3PO_4$	**10361-65-6**

Type of data: cloud points

T/K = 293.15

w_A	0.8516	0.8672	0.8685	0.8707	0.8724	0.8725	0.8682	0.8587	0.8428
w_B	0.0049	0.0066	0.0084	0.0102	0.0135	0.0188	0.0314	0.0499	0.0763
w_C	0.1435	0.1262	0.1231	0.1191	0.1141	0.1087	0.1004	0.0914	0.0809
w_A	0.8211	0.7988	0.7760	0.7520	0.7312	0.7086	0.6674		
w_B	0.1092	0.1420	0.1743	0.2063	0.2327	0.2605	0.3086		
w_C	0.0697	0.0592	0.0497	0.0417	0.0361	0.0309	0.0240		

T/K = 298.15

w_A	0.8644	0.8717	0.8730	0.8746	0.8753	0.8736	0.8643	0.8432	0.8192
w_B	0.0051	0.0075	0.0097	0.0120	0.0165	0.0250	0.0438	0.0781	0.1146
w_C	0.1305	0.1208	0.1173	0.1134	0.1082	0.1014	0.0919	0.0787	0.0662
w_A	0.7911	0.7620	0.7301	0.7004	0.6798				
w_B	0.1546	0.1942	0.2347	0.2706	0.2946				
w_C	0.0543	0.0438	0.0352	0.0290	0.0256				

T/K = 303.15

w_A	0.8366	0.8644	0.8724	0.8785	0.8760	0.8739	0.8688	0.8620	0.8528
w_B	0.0063	0.0074	0.0089	0.0195	0.0246	0.0308	0.0407	0.0524	0.0670
w_C	0.1571	0.1282	0.1187	0.1020	0.0994	0.0953	0.0905	0.0856	0.0802
w_A	0.8411	0.8245	0.8077	0.7878	0.7675	0.7468	0.7257	0.7050	0.6912
w_B	0.0847	0.1092	0.1332	0.1604	0.1877	0.2143	0.2403	0.2651	0.2814
w_C	0.0742	0.0663	0.0591	0.0518	0.0448	0.0389	0.0340	0.0299	0.0274

T/K = 308.15

w_A	0.8375	0.8499	0.8662	0.8734	0.8792	0.8797	0.8747	0.8665	0.8532
w_B	0.0051	0.0070	0.0081	0.0095	0.0128	0.0176	0.0356	0.0499	0.0702
w_C	0.1574	0.1431	0.1257	0.1171	0.1080	0.1027	0.0897	0.0836	0.0766
w_A	0.8385	0.8217	0.8013	0.7830	0.7624	0.7373	0.7125	0.6787	0.6612
w_B	0.0916	0.1156	0.1435	0.1681	0.1948	0.2262	0.2562	0.2954	0.3151
w_C	0.0699	0.0627	0.0552	0.0489	0.0428	0.0365	0.0313	0.0259	0.0237

T/K = 313.15

w_A	0.8467	0.8663	0.8768	0.8809	0.8823	0.8826	0.8831	0.8807	0.8754
w_B	0.0050	0.0064	0.0107	0.0117	0.0143	0.0182	0.0212	0.0286	0.0388
w_C	0.1483	0.1273	0.1125	0.1074	0.1034	0.0992	0.0957	0.0907	0.0858
w_A	0.8676	0.8563	0.8429	0.8264	0.8059	0.7842	0.7622	0.7370	0.7110
w_B	0.0520	0.0690	0.0883	0.1110	0.1388	0.1672	0.1958	0.2269	0.2580
w_C	0.0804	0.0747	0.0688	0.0626	0.0553	0.0486	0.0420	0.0361	0.0310

continued

continued

T/K = 318.15

w_A	0.8556	0.8791	0.8851	0.8878	0.8855	0.8778	0.8668	0.8523	0.8343
w_B	0.0050	0.0063	0.0083	0.0164	0.0265	0.0409	0.0582	0.0788	0.1035
w_C	0.1394	0.1146	0.1066	0.0958	0.0880	0.0813	0.0750	0.0689	0.0622
w_A	0.8137	0.7911	0.7689	0.7459	0.7194	0.6953	0.6540		
w_B	0.1310	0.1604	0.1886	0.2172	0.2490	0.2772	0.3238		
w_C	0.0553	0.0485	0.0425	0.0369	0.0316	0.0275	0.0222		

Polymer (B): **poly(ethylene oxide)** **2011LEM**
Characterization: M_w/g.mol^{-1} = 1500, Vetec, Brazil
Solvent (A): **water** **H_2O** **7732-18-5**
Salt (C): **ammonium sulfate** **$(NH_4)_2SO_4$** **7783-20-2**

Type of data: coexistence data (tie lines)

Total system			Top phase			Bottom phase		
w_A	w_B	w_C	w_A	w_B	w_C	w_A	w_B	w_C
T/K = 283.15								
0.7034	0.1723	0.1243	0.6609	0.2949	0.0442	0.7468	0.0387	0.2145
0.6512	0.2056	0.1433	0.5973	0.3721	0.0306	0.7081	0.0304	0.2616
0.5955	0.2380	0.1664	0.5132	0.4622	0.0246	0.6582	0.0286	0.3132
0.5342	0.2806	0.1852	0.4530	0.5284	0.0186	0.5948	0.0335	0.3717
0.5027	0.2973	0.1999	0.4314	0.5550	0.0136	0.5609	0.0358	0.4034
T/K = 298.15								
0.7138	0.1678	0.1184	0.6327	0.3160	0.0513	0.7622	0.0525	0.1853
0.6617	0.2043	0.1340	0.5579	0.4080	0.0341	0.7285	0.0380	0.2336
0.5970	0.2518	0.1512	0.5002	0.4760	0.0238	0.6736	0.0407	0.2856
0.5498	0.2780	0.1722	0.4221	0.5604	0.0176	0.6271	0.0484	0.3245
0.5096	0.3011	0.1893	0.3880	0.5988	0.0132	0.5978	0.0478	0.3544
T/K = 313.15								
0.7039	0.1720	0.1242	0.6010	0.3590	0.0400	0.7683	0.0320	0.1997
0.6512	0.2061	0.1427	0.5188	0.4541	0.0271	0.7229	0.0362	0.2409
0.5949	0.2382	0.1669	0.4577	0.5227	0.0196	0.6818	0.0369	0.2813
0.5338	0.2814	0.1848	0.4031	0.5845	0.0123	0.6322	0.0251	0.3427
0.5027	0.2976	0.1997	0.3655	0.6254	0.0091	0.5839	0.0392	0.3770

Polymer (B):	**poly(ethylene oxide)**		**2008SA1**
Characterization:	M_n/g.mol^{-1} = 4000, Merck, Darmstadt, Germany		
Solvent (A):	**water**	H_2O	**7732-18-5**
Salt (C):	**disodium phosphate**	Na_2HPO_4	**7558-79-4**

Type of data: cloud points

T/K = 298.15

w_A	0.9031	0.9008	0.8981	0.8951	0.8910	0.8857	0.8799	0.8734	0.8664
w_B	0.0080	0.0130	0.0189	0.0249	0.0320	0.0403	0.0492	0.0588	0.0688
w_C	0.0889	0.0862	0.0830	0.0800	0.0770	0.0740	0.0709	0.0678	0.0648
w_A	0.8589	0.8513	0.8436	0.8357	0.8337	0.8281	0.8233	0.8216	0.8223
w_B	0.0795	0.0903	0.1013	0.1120	0.1158	0.1229	0.1280	0.1310	0.1311
w_C	0.0616	0.0584	0.0551	0.0523	0.0505	0.0490	0.0487	0.0474	0.0466
w_A	0.8157	0.8148	0.8096	0.8019	0.7952	0.7894	0.7852	0.7823	0.7821
w_B	0.1393	0.1406	0.1478	0.1577	0.1661	0.1734	0.1787	0.1818	0.1825
w_C	0.0450	0.0446	0.0426	0.0404	0.0387	0.0372	0.0361	0.0359	0.0354
w_A	0.7796	0.7711	0.7578	0.7435	0.7283	0.7114	0.6927	0.6742	0.6574
w_B	0.1856	0.1956	0.2120	0.2289	0.2468	0.2664	0.2877	0.3086	0.3277
w_C	0.0348	0.0333	0.0302	0.0276	0.0249	0.0222	0.0196	0.0172	0.0149
w_A	0.6358	0.6033	0.5324						
w_B	0.3514	0.3867	0.4620						
w_C	0.0128	0.0100	0.0056						

T/K = 303.15

w_A	0.8955	0.9024	0.9042	0.9025	0.8992	0.8946	0.8880	0.8811	0.8738
w_B	0.0029	0.0054	0.0101	0.0162	0.0234	0.0316	0.0417	0.0518	0.0621
w_C	0.1016	0.0922	0.0857	0.0813	0.0774	0.0738	0.0703	0.0671	0.0641
w_A	0.8665	0.8589	0.8559	0.8513	0.8504	0.8439	0.8440	0.8376	0.8291
w_B	0.0724	0.0829	0.0873	0.0934	0.0951	0.1035	0.1038	0.1126	0.1235
w_C	0.0611	0.0582	0.0568	0.0553	0.0545	0.0526	0.0522	0.0498	0.0474
w_A	0.8283	0.8218	0.8181	0.8137	0.8063	0.7931	0.7779	0.7575	0.7346
w_B	0.1252	0.1333	0.1384	0.1437	0.1536	0.1702	0.1889	0.2132	0.2400
w_C	0.0465	0.0449	0.0435	0.0426	0.0401	0.0367	0.0332	0.0293	0.0254
w_A	0.7072	0.6743	0.6333	0.5804	0.5273				
w_B	0.2717	0.3094	0.3546	0.4118	0.4683				
w_C	0.0211	0.0163	0.0121	0.0078	0.0044				

T/K = 308.15

w_A	0.9067	0.9039	0.8987	0.8913	0.8818	0.8711	0.8682	0.8629	0.8604
w_B	0.0097	0.0211	0.0301	0.0419	0.0551	0.0697	0.0751	0.0818	0.0846
w_C	0.0836	0.0750	0.0712	0.0668	0.0631	0.0592	0.0567	0.0553	0.0550

continued

continued

w_A	0.8579	0.8516	0.8496	0.8454	0.8391	0.8374	0.8360	0.8327	0.8311
w_B	0.0889	0.0968	0.0991	0.1056	0.1127	0.1155	0.1173	0.1210	0.1232
w_C	0.0532	0.0516	0.0513	0.0490	0.0482	0.0471	0.0467	0.0463	0.0457
w_A	0.8295	0.8274	0.8231	0.8200	0.8119	0.8105	0.8087	0.8026	0.7945
w_B	0.1261	0.1285	0.1335	0.1386	0.1475	0.1498	0.1529	0.1595	0.1704
w_C	0.0444	0.0441	0.0434	0.0414	0.0406	0.0397	0.0384	0.0379	0.0351
w_A	0.7884	0.7773	0.7729	0.7569	0.7402	0.7370	0.7197	0.7146	0.6956
w_B	0.1767	0.1909	0.1954	0.2143	0.2338	0.2385	0.2573	0.2639	0.2847
w_C	0.0349	0.0318	0.0317	0.0288	0.0260	0.0245	0.0230	0.0215	0.0197
w_A	0.6894	0.6669	0.6619	0.6497	0.6332	0.6173	0.6066	0.5814	0.5544
w_B	0.2923	0.3170	0.3228	0.3361	0.3543	0.3713	0.3830	0.4100	0.4387
w_C	0.0183	0.0161	0.0153	0.0142	0.0125	0.0114	0.0104	0.0086	0.0069

T/K = 313.15

w_A	0.9103	0.9060	0.9013	0.8954	0.8870	0.8776	0.8687	0.8592	0.8567
w_B	0.0133	0.0223	0.0307	0.0400	0.0517	0.0643	0.0764	0.0895	0.0922
w_C	0.0764	0.0717	0.0680	0.0646	0.0613	0.0581	0.0549	0.0513	0.0511
w_A	0.8549	0.8500	0.8466	0.8434	0.8369	0.8357	0.8271	0.8275	0.8174
w_B	0.0954	0.1017	0.1052	0.1104	0.1177	0.1202	0.1303	0.1307	0.1425
w_C	0.0497	0.0483	0.0482	0.0462	0.0454	0.0441	0.0426	0.0418	0.0401
w_A	0.8172	0.8082	0.8054	0.8021	0.7948	0.7930	0.7925	0.7863	0.7801
w_B	0.1436	0.1540	0.1581	0.1614	0.1707	0.1728	0.1737	0.1806	0.1879
w_C	0.0392	0.0378	0.0365	0.0365	0.0345	0.0342	0.0338	0.0331	0.0320
w_A	0.7791	0.7773	0.7736	0.7591	0.7557	0.7383	0.7300	0.7138	0.7001
w_B	0.1894	0.1918	0.1956	0.2132	0.2170	0.2372	0.2468	0.2650	0.2810
w_C	0.0315	0.0309	0.0308	0.0277	0.0273	0.0245	0.0232	0.0212	0.0189
w_A	0.6870	0.6575	0.6551	0.6295	0.6045	0.6017	0.5432		
w_B	0.2951	0.3275	0.3304	0.3583	0.3855	0.3883	0.4504		
w_C	0.0179	0.0150	0.0145	0.0122	0.0100	0.0100	0.0064		

T/K = 318.15

w_A	0.9130	0.9138	0.9120	0.9069	0.9012	0.8943	0.8874	0.8837	0.8808
w_B	0.0059	0.0119	0.0181	0.0269	0.0358	0.0454	0.0549	0.0599	0.0637
w_C	0.0811	0.0743	0.0699	0.0662	0.0630	0.0603	0.0577	0.0564	0.0555
w_A	0.8791	0.8769	0.8746	0.8731	0.8684	0.8674	0.8632	0.8595	0.8577
w_B	0.0658	0.0687	0.0715	0.0737	0.0797	0.0810	0.0864	0.0913	0.0936
w_C	0.0551	0.0544	0.0539	0.0532	0.0519	0.0516	0.0504	0.0492	0.0487
w_A	0.8554	0.8527	0.8516	0.8452	0.8375	0.8292	0.8195	0.8090	0.7968
w_B	0.0965	0.1000	0.1013	0.1097	0.1193	0.1297	0.1418	0.1546	0.1693
w_C	0.0481	0.0473	0.0471	0.0451	0.0432	0.0411	0.0387	0.0364	0.0339

continued

continued

w_A	0.7826	0.7760	0.7662	0.7572	0.7474	0.7247	0.7152	0.7001	0.6870
w_B	0.1862	0.1947	0.2053	0.2167	0.2269	0.2528	0.2640	0.2803	0.2954
w_C	0.0312	0.0293	0.0285	0.0261	0.0257	0.0225	0.0208	0.0196	0.0176
w_A	0.6703	0.6538	0.6444	0.6201	0.6142	0.5981	0.5724	0.5368	
w_B	0.3133	0.3316	0.3417	0.3681	0.3747	0.3917	0.4193	0.4571	
w_C	0.0164	0.0146	0.0139	0.0118	0.0111	0.0102	0.0083	0.0061	

Polymer (B): **poly(ethylene oxide)** **2008SA1**
Characterization: M_n/g.mol^{-1} = 4000, Merck, Darmstadt, Germany
Solvent (A): **water** H_2O **7732-18-5**
Salt (C): **monosodium phosphate** NaH_2PO_4 **7558-80-7**

Type of data: cloud points

T/K = 298.15

w_A	0.8372	0.8352	0.8327	0.8286	0.8228	0.8153	0.8064	0.7965	0.7907
w_B	0.0088	0.0169	0.0254	0.0354	0.0472	0.0612	0.0771	0.0939	0.1048
w_C	0.1540	0.1479	0.1419	0.1360	0.1300	0.1235	0.1165	0.1096	0.1045
w_A	0.7869	0.7848	0.7777	0.7772	0.7704	0.7669	0.7611	0.7532	0.7500
w_B	0.1100	0.1146	0.1257	0.1265	0.1380	0.1424	0.1532	0.1635	0.1697
w_C	0.1031	0.1006	0.0966	0.0963	0.0916	0.0907	0.0857	0.0833	0.0803
w_A	0.7366	0.7218	0.7057	0.6852	0.6621	0.6391	0.6167	0.5915	0.5780
w_B	0.1887	0.2094	0.2325	0.2588	0.2894	0.3178	0.3462	0.3764	0.3939
w_C	0.0747	0.0688	0.0618	0.0560	0.0485	0.0431	0.0371	0.0321	0.0281

T/K = 303.15

w_A	0.8329	0.8419	0.8407	0.8382	0.8340	0.8280	0.8201	0.8111	0.8025
w_B	0.0051	0.0092	0.0165	0.0253	0.0353	0.0478	0.0616	0.0764	0.0906
w_C	0.1620	0.1489	0.1428	0.1365	0.1307	0.1242	0.1183	0.1125	0.1069
w_A	0.7944	0.7844	0.7768	0.7671	0.7558	0.7467	0.7387	0.7254	0.7115
w_B	0.1042	0.1199	0.1322	0.1472	0.1638	0.1770	0.1881	0.2062	0.2250
w_C	0.1014	0.0957	0.0910	0.0857	0.0804	0.0763	0.0732	0.0684	0.0635
w_A	0.6958	0.6788	0.6570	0.6331	0.6081	0.5772	0.5396	0.5174	0.4723
w_B	0.2461	0.2684	0.2957	0.3253	0.3564	0.3931	0.4366	0.4626	0.5158
w_C	0.0581	0.0528	0.0473	0.0416	0.0355	0.0297	0.0238	0.0200	0.0119

T/K = 308.15

w_A	0.8450	0.8458	0.8452	0.8420	0.8377	0.8325	0.8266	0.8186	0.8111
w_B	0.0081	0.0133	0.0200	0.0290	0.0382	0.0483	0.0587	0.0719	0.0843
w_C	0.1469	0.1409	0.1348	0.1290	0.1241	0.1192	0.1147	0.1095	0.1046
w_A	0.8001	0.7913	0.7830	0.7741	0.7654	0.7632	0.7553	0.7520	0.7428
w_B	0.1012	0.1149	0.1277	0.1413	0.1540	0.1572	0.1686	0.1730	0.1859
w_C	0.0987	0.0938	0.0893	0.0846	0.0806	0.0796	0.0761	0.0750	0.0713

continued

continued

w_A	0.7412	0.7297	0.7151	0.6996	0.6813	0.6587	0.6353	0.6101	0.5785
w_B	0.1881	0.2036	0.2229	0.2428	0.2664	0.2942	0.3233	0.3538	0.3914
w_C	0.0707	0.0667	0.0620	0.0576	0.0523	0.0471	0.0414	0.0361	0.0301
w_A	0.5437	0.5053	0.4831						
w_B	0.4329	0.4802	0.5091						
w_C	0.0234	0.0145	0.0078						

T/K = 313.15

w_A	0.8486	0.8531	0.8507	0.8451	0.8357	0.8248	0.8129	0.8027	0.8019
w_B	0.0053	0.0123	0.0222	0.0350	0.0522	0.0697	0.0882	0.1026	0.1053
w_C	0.1461	0.1346	0.1271	0.1199	0.1121	0.1055	0.0989	0.0947	0.0928
w_A	0.7941	0.7887	0.7878	0.7796	0.7765	0.7711	0.7617	0.7620	0.7496
w_B	0.1155	0.1244	0.1259	0.1372	0.1433	0.1496	0.1632	0.1637	0.1797
w_C	0.0904	0.0869	0.0863	0.0832	0.0802	0.0793	0.0751	0.0743	0.0707
w_A	0.7493	0.7362	0.7352	0.7204	0.7027	0.6824	0.6569	0.6299	0.6009
w_B	0.1815	0.1977	0.2006	0.2186	0.2412	0.2672	0.2982	0.3310	0.3657
w_C	0.0692	0.0661	0.0642	0.0610	0.0561	0.0504	0.0449	0.0391	0.0334
w_A	0.5690	0.5374	0.4846						
w_B	0.4029	0.4397	0.4991						
w_C	0.0281	0.0229	0.0163						

T/K = 318.15

w_A	0.8390	0.8579	0.8568	0.8533	0.8480	0.8402	0.8281	0.8165	0.8046
w_B	0.0053	0.0104	0.0183	0.0284	0.0395	0.0532	0.0716	0.0889	0.1065
w_C	0.1557	0.1317	0.1249	0.1183	0.1125	0.1066	0.1003	0.0946	0.0889
w_A	0.7919	0.7787	0.7665	0.7635	0.7554	0.7446	0.7434	0.7298	0.7153
w_B	0.1253	0.1442	0.1611	0.1651	0.1763	0.1900	0.1921	0.2098	0.2283
w_C	0.0828	0.0771	0.0724	0.0714	0.0683	0.0654	0.0645	0.0604	0.0564
w_A	0.6977	0.6792	0.6569	0.6328	0.6061	0.5778	0.5556	0.5264	0.4916
w_B	0.2504	0.2734	0.3003	0.3291	0.3605	0.3934	0.4200	0.4535	0.4916
w_C	0.0519	0.0474	0.0428	0.0381	0.0334	0.0288	0.0244	0.0201	0.0168

Polymer (B): **poly(ethylene oxide)** **2011LEM**
Characterization: M_w/g.mol^{-1} = 1500, Vetec, Brazil
Solvent (A): **water** H_2O **7732-18-5**
Salt (C): **potassium hydroxide** **KOH** **1310-58-3**

Type of data: coexistence data (tie lines)

continued

continued

Total system			Top phase			Bottom phase		
w_A	w_B	w_C	w_A	w_B	w_C	w_A	w_B	w_C
T/K = 283.15								
0.7232	0.1543	0.1225	0.6806	0.2583	0.0611	0.7859	0.0162	0.1979
0.6981	0.1690	0.1329	0.6184	0.3373	0.0443	0.7676	0.0092	0.2232
0.6722	0.1902	0.1376	0.5825	0.3811	0.0364	0.7484	0.0079	0.2437
0.6313	0.2051	0.1636	0.5338	0.4370	0.0292	0.7250	0.0072	0.2678
0.6086	0.2250	0.1664	0.4980	0.4735	0.0285	0.7064	0.0005	0.2931
T/K = 298.15								
0.7463	0.1382	0.1155	0.6669	0.2825	0.0506	0.7957	0.0561	0.1482
0.7087	0.1634	0.1279	0.6182	0.3482	0.0336	0.7743	0.0394	0.1864
0.6906	0.1770	0.1324	0.6101	0.3594	0.0305	0.7577	0.0358	0.2064
0.6694	0.1895	0.1411	0.5692	0.4068	0.0240	0.7424	0.0026	0.2550
0.6436	0.2041	0.1523	0.5313	0.4488	0.0199	0.7103	0.0005	0.2892
T/K = 313.15								
0.7294	0.1454	0.1252	0.6475	0.3139	0.0385	0.7855	0.0128	0.2017
0.6918	0.1703	0.1379	0.5898	0.3835	0.0267	0.7608	0.0089	0.2302
0.6831	0.1766	0.1404	0.5623	0.4148	0.0229	0.7580	0.0073	0.2347
0.6509	0.1959	0.1533	0.5355	0.4444	0.0202	0.7373	0.0048	0.2579
0.6213	0.2138	0.1649	0.5068	0.4771	0.0161	0.7092	0.0022	0.2886

Polymer (B): **poly(ethylene oxide)** **2008SA1**
Characterization: M_n/g.mol^{-1} = 4000, Merck, Darmstadt, Germany
Solvent (A): **water** H_2O **7732-18-5**
Salt (C): **sodium phosphate** Na_3PO_4 **7601-54-9**

Type of data: cloud points

T/K = 298.15

w_A	0.9180	0.9161	0.9136	0.9099	0.9047	0.8970	0.8880	0.8781	0.8682
w_B	0.0138	0.0179	0.0224	0.0284	0.0363	0.0473	0.0596	0.0730	0.0863
w_C	0.0682	0.0660	0.0640	0.0617	0.0590	0.0557	0.0524	0.0489	0.0455
w_A	0.8571	0.8476	0.8471	0.8433	0.8349	0.8252	0.8149	0.8071	0.7958
w_B	0.1008	0.1134	0.1140	0.1190	0.1297	0.1418	0.1545	0.1640	0.1776
w_C	0.0421	0.0390	0.0389	0.0377	0.0354	0.0330	0.0306	0.0289	0.0266
w_A	0.7818	0.7675	0.7521	0.7344	0.7131	0.6959	0.6680	0.6381	0.5664
w_B	0.1940	0.2105	0.2283	0.2486	0.2722	0.2914	0.3207	0.3544	0.4306
w_C	0.0242	0.0220	0.0196	0.0170	0.0147	0.0127	0.0113	0.0075	0.0030

continued

continued

T/K = 303.15

w_A	0.9223	0.9217	0.9151	0.9063	0.8952	0.8844	0.8740	0.8608	0.8489
w_B	0.0057	0.0121	0.0243	0.0375	0.0525	0.0669	0.0806	0.0975	0.1129
w_C	0.0720	0.0662	0.0606	0.0562	0.0523	0.0487	0.0454	0.0417	0.0382
w_A	0.8360	0.8237	0.8116	0.7992	0.7946	0.7866	0.7844	0.7694	0.7541
w_B	0.1291	0.1442	0.1589	0.1738	0.1804	0.1885	0.1920	0.2094	0.2269
w_C	0.0349	0.0321	0.0295	0.0270	0.0250	0.0249	0.0236	0.0212	0.0190
w_A	0.7406	0.7234	0.7066	0.6857	0.6626	0.6431	0.6035	0.5822	0.5010
w_B	0.2424	0.2615	0.2800	0.3029	0.3279	0.3492	0.3903	0.4143	0.4975
w_C	0.0170	0.0151	0.0134	0.0114	0.0095	0.0077	0.0062	0.0035	0.0015

T/K = 308.15

w_A	0.9218	0.9241	0.9233	0.9197	0.9142	0.9069	0.8987	0.8894	0.8801
w_B	0.0040	0.0083	0.0132	0.0206	0.0293	0.0397	0.0508	0.0629	0.0750
w_C	0.0742	0.0676	0.0635	0.0597	0.0565	0.0534	0.0505	0.0477	0.0449
w_A	0.8703	0.8612	0.8552	0.8506	0.8472	0.8405	0.8392	0.8311	0.8296
w_B	0.0875	0.0992	0.1068	0.1126	0.1169	0.1252	0.1268	0.1367	0.1386
w_C	0.0422	0.0396	0.0380	0.0368	0.0359	0.0343	0.0340	0.0322	0.0318
w_A	0.8218	0.8196	0.8122	0.8083	0.8022	0.7967	0.7931	0.7823	0.7658
w_B	0.1480	0.1508	0.1595	0.1641	0.1714	0.1778	0.1824	0.1946	0.2136
w_C	0.0302	0.0296	0.0283	0.0276	0.0264	0.0255	0.0245	0.0231	0.0206
w_A	0.7463	0.7249	0.7150	0.7038	0.6870	0.6809	0.6595	0.6523	0.6415
w_B	0.2356	0.2595	0.2700	0.2829	0.3005	0.3078	0.3309	0.3381	0.3504
w_C	0.0181	0.0156	0.0150	0.0133	0.0125	0.0113	0.0096	0.0096	0.0081
w_A	0.6265	0.5828	0.5296						
w_B	0.3657	0.4118	0.4669						
w_C	0.0078	0.0054	0.0035						

T/K = 313.15

w_A	0.9210	0.9274	0.9260	0.9216	0.9150	0.9088	0.9008	0.8921	0.8824
w_B	0.0025	0.0069	0.0137	0.0219	0.0315	0.0402	0.0508	0.0620	0.0742
w_C	0.0765	0.0657	0.0603	0.0565	0.0535	0.0510	0.0484	0.0459	0.0434
w_A	0.8729	0.8632	0.8543	0.8494	0.8449	0.8376	0.8357	0.8271	0.8264
w_B	0.0862	0.0983	0.1095	0.1158	0.1211	0.1300	0.1322	0.1427	0.1435
w_C	0.0409	0.0385	0.0362	0.0348	0.0340	0.0324	0.0321	0.0302	0.0301
w_A	0.8165	0.8158	0.8071	0.8020	0.7971	0.7880	0.7862	0.7785	0.7704
w_B	0.1552	0.1562	0.1664	0.1723	0.1779	0.1886	0.1906	0.1995	0.2087
w_C	0.0283	0.0280	0.0265	0.0257	0.0250	0.0234	0.0232	0.0220	0.0209
w_A	0.7517	0.7287	0.7209	0.7048	0.6968	0.6787	0.6639	0.6527	0.6428
w_B	0.2299	0.2553	0.2641	0.2816	0.2906	0.3101	0.3261	0.3381	0.3488
w_C	0.0184	0.0160	0.0150	0.0136	0.0126	0.0112	0.0100	0.0092	0.0084

continued

continued

w_A	0.6253	0.5844	0.5462
w_B	0.3672	0.4103	0.4497
w_C	0.0075	0.0053	0.0041

T/K = 318.15

w_A	0.9290	0.9293	0.9242	0.9172	0.9065	0.8967	0.8873	0.8762	0.8662
w_B	0.0051	0.0127	0.0219	0.0327	0.0464	0.0589	0.0708	0.0845	0.0969
w_C	0.0659	0.0580	0.0539	0.0501	0.0471	0.0444	0.0419	0.0393	0.0369
w_A	0.8546	0.8437	0.8324	0.8205	0.8117	0.8087	0.7978	0.7957	0.7850
w_B	0.1110	0.1243	0.1379	0.1519	0.1621	0.1656	0.1782	0.1807	0.1928
w_C	0.0344	0.0320	0.0297	0.0276	0.0262	0.0257	0.0240	0.0236	0.0222
w_A	0.7796	0.7725	0.7626	0.7410	0.7334	0.7180	0.7049	0.6946	0.6736
w_B	0.1989	0.2071	0.2181	0.2421	0.2513	0.2675	0.2824	0.2929	0.3160
w_C	0.0215	0.0204	0.0193	0.0169	0.0153	0.0145	0.0127	0.0125	0.0104
w_A	0.6706	0.6505	0.6373	0.6339	0.5999	0.5604	0.5064		
w_B	0.3192	0.3406	0.3549	0.3584	0.3945	0.4356	0.4912		
w_C	0.0102	0.0089	0.0078	0.0077	0.0056	0.0040	0.0024		

Polymer (B): **poly(ethylene oxide)** **2011LEM**
Characterization: M_w/g.mol^{-1} = 1500, Vetec, Brazil
Solvent (A): **water** H_2O **7732-18-5**
Salt (C): **zinc sulfate** $ZnSO_4$ **7733-02-0**

Type of data: coexistence data (tie lines)

Total system			Top phase			Bottom phase		
w_A	w_B	w_C	w_A	w_B	w_C	w_A	w_B	w_C
T/K = 283.15								
0.6802	0.1823	0.1375	0.6155	0.3672	0.0173	0.7417	0.0010	0.2573
0.6632	0.1873	0.1495	0.5961	0.3906	0.0133	0.7266	0.0045	0.2689
0.6456	0.1928	0.1616	0.5736	0.4181	0.0083	0.7091	0.0012	0.2897
0.6291	0.1973	0.1736	0.5483	0.4463	0.0054	0.6891	0.0120	0.2989
0.6120	0.2027	0.1853	0.5166	0.4803	0.0031	0.6770	0.0121	0.3109
T/K = 298.15								
0.6656	0.2548	0.0796	0.6253	0.3527	0.0220	0.7684	0.0023	0.2293
0.6486	0.2667	0.0847	0.6022	0.3808	0.0170	0.7513	0.0106	0.2381
0.6314	0.2791	0.0895	0.5830	0.4034	0.0136	0.7362	0.0086	0.2552
0.6145	0.2910	0.0945	0.5614	0.4287	0.0099	0.7238	0.0083	0.2679
0.5973	0.3032	0.0995	0.5383	0.4540	0.0077	0.7082	0.0166	0.2752
0.5807	0.3148	0.1045	0.5222	0.4721	0.0057	0.6933	0.0218	0.2849

continued

continued

T/K = 313.15

0.6976	0.1770	0.1254	0.6172	0.3645	0.0183	0.7599	0.0053	0.2348
0.6807	0.1819	0.1374	0.5924	0.3946	0.0130	0.7513	0.0010	0.2477
0.6634	0.1870	0.1496	0.5663	0.4250	0.0087	0.7191	0.0095	0.2714
0.6454	0.1928	0.1618	0.5428	0.4518	0.0054	0.7180	0.0014	0.2806
0.6286	0.1979	0.1735	0.5160	0.4807	0.0033	0.7076	0.0056	0.2868

Polymer (B): **poly(ethylene oxide-*co*-propylene oxide)** **2004PER**

Characterization: M_n/g.mol^{-1} = 3900, 50.0 mol% ethylene oxide, Ucon 50-HB-5100, Union Carbide Corp., NY

Solvent (A): **water** **H_2O** **7732-18-5**

Salt (C): **ammonium sulfate** **$(NH_4)_2SO_4$** **7783-20-2**

Type of data: coexistence data (tie lines)

T/K = 295.15

Total system			Top phase			Bottom phase		
w_A	w_B	w_C	w_A	w_B	w_C	w_A	w_B	w_C
0.8344	0.1001	0.0655	0.7044	0.2652	0.0304	0.8949	0.0206	0.0845
0.7806	0.1614	0.0580	0.6672	0.3072	0.0256	0.8930	0.0134	0.0936
0.7684	0.1696	0.0620	0.6314	0.3472	0.0214	0.8889	0.0091	0.1020
0.7411	0.1891	0.0698	0.5865	0.3967	0.0168	0.8781	0.0038	0.1181

Plait-point composition: w_A = 0.8388 + w_B = 0.1080 + w_C = 0.0532

T/K = 303.15

Total system			Top phase			Bottom phase		
w_A	w_B	w_C	w_A	w_B	w_C	w_A	w_B	w_C
0.8397	0.1203	0.0400	0.8898	0.0608	0.0494	0.7746	0.1966	0.0288
0.8344	0.1206	0.0450	0.9107	0.0289	0.0604	0.7078	0.2700	0.0222
0.8295	0.1205	0.0500	0.9132	0.0196	0.0672	0.6665	0.3151	0.0184
0.8245	0.1203	0.0552	0.9127	0.0130	0.0743	0.6348	0.3496	0.0156

Plait-point composition: w_A = 0.8443 + w_B = 0.1180 + w_C = 0.0377

continued

continued

T/K = 313.15

Total system			Top phase			Bottom phase		
w_A	w_B	w_C	w_A	w_B	w_C	w_A	w_B	w_C
0.8734	0.1009	0.0257	0.9197	0.0494	0.0309	0.7104	0.2772	0.0124
0.8650	0.1000	0.0350	0.9378	0.0192	0.0430	0.6064	0.3841	0.0095
0.8558	0.0995	0.0447	0.9345	0.0105	0.0550	0.5351	0.4573	0.0076
0.8447	0.1002	0.0551	0.9258	0.0060	0.0682	0.4834	0.5103	0.0063

Plait-point composition: $w_A = 0.8414 + w_B = 0.1400 + w_C = 0.0186$

Polymer (B): **poly(ethylene oxide-*co*-propylene oxide)** **2010SIL**
Characterization: M_n/g.mol^{-1} = 3900, 50.0 mol% ethylene oxide, Ucon 50-HB-5100, Union Carbide Corp., NY
Solvent (A): **water** **H_2O** **7732-18-5**
Salt (C): **dipotassium phosphate** **K_2HPO_4** **7758-11-4**

Type of data: coexistence data (tie lines)

T/K = 296.15

Total system			Top phase			Bottom phase		
w_A	w_B	w_C	w_A	w_B	w_C	w_A	w_B	w_C
0.7818	0.1802	0.0380	0.7380	0.2370	0.0250	0.9160	0.0110	0.0730
0.7602	0.1998	0.0400	0.6940	0.2860	0.0200	0.9140	0.0030	0.0830
0.7380	0.2200	0.0420	0.6590	0.3240	0.0170	0.9060	0.0010	0.0930

Polymer (B): **poly(ethylene oxide-*co*-propylene oxide)** **2011NAS**
Characterization: M_n/g.mol^{-1} = 2000, 50.0 mol% ethylene oxide, Ucon 50-HB-2000, Dow Chemical, Midland, MI
Solvent (A): **water** **H_2O** **7732-18-5**
Salt (C): **dipotassium phosphate/monopotassium phosphate**
K_2HPO_4/H_2KPO_4 **7758-11-4/7778-77-0**

Comments: The mixture of H_2KPO_4/K_2HPO_4 was chosen in a ratio of
1:0.7 that provides a pH = 6,
1:1.91 that provides a pH = 7,
1:12.04 that provides a pH = 8.

continued

continued

Type of data: coexistence data (tie lines)

Total system			Top phase			Bottom phase		
w_A	w_B	w_C	w_A	w_B	w_C	w_A	w_B	w_C
T/K = 277.15	pH = 6.0							
0.7724	0.1601	0.0675	0.6854	0.2895	0.0251	0.8627	0.0402	0.0971
0.7479	0.1800	0.0721	0.6731	0.3028	0.0241	0.8352	0.0523	0.1125
0.6977	0.2158	0.0865	0.6189	0.3651	0.0160	0.8211	0.0237	0.1552
0.6377	0.2573	0.1050	0.5509	0.4373	0.0118	0.7524	0.0469	0.2007
T/K = 298.15	pH = 6.0							
0.7901	0.1499	0.0600	0.6732	0.3038	0.0230	0.8766	0.0522	0.0712
0.7479	0.1801	0.0720	0.6052	0.3781	0.0167	0.8762	0.0211	0.1027
0.6977	0.2158	0.0865	0.5236	0.4649	0.0115	0.8350	0.0364	0.1286
0.6377	0.2573	0.1050	0.4421	0.5530	0.0049	0.7938	0.0454	0.1608
T/K = 313.15	pH = 6.0							
0.7907	0.1495	0.0598	0.5236	0.4663	0.0101	0.9134	0.0154	0.0712
0.7482	0.1798	0.0720	0.5101	0.4819	0.0080	0.8763	0.0322	0.0915
0.6976	0.2159	0.0865	0.4421	0.5503	0.0076	0.8624	0.0223	0.1153
0.6376	0.2573	0.1051	0.3469	0.6509	0.0022	0.8348	0.0177	0.1475
T/K = 277.15	pH = 7.0							
0.7725	0.1600	0.0675	0.6868	0.2895	0.0237	0.8489	0.0393	0.1118
0.7480	0.1800	0.0720	0.6732	0.3059	0.0209	0.8350	0.0322	0.1328
0.6977	0.2159	0.0864	0.6324	0.3502	0.0174	0.7800	0.0431	0.1769
0.6377	0.2574	0.1049	0.5508	0.4370	0.0122	0.7110	0.0456	0.2434
T/K = 298.15	pH = 7.0							
0.7899	0.1500	0.0601	0.6732	0.3066	0.0202	0.8764	0.0342	0.0894
0.7481	0.1799	0.0720	0.6053	0.3794	0.0153	0.8624	0.0181	0.1195
0.6983	0.2154	0.0863	0.5236	0.4656	0.0108	0.8348	0.0170	0.1482
0.6373	0.2588	0.1039	0.4421	0.5537	0.0042	0.7933	0.0102	0.1965
T/K = 313.15	pH = 7.0							
0.7900	0.1500	0.0600	0.5101	0.4819	0.0080	0.9132	0.0016	0.0852
0.7491	0.1791	0.0718	0.4693	0.5231	0.0076	0.8625	0.0299	0.1076
0.6977	0.2157	0.0866	0.3876	0.6065	0.0059	0.8349	0.0281	0.1370
0.6376	0.2587	0.1037	0.3604	0.6337	0.0059	0.7935	0.0261	0.1804
T/K = 277.15	pH = 8.0							
0.7724	0.1600	0.0676	0.6731	0.3007	0.0262	0.8486	0.0172	0.1342
0.7484	0.1796	0.0720	0.6731	0.3042	0.0227	0.8346	0.0067	0.1587
0.6976	0.2158	0.0866	0.6052	0.3781	0.0167	0.7795	0.0086	0.2119
0.6372	0.2590	0.1038	0.5100	0.4792	0.0108	0.7268	0.0025	0.2707

continued

continued

T/K = 298.15 pH = 8.0

0.7878	0.1458	0.0664	0.6324	0.3498	0.0178	0.8760	0.0094	0.1146
0.7480	0.1798	0.0722	0.5508	0.4356	0.0136	0.8349	0.0281	0.1370
0.6976	0.2158	0.0866	0.5101	0.4802	0.0097	0.8208	0.0058	0.1734
0.6374	0.2588	0.1038	0.4013	0.5954	0.0033	0.7795	0.0072	0.2133

Polymer (B): **poly(ethylene oxide-*co*-propylene oxide)** **2010SIL**
Characterization: M_n/g.mol^{-1} = 3900, 50.0 mol% ethylene oxide,
Ucon 50-HB-5100, Union Carbide Corp., NY
Solvent (A): **water** **H_2O** **7732-18-5**
Salt (C): **dipotassium phosphate/monopotassium phosphate**
K_2HPO_4/H_2KPO_4 **7758-11-4/7778-77-0**

Comments: The mixture of K_2HPO_4/H_2KPO_4 was chosen in a ratio that provides a pH = 7.

Type of data: coexistence data (tie lines)

T/K = 296.15

Total system			Top phase			Bottom phase		
w_A	w_B	w_C	w_A	w_B	w_C	w_A	w_B	w_C
0.8169	0.1301	0.0530	0.7440	0.2270	0.0290	0.9030	0.0170	0.0800
0.8029	0.1401	0.0570	0.7090	0.2670	0.0240	0.9030	0.0060	0.0910
0.7888	0.1502	0.0610	0.6810	0.2990	0.0200	0.8970	0.0020	0.1010

Polymer (B): **poly(ethylene oxide-*co*-propylene oxide)** **2010SIL**
Characterization: M_n/g.mol^{-1} = 3900, 50.0 mol% ethylene oxide,
Ucon 50-HB-5100, Union Carbide Corp., NY
Solvent (A): **water** **H_2O** **7732-18-5**
Salt (C): **disodium phosphate** **Na_2HPO_4** **7558-79-4**

Type of data: coexistence data (tie lines)

T/K = 296.15

Total system			Top phase			Bottom phase		
w_A	w_B	w_C	w_A	w_B	w_C	w_A	w_B	w_C
0.8159	0.1491	0.0350	0.7440	0.2370	0.0190	0.9270	0.0140	0.0590
0.8009	0.1601	0.0390	0.7040	0.2820	0.0140	0.9280	0.0040	0.0680
0.7840	0.1730	0.0430	0.6730	0.3150	0.0120	0.9210	0.0010	0.0780

Polymer (B): **poly(ethylene oxide-*co*-propylene oxide)** **2010SIL**
Characterization: M_n/g.mol^{-1} = 3900, 50.0 mol% ethylene oxide, Ucon 50-HB-5100, Union Carbide Corp., NY
Solvent (A): **water** **H_2O** **7732-18-5**
Salt (C): **disodium phosphate/monosodium phosphate** **Na_2HPO_4/H_2NaPO_4** **7558-79-4/7558-80-7**

Comments: The mixture of Na_2HPO_4/H_2NaPO_4 was chosen in a ratio that provides a pH = 7.

Type of data: coexistence data (tie lines)

T/K = 296.15

Total system			Top phase			Bottom phase		
w_A	w_B	w_C	w_A	w_B	w_C	w_A	w_B	w_C
0.8012	0.1597	0.0391	0.7190	0.2630	0.0180	0.9230	0.0090	0.0680
0.7846	0.1724	0.0430	0.6880	0.2970	0.0150	0.9200	0.0020	0.0780
0.7671	0.1859	0.0470	0.6520	0.3360	0.0120	0.9126	0.0004	0.0870

Polymer (B): **poly(ethylene oxide-*co*-propylene oxide)** **2008SI2**
Characterization: M_n/g.mol^{-1} = 3900, 50.0 mol% ethylene oxide, Ucon 50-HB-5100, Union Carbide Corp., NY
Solvent (A): **water** **H_2O** **7732-18-5**
Salt (C): **lithium sulfate** **Li_2SO_4** **10377-48-7**

Type of data: coexistence data (tie lines)

T/K = 296.15

Total system			Top phase			Bottom phase		
w_A	w_B	w_C	w_A	w_B	w_C	w_A	w_B	w_C
0.8230	0.1200	0.0570	0.7003	0.2721	0.0276	0.9151	0.0058	0.0791
0.8350	0.1100	0.0550	0.7266	0.2427	0.0307	0.9144	0.0129	0.0727
0.8100	0.1300	0.0600	0.6848	0.2894	0.0258	0.9108	0.0017	0.0875

Polymer (B): **poly(ethylene oxide-*co*-propylene oxide)** **2010SIL**
Characterization: M_n/g.mol^{-1} = 3900, 50.0 mol% ethylene oxide, Ucon 50-HB-5100, Union Carbide Corp., NY
Solvent (A): **water** **H_2O** **7732-18-5**
Salt (C): **monopotassium phosphate** **KH_2PO_4** **7778-77-0**

Type of data: coexistence data (tie lines)

continued

continued

T/K = 296.15

Total system w_A	w_B	w_C	Top phase w_A	w_B	w_C	Bottom phase w_A	w_B	w_C
0.7800	0.1500	0.0700	0.6960	0.2640	0.0400	0.8790	0.0150	0.1060
0.7520	0.1720	0.0760	0.6500	0.3170	0.0330	0.8710	0.0030	0.1260
0.7251	0.1899	0.0850	0.6030	0.3690	0.0280	0.8537	0.0003	0.1460

Polymer (B): **poly(ethylene oxide-*co*-propylene oxide)** **2010SIL**
Characterization: M_n/g.mol^{-1} = 3900, 50.0 mol% ethylene oxide, Ucon 50-HB-5100, Union Carbide Corp., NY
Solvent (A): **water** **H_2O** **7732-18-5**
Salt (C): **monosodium phosphate** **NaH_2PO_4** **7558-80-7**

Type of data: coexistence data (tie lines)

T/K = 296.15

Total system w_A	w_B	w_C	Top phase w_A	w_B	w_C	Bottom phase w_A	w_B	w_C
0.7750	0.1600	0.0650	0.6760	0.2920	0.0320	0.8920	0.0050	0.1030
0.7475	0.1826	0.0699	0.6320	0.3400	0.0280	0.8800	0.0010	0.1190
0.7200	0.2000	0.0800	0.5850	0.3920	0.0230	0.8600	0.00003	0.1400

Polymer (B): **poly(ethylene oxide-*co*-propylene oxide)** **2011NAS**
Characterization: M_n/g.mol^{-1} = 2000, 50.0 mol% ethylene oxide, Ucon 50-HB-2000, Dow Chemical, Midland, MI
Solvent (A): **water** **H_2O** **7732-18-5**
Salt (C): **sodium citrate** **$Na_3C_6H_5O_7$** **68-04-2**

Comments: The pH-value was adjusted by the addition of the citric acid.

Type of data: coexistence data (tie lines)

Total system w_A	w_B	w_C	Top phase w_A	w_B	w_C	Bottom phase w_A	w_B	w_C
T/K = 277.15		pH = 6.0						
0.7898	0.1499	0.0603	0.6844	0.2924	0.0232	0.9042	0.0001	0.0957
0.7478	0.1803	0.0719	0.5800	0.4058	0.0142	0.8752	0.0061	0.1187
0.6985	0.2153	0.0862	0.4918	0.4992	0.0090	0.8409	0.0057	0.1534
0.6370	0.2592	0.1038	0.4214	0.5683	0.0103	0.8006	0.0105	0.1889

continued

continued

T/K = 277.15	pH = 8.0							
0.7898	0.1501	0.0601	0.7135	0.2629	0.0236	0.8734	0.0439	0.0827
0.7485	0.1796	0.0719	0.6667	0.3172	0.0161	0.8476	0.0377	0.1147
0.6980	0.2158	0.0862	0.5926	0.3967	0.0107	0.8259	0.0203	0.1538
0.6371	0.2591	0.1038	0.5904	0.4032	0.0064	0.7652	0.0346	0.2002
T/K = 298.15	pH = 7.0							
0.7899	0.1504	0.0597	0.6904	0.2865	0.0231	0.8902	0.0180	0.0918
0.7469	0.1814	0.0717	0.6077	0.3763	0.0160	0.8764	0.0086	0.1150
0.6974	0.2162	0.0864	0.5351	0.4540	0.0109	0.8422	0.0068	0.1510
0.6370	0.2587	0.1043	0.4589	0.5333	0.0078	0.8003	0.0153	0.1844
T/K = 298.15	pH = 8.0							
0.7898	0.1499	0.0603	0.6808	0.3034	0.0158	0.9228	0.0005	0.0767
0.7481	0.1800	0.0719	0.5922	0.3977	0.0101	0.8956	0.0066	0.0978
0.6976	0.2162	0.0862	0.4905	0.5034	0.0061	0.8610	0.0111	0.1279
0.6372	0.2590	0.1038	0.4031	0.5944	0.0025	0.8262	0.0194	0.1544

Polymer (B): **poly(ethylene oxide-*co*-propylene oxide)** **2009TUB**
Characterization: M_n/g.mol^{-1} = 3900, 50.0 mol% ethylene oxide, Ucon 50-HB-5100, Union Carbide Corp., NY
Solvent (A): **water** H_2O **7732-18-5**
Salt (C): **sodium citrate** $Na_3C_6H_5O_7$ **68-04-2**

Type of data: coexistence data (tie lines)

Total system			Top phase			Bottom phase		
w_A	w_B	w_C	w_A	w_B	w_C	w_A	w_B	w_C
T/K = 278.15	pH = 5.20							
0.8307	0.0896	0.0797	0.7447	0.2231	0.0322	0.8852	0.0051	0.1097
0.8203	0.0898	0.0899	0.7131	0.2588	0.0281	0.8749	0.0039	0.1212
0.8094	0.0900	0.1006	0.6870	0.2882	0.0248	0.8629	0.0034	0.1337
0.8003	0.0898	0.1099	0.6682	0.3096	0.0222	0.8524	0.0032	0.1444
T/K = 293.15	pH = 5.20							
0.8365	0.1065	0.0570	0.7528	0.2140	0.0332	0.8924	0.0348	0.0728
0.8299	0.1102	0.0599	0.7195	0.2521	0.0284	0.8961	0.0252	0.0787
0.7670	0.1708	0.0622	0.6471	0.3340	0.0189	0.8894	0.0042	0.1064
0.7559	0.1787	0.0654	0.6283	0.3553	0.0164	0.8841	0.0021	0.1138
T/K = 313.15	pH = 5.20							
0.7662	0.1715	0.0623	0.5506	0.4375	0.0119	0.8990	0.0078	0.0932
0.7571	0.1781	0.0648	0.5260	0.4643	0.0097	0.8994	0.0019	0.0987
0.7342	0.1949	0.0709	0.4700	0.5251	0.0049	0.8876	0.0009	0.1115

continued

continued

T/K = 278.15 pH = 8.20

0.8411	0.0944	0.0645	0.7532	0.2273	0.0195	0.9029	0.0009	0.0962
0.8315	0.0937	0.0748	0.7172	0.2678	0.0150	0.8910	0.0019	0.1071
0.8213	0.0940	0.0847	0.6849	0.3039	0.0112	0.8801	0.0019	0.1180
0.8122	0.0937	0.0941	0.6536	0.3390	0.0074	0.8704	0.0018	0.1278

T/K = 293.15 pH = 8.20

0.8080	0.1488	0.0432	0.7139	0.2645	0.0216	0.9214	0.0093	0.0693
0.7918	0.1633	0.0449	0.6870	0.2937	0.0193	0.9181	0.0061	0.0758
0.7730	0.1770	0.0500	0.6504	0.3333	0.0163	0.9093	0.0033	0.0874
0.7533	0.1918	0.0549	0.6235	0.3624	0.0141	0.8976	0.0022	0.1002

T/K = 313.15 pH = 8.20

0.8216	0.1585	0.0199	0.7218	0.2695	0.0087	0.9434	0.0231	0.0335
0.8141	0.1639	0.0220	0.7038	0.2880	0.0082	0.9442	0.0175	0.0383
0.7715	0.2021	0.0264	0.5789	0.4169	0.0042	0.9493	0.0004	0.0503
0.7351	0.2336	0.0313	0.5353	0.4606	0.0041	0.9309	0.0112	0.0579

Polymer (B): **poly(ethylene oxide-*co*-propylene oxide)** **2008SI2**
Characterization: M_n/g.mol^{-1} = 3900, 50.0 mol% ethylene oxide, Ucon 50-HB-5100, Union Carbide Corp., NY
Solvent (A): **water** **H_2O** **7732-18-5**
Salt (C): **sodium sulfate** **Na_2SO_4** **7757-82-6**

Type of data: coexistence data (tie lines)

T/K = 296.15

Total system			Top phase			Bottom phase		
w_A	w_B	w_C	w_A	w_B	w_C	w_A	w_B	w_C
0.7911	0.1676	0.0413	0.6868	0.2970	0.0162	0.9246	0.0021	0.0733
0.7709	0.1751	0.0540	0.6183	0.3712	0.0105	0.9071	0.0001	0.0928
0.7800	0.1700	0.0500	0.6579	0.3285	0.0136	0.9108	0.0001	0.0891

Polymer (B): **poly(ethylene oxide-*b*-propylene oxide-*b*-ethylene oxide)** **2010LEM**
Characterization: M_w/g.mol^{-1} = 1900, (EO)11-(PO)16-(EO)11, L35, Aldrich Chem. Co., Inc., Milwaukee, WI
Solvent (A): **water** **H_2O** **7732-18-5**
Salt (C): **ammonium citrate** **$(NH_4)_3C_6H_5O_7$** **3458-72-8**

continued

continued

Type of data: coexistence data (tie lines)

Total system			Top phase			Bottom phase		
w_A	w_B	w_C	w_A	w_B	w_C	w_A	w_B	w_C
T/K = 283.15								
0.6378	0.1994	0.1628	0.5866	0.3321	0.0813	0.7105	0.0093	0.2802
0.6044	0.2280	0.1676	0.5377	0.3945	0.0678	0.6897	0.0040	0.3063
0.5603	0.2573	0.1824	0.4771	0.4677	0.0552	0.6542	0.0013	0.3445
0.5340	0.2794	0.1866	0.4482	0.5025	0.0493	0.6371	0.0008	0.3621
0.4965	0.3008	0.2027	0.3985	0.5596	0.0419	0.6035	0.0004	0.3961
T/K = 298.15								
0.6258	0.2326	0.1416	0.5346	0.3906	0.0748	0.7410	0.0081	0.2509
0.5926	0.2590	0.1484	0.4828	0.4554	0.0618	0.7215	0.0033	0.2752
0.5585	0.2850	0.1565	0.4357	0.5118	0.0525	0.6836	0.0035	0.3129
0.5248	0.3120	0.1632	0.3996	0.5551	0.0453	0.6682	0.0039	0.3279
0.4911	0.3394	0.1695	0.3643	0.5951	0.0406	0.6282	0.0072	0.3646
T/K = 313.15								
0.6380	0.1992	0.1628	0.4689	0.4884	0.0427	0.7526	0.0028	0.2446
0.6014	0.2302	0.1684	0.4109	0.5489	0.0402	0.7301	0.0007	0.2692
0.5595	0.2579	0.1826	0.3631	0.6038	0.0331	0.6923	0.0005	0.3072
0.5338	0.2795	0.1867	0.3212	0.6490	0.0298	0.6686	0.0002	0.3312
0.4963	0.3007	0.2030	0.2919	0.6810	0.0271	0.6453	0.0004	0.3543

Polymer (B): **poly(ethylene oxide-*b*-propylene oxide-*b*-ethylene oxide)** **2011AND**

Characterization: M_w/g.mol^{-1} = 2930, M_w/M_n = 1.09, 40.0 wt% ethylene oxide, (EO)13(PO)30(EO)13, L64, Aldrich Chem. Co., Inc., Milwaukee, WI

Solvent (A): **water** **H_2O** **7732-18-5**

Salt (C): **ammonium citrate** **$(NH_4)_3C_6H_5O_7$** **3458-72-8**

Type of data: coexistence data (tie lines)

Total system			Top phase			Bottom phase		
w_A	w_B	w_C	w_A	w_B	w_C	w_A	w_B	w_C
T/K = 278.15								
0.5803	0.1273	0.2924	0.5471	0.1976	0.2554	0.6225	0.0028	0.3747
0.5630	0.1391	0.2979	0.5044	0.2683	0.2273	0.6091	0.0065	0.3845
0.5299	0.1620	0.3081	0.4338	0.3879	0.1783	0.5838	0.0050	0.4113
0.5127	0.1745	0.3128	0.4063	0.4210	0.1727	0.5536	0.0108	0.4355

continued

continued

T/K = 288.15								
0.5800	0.1244	0.2956	0.5061	0.2718	0.2221	0.6396	0.0031	0.3573
0.5648	0.1377	0.2975	0.4777	0.3131	0.2092	0.6285	0.0028	0.3688
0.5483	0.1492	0.3025	0.4454	0.3599	0.1947	0.6010	0.0080	0.3911
0.5338	0.1588	0.3074	0.4143	0.4128	0.1729	0.5922	0.0040	0.4038
T/K = 298.15								
0.5995	0.1119	0.2886	0.5676	0.1604	0.2719	0.6417	0.0040	0.3542
0.5800	0.1248	0.2952	0.5145	0.2480	0.2375	0.6345	0.0026	0.3629
0.5653	0.1369	0.2978	0.4665	0.3317	0.2018	0.6203	0.0014	0.3782
0.5483	0.1488	0.3029	0.4294	0.3825	0.1881	0.6085	0.0041	0.3874

Polymer (B): **poly(ethylene oxide-*b*-propylene oxide-*b*-ethylene oxide)** **2010RO1**
Characterization: M_w/g.mol^{-1} = 8400, (EO)80-(PO)30-(EO)80, F68, Aldrich Chem. Co., Inc., Milwaukee, WI
Solvent (A): **water** **H_2O** **7732-18-5**
Salt (C): **ammonium citrate** **$(NH_4)_3C_6H_5O_7$** **3458-72-8**

Type of data: coexistence data (tie lines)

Total system			Top phase			Bottom phase		
w_A	w_B	w_C	w_A	w_B	w_C	w_A	w_B	w_C
T/K = 283.15								
0.6927	0.1337	0.1737	0.6591	0.2361	0.1048	0.7453	0.0100	0.2447
0.6745	0.1464	0.1791	0.6381	0.2590	0.1029	0.7295	0.0068	0.2637
0.6576	0.1588	0.1836	0.6194	0.2820	0.0986	0.7148	0.0037	0.2815
0.6407	0.1713	0.1880	0.6090	0.2945	0.0965	0.7010	0.0004	0.2986
0.6233	0.1839	0.1928	0.5878	0.3174	0.0948	0.6728	0.0035	0.3237
T/K = 298.15								
0.6918	0.1341	0.1741	0.6399	0.2609	0.0992	0.7447	0.0207	0.2346
0.6748	0.1466	0.1786	0.6168	0.2895	0.0937	0.7343	0.0192	0.2465
0.6579	0.1588	0.1832	0.6047	0.3033	0.0920	0.7219	0.0176	0.2605
0.6422	0.1707	0.1871	0.5848	0.3245	0.0907	0.7055	0.0219	0.2726
0.6240	0.1835	0.1925	0.5715	0.3389	0.0896	0.6895	0.0183	0.2922
T/K = 313.15								
0.7111	0.1191	0.1698	0.6560	0.2303	0.1137	0.7679	0.0031	0.2290
0.6930	0.1327	0.1743	0.6212	0.2778	0.1010	0.7570	0.0051	0.2379
0.6733	0.1469	0.1798	0.5953	0.3102	0.0945	0.7426	0.0044	0.2530
0.6545	0.1608	0.1847	0.5731	0.3388	0.0881	0.7169	0.0091	0.2740

Polymer (B): **poly(ethylene oxide-*b*-propylene oxide-*b*-ethylene oxide)** **2010LEM**
Characterization: M_w/g.mol^{-1} = 1900, (EO)11-(PO)16-(EO)11, L35, Aldrich Chem. Co., Inc., Milwaukee, WI
Solvent (A): **water** **H_2O** **7732-18-5**
Salt (C): **ammonium sulfate** **$(NH_4)_2SO_4$** **7783-20-2**

Type of data: coexistence data (tie lines)

Total system			Top phase			Bottom phase		
w_A	w_B	w_C	w_A	w_B	w_C	w_A	w_B	w_C
T/K = 283.15								
0.7155	0.2040	0.0805	0.6414	0.3226	0.0360	0.8161	0.0260	0.1579
0.6726	0.2412	0.0862	0.5764	0.3981	0.0255	0.7948	0.0181	0.1871
0.6359	0.2710	0.0931	0.5265	0.4527	0.0208	0.7739	0.0194	0.2067
0.5942	0.2992	0.1066	0.4619	0.5230	0.0151	0.7403	0.0233	0.2364
0.5586	0.3248	0.1166	0.4130	0.5749	0.0121	0.7184	0.0238	0.2578
T/K = 298.15								
0.6860	0.2520	0.0620	0.5900	0.3786	0.0314	0.8411	0.0294	0.1295
0.6377	0.2954	0.0669	0.5183	0.4598	0.0219	0.8264	0.0231	0.1505
0.5882	0.3398	0.0720	0.4697	0.5138	0.0165	0.8010	0.0284	0.1706
0.5393	0.3837	0.0770	0.4163	0.5716	0.0121	0.7721	0.0268	0.2011
0.4904	0.4275	0.0821	0.3742	0.6164	0.0094	0.7424	0.0340	0.2236
T/K = 313.15								
0.7154	0.2042	0.0804	0.4689	0.5140	0.0171	0.8573	0.0126	0.1301
0.6726	0.2412	0.0862	0.4202	0.5669	0.0129	0.8385	0.0108	0.1507
0.6359	0.2703	0.0938	0.4245	0.5623	0.0132	0.8151	0.0142	0.1707
0.5941	0.2992	0.1067	0.3854	0.6047	0.0099	0.7826	0.0265	0.1909
0.5584	0.3252	0.1164	0.3479	0.6449	0.0072	0.7569	0.0160	0.2271

Polymer (B): **poly(ethylene oxide-*b*-propylene oxide-*b*-ethylene oxide)** **2010MA2**
Characterization: M_w/g.mol^{-1} = 8460, M_w/M_n = 1.42, 80 wt% ethylene oxide, (EO)80-(PO)30-(EO)80, F68, Sigma Chemical Co., Inc., St. Louis, MO
Solvent (A): **water** **H_2O** **7732-18-5**
Salt (C): **ammonium sulfate** **$(NH_4)_2SO_4$** **7783-20-2**

Type of data: coexistence data (tie lines)

continued

continued

Total system			Top phase			Bottom phase		
w_A	w_B	w_C	w_A	w_B	w_C	w_A	w_B	w_C
T/K = 278.15								
0.8035	0.1029	0.0936	0.7514	0.1897	0.0589	0.8556	0.0161	0.1283
0.7852	0.1185	0.0963	0.7191	0.2296	0.0513	0.8513	0.0075	0.1412
0.7656	0.1365	0.0979	0.6886	0.2672	0.0442	0.8427	0.0058	0.1515
0.7325	0.1658	0.1017	0.6441	0.3191	0.0368	0.8209	0.0125	0.1666
0.7085	0.1777	0.1138	0.6162	0.3501	0.0337	0.8009	0.0053	0.1938
T/K = 288.15								
0.8110	0.0925	0.0965	0.7603	0.1820	0.0577	0.8617	0.0030	0.1353
0.7847	0.1234	0.0919	0.7080	0.2454	0.0466	0.8614	0.0014	0.1372
0.7642	0.1388	0.0970	0.6804	0.2765	0.0431	0.8481	0.0010	0.1509
0.7418	0.1452	0.1130	0.6675	0.2904	0.0421	0.8161	0.0000	0.1839
0.6972	0.1916	0.1112	0.5866	0.3832	0.0302	0.8079	0.0000	0.1921
T/K = 298.15								
0.8247	0.0951	0.0802	0.7652	0.1773	0.0575	0.8840	0.0130	0.1030
0.7899	0.1306	0.0795	0.7061	0.2449	0.0490	0.8736	0.0163	0.1101
0.7550	0.1578	0.0872	0.6483	0.3146	0.0371	0.8618	0.0009	0.1373
0.7290	0.1691	0.1019	0.6263	0.3379	0.0358	0.8316	0.0004	0.1680
0.7029	0.1864	0.1107	0.5912	0.3723	0.0365	0.8146	0.0004	0.1850

Polymer (B): **poly(ethylene oxide-*b*-propylene oxide-*b*-ethylene oxide)** **2004HAR**

Characterization: M_n/g.mol^{-1} = 2100, 20.0 wt% ethylene oxide, L62, (EO)6(PO)30(EO)6, ICI Surfactants, Cleveland, UK

Solvent (A): **water** **H_2O** **7732-18-5**

Salt (C): **dipotassium phosphate/monopotassium phosphate K_2HPO_4/H_2KPO_4** **7758-11-4/7778-77-0**

Type of data: cloud points

T/K = 283.15 pH = 5.0

w_A	0.611	0.626	0.650	0.669	0.691	0.709	0.725	0.753	0.764
w_B	0.319	0.300	0.272	0.246	0.215	0.192	0.170	0.133	0.119
w_C	0.070	0.074	0.078	0.085	0.094	0.099	0.105	0.114	0.117

w_A	0.775
w_B	0.098
w_C	0.127

continued

continued

T/K = 283.15 pH = 7.0

w_A	0.574	0.596	0.620	0.640	0.664	0.684	0.703	0.740	0.756
w_B	0.381	0.357	0.328	0.307	0.277	0.254	0.230	0.182	0.162
w_C	0.045	0.047	0.052	0.053	0.059	0.062	0.067	0.078	0.082
w_A	0.774								
w_B	0.135								
w_C	0.091								

T/K = 298.15 pH = 5.0

w_A	0.727	0.746	0.761	0.777	0.797	0.808	0.822	0.837	0.851
w_B	0.195	0.172	0.158	0.142	0.120	0.104	0.089	0.073	0.057
w_C	0.078	0.082	0.081	0.081	0.083	0.088	0.089	0.090	0.092

T/K = 298.15 pH = 7.0

w_A	0.699	0.739	0.758	0.780	0.816	0.835	0.844	0.870	0.881
w_B	0.240	0.195	0.180	0.154	0.116	0.096	0.082	0.056	0.036
w_C	0.061	0.066	0.062	0.066	0.068	0.069	0.074	0.074	0.083

Type of data: coexistence data (tie lines)

Total system			Top phase			Bottom phase		
w_A	w_B	w_C	w_A	w_B	w_C	w_A	w_B	w_C
T/K = 283.15	pH = 5.0							
0.699	0.182	0.119	0.440	0.522	0.038	0.783	0.065	0.152
0.676	0.194	0.130	0.423	0.541	0.036	0.777	0.048	0.175
T/K = 283.15	pH = 7.0							
0.702	0.210	0.088	0.533	0.433	0.034	0.791	0.094	0.115
0.673	0.225	0.102	0.498	0.473	0.029	0.789	0.062	0.149
0.645	0.237	0.118	0.477	0.497	0.026	0.769	0.048	0.183
0.615	0.254	0.131	0.426	0.549	0.025	0.737	0.068	0.195
T/K = 298.15	pH = 5.0							
0.776	0.118	0.106	0.571	0.365	0.064	0.850	0.027	0.123
0.740	0.140	0.120	0.521	0.416	0.063	0.835	0.019	0.146
0.700	0.159	0.141	0.474	0.465	0.061	0.808	0.020	0.172
T/K = 298.15	pH = 7.0							
0.780	0.137	0.083	0.554	0.392	0.054	0.880	0.025	0.095
0.760	0.140	0.100	0.519	0.433	0.048	0.862	0.022	0.116
0.724	0.162	0.114	0.494	0.454	0.052	0.841	0.016	0.143
0.694	0.181	0.125	0.460	0.490	0.050	0.821	0.012	0.167

Polymer (B): **poly(ethylene oxide-*b*-propylene oxide-*b*-ethylene oxide)** **2004HAR**
Characterization: M_w/g.mol^{-1} = 2900, 40.0 wt% ethylene oxide, L64, (EO)13(PO)30(EO)13, Aldrich Chem. Co., Inc., Milwaukee, WI
Solvent (A): **water** **H_2O** **7732-18-5**
Salt (C): **dipotassium phosphate/monopotassium phosphate** **K_2HPO_4/H_2KPO_4** **7758-11-4/7778-77-0**

Type of data: cloud points

T/K = 283.15 pH = 5.0

w_A	0.613	0.629	0.656	0.681	0.699	0.708	0.727	0.753	0.764
w_B	0.313	0.293	0.255	0.220	0.199	0.187	0.159	0.117	0.101
w_C	0.074	0.078	0.089	0.099	0.102	0.105	0.114	0.130	0.135
w_A	0.768								
w_B	0.085								
w_C	0.147								

T/K = 283.15 pH = 7.0

w_A	0.576	0.606	0.620	0.642	0.662	0.710	0.704	0.740	0.756
w_B	0.382	0.348	0.328	0.301	0.277	0.213	0.219	0.178	0.155
w_C	0.042	0.046	0.052	0.057	0.061	0.077	0.077	0.082	0.089
w_A	0.771								
w_B	0.134								
w_C	0.095								

T/K = 298.15 pH = 5.0

w_A	0.746	0.771	0.783	0.794	0.803	0.812	0.820	0.828	0.834
w_B	0.147	0.116	0.099	0.083	0.071	0.060	0.052	0.041	0.032
w_C	0.107	0.113	0.118	0.123	0.126	0.128	0.128	0.131	0.134

T/K = 298.15 pH = 7.0

w_A	0.699	0.718	0.736	0.753	0.769	0.786	0.802	0.817	0.833
w_B	0.233	0.211	0.189	0.165	0.144	0.124	0.105	0.085	0.068
w_C	0.068	0.071	0.075	0.082	0.087	0.090	0.093	0.098	0.099
w_A	0.846	0.855							
w_B	0.048	0.026							
w_C	0.106	0.119							

Type of data: coexistence data (tie lines)

Total system			Top phase			Bottom phase		
w_A	w_B	w_C	w_A	w_B	w_C	w_A	w_B	w_C
T/K = 283.15	pH = 5.0							
0.708	0.166	0.126	0.523	0.421	0.056	0.768	0.085	0.147
0.683	0.179	0.138	0.514	0.441	0.045	0.780	0.037	0.183

continued

continued

T/K = 283.15	pH = 7.0							
0.725	0.175	0.100	0.576	0.384	0.040	0.794	0.079	0.127
0.696	0.190	0.114	0.518	0.455	0.027	0.808	0.020	0.172
0.668	0.204	0.128	0.507	0.461	0.032	0.774	0.026	0.200
0.639	0.219	0.142	0.474	0.499	0.027	0.753	0.012	0.235
T/K = 298.15	pH = 5.0							
0.750	0.115	0.135	0.520	0.420	0.060	0.829	0.014	0.157
0.730	0.125	0.145	0.488	0.457	0.055	0.815	0.008	0.177
0.710	0.135	0.155	0.467	0.483	0.050	0.786	0.010	0.204
0.692	0.142	0.166	0.437	0.514	0.049	0.777	0.006	0.217
T/K = 298.15	pH = 7.0							
0.757	0.128	0.115	0.566	0.379	0.055	0.832	0.028	0.140
0.744	0.126	0.130	0.518	0.431	0.051	0.823	0.013	0.164
0.713	0.145	0.142	0.461	0.493	0.046	0.810	0.011	0.179
0.692	0.149	0.159	0.420	0.533	0.047	0.792	0.008	0.200

Polymer (B): **poly(ethylene oxide-*b*-propylene oxide-*b*-ethylene oxide)** **2005SIL**

Characterization: M_n/g.mol^{-1} = 1706, M_w/g.mol^{-1} = 1945, 50.0 wt% ethylene oxide, (EO)11-(PO)16-(EO)11, L35, Aldrich Chem. Co., Inc., Milwaukee, WI

Solvent (A): **water** **H_2O** **7732-18-5**

Salt (C): **dipotassium phosphate/monopotassium phosphate** **K_2HPO_4/H_2KPO_4** **7758-11-4/7778-77-0**

Comments: The mixture of K_2HPO_4/H_2KPO_4 was chosen in a ratio that provides a pH = 7.

Type of data: coexistence data (tie lines)

Total system			Top phase			Bottom phase		
w_A	w_B	w_C	w_A	w_B	w_C	w_A	w_B	w_C
T/K = 283.15								
0.7690	0.1311	0.0999	0.6913	0.3041	0.0046	0.8217	0.0504	0.1279
0.6764	0.1836	0.1400	0.5989	0.4010	0.0001	0.7414	0.0435	0.2151
0.6001	0.2300	0.1699	0.4785	0.5214	0.0001	0.6652	0.0455	0.2893
0.5078	0.2822	0.2100	0.3575	0.6424	0.0001	0.5977	0.0477	0.3546
T/K = 298.15								
0.7839	0.1278	0.0883	0.6500	0.3497	0.0003	0.8563	0.0260	0.1177
0.7087	0.1715	0.1198	0.5144	0.4854	0.0002	0.8160	0.0274	0.1566
0.6501	0.2007	0.1492	0.4334	0.5665	0.0001	0.7531	0.0271	0.2198
0.5903	0.2401	0.1696	0.3699	0.6300	0.0001	0.7286	0.0304	0.2410

continued

continued

T/K = 313.15

0.7394	0.2107	0.0499	0.6758	0.3225	0.0017	0.8802	0.0312	0.0886
0.6286	0.3015	0.0699	0.4689	0.5310	0.0001	0.8301	0.0164	0.1535
0.5320	0.3803	0.0877	0.3616	0.6382	0.0002	0.7743	0.0210	0.2047
0.4195	0.4706	0.1099	0.2721	0.7277	0.0002	0.6768	0.0271	0.2961

Polymer (B): **poly(ethylene oxide-*b*-propylene oxide-*b*-ethylene oxide)** **2005SIL**

Characterization: M_n/g.mol^{-1} = 5960, M_w/g.mol^{-1} = 8460, 80.0 wt% ethylene oxide, (EO)80-(PO)30-(EO)80, F68, Aldrich Chem. Co., Inc., Milwaukee, WI

Solvent (A): **water** **H_2O** **7732-18-5**

Salt (C): **dipotassium phosphate/monopotassium phosphate** **K_2HPO_4/H_2KPO_4** **7758-11-4/7778-77-0**

Comments: The mixture of K_2HPO_4/H_2KPO_4 was chosen in a ratio that provides a pH = 7.

Type of data: coexistence data (tie lines)

Total system			Top phase			Bottom phase		
w_A	w_B	w_C	w_A	w_B	w_C	w_A	w_B	w_C
T/K = 283.15								
0.7777	0.1425	0.0798	0.7545	0.2278	0.0177	0.8550	0.0164	0.1286
0.7300	0.1702	0.0998	0.6659	0.3269	0.0072	0.8045	0.0257	0.1698
0.6889	0.2012	0.1099	0.6138	0.3782	0.0080	0.7835	0.0304	0.1861
0.6532	0.2189	0.1279	0.5718	0.4176	0.0106	0.7430	0.0410	0.2160
T/K = 298.15								
0.7898	0.1402	0.0700	0.7123	0.2848	0.0029	0.8727	0.0353	0.0920
0.7499	0.1702	0.0799	0.6575	0.3408	0.0017	0.8516	0.0233	0.1251
0.7203	0.1899	0.0898	0.6221	0.3730	0.0049	0.8211	0.0260	0.1529
0.6694	0.2207	0.1099	0.5740	0.4103	0.0157	0.7847	0.0254	0.1899
T/K = 313.15								
0.8101	0.1299	0.0600	0.7435	0.2511	0.0054	0.8854	0.0306	0.0840
0.7602	0.1599	0.0799	0.6443	0.3553	0.0004	0.8590	0.0211	0.1199
0.7201	0.1900	0.0899	0.6090	0.3904	0.0006	0.8464	0.0192	0.1344
0.6910	0.2093	0.0997	0.5712	0.4283	0.0005	0.8124	0.0237	0.1639

Polymer (B):	**poly(ethylene oxide-*b*-propylene oxide-*b*-ethylene oxide)**	**2004HAR**
Characterization:	M_w/g.mol^{-1} = 8400, 80.0 wt% ethylene oxide, F68, (EO)76-(PO)30(EO)76, Aldrich Chem. Co., Inc., Milwaukee, WI	
Solvent (A):	**water** H_2O	**7732-18-5**
Salt (C):	**dipotassium phosphate/monopotassium phosphate** K_2HPO_4/H_2KPO_4	**7758-11-4/7778-77-0**

Type of data: cloud points

T/K = 283.15 pH = 5.0

w_A	0.665	0.685	0.705	0.716	0.750	0.765	0.776
w_B	0.262	0.236	0.205	0.199	0.143	0.116	0.086
w_C	0.073	0.079	0.090	0.085	0.107	0.119	0.138

T/K = 283.15 pH = 7.0

w_A	0.682	0.701	0.742	0.759	0.776
w_B	0.265	0.247	0.191	0.156	0.144
w_C	0.053	0.052	0.067	0.085	0.080

T/K = 298.15 pH = 5.0

w_A	0.746	0.771	0.783	0.794	0.803	0.812	0.820	0.828	0.834
w_B	0.147	0.116	0.099	0.083	0.071	0.060	0.052	0.041	0.032
w_C	0.107	0.113	0.118	0.123	0.126	0.128	0.128	0.131	0.134

T/K = 298.15 pH = 7.0

w_A	0.699	0.734	0.756	0.775	0.790	0.803	0.818	0.821	0.841
w_B	0.242	0.197	0.175	0.149	0.130	0.105	0.086	0.062	0.049
w_C	0.059	0.069	0.069	0.076	0.080	0.092	0.096	0.117	0.110
w_A	0.834								
w_B	0.044								
w_C	0.122								

Type of data: coexistence data (tie lines)

Total system			Top phase			Bottom phase		
w_A	w_B	w_C	w_A	w_B	w_C	w_A	w_B	w_C
T/K = 283.15	pH = 7.0							
0.756	0.149	0.095	0.720	0.231	0.049	0.795	0.074	0.131
0.728	0.164	0.108	0.683	0.270	0.047	0.802	0.009	0.189
0.703	0.177	0.120	0.643	0.324	0.033	0.760	0.017	0.223
0.671	0.193	0.136	0.611	0.358	0.031	0.741	0.012	0.247
T/K = 298.15	pH = 7.0							
0.726	0.178	0.096	0.671	0.280	0.049	0.797	0.024	0.179
0.694	0.196	0.110	0.639	0.315	0.046	0.768	0.022	0.210
0.665	0.210	0.125	0.621	0.337	0.042	0.729	0.017	0.254

Polymer (B): **poly(ethylene oxide-*b*-propylene oxide-*b*-ethylene oxide)** **2010LEM**
Characterization: M_w/g.mol^{-1} = 1900, (EO)11-(PO)16-(EO)11, L35, Aldrich Chem. Co., Inc., Milwaukee, WI
Solvent (A): **water** H_2O **7732-18-5**
Salt (C): **disodium succinate** $C_4H_4Na_2O_4$ **150-90-3**

Type of data: coexistence data (tie lines)

Total system			Top phase			Bottom phase		
w_A	w_B	w_C	w_A	w_B	w_C	w_A	w_B	w_C
T/K = 283.15								
0.6354	0.2915	0.0731	0.5408	0.4228	0.0364	0.8368	0.0092	0.1540
0.5962	0.3290	0.0748	0.4632	0.5075	0.0293	0.8226	0.0068	0.1706
0.5601	0.3582	0.0817	0.4119	0.5655	0.0226	0.8092	0.0020	0.1888
0.5186	0.3948	0.0866	0.3759	0.6069	0.0172	0.7880	0.0019	0.2101
0.4946	0.4154	0.0900	0.3632	0.6202	0.0166	0.7825	0.0001	0.2174
T/K = 298.15								
0.6240	0.2997	0.0763	0.4759	0.4911	0.0330	0.8182	0.0297	0.1521
0.5879	0.3316	0.0805	0.4391	0.5374	0.0235	0.7926	0.0278	0.1796
0.5529	0.3617	0.0854	0.3585	0.6221	0.0194	0.7888	0.0271	0.1841
0.5133	0.3974	0.0893	0.3167	0.6680	0.0153	0.7841	0.0244	0.1915
0.4863	0.4179	0.0958	0.2995	0.6882	0.0123	0.7696	0.0207	0.2097
T/K = 313.15								
0.6265	0.2983	0.0752	0.4272	0.5541	0.0187	0.8488	0.0182	0.1330
0.5822	0.3373	0.0805	0.3700	0.6165	0.0135	0.8258	0.0147	0.1595
0.5552	0.3603	0.0845	0.3267	0.6629	0.0104	0.8184	0.0107	0.1709
0.5078	0.4025	0.0897	0.2802	0.7128	0.0070	0.8002	0.0057	0.1941
0.4752	0.4280	0.0968	0.2303	0.7654	0.0043	0.7817	0.0012	0.2171

Polymer (B): **poly(ethylene oxide-*b*-propylene oxide-*b*-ethylene oxide)** **2011AND**
Characterization: M_w/g.mol^{-1} = 2930, M_w/M_n = 1.09, 40.0 wt% ethylene oxide, (EO)13(PO)30(EO)13, L64, Aldrich Chem. Co., Inc., Milwaukee, WI
Solvent (A): **water** H_2O **7732-18-5**
Salt (C): **disodium succinate** $C_4H_4Na_2O_4$ **150-90-3**

Type of data: coexistence data (tie lines)

continued

continued

Total system			Top phase			Bottom phase		
w_A	w_B	w_C	w_A	w_B	w_C	w_A	w_B	w_C
T/K = 278.15								
0.7022	0.2050	0.0928	0.4852	0.4772	0.0376	0.8546	0.0096	0.1358
0.6860	0.2200	0.0940	0.4620	0.5014	0.0366	0.8539	0.0026	0.1435
0.6697	0.2349	0.0954	0.4465	0.5173	0.0362	0.8475	0.0044	0.1482
0.6213	0.2788	0.0999	0.4120	0.5536	0.0344	0.8177	0.0022	0.1801
T/K = 288.15								
0.7522	0.1593	0.0885	0.5702	0.3852	0.0446	0.8807	0.0023	0.1170
0.7304	0.1792	0.0904	0.5129	0.4476	0.0395	0.8698	0.0027	0.1275
0.7023	0.2048	0.0929	0.4637	0.5019	0.0344	0.8567	0.0022	0.1411
0.6697	0.2349	0.0954	0.4281	0.5398	0.0321	0.8456	0.0033	0.1511
T/K = 298.15								
0.7905	0.1245	0.0850	0.7604	0.1628	0.0768	0.8889	0.0015	0.1096
0.7735	0.1400	0.0865	0.6606	0.2726	0.0667	0.8787	0.0068	0.1145
0.7521	0.1594	0.0885	0.5997	0.3430	0.0573	0.8758	0.0052	0.1190
0.7303	0.1792	0.0905	0.5805	0.3651	0.0544	0.8710	0.0069	0.1221

Polymer (B): **poly(ethylene oxide-*b*-propylene oxide-*b*-ethylene oxide)** **2010RO1**

Characterization: M_w/g.mol^{-1} = 8400, (EO)80-(PO)30-(EO)80, F68, Aldrich Chem. Co., Inc., Milwaukee, WI

Solvent (A): **water** **H_2O** **7732-18-5**

Salt (C): **disodium succinate** **$C_4H_4Na_2O_4$** **150-90-3**

Type of data: coexistence data (tie lines)

Total system			Top phase			Bottom phase		
w_A	w_B	w_C	w_A	w_B	w_C	w_A	w_B	w_C
T/K = 283.15								
0.7282	0.1846	0.0872	0.6673	0.2695	0.0632	0.8431	0.0117	0.1452
0.7024	0.2065	0.0911	0.6401	0.3000	0.0599	0.8322	0.0162	0.1516
0.6895	0.2174	0.0931	0.6290	0.3116	0.0594	0.8234	0.0160	0.1606
0.6768	0.2282	0.0951	0.6140	0.3255	0.0605	0.8171	0.0001	0.1828
T/K = 298.15								
0.7511	0.1651	0.0838	0.6973	0.2344	0.0683	0.8616	0.0129	0.1255
0.7412	0.1736	0.0852	0.6717	0.2662	0.0621	0.8621	0.0044	0.1335
0.7213	0.1904	0.0882	0.6340	0.3121	0.0539	0.8504	0.0048	0.1448
0.7005	0.2099	0.0896	0.6157	0.3338	0.0505	0.8464	0.0039	0.1497

continued

continued

T/K = 313.15								
0.7893	0.1326	0.0781	0.6900	0.2584	0.0516	0.8890	0.0015	0.1095
0.7802	0.1406	0.0792	0.6718	0.2790	0.0492	0.8842	0.0035	0.1123
0.7700	0.1494	0.0806	0.6504	0.3038	0.0458	0.8826	0.0001	0.1173
0.7605	0.1572	0.0822	0.6302	0.3262	0.0436	0.8775	0.0005	0.1220

Polymer (B): **poly(ethylene oxide-*b*-propylene oxide-*b*-ethylene oxide)** **2009MAR**

Characterization: M_w/g.mol^{-1} = 1900, 50.0 wt% ethylene oxide, (EO)11-(PO)16-(EO)11, L35, Sigma Chemical Co., Inc., St. Louis, MO

Solvent (A): **water** **H_2O** **7732-18-5**

Salt (C): **disodium (±) tartrate** **$C_4H_4Na_2O_6$** **51307-92-7**

Type of data: coexistence data (tie lines)

Total system			Top phase			Bottom phase		
w_A	w_B	w_C	w_A	w_B	w_C	w_A	w_B	w_C
T/K = 283.15								
0.7179	0.2000	0.0821	0.6880	0.2672	0.0448	0.7667	0.0651	0.1682
0.6768	0.2211	0.1021	0.5958	0.3785	0.0257	0.7634	0.0288	0.2078
0.6340	0.2528	0.1132	0.5552	0.4259	0.0189	0.7286	0.0241	0.2473
0.5967	0.2820	0.1213	0.5173	0.4678	0.0149	0.7008	0.0233	0.2759
0.5602	0.3059	0.1339	0.4599	0.5283	0.0118	0.6633	0.0451	0.2916
T/K = 298.15								
0.5946	0.3599	0.0455	0.5640	0.4106	0.0254	0.7954	0.0302	0.1744
0.5474	0.3992	0.0534	0.5066	0.4757	0.0177	0.7625	0.0313	0.2044
0.4933	0.4504	0.0563	0.4609	0.5267	0.0124	0.7256	0.0455	0.2289
0.4401	0.4966	0.0633	0.4149	0.5739	0.0112	0.7037	0.0424	0.2539
0.3879	0.5462	0.0659	0.3119	0.6825	0.0056	0.6359	0.0582	0.3059
T/K = 313.15								
0.6429	0.2578	0.0993	0.4528	0.5320	0.0152	0.8033	0.0203	0.1764
0.6326	0.2558	0.1116	0.4085	0.5802	0.0113	0.7755	0.0363	0.1882
0.5896	0.2902	0.1202	0.3729	0.6186	0.0085	0.7531	0.0312	0.2157
0.5465	0.3269	0.1266	0.3149	0.6789	0.0062	0.7175	0.0326	0.2499
0.4689	0.3889	0.1422	0.2901	0.7063	0.0036	0.6482	0.0610	0.2908

Polymer (B):	**poly(ethylene oxide-*b*-propylene oxide-*b*-ethylene oxide)**	**2011AND**
Characterization:	M_w/g.mol^{-1} = 2930, M_w/M_n = 1.09, 40.0 wt% ethylene oxide, (EO)13(PO)30(EO)13, L64, Aldrich Chem. Co., Inc., Milwaukee, WI	
Solvent (A):	**water** H_2O	**7732-18-5**
Salt (C):	**disodium tartrate** $C_4H_4Na_2O_6$	**868-18-8**

Type of data: coexistence data (tie lines)

Total system			Top phase			Bottom phase		
w_A	w_B	w_C	w_A	w_B	w_C	w_A	w_B	w_C
T/K = 278.15								
0.7395	0.1782	0.0823	0.6819	0.2709	0.0471	0.8323	0.0081	0.1595
0.7082	0.2092	0.0826	0.6362	0.3280	0.0358	0.8238	0.0009	0.1753
0.6878	0.2275	0.0847	0.6110	0.3548	0.0342	0.8137	0.0033	0.1830
0.6556	0.2618	0.0826	0.5790	0.3908	0.0301	0.8023	0.0048	0.1929
0.6242	0.2887	0.0871	0.5463	0.4274	0.0263	0.7883	0.0003	0.2114
T/K = 288.15								
0.7397	0.1784	0.0819	0.6720	0.2653	0.0627	0.8525	0.0268	0.1207
0.7084	0.2092	0.0824	0.5779	0.3820	0.0401	0.8465	0.0189	0.1346
0.6871	0.2285	0.0844	0.5437	0.4226	0.0337	0.8445	0.0111	0.1444
0.6564	0.2614	0.0822	0.5197	0.4506	0.0296	0.8385	0.0077	0.1538
0.6416	0.2746	0.0838	0.4887	0.4854	0.0259	0.8332	0.0051	0.1617
T/K = 298.15								
0.6957	0.2320	0.0723	0.6226	0.3228	0.0547	0.8556	0.0351	0.1093
0.6774	0.2483	0.0743	0.5612	0.3927	0.0461	0.8545	0.0265	0.1190
0.6591	0.2646	0.0763	0.5296	0.4303	0.0401	0.8444	0.0270	0.1286
0.6408	0.2809	0.0783	0.4756	0.4926	0.0318	0.8351	0.0317	0.1332
0.6225	0.2972	0.0803	0.4546	0.5161	0.0294	0.8274	0.0342	0.1384

Polymer (B):	**poly(ethylene oxide-*b*-propylene oxide-*b*-ethylene oxide)**	**2010RO1**
Characterization:	M_w/g.mol^{-1} = 8400, (EO)80-(PO)30-(EO)80, F68, Aldrich Chem. Co., Inc., Milwaukee, WI	
Solvent (A):	**water** H_2O	**7732-18-5**
Salt (C):	**disodium tartrate** $C_4H_4Na_2O_6$	**868-18-8**

Type of data: coexistence data (tie lines)

continued

continued

Total system			Top phase			Bottom phase		
w_A	w_B	w_C	w_A	w_B	w_C	w_A	w_B	w_C
T/K = 283.15								
0.7843	0.1235	0.0922	0.7354	0.2098	0.0548	0.8516	0.0154	0.1330
0.7626	0.1444	0.0929	0.7117	0.2374	0.0509	0.8432	0.0084	0.1484
0.7460	0.1591	0.0949	0.6838	0.2747	0.0415	0.8312	0.0089	0.1599
0.7195	0.1799	0.1006	0.6499	0.3155	0.0346	0.8132	0.0103	0.1765
0.6937	0.2004	0.1058	0.6179	0.3509	0.0312	0.7930	0.0088	0.1982
T/K = 298.15								
0.7969	0.1206	0.0825	0.7450	0.1897	0.0653	0.8686	0.0149	0.1165
0.7795	0.1353	0.0852	0.7076	0.2393	0.0531	0.8608	0.0083	0.1309
0.7620	0.1498	0.0883	0.6666	0.2880	0.0454	0.8584	0.0024	0.1392
0.7442	0.1646	0.0912	0.6436	0.3156	0.0408	0.8482	0.0035	0.1483
0.7259	0.1789	0.0952	0.6059	0.3577	0.0364	0.8330	0.0030	0.1640
T/K = 313.15								
0.8390	0.0795	0.0815	0.7537	0.1869	0.0594	0.8684	0.0355	0.0961
0.8532	0.0562	0.0906	0.7267	0.2210	0.0523	0.8880	0.0045	0.1075
0.8298	0.0810	0.0892	0.6960	0.2575	0.0465	0.8863	0.0001	0.1136
0.8074	0.1021	0.0906	0.6632	0.2948	0.0420	0.8766	0.0001	0.1233
0.7839	0.1239	0.0922	0.6409	0.3211	0.0380	0.8667	0.0001	0.1332

Polymer (B): **poly(ethylene oxide-*b*-propylene oxide-*b*-ethylene oxide)** **2006SIL**

Characterization: M_n/g.mol^{-1} = 1706, M_w/g.mol^{-1} = 1945, 50.0 wt% ethylene oxide, (EO)11-(PO)16-(EO)11, L35, Aldrich Chem. Co., Inc., Milwaukee, WI

Solvent (A): **water** H_2O **7732-18-5**

Salt (C): **lithium sulfate** Li_2SO_4 **10377-48-7**

Type of data: coexistence data (tie lines)

Total system			Top phase			Bottom phase		
w_A	w_B	w_C	w_A	w_B	w_C	w_A	w_B	w_C
T/K = 298.15								
0.7874	0.0778	0.1348	0.5436	0.4448	0.0116	0.8218	0.0292	0.1490
0.7689	0.0903	0.1408	0.5304	0.4587	0.0109	0.8101	0.0262	0.1637
0.7535	0.0999	0.1466	0.4935	0.4985	0.0080	0.8028	0.0218	0.1754
0.7251	0.1179	0.1570	0.4525	0.5425	0.0050	0.7950	0.0095	0.1955
0.6906	0.1401	0.1693	0.4061	0.5892	0.0047	0.7788	0.0019	0.2193

continued

continued

T/K = 313.15								
0.8024	0.1266	0.0710	0.5473	0.4361	0.0166	0.8890	0.0214	0.0896
0.7879	0.1385	0.0736	0.5212	0.4641	0.0147	0.8857	0.0198	0.0945
0.7684	0.1548	0.0768	0.4984	0.4895	0.0121	0.8792	0.0181	0.1027
0.7166	0.1974	0.0860	0.4827	0.5058	0.0115	0.8692	0.0010	0.1298
0.6850	0.2234	0.0916	0.4695	0.5193	0.0112	0.8520	0.0004	0.1476

Polymer (B): **poly(ethylene oxide-*b*-propylene oxide-*b*-ethylene oxide)** **2009ROD**

Characterization: M_w/g.mol^{-1} = 2900, 40.0 wt% ethylene oxide, L64, (EO)13(PO)30(EO)13, Aldrich Chem. Co., Inc., Milwaukee, WI

Solvent (A): **water** H_2O **7732-18-5**

Salt (C): **lithium sulfate** Li_2SO_4 **10377-48-7**

Type of data: coexistence data (tie lines)

Total system			Top phase			Bottom phase		
w_A	w_B	w_C	w_A	w_B	w_C	w_A	w_B	w_C
T/K = 278.15								
0.7108	0.2257	0.0635	0.6216	0.3361	0.0423	0.8623	0.0292	0.1085
0.6884	0.2461	0.0655	0.5632	0.4042	0.0326	0.8678	0.0175	0.1147
0.6661	0.2663	0.0676	0.5482	0.4168	0.0350	0.8638	0.0097	0.1265
0.6435	0.2873	0.0693	0.5195	0.4488	0.0317	0.8598	0.0035	0.1366
T/K = 288.15								
0.7095	0.2273	0.0631	0.5994	0.3591	0.0415	0.8623	0.0420	0.0957
0.6879	0.2463	0.0658	0.5746	0.3872	0.0382	0.8678	0.0259	0.1063
0.6642	0.2684	0.0674	0.5196	0.4497	0.0307	0.8638	0.0246	0.1116
0.6425	0.2880	0.0695	0.4765	0.4990	0.0245	0.8598	0.0194	0.1208
T/K = 298.15								
0.7780	0.1645	0.0575	0.7153	0.2340	0.0507	0.8996	0.0209	0.0795
0.7552	0.1855	0.0593	0.6378	0.3208	0.0414	0.9039	0.0112	0.0848
0.7334	0.2049	0.0616	0.6023	0.3577	0.0401	0.8994	0.0056	0.0950
0.7110	0.2262	0.0628	0.5500	0.4172	0.0327	0.8957	0.0041	0.1001

Polymer (B): **poly(ethylene oxide-*b*-propylene oxide-*b*-ethylene oxide)** **2010MA2**

Characterization: M_w/g.mol^{-1} = 8460, M_w/M_n = 1.42, 80 wt% ethylene oxide, (EO)80-(PO)30-(EO)80, F68

Solvent (A): **water** H_2O **7732-18-5**

Salt (C): **lithium sulfate** Li_2SO_4 **10377-48-7**

continued

continued

Type of data: coexistence data (tie lines)

Total system			Top phase			Bottom phase		
w_A	w_B	w_C	w_A	w_B	w_C	w_A	w_B	w_C
T/K = 278.15								
0.7427	0.1751	0.0822	0.6370	0.3258	0.0372	0.8485	0.0243	0.1272
0.7269	0.1875	0.0856	0.6070	0.3580	0.0350	0.8469	0.0170	0.1361
0.7085	0.2070	0.0845	0.5721	0.3995	0.0284	0.8449	0.0144	0.1407
0.6983	0.2105	0.0912	0.5594	0.4101	0.0305	0.8381	0.0099	0.1520
0.6888	0.2159	0.0953	0.5466	0.4244	0.0290	0.8309	0.0075	0.1616
T/K = 288.15								
0.8032	0.1177	0.0791	0.7261	0.2231	0.0509	0.8804	0.0123	0.1072
0.7804	0.1403	0.0793	0.6853	0.2711	0.0435	0.8755	0.0094	0.1151
0.7626	0.1541	0.0833	0.6584	0.3012	0.0404	0.8667	0.0071	0.1263
0.7428	0.1735	0.0837	0.6223	0.3419	0.0359	0.8633	0.0051	0.1316
0.7298	0.1833	0.0869	0.6025	0.3627	0.0348	0.8569	0.0040	0.1391
T/K = 298.15								
0.7589	0.1527	0.0884	0.6537	0.3003	0.0460	0.8641	0.0051	0.1308
0.7091	0.1841	0.1068	0.5906	0.3645	0.0449	0.8275	0.0037	0.1688
0.6877	0.1976	0.1147	0.5634	0.3933	0.0433	0.9794	0.0020	0.1860
0.6588	0.2183	0.1229	0.5266	0.4320	0.0414	0.7908	0.0047	0.2045
0.6408	0.2263	0.1329	0.5069	0.4523	0.0408	0.7749	0.0004	0.2250

Polymer (B): **poly(ethylene oxide-*b*-propylene oxide-*b*-ethylene oxide)** **2010MAH**
Characterization: M_w/g.mol^{-1} = 2036, (EO)2.5(PO)31(EO)2.5, Aldrich Chem. Co., Inc., Milwaukee, WI
Solvent (A): **water** **H_2O** **7732-18-5**
Salt (C): **magnesium chloride** **$MgCl_2$** **7786-30-3**

Type of data: cloud points (LCST-behavior)

w_B = 0.01 in the binary aqueous solution was kept constant

c_C/(mol/l)	0.00	0.10	0.20	0.30	0.40
T/K	300.45	299.95	297.85	296.75	294.85

Polymer (B): **poly(ethylene oxide-*b*-propylene oxide-*b*-ethylene oxide)** **2010MAH**
Characterization: M_w/g.mol^{-1} = 2900, 40.0 wt% ethylene oxide, L64, (EO)13(PO)30(EO)13, Aldrich Chem. Co., Inc., Milwaukee, WI
Solvent (A): **water** **H_2O** **7732-18-5**
Salt (C): **magnesium chloride** **$MgCl_2$** **7786-30-3**

continued

continued

Type of data: cloud points (LCST-behavior)

w_B = 0.01 in the binary aqueous solution was kept constant

c_C/(mol/l)	0.00	0.10	0.20	0.30	0.40	0.50
T/K	330.25	328.05	325.15	322.55	320.65	319.85

Polymer (B): **poly(ethylene oxide-*b*-propylene oxide-*b*-ethylene oxide)** **2006SIL**

Characterization: M_n/g.mol^{-1} = 1706, M_w/g.mol^{-1} = 1945, 50.0 wt% ethylene oxide, (EO)11-(PO)16-(EO)11, L35, Aldrich Chem. Co., Inc., Milwaukee, WI

Solvent (A): **water** **H_2O** **7732-18-5**

Salt (C): **magnesium sulfate** **$MgSO_4$** **7487-88-9**

Type of data: coexistence data (tie lines)

Total system			Top phase			Bottom phase		
w_A	w_B	w_C	w_A	w_B	w_C	w_A	w_B	w_C
T/K = 283.15								
0.5769	0.3918	0.0313	0.5382	0.4574	0.0044	0.7618	0.0577	0.1805
0.5215	0.4409	0.0376	0.4860	0.5114	0.0026	0.7396	0.0512	0.2092
0.4865	0.4719	0.0416	0.4341	0.5645	0.0014	0.7265	0.0397	0.2338
0.4365	0.5164	0.0471	0.3728	0.6267	0.0005	0.7086	0.0351	0.2563
0.3931	0.5548	0.0521	0.3266	0.6733	0.0001	0.6967	0.0371	0.2662
T/K = 298.15								
0.6169	0.3500	0.0331	0.5723	0.4195	0.0082	0.8045	0.0483	0.1472
0.5711	0.3908	0.0381	0.5203	0.4739	0.0058	0.7869	0.0476	0.1655
0.5254	0.4315	0.0431	0.4640	0.5318	0.0042	0.7653	0.0401	0.1946
0.4796	0.4723	0.0481	0.4160	0.5815	0.0025	0.7488	0.0312	0.2200
0.4338	0.5131	0.0531	0.3665	0.6323	0.0012	0.7423	0.0196	0.2381
T/K = 313.15								
0.6241	0.3486	0.0273	0.5614	0.4334	0.0052	0.8656	0.0335	0.1009
0.5683	0.3994	0.0323	0.4928	0.5042	0.0030	0.8443	0.0312	0.1245
0.5125	0.4502	0.0373	0.4361	0.5618	0.0021	0.8206	0.0289	0.1505
0.4567	0.5010	0.0423	0.3760	0.6227	0.0013	0.7927	0.0223	0.1850
0.4009	0.5518	0.0473	0.3123	0.6872	0.0005	0.7717	0.0170	0.2113

Polymer (B): **poly(ethylene oxide-*b*-propylene oxide-*b*-ethylene oxide)** **2009ROD**
Characterization: M_w/g.mol^{-1} = 2900, 40.0 wt% ethylene oxide, L64, (EO)13(PO)30(EO)13, Aldrich Chem. Co., Inc., Milwaukee, WI
Solvent (A): **water** **H_2O** **7732-18-5**
Salt (C): **magnesium sulfate** **$MgSO_4$** **7487-88-9**

Type of data: coexistence data (tie lines)

Total system			Top phase			Bottom phase		
w_A	w_B	w_C	w_A	w_B	w_C	w_A	w_B	w_C
T/K = 278.15								
0.7005	0.2528	0.0467	0.6340	0.3540	0.0120	0.8495	0.0049	0.1455
0.6702	0.2788	0.0510	0.5962	0.3951	0.0087	0.8397	0.0021	0.1582
0.6415	0.3015	0.0570	0.5504	0.4432	0.0064	0.8282	0.0046	0.1673
0.6202	0.3157	0.0641	0.5080	0.4861	0.0059	0.8164	0.0052	0.1784
0.5889	0.3404	0.0707	0.4664	0.5300	0.0035	0.8043	0.0067	0.1890
T/K = 288.15								
0.7519	0.2030	0.0451	0.6550	0.3256	0.0194	0.8972	0.0088	0.0940
0.6996	0.2532	0.0472	0.5871	0.4005	0.0124	0.8818	0.0045	0.1137
0.6701	0.2787	0.0512	0.5387	0.4515	0.0098	0.8709	0.0002	0.1290
0.6434	0.2997	0.0569	0.5019	0.4910	0.0071	0.8593	0.0016	0.1391
0.6204	0.3158	0.0638	0.4632	0.5313	0.0054	0.8453	0.0033	0.1514
T/K = 298.15								
0.7350	0.2187	0.0463	0.6198	0.3515	0.0287	0.8835	0.0556	0.0609
0.6950	0.2556	0.0494	0.5222	0.4558	0.0220	0.8766	0.0515	0.0718
0.6560	0.2910	0.0530	0.4712	0.5087	0.0201	0.8683	0.0497	0.0820
0.6149	0.3274	0.0577	0.4307	0.5503	0.0191	0.8517	0.0530	0.0953
0.5702	0.3702	0.0596	0.3999	0.5819	0.0182	0.8381	0.0516	0.1103

Polymer (B): **poly(ethylene oxide-*b*-propylene oxide-*b*-ethylene oxide)** **2006SIL**
Characterization: M_n/g.mol^{-1} = 5960, M_w/g.mol^{-1} = 8460, 80.0 wt% ethylene oxide, (EO)80-(PO)30-(EO)80, F68, Aldrich Chem. Co., Inc., Milwaukee, WI
Solvent (A): **water** **H_2O** **7732-18-5**
Salt (C): **magnesium sulfate** **$MgSO_4$** **7487-88-9**

Type of data: coexistence data (tie lines)

continued

continued

Total system			Top phase			Bottom phase		
w_A	w_B	w_C	w_A	w_B	w_C	w_A	w_B	w_C
T/K = 283.15								
0.7822	0.1720	0.0458	0.7220	0.2596	0.0184	0.8585	0.0587	0.0828
0.7531	0.1948	0.0521	0.6809	0.3047	0.0144	0.8449	0.0583	0.0968
0.7299	0.2130	0.0571	0.6599	0.3288	0.0113	0.8242	0.0521	0.1237
0.7067	0.2312	0.0621	0.6407	0.3502	0.0091	0.8069	0.0443	0.1488
0.6835	0.2494	0.0671	0.6203	0.3713	0.0084	0.7925	0.0384	0.1691
T/K = 298.15								
0.8237	0.1322	0.0441	0.7742	0.2071	0.0187	0.8956	0.0267	0.0777
0.8024	0.1498	0.0478	0.7321	0.2519	0.0160	0.8880	0.0188	0.0932
0.7618	0.1832	0.0550	0.6908	0.2959	0.0133	0.8661	0.0156	0.1183
0.7357	0.2047	0.0596	0.6578	0.3322	0.0100	0.8546	0.0112	0.1342
0.7175	0.2197	0.0628	0.6240	0.3676	0.0084	0.8526	0.0042	0.1432
T/K = 313.15								
0.7903	0.1694	0.0403	0.7201	0.2633	0.0166	0.8690	0.0638	0.0672
0.7602	0.1945	0.0453	0.6729	0.3137	0.0134	0.8550	0.0636	0.0814
0.7434	0.2085	0.0481	0.6592	0.3290	0.0118	0.8477	0.0602	0.0921
0.7242	0.2245	0.0513	0.6432	0.3461	0.0107	0.8336	0.0589	0.1075
0.6794	0.2618	0.0588	0.5790	0.4109	0.0101	0.8164	0.0586	0.1250

Polymer (B): **poly(ethylene oxide-*b*-propylene oxide-*b*-ethylene oxide)** **2010VIR**

Characterization: M_w/g.mol^{-1} = 1900, (EO)11-(PO)16-(EO)11, L35, Aldrich Chem. Co., Inc., Milwaukee, WI

Solvent (A): **water** H_2O **7732-18-5**

Salt (C): **potassium hydroxide** **KOH** **1310-58-3**

Type of data: coexistence data (tie lines)

Total system			Top phase			Bottom phase		
w_A	w_B	w_C	w_A	w_B	w_C	w_A	w_B	w_C
T/K = 283.15								
0.6834	0.2131	0.1035	0.6019	0.3409	0.0572	0.7549	0.0961	0.1490
0.6335	0.2513	0.1152	0.5467	0.4092	0.0441	0.7146	0.1101	0.1753
0.5262	0.3434	0.1304	0.4184	0.5566	0.0250	0.6587	0.1032	0.2381
0.4719	0.3839	0.1442	0.3650	0.6137	0.0213	0.6166	0.1182	0.2652
0.3843	0.4540	0.1617	0.3203	0.6602	0.0195	0.5480	0.1143	0.3377

continued

continued

T/K = 293.15								
0.5880	0.3028	0.1092	0.4710	0.4813	0.0477	0.7000	0.1335	0.1665
0.5504	0.3306	0.1190	0.4293	0.5376	0.0331	0.6773	0.1326	0.1901
0.5082	0.3613	0.1305	0.3936	0.5750	0.0314	0.6280	0.1664	0.2056
0.4642	0.3891	0.1467	0.3013	0.6815	0.0172	0.6008	0.1655	0.2337
0.3887	0.4418	0.1695	0.2595	0.7165	0.0240	0.5420	0.1619	0.2961
0.3541	0.4632	0.1827	0.2025	0.7824	0.0151	0.5153	0.1459	0.3388
T/K = 303.15								
0.5535	0.3621	0.0844	0.3750	0.5904	0.0346	0.7238	0.1412	0.1350
0.4719	0.4271	0.1010	0.2964	0.6753	0.0283	0.6756	0.1418	0.1826
0.4196	0.4700	0.1104	0.2459	0.7273	0.0268	0.6475	0.1506	0.2019
0.3717	0.5101	0.1182	0.1934	0.7830	0.0236	0.6083	0.1549	0.2368
0.3457	0.5284	0.1259	0.1753	0.8018	0.0229	0.5912	0.1531	0.2557
0.3137	0.5556	0.1307	0.1391	0.8406	0.0203	0.5762	0.1499	0.2739
T/K = 313.15								
0.5784	0.3179	0.1037	0.3526	0.6071	0.0403	0.7661	0.0772	0.1567
0.5630	0.3266	0.1104	0.3202	0.6427	0.0371	0.7474	0.0843	0.1683
0.5368	0.3482	0.1150	0.2897	0.6781	0.0322	0.7216	0.0963	0.1821
0.5159	0.3631	0.1210	0.2625	0.7067	0.0308	0.7293	0.0790	0.1917
0.4896	0.3827	0.1277	0.2116	0.7641	0.0243	0.7159	0.0698	0.2143
0.4500	0.4082	0.1418	0.1740	0.8065	0.0195	0.6682	0.0777	0.2541
0.3657	0.4708	0.1635	0.1368	0.8442	0.0190	0.6450	0.0761	0.2789

Polymer (B): **poly(ethylene oxide-*b*-propylene oxide-*b*-ethylene oxide)** **2010LEM**

Characterization: M_w/g.mol^{-1} = 1900, (EO)11-(PO)16-(EO)11, L35, Aldrich Chem. Co., Inc., Milwaukee, WI

Solvent (A): **water** **H_2O** **7732-18-5**

Salt (C): **sodium acetate** **$C_2H_3NaO_2$** **127-09-3**

Type of data: coexistence data (tie lines)

Total system			Top phase			Bottom phase		
w_A	w_B	w_C	w_A	w_B	w_C	w_A	w_B	w_C
T/K = 283.15								
0.6780	0.1373	0.1847	0.6121	0.2321	0.1558	0.7635	0.0043	0.2322
0.6149	0.2089	0.1762	0.5483	0.3025	0.1492	0.7460	0.0029	0.2511
0.5626	0.2643	0.1731	0.4860	0.3809	0.1331	0.7230	0.0001	0.2769
0.5054	0.3243	0.1703	0.3973	0.5113	0.0914	0.6916	0.0001	0.3083
0.4734	0.3573	0.1693	0.3336	0.5967	0.0697	0.6861	0.0001	0.3138

continued

continued

T/K = 298.15								
0.6427	0.1955	0.1618	0.5969	0.2605	0.1426	0.7641	0.0091	0.2268
0.6125	0.2207	0.1668	0.4837	0.4044	0.1119	0.7529	0.0065	0.2406
0.5823	0.2459	0.1718	0.3931	0.5177	0.0892	0.7359	0.0069	0.2572
0.5521	0.2711	0.1768	0.2911	0.6431	0.0658	0.7196	0.0056	0.2748
0.5219	0.2963	0.1818	0.2423	0.7076	0.0501	0.7048	0.0067	0.2885
T/K = 313.15								
0.5972	0.3006	0.1022	0.5030	0.4239	0.0731	0.8064	0.0198	0.1738
0.5673	0.3311	0.1016	0.4429	0.4946	0.0625	0.7971	0.0201	0.1828
0.5055	0.3243	0.1703	0.2059	0.7784	0.0157	0.6970	0.0246	0.2784
0.4735	0.3573	0.1693	0.1624	0.8312	0.0064	0.6799	0.0314	0.2887
0.4091	0.5036	0.0872	0.3038	0.6611	0.0351	0.7521	0.0195	0.2284

Polymer (B): **poly(ethylene oxide-*b*-propylene oxide-*b*-ethylene oxide)** **2010LEM**

Characterization: M_w/g.mol^{-1} = 1900, (EO)11-(PO)16-(EO)11, L35, Aldrich Chem. Co., Inc., Milwaukee, WI

Solvent (A): **water** **H_2O** **7732-18-5**

Salt (C): **sodium carbonate** **Na_2CO_3** **497-19-8**

Type of data: coexistence data (tie lines)

Total system			Top phase			Bottom phase		
w_A	w_B	w_C	w_A	w_B	w_C	w_A	w_B	w_C
T/K = 283.15								
0.7547	0.1902	0.0551	0.6808	0.2938	0.0254	0.8539	0.0254	0.1207
0.7079	0.2254	0.0667	0.6104	0.3739	0.0157	0.8158	0.0370	0.1472
0.6632	0.2608	0.0760	0.5506	0.4382	0.0112	0.7873	0.0322	0.1805
0.6170	0.2944	0.0886	0.4928	0.4990	0.0082	0.7510	0.0368	0.2122
0.5745	0.3216	0.1039	0.4294	0.5655	0.0051	0.7162	0.0333	0.2505
T/K = 298.15								
0.7547	0.1902	0.0551	0.6111	0.3695	0.0194	0.8685	0.0338	0.0977
0.7024	0.2440	0.0536	0.5745	0.4088	0.0167	0.8578	0.0255	0.1167
0.6423	0.2991	0.0586	0.4996	0.4904	0.0100	0.8366	0.0214	0.1420
0.5822	0.3542	0.0636	0.4396	0.5534	0.0070	0.8012	0.0286	0.1702
0.5221	0.4093	0.0686	0.3858	0.6091	0.0051	0.7685	0.0288	0.2027
T/K = 313.15								
0.7547	0.1902	0.0551	0.5659	0.4207	0.0134	0.8898	0.0149	0.0953
0.7078	0.2255	0.0667	0.4717	0.5194	0.0089	0.8679	0.0137	0.1184
0.6631	0.2609	0.0760	0.4029	0.5917	0.0054	0.8341	0.0265	0.1394
0.6169	0.2945	0.0886	0.3522	0.6437	0.0041	0.7995	0.0266	0.1739
0.5746	0.3215	0.1039	0.3485	0.6475	0.0040	0.7602	0.0332	0.2066

Polymer (B): **poly(ethylene oxide-*b*-propylene oxide-*b*-ethylene oxide)** **2010RO1**
Characterization: M_w/g.mol^{-1} = 8400, (EO)80-(PO)30-(EO)80, F68, Aldrich Chem. Co., Inc., Milwaukee, WI
Solvent (A): **water** **H_2O** **7732-18-5**
Salt (C): **sodium carbonate** **Na_2CO_3** **497-19-8**

Type of data: coexistence data (tie lines)

Total system			Top phase			Bottom phase		
w_A	w_B	w_C	w_A	w_B	w_C	w_A	w_B	w_C
T/K = 283.15								
0.8334	0.1216	0.0449	0.7631	0.2110	0.0259	0.9217	0.0007	0.0776
0.8130	0.1399	0.0471	0.7281	0.2498	0.0221	0.9142	0.0001	0.0857
0.7917	0.1594	0.0489	0.6961	0.2850	0.0189	0.9016	0.0001	0.0983
0.7713	0.1775	0.0512	0.6717	0.3121	0.0162	0.8974	0.0001	0.1025
0.7503	0.1963	0.0533	0.6465	0.3396	0.0139	0.8810	0.0001	0.1189
T/K = 298.15								
0.8362	0.1277	0.0361	0.7972	0.1724	0.0304	0.9275	0.0178	0.0547
0.8177	0.1448	0.0375	0.7437	0.2334	0.0229	0.9336	0.0037	0.0627
0.7916	0.1683	0.0401	0.6848	0.2974	0.0178	0.9288	0.0020	0.0692
0.7651	0.1925	0.0424	0.6438	0.3417	0.0145	0.9188	0.0025	0.0787
0.7389	0.2160	0.0451	0.6112	0.3774	0.0114	0.9105	0.0001	0.0894
T/K = 313.15								
0.8883	0.0715	0.0402	0.9508	0.0001	0.0491	0.7624	0.2162	0.0214
0.8751	0.0835	0.0414	0.7281	0.2526	0.0193	0.9455	0.0001	0.0544
0.8622	0.0951	0.0427	0.7009	0.2813	0.0178	0.9420	0.0003	0.0577
0.8491	0.1069	0.0440	0.6759	0.3086	0.0155	0.9383	0.0002	0.0615

Polymer (B): **poly(ethylene oxide-*b*-propylene oxide-*b*-ethylene oxide)** **2010MAH**
Characterization: M_w/g.mol^{-1} = 2036, (EO)2.5(PO)31(EO)2.5, Aldrich Chem. Co., Inc., Milwaukee, WI
Solvent (A): **water** **H_2O** **7732-18-5**
Salt (C): **sodium chloride** **NaCl** **7647-14-5**

Type of data: cloud points (LCST-behavior)

w_B = 0.01 in the binary aqueous solution was kept constant

c_C/(mol/l)	0.00	0.10	0.20	0.30	0.40
T/K	300.45	299.35	296.55	294.15	291.15

Polymer (B): **poly(ethylene oxide-*b*-propylene oxide-*b*-ethylene oxide)** **2010MAH**

Characterization: M_w/g.mol^{-1} = 2900, 40.0 wt% ethylene oxide, L64, (EO)13(PO)30(EO)13, Aldrich Chem. Co., Inc., Milwaukee, WI

Solvent (A): **water** **H_2O** **7732-18-5**

Salt (C): **sodium chloride** **NaCl** **7647-14-5**

Type of data: cloud points (LCST-behavior)

w_B = 0.01 in the binary aqueous solution was kept constant

c_C/(mol/l)	0.00	0.10	0.20	0.30	0.40	0.50
T/K	330.25	327.75	324.75	321.75	318.85	314.55

Polymer (B): **poly(ethylene oxide-*b*-propylene oxide-*b*-ethylene oxide)** **2009MAR**

Characterization: M_w/g.mol^{-1} = 1900, 50.0 wt% ethylene oxide, (EO)11-(PO)16-(EO)11, L35, Sigma Chemical Co., Inc., St. Louis, MO

Solvent (A): **water** **H_2O** **7732-18-5**

Salt (C): **sodium citrate** **$Na_3C_6H_5O_7$** **68-04-2**

Type of data: coexistence data (tie lines)

Total system			Top phase			Bottom phase		
w_A	w_B	w_C	w_A	w_B	w_C	w_A	w_B	w_C
T/K = 283.15								
0.7221	0.1868	0.0910	0.6205	0.3474	0.0321	0.7829	0.0419	0.1752
0.6834	0.2197	0.0968	0.6011	0.3760	0.0229	0.7580	0.0377	0.2043
0.6453	0.2467	0.1079	0.5568	0.4275	0.0157	0.7159	0.0582	0.2259
0.6052	0.2740	0.1206	0.5011	0.4869	0.0120	0.6775	0.0628	0.2597
0.5661	0.3043	0.1294	0.4554	0.5360	0.0086	0.6518	0.0699	0.2783
T/K = 298.15								
0.6767	0.2732	0.0511	0.6214	0.3494	0.0292	0.8229	0.0243	0.1528
0.6332	0.3106	0.0562	0.5621	0.4168	0.0211	0.8043	0.0233	0.1724
0.5907	0.3482	0.0611	0.5152	0.4693	0.0155	0.7755	0.0243	0.2002
0.5481	0.3857	0.0662	0.4639	0.5251	0.0110	0.7472	0.0225	0.2303
0.5057	0.4231	0.0712	0.4169	0.5756	0.0075	0.7152	0.0332	0.2516
T/K = 313.15								
0.7223	0.1867	0.0910	0.5294	0.4539	0.0167	0.8207	0.0268	0.1525
0.6832	0.2201	0.0967	0.4787	0.5088	0.0125	0.8022	0.0281	0.1697
0.6454	0.2468	0.1078	0.4329	0.5578	0.0093	0.7780	0.0317	0.1903
0.6056	0.2738	0.1206	0.3953	0.5980	0.0067	0.7394	0.0443	0.2163
0.5665	0.3041	0.1294	0.3627	0.6324	0.0049	0.7062	0.0518	0.2420

Polymer (B): **poly(ethylene oxide-*b*-propylene oxide-*b*-ethylene oxide)** **2011AND**

Characterization: M_w/g.mol^{-1} = 2930, M_w/M_n = 1.09, 40.0 wt% ethylene oxide, (EO)13(PO)30(EO)13, L64, Aldrich Chem. Co., Inc., Milwaukee, WI

Solvent (A): **water** H_2O **7732-18-5**

Salt (C): **sodium citrate** $Na_3C_6H_5O_7$ **68-04-2**

Type of data: coexistence data (tie lines)

Total system			Top phase			Bottom phase		
w_A	w_B	w_C	w_A	w_B	w_C	w_A	w_B	w_C
T/K = 278.15								
0.7370	0.1913	0.0717	0.6822	0.2795	0.0383	0.8471	0.0097	0.1432
0.7066	0.2209	0.0725	0.6372	0.3342	0.0286	0.8302	0.0152	0.1545
0.6908	0.2237	0.0855	0.5982	0.3783	0.0235	0.8171	0.0061	0.1768
0.6528	0.2562	0.0910	0.5393	0.4419	0.0187	0.8021	0.0096	0.1883
0.6250	0.2821	0.0929	0.5113	0.4725	0.0162	0.7904	0.0109	0.1987
T/K = 288.15								
0.7376	0.1919	0.0705	0.6247	0.3398	0.0355	0.8740	0.0149	0.1110
0.7054	0.2219	0.0727	0.5879	0.3809	0.0312	0.8575	0.0131	0.1294
0.6908	0.2249	0.0843	0.5346	0.4421	0.0234	0.8500	0.0021	0.1479
0.6527	0.2574	0.0899	0.4895	0.4925	0.0181	0.8341	0.0014	0.1645
0.6246	0.2833	0.0921	0.4721	0.5114	0.0165	0.8206	0.0003	0.1792
T/K = 298.15								
0.7971	0.1153	0.0876	0.5938	0.3715	0.0347	0.8806	0.0128	0.1066
0.7596	0.1506	0.0898	0.5417	0.4299	0.0284	0.8691	0.0137	0.1172
0.6845	0.2416	0.0739	0.5187	0.4476	0.0337	0.8613	0.0033	0.1354
0.6484	0.2756	0.0760	0.4604	0.5143	0.0253	0.8443	0.0008	0.1549
0.6121	0.3089	0.0790	0.4175	0.5631	0.0193	0.8313	0.0035	0.1652

Polymer (B): **poly(ethylene oxide-*b*-propylene oxide-*b*-ethylene oxide)** **2010RO1**

Characterization: M_w/g.mol^{-1} = 8400, (EO)80-(PO)30-(EO)80, F68, Aldrich Chem. Co., Inc., Milwaukee, WI

Solvent (A): **water** H_2O **7732-18-5**

Salt (C): **sodium citrate** $Na_3C_6H_5O_7$ **68-04-2**

Type of data: coexistence data (tie lines)

continued

continued

Total system			Top phase			Bottom phase		
w_A	w_B	w_C	w_A	w_B	w_C	w_A	w_B	w_C
T/K = 283.15								
0.7843	0.1363	0.0794	0.7261	0.2284	0.0455	0.8581	0.0113	0.1306
0.7631	0.1535	0.0834	0.6954	0.2660	0.0386	0.8476	0.0001	0.1523
0.7469	0.1666	0.0866	0.6751	0.2908	0.0341	0.8382	0.0001	0.1617
0.7252	0.1830	0.0918	0.6474	0.3237	0.0289	0.8194	0.0001	0.1805
0.7028	0.1997	0.0975	0.6231	0.3514	0.0255	0.8038	0.0001	0.1961
T/K = 298.15								
0.7869	0.1455	0.0677	0.7328	0.2196	0.0476	0.8788	0.0100	0.1112
0.7674	0.1629	0.0697	0.6955	0.2645	0.0400	0.8722	0.0079	0.1199
0.7476	0.1796	0.0727	0.6609	0.3057	0.0334	0.8640	0.0019	0.1341
0.7275	0.1967	0.0758	0.6383	0.3319	0.0298	0.8538	0.0016	0.1446
0.7077	0.2137	0.0786	0.6103	0.3640	0.0257	0.8378	0.0029	0.1593
T/K = 313.15								
0.8431	0.0869	0.0700	0.7774	0.1691	0.0535	0.8937	0.0188	0.0875
0.8341	0.0932	0.0727	0.7324	0.2255	0.0421	0.9016	0.0008	0.0976
0.8261	0.0993	0.0746	0.7040	0.2580	0.0380	0.8977	0.0001	0.1022
0.8061	0.1147	0.0792	0.6695	0.2989	0.0316	0.8874	0.0001	0.1125
0.7840	0.1367	0.0793	0.6476	0.3242	0.0282	0.8788	0.0001	0.1211

Polymer (B): **poly(ethylene oxide-*b*-propylene oxide-*b*-ethylene oxide)** **2010VIR**

Characterization: M_w/g.mol^{-1} = 1900, (EO)11-(PO)16-(EO)11, L35, Aldrich Chem. Co., Inc., Milwaukee, WI

Solvent (A): **water** **H_2O** **7732-18-5**

Salt (C): **sodium hydroxide** **NaOH** **1310-73-2**

Type of data: coexistence data (tie lines)

Total system			Top phase			Bottom phase		
w_A	w_B	w_C	w_A	w_B	w_C	w_A	w_B	w_C
T/K = 283.15								
0.4376	0.5059	0.0565	0.3089	0.6830	0.0081	0.6613	0.2031	0.1356
0.4053	0.5252	0.0695	0.2797	0.7134	0.0069	0.6123	0.2169	0.1708
0.3870	0.5345	0.0785	0.2441	0.7507	0.0052	0.5983	0.2067	0.1950
0.3528	0.5537	0.0935	0.2190	0.7769	0.0041	0.5503	0.2230	0.2267

continued

continued

T/K = 293.15								
0.5903	0.3506	0.0591	0.4558	0.5194	0.0248	0.7274	0.1784	0.0942
0.5368	0.3898	0.0734	0.3380	0.6463	0.0157	0.6968	0.1822	0.1210
0.4725	0.4391	0.0884	0.2449	0.7470	0.0081	0.6571	0.1803	0.1626
0.4073	0.4883	0.1044	0.1520	0.8440	0.0040	0.6286	0.1830	0.1884
T/K = 303.15								
0.6541	0.2930	0.0529	0.4854	0.4933	0.0213	0.7476	0.1775	0.0749
0.5546	0.3825	0.0629	0.3033	0.6873	0.0094	0.7147	0.1836	0.1017
0.4340	0.4913	0.0747	0.2111	0.7843	0.0046	0.6848	0.1658	0.1494
0.3354	0.5827	0.0819	0.1602	0.8366	0.0032	0.6334	0.1704	0.1962
T/K = 313.15								
0.5098	0.4473	0.0429	0.4380	0.5463	0.0157	0.6768	0.2022	0.1210
0.4375	0.5142	0.0483	0.2785	0.7172	0.0043	0.6714	0.1948	0.1338
0.3228	0.6212	0.0560	0.1754	0.8216	0.0030	0.6440	0.2070	0.1490
0.2490	0.6885	0.0625	0.1255	0.8739	0.0006	0.6096	0.1993	0.1911

Polymer (B): **poly(ethylene oxide-*b*-propylene oxide-*b*-ethylene oxide)** **2009MAR**

Characterization: M_w/g.mol^{-1} = 1900, 50.0 wt% ethylene oxide, (EO)11-(PO)16-(EO)11, L35, Sigma Chemical Co., Inc., St. Louis, MO

Solvent (A): **water** **H_2O** **7732-18-5**

Salt (C): **sodium nitrate** **$NaNO_3$** **7631-99-4**

Type of data: coexistence data (tie lines)

Total system			Top phase			Bottom phase		
w_A	w_B	w_C	w_A	w_B	w_C	w_A	w_B	w_C
T/K = 283.15								
0.5376	0.2111	0.2513	0.4565	0.3856	0.1579	0.6451	0.0208	0.3341
0.5135	0.2304	0.2561	0.4166	0.4314	0.1520	0.6366	0.0079	0.3555
0.4873	0.2505	0.2622	0.3739	0.4822	0.1439	0.6093	0.0108	0.3799
0.4638	0.2684	0.2678	0.3483	0.5117	0.1400	0.6000	0.0082	0.3918
0.4399	0.2867	0.2734	0.3284	0.5336	0.1380	0.5764	0.0001	0.4235
T/K = 298.15								
0.5624	0.1916	0.2460	0.4109	0.4488	0.1403	0.6758	0.0171	0.3091
0.5328	0.2128	0.2544	0.3810	0.4871	0.1319	0.6543	0.0171	0.3286
0.5121	0.2312	0.2567	0.3426	0.5311	0.1263	0.6420	0.0130	0.3450
0.4875	0.2514	0.2611	0.3063	0.5735	0.1202	0.6268	0.0101	0.3631
0.4631	0.2714	0.2655	0.2823	0.6006	0.1171	0.5901	0.0320	0.3779

continued

continued

T/K = 313.15

0.5598	0.1919	0.2483	0.4960	0.3394	0.1728	0.6733	0.0100	0.3167
0.5363	0.2125	0.2512	0.4423	0.3913	0.1664	0.6506	0.0222	0.3272
0.5174	0.2297	0.2529	0.4166	0.4306	0.1528	0.6476	0.0107	0.3417
0.4869	0.2510	0.2621	0.3739	0.4826	0.1435	0.6196	0.0004	0.3800
0.4621	0.2710	0.2669	0.3286	0.5346	0.1368	0.5885	0.0174	0.3941

Polymer (B): **poly(ethylene oxide-*b*-propylene oxide-*b*-ethylene oxide)** **2010MAH**

Characterization: M_w/g.mol^{-1} = 2036, (EO)2.5(PO)31(EO)2.5, Aldrich Chem. Co., Inc., Milwaukee, WI

Solvent (A): **water** **H_2O** **7732-18-5**

Salt (C): **sodium phosphate** **Na_3PO_4** **7601-54-9**

Type of data: cloud points (LCST-behavior)

w_B = 0.01 in the binary aqueous solution was kept constant

c_C/(mol/l)	0.00	0.10	0.20	0.30	0.40
T/K	300.45	297.65	294.75	292.65	289.55

Polymer (B): **poly(ethylene oxide-*b*-propylene oxide-*b*-ethylene oxide)** **2010MAH**

Characterization: M_w/g.mol^{-1} = 2900, 40.0 wt% ethylene oxide, L64, (EO)13(PO)30(EO)13, Aldrich Chem. Co., Inc., Milwaukee, WI

Solvent (A): **water** **H_2O** **7732-18-5**

Salt (C): **sodium phosphate** **Na_3PO_4** **7601-54-9**

Type of data: cloud points (LCST-behavior)

w_B = 0.01 in the binary aqueous solution was kept constant

c_C/(mol/l)	0.00	0.10	0.20	0.30	0.40	0.50
T/K	330.25	321.85	315.85	313.95	311.15	305.35

Polymer (B): **poly(ethylene oxide-*b*-propylene oxide-*b*-ethylene oxide)** **2006SIL**

Characterization: M_n/g.mol^{-1} = 1706, M_w/g.mol^{-1} = 1945, 50.0 wt% ethylene oxide, (EO)11-(PO)16-(EO)11, L35, Aldrich Chem. Co., Inc., Milwaukee, WI

Solvent (A): **water** **H_2O** **7732-18-5**

Salt (C): **sodium sulfate** **Na_2SO_4** **7757-82-6**

Type of data: coexistence data (tie lines)

continued

continued

Total system			Top phase			Bottom phase		
w_A	w_B	w_C	w_A	w_B	w_C	w_A	w_B	w_C
T/K = 298.15								
0.7458	0.1336	0.1206	0.5227	0.4691	0.0082	0.8166	0.0259	0.1575
0.7195	0.1506	0.1299	0.4849	0.5085	0.0066	0.8063	0.0191	0.1746
0.6766	0.1769	0.1465	0.4454	0.5508	0.0038	0.7798	0.0130	0.2072
0.6614	0.1855	0.1531	0.4198	0.5780	0.0022	0.7711	0.0100	0.2189
0.6210	0.2109	0.1681	0.4026	0.5959	0.0015	0.7370	0.0077	0.2553
T/K = 313.15								
0.7749	0.1622	0.0629	0.5228	0.4656	0.0116	0.8910	0.0243	0.0847
0.7525	0.1818	0.0657	0.5136	0.4755	0.0109	0.8894	0.0192	0.0914
0.7428	0.1902	0.0670	0.4701	0.5216	0.0083	0.8855	0.0180	0.0965
0.7084	0.2203	0.0713	0.4044	0.5889	0.0067	0.8774	0.0166	0.1060
0.6609	0.2618	0.0773	0.3933	0.6006	0.0061	0.8693	0.0009	0.1298

Polymer (B): **poly(ethylene oxide-*b*-propylene oxide-*b*-ethylene oxide)** **2010MAH**
Characterization: M_w/g.mol^{-1} = 2036, (EO)2.5(PO)31(EO)2.5, Aldrich Chem. Co., Inc., Milwaukee, WI
Solvent (A): **water** **H_2O** **7732-18-5**
Salt (C): **sodium sulfate** **Na_2SO_4** **7757-82-6**

Type of data: cloud points (LCST-behavior)

w_B = 0.01 in the binary aqueous solution was kept constant

c_C/(mol/l)	0.00	0.10	0.20	0.30	0.40
T/K	300.45	299.35	296.85	293.55	290.75

Polymer (B): **poly(ethylene oxide-*b*-propylene oxide-*b*-ethylene oxide)** **2010MAH**
Characterization: M_w/g.mol^{-1} = 2900, 40.0 wt% ethylene oxide, L64, (EO)13(PO)30(EO)13, Aldrich Chem. Co., Inc., Milwaukee, WI
Solvent (A): **water** **H_2O** **7732-18-5**
Salt (C): **sodium sulfate** **Na_2SO_4** **7757-82-6**

Type of data: cloud points (LCST-behavior)

w_B = 0.01 in the binary aqueous solution was kept constant

c_C/(mol/l)	0.00	0.10	0.20	0.30	0.40	0.50
T/K	330.25	323.25	319.15	315.95	312.75	309.35

Polymer (B): **poly(ethylene oxide-*b*-propylene oxide-*b*-ethylene oxide)** **2009ROD**
Characterization: M_w/g.mol^{-1} = 2900, 40.0 wt% ethylene oxide, L64, (EO)13(PO)30(EO)13, Aldrich Chem. Co., Inc., Milwaukee, WI
Solvent (A): **water** **H_2O** **7732-18-5**
Salt (C): **sodium sulfate** **Na_2SO_4** **7757-82-6**

Type of data: coexistence data (tie lines)

Total system			Top phase			Bottom phase		
w_A	w_B	w_C	w_A	w_B	w_C	w_A	w_B	w_C
T/K = 278.15								
0.7489	0.1993	0.0518	0.6820	0.2992	0.0188	0.8759	0.0104	0.1137
0.7134	0.2319	0.0547	0.6340	0.3511	0.0150	0.8617	0.0075	0.1309
0.6776	0.2646	0.0578	0.5811	0.4068	0.0122	0.8509	0.0107	0.1383
0.6425	0.2972	0.0603	0.5311	0.4591	0.0098	0.8409	0.0045	0.1546
0.6052	0.3315	0.0634	0.4919	0.4992	0.0089	0.8229	0.0084	0.1687
0.5744	0.3594	0.0662	0.4715	0.5207	0.0078	0.8057	0.0068	0.1875
T/K = 288.15								
0.7488	0.1995	0.0517	0.6719	0.2950	0.0331	0.9101	0.0002	0.0897
0.7125	0.2329	0.0545	0.5946	0.3832	0.0222	0.8938	0.0001	0.1062
0.6782	0.2642	0.0575	0.5422	0.4407	0.0170	0.8823	0.0001	0.1176
0.6416	0.2978	0.0606	0.4968	0.4899	0.0133	0.8662	0.0002	0.1336
0.6068	0.3299	0.0633	0.4668	0.5226	0.0107	0.8499	0.0001	0.1501
T/K = 298.15								
0.7484	0.1997	0.0519	0.6259	0.3431	0.0310	0.9020	0.0121	0.0858
0.7247	0.2214	0.0539	0.5605	0.4173	0.0223	0.8985	0.0055	0.0960
0.7001	0.2429	0.0570	0.5075	0.4749	0.0176	0.8908	0.0047	0.1046
0.6742	0.2659	0.0598	0.4884	0.4983	0.0133	0.8808	0.0055	0.1137
0.6522	0.2848	0.0630	0.4469	0.5412	0.0119	0.8713	0.0037	0.1250

Polymer (B): **poly(ethylene oxide-*b*-propylene oxide-*b*-ethylene oxide)** **2010MA2**
Characterization: M_w/g.mol^{-1} = 8460, M_w/M_n = 1.42, 80 wt% ethylene oxide, (EO)80-(PO)30-(EO)80, F68, Sigma Chemical Co., Inc., St. Louis, MO
Solvent (A): **water** **H_2O** **7732-18-5**
Salt (C): **sodium sulfate** **Na_2SO_4** **7757-82-6**

Type of data: coexistence data (tie lines)

continued

continued

Total system			Top phase			Bottom phase		
w_A	w_B	w_C	w_A	w_B	w_C	w_A	w_B	w_C
T/K = 278.15								
0.8203	0.1091	0.0706	0.7580	0.2043	0.0377	0.8826	0.0140	0.1034
0.7938	0.1294	0.0768	0.7155	0.2588	0.0257	0.8720	0.0001	0.1279
0.7623	0.1583	0.0794	0.6633	0.3166	0.0201	0.8613	0.0001	0.1386
0.7463	0.1741	0.0796	0.6415	0.3424	0.0161	0.8511	0.0058	0.1431
0.7149	0.1882	0.0969	0.6098	0.3762	0.0140	0.8200	0.0001	0.1799
T/K = 288.15								
0.8086	0.1252	0.0662	0.7309	0.2381	0.0310	0.8863	0.0123	0.1014
0.7793	0.1519	0.0688	0.6853	0.2904	0.0243	0.8733	0.0134	0.1133
0.7491	0.1758	0.0751	0.6409	0.3394	0.0197	0.8572	0.0123	0.1305
0.7262	0.1893	0.0845	0.6118	0.3712	0.0170	0.8407	0.0073	0.1520
0.6994	0.2113	0.0893	0.5761	0.4093	0.0145	0.8225	0.0133	0.1642
T/K = 298.15								
0.8180	0.1239	0.0581	0.7320	0.2428	0.0252	0.9038	0.0051	0.0911
0.8076	0.1290	0.0634	0.7267	0.2565	0.0268	0.8985	0.0015	0.1000
0.7768	0.1535	0.0697	0.6677	0.3069	0.0254	0.8858	0.0001	0.1141
0.7445	0.1785	0.0770	0.6279	0.3556	0.0165	0.8610	0.0015	0.1375
0.7261	0.1870	0.0869	0.6086	0.3740	0.0174	0.8436	0.0001	0.1563

Polymer (B): **poly(ethylene oxide-*b*-propylene oxide-*b*-ethylene oxide)** **2009ROD**

Characterization: M_w/g.mol^{-1} = 2900, 40.0 wt% ethylene oxide, L64, (EO)13(PO)30(EO)13, Aldrich Chem. Co., Inc., Milwaukee, WI

Solvent (A): **water** **H_2O** **7732-18-5**

Salt (C): **zinc sulfate** **$ZnSO_4$** **7733-02-0**

Type of data: coexistence data (tie lines)

Total system			Top phase			Bottom phase		
w_A	w_B	w_C	w_A	w_B	w_C	w_A	w_B	w_C
T/K = 278.15								
0.7297	0.1975	0.0728	0.6530	0.3333	0.0137	0.8335	0.0027	0.1638
0.6990	0.2233	0.0777	0.6242	0.3646	0.0112	0.8094	0.0050	0.1855
0.6677	0.2494	0.0829	0.5886	0.4027	0.0087	0.7901	0.0090	0.2009
0.6371	0.2749	0.0880	0.5451	0.4482	0.0067	0.7770	0.0117	0.2113
0.6165	0.2910	0.0925	0.5141	0.4807	0.0051	0.7671	0.0134	0.2195

continued

continued

T/K = 288.15								
0.7602	0.1716	0.0682	0.6433	0.3343	0.0224	0.8731	0.0064	0.1205
0.7292	0.1978	0.0730	0.6033	0.3804	0.0163	0.8634	0.0003	0.1363
0.6982	0.2236	0.0782	0.5542	0.4321	0.0137	0.8497	0.0003	0.1500
0.6673	0.2496	0.0831	0.5298	0.4591	0.0110	0.8343	0.0003	0.1654
0.6382	0.2742	0.0876	0.4969	0.4947	0.0084	0.8213	0.0010	0.1777
T/K = 298.15								
0.7605	0.1712	0.0683	0.6264	0.3414	0.0321	0.8849	0.0029	0.1122
0.7294	0.1975	0.0731	0.5605	0.4195	0.0200	0.8731	0.0036	0.1234
0.6987	0.2228	0.0785	0.5146	0.4706	0.0147	0.8585	0.0079	0.1336
0.6677	0.2494	0.0829	0.4819	0.5054	0.0127	0.8452	0.0031	0.1517
0.6552	0.2581	0.0867	0.4690	0.5198	0.0112	0.8348	0.0078	0.1574

Polymer (B): **poly(ethylene oxide-*b*-propylene oxide-*b*-ethylene oxide)** **2010MA2**

Characterization: M_w/g.mol^{-1} = 8460, M_w/M_n = 1.42, 80 wt% ethylene oxide, F68 Sigma Chemical Co., Inc., St. Louis, MO

Solvent (A): **water** H_2O **7732-18-5**

Salt (C): **zinc sulfate** $ZnSO_4$ **7733-02-0**

Type of data: coexistence data (tie lines)

Total system			Top phase			Bottom phase		
w_A	w_B	w_C	w_A	w_B	w_C	w_A	w_B	w_C
T/K = 278.15								
0.8040	0.1109	0.0851	0.7413	0.2177	0.0410	0.8667	0.0040	0.1293
0.7818	0.1296	0.0886	0.7161	0.2546	0.0293	0.8475	0.0046	0.1479
0.7675	0.1374	0.0951	0.6995	0.2735	0.0270	0.8355	0.0013	0.1632
0.7445	0.1616	0.0939	0.6692	0.3102	0.0206	0.8198	0.0130	0.1672
0.7256	0.1662	0.1082	0.6517	0.3323	0.0160	0.7995	0.0001	0.2004
T/K = 288.15								
0.8334	0.0852	0.0814	0.7873	0.1618	0.0509	0.8795	0.0085	0.1120
0.7963	0.1163	0.0874	0.7343	0.2321	0.0336	0.8632	0.0004	0.1413
0.7749	0.1391	0.0860	0.7033	0.2704	0.0264	0.8466	0.0077	0.1456
0.7533	0.1524	0.0943	0.6738	0.3043	0.0218	0.8339	0.0004	0.1668
0.7262	0.1701	0.1037	0.6435	0.3398	0.0167	0.8178	0.0004	0.1905
T/K = 298.15								
0.8357	0.0931	0.0712	0.7901	0.1703	0.0396	0.8814	0.0158	0.1028
0.8151	0.1061	0.0788	0.7513	0.2115	0.0372	0.8789	0.0007	0.1204
0.7853	0.1343	0.0804	0.7027	0.2684	0.0289	0.8680	0.0001	0.1319
0.7605	0.1546	0.0849	0.6681	0.3092	0.0227	0.8529	0.0001	0.1470
0.7396	0.1706	0.0898	0.6401	0.3410	0.0189	0.8392	0.0001	0.1607

Polymer (B): **poly(propylene glycol)** **2010WUC**
Characterization: M_n/g.mol^{-1} = 400, Alfa Aesar China Co., Tianjin, China
Solvent (A): **water** H_2O **7732-18-5**
Salt (C): **1-allyl-3-methylimidazolium chloride** $C_7H_{11}ClN_2$ **65039-10-3**

Type of data: cloud points

T/K = 298.15

w_A	0.4907	0.5256	0.5590	0.5840	0.5928	0.6023	0.6105	0.6174	0.6217
w_B	0.0934	0.1040	0.1196	0.1435	0.1559	0.1611	0.1745	0.1975	0.2167
w_C	0.4159	0.3704	0.3214	0.2725	0.2513	0.2366	0.2150	0.1851	0.1616
w_A	0.6211	0.6199	0.6146	0.6086	0.6011	0.5925	0.5841		
w_B	0.2340	0.2552	0.2833	0.3027	0.3221	0.3416	0.3588		
w_C	0.1449	0.1249	0.1021	0.0887	0.0768	0.0659	0.0571		

Type of data: coexistence data (tie lines)

T/K = 298.15

Total system			Top phase			Bottom phase		
w_A	w_B	w_C	w_A	w_B	w_C	w_A	w_B	w_C
0.5562	0.2606	0.1832	0.2681	0.6821	0.0498	0.6143	0.1734	0.2123
0.5473	0.2711	0.1816	0.2535	0.6991	0.0474	0.6185	0.1619	0.2196
0.5210	0.2784	0.2006	0.1936	0.7665	0.0399	0.6162	0.1354	0.2484
0.4912	0.2803	0.2285	0.1017	0.8701	0.0282	0.6131	0.0973	0.2896
0.5035	0.2829	0.2136	0.1450	0.8229	0.0321	0.6190	0.1089	0.2721

Polymer (B): **poly(propylene glycol)** **2010WUC**
Characterization: M_n/g.mol^{-1} = 1000, Alfa Aesar China Co., Tianjin, China
Solvent (A): **water** H_2O **7732-18-5**
Salt (C): **1-allyl-3-methylimidazolium chloride** $C_7H_{11}ClN_2$ **65039-10-3**

Type of data: cloud points

T/K = 298.15

w_A	0.9143	0.8973	0.8892	0.8810	0.8724	0.8643	0.8525	0.8440	0.8358
w_B	0.0465	0.0408	0.0360	0.0322	0.0281	0.0247	0.0214	0.0192	0.0174
w_C	0.0392	0.0619	0.0748	0.0868	0.0995	0.1110	0.1261	0.1368	0.1468
w_A	0.8327	0.8232	0.8169	0.8120	0.8076	0.8021	0.7969	0.7913	0.7850
w_B	0.0159	0.0141	0.0130	0.0121	0.0113	0.0104	0.0097	0.0088	0.0082
w_C	0.1514	0.1627	0.1701	0.1759	0.1811	0.1875	0.1934	0.1999	0.2068

continued

continued

w_A	0.7814	0.7528
w_B	0.0076	0.0063
w_C	0.2110	0.2409

Polymer (B): **poly(propylene glycol)** **2011ZH1**
Characterization: M_n/g.mol^{-1} = 400, Aladdin Reagent Co., Ltd., Shanghai, China
Solvent (A): **water** H_2O **7732-18-5**
Salt (C): **ammonium sulfate** $(NH_4)_2SO_4$ **7783-20-2**

Type of data: cloud points

T/K = 298.15

w_A	0.3763	0.4171	0.4349	0.4613	0.4946	0.5191	0.5899	0.6088	0.6243
w_B	0.6192	0.5753	0.5559	0.5265	0.4883	0.4598	0.3724	0.3483	0.3280
w_C	0.0045	0.0076	0.0092	0.0122	0.0171	0.0211	0.0377	0.0429	0.0477
w_A	0.6380	0.6505	0.6619	0.6828	0.7367	0.7824	0.8007	0.8107	0.8118
w_B	0.3104	0.2944	0.2795	0.2507	0.1765	0.1041	0.0727	0.0514	0.0348
w_C	0.0516	0.0551	0.0586	0.0665	0.0868	0.1135	0.1266	0.1379	0.1534
w_A	0.8112	0.8080	0.7892						
w_B	0.0280	0.0186	0.0112						
w_C	0.1608	0.1734	0.1996						

Type of data: coexistence data (tie lines)

T/K = 298.15

Total system			Top phase			Bottom phase		
w_A	w_B	w_C	w_A	w_B	w_C	w_A	w_B	w_C
0.7190	0.1807	0.1003	0.5330	0.4424	0.0246	0.8104	0.0518	0.1378
0.6902	0.2001	0.1097	0.4990	0.4835	0.0175	0.8117	0.0216	0.1667
0.6717	0.2092	0.1191	0.4786	0.5068	0.0146	0.8005	0.0125	0.1870
0.6446	0.2253	0.1301	0.4513	0.5369	0.0118	0.7804	0.0048	0.2148

Polymer (B): **poly(propylene glycol)** **2004SA2**
Characterization: M_n/g.mol^{-1} = 405, Aldrich Chem. Co., Inc., Milwaukee, WI
Solvent (A): **water** H_2O **7732-18-5**
Salt (C): **ammonium sulfate** $(NH_4)_2SO_4$ **7783-20-2**

Type of data: coexistence data (tie lines)

T/K = 298.15

continued

continued

Total system			Top phase			Bottom phase		
w_A	w_B	w_C	w_A	w_B	w_C	w_A	w_B	w_C
0.594	0.372	0.034	0.390	0.605	0.005	0.747	0.193	0.060
0.617	0.340	0.044	0.359	0.637	0.004	0.772	0.158	0.070
0.579	0.374	0.047	0.325	0.671	0.004	0.802	0.113	0.085
0.586	0.350	0.064	0.317	0.681	0.002	0.818	0.064	0.118
0.551	0.372	0.077	0.291	0.707	0.002	0.823	0.027	0.150
0.550	0.360	0.090	0.271	0.729	0.002	0.814	0.019	0.167
0.515	0.376	0.109	0.229	0.770	0.001	0.792	0.000	0.208
0.470	0.390	0.140	0.227	0.772	0.001	0.726	0.000	0.274

Polymer (B): **poly(propylene glycol)** **2010SA1**
Characterization: M_n/g.mol^{-1} = 425, Aldrich Chem. Co., Inc., Milwaukee, WI
Solvent (A): **water** H_2O **7732-18-5**
Salt (C): **ammonium sulfate** $(NH_4)_2SO_4$ **7783-20-2**

Type of data: coexistence data (tie lines)

T/K = 298.15 pH = 7.6

Top phase			Bottom phase		
w_A	w_B	w_C	w_A	w_B	w_C
0.359	0.637	0.0040	0.772	0.158	0.070
0.317	0.681	0.0020	0.818	0.064	0.118
0.269	0.729	0.0020	0.814	0.019	0.167
0.229	0.770	0.0010	0.792	0.000	0.208

Polymer (B): **poly(propylene glycol)** **2010WUC**
Characterization: M_n/g.mol^{-1} = 400, Alfa Aesar China Co., Tianjin, China
Solvent (A): **water** H_2O **7732-18-5**
Salt (C): **1-butyl-3-methylimidazolium acetate**
$C_{10}H_{18}N_2O_2$ **284049-75-8**

Type of data: cloud points

T/K = 298.15

w_A	0.4276	0.4597	0.4991	0.5255	0.5493	0.5674	0.5766	0.6087	0.5923
w_B	0.1323	0.1405	0.1542	0.1685	0.1854	0.2037	0.2297	0.2453	0.2742
w_C	0.4401	0.3998	0.3467	0.3060	0.2653	0.2289	0.1937	0.1460	0.1335

continued

continued

w_A	0.5983	0.5799	0.5658	0.5468	0.5052	0.4773
w_B	0.2960	0.3262	0.3600	0.3968	0.4422	0.4910
w_C	0.1057	0.0939	0.0742	0.0564	0.0526	0.0317

Type of data: coexistence data (tie lines)

T/K = 298.15

Total system			Top phase			Bottom phase		
w_A	w_B	w_C	w_A	w_B	w_C	w_A	w_B	w_C
0.5356	0.2304	0.2340	0.3353	0.5593	0.1054	0.5942	0.1350	0.2708
0.4813	0.2981	0.2206	0.2898	0.6191	0.0911	0.5884	0.1143	0.2973
0.4473	0.3170	0.2357	0.2389	0.6755	0.0856	0.5828	0.0869	0.3303
0.4191	0.3347	0.2462	0.2099	0.7065	0.0836	0.5620	0.0811	0.3569
0.3949	0.3562	0.2489	0.1961	0.7212	0.0827	0.5508	0.0706	0.3786

Polymer (B): **poly(propylene glycol)** **2010WUC**
Characterization: M_n/g.mol^{-1} = 400, Alfa Aesar China Co., Tianjin, China
Solvent (A): **water** **H_2O** **7732-18-5**
Salt (C): **1-butyl-3-methylimidazolium chloride**
$C_8H_{15}ClN_2$ **79917-90-1**

Type of data: cloud points

T/K = 298.15

w_A	0.2622	0.2844	0.2985	0.3173	0.3383	0.3558	0.3681	0.3855	0.3930
w_B	0.2110	0.2229	0.2329	0.2398	0.2521	0.2679	0.2796	0.2944	0.3145
w_C	0.5268	0.4927	0.4686	0.4429	0.4096	0.3763	0.3523	0.3201	0.2925
w_A	0.3977	0.4038	0.4068	0.4046	0.3974	0.3888	0.3814	0.3562	
w_B	0.3359	0.3575	0.3781	0.4000	0.4228	0.4450	0.4677	0.5112	
w_C	0.2664	0.2387	0.2151	0.1954	0.1798	0.1662	0.1509	0.1326	

Type of data: coexistence data (tie lines)

T/K = 298.15

Total system			Top phase			Bottom phase		
w_A	w_B	w_C	w_A	w_B	w_C	w_A	w_B	w_C
0.2533	0.4422	0.3045	0.1864	0.6231	0.1905	0.3693	0.1401	0.4906
0.2438	0.3785	0.3777	0.1403	0.6721	0.1876	0.3326	0.1208	0.5466
0.3087	0.3574	0.3339	0.2113	0.5963	0.1924	0.3940	0.1546	0.4514
0.2867	0.3632	0.3501	0.1927	0.6162	0.1911	0.3756	0.1373	0.4871
0.2622	0.3774	0.3604	0.1658	0.6467	0.1875	0.3517	0.1225	0.5258

Polymer (B): **poly(propylene glycol)** **2010WUC**
Characterization: M_n/g.mol^{-1} = 1000, Alfa Aesar China Co., Tianjin, China
Solvent (A): **water** H_2O **7732-18-5**
Salt (C): **1-butyl-3-methylimidazolium chloride** $C_8H_{15}ClN_2$ **79917-90-1**

Type of data: cloud points

T/K = 298.15

w_A	0.8759	0.8520	0.8459	0.8342	0.8245	0.8122	0.8034	0.7944	0.7867
w_B	0.0414	0.0334	0.0302	0.0271	0.0242	0.0210	0.0190	0.0173	0.0160
w_C	0.0827	0.1146	0.1239	0.1387	0.1513	0.1668	0.1776	0.1883	0.1973
w_A	0.7808	0.7723	0.7644	0.7566	0.7511	0.7454	0.7391	0.7318	
w_B	0.0148	0.0132	0.0120	0.0109	0.0100	0.0094	0.0085	0.0079	
w_C	0.2044	0.2145	0.2236	0.2325	0.2389	0.2452	0.2524	0.2603	

Polymer (B): **poly(propylene glycol)** **2011SAD**
Characterization: M_n/g.mol^{-1} = 450, Fluka AG, Buchs, Switzerland
Solvent (A): **water** H_2O **7732-18-5**
Salt (C): **diammonium phosphate** $(NH_4)_2HPO_4$ **7783-28-0**

Type of data: cloud points

T/K = 303.15

w_A	0.8508	0.8518	0.8532	0.8550	0.8565	0.8562	0.8546	0.8533	0.8510
w_B	0.0216	0.0242	0.0268	0.0298	0.0334	0.0374	0.0421	0.0473	0.0531
w_C	0.1276	0.1240	0.1200	0.1152	0.1101	0.1064	0.1033	0.0994	0.0959
w_A	0.8481	0.8457	0.8424	0.8399	0.8367	0.8341	0.8321	0.8301	0.8279
w_B	0.0592	0.0651	0.0716	0.0770	0.0824	0.0870	0.0903	0.0939	0.0977
w_C	0.0927	0.0892	0.0860	0.0831	0.0809	0.0789	0.0776	0.0760	0.0744
w_A	0.8253	0.8227	0.8197	0.8182	0.8165	0.8148	0.8129	0.8110	0.8091
w_B	0.1019	0.1065	0.1115	0.1141	0.1170	0.1199	0.1229	0.1260	0.1291
w_C	0.0728	0.0708	0.0688	0.0677	0.0665	0.0653	0.0642	0.0630	0.0618
w_A	0.8069	0.8047	0.8021	0.7995	0.7968	0.7939	0.7908	0.7896	0.7876
w_B	0.1327	0.1363	0.1403	0.1444	0.1487	0.1534	0.1582	0.1612	0.1632
w_C	0.0604	0.0590	0.0576	0.0561	0.0545	0.0527	0.0510	0.0492	0.0492
w_A	0.7843	0.7822	0.7774	0.7750	0.7734	0.7658	0.7650	0.7587	0.7491
w_B	0.1685	0.1707	0.1769	0.1803	0.1831	0.1917	0.1931	0.2007	0.2121
w_C	0.0472	0.0471	0.0457	0.0447	0.0435	0.0425	0.0419	0.0406	0.0388
w_A	0.7420	0.7328	0.7225	0.7182	0.7134	0.7063	0.7011	0.6932	0.6850
w_B	0.2207	0.2313	0.2428	0.2479	0.2535	0.2614	0.2675	0.2763	0.2853
w_C	0.0373	0.0359	0.0347	0.0339	0.0331	0.0323	0.0314	0.0305	0.0297
w_A	0.6762	0.6695	0.6595	0.6488	0.6374	0.6251	0.6117	0.5971	0.5811
w_B	0.2949	0.3027	0.3137	0.3254	0.3378	0.3513	0.3660	0.3822	0.4000
w_C	0.0289	0.0278	0.0268	0.0258	0.0248	0.0236	0.0223	0.0207	0.0189

continued

continued

w_A	0.5651	0.5448	0.5242	0.5010	0.4673	0.4277	0.3603		
w_B	0.4177	0.4398	0.4625	0.4879	0.5240	0.5665	0.6370		
w_C	0.0172	0.0154	0.0133	0.0111	0.0087	0.0058	0.0027		

T/K = 308.15

w_A	0.8574	0.8670	0.8683	0.8710	0.8681	0.8617	0.8596	0.8564	0.8482
w_B	0.0214	0.0422	0.0524	0.0576	0.0686	0.0790	0.0848	0.0928	0.1040
w_C	0.1212	0.0908	0.0793	0.0714	0.0633	0.0593	0.0556	0.0508	0.0478
w_A	0.8432	0.8315	0.8200	0.8077	0.7933	0.7841	0.7774	0.7626	0.7585
w_B	0.1137	0.1292	0.1450	0.1610	0.1790	0.1896	0.1973	0.2134	0.2179
w_C	0.0431	0.0393	0.0350	0.0313	0.0277	0.0263	0.0253	0.0240	0.0236
w_A	0.7541	0.7426	0.7394	0.7251	0.7103	0.6997	0.6829	0.6602	0.6416
w_B	0.2235	0.2353	0.2396	0.2541	0.2710	0.2826	0.3003	0.3237	0.3434
w_C	0.0224	0.0221	0.0210	0.0208	0.0187	0.0177	0.0168	0.0161	0.0150
w_A	0.6210	0.5877	0.5541	0.4867					
w_B	0.3653	0.3997	0.4351	0.5069					
w_C	0.0137	0.0126	0.0108	0.0064					

T/K = 313.15

w_A	0.8935	0.8920	0.8941	0.8978	0.8984	0.8989	0.8969	0.8939	0.8901
w_B	0.0118	0.0175	0.0228	0.0277	0.0328	0.0376	0.0442	0.0521	0.0610
w_C	0.0947	0.0905	0.0831	0.0745	0.0688	0.0635	0.0589	0.0540	0.0489
w_A	0.8850	0.8796	0.8737	0.8695	0.8647	0.8588	0.8502	0.8470	0.8454
w_B	0.0691	0.0775	0.0873	0.0935	0.1007	0.1091	0.1206	0.1249	0.1271
w_C	0.0459	0.0429	0.0390	0.0370	0.0346	0.0321	0.0292	0.0281	0.0275
w_A	0.8404	0.8399	0.8349	0.8313	0.8305	0.8250	0.8208	0.8192	0.8161
w_B	0.1334	0.1345	0.1405	0.1450	0.1461	0.1527	0.1575	0.1599	0.1638
w_C	0.0262	0.0256	0.0246	0.0237	0.0234	0.0223	0.0217	0.0209	0.0201
w_A	0.8127	0.8080	0.8058	0.8036	0.7996	0.7944	0.7874	0.7813	0.7721
w_B	0.1678	0.1729	0.1754	0.1784	0.1833	0.1892	0.1971	0.2042	0.2144
w_C	0.0195	0.0191	0.0188	0.0180	0.0171	0.0164	0.0155	0.0145	0.0135
w_A	0.7626	0.7628	0.7526	0.7423	0.7203	0.6940	0.6388	0.5354	0.4077
w_B	0.2249	0.2254	0.2360	0.2474	0.2705	0.2979	0.3540	0.4585	0.5893
w_C	0.0125	0.0118	0.0114	0.0103	0.0092	0.0081	0.0072	0.0061	0.0030

T/K = 318.15

w_A	0.8876	0.8903	0.8928	0.8950	0.8978	0.9001	0.9024	0.9039	0.9054
w_B	0.0169	0.0188	0.0204	0.0221	0.0239	0.0260	0.0282	0.0309	0.0336
w_C	0.0955	0.0909	0.0868	0.0829	0.0783	0.0739	0.0694	0.0652	0.0610
w_A	0.9058	0.9057	0.9055	0.9050	0.9048	0.9036	0.9026	0.9016	0.9002
w_B	0.0363	0.0387	0.0415	0.0441	0.0466	0.0502	0.0531	0.0562	0.0598
w_C	0.0579	0.0556	0.0530	0.0509	0.0486	0.0462	0.0443	0.0422	0.0400

continued

continued

w_A	0.8993	0.8981	0.8970	0.8958	0.8939	0.8924	0.8899	0.8872	0.8841
w_B	0.0619	0.0649	0.0673	0.0700	0.0735	0.0767	0.0809	0.0854	0.0904
w_C	0.0388	0.0370	0.0357	0.0342	0.0326	0.0309	0.0292	0.0274	0.0255
w_A	0.8806	0.8782	0.8757	0.8765	0.8732	0.8726	0.8701	0.8688	0.8667
w_B	0.0960	0.0995	0.1033	0.1037	0.1073	0.1090	0.1117	0.1142	0.1165
w_C	0.0234	0.0223	0.0210	0.0198	0.0195	0.0184	0.0182	0.0170	0.0168
w_A	0.8643	0.8616	0.8593	0.8566	0.8559	0.8533	0.8496	0.8472	0.8462
w_B	0.1199	0.1232	0.1261	0.1294	0.1309	0.1336	0.1379	0.1412	0.1426
w_C	0.0158	0.0152	0.0146	0.0140	0.0132	0.0131	0.0125	0.0116	0.0112
w_A	0.8423	0.8372	0.8356	0.8315	0.8264	0.8199	0.8179	0.8125	0.8004
w_B	0.1469	0.1528	0.1553	0.1594	0.1655	0.1732	0.1753	0.1817	0.1949
w_C	0.0108	0.0100	0.0091	0.0091	0.0081	0.0069	0.0068	0.0058	0.0047
w_A	0.7876	0.7840	0.7510						
w_B	0.2084	0.2128	0.2475						
w_C	0.0040	0.0032	0.0015						

Polymer (B): **poly(propylene glycol)** **2010XI1**
Characterization: M_w/g.mol^{-1} = 400, ρ = 1.004 g/cm^3 (298 K), Aladdin Reagent Co., Ltd., Shanghai, China
Solvent (A): **water** H_2O **7732-18-5**
Salt (C): **dipotassium oxalate** $C_2K_2O_4$ **583-52-8**

Type of data: cloud points

T/K = 298.15

w_A	0.3214	0.3998	0.4442	0.4594	0.4916	0.5134	0.5371	0.5705	0.5920
w_B	0.6760	0.5939	0.5459	0.5291	0.4929	0.4680	0.4408	0.4010	0.3747
w_C	0.0026	0.0063	0.0099	0.0115	0.0155	0.0186	0.0221	0.0285	0.0333
w_A	0.6528	0.7075	0.7430	0.7738	0.7946	0.8096	0.8199	0.8243	0.8244
w_B	0.3004	0.2315	0.1848	0.1406	0.1074	0.0813	0.0559	0.0414	0.0319
w_C	0.0468	0.0610	0.0722	0.0856	0.0980	0.1091	0.1242	0.1343	0.1437
w_A	0.8237	0.8146							
w_B	0.0239	0.0160							
w_C	0.1524	0.1694							

T/K = 308.15

w_A	0.2606	0.3448	0.4056	0.4535	0.5114	0.5655	0.6167	0.6562	0.6788
w_B	0.7388	0.6531	0.5900	0.5388	0.4760	0.4171	0.3599	0.3159	0.2905
w_C	0.0006	0.0021	0.0044	0.0077	0.0126	0.0174	0.0234	0.0279	0.0307
w_A	0.7186	0.7753	0.8037	0.8112	0.8259	0.8393	0.8534	0.8568	0.8569
w_B	0.2440	0.1759	0.1392	0.1280	0.1062	0.0838	0.0567	0.0446	0.0357
w_C	0.0374	0.0488	0.0571	0.0608	0.0679	0.0769	0.0899	0.0986	0.1074

continued

continued

w_A	0.8542	0.8463	0.8441
w_B	0.0228	0.0148	0.0126
w_C	0.1230	0.1389	0.1433

T/K = 318.15

w_A	0.3488	0.3925	0.4509	0.5557	0.6109	0.6530	0.7620	0.7844	0.8475
w_B	0.6497	0.6050	0.5451	0.4371	0.3800	0.3363	0.2192	0.1942	0.1181
w_C	0.0015	0.0025	0.0040	0.0072	0.0091	0.0107	0.0188	0.0214	0.0344
w_A	0.8656	0.8719	0.8833	0.8856	0.8886	0.8873	0.8835	0.8800	0.8757
w_B	0.0903	0.0801	0.0602	0.0484	0.0384	0.0301	0.0269	0.0215	0.0172
w_C	0.0441	0.0480	0.0565	0.0660	0.0730	0.0826	0.0896	0.0985	0.1071
w_A	0.8718	0.8672							
w_B	0.0141	0.0118							
w_C	0.1141	0.1210							

Type of data: coexistence data (tie lines)

Total system			Top phase			Bottom phase		
w_A	w_B	w_C	w_A	w_B	w_C	w_A	w_B	w_C
T/K = 298.15								
0.7734	0.1022	0.1244	0.3398	0.6573	0.0029	0.8242	0.0382	0.1376
0.7296	0.1407	0.1297	0.3075	0.6902	0.0023	0.8217	0.0219	0.1564
0.7247	0.1406	0.1347	0.2967	0.7012	0.0021	0.8192	0.0190	0.1618
T/K = 308.15								
0.6359	0.2993	0.0648	0.4090	0.5869	0.0041	0.8538	0.0251	0.1211
0.5996	0.3306	0.0698	0.3718	0.6255	0.0027	0.8467	0.0151	0.1382
0.5756	0.3490	0.0754	0.3328	0.6653	0.0019	0.8363	0.0099	0.1538
T/K = 318.15								
0.7705	0.2038	0.0257	0.4742	0.5216	0.0042	0.8350	0.1346	0.0304
0.7493	0.2250	0.0257	0.4560	0.5401	0.0039	0.8351	0.1323	0.0326
0.7184	0.2506	0.0310	0.3626	0.6353	0.0021	0.8615	0.0953	0.0432

Polymer (B):	**poly(propylene glycol)**		**2010XI2**
Characterization:	M_w/g.mol^{-1} = 400, ρ = 1.004 g/cm^3 (298 K), Aladdin Reagent Co., Ltd., Shanghai, China		
Solvent (A):	**water**	**H_2O**	**7732-18-5**
Salt (C):	**dipotassium phosphate**	**K_2HPO_4**	**7758-11-4**

Type of data: cloud points

continued

continued

T/K = 298.15

w_A	0.3249	0.3914	0.4181	0.4588	0.4856	0.5145	0.5487	0.6049	0.6284
w_B	0.6737	0.6052	0.5778	0.5345	0.5056	0.4740	0.4355	0.3693	0.3400
w_C	0.0014	0.0034	0.0041	0.0067	0.0088	0.0115	0.0158	0.0258	0.0316
w_A	0.6443	0.6622	0.6876	0.7103	0.7413	0.7658	0.7865	0.8016	0.8126
w_B	0.3198	0.2962	0.2615	0.2287	0.1828	0.1452	0.1120	0.0847	0.0640
w_C	0.0359	0.0416	0.0509	0.0610	0.0759	0.0890	0.1015	0.1137	0.1234
w_A	0.8163	0.8199	0.8158	0.8133	0.8110	0.8091			
w_B	0.0441	0.0258	0.0169	0.0143	0.0122	0.0107			
w_C	0.1396	0.1543	0.1673	0.1724	0.1768	0.1802			

Type of data: coexistence data (tie lines)

T/K = 298.15

Total system			Top phase			Bottom phase		
w_A	w_B	w_C	w_A	w_B	w_C	w_A	w_B	w_C
0.7200	0.2001	0.0799	0.6498	0.3125	0.0377	0.8223	0.0300	0.1477
0.6809	0.2488	0.0703	0.6213	0.3485	0.0302	0.8168	0.0150	0.1682
0.6696	0.2489	0.0815	0.5969	0.3801	0.0230	0.8027	0.0071	0.1902
0.6296	0.2999	0.0705	0.5717	0.4087	0.0196	0.7848	0.0036	0.2116

Polymer (B): **poly(propylene glycol)** **2010XI1**
Characterization: M_w/g.mol^{-1} = 400, ρ = 1.004 g/cm^3 (298 K), Aladdin Reagent Co., Ltd., Shanghai, China
Solvent (A): **water** **H_2O** **7732-18-5**
Salt (C): **dipotassium tartrate** **$C_4H_4K_2O_6$** **6100-19-2**

Type of data: cloud points

T/K = 298.15

w_A	0.2632	0.3334	0.4020	0.4332	0.4688	0.5015	0.5417	0.5960	0.7099
w_B	0.7350	0.6627	0.5890	0.5540	0.5135	0.4750	0.4256	0.3562	0.2000
w_C	0.0018	0.0039	0.0090	0.0128	0.0177	0.0235	0.0327	0.0478	0.0901
w_A	0.7461	0.7652	0.7747	0.7810	0.7808	0.7825	0.7848	0.7814	0.7790
w_B	0.1374	0.1040	0.0856	0.0648	0.0595	0.0503	0.0403	0.0300	0.0237
w_C	0.1165	0.1308	0.1397	0.1542	0.1597	0.1672	0.1749	0.1886	0.1973
w_A	0.7731	0.7611	0.7554						
w_B	0.0180	0.0138	0.0105						
w_C	0.2089	0.2251	0.2341						

continued

continued

T/K = 308.15

w_A	0.2182	0.3366	0.4370	0.5320	0.5741	0.6498	0.7117	0.7390	0.7732
w_B	0.7807	0.6601	0.5534	0.4468	0.3980	0.3104	0.2353	0.2003	0.1527
w_C	0.0011	0.0033	0.0096	0.0212	0.0279	0.0398	0.0530	0.0607	0.0741
w_A	0.7858	0.7957	0.8033	0.8135	0.8177	0.8188	0.8205	0.8187	0.8123
w_B	0.1343	0.1179	0.0996	0.0845	0.0657	0.0588	0.0491	0.0412	0.0319
w_C	0.0799	0.0864	0.0971	0.1020	0.1166	0.1224	0.1304	0.1401	0.1558
w_A	0.8120	0.8017	0.7875						
w_B	0.0243	0.0171	0.0112						
w_C	0.1637	0.1812	0.2013						

T/K = 318.15

w_A	0.3225	0.3489	0.3727	0.4004	0.4293	0.4674	0.5134	0.5626	0.6540
w_B	0.6764	0.6495	0.6253	0.5969	0.5671	0.5273	0.4784	0.4257	0.3272
w_C	0.0011	0.0016	0.0020	0.0027	0.0036	0.0053	0.0082	0.0117	0.0188
w_A	0.7111	0.7567	0.8072	0.8324	0.8520	0.8565	0.8605	0.8602	0.8586
w_B	0.2641	0.2093	0.1472	0.1094	0.0789	0.0661	0.0521	0.0466	0.0348
w_C	0.0248	0.0340	0.0456	0.0582	0.0691	0.0774	0.0874	0.0932	0.1066
w_A	0.8546	0.8453							
w_B	0.0268	0.0203							
w_C	0.1186	0.1344							

Type of data: coexistence data (tie lines)

Total system			Top phase			Bottom phase		
w_A	w_B	w_C	w_A	w_B	w_C	w_A	w_B	w_C
T/K = 298.15								
0.7050	0.1546	0.1404	0.3858	0.6065	0.0077	0.7841	0.0399	0.1760
0.6746	0.1805	0.1449	0.3560	0.6393	0.0047	0.7836	0.0288	0.1876
0.6102	0.2390	0.1508	0.2813	0.7170	0.0017	0.7684	0.0105	0.2211
T/K = 308.15								
0.6634	0.2560	0.0806	0.3877	0.6058	0.0065	0.8200	0.0558	0.1242
0.5950	0.3155	0.0895	0.3623	0.6327	0.0050	0.8154	0.0270	0.1576
0.5487	0.3513	0.1000	0.3182	0.6789	0.0029	0.7898	0.0123	0.1979
T/K = 318.15								
0.7573	0.2017	0.0410	0.3873	0.6104	0.0023	0.8212	0.1324	0.0464
0.7184	0.2317	0.0499	0.3706	0.6275	0.0019	0.8524	0.0803	0.0673
0.6844	0.2606	0.0550	0.3591	0.6391	0.0018	0.8563	0.0620	0.0817

Polymer (B):	**poly(propylene glycol)**		**2009SAD**
Characterization:	M_n/g.mol^{-1} = 400, Fluka AG, Buchs, Switzerland		
Solvent (A):	**water**	H_2O	**7732-18-5**
Salt (C):	**disodium phosphate**	Na_2HPO_4	**7558-79-4**

Type of data: cloud points

T/K = 298.15

w_A	0.3858	0.4081	0.4819	0.6980	0.7139	0.7259	0.7424	0.7556	0.7688
w_B	0.6123	0.5891	0.5131	0.2701	0.2506	0.2349	0.2148	0.1976	0.1807
w_C	0.0019	0.0028	0.0050	0.0319	0.0355	0.0392	0.0428	0.0468	0.0505
w_A	0.7828	0.7951	0.8147	0.8261	0.8351	0.8440	0.8511	0.8548	0.8576
w_B	0.1631	0.1470	0.1231	0.1060	0.0920	0.0781	0.0660	0.0582	0.0526
w_C	0.0541	0.0579	0.0622	0.0679	0.0729	0.0779	0.0829	0.0870	0.0898

T/K = 303.15

w_A	0.3185	0.3781	0.4216	0.4367	0.4692	0.4866	0.5015	0.5261	0.5331
w_B	0.6810	0.6205	0.5753	0.5601	0.5260	0.5071	0.4916	0.4647	0.4575
w_C	0.0005	0.0014	0.0031	0.0032	0.0048	0.0063	0.0069	0.0092	0.0094
w_A	0.5626	0.5640	0.5945	0.5966	0.6256	0.6322	0.6599	0.6719	0.6841
w_B	0.4255	0.4240	0.3908	0.3882	0.3562	0.3486	0.3187	0.3046	0.2909
w_C	0.0119	0.0120	0.0147	0.0152	0.0182	0.0192	0.0214	0.0235	0.0250
w_A	0.6913	0.6944	0.7124	0.7388	0.7760	0.8149	0.8318	0.8369	0.8499
w_B	0.2833	0.2799	0.2605	0.2316	0.1909	0.1436	0.1204	0.1113	0.0945
w_C	0.0254	0.0257	0.0271	0.0296	0.0331	0.0415	0.0478	0.0518	0.0556
w_A	0.8633	0.8781	0.8786	0.8789					
w_B	0.0713	0.0436	0.0277	0.0165					
w_C	0.0654	0.0783	0.0937	0.1046					

T/K = 308.15

w_A	0.3606	0.3943	0.4555	0.5046	0.5559	0.6072	0.6662	0.7268	0.7745
w_B	0.6383	0.6034	0.5403	0.4889	0.4354	0.3810	0.3191	0.2564	0.2042
w_C	0.0011	0.0023	0.0042	0.0065	0.0087	0.0118	0.0147	0.0168	0.0213
w_A	0.8098	0.8113	0.8307	0.8303	0.8434	0.8561	0.8677	0.8764	0.8852
w_B	0.1641	0.1624	0.1395	0.1384	0.1225	0.1050	0.0879	0.0720	0.0550
w_C	0.0261	0.0263	0.0298	0.0313	0.0341	0.0389	0.0444	0.0516	0.0598
w_A	0.8920	0.8793	0.8737						
w_B	0.0339	0.0200	0.0103						
w_C	0.0741	0.1007	0.1160						

T/K = 313.15

w_A	0.3884	0.4288	0.4744	0.5392	0.5715	0.6152	0.6220	0.6735	0.7073
w_B	0.6100	0.5688	0.5220	0.4557	0.4226	0.3781	0.3712	0.3186	0.2845
w_C	0.0016	0.0024	0.0036	0.0051	0.0059	0.0067	0.0068	0.0079	0.0082

continued

continued

w_A	0.7415	0.7844	0.8120	0.8370	0.8552	0.8673	0.8662	0.8750	0.8802
w_B	0.2488	0.2017	0.1703	0.1418	0.1186	0.1026	0.1024	0.0907	0.0800
w_C	0.0097	0.0139	0.0177	0.0212	0.0262	0.0301	0.0314	0.0343	0.0398
w_A	0.8957	0.8984	0.8992	0.8915	0.8732				
w_B	0.0581	0.0457	0.0341	0.0257	0.0136				
w_C	0.0462	0.0559	0.0667	0.0828	0.1132				

T/K = 318.15

w_A	0.3299	0.3327	0.3466	0.3802	0.3963	0.4181	0.4256	0.4474	0.5124
w_B	0.6695	0.6666	0.6526	0.6188	0.6025	0.5806	0.5730	0.5511	0.4857
w_C	0.0006	0.0007	0.0008	0.0010	0.0012	0.0013	0.0014	0.0015	0.0019
w_A	0.5963	0.6252	0.6993	0.7225	0.7308	0.7382	0.7527	0.7636	0.7700
w_B	0.4018	0.3728	0.2985	0.2750	0.2665	0.2588	0.2439	0.2330	0.2261
w_C	0.0019	0.0020	0.0022	0.0025	0.0027	0.0030	0.0034	0.0034	0.0039
w_A	0.8116	0.8443	0.8684	0.8807	0.8810	0.8875	0.8949	0.8947	0.9033
w_B	0.1822	0.1445	0.1154	0.0991	0.0986	0.0891	0.0790	0.0772	0.0628
w_C	0.0062	0.0112	0.0162	0.0202	0.0204	0.0234	0.0261	0.0281	0.0339
w_A	0.9055	0.9091	0.9067	0.8919	0.8754				
w_B	0.0517	0.0376	0.0270	0.0167	0.0103				
w_C	0.0428	0.0533	0.0663	0.0914	0.1143				

Type of data: coexistence data (tie lines)

Total system			Top phase			Bottom phase		
w_A	w_B	w_C	w_A	w_B	w_C	w_A	w_B	w_C
T/K = 298.15								
0.6807	0.2807	0.0386	0.5074	0.4846	0.0080	0.8101	0.1310	0.0589
0.6715	0.2865	0.0420	0.4866	0.5074	0.0060	0.8151	0.1182	0.0667
0.6457	0.3092	0.0451	0.4455	0.5499	0.0046	0.8550	0.0557	0.0893
0.6311	0.3205	0.0484	0.4368	0.5588	0.0044	0.8508	0.0481	0.1011
0.6066	0.3413	0.0521	0.4063	0.5909	0.0028	0.8481	0.0390	0.1129
T/K = 303.15								
0.6478	0.3246	0.0276	0.4020	0.5956	0.0024	0.8206	0.1313	0.0481
0.6170	0.3515	0.0315	0.3881	0.6100	0.0019	0.8515	0.0844	0.0641
0.5896	0.3754	0.0350	0.3815	0.6162	0.0023	0.8703	0.0463	0.0834
0.5619	0.3995	0.0386	0.3403	0.6579	0.0018	0.8673	0.0416	0.0911
0.5410	0.4168	0.0422	0.3305	0.6684	0.0011	0.8648	0.0338	0.1014
T/K = 308.15								
0.6356	0.3441	0.0203	0.3119	0.6870	0.0011	0.8148	0.1551	0.0301
0.5806	0.3930	0.0264	0.2877	0.7113	0.0010	0.8752	0.0706	0.0542
0.5396	0.4281	0.0323	0.2769	0.7226	0.0005	0.8763	0.0528	0.0709
0.4850	0.4768	0.0382	0.2487	0.7507	0.0006	0.8643	0.0371	0.0986
0.4565	0.4990	0.0445	0.2314	0.7680	0.0006	0.8517	0.0282	0.1201

continued

continued

T/K = 313.15								
0.6334	0.3535	0.0131	0.3265	0.6722	0.0013	0.8186	0.1616	0.0198
0.5836	0.3954	0.0210	0.2963	0.7015	0.0022	0.8869	0.0708	0.0423
0.5367	0.4355	0.0278	0.2695	0.7288	0.0017	0.8926	0.0437	0.0637
0.5043	0.4622	0.0335	0.2410	0.7577	0.0013	0.8808	0.0407	0.0785
0.4762	0.4852	0.0386	0.2164	0.7825	0.0011	0.8757	0.0285	0.0958
T/K = 318.15								
0.6531	0.3319	0.0150	0.2527	0.7467	0.0006	0.8733	0.1040	0.0227
0.6106	0.3699	0.0195	0.2434	0.7557	0.0009	0.9006	0.0642	0.0352
0.5875	0.3900	0.0225	0.2311	0.7683	0.0006	0.8998	0.0574	0.0428
0.5333	0.4380	0.0287	0.2222	0.7772	0.0006	0.9023	0.0349	0.0628
0.4829	0.4829	0.0342	0.1987	0.8007	0.0006	0.8876	0.0288	0.0836

Polymer (B): **poly(propylene glycol)** **2011PAT**
Characterization: M_w/g.mol^{-1} = 400, Sigma Chemical Co., Inc., St. Louis, MO
Solvent (A): **water** H_2O **7732-18-5**
Salt (C): **disodium succinate** $C_4H_4Na_2O_4$ **150-90-3**

Type of data: coexistence data (tie lines)

Total system			Top phase			Bottom phase		
w_A	w_B	w_C	w_A	w_B	w_C	w_A	w_B	w_C
T/K = 298.15								
0.4459	0.5375	0.0166	0.3863	0.6098	0.0039	0.7901	0.1583	0.0516
0.3985	0.5834	0.0181	0.2739	0.7241	0.0020	0.8195	0.1055	0.0750
0.3586	0.6218	0.0196	0.2396	0.7593	0.0011	0.8299	0.0734	0.0967
0.3168	0.6621	0.0211	0.2158	0.7834	0.0008	0.8281	0.0461	0.1258
0.3127	0.6648	0.0225	0.2062	0.7931	0.0007	0.8287	0.0369	0.1344
T/K = 313.15								
0.4896	0.4954	0.0150	0.2185	0.7809	0.0006	0.8527	0.1116	0.0357
0.4461	0.5373	0.0166	0.1983	0.8011	0.0006	0.8618	0.0914	0.0468
0.3989	0.5829	0.0182	0.1892	0.8103	0.0005	0.8682	0.0705	0.0613
0.3184	0.6605	0.0211	0.1532	0.8464	0.0004	0.8645	0.0316	0.1039

Polymer (B): **poly(propylene glycol)** **2011PAT**
Characterization: M_w/g.mol^{-1} = 400, Sigma Chemical Co., Inc., St. Louis, MO
Solvent (A): **water** H_2O **7732-18-5**
Salt (C): **disodium tartrate** $C_4H_4Na_2O_6$ **868-18-8**

Type of data: coexistence data (tie lines)

continued

continued

Total system			Top phase			Bottom phase		
w_A	w_B	w_C	w_A	w_B	w_C	w_A	w_B	w_C
T/K = 283.15								
0.5757	0.3465	0.0778	0.4668	0.5220	0.0112	0.7851	0.0055	0.2094
0.5359	0.3816	0.0825	0.4294	0.5624	0.0082	0.7533	0.0208	0.2259
0.4936	0.4186	0.0878	0.3769	0.6196	0.0036	0.7357	0.0197	0.2446
0.4504	0.4570	0.0926	0.3181	0.6795	0.0024	0.7300	0.0156	0.2544
0.4100	0.4925	0.0975	0.2426	0.7557	0.0017	0.7102	0.0139	0.2759
T/K = 298.15								
0.6604	0.2722	0.0674	0.3994	0.5952	0.0054	0.8157	0.0873	0.0969
0.6184	0.3090	0.0725	0.3754	0.6208	0.0038	0.8226	0.0572	0.1202
0.5744	0.3482	0.0774	0.3510	0.6464	0.0025	0.8270	0.0239	0.1491
0.5343	0.3831	0.0826	0.3090	0.6897	0.0013	0.8166	0.0148	0.1686
0.4919	0.4201	0.0881	0.2782	0.7213	0.0005	0.7970	0.0001	0.2029
T/K = 313.15								
0.6178	0.3093	0.0729	0.2199	0.7797	0.0004	0.8476	0.0325	0.1199
0.5748	0.3474	0.0778	0.1992	0.8005	0.0003	0.8381	0.0296	0.1323
0.5377	0.3803	0.0820	0.2090	0.7907	0.0003	0.8322	0.0227	0.1451
0.4930	0.4195	0.0875	0.1753	0.8245	0.0002	0.8151	0.0103	0.1746
0.4519	0.4555	0.0926	0.1644	0.8354	0.0002	0.7846	0.0119	0.2035

Polymer (B): **poly(propylene glycol)** **2011ZH1**
Characterization: M_n/g.mol^{-1} = 400, Aladdin Reagent Co., Ltd., Shanghai, China
Solvent (A): **water** H_2O **7732-18-5**
Salt (C): **magnesium sulfate** $MgSO_4$ **7487-88-9**

Type of data: cloud points

T/K = 298.15

w_A	0.5330	0.5499	0.5619	0.5921	0.6105	0.6381	0.6579	0.6709	0.6951
w_B	0.4615	0.4440	0.4310	0.3979	0.3761	0.3441	0.3207	0.3048	0.2726
w_C	0.0055	0.0061	0.0071	0.0100	0.0134	0.0178	0.0214	0.0243	0.0323
w_A	0.7127	0.7347	0.7557	0.7671	0.7807	0.7958	0.8065	0.8143	0.8239
w_B	0.2487	0.2162	0.1836	0.1658	0.1430	0.1175	0.0985	0.0839	0.0613
w_C	0.0386	0.0491	0.0607	0.0671	0.0763	0.0867	0.0950	0.1018	0.1148
w_A	0.8277	0.8310	0.8314	0.8318	0.8312	0.8293			
w_B	0.0497	0.0401	0.0350	0.0312	0.0269	0.0227			
w_C	0.1226	0.1289	0.1336	0.1370	0.1419	0.1480			

continued

continued

Type of data: coexistence data (tie lines)

T/K = 298.15

Total system			Top phase			Bottom phase		
w_A	w_B	w_C	w_A	w_B	w_C	w_A	w_B	w_C
0.7500	0.1850	0.0650	0.7004	0.2653	0.0343	0.8178	0.0760	0.1062
0.7393	0.1909	0.0698	0.6573	0.3216	0.0211	0.8267	0.0514	0.1219
0.7294	0.2008	0.0698	0.6357	0.3467	0.0176	0.8311	0.0408	0.1281
0.6968	0.2239	0.0793	0.5727	0.4189	0.0084	0.8285	0.0209	0.1506

Polymer (B): **poly(propylene glycol)** **2010SA1**
Characterization: M_n/g.mol^{-1} = 425, Aldrich Chem. Co., Inc., Milwaukee, WI
Solvent (A): **water** **H_2O** **7732-18-5**
Salt (C): **magnesium sulfate** **$MgSO_4$** **7487-88-9**

Type of data: coexistence data (tie lines)

T/K = 298.15 pH = 7.3

Top phase			Bottom phase		
w_A	w_B	w_C	w_A	w_B	w_C
0.683	0.289	0.0280	0.784	0.149	0.067
0.523	0.469	0.0080	0.815	0.071	0.114
0.399	0.599	0.0020	0.814	0.036	0.149
0.198	0.802	0.0006	0.761	0.021	0.217

Polymer (B): **poly(propylene glycol)** **2008SA2**
Characterization: M_n/g.mol^{-1} = 400, Fluka AG, Buchs, Switzerland
Solvent (A): **water** **H_2O** **7732-18-5**
Salt (C): **monosodium phosphate** **NaH_2PO_4** **7558-80-7**

Type of data: cloud points

T/K = 298.15

w_A	0.2566	0.3460	0.3726	0.4036	0.4166	0.4362	0.4625	0.4702	0.5067
w_B	0.7423	0.6510	0.6218	0.5885	0.5746	0.5539	0.5246	0.5159	0.4751
w_C	0.0011	0.0030	0.0056	0.0079	0.0088	0.0099	0.0129	0.0139	0.0182
w_A	0.5145	0.5569	0.6059	0.6523	0.6935	0.7348	0.7602	0.7737	0.7791
w_B	0.4668	0.4182	0.3627	0.3089	0.2617	0.2145	0.1844	0.1667	0.1585
w_C	0.0187	0.0249	0.0314	0.0388	0.0448	0.0507	0.0554	0.0596	0.0624

continued

continued

w_A	0.7939	0.8098	0.8191	0.8274	0.8336	0.8374	0.8329	0.8291	0.7970
w_B	0.1360	0.1134	0.0958	0.0790	0.0635	0.0486	0.0351	0.0239	0.0196
w_C	0.0701	0.0768	0.0851	0.0936	0.1029	0.1140	0.1320	0.1470	0.1834

T/K = 303.15

w_A	0.2942	0.3171	0.3623	0.4080	0.4548	0.4998	0.5342	0.5363	0.5489
w_B	0.7040	0.6801	0.6331	0.5849	0.5349	0.4863	0.4485	0.4461	0.4323
w_C	0.0018	0.0028	0.0046	0.0071	0.0103	0.0139	0.0173	0.0176	0.0188
w_A	0.5653	0.5711	0.5880	0.6048	0.6102	0.6327	0.6329	0.6544	0.6570
w_B	0.4144	0.4084	0.3898	0.3717	0.3657	0.3412	0.3407	0.3175	0.3142
w_C	0.0203	0.0205	0.0222	0.0235	0.0241	0.0261	0.0264	0.0281	0.0288
w_A	0.6781	0.6834	0.7093	0.7087	0.7386	0.7619	0.7836	0.7936	0.8001
w_B	0.2914	0.2857	0.2583	0.2579	0.2253	0.1958	0.1692	0.1541	0.1467
w_C	0.0305	0.0309	0.0324	0.0334	0.0361	0.0423	0.0472	0.0523	0.0532
w_A	0.7997	0.8037	0.8086	0.8102	0.8166	0.8162	0.8253	0.8289	0.8288
w_B	0.1449	0.1383	0.1310	0.1293	0.1203	0.1202	0.1079	0.1021	0.1001
w_C	0.0554	0.0580	0.0604	0.0605	0.0631	0.0636	0.0668	0.0690	0.0711
w_A	0.8359	0.8420	0.8444	0.8489	0.8518	0.8552	0.8502	0.8508	0.8261
w_B	0.0887	0.0768	0.0692	0.0611	0.0511	0.0408	0.0320	0.0232	0.0159
w_C	0.0754	0.0812	0.0864	0.0900	0.0971	0.1040	0.1178	0.1260	0.1580

T/K = 308.15

w_A	0.2877	0.3279	0.3864	0.4447	0.5025	0.5624	0.5818	0.5977	0.6216
w_B	0.7113	0.6697	0.6087	0.5477	0.4867	0.4237	0.4030	0.3863	0.3616
w_C	0.0010	0.0024	0.0049	0.0076	0.0108	0.0139	0.0152	0.0160	0.0168
w_A	0.6227	0.6494	0.6728	0.6809	0.7095	0.7111	0.7429	0.7436	0.7674
w_B	0.3602	0.3319	0.3074	0.2984	0.2674	0.2657	0.2311	0.2295	0.2015
w_C	0.0171	0.0187	0.0198	0.0207	0.0231	0.0232	0.0260	0.0269	0.0311
w_A	0.7695	0.7865	0.7938	0.8091	0.8136	0.8158	0.8208	0.8285	0.8269
w_B	0.1984	0.1776	0.1700	0.1513	0.1437	0.1415	0.1335	0.1234	0.1230
w_C	0.0321	0.0359	0.0362	0.0396	0.0427	0.0427	0.0457	0.0481	0.0501
w_A	0.8369	0.8350	0.8430	0.8418	0.8481	0.8511	0.8571	0.8607	0.8594
w_B	0.1112	0.1099	0.0999	0.0980	0.0900	0.0827	0.0727	0.0629	0.0559
w_C	0.0519	0.0551	0.0571	0.0602	0.0619	0.0662	0.0702	0.0764	0.0847
w_A	0.8599	0.8616	0.8542	0.8475	0.8385				
w_B	0.0478	0.0383	0.0332	0.0268	0.0216				
w_C	0.0923	0.1001	0.1126	0.1257	0.1399				

T/K = 313.15

w_A	0.3160	0.4055	0.4580	0.4801	0.4936	0.5523	0.5649	0.6234	0.6423
w_B	0.6825	0.5904	0.5365	0.5137	0.4998	0.4398	0.4267	0.3671	0.3475
w_C	0.0015	0.0041	0.0055	0.0062	0.0066	0.0079	0.0084	0.0095	0.0102

continued

continued

w_A	0.6892	0.6989	0.7332	0.7357	0.7623	0.7695	0.7858	0.7915	0.8025
w_B	0.2999	0.2892	0.2534	0.2498	0.2208	0.2134	0.1938	0.1870	0.1730
w_C	0.0109	0.0119	0.0134	0.0145	0.0169	0.0171	0.0204	0.0215	0.0245
w_A	0.8022	0.8182	0.8300	0.8294	0.8412	0.8416	0.8483	0.8571	0.8621
w_B	0.1722	0.1532	0.1379	0.1375	0.1224	0.1200	0.1068	0.0925	0.0808
w_C	0.0256	0.0286	0.0321	0.0331	0.0364	0.0384	0.0449	0.0504	0.0571
w_A	0.8618	0.8632	0.8631	0.8603	0.8478	0.8272	0.8124		
w_B	0.0741	0.0657	0.0577	0.0502	0.0383	0.0294	0.0146		
w_C	0.0641	0.0711	0.0792	0.0895	0.1139	0.1434	0.1730		

T/K = 318.15

w_A	0.3227	0.5192	0.6569	0.7479	0.7846	0.8117	0.8246	0.8398	0.8411
w_B	0.6759	0.4787	0.3403	0.2481	0.2087	0.1783	0.1649	0.1462	0.1447
w_C	0.0014	0.0021	0.0028	0.0040	0.0067	0.0100	0.0105	0.0140	0.0142
w_A	0.8543	0.8654	0.8740	0.8821	0.8887	0.8905	0.8919	0.8922	0.8961
w_B	0.1285	0.1125	0.1003	0.0881	0.0769	0.0750	0.0686	0.0676	0.0578
w_C	0.0172	0.0221	0.0257	0.0298	0.0344	0.0345	0.0395	0.0402	0.0461
w_A	0.8998	0.8979	0.8856	0.8574	0.8175				
w_B	0.0472	0.0358	0.0230	0.0180	0.0088				
w_C	0.0530	0.0663	0.0914	0.1246	0.1737				

Type of data: coexistence data (tie lines)

Total system			Top phase			Bottom phase		
w_A	w_B	w_C	w_A	w_B	w_C	w_A	w_B	w_C
T/K = 298.15								
0.6627	0.2745	0.0628	0.3764	0.6155	0.0081	0.8100	0.0956	0.0944
0.6463	0.2819	0.0718	0.3470	0.6473	0.0057	0.8260	0.0609	0.1131
0.6043	0.3133	0.0824	0.3194	0.6762	0.0044	0.8216	0.0369	0.1415
0.5882	0.3237	0.0881	0.2776	0.7190	0.0034	0.8159	0.0277	0.1564
0.5565	0.3488	0.0947	0.2470	0.7493	0.0037	0.8043	0.0228	0.1729
T/K = 303.15								
0.6177	0.3489	0.0334	0.3663	0.6272	0.0065	0.7523	0.2031	0.0446
0.5807	0.3798	0.0395	0.3479	0.6469	0.0052	0.8138	0.1135	0.0727
0.5582	0.3979	0.0439	0.3165	0.6789	0.0046	0.8285	0.0849	0.0866
0.5241	0.4255	0.0504	0.2798	0.7160	0.0042	0.8457	0.0414	0.1129
0.4955	0.4481	0.0564	0.2481	0.7513	0.0006	0.8369	0.0316	0.1315

continued

continued

T/K = 308.15

0.6087	0.3683	0.0230	0.2815	0.7155	0.0030	0.7910	0.1738	0.0352
0.5726	0.3996	0.0278	0.2655	0.7324	0.0021	0.8309	0.1180	0.0511
0.5472	0.4206	0.0322	0.2545	0.7445	0.0010	0.8522	0.0818	0.0660
0.5181	0.4440	0.0379	0.2428	0.7560	0.0012	0.8523	0.0669	0.0808
0.4922	0.4654	0.0424	0.2298	0.7674	0.0028	0.8595	0.0374	0.1031

T/K = 313.15

0.6559	0.3255	0.0186	0.2550	0.7435	0.0015	0.7771	0.2001	0.0228
0.6096	0.3644	0.0260	0.2395	0.7583	0.0022	0.8366	0.1218	0.0416
0.5682	0.4011	0.0307	0.2083	0.7909	0.0008	0.8558	0.0900	0.0542
0.5263	0.4386	0.0351	0.1864	0.8129	0.0007	0.8676	0.0596	0.0728
0.4915	0.4679	0.0406	0.1696	0.8295	0.0009	0.8573	0.0578	0.0849

T/K = 318.15

0.6671	0.3108	0.0221	0.2128	0.7870	0.0002	0.8533	0.1162	0.0305
0.6363	0.3347	0.0290	0.2030	0.7962	0.0008	0.8750	0.0806	0.0444
0.5963	0.3678	0.0359	0.1982	0.8005	0.0013	0.8770	0.0631	0.0599
0.5623	0.3954	0.0423	0.1833	0.8155	0.0012	0.8801	0.0424	0.0775
0.5226	0.4288	0.0486	0.1681	0.8310	0.0009	0.8802	0.0220	0.0978

Polymer (B): **poly(propylene glycol)** **2011ZH1**
Characterization: M_n/g.mol^{-1} = 400, Aladdin Reagent Co., Ltd., Shanghai, China
Solvent (A): **water** H_2O **7732-18-5**
Salt (C): **potassium acetate** $C_2H_3KO_2$ **127-08-2**

Type of data: cloud points

T/K = 298.15

w_A	0.1024	0.1455	0.2220	0.2892	0.4010	0.5283	0.5898	0.6211	0.6672
w_B	0.8947	0.8509	0.7716	0.6981	0.5741	0.4300	0.3578	0.3207	0.2639
w_C	0.0029	0.0036	0.0064	0.0127	0.0249	0.0417	0.0524	0.0582	0.0689
w_A	0.6907	0.7165	0.7275	0.7517	0.7569	0.7888	0.7896	0.7845	0.7797
w_B	0.2324	0.1966	0.1803	0.1407	0.1312	0.0613	0.0430	0.0317	0.0219
w_C	0.0769	0.0869	0.0922	0.1076	0.1119	0.1499	0.1674	0.1838	0.1984
w_A	0.7679	0.7546							
w_B	0.0158	0.0109							
w_C	0.2163	0.2345							

T/K = 308.15

w_A	0.1291	0.1643	0.4009	0.4330	0.4600	0.4980	0.6042	0.7192	0.7399
w_B	0.8694	0.8330	0.5842	0.5499	0.5210	0.4796	0.3644	0.2311	0.2065
w_C	0.0015	0.0027	0.0149	0.0171	0.0190	0.0224	0.0314	0.0497	0.0536

continued

continued

w_A	0.7586	0.7978	0.8069	0.8178	0.8172	0.8208	0.8236	0.8242	0.8243
w_B	0.1821	0.1245	0.1087	0.0907	0.0903	0.0814	0.0614	0.0501	0.0378
w_C	0.0593	0.0777	0.0844	0.0915	0.0925	0.0978	0.1150	0.1257	0.1379
w_A	0.8210	0.8182							
w_B	0.0332	0.0295							
w_C	0.1458	0.1523							

T/K = 318.15

w_A	0.3454	0.5027	0.5449	0.5796	0.6872	0.7085	0.7447	0.7600	0.7932
w_B	0.6532	0.4903	0.4469	0.4105	0.2946	0.2718	0.2317	0.2143	0.1747
w_C	0.0014	0.0070	0.0082	0.0099	0.0182	0.0197	0.0236	0.0257	0.0321
w_A	0.8043	0.8208	0.8523	0.8612	0.8679	0.8653	0.8619	0.8597	0.8492
w_B	0.1610	0.1403	0.0929	0.0761	0.0420	0.0365	0.0296	0.0268	0.0201
w_C	0.0347	0.0389	0.0548	0.0627	0.0901	0.0982	0.1085	0.1135	0.1307
w_A	0.8391	0.8307	0.8270						
w_B	0.0154	0.0135	0.0119						
w_C	0.1455	0.1558	0.1611						

Type of data: coexistence data (tie lines)

Total system			Top phase			Bottom phase		
w_A	w_B	w_C	w_A	w_B	w_C	w_A	w_B	w_C
T/K = 298.15								
0.6842	0.1808	0.1350	0.2092	0.7826	0.0082	0.7752	0.0633	0.1615
0.6140	0.2456	0.1404	0.1786	0.8156	0.0058	0.7689	0.0356	0.1955
0.5710	0.2763	0.1527	0.1646	0.8300	0.0054	0.7597	0.0200	0.2203
T/K = 308.15								
0.6864	0.2434	0.0702	0.2543	0.7382	0.0075	0.8113	0.1001	0.0886
0.6598	0.2602	0.0800	0.2162	0.7790	0.0048	0.8270	0.0650	0.1080
0.6243	0.2805	0.0952	0.1722	0.8250	0.0028	0.8225	0.0407	0.1368
T/K = 318.15								
0.6967	0.2704	0.0329	0.4643	0.5302	0.0055	0.8443	0.1056	0.0501
0.6776	0.2804	0.0420	0.3720	0.6255	0.0025	0.8565	0.0788	0.0647
0.6406	0.3096	0.0498	0.2901	0.7089	0.0010	0.8639	0.0544	0.0817

Polymer (B):	**poly(propylene glycol)**		**2010XI2**
Characterization:	M_w/g.mol^{-1} = 400, ρ = 1.004 g/cm^3 (298 K), Aladdin Reagent Co., Ltd., Shanghai, China		
Solvent (A):	**water**	**H_2O**	**7732-18-5**
Salt (C):	**potassium carbonate**	**K_2CO_3**	**584-08-7**

Type of data: cloud points

T/K = 298.15

w_A	0.3260	0.3809	0.4048	0.4455	0.4810	0.5555	0.5035	0.5264	0.5889
w_B	0.6721	0.6152	0.5904	0.5473	0.5089	0.4259	0.4835	0.4586	0.3870
w_C	0.0019	0.0039	0.0048	0.0072	0.0101	0.0186	0.0130	0.0150	0.0241
w_A	0.6213	0.6539	0.6850	0.7136	0.7612	0.7977	0.8253	0.8375	0.8459
w_B	0.3477	0.3081	0.2693	0.2332	0.1704	0.1221	0.0795	0.0544	0.0352
w_C	0.0310	0.0380	0.0457	0.0532	0.0684	0.0802	0.0952	0.1081	0.1189
w_A	0.8466	0.8446	0.8399	0.8372					
w_B	0.0271	0.0166	0.0102	0.0084					
w_C	0.1263	0.1388	0.1499	0.1544					

Type of data: coexistence data (tie lines)

T/K = 298.15

Total system			Top phase			Bottom phase		
w_A	w_B	w_C	w_A	w_B	w_C	w_A	w_B	w_C
0.7296	0.2003	0.0701	0.5928	0.3828	0.0244	0.8240	0.0754	0.1006
0.7202	0.1996	0.0802	0.5386	0.4452	0.0162	0.8404	0.0418	0.1178
0.6897	0.2402	0.0701	0.5256	0.4605	0.0139	0.8419	0.0352	0.1229
0.6785	0.2412	0.0803	0.4907	0.4982	0.0111	0.8433	0.0206	0.1361

Polymer (B):	**poly(propylene glycol)**		**2011ZH1**
Characterization:	M_n/g.mol^{-1} = 400, Aladdin Reagent Co., Ltd., Shanghai, China		
Solvent (A):	**water**	**H_2O**	**7732-18-5**
Salt (C):	**potassium chloride**	**KCl**	**7447-40-7**

Type of data: cloud points

T/K = 298.15

w_A	0.0852	0.1339	0.2030	0.2451	0.2751	0.3332	0.3990	0.4801	0.5255
w_B	0.9112	0.8601	0.7875	0.7411	0.7081	0.6417	0.5653	0.4705	0.4173
w_C	0.0036	0.0060	0.0095	0.0138	0.0168	0.0251	0.0357	0.0494	0.0572
w_A	0.5849	0.6229	0.6596	0.6804	0.6904	0.7128	0.7258	0.7413	0.7466
w_B	0.3465	0.3005	0.2506	0.2223	0.2089	0.1700	0.1505	0.1225	0.1038
w_C	0.0686	0.0766	0.0898	0.0973	0.1007	0.1172	0.1237	0.1362	0.1496

continued

continued

w_A	0.7558	0.7598	0.7579	0.7594	0.7524	0.7446
w_B	0.0773	0.0580	0.0431	0.0347	0.0219	0.0099
w_C	0.1669	0.1822	0.1990	0.2059	0.2257	0.2455

Type of data: coexistence data (tie lines)

T/K = 298.15

Total system			Top phase			Bottom phase		
w_A	w_B	w_C	w_A	w_B	w_C	w_A	w_B	w_C
0.6673	0.2027	0.1300	0.2030	0.7875	0.0095	0.7539	0.0911	0.1550
0.6868	0.1730	0.1402	0.1890	0.8026	0.0084	0.7543	0.0884	0.1573
0.6603	0.1995	0.1402	0.1479	0.8459	0.0062	0.7534	0.0836	0.1630

Polymer (B): **poly(propylene glycol)** **2010XI1**
Characterization: M_w/g.mol^{-1} = 400, ρ = 1.004 g/cm^3 (298 K), Aladdin Reagent Co., Ltd., Shanghai, China
Solvent (A): **water** **H_2O** **7732-18-5**
Salt (C): **potassium citrate** **$K_3C_6H_5O_7$** **866-84-2**

Type of data: cloud points

T/K = 298.15

w_A	0.2217	0.3047	0.3544	0.4344	0.4596	0.4866	0.5400	0.5844	0.6660
w_B	0.7779	0.6938	0.6425	0.5548	0.5272	0.4959	0.4307	0.3748	0.2701
w_C	0.0004	0.0015	0.0031	0.0108	0.0132	0.0175	0.0293	0.0408	0.0639
w_A	0.6894	0.7113	0.7389	0.7511	0.7645	0.7735	0.7820	0.7851	0.7875
w_B	0.2367	0.2065	0.1605	0.1427	0.1187	0.1016	0.0782	0.0701	0.0555
w_C	0.0739	0.0822	0.1006	0.1062	0.1168	0.1249	0.1398	0.1448	0.1570
w_A	0.7869	0.7868	0.7855	0.7799	0.7784	0.7704	0.7663		
w_B	0.0464	0.0365	0.0301	0.0249	0.0211	0.0159	0.0107		
w_C	0.1667	0.1767	0.1844	0.1952	0.2005	0.2137	0.2230		

T/K = 308.15

w_A	0.3353	0.3960	0.4320	0.4680	0.5107	0.5537	0.5849	0.6186	0.6371
w_B	0.6632	0.6003	0.5619	0.5230	0.4751	0.4261	0.3906	0.3503	0.3286
w_C	0.0015	0.0037	0.0061	0.0090	0.0142	0.0202	0.0245	0.0311	0.0343
w_A	0.6500	0.7399	0.7882	0.7911	0.8040	0.8138	0.8238	0.8292	0.8288
w_B	0.3126	0.2010	0.1364	0.1298	0.1078	0.0928	0.0692	0.0577	0.0463
w_C	0.0374	0.0591	0.0754	0.0791	0.0882	0.0934	0.1070	0.1131	0.1249
w_A	0.8295	0.8219	0.8194	0.8114	0.8049	0.7967			
w_B	0.0374	0.0297	0.0211	0.0158	0.0128	0.0099			
w_C	0.1331	0.1484	0.1595	0.1728	0.1823	0.1934			

continued

continued

T/K = 318.15

w_A	0.3556	0.4191	0.4903	0.5470	0.6023	0.6473	0.6807	0.7386	0.7639
w_B	0.6437	0.5785	0.5041	0.4441	0.3843	0.3364	0.3004	0.2368	0.2077
w_C	0.0007	0.0024	0.0056	0.0089	0.0134	0.0163	0.0189	0.0246	0.0284
w_A	0.8249	0.8442	0.8584	0.8631	0.8669	0.8676	0.8725	0.8676	0.8653
w_B	0.1339	0.1069	0.0864	0.0773	0.0710	0.0644	0.0471	0.0397	0.0323
w_C	0.0412	0.0489	0.0552	0.0596	0.0621	0.0680	0.0804	0.0927	0.1024
w_A	0.8657	0.8596	0.8560	0.8502	0.8471	0.8423			
w_B	0.0238	0.0226	0.0192	0.0169	0.0151	0.0131			
w_C	0.1105	0.1178	0.1248	0.1329	0.1378	0.1446			

Type of data: coexistence data (tie lines)

Total system			Top phase			Bottom phase		
w_A	w_B	w_C	w_A	w_B	w_C	w_A	w_B	w_C
T/K = 298.15								
0.6454	0.2791	0.0755	0.4665	0.5179	0.0156	0.7432	0.1483	0.1085
0.6146	0.3083	0.0771	0.4374	0.5509	0.0117	0.7645	0.1107	0.1248
0.5752	0.3492	0.0756	0.3840	0.6090	0.0070	0.7796	0.0711	0.1493
T/K = 308.15								
0.6797	0.2402	0.0801	0.4319	0.5609	0.0072	0.8241	0.0557	0.1202
0.6008	0.3095	0.0897	0.3677	0.6287	0.0036	0.8126	0.0180	0.1694
0.5658	0.3344	0.0998	0.3345	0.6639	0.0016	0.7977	0.0100	0.1923
T/K = 318.15								
0.7826	0.1811	0.0363	0.4701	0.5244	0.0055	0.8265	0.1325	0.0410
0.7301	0.2291	0.0408	0.4017	0.5961	0.0022	0.8627	0.0793	0.0580
0.7062	0.2508	0.0430	0.3876	0.6103	0.0021	0.8666	0.0683	0.0651

Polymer (B): **poly(propylene glycol)** **2008ZA1**
Characterization: M_n/g.mol^{-1} = 403, Aldrich Chem. Co., Inc., Milwaukee, WI
Solvent (A): **water** H_2O **7732-18-5**
Salt (C): **potassium citrate** $K_3C_6H_5O_7$ **866-84-2**

Type of data: cloud points

T/K = 293.15

w_A	0.3846	0.4121	0.4922	0.5165	0.5442	0.5811	0.6185	0.6565	0.7048
w_B	0.6088	0.5792	0.4900	0.4606	0.4272	0.3811	0.3324	0.2785	0.2090
w_C	0.0066	0.0087	0.0178	0.0229	0.0286	0.0378	0.0491	0.0650	0.0862

continued

continued

T/K = 298.15

w_A	0.3256	0.3355	0.3753	0.4139	0.4715	0.4988	0.5314	0.5605	0.5907
w_B	0.6718	0.6614	0.6200	0.5788	0.5158	0.4852	0.4475	0.4133	0.3765
w_C	0.0026	0.0031	0.0047	0.0073	0.0127	0.0160	0.0211	0.0262	0.0328
w_A	0.6192	0.6427	0.6688	0.6872	0.7056	0.7536	0.7583	0.7610	0.7696
w_B	0.3418	0.3122	0.2805	0.2567	0.2334	0.1667	0.1615	0.1576	0.1458
w_C	0.0390	0.0451	0.0507	0.0561	0.0610	0.0797	0.0802	0.0814	0.0846
w_A	0.7744	0.7797	0.7901						
w_B	0.1359	0.1283	0.1117						
w_C	0.0897	0.0920	0.0982						

T/K = 308.15

w_A	0.3070	0.3385	0.3791	0.4444	0.4936	0.5427	0.5991	0.6621	0.7056
w_B	0.6911	0.6590	0.6167	0.5485	0.4967	0.4437	0.3828	0.3142	0.2656
w_C	0.0019	0.0025	0.0042	0.0071	0.0097	0.0136	0.0181	0.0237	0.0288
w_A	0.7408	0.7602	0.7824	0.8006	0.8247				
w_B	0.2241	0.2004	0.1711	0.1445	0.1055				
w_C	0.0351	0.0394	0.0465	0.0549	0.0698				

T/K = 318.15

w_A	0.5024	0.5481	0.6456	0.6900	0.7304	0.7643	0.7883	0.8091	0.8221
w_B	0.4944	0.4485	0.3511	0.3059	0.2640	0.2257	0.1982	0.1728	0.1560
w_C	0.0032	0.0034	0.0033	0.0041	0.0056	0.0100	0.0135	0.0181	0.0219
w_A	0.8321	0.8449	0.8533	0.8582	0.8666				
w_B	0.1410	0.1203	0.1067	0.0976	0.0783				
w_C	0.0269	0.0348	0.0400	0.0442	0.0551				

Type of data: coexistence data (tie lines)

Top phase			Bottom phase		
w_A	w_B	w_C	w_A	w_B	w_C
T/K = 293.15					
0.4113	0.5757	0.0130	0.7568	0.1037	0.1395
0.3587	0.6373	0.0040	0.7555	0.0931	0.1514
0.2828	0.7153	0.0019	0.7407	0.0623	0.1970
0.2334	0.7657	0.0009	0.7234	0.0511	0.2255
T/K = 298.15					
0.4176	0.5735	0.0089	0.7927	0.1017	0.1056
0.3903	0.6034	0.0063	0.7878	0.0830	0.1292
0.3428	0.6546	0.0026	0.7866	0.0579	0.1555
0.2928	0.7053	0.0019	0.7699	0.0438	0.1863
0.1578	0.8417	0.0005	0.6893	0.0054	0.3053

continued

continued

T/K = 308.15

0.3178	0.6810	0.0012	0.8140	0.0755	0.1105
0.2572	0.7421	0.0007	0.8015	0.0523	0.1462
0.2141	0.7854	0.0005	0.7896	0.0501	0.1603
0.1334	0.8666	ß.0004	0.7646	0.0535	0.1819
0.0871	0.9128	0.0001	0.7493	0.0359	0.2148

T/K = 318.15

0.2716	0.7277	0.0007	0.8234	0.1504	0.0262
0.2495	0.7502	0.0003	0.8517	0.1093	0.0390
0.2409	0.7578	0.0013	0.8594	0.0948	0.0458
0.2191	0.7805	0.0004	0.8637	0.0741	0.0622
0.2086	0.7909	0.0005	0.8635	0.0710	0.0655

Polymer (B): **poly(propylene glycol)** **2010XI2**
Characterization: M_w/g.mol^{-1} = 400, ρ = 1.004 g/cm^3 (298 K), Aladdin Reagent Co., Ltd., Shanghai, China
Solvent (A): **water** H_2O **7732-18-5**
Salt (C): **potassium phosphate** K_3PO_4 **7778-53-2**

Type of data: cloud points

T/K = 298.15

w_A	0.3881	0.4280	0.4665	0.4978	0.5208	0.5627	0.5838	0.6043	0.6217
w_B	0.6099	0.5680	0.5282	0.4946	0.4691	0.4227	0.3984	0.3736	0.3526
w_C	0.0020	0.0040	0.0053	0.0076	0.0101	0.0146	0.0178	0.0221	0.0257
w_A	0.6412	0.6673	0.6927	0.7042	0.7217	0.7326	0.7479	0.7611	0.7902
w_B	0.3289	0.2957	0.2623	0.2470	0.2236	0.2075	0.1860	0.1675	0.1259
w_C	0.0299	0.0370	0.0450	0.0488	0.0547	0.0599	0.0661	0.0714	0.0839
w_A	0.8002	0.8257	0.8321	0.8412	0.8421	0.8386	0.8356		
w_B	0.1113	0.0693	0.0537	0.0342	0.0207	0.0138	0.0102		
w_C	0.0885	0.1050	0.1142	0.1246	0.1372	0.1476	0.1542		

Type of data: coexistence data (tie lines)

T/K = 298.15

Total system			Top phase			Bottom phase		
w_A	w_B	w_C	w_A	w_B	w_C	w_A	w_B	w_C
0.7194	0.1997	0.0809	0.5945	0.3865	0.0190	0.8413	0.0162	0.1425
0.6791	0.2506	0.0703	0.5829	0.4004	0.0167	0.8351	0.0120	0.1529
0.6707	0.2495	0.0798	0.5616	0.4237	0.0147	0.8234	0.0084	0.1682

Polymer (B): **poly(propylene glycol)** **2011PAT**
Characterization: M_w/g.mol^{-1} = 400, Sigma Chemical Co., Inc., St. Louis, MO
Solvent (A): **water** **H_2O** **7732-18-5**
Salt (C): **sodium acetate** **$C_2H_3NaO_2$** **127-09-3**

Type of data: coexistence data (tie lines)

Total system			Top phase			Bottom phase		
w_A	w_B	w_C	w_A	w_B	w_C	w_A	w_B	w_C
T/K = 283.15								
0.6325	0.2843	0.0832	0.2896	0.7059	0.0046	0.7690	0.1171	0.1139
0.5948	0.3170	0.0882	0.2874	0.7079	0.0047	0.7699	0.0957	0.1345
0.5573	0.3495	0.0932	0.1846	0.8108	0.0046	0.7622	0.0974	0.1404
0.5204	0.3816	0.0980	0.1717	0.8235	0.0048	0.7721	0.0649	0.1630
T/K = 298.15								
0.6336	0.2833	0.0832	0.1666	0.8318	0.0016	0.8291	0.0518	0.1191
0.5965	0.3155	0.0879	0.1601	0.8381	0.0018	0.8291	0.0373	0.1335
0.5582	0.3488	0.0930	0.1579	0.8401	0.0020	0.8285	0.0331	0.1384
0.5213	0.3809	0.0978	0.1456	0.8522	0.0022	0.8173	0.0149	0.1678
0.4844	0.4121	0.1035	0.1365	0.8612	0.0022	0.8025	0.0081	0.1894

Polymer (B): **poly(propylene glycol)** **2005ZA1**
Characterization: M_n/g.mol^{-1} = 400, Aldrich Chem. Co., Inc., Milwaukee, WI
Solvent (A): **water** **H_2O** **7732-18-5**
Salt (C): **sodium carbonate** **Na_2CO_3** **497-19-8**

Type of data: cloud points

T/K = 298.15

w_A	0.8198	0.8049	0.7887	0.7782	0.7559	0.7369	0.7267	0.7172	0.7009
w_B	0.1336	0.1524	0.1710	0.1855	0.2098	0.2323	0.2431	0.2547	0.2729
w_C	0.0466	0.0427	0.0403	0.0363	0.0343	0.0308	0.0302	0.0281	0.0262
w_A	0.6657	0.6633	0.6366	0.6276	0.6112	0.5887	0.5622	0.5155	0.4791
w_B	0.3123	0.3152	0.3450	0.3551	0.3732	0.3989	0.4269	0.4768	0.5155
w_C	0.0220	0.0215	0.0184	0.0173	0.0156	0.0124	0.0109	0.0077	0.0054
w_A	0.4398	0.4068							
w_B	0.5563	0.5911							
w_C	0.0039	0.0021							

Type of data: coexistence data (tie lines)

continued

continued

T/K = 298.15

Top phase			Bottom phase		
w_A	w_B	w_C	w_A	w_B	w_C
0.5284	0.4626	0.0090	0.8189	0.1300	0.0511
0.4324	0.5625	0.0051	0.8587	0.0602	0.0811
0.3495	0.6479	0.0026	0.8659	0.0324	0.1017
0.2705	0.7284	0.0011	0.8496	0.0217	0.1287
0.2278	0.7712	0.0010	0.8381	0.0043	0.1576
0.2105	0.7886	0.0009	0.8246	0.0090	0.1664
0.0990	0.9010	0.0000	0.6829	0.0000	0.3171

Polymer (B): **poly(propylene glycol)** **2011PAT**
Characterization: M_w/g.mol^{-1} = 400, Sigma Chemical Co., Inc., St. Louis, MO
Solvent (A): **water** **H_2O** **7732-18-5**
Salt (C): **sodium citrate** **$Na_3C_6H_5O_7$** **68-04-2**

Type of data: coexistence data (tie lines)

Total system			Top phase			Bottom phase		
w_A	w_B	w_C	w_A	w_B	w_C	w_A	w_B	w_C
T/K = 283.15								
0.6219	0.2923	0.0858	0.4880	0.5006	0.0115	0.7680	0.0420	0.1899
0.5939	0.3176	0.0885	0.4425	0.5502	0.0073	0.7587	0.0388	0.2025
0.5572	0.3482	0.0946	0.4022	0.5934	0.0044	0.7511	0.0387	0.2102
0.5298	0.3713	0.0989	0.3595	0.6385	0.0020	0.7497	0.0351	0.2152
0.5033	0.3945	0.1022	0.3467	0.6524	0.0009	0.7224	0.0340	0.2436
T/K = 298.15								
0.6299	0.3019	0.0682	0.4057	0.5887	0.0056	0.8088	0.0660	0.1252
0.5864	0.3405	0.0731	0.3571	0.6398	0.0030	0.8089	0.0397	0.1514
0.5422	0.3793	0.0784	0.3288	0.6692	0.0020	0.8003	0.0122	0.1875
0.4992	0.4174	0.0834	0.2787	0.7203	0.0011	0.7879	0.0128	0.1993
0.4557	0.4561	0.0882	0.2586	0.7409	0.0005	0.7641	0.0135	0.2225
T/K = 313.15								
0.6196	0.2954	0.0850	0.2315	0.7681	0.0005	0.8361	0.0277	0.1362
0.5887	0.3214	0.0899	0.2265	0.7731	0.0005	0.8324	0.0189	0.1487
0.5575	0.3475	0.0949	0.1712	0.8284	0.0005	0.8234	0.0176	0.1589
0.5208	0.3793	0.0999	0.1935	0.8061	0.0005	0.8057	0.0124	0.1818
0.4955	0.3996	0.1049	0.1641	0.8355	0.0005	0.7911	0.0103	0.1987

Polymer (B): **poly(propylene glycol)** **2005SAL**
Characterization: M_n/g.mol^{-1} = 405, Aldrich Chem. Co., Inc., Milwaukee, WI
Solvent (A): **water** **H_2O** **7732-18-5**
Salt (C): **sodium citrate** **$Na_3C_6H_5O_7$** **68-04-2**

Type of data: coexistence data (tie lines)

T/K = 298.15

Total system			Top phase			Bottom phase		
w_A	w_B	w_C	w_A	w_B	w_C	w_A	w_B	w_C
0.673	0.244	0.083	0.615	0.350	0.035	0.789	0.051	0.160
0.632	0.277	0.091	0.509	0.481	0.010	0.788	0.021	0.191
0.598	0.300	0.102	0.452	0.541	0.007	0.769	0.021	0.210
0.580	0.311	0.109	0.414	0.581	0.005	0.765	0.015	0.220
0.561	0.326	0.113	0.376	0.620	0.004	0.762	0.008	0.230
0.534	0.343	0.123	0.345	0.650	0.005	0.745	0.010	0.245
0.509	0.363	0.128	0.292	0.702	0.006	0.736	0.007	0.257
0.498	0.366	0.136	0.270	0.724	0.006	0.722	0.006	0.272
0.481	0.380	0.139	0.251	0.744	0.005	0.714	0.005	0.281

Polymer (B): **poly(propylene glycol)** **2005SAL**
Characterization: M_n/g.mol^{-1} = 725, M_w/M_n = 1.05,
Merck, Darmstadt, Germany
Solvent (A): **water** **H_2O** **7732-18-5**
Salt (C): **sodium citrate** **$Na_3C_6H_5O_7$** **68-04-2**

Type of data: coexistence data (tie lines)

T/K = 298.15

Total system			Top phase			Bottom phase		
w_A	w_B	w_C	w_A	w_B	w_C	w_A	w_B	w_C
0.533	0.437	0.030	0.341	0.652	0.007	0.812	0.119	0.069
0.523	0.440	0.037	0.303	0.691	0.006	0.834	0.081	0.085
0.528	0.426	0.046	0.275	0.720	0.005	0.869	0.030	0.101
0.504	0.446	0.050	0.235	0.760	0.005	0.858	0.032	0.110
0.501	0.440	0.059	0.194	0.801	0.005	0.854	0.025	0.121
0.493	0.437	0.070	0.145	0.851	0.004	0.828	0.031	0.141
0.480	0.435	0.085	0.116	0.880	0.004	0.822	0.010	0.168
0.480	0.410	0.110	0.066	0.931	0.003	0.800	0.010	0.190

Polymer (B):	**poly(propylene glycol)**		**2005ZA1**
Characterization:	M_n/g.mol^{-1} = 400, Aldrich Chem. Co., Inc., Milwaukee, WI		
Solvent (A):	**water**	**H_2O**	**7732-18-5**
Salt (C):	**sodium nitrate**	**$NaNO_3$**	**7631-99-4**

Type of data: cloud points

T/K = 298.15

w_A	0.6905	0.6885	0.6839	0.6795	0.6734	0.6638	0.6420	0.6227	0.6018
w_B	0.1045	0.1147	0.1268	0.1366	0.1494	0.1686	0.1975	0.2247	0.2537
w_C	0.2050	0.1968	0.1893	0.1839	0.1772	0.1676	0.1605	0.1526	0.1445
w_A	0.5813	0.5590	0.5337	0.4759	0.4343	0.3871	0.3366	0.2941	0.2549
w_B	0.2810	0.3104	0.3413	0.4142	0.4670	0.5283	0.5916	0.6443	0.6910
w_C	0.1377	0.1306	0.1250	0.1099	0.0987	0.0846	0.0718	0.0616	0.0541
w_A	0.2336	0.1804							
w_B	0.7173	0.7817							
w_C	0.0491	0.0379							

Type of data: coexistence data (tie lines)

T/K = 298.15

Top phase			Bottom phase		
w_A	w_B	w_C	w_A	w_B	w_C
0.2875	0.6575	0.0550	0.6748	0.1200	0.2052
0.2218	0.7328	0.0454	0.6862	0.0396	0.2742
0.1713	0.7866	0.0421	0.6696	0.0000	0.3304
0.1442	0.8118	0.0440	0.6051	0.0000	0.3949
0.1112	0.8448	0.0440	0.5300	0.0000	0.4700

Polymer (B):	**poly(propylene glycol)**		**2009SAD**
Characterization:	M_n/g.mol^{-1} = 400, Fluka AG, Buchs, Switzerland		
Solvent (A):	**water**	**H_2O**	**7732-18-5**
Salt (C):	**sodium phosphate**	**Na_3PO_4**	**7601-54-9**

Type of data: cloud points

T/K = 298.15

w_A	0.3530	0.4039	0.4581	0.4872	0.5259	0.5646	0.5961	0.6260	0.6533
w_B	0.6465	0.5948	0.5400	0.5097	0.4694	0.4285	0.3948	0.3620	0.3316
w_C	0.0005	0.0013	0.0019	0.0031	0.0047	0.0069	0.0091	0.0120	0.0151
w_A	0.6743	0.6772	0.6790	0.6895	0.6971	0.7000	0.7115	0.7218	0.7322
w_B	0.3078	0.3041	0.3019	0.2898	0.2811	0.2773	0.2637	0.2511	0.2383
w_C	0.0179	0.0187	0.0191	0.0207	0.0218	0.0227	0.0248	0.0271	0.0295

continued

continued

w_A	0.7437	0.7551	0.7667	0.7792	0.7906	0.8019	0.8110	0.8175	0.8237
w_B	0.2240	0.2097	0.1950	0.1791	0.1644	0.1501	0.1383	0.1296	0.1215
w_C	0.0323	0.0352	0.0383	0.0417	0.0450	0.0480	0.0507	0.0529	0.0548
w_A	0.8314								
w_B	0.1124								
w_C	0.0562								

T/K = 303.15

w_A	0.3558	0.3672	0.4540	0.4714	0.4932	0.4986	0.5235	0.5267	0.5459
w_B	0.6439	0.6319	0.5441	0.5263	0.5031	0.4975	0.4713	0.4679	0.4477
w_C	0.0003	0.0009	0.0019	0.0023	0.0037	0.0039	0.0052	0.0054	0.0064
w_A	0.5544	0.5750	0.5789	0.5995	0.6230	0.6239	0.6445	0.6448	0.6637
w_B	0.4386	0.4167	0.4124	0.3903	0.3647	0.3637	0.3409	0.3405	0.3193
w_C	0.0070	0.0083	0.0087	0.0102	0.0123	0.0124	0.0146	0.0147	0.0170
w_A	0.6672	0.6714	0.6820	0.6860	0.6997	0.7035	0.7146	0.7298	0.7392
w_B	0.3156	0.3106	0.2986	0.2942	0.2787	0.2744	0.2618	0.2443	0.2348
w_C	0.0172	0.0180	0.0194	0.0198	0.0216	0.0221	0.0236	0.0259	0.0260
w_A	0.7569	0.7900	0.8145	0.8317	0.8469	0.8594	0.8709	0.8742	0.8791
w_B	0.2142	0.1782	0.1492	0.1270	0.1070	0.0898	0.0733	0.0654	0.0582
w_C	0.0289	0.0318	0.0363	0.0413	0.0461	0.0508	0.0558	0.0604	0.0627

T/K = 308.15

w_A	0.3780	0.4206	0.4413	0.4687	0.4984	0.5222	0.5488	0.5762	0.6102
w_B	0.6210	0.5777	0.5562	0.5280	0.4973	0.4724	0.4446	0.4159	0.3802
w_C	0.0010	0.0017	0.0025	0.0033	0.0043	0.0054	0.0066	0.0079	0.0096
w_A	0.6377	0.6653	0.6913	0.7026	0.7211	0.7239	0.7494	0.7795	0.8016
w_B	0.3513	0.3224	0.2952	0.2835	0.2646	0.2616	0.2350	0.2025	0.1784
w_C	0.0110	0.0123	0.0135	0.0139	0.0143	0.0145	0.0156	0.0180	0.0200
w_A	0.8106	0.8203	0.8344	0.8475	0.8634	0.8760	0.8838	0.8921	0.8915
w_B	0.1678	0.1560	0.1382	0.1216	0.1016	0.0845	0.0718	0.0573	0.0486
w_C	0.0216	0.0237	0.0274	0.0309	0.0350	0.0395	0.0444	0.0506	0.0599

T/K = 313.15

w_A	0.3636	0.3891	0.4317	0.4761	0.5189	0.5748	0.6412	0.6914	0.7321
w_B	0.6357	0.6096	0.5663	0.5210	0.4772	0.4202	0.3528	0.3017	0.2597
w_C	0.0007	0.0013	0.0020	0.0029	0.0039	0.0050	0.0060	0.0069	0.0082
w_A	0.7600	0.7825	0.8131	0.8402	0.8542	0.8610	0.8677	0.8700	0.8792
w_B	0.2298	0.2056	0.1721	0.1410	0.1251	0.1155	0.1080	0.1040	0.0919
w_C	0.0102	0.0119	0.0148	0.0188	0.0207	0.0235	0.0243	0.0260	0.0289
w_A	0.8876	0.8966	0.9059	0.9076	0.9100				
w_B	0.0792	0.0652	0.0502	0.0391	0.0260				
w_C	0.0332	0.0382	0.0439	0.0533	0.0640				

continued

continued

T/K = 318.15

w_A	0.2932	0.3528	0.3541	0.3881	0.4035	0.4253	0.4373	0.4911	0.5165
w_B	0.7066	0.6467	0.6452	0.6111	0.5956	0.5736	0.5614	0.5075	0.4819
w_C	0.0002	0.0005	0.0007	0.0008	0.0009	0.0011	0.0013	0.0014	0.0016
w_A	0.5465	0.5842	0.6820	0.6866	0.7227	0.7329	0.7567	0.7650	0.7749
w_B	0.4519	0.4142	0.3163	0.3116	0.2752	0.2648	0.2403	0.2317	0.2215
w_C	0.0016	0.0016	0.0017	0.0018	0.0021	0.0023	0.0030	0.0033	0.0036
w_A	0.7792	0.7883	0.7966	0.7993	0.8075	0.8196	0.8237	0.8408	0.8447
w_B	0.2164	0.2070	0.1979	0.1952	0.1865	0.1729	0.1687	0.1495	0.1443
w_C	0.0044	0.0047	0.0055	0.0055	0.0060	0.0075	0.0076	0.0097	0.0110
w_A	0.8494	0.8583	0.8593	0.8655	0.8711	0.8762	0.8799	0.8835	0.8903
w_B	0.1392	0.1290	0.1278	0.1202	0.1141	0.1078	0.1028	0.0989	0.0894
w_C	0.0114	0.0127	0.0129	0.0143	0.0148	0.0160	0.0173	0.0176	0.0203
w_A	0.8953	0.8995	0.9033	0.9054	0.9074	0.9105	0.9085	0.9063	0.9153
w_B	0.0814	0.0739	0.0696	0.0624	0.0574	0.0482	0.0423	0.0348	0.0185
w_C	0.0233	0.0266	0.0271	0.0322	0.0352	0.0413	0.0492	0.0589	0.0662

Type of data: coexistence data (tie lines)

Total system			Top phase			Bottom phase		
w_A	w_B	w_C	w_A	w_B	w_C	w_A	w_B	w_C
T/K = 298.15								
0.7041	0.2677	0.0282	0.5893	0.4014	0.0093	0.7935	0.1620	0.0445
0.6953	0.2754	0.0293	0.5784	0.4134	0.0082	0.8180	0.1305	0.0515
0.6913	0.2776	0.0311	0.5612	0.4307	0.0081	0.8374	0.1041	0.0585
0.6816	0.2856	0.0328	0.5305	0.4642	0.0053	0.8507	0.0846	0.0647
0.6580	0.3093	0.0327	0.5255	0.4692	0.0053	0.8664	0.0552	0.0784
T/K = 303.15								
0.6825	0.2930	0.0245	0.4525	0.5452	0.0023	0.8372	0.1229	0.0399
0.6549	0.3175	0.0276	0.4425	0.5553	0.0022	0.8622	0.0875	0.0503
0.6256	0.3436	0.0308	0.4363	0.5610	0.0027	0.8692	0.0639	0.0669
0.6001	0.3657	0.0342	0.4227	0.5743	0.0030	0.8683	0.0457	0.0860
0.5772	0.3854	0.0374	0.4022	0.5953	0.0025	0.8827	0.0219	0.0954
T/K = 308.15								
0.6540	0.3294	0.0166	0.3339	0.6653	0.0008	0.8294	0.1450	0.0256
0.6347	0.3458	0.0195	0.3307	0.6683	0.0010	0.8651	0.0999	0.0350
0.6132	0.3645	0.0223	0.3247	0.6746	0.0007	0.8747	0.0847	0.0406
0.5940	0.3812	0.0248	0.3219	0.6770	0.0011	0.8852	0.0636	0.0512
0.5722	0.4000	0.0278	0.3063	0.6932	0.0005	0.8851	0.0555	0.0594

continued

continued

T/K = 313.15

0.6117	0.3780	0.0103	0.3163	0.6832	0.0005	0.8325	0.1499	0.0176
0.5988	0.3882	0.0130	0.3101	0.6892	0.0007	0.8558	0.1197	0.0245
0.5755	0.4091	0.0154	0.2976	0.7018	0.0006	0.8777	0.0911	0.0312
0.5421	0.4397	0.0182	0.2769	0.7226	0.0005	0.8980	0.0593	0.0427
0.5324	0.4457	0.0219	0.2673	0.7321	0.0006	0.9024	0.0458	0.0518

T/K = 318.15

0.6862	0.3041	0.0097	0.2680	0.7312	0.0008	0.8449	0.1421	0.0130
0.6415	0.3447	0.0138	0.2496	0.7498	0.0006	0.8830	0.0940	0.0230
0.5949	0.3863	0.0188	0.2319	0.7675	0.0006	0.8973	0.0673	0.0354
0.5487	0.4271	0.0242	0.2110	0.7884	0.0006	0.9018	0.0470	0.0512
0.5058	0.4660	0.0282	0.1775	0.8220	0.0005	0.9014	0.0341	0.0645

Polymer (B): **poly(propylene glycol)** **2005ZA1**
Characterization: M_n/g.mol^{-1} = 400, Aldrich Chem. Co., Inc., Milwaukee, WI
Solvent (A): **water** **H_2O** **7732-18-5**
Salt (C): **sodium sulfate** **Na_2SO_4** **7757-82-6**

Type of data: cloud points

T/K = 298.15

w_A	0.8217	0.8138	0.8028	0.7894	0.7770	0.7610	0.7495	0.7464	0.7445
w_B	0.1098	0.1228	0.1362	0.1538	0.1695	0.1890	0.2029	0.2070	0.2099
w_C	0.0685	0.0634	0.0610	0.0568	0.0535	0.0500	0.0476	0.0466	0.0456
w_A	0.7192	0.6924	0.6639	0.6345	0.6044	0.5676	0.5316	0.4950	0.4646
w_B	0.2402	0.2724	0.3064	0.3415	0.3762	0.4175	0.4572	0.4967	0.5291
w_C	0.0406	0.0352	0.0297	0.0240	0.0194	0.0149	0.0112	0.0083	0.0063
w_A	0.4404	0.3429							
w_B	0.5546	0.6552							
w_C	0.0050	0.0019							

Type of data: coexistence data (tie lines)

T/K = 298.15

Top phase			Bottom phase		
w_A	w_B	w_C	w_A	w_B	w_C
0.5580	0.4276	0.0144	0.8058	0.1220	0.0722
0.4612	0.5336	0.0052	0.8346	0.0665	0.0989
0.3733	0.6242	0.0025	0.8314	0.0474	0.1212
0.3049	0.6932	0.0019	0.8284	0.0174	0.1542
0.2445	0.7546	0.0009	0.8041	0.0018	0.1941
0.1185	0.8815	0.0000	0.6599	0.0000	0.3401

Polymer (B): **poly(propylene glycol)** **2004SA2**
Characterization: M_n/g.mol^{-1} = 405, Aldrich Chem. Co., Inc., Milwaukee, WI
Solvent (A): **water** **H_2O** **7732-18-5**
Salt (C): **sodium sulfate** **Na_2SO_4** **7757-82-6**

Type of data: coexistence data (tie lines)

T/K = 298.15

Total system			Top phase			Bottom phase		
w_A	w_B	w_C	w_A	w_B	w_C	w_A	w_B	w_C
0.638	0.333	0.029	0.440	0.555	0.005	0.798	0.151	0.051
0.657	0.312	0.032	0.469	0.528	0.003	0.757	0.198	0.045
0.646	0.301	0.053	0.393	0.604	0.003	0.835	0.077	0.088
0.604	0.344	0.052	0.356	0.642	0.002	0.840	0.059	0.101
0.597	0.324	0.080	0.315	0.683	0.002	0.841	0.014	0.145
0.569	0.343	0.088	0.289	0.709	0.002	0.831	0.008	0.161
0.524	0.357	0.119	0.240	0.759	0.001	0.783	0.001	0.216
0.460	0.390	0.150	0.216	0.783	0.001	0.705	0.000	0.295

Polymer (B): **poly(propylene glycol)** **2010SA1**
Characterization: M_n/g.mol^{-1} = 425, Aldrich Chem. Co., Inc., Milwaukee, WI
Solvent (A): **water** **H_2O** **7732-18-5**
Salt (C): **sodium sulfate** **Na_2SO_4** **7757-82-6**

Type of data: coexistence data (tie lines)

T/K = 298.15 pH = 7.6

Top phase			Bottom phase		
w_A	w_B	w_C	w_A	w_B	w_C
0.469	0.528	0.0030	0.757	0.198	0.045
0.393	0.604	0.0030	0.835	0.077	0.088
0.356	0.642	0.0020	0.840	0.059	0.101
0.315	0.683	0.0020	0.841	0.014	0.145

Polymer (B): **poly(*N*-vinylcaprolactam)** **2008FO3**
Characterization: M_n/g.mol^{-1} = 1150, synthesized in the laboratory
Solvent (A): **water** **H_2O** **7732-18-5**
Salt (C): **monoammonium phosphate** **$(NH_4)H_2PO_4$** **7722-76-1**

Type of data: cloud points

continued

continued

T/K = 298.15

w_A	0.9271	0.9267	0.9269	0.9203	0.9099	0.9029	0.8988	0.9271	0.9267
w_B	0.0446	0.0387	0.0333	0.0285	0.0277	0.0274	0.0274	0.0446	0.0387
w_C	0.0283	0.0346	0.0398	0.0512	0.0624	0.0697	0.0738	0.0283	0.0346

Polymer (B): **poly(*N*-vinylcaprolactam)** **2008FO3**
Characterization: M_n/g.mol^{-1} = 1150, synthesized in the laboratory
Solvent (A): **water** H_2O **7732-18-5**
Salt (C): **monopotassium phosphate** KH_2PO_4 **7778-77-0**

Type of data: cloud points

T/K = 298.15

w_A	0.9261	0.9307	0.9295	0.9215	0.9153	0.9125	0.9101	0.9261	0.9307
w_B	0.0453	0.0338	0.0266	0.0266	0.0264	0.0260	0.0260	0.0453	0.0338
w_C	0.0286	0.0355	0.0439	0.0519	0.0583	0.0615	0.0639	0.0286	0.0355

Polymer (B): **poly(*N*-vinylcaprolactam)** **2008FO3**
Characterization: M_n/g.mol^{-1} = 1150, synthesized in the laboratory
Solvent (A): **water** H_2O **7732-18-5**
Salt (C): **monosodium carbonate** $NaHCO_3$ **144-55-8**

Type of data: cloud points

T/K = 298.15

w_A	0.9433	0.9485	0.9534	0.9448	0.9379	0.9310
w_B	0.0425	0.0338	0.0237	0.0235	0.0230	0.0240
w_C	0.0142	0.0177	0.0229	0.0317	0.0391	0.0450

Polymer (B): **poly(*N*-vinylcaprolactam)** **2008FO3**
Characterization: M_n/g.mol^{-1} = 1150, synthesized in the laboratory
Solvent (A): **water** H_2O **7732-18-5**
Salt (C): **sodium chloride** $NaCl$ **7647-14-5**

Type of data: cloud points

T/K = 298.15

w_A	0.9498	0.9480	0.9459	0.9420	0.9382	0.9342	0.9219	0.9078	0.8966
w_B	0.0222	0.0210	0.0206	0.0190	0.0178	0.0158	0.0128	0.0126	0.0123
w_C	0.0280	0.0310	0.0335	0.0390	0.0440	0.0500	0.0653	0.0796	0.0911

w_A	0.8928	0.8880
w_B	0.0121	0.0121
w_C	0.0951	0.0999

Polymer (B):	**poly(1-vinyl-2-pyrrolidinone)**		**2008FO2**
Characterization:	M_w/g.mol^{-1} = 24000, Aldrich Chem. Co., Inc., Milwaukee, WI		
Solvent (A):	**water**	**H_2O**	**7732-18-5**
Salt (C):	**diammonium phosphate/monoammonium phosphate**		
	$(NH_4)_2HPO_4/(NH_4)H_2PO_4$		**7783-28-0/7722-76-1**

Type of data: cloud points

pH = 5.0 (prepared by a mixture of dibasic and monobasic ammonium phosphate)

T/K = 308.15

w_A	0.7210	0.7657	0.8008	0.8161	0.8244	0.8412	0.8440	0.8453	0.8439
w_B	0.1319	0.1329	0.1352	0.1368	0.1381	0.1438	0.1496	0.1507	0.1525
w_C	0.1471	0.1014	0.0640	0.0471	0.0375	0.0150	0.0064	0.0040	0.0036

w_A	0.8428	0.8393	0.8311
w_B	0.1542	0.1593	0.1679
w_C	0.0030	0.0014	0.0010

T/K = 318.15

w_A	0.7799	0.8536	0.8743	0.8767	0.8784	0.8804	0.8671
w_B	0.0900	0.0950	0.1000	0.1028	0.1052	0.1089	0.1319
w_C	0.1301	0.0514	0.0257	0.0205	0.0164	0.0107	0.0010

pH = 7.0 (prepared by a mixture of dibasic and monobasic ammonium phosphate)

T/K = 298.15

w_A	0.8562	0.8590	0.8588	0.8614	0.8625	0.8591	0.8523	0.8386	0.8194
w_B	0.1238	0.1270	0.1262	0.1285	0.1300	0.1359	0.1450	0.1600	0.1801
w_C	0.0200	0.0140	0.0150	0.0101	0.0075	0.0050	0.0027	0.0014	0.0005

T/K = 308.15

w_A	0.8877	0.8878	0.8878	0.8874	0.8891	0.8899	0.8856	0.8754	0.8724
w_B	0.0921	0.0931	0.0937	0.0944	0.0968	0.1063	0.1125	0.1238	0.1271
w_C	0.0202	0.0191	0.0185	0.0182	0.0141	0.0038	0.0019	0.0008	0.0005

T/K = 318.15

w_A	0.8983	0.9020	0.9045	0.9042	0.9036	0.9034	0.9030	0.9033	0.9028
w_B	0.0817	0.0858	0.0865	0.0873	0.0889	0.0921	0.0930	0.0937	0.0944
w_C	0.0200	0.0122	0.0090	0.0085	0.0075	0.0045	0.0040	0.0030	0.0028

w_A	0.9012	0.8927	0.8868
w_B	0.0968	0.1063	0.1125
w_C	0.0020	0.0010	0.0007

Type of data: coexistence data (tie lines)

continued

continued

Top phase w_A	w_B	w_C	Bottom phase w_A	w_B	w_C
T/K = 298.15	pH = 5				
0.7091	0.1402	0.1507	0.8058	0.0041	0.1901
0.7274	0.1201	0.1525	0.8108	0.0042	0.1850
0.7453	0.1005	0.1542	0.8152	0.0049	0.1799
T/K = 308.15	pH = 5				
0.7282	0.1401	0.1317	0.8346	0.0014	0.1640
0.7437	0.1222	0.1341	0.8389	0.0031	0.1580
0.7670	0.0998	0.1332	0.8438	0.0042	0.1520
T/K = 318.15	pH = 5				
0.7741	0.1368	0.0891	0.8665	0.0031	0.1304
0.8277	0.0801	0.0922	0.8800	0.0099	0.1101
0.8062	0.0997	0.0941	0.8716	0.0079	0.1205
T/K = 298.15	pH = 7				
0.8572	0.0166	0.1262	0.8194	0.0005	0.1801
0.8577	0.0141	0.1282	0.8289	0.0009	0.1702
0.8606	0.0123	0.1271	0.8386	0.0014	0.1600
T/K = 308.15	pH = 7				
0.8868	0.0193	0.0939	0.8717	0.0011	0.1272
0.8881	0.0160	0.0959	0.8781	0.0013	0.1206
0.8893	0.0126	0.0981	0.8822	0.0018	0.1160
T/K = 318.15	pH = 7				
0.8950	0.0013	0.1037	0.9006	0.0152	0.0842
0.8907	0.0007	0.1086	0.8998	0.0165	0.0837
0.8970	0.0022	0.1008	0.9018	0.0131	0.0851

Comments: The mixture of $(NH_4)_2HPO_4/(NH_4)H_2PO_4$ was chosen in a ratio that provides a constant pH-value in the feed phase. The ratio of $(NH_4)_2HPO_4/(NH_4)H_2PO_4$ in the coexisting phases may be different but was not measured. Only the total salt concentration in each phase was determined.

Polymer (B): **poly(1-vinyl-2-pyrrolidinone)** **2009ZA1**
Characterization: M_n/g.mol^{-1} = 3500, PVP-K12, Acros Organics, New Jersey
Solvent (A): **water** H_2O **7732-18-5**
Salt (C): **dipotassium oxalate** $C_2K_2O_4$ **583-52-8**

Type of data: cloud points

continued

continued

T/K = 298.15

w_A	0.5637	0.5906	0.6068	0.6284	0.6666	0.7055	0.7369
w_B	0.3934	0.3583	0.3358	0.3064	0.2513	0.1944	0.1468
w_C	0.0429	0.0511	0.0574	0.0652	0.0821	0.1001	0.1163

T/K = 308.15

w_A	0.5318	0.5358	0.5432	0.5536	0.5753	0.5970	0.6349	0.6603	0.6873
w_B	0.4378	0.4330	0.4241	0.4115	0.3849	0.3576	0.3085	0.2745	0.2373
w_C	0.0304	0.0312	0.0327	0.0349	0.0398	0.0454	0.0566	0.0652	0.0754

w_A	0.7142	0.7419
w_B	0.1999	0.1612
w_C	0.0859	0.0969

T/K = 318.15

w_A	0.5406	0.5731	0.5973	0.6258	0.6557	0.6825	0.7123	0.7370	0.7569
w_B	0.4278	0.3887	0.3589	0.3228	0.2839	0.2480	0.2077	0.1744	0.1465
w_C	0.0316	0.0382	0.0438	0.0514	0.0604	0.0695	0.0800	0.0886	0.0966

w_A	0.7762
w_B	0.1188
w_C	0.1050

Type of data: coexistence data (tie lines)

Top phase			Bottom phase		
w_A	w_B	w_C	w_A	w_B	w_C
T/K = 298.15					
0.5474	0.4146	0.0380	0.7663	0.0812	0.1525
0.5095	0.4614	0.0291	0.7700	0.0558	0.1742
0.4761	0.5020	0.0219	0.7582	0.0409	0.2009
0.4550	0.5262	0.0188	0.7453	0.0347	0.2200
0.4246	0.5608	0.0146	0.7300	0.0327	0.2373
0.4054	0.5825	0.0121	0.7147	0.0304	0.2549
T/K = 308.15					
0.5568	0.4089	0.0343	0.7909	0.0519	0.1572
0.5217	0.4519	0.0264	0.7858	0.0343	0.1799
0.4915	0.4873	0.0212	0.7782	0.0213	0.2005
0.4690	0.5127	0.0183	0.7628	0.0186	0.2186
0.4499	0.5339	0.0162	0.7515	0.0176	0.2309
0.4304	0.5557	0.0139	0.7364	0.0164	0.2472

continued

continued

T/K = 318.15

0.5648	0.3986	0.0366	0.7993	0.0554	0.1453
0.5087	0.4676	0.0237	0.7914	0.0286	0.1800
0.4831	0.4960	0.0209	0.7808	0.0198	0.1994
0.4555	0.5276	0.0169	0.7687	0.0144	0.2169
0.4305	0.5552	0.0143	0.7561	0.0086	0.2353
0.4138	0.5737	0.0125	0.7440	0.0048	0.2512

Polymer (B): **poly(1-vinyl-2-pyrrolidinone)** **2007FOR, 2008FO1**
Characterization: M_w/g.mol^{-1} = 24000, Aldrich Chem. Co., Inc., Milwaukee, WI
Solvent (A): **water** **H_2O** **7732-18-5**
Salt (C): **dipotassium phosphate/monopotassium phosphate**
K_2HPO_4/H_2KPO_4 **7758-11-4/7778-77-0**

Type of data: cloud points

pH = 5.0 (prepared by a mixture of dibasic and monobasic potassium phosphate)

T/K = 298.15

w_A	0.7029	0.7232	0.7279	0.7555	0.7858	0.7961	0.8079	0.8173	0.8194
w_B	0.1490	0.1261	0.1210	0.0910	0.0571	0.0446	0.0300	0.0150	0.0105
w_C	0.1481	0.1507	0.1511	0.1535	0.1571	0.1593	0.1621	0.1677	0.1701
w_A	0.8213	0.8199	0.8146	0.8078	0.7800				
w_B	0.0041	0.0011	0.0002	0.0002	0.0000				
w_C	0.1746	0.1790	0.1852	0.1920	0.2200				

T/K = 308.15

w_A	0.7139	0.7657	0.8161	0.8244	0.8413	0.8440	0.8444	0.8439	0.8410
w_B	0.1471	0.1014	0.0471	0.0375	0.0149	0.0064	0.0049	0.0036	0.0020
w_C	0.1390	0.1329	0.1368	0.1381	0.1438	0.1496	0.1507	0.1525	0.1570
w_A	0.8389	0.8240	0.8137	0.7995	0.7909	0.7794			
w_B	0.0018	0.0014	0.0013	0.0004	0.0002	0.0001			
w_C	0.1593	0.1746	0.1850	0.2001	0.2089	0.2205			

T/K = 318.15

w_A	0.7659	0.8609	0.8742	0.8764	0.8770	0.8789	0.8804	0.8760	0.8630
w_B	0.1471	0.0440	0.0257	0.0211	0.0202	0.0140	0.0107	0.0040	0.0002
w_C	0.0870	0.0951	0.1001	0.1025	0.1028	0.1071	0.1089	0.1200	0.1368

pH = 7.0 (prepared by a mixture of dibasic and monobasic potassium phosphate)

T/K = 298.15

w_A	0.8530	0.8570	0.8595	0.8605	0.8590	0.8522	0.8429	0.8382	0.7802
w_B	0.0199	0.0150	0.0102	0.0060	0.0051	0.0027	0.0020	0.0014	0.0003
w_C	0.1271	0.1280	0.1303	0.1335	0.1359	0.1451	0.1551	0.1604	0.2195

continued

continued

w_A	0.7768	0.7748	0.7714	0.7697
w_B	0.0002	0.0001	0.0001	0.0000
w_C	0.2230	0.2251	0.2285	0.2303

T/K = 308.15

w_A	0.8832	0.8867	0.8884	0.8925	0.8902	0.8855	0.8752	0.8723	0.8691
w_B	0.0201	0.0160	0.0142	0.0070	0.0035	0.0020	0.0010	0.0007	0.0006
w_C	0.0967	0.0973	0.0974	0.1005	0.1063	0.1125	0.1238	0.1270	0.1303

w_A	0.8660	0.8215
w_B	0.0005	0.0001
w_C	0.1335	0.1784

T/K = 318.15

w_A	0.8982	0.9016	0.9049	0.9046	0.9051	0.9042	0.9041	0.9040	0.9036
w_B	0.0201	0.0162	0.0110	0.0102	0.0089	0.0085	0.0080	0.0075	0.0075
w_C	0.0817	0.0822	0.0841	0.0852	0.0860	0.0873	0.0879	0.0885	0.0889

w_A	0.9034	0.9032	0.9026	0.9023	0.8978	0.8927	0.8868	0.8758	0.8726
w_B	0.0045	0.0043	0.0039	0.0033	0.0021	0.0010	0.0007	0.0004	0.0003
w_C	0.0921	0.0925	0.0935	0.0944	0.1001	0.1063	0.1125	0.1238	0.1271

Type of data: coexistence data (tie lines)

Top phase			Bottom phase		
w_A	w_B	w_C	w_A	w_B	w_C
T/K = 298.15	pH = 5				
0.7188	0.1313	0.1499	0.7901	0.0001	0.2098
0.7473	0.1002	0.1525	0.7998	0.0002	0.2000
0.7657	0.0801	0.1542	0.8112	0.0003	0.1885
T/K = 308.15	pH = 5				
0.7478	0.1200	0.1322	0.8077	0.0003	0.1920
0.7860	0.0800	0.1340	0.8308	0.0013	0.1679
0.8008	0.0640	0.1352	0.8366	0.0013	0.1621
T/K = 318.15	pH = 5				
0.7820	0.1300	0.0880	0.8079	0.0001	0.1920
0.8105	0.1004	0.0891	0.8502	0.0002	0.1496
0.8301	0.0799	0.0900	0.8671	0.0010	0.1319
T/K = 298.15	pH = 7				
0.8555	0.0170	0.1275	0.7856	0.0004	0.2140
0.8577	0.0131	0.1292	0.8095	0.0007	0.1898
0.8600	0.0075	0.1325	0.8189	0.0009	0.1802

continued

continued

T/K = 308.15 pH = 7

0.8850	0.0180	0.0970	0.8402	0.0002	0.1596
0.8905	0.0120	0.0975	0.8548	0.0002	0.1450
0.8920	0.0100	0.0980	0.8638	0.0004	0.1358

T/K = 318.15 pH = 7

0.9011	0.0170	0.0819	0.8499	0.0001	0.1500
0.9020	0.0155	0.0825	0.8645	0.0001	0.1354
0.9030	0.0140	0.0830	0.8696	0.0002	0.1302

Comments: The mixture of K_2HPO_4/H_2KPO_4 was chosen in a ratio that provides a constant pH-value in the feed phase. The ratio of K_2HPO_4/H_2KPO_4 in the coexisting phases may be different but was not measured. Only the total salt concentration in each phase was determined.

Polymer (B): **poly(1-vinyl-2-pyrrolidinone)** **2009ZA1**
Characterization: M_n/g.mol^{-1} = 3500, PVP-K12, Acros Organics, New Jersey
Solvent (A): **water** H_2O **7732-18-5**
Salt (C): **dipotassium tartrate** $C_4H_4K_2O_6$ **6100-19-2**

Type of data: cloud points

T/K = 298.15

w_A	0.5529	0.5666	0.5823	0.5934	0.6030	0.6108	0.6188	0.6386	0.6612
w_B	0.3961	0.3777	0.3551	0.3387	0.3249	0.3123	0.2994	0.2692	0.2334
w_C	0.0510	0.0557	0.0626	0.0679	0.0721	0.0769	0.0818	0.0922	0.1054

w_A	0.6699	0.6842	0.7021	0.7216
w_B	0.2189	0.1947	0.1650	0.1326
w_C	0.1112	0.1211	0.1329	0.1458

T/K = 308.15

w_A	0.5873	0.5968	0.6089	0.6373	0.6623	0.6855	0.7046	0.7199	0.7335
w_B	0.3484	0.3344	0.3174	0.2746	0.2347	0.1982	0.1671	0.1433	0.1213
w_C	0.0643	0.0688	0.0737	0.0881	0.1030	0.1163	0.1283	0.1368	0.1452

T/K = 318.15

w_A	0.5649	0.5803	0.6035	0.6258	0.6475	0.6733	0.6945	0.7145	0.7321
w_B	0.3806	0.3596	0.3269	0.2949	0.2625	0.2233	0.1907	0.1602	0.1341
w_C	0.0545	0.0601	0.0696	0.0793	0.0900	0.1034	0.1148	0.1253	0.1338

w_A	0.7457	0.7293
w_B	0.1130	0.0962
w_C	0.1413	0.1745

continued

continued

Type of data: coexistence data (tie lines)

Top phase			Bottom phase		
w_A	w_B	w_C	w_A	w_B	w_C
T/K = 298.15					
0.5649	0.3772	0.0579	0.7431	0.2311	0.0258
0.5145	0.4455	0.0400	0.7238	0.2593	0.0169
0.4820	0.4875	0.0305	0.7049	0.2843	0.0108
0.4501	0.5269	0.0230	0.6794	0.3127	0.0079
0.4274	0.5545	0.0181	0.6583	0.3356	0.0061
T/K = 308.15					
0.5273	0.4302	0.0425	0.7509	0.0277	0.2214
0.5145	0.4470	0.0385	0.7430	0.0200	0.2370
0.4878	0.4811	0.0311	0.7282	0.0176	0.2542
0.4684	0.5046	0.0270	0.7132	0.0124	0.2744
0.4604	0.5146	0.0250	0.7062	0.0115	0.2823
T/K = 318.15					
0.5310	0.4273	0.0417	0.7498	0.0421	0.2081
0.5174	0.4442	0.0384	0.7448	0.0296	0.2256
0.4985	0.4682	0.0333	0.7348	0.0197	0.2455
0.4767	0.4941	0.0292	0.7276	0.0141	0.2583
0.4630	0.5114	0.0256	0.7174	0.0097	0.2729
0.4504	0.5258	0.0238	0.7100	0.0084	0.2816

Polymer (B): **poly(1-vinyl-2-pyrrolidinone)** **2005ZA2**
Characterization: M_w/g.mol^{-1} = 10000, Aldrich Chem. Co., Inc., Milwaukee, WI
Solvent (A): **water** H_2O **7732-18-5**
Salt (C): **disodium phosphate** Na_2HPO_4 **7558-79-4**

Type of data: cloud points

T/K = 298.15

w_A	0.6216	0.6464	0.6648	0.6991	0.7191	0.7374	0.7602	0.7838	0.8029
w_B	0.3508	0.3214	0.2990	0.2574	0.2323	0.2090	0.1803	0.1511	0.1276
w_C	0.0276	0.0322	0.0362	0.0435	0.0486	0.0536	0.0595	0.0651	0.0695

T/K = 308.15

w_A	0.6194	0.6454	0.6783	0.6931	0.7215	0.7399	0.7513	0.7709	0.7840
w_B	0.3528	0.3222	0.2816	0.2646	0.2297	0.2067	0.1926	0.1688	0.1527
w_C	0.0278	0.0324	0.0401	0.0423	0.0488	0.0534	0.0561	0.0603	0.0633

continued

continued

w_A	0.7921	0.8082
w_B	0.1429	0.1240
w_C	0.0650	0.0678

T/K = 318.15

w_A	0.6122	0.6340	0.6468	0.6661	0.6908	0.7325	0.7504	0.7676	0.7830
w_B	0.3595	0.3349	0.3195	0.2967	0.2675	0.2169	0.1953	0.1749	0.1568
w_C	0.0283	0.0311	0.0337	0.0372	0.0417	0.0506	0.0543	0.0575	0.0602

w_A	0.7913	0.8030	0.8169
w_B	0.1469	0.1333	0.1174
w_C	0.0618	0.0637	0.0657

T/K = 328.15

w_A	0.5480	0.5667	0.5803	0.5929	0.6150	0.6398	0.6680	0.7013	0.7324
w_B	0.4321	0.4115	0.3950	0.3805	0.3554	0.3270	0.2942	0.2556	0.2188
w_C	0.0199	0.0218	0.0247	0.0266	0.0296	0.0332	0.0378	0.0431	0.0488

w_A	0.7616	0.7824	0.7973
w_B	0.1848	0.1601	0.1435
w_C	0.0536	0.0575	0.0592

Type of data: coexistence data (tie lines)

Top phase			Bottom phase		
w_A	w_B	w_C	w_A	w_B	w_C
T/K = 298.15					
0.6972	0.2608	0.0420	0.8266	0.0817	0.0917
0.6772	0.2837	0.0391	0.8308	0.0675	0.1017
0.6402	0.3297	0.0301	0.8375	0.0468	0.1157
0.6219	0.3500	0.0281	0.8331	0.0428	0.1241
0.6065	0.3651	0.0284	0.8245	0.0441	0.1314
0.5695	0.4087	0.0218	0.8108	0.0361	0.1531
T/K = 308.15					
0.6568	0.3076	0.0356	0.8352	0.0678	0.0970
0.6179	0.3531	0.0290	0.8340	0.0482	0.1178
0.5641	0.4144	0.0215	0.8264	0.0404	0.1332
0.5066	0.4783	0.0151	0.8101	0.0338	0.1561
T/K = 318.15					
0.8265	0.0947	0.0788	0.6721	0.2902	0.0377
0.8298	0.0793	0.0909	0.6410	0.3250	0.0340
0.5928	0.3830	0.0242	0.8360	0.0573	0.1067
0.5630	0.4160	0.0210	0.8270	0.0533	0.1197
0.5208	0.4640	0.0152	0.8218	0.0400	0.1382

continued

continued

T/K = 328.15

0.8354	0.0834	0.0812	0.6677	0.2967	0.0356
0.8406	0.0603	0.0991	0.6101	0.3623	0.0276
0.5690	0.4094	0.0216	0.8416	0.0361	0.1223
0.5240	0.4586	0.0174	0.8328	0.0272	0.1400
0.4826	0.5024	0.0150	0.8225	0.0229	0.1546

Polymer (B): **poly(1-vinyl-2-pyrrolidinone)** **2005SAD**
Characterization: M_w/g.mol^{-1} = 10000, Aldrich Chem. Co., Inc., Milwaukee, WI
Solvent (A): **water** **H_2O** **7732-18-5**
Salt (C): **disodium succinate** **$C_4H_4Na_2O_4$** **150-90-3**

Type of data: cloud points

T/K = 303.15

w_A	0.5617	0.5691	0.5753	0.5836	0.6106	0.6236	0.6387	0.6542	0.6703
w_B	0.3653	0.3542	0.3461	0.3352	0.2995	0.2819	0.2618	0.2410	0.2195
w_C	0.0730	0.0767	0.0786	0.0812	0.0899	0.0945	0.0995	0.1048	0.1102

w_A	0.6858	0.7006	0.7142	0.7393
w_B	0.1987	0.1790	0.1613	0.1296
w_C	0.1155	0.1204	0.1245	0.1311

Type of data: coexistence data (tie lines)

T/K = 303.15

Top phase			Bottom phase		
w_A	w_B	w_C	w_A	w_B	w_C
0.7509	0.0885	0.1606	0.5834	0.3387	0.0779
0.7549	0.0850	0.1601	0.5716	0.3500	0.0784
0.5035	0.4408	0.0557	0.7499	0.0500	0.2001
0.4240	0.5380	0.0380	0.7399	0.0237	0.2364
0.3730	0.5990	0.0280	0.7104	0.0180	0.2716

Polymer (B): **poly(1-vinyl-2-pyrrolidinone)** **2006SAL**
Characterization: M_w/g.mol^{-1} = 10000, Aldrich Chem. Co., Inc., Milwaukee, WI
Solvent (A): **water** **H_2O** **7732-18-5**
Salt (C): **magnesium sulfate** **$MgSO_4$** **7487-88-9**

Type of data: coexistence data (tie lines)

continued

continued

Top phase			Bottom phase		
w_A	w_B	w_C	w_A	w_B	w_C
T/K = 298.15					
0.670	0.284	0.047	0.818	0.110	0.072
0.630	0.330	0.044	0.814	0.104	0.082
0.560	0.410	0.035	0.811	0.086	0.103
0.535	0.436	0.031	0.810	0.058	0.132
0.530	0.447	0.028	0.801	0.055	0.144
T/K = 303.15					
0.655	0.307	0.043	0.830	0.098	0.078
0.624	0.340	0.041	0.830	0.090	0.087
0.550	0.425	0.031	0.820	0.074	0.113
0.520	0.458	0.026	0.815	0.051	0.138
0.500	0.485	0.025	0.810	0.031	0.162
T/K = 308.15					
0.650	0.315	0.041	0.830	0.117	0.096
0.615	0.362	0.031	0.825	0.110	0.102
0.550	0.430	0.026	0.823	0.094	0.121
0.510	0.471	0.025	0.820	0.069	0.146
0.486	0.490	0.024	0.816	0.026	0.184

Polymer (B): **poly(1-vinyl-2-pyrrolidinone)** **2006SA1**
Characterization: M_w/g.mol^{-1} = 10000, Aldrich Chem. Co., Inc., Milwaukee, WI
Solvent (A): **water** **H_2O** **7732-18-5**
Salt (C): **potassium citrate** **$K_3C_6H_5O_7$** **866-84-2**

Type of data: cloud points

T/K = 298.15

w_A	0.5567	0.5687	0.5901	0.6023	0.6168	0.6341	0.6535	0.6732	0.6911
w_B	0.4003	0.3855	0.3563	0.3405	0.3209	0.2968	0.2699	0.2420	0.2163
w_C	0.0430	0.0458	0.0536	0.0572	0.0623	0.0691	0.0766	0.0848	0.0926
w_A	0.7087	0.7206	0.7479	0.7647					
w_B	0.1906	0.1736	0.1300	0.0950					
w_C	0.1007	0.1058	0.1221	0.1403					

T/K = 308.15

w_A	0.5649	0.5714	0.5848	0.5972	0.6133	0.6328	0.6534	0.6742	0.6934
w_B	0.3930	0.3846	0.3669	0.3506	0.3296	0.3037	0.2756	0.2462	0.2193
w_C	0.0421	0.0440	0.0483	0.0522	0.0571	0.0635	0.0710	0.0796	0.0873

continued

continued

w_A	0.7109	0.7227	0.7554	0.7715
w_B	0.1952	0.1781	0.1278	0.0980
w_C	0.0939	0.0992	0.1168	0.1305

T/K = 318.15

w_A	0.5738	0.5888	0.5937	0.6044	0.6181	0.6365	0.6515	0.6644	0.6800
w_B	0.3849	0.3656	0.3593	0.3459	0.3279	0.3037	0.2836	0.2663	0.2448
w_C	0.0413	0.0456	0.0470	0.0497	0.0540	0.0598	0.0649	0.0693	0.0752

w_A	0.6950	0.7065	0.7193	0.7410	0.7625	0.7772	0.7843
w_B	0.2242	0.2085	0.1919	0.1598	0.1265	0.1010	0.0852
w_C	0.0808	0.0850	0.0888	0.0992	0.1110	0.1218	0.1305

Type of data: coexistence data (tie lines)

Top phase			Bottom phase		
w_A	w_B	w_C	w_A	w_B	w_C
T/K = 298.15					
0.6749	0.2406	0.0845	0.7443	0.1260	0.1297
0.6466	0.2795	0.0739	0.7589	0.0825	0.1586
0.5771	0.3745	0.0484	0.7596	0.0421	0.1983
0.5529	0.4046	0.0425	0.7483	0.0353	0.2164
0.5231	0.4420	0.0349	0.7408	0.0261	0.2331
0.4650	0.5133	0.0217	0.6977	0.0193	0.2830
T/K = 308.15					
0.6424	0.2888	0.0688	0.7668	0.0963	0.1369
0.6147	0.3276	0.0577	0.7687	0.0784	0.1529
0.5586	0.3968	0.0446	0.7629	0.0516	0.1855
0.5472	0.4140	0.0388	0.7600	0.0493	0.1907
0.5052	0.4631	0.0317	0.7500	0.0347	0.2153
0.4370	0.5448	0.0182	0.7032	0.0290	0.2678
T/K = 318.15					
0.6253	0.3161	0.0586	0.7807	0.0641	0.1552
0.5781	0.3811	0.0408	0.7789	0.0447	0.1764
0.5222	0.4477	0.0301	0.7657	0.0328	0.2015
0.5164	0.4553	0.0283	0.7652	0.0315	0.2033
0.4881	0.4828	0.0291	0.7575	0.0240	0.2185

Polymer (B):	**poly(1-vinyl-2-pyrrolidinone)**		**2005SAD**
Characterization:	M_w/g.mol^{-1} = 10000, Aldrich Chem. Co., Inc., Milwaukee, WI		
Solvent (A):	**water**	**H_2O**	**7732-18-5**
Salt (C):	**sodium carbonate**	**Na_2CO_3**	**497-19-8**

Type of data: cloud points

T/K = 303.15

w_A	0.6310	0.6365	0.6474	0.6654	0.6861	0.7108	0.7377	0.7672	0.7922
w_B	0.3465	0.3398	0.3272	0.3067	0.2826	0.2538	0.2221	0.1874	0.1581
w_C	0.0225	0.0237	0.0254	0.0279	0.0313	0.0354	0.0402	0.0454	0.0497

w_A	0.8139	0.8325	0.8503
w_B	0.1327	0.1114	0.0911
w_C	0.0534	0.0561	0.0586

Type of data: coexistence data (tie lines)

T/K = 303.15

Top phase			Bottom phase		
w_A	w_B	w_C	w_A	w_B	w_C
0.8726	0.0473	0.0801	0.6475	0.3274	0.0251
0.8809	0.0274	0.0917	0.6079	0.3726	0.0195
0.5614	0.4236	0.0150	0.8705	0.0200	0.1095
0.5219	0.4636	0.0145	0.8619	0.0052	0.1329
0.5217	0.4645	0.0138	0.8600	0.0059	0.1341
0.4845	0.5036	0.0119	0.8469	0.0050	0.1481

Polymer (B):	**poly(1-vinyl-2-pyrrolidinone)**		**2006SA2**
Characterization:	M_w/g.mol^{-1} = 10000, Aldrich Chem. Co., Inc., Milwaukee, WI		
Solvent (A):	**water**	**H_2O**	**7732-18-5**
Salt (C):	**sodium citrate**	**$Na_3C_6H_5O_7$**	**68-04-2**

Type of data: cloud points

T/K = 298.15

w_A	0.5985	0.6029	0.6187	0.6356	0.6605	0.6823	0.7052	0.7101	0.7323
w_B	0.3574	0.3502	0.3304	0.3078	0.2753	0.2448	0.2131	0.2058	0.1760
w_C	0.0441	0.0469	0.0509	0.0566	0.0642	0.0729	0.0817	0.0841	0.0917

w_A	0.7520	0.7703	0.7887
w_B	0.1492	0.1250	0.1017
w_C	0.0988	0.1047	0.1096

continued

continued

T/K = 308.15

w_A	0.5924	0.6068	0.6286	0.6367	0.6590	0.6840	0.7051	0.7284	0.7515
w_B	0.3632	0.3452	0.3188	0.3069	0.2777	0.2446	0.2155	0.1845	0.1538
w_C	0.0444	0.0480	0.0526	0.0564	0.0633	0.0714	0.0794	0.0871	0.0947
w_A	0.7715	0.7736	0.7794	0.7883					
w_B	0.1278	0.1252	0.1184	0.1070					
w_C	0.1007	0.1012	0.1022	0.1047					

T/K = 318.15

w_A	0.6032	0.6126	0.6258	0.6394	0.6595	0.6848	0.7113	0.7366	0.7398
w_B	0.3511	0.3392	0.3235	0.3052	0.2792	0.2466	0.2125	0.1798	0.1758
w_C	0.0457	0.0482	0.0507	0.0554	0.0613	0.0686	0.0762	0.0836	0.0844
w_A	0.7582	0.7681	0.7797	0.7920	0.7984				
w_B	0.1516	0.1390	0.1248	0.1103	0.1019				
w_C	0.0902	0.0929	0.0955	0.0977	0.0997				

T/K = 328.15

w_A	0.5908	0.5995	0.6196	0.6340	0.6519	0.6739	0.6985	0.7236	0.7430
w_B	0.3669	0.3564	0.3326	0.3153	0.2921	0.2645	0.2332	0.2010	0.1763
w_C	0.0423	0.0441	0.0478	0.0507	0.0560	0.0616	0.0683	0.0754	0.0807
w_A	0.7705	0.7960							
w_B	0.1410	0.1090							
w_C	0.0885	0.0950							

Type of data: coexistence data (tie lines)

Top phase			Bottom phase		
w_A	w_B	w_C	w_A	w_B	w_C
T/K = 298.15					
0.6305	0.3152	0.0543	0.7874	0.0545	0.1581
0.6042	0.3490	0.0468	0.7878	0.0375	0.1747
0.5740	0.3862	0.0398	0.7818	0.0330	0.1852
0.5255	0.4436	0.0309	0.7637	0.0200	0.2163
0.4541	0.5271	0.0188	0.7234	0.0116	0.2650
T/K = 308.15					
0.6055	0.3480	0.0465	0.7979	0.0462	0.1559
0.5824	0.3741	0.0435	0.7917	0.0309	0.1774
0.5241	0.4447	0.0312	0.7734	0.0204	0.2062
0.4659	0.5152	0.0189	0.7426	0.0098	0.2476

continued

continued

T/K = 318.15

0.5931	0.3630	0.0439	0.8049	0.0404	0.1547
0.5609	0.4010	0.0381	0.8007	0.0228	0.1765
0.5512	0.4114	0.0374	0.7967	0.0252	0.1781
0.5157	0.4567	0.0276	0.7855	0.0241	0.1904

T/K = 328.15

0.8172	0.0616	0.1212	0.6750	0.2635	0.0615
0.8205	0.0400	0.1395	0.6204	0.3309	0.0487
0.5789	0.3807	0.0404	0.8131	0.0356	0.1513
0.5493	0.4157	0.0350	0.8024	0.0309	0.1667
0.4857	0.4921	0.0222	0.7766	0.0264	0.1970

Polymer (B): **poly(1-vinyl-2-pyrrolidinone)** **2005ZA2**
Characterization: M_w/g.mol^{-1} = 10000, Aldrich Chem. Co., Inc., Milwaukee, WI
Solvent (A): **water** **H_2O** **7732-18-5**
Salt (C): **sodium phosphate** **Na_3PO_4** **7601-54-9**

Type of data: cloud points

T/K = 298.15

w_A	0.6229	0.6243	0.6378	0.6435	0.6494	0.6561	0.6699	0.6722	0.6930
w_B	0.3617	0.3602	0.3450	0.3385	0.3318	0.3241	0.3083	0.3056	0.2816
w_C	0.0154	0.0155	0.0172	0.0180	0.0188	0.0198	0.0218	0.0222	0.0254
w_A	0.6945	0.7230	0.7483	0.7752	0.7900	0.8039	0.8152		
w_B	0.2798	0.2466	0.2161	0.1841	0.1657	0.1487	0.1350		
w_C	0.0257	0.0304	0.0356	0.0407	0.0443	0.0474	0.0498		

T/K = 308.15

w_A	0.6192	0.6268	0.6423	0.6653	0.6910	0.7187	0.7437	0.7665	0.7863
w_B	0.3651	0.3565	0.3390	0.3132	0.2833	0.2512	0.2218	0.1949	0.1710
w_C	0.0157	0.0167	0.0187	0.0215	0.0257	0.0301	0.0345	0.0386	0.0427
w_A	0.8062	0.8233	0.8374						
w_B	0.1474	0.1274	0.1110						
w_C	0.0464	0.0493	0.0516						

T/K = 318.15

w_A	0.6130	0.6166	0.6260	0.6430	0.6652	0.6889	0.7106	0.7370	0.7593
w_B	0.3711	0.3671	0.3567	0.3378	0.3128	0.2861	0.2611	0.2304	0.2043
w_C	0.0159	0.0163	0.0173	0.0192	0.0220	0.0250	0.0283	0.0326	0.0364
w_A	0.7821	0.7938	0.8038						
w_B	0.1777	0.1641	0.1525						
w_C	0.0402	0.0421	0.0437						

continued

continued

T/K = 328.15

w_A	0.5974	0.6059	0.6165	0.6300	0.6501	0.6737	0.7017	0.7276	0.7526
w_B	0.3881	0.3786	0.3671	0.3521	0.3300	0.3040	0.2727	0.2429	0.2140
w_C	0.0145	0.0155	0.0164	0.0179	0.0199	0.0223	0.0256	0.0295	0.0334

w_A	0.7773	0.7974	0.8075
w_B	0.1860	0.1627	0.1510
w_C	0.0367	0.0399	0.0415

Type of data: coexistence data (tie lines)

Top phase			Bottom phase		
w_A	w_B	w_C	w_A	w_B	w_C
T/K = 298.15					
0.7275	0.2408	0.0317	0.8850	0.0400	0.0750
0.6678	0.3109	0.0213	0.8978	0.0159	0.0863
0.6453	0.3362	0.0185	0.8967	0.0120	0.0913
0.6385	0.3432	0.0183	0.8967	0.0059	0.0974
0.6037	0.3831	0.0132	0.8883	0.0060	0.1057
T/K = 308.15					
0.6805	0.2960	0.0235	0.8836	0.0447	0.0717
0.6574	0.3216	0.0210	0.8996	0.0150	0.0854
0.6147	0.3697	0.0156	0.8990	0.0070	0.0940
0.5959	0.3903	0.0138	0.8962	0.0020	0.1018
0.5734	0.4147	0.0119	0.8917	0.0013	0.1070
0.5412	0.4490	0.0098	0.8806	0.0010	0.1184
T/K = 318.15					
0.8586	0.0853	0.0561	0.7297	0.2388	0.0315
0.8722	0.0650	0.0628	0.6990	0.2745	0.0265
0.8910	0.0370	0.0720	0.6623	0.3160	0.0217
0.6313	0.3500	0.0187	0.8954	0.0188	0.0858
0.5945	0.3904	0.0151	0.8994	0.0010	0.0996
T/K = 328.15					
0.8783	0.0545	0.0672	0.6625	0.3162	0.0213
0.8907	0.0291	0.0802	0.6216	0.3611	0.0173
0.5838	0.4014	0.0148	0.8949	0.0100	0.0951
0.5464	0.4398	0.0138	0.8853	0.0070	0.1077
0.5186	0.4689	0.0125	0.8716	0.0050	0.1234

Polymer (B): **poly(1-vinyl-2-pyrrolidinone)** **2007FED**
Characterization: M_n/g.mol^{-1} = 3880, PVPK17,
BASF AG, Ludwigshafen, Germany
Solvent (A): **water** H_2O **7732-18-5**
Salt (C): **sodium sulfate** Na_2SO_4 **7757-82-6**

Type of data: cloud points

T/K = 298.15

w_A	0.8484	0.8578	0.8631	0.8539	0.8135	0.8370	0.8701	0.8480	0.8777
w_B	0.0012	0.0018	0.0020	0.0022	0.0029	0.0035	0.0036	0.0042	0.0048
w_C	0.1504	0.1404	0.1349	0.1439	0.1836	0.1595	0.1263	0.1478	0.1175
w_A	0.8805	0.8776	0.8776	0.8748	0.8709	0.8709	0.8673	0.8629	0.8585
w_B	0.0048	0.0182	0.0242	0.0279	0.0336	0.0343	0.0380	0.0444	0.0497
w_C	0.1147	0.1042	0.0982	0.0973	0.0955	0.0948	0.0947	0.0927	0.0918
w_A	0.8542	0.8521	0.8474	0.8455	0.8417	0.8371	0.8301	0.8295	0.8242
w_B	0.0570	0.0584	0.0655	0.0673	0.0712	0.0782	0.0836	0.0878	0.0949
w_C	0.0888	0.0895	0.0871	0.0872	0.0871	0.0847	0.0863	0.0827	0.0809
w_A	0.8168	0.8002	0.7935	0.7903	0.7014	0.6442			
w_B	0.1074	0.1302	0.1371	0.1421	0.2531	0.3218			
w_C	0.0758	0.0696	0.0694	0.0676	0.0455	0.0340			

T/K = 338.15

w_A	0.8107	0.8491	0.8312	0.8596	0.8412	0.8872	0.8771	0.8779	0.8483
w_B	0.0007	0.0011	0.0012	0.0012	0.0015	0.0021	0.0024	0.0031	0.0032
w_C	0.1886	0.1498	0.1676	0.1392	0.1573	0.1107	0.1205	0.1190	0.1485
w_A	0.8964	0.8964	0.8787	0.8953	0.8953	0.8880	0.8833	0.8697	0.8614
w_B	0.0035	0.0035	0.0036	0.0036	0.0036	0.0214	0.0276	0.0484	0.0532
w_C	0.1001	0.1001	0.1177	0.1011	0.1011	0.0906	0.0891	0.0819	0.0854
w_A	0.8630	0.8525	0.8437	0.8411	0.8306	0.8276	0.8172	0.8181	0.8050
w_B	0.0618	0.0724	0.0814	0.0858	0.0932	0.0972	0.1086	0.1104	0.1243
w_C	0.0752	0.0751	0.0749	0.0731	0.0762	0.0752	0.0742	0.0715	0.0707
w_A	0.7983	0.7900	0.7803	0.7848	0.7745	0.7680	0.7551	0.7542	0.7256
w_B	0.1329	0.1437	0.1567	0.1591	0.1657	0.1719	0.1917	0.1922	0.2241
w_C	0.0688	0.0663	0.0630	0.0561	0.0598	0.0601	0.0532	0.0536	0.0503
w_A	0.7229	0.6944	0.6896	0.6538	0.6230				
w_B	0.2298	0.2617	0.2676	0.3131	0.3456				
w_C	0.0473	0.0439	0.0428	0.0331	0.0314				

Type of data: coexistence data (tie lines)

continued

continued

Total system			Top phase			Bottom phase		
w_A	w_B	w_C	w_A	w_B	w_C	w_A	w_B	w_C
T/K = 298.15								
0.8286	0.0644	0.1070	0.7770	0.1551	0.0679	0.8640	0.0111	0.1249
0.8132	0.0560	0.1308	0.6644	0.2672	0.0684	0.8436	0.0144	0.1420
0.7894	0.0884	0.1222	0.5500	0.4308	0.0192	0.8385	0.0156	0.1459
0.7931	0.0601	0.1468	0.4949	0.4764	0.0287	0.8264	0.0102	0.1634
T/K = 338.15								
0.8137	0.0632	0.1231	0.5847	0.3748	0.0405	0.8516	0.0101	0.1383
0.7875	0.0768	0.1357	0.5733	0.3942	0.0325	0.8408	0.0068	0.1524
0.7877	0.0820	0.1303	0.5527	0.4172	0.0301	0.8398	0.0073	0.1529
0.7978	0.0997	0.1025	0.5421	0.4325	0.0254	0.8440	0.0338	0.1222
0.7561	0.0724	0.1715	0.4419	0.5476	0.0105	0.7990	0.0051	0.1959

Polymer (B): **poly(1-vinyl-2-pyrrolidinone)** **2005SAD**
Characterization: M_w/g.mol^{-1} = 10000, Aldrich Chem. Co., Inc., Milwaukee, WI
Solvent (A): **water** H_2O **7732-18-5**
Salt (C): **sodium sulfate** Na_2SO_4 **7757-82-6**

Type of data: cloud points

T/K = 303.15

w_A	0.6047	0.6243	0.6376	0.6610	0.6995	0.7047	0.7208	0.7232	0.7434
w_B	0.3675	0.3439	0.3283	0.2997	0.2527	0.2461	0.2265	0.2226	0.1974
w_C	0.0278	0.0318	0.0341	0.0393	0.0478	0.0492	0.0527	0.0542	0.0592
w_A	0.7686	0.7915	0.8087	0.8208					
w_B	0.1659	0.1375	0.1167	0.1019					
w_C	0.0655	0.0710	0.0746	0.0773					

Type of data: coexistence data (tie lines)

T/K = 303.15

Top phase			Bottom phase		
w_A	w_B	w_C	w_A	w_B	w_C
0.6981	0.2528	0.0491	0.8380	0.0750	0.0870
0.6609	0.2991	0.0400	0.8525	0.0455	0.1020
0.6132	0.3568	0.0300	0.8435	0.0317	0.1248
0.5695	0.4065	0.0240	0.8301	0.0178	0.1521
0.5093	0.4704	0.0203	0.8044	0.0166	0.1790

Polymer (B): **poly(1-vinyl-2-pyrrolidinone)** **2007FED**
Characterization: M_n/g.mol^{-1} = 17750, PVPK30,
BASF AG, Ludwigshafen, Germany
Solvent (A): **water** H_2O **7732-18-5**
Salt (C): **sodium sulfate** Na_2SO_4 **7757-82-6**

Type of data: cloud points

T/K = 298.15

w_A	0.8102	0.8242	0.8370	0.8397	0.8737	0.8549	0.8668	0.8835	0.8993
w_B	0.0003	0.0004	0.0005	0.0008	0.0012	0.0012	0.0013	0.0018	0.0022
w_C	0.1895	0.1754	0.1625	0.1595	0.1251	0.1439	0.1319	0.1147	0.0985
w_A	0.8668	0.8583	0.8452	0.8157	0.8005	0.7748	0.7598	0.7426	0.7194
w_B	0.0637	0.0772	0.0918	0.1296	0.1503	0.1778	0.1946	0.2147	0.2414
w_C	0.0695	0.0645	0.0630	0.0547	0.0492	0.0474	0.0456	0.0427	0.0392
w_A	0.6906	0.6414							
w_B	0.2725	0.3255							
w_C	0.0369	0.0331							

T/K = 338.15

w_A	0.8102	0.8242	0.8370	0.8397	0.8737	0.8549	0.8668	0.8835	0.8993
w_B	0.0003	0.0004	0.0005	0.0008	0.0012	0.0012	0.0013	0.0018	0.0022
w_C	0.1895	0.1754	0.1625	0.1595	0.1251	0.1439	0.1319	0.1147	0.0985
w_A	0.8668	0.8583	0.8452	0.8157	0.8005	0.7748	0.7598	0.7426	0.7194
w_B	0.0637	0.0772	0.0918	0.1296	0.1503	0.1778	0.1946	0.2147	0.2414
w_C	0.0695	0.0645	0.0630	0.0547	0.0492	0.0474	0.0456	0.0427	0.0392
w_A	0.6906	0.6414							
w_B	0.2725	0.3255							
w_C	0.0369	0.0331							

Type of data: coexistence data (tie lines)

Total system			Top phase			Bottom phase		
w_A	w_B	w_C	w_A	w_B	w_C	w_A	w_B	w_C
T/K = 298.15								
0.8560	0.0546	0.0894	0.6746	0.2936	0.0318	0.8881	0.0129	0.0990
0.8480	0.0487	0.1033	0.6124	0.3504	0.0372	0.8755	0.0102	0.1143
0.8182	0.0614	0.1204	0.6075	0.3603	0.0322	0.8604	0.0044	0.1352
0.8384	0.0419	0.1197	0.5683	0.3973	0.0344	0.8653	0.0054	0.1293
0.8015	0.0522	0.1463	0.5591	0.4186	0.0223	0.8366	0.0025	0.1609
T/K = 338.15								
0.8497	0.0505	0.0998	0.6035	0.3680	0.0285	0.8874	0.0055	0.1071
0.8437	0.0395	0.1168	0.5867	0.3784	0.0349	0.8678	0.0060	0.1262

Polymer (B):	**poly(1-vinyl-2-pyrrolidinone)**		**2007FED**
Characterization:	M_n/g.mol^{-1} = 138600, PVPK90, BASF AG, Ludwigshafen, Germany		
Solvent (A):	**water**	H_2O	**7732-18-5**
Salt (C):	**sodium sulfate**	Na_2SO_4	**7757-82-6**

Type of data: cloud points

T/K = 298.15

w_A	0.8008	0.8288	0.8478	0.8771	0.9124	0.9134	0.9190	0.9167	0.9117
w_B	0.0005	0.0006	0.0008	0.0009	0.0098	0.0183	0.0195	0.0222	0.0232
w_C	0.1987	0.1706	0.1514	0.1220	0.0778	0.0683	0.0615	0.0611	0.0651
w_A	0.9093	0.8923	0.8572	0.8499	0.8421	0.8345	0.8294		
w_B	0.0308	0.0482	0.0852	0.0933	0.1018	0.1105	0.1168		
w_C	0.0599	0.0595	0.0576	0.0568	0.0561	0.0550	0.0538		

T/K = 338.15

w_A	0.7634	0.7220	0.8005	0.8264	0.8429	0.8848	0.9224	0.9099	0.9001
w_B	0.0001	0.0001	0.0001	0.0001	0.0001	0.0002	0.0140	0.0363	0.0413
w_C	0.2365	0.2779	0.1994	0.1735	0.1570	0.1150	0.0636	0.0538	0.0586
w_A	0.9029	0.8970	0.8882	0.8889	0.8761	0.8654	0.8603	0.8533	0.8537
w_B	0.0421	0.0500	0.0597	0.0599	0.0738	0.0866	0.0940	0.0998	0.1007
w_C	0.0550	0.0530	0.0521	0.0512	0.0501	0.0480	0.0457	0.0469	0.0456
w_A	0.8441	0.8341	0.8328	0.8249					
w_B	0.1118	0.1231	0.1247	0.1324					
w_C	0.0441	0.0428	0.0425	0.0427					

Type of data: coexistence data (tie lines)

Total system			Top phase			Bottom phase		
w_A	w_B	w_C	w_A	w_B	w_C	w_A	w_B	w_C
T/K = 298.15								
0.8486	0.0572	0.0942	0.6734	0.2954	0.0312	0.8887	0.0016	0.1097
0.8778	0.0398	0.0824	0.7423	0.2222	0.0355	0.9083	0.0025	0.0892
0.8480	0.0396	0.1124	0.6456	0.3258	0.0286	0.8756	0.0013	0.1231
T/K = 338.15								
0.8940	0.0547	0.0513	0.8727	0.0760	0.0513	0.9246	0.0166	0.0588
0.8677	0.0724	0.0599	0.7537	0.2071	0.0392	0.9178	0.0070	0.0752
0.8527	0.0392	0.1081	0.6610	0.2880	0.0510	0.8842	0.0012	0.1146
0.8377	0.0470	0.1153	0.4641	0.5163	0.0196	0.8689	0.0046	0.1265

3.4. Table of systems where ternary LLE data were published only in graphical form as phase diagrams or related figures

Polymer (B)	Second and third component	Ref.
Agarose		
	maltodextrin and water	2010FRI
Amylopectin		
	β-lactoglobulin and water	2008QUI
Amylose		
	xanthan and water	2004MAN
N-Carboxyethylchitosan		
	gemini surfactant and water	2010PEN
Casein		
	guar gum and water	2007AN1
	sodium alginate and water	2007AN2
Cellulose		
	N-methylmorpholine-*N*-oxide and water	2010GOL
Cellulose acetate		
(39.8 % acetyl content)	1-butyl-3-methylimidazolium methyl sulfate and water	2010XIN
(39.8 % acetyl content)	1-butyl-3-methylimidazolium thiocyanate and and water	2010XIN
	1-methyl-2-pyrrolidinone and water	2010XIN
	2-propanone and water	2010XIN
Chitosan-*g*-poly(*N*-isopropyl-acrylamide)		
	sodium chloride and water	2010BAO
	sodium iodide and water	2010BAO

Polymer (B)	**Second and third component**	**Ref.**
Dextran		
	bovine serum albumin and water	2006ANT
	ficoll and water	2008MAD
	ficoll and water	2010MAD
	gelatine and water	2004ANT
	gelatine and water	2005LUN
	guar-*g*-poly(1-vinyl-2-pyrrolidinone) and water	2009KAU
	hydroxypropylstarch and water	2008MAD
	hydroxypropylstarch and water	2010MAD
	poly(ethylene glycol) and water	2008MAD
	poly(ethylene glycol) and water	2004HOP
	poly(ethylene glycol) and water	2004KAV
	poly(ethylene glycol) and water	2005MOO
	poly(ethylene glycol) and water	2006EDA
	poly(ethylene glycol) and water	2010MAD
	poly(ethylene oxide) and water	2003EDE
	poly(ethylene oxide) and water	2005OL1
	poly(ethylene oxide-*co*-propylene oxide) and water	2005OL1
	poly(ethylene oxide-*co*-propylene oxide) and water	2008MAD
	poly(ethylene oxide-*co*-propylene oxide) and water	2010MAD
	poly(*N*-isopropylacrylamide-*co*-butyl acrylate-*co*-chlorophyllin sodium copper salt) and water	2007KON
	poly(*N*-isopropylacrylamide-*co*-butyl acrylate-*co*-chlorophyllin sodium copper salt) and water	2009SHA
	poly(1-vinyl-2-pyrrolidinone) and water	2008AHM
	soy protein and water	2008LIX
	whey protein and water	2007SCH
Ethyl(hydroxyethyl)cellulose		
	1-butanol and water	1996THU
	dodecyl trimethylammonium bromide and water	2005OL2
	ethanol and water	1996THU
	guanidine hydrochloride and water	1996THU
	guanidine hydrothiocyanate and water	1996THU
	hexadecyl trimethylammonium bromide and water	2005OL2
	1-hexanol and water	1995THU
	1-hexanol and water	1996THU
	hexyl trimethylammonium bromide and water	2005OL2
	methanol and water	1996THU
	octa(ethylene glycol) mono dodecyl ether and water	2005OL2
	octyl-β-D-glucopyranoside and water	2005OL2
	octyl trimethylammonium bromide and water	2005OL2
	1-pentanol and water	1996THU
	1-propanol and water	1996THU

Polymer (B)	Second and third component	Ref.
Ethyl(hydroxyethyl)cellulose		
	sodium bromide and water	1996THU
	sodium chloride and water	1996THU
	sodium chloride and water	2004OLS
	sodium dodecyl sulfate and deuterium oxide	2005KJ2
	sodium dodecyl sulfate and water	2005KJ2
	sodium dodecyl sulfate and water	2005OL2
	sodium dodecyl (diethylene oxide) sulfate and water	2005OL2
	sodium octyl sulfate and water	2005OL2
	sodium thiocyanate and water	1996THU
	Triton X-100 and water	2005OL2
	urea and water	1996THU
Ficoll		
	dextran and water	2008MAD
	dextran and water	2010MAD
	hydroxypropylstarch and water	2008MAD
	hydroxypropylstarch and water	2010MAD
	poly(ethylene glycol) and water	2008MAD
	poly(ethylene glycol) and water	2010MAD
	poly(ethylene oxide-*co*-propylene oxide) and water	2008MAD
	poly(ethylene oxide-*co*-propylene oxide) and water	2010MAD
Gelatine		
	tert-butanol and water	2005GUP
	dextran and water	2004ANT
	dextran and water	2005LUN
	ethanol and water	2005GUP
	maltodextrin and water	2010FRI
	methanol and water	2005GUP
	pectin and water	2007PL1
	1-propanol and water	2005GUP
	sodium chloride and water	2005GUP
Guar gum		
	casein and water	2007AN1
Hepta(ethylene glycol) monododecyl ether		
	n-dodecane and water	2008MIY

Polymer (B)	Second and third component	Ref.
Hepta(ethylene glycol) monotetradecyl ether	penta(ethylene glycol) monodecyl ether and water	2007IMA
	penta(ethylene glycol) monotetradecyl ether and water	2007IMA
Hexa(ethylene glycol) monodecyl ether	urea and water	2011BIA
Hexa(ethylene glycol) monododecyl ether	n-dodecane and water	2008MIY
	poly(ethylene glycol) and water	2005HO1
	poly(ethylene glycol) and water	2005HO2
	urea and water	2011BIA
Hydroxyethylcellulose	Triton-X and water	2006ZH2
Hydroxyethylcellulose (modified)	sodium dodecyl sulfate and water	2004ZHO
	Triton-X and water	2006ZH2
	Triton-X and water	2007ZHA
Hydroxyethylcellulose-*g*-poly(*N*-isopropylacrylamide)	sodium chloride and water	2011PHA
Hydroxypropylcellulose	methylcellulose and water	2011FET
	poly(ethylene glycol) and water	2010VSH
	poly(maleic acid-*co*-acrylic acid) and water	2005BUM
	poly(maleic acid-*co*-styrene) and water	2005BUM
	poly(maleic acid-*co*-vinyl acetate) and water	2005BUM
Hydroxypropylstarch	dextran and water	2008MAD
	dextran and water	2010MAD
	ficoll and water	2008MAD
	ficoll and water	2010MAD

Polymer (B)	Second and third component	Ref.
Hydroxypropylstarch		
	poly(ethylene glycol) and water	2008MAD
	poly(ethylene glycol) and water	2010MAD
	poly(ethylene oxide-*co*-propylene oxide) and water	2008MAD
	poly(ethylene oxide-*co*-propylene oxide) and water	2010MAD
Lysozyme		
	magnesium chloride and water	2007WEN
	sodium chloride and water	2007WEN
Maltodextrin		
	agarose and water	2010FRI
	gelatine and water	2010FRI
Methylcellulose		
	hydroxypropylcellulose and water	2011FET
	sodium chloride and water	2011VIL
	sodium chloride and water	2005XUY
	sodium iodide and water	2005XUY
Methyl(hydroxypropyl)cellulose		
	sodium dodecyl sulfate and water	2010KA1
	surfactants and water	2010SAR
Nitrocellulose		
	2-(2-ethoxyethoxy)-ethanol and water	2008MIK
Octa(ethylene glycol) monododecyl ether		
	urea and water	2011BIA
Pectin		
	gelatine and water	2007PL1
	methanol and water	2006THO
Penta(ethylene glycol) monodecyl ether		
	hepta(ethylene glycol) monotetradecyl ether and water	2007IMA

Polymer (B)	Second and third component	Ref.
Penta(ethylene glycol) monododecyl ether		
	n-dodecane and water	2008MIY
	1-dodecanol and water	2007MIY
Penta(ethylene glycol) monooctyl ether		
	urea and water	2011BIA
Penta(ethylene glycol) monotetradecyl ether		
	hepta(ethylene glycol) monotetradecyl ether and water	2007IMA
Poly(acrylamide)		
	ammonium sulfate and water	2010CH5
	1,4-dioxane and water	2006DAL
	poly(ethylene glycol) and water	2010LUE
Poly(acrylamide-*co*-*N*-benzyl-acrylamide)		
	poly(ethylene glycol) and water	2003ABU
Poly(acrylamide-*co*-*N*-isopropyl-acrylamide)		
	1,4-dioxane and water	2006DAL
Poly(acrylamide-*co*-4-methoxy-styrene)		
	poly(ethylene glycol) and water	2003ABU
Poly(acrylamide-*co*-*N*-phenyl-acrylamide)		
	poly(ethylene glycol) and water	2003ABU
Poly(acrylamide-*co*-sodium acrylate)		
	polystyrenesulfonate and water	2003HEL

Polymer (B)	Second and third component	Ref.
Poly(acrylic acid)		
	2,6-lutidine and water	2001TOK
	poly(allylamine) hydrochloride and water	2010CHO
	poly(*N,N*-diallyl-*N,N*-dimethylammonium chloride) and water	2009LIT
	poly(*N,N*-diallyl-*N,N*-dimethylammonium chloride) and water	2010LIT
	poly(1,2-dimethyl-5-vinylpyridinium methylsulfate) and water	2009LIT
	poly(ethylene glycol) and water	2008SAR
	poly(*N*-ethyl-4-vinylpyridinium) bromide and water	2006IZU
	poly(2-hydroxyethyl acrylate-*co*-butyl acrylate) and water	2007MUN
Poly(acrylonitrile-*co*-acrylic acid)		
	N,N-dimethylformamide and water	2011AGR
Poly(acrylonitrile-*co*-itaconic acid)		
	N,N-dimethylformamide and water	2008TAN
	dimethylsulfoxide and water	2009TAN
Poly(acrylonitrile-*co*-methyl acrylate-*co*-itaconic acid)		
	dimethylsulfoxide and water	2009DON
Poly(*N*-acryloyl-L-aspartic acid *N*′-propylamide)		
	ammonium acetate and water	2004KUR
Poly(*N*-acryloyl-L-glutamic acid *N*′-ethylamide)		
	ammonium acetate and water	2004KUR
Poly(*N*-acryloyl-L-glutamic acid *N*′-methylamide)		
	ammonium acetate and water	2004KUR
Poly(*N*-acryloyl-L-glutamic acid *N*′-propylamide)		
	ammonium acetate and water	2004KUR

Polymer (B)	Second and third component	Ref.
Poly(allylamine) hydrochloride		
	poly(acrylic acid) and water	2010CHO
Poly(*N-tert*-butylacrylamide-*co*-acrylamide)		
	methanol and water	2009MAH
Poly(butyl acrylate-*b*-acrylic acid)		
	tetrahydrofuran and water	2008CRI
Poly(*tert*-butyl acrylate)-*g*-poly[oligo(ethylene glycol) methyl ether methacrylate)]		
	sodium chloride and water	2011SON
	sodium sulfate and water	2011SON
	sodium thiocyanate and water	2011SON
Poly(diacetone acrylamide-*co*-acrylamide)		
	guanidine hydrochloride and water	1975TAY
	potassium hydroxide and water	1975TAY
	sodium benzoate and water	1975TAY
	sodium bromide and water	1975TAY
	sodium chloride and water	1975TAY
	tetrabutylammonium bromide and water	1975TAY
	tetraethylammonium bromide and water	1975TAY
	tetramethylammonium bromide and water	1975TAY
	tetrapropylammonium bromide and water	1975TAY
	urea and water	1975TAY
Poly(diallyaminoethanoate-*co*-dimethylsulfoxide)		
	poly(ethylene glycol) and water	2004WAZ
Poly(*N,N*-diallyl-*N,N*-dimethyl-ammonium chloride)		
	cetyltrimethylammonium bromide and water	2010KAL
	isobutyric acid and water	2011WA3
	poly(acrylic acid) and water	2009LIT
	poly(acrylic acid) and water	2010LIT

Polymer (B)	**Second and third component**	**Ref.**
Poly(*N,N*-diethylacrylamide)		
	tert-butanol and water	2004PAN
	N,N-dimethylformamide and water	2004PAN
	1,4-dioxane and water	2004PAN
	ethanol and water	2004PAN
	methanol and water	2004PAN
	2-methyl-1-propanol and water	2004PAN
	poly(*N*-isopropylacrylamide) and water	2003MAO
	1-propanol and water	2004PAN
	2-propanol and water	2004PAN
	sodium chloride and water	2008FAN
Poly(*N,N*-diethylacrylamide-*co*-acrylic acid)		
	sodium chloride and water	2008FAN
Poly(*N,N*-diethylacrylamide-*co*-*N*-acryloxysuccinimide)		
	magnesium sulfate and water	2010ZH2
	potassium chloride and water	2010ZH2
	sodium chloride and water	2010ZH2
	sodium phosphate and water	2010ZH2
Poly[di(ethylene glycol) methacrylate]		
	sodium chloride and water	2010LUZ
	sodium sulfate and water	2010LUZ
Poly[di(ethylene glycol) methacrylate-*co*-oligo(ethylene glycol) methacrylate]		
	sodium chloride and water	2010LUZ
	sodium sulfate and water	2010LUZ
Poly(*N,N*-dimethylacrylamide)		
	ammonium sulfate and water	2011FIS
	1,4-dioxane and water	2004PAG
	2-propanone and water	2004PAG
	potassium carbonate and water	2011FIS
	sodium hydroxide and water	2011FIS
	sodium phosphate and water	2011FIS

Polymer (B)	**Second and third component**	**Ref.**
Poly(*N,N*-dimethylacrylamide-*co*-2-hydroxyethyl methacrylate)		
	β-cyclodextrin and water	1998GOS
Poly[2-(*N,N*-dimethylamino)ethyl methacrylate]		
	cyclodextrin and water	2010ZH3
	sodium dodecyl sulfate and water	2007PL2
Poly[2-(*N,N*-dimethylamino)ethyl methacrylate-*co*-acrylic acid-*co*-butyl methacrylate]		
	poly(*N*-isopropylacrylamide-*co*-butyl acrylate-*co*-chlorophyllin sodium copper salt) and water	2009NIN
Poly[2-(*N,N*-dimethylamino)ethyl methacrylate-*b*-*tert*-butyl methacrylate-*b*-methyl methacrylate]		
	poly(ethylene glycol) and water	2008QIN
Poly(1,2-dimethyl-5-vinylpyridinium methylsulfate)		
	poly(acrylic acid) and water	2009LIT
Poly(divinyl ether-*alt*-maleic anhydride)		
	poly(*N*-ethyl-4-vinylpyridinium) bromide and water	2004VOL
	poly(*N*-ethyl-4-vinylpyridinium) bromide and water	2005IZU
	poly(methacrylic acid) and water	2004VOL
	poly(methacrylic acid) and water	2005IZU
	sodium chloride and water	2004VOL
Polyesterurethane		
	dimethylsulfoxide and water	2005HEI
Poly(ether amine)		
	sodium chloride and water	2010YUB

Polymer (B)	**Second and third component**	**Ref.**
Polyethersulfone		
	N,N-dimethylacetamide and water	2007BAR
	1-methyl-2-pyrrolidinone and water	2007BAR
Poly[*N*-(2-ethoxyethyl)acryl-amide]		
	methanol and water	2009MA3
Poly[*N*-(2-ethoxyethyl)acryl-amide-*co*-*N*-isopropylacrylamide]		
	methanol and water	2009MA3
Poly[*N*-(3-ethoxypropyl)acryl-amide]		
	ethanol and water	2005UGU
	sodium chloride and water	2005UGU
Poly(*N*-ethylacrylamide)		
	potassium chloride and water	2003XUE
	sodium dodecyl sulfate and water	2003XUE
Poly(ethylene carbonate)		
	dimethylsulfoxide and water	2010LI3
	1-methyl-2-pyrrolidinone and water	2010LI3
Poly(ethylene glycol)		
	ammonium carbamate and water	2007DAL
	ammonium sulfate and water	2005HEY
	ammonium sulfate and water	2009IMA
	ammonium sulfate and water	2010DIP
	dextran and water	2004HOP
	dextran and water	2004KAV
	dextran and water	2005MOO
	dextran and water	2006EDA
	dextran and water	2008MAD
	dextran and water	2010MAD
	cocamidopropyl betaine and water	2009YOU
	diammonium phosphate and water	2009FO1
	dipotassium phosphate and water	2003ELI
	dipotassium phosphate and water	2007FAR
	disodium phosphate and water	2003ELI
	disodium phosphate and water	2005HEY

Polymer (B)	Second and third component	Ref.
Poly(ethylene glycol)		
	ficoll and water	2008MAD
	ficoll and water	2010MAD
	hexa(ethylene glycol) monododecyl ether and water	2005HO1
	hexa(ethylene glycol) monododecyl ether and water	2005HO2
	hydroxypropylcellulose and water	2010VSH
	hydroxypropylstarch and water	2008MAD
	hydroxypropylstarch and water	2010MAD
	isobutyric acid and water	2007NOR
	isobutyric acid and water	2008NOR
	magnesium sulfate and water	2005HEY
	monopotassium phosphate and water	2003ELI
	monosodium phosphate and water	2003ELI
	monosodium sulfate and water	2004CHE
	2H,3H-perfluoropentane + water	2005PAU
	2H,3H-perfluoropentane + water	2008COT
	poly(acrylamide) and water	2010LUE
	poly(acrylamide-*co*-*N*-benzylacrylamide) and water	2003ABU
	poly(acrylamide-*co*-4-methoxystyrene) and water	2003ABU
	poly(acrylamide-*co*-*N*-phenylacrylamide) and water	2003ABU
	poly(acrylic acid) and water	2008SAR
	poly(diallyaminoethanoate-*co*-dimethylsulfoxide) and water	2004WAZ
	poly[2-(*N,N*-dimethylamino)ethyl methacrylate-*b*-*tert*-butyl methacrylate-*b*-methyl methacrylate] and water	2008QIN
	poly(ethylene oxide-*co*-propylene oxide) and water	2008MAD
	poly(ethylene oxide-*co*-propylene oxide) and water	2010MAD
	poly(*N*-isopropylacrylamide) and water	2004HUE
	poly(sodium acrylate) and water	2008JOH
	poly(1-vinyl-2-pyrrolidinone) and water	2004INA
	potassium citrate and water	2010LUY
	potassium phosphate and water	2006WON
	potassium sulfate and water	2005HEY
	sodium carbonate and water	2005HEY
	sodium citrate and water	2007POR
	sodium citrate and water	2009SIL
	sodium citrate and water	2011LIX
	sodium hydroxide and water	2005HEY
	sodium phosphate and water	2005HEY
	sodium sulfate and water	2005HEY
	sodium sulfate and water	2009AN1
	sodium sulfate and water	2009IMA
	Triton-X and water	2011NAQ

Polymer (B)	Second and third component	Ref.
Poly(ethylene glycol)		
	Tween surfactant and water	2004MAH
	urate oxidase and water	2004VIV
	zinc sulfate and water	2005HEY
Poly(ethylene glycol-*b*-*N*-acryloyl-2,2-dimethyl-1,3-oxazolidine)		
	sodium chloride and water	2011CUI
Poly(ethylene glycol) dimethyl ether		
	ammonium sulfate and water	2010DIP
	isobutyric acid and water	2008NOR
Poly(ethylene glycol) mono-dodecyl ether (Brij-35)		
	ammonium sulfate and water	2011BAT
	cobalt sulfate and water	2011BAT
	copper sulfate and water	2011BAT
	lithium sulfate and water	2011BAT
	potassium carbonate and water	2011BAT
	potassium nitrate and water	2011BAT
	potassium phosphate and water	2011BAT
	potassum sulfate and water	2011BAT
	sodium bromide and water	2011BAT
	sodium chloride and water	2011BAT
	sodium fluoride and water	2011BAT
	sodium sulfate and water	2011BAT
	zinc sulfate and water	2011BAT
Poly(ethylene glycol) mono-hexadecyl ether		
	alkyl triphenyl phosphonium bromide and water	2005PRA
Poly(ethylene glycol) monooleyl ether		
	alkyl triphenyl phosphonium bromide and water	2005PRA

Polymer (B)	Second and third component	Ref.
Poly(ethylene glycol) p-octyl-phenyl ether		
	cyclohexane and water	2009CHE2
	magnesium sulfate and water	2010SA2
	magnesium sulfate and water	2011SAL1
	poly(ethylene glycol) and water	2011NAQ
	sodium citrate and water	2010SA2
	sodium citrate and water	2011SAL1
Poly(ethylene oxide)		
	dextran and water	2003EDE
	dextran and water	2005OL1
	isobutyric acid and water	2011RE2
	poly(*N*-vinylcaprolactam) and water	2003VE1
	sodium dodecyl sulfate and water	2005DEN
Poly(ethylene oxide-*b*-isoprene)		
	oligo(ethylene oxide) monododecyl ether and water	2004KUN
Poly(ethylene oxide-*co*-propylene oxide)		
	ammonium sulfate and water	2010DIP
	dextran and water	2005OL1
	dextran and water	2008MAD
	dextran and water	2010MAD
	ficoll and water	2008MAD
	ficoll and water	2010MAD
	hydroxypropylstarch and water	2008MAD
	hydroxypropylstarch and water	2010MAD
	poly(ethylene glycol) and water	2008MAD
	poly(ethylene glycol) and water	2010MAD
	potassium phosphate and water	2010DIP
	sodium carbonate and water	2010DIP
Poly(ethylene oxide-*b*-propylene oxide-*b*-ethylene oxide)		
	acetic acid and water	2008SHA
	ammonium carbamate and water	2007OLI
	1-butanol and water	2007BHA
	1-butanol and water	2008SHA
	calcium chloride and water	2008SHA
	cetyltrimethylammonium bromide and water	2010PAT
	decyltrimethylammonium bromide and water	2010PAT

Polymer (B)	Second and third component	Ref.
Poly(ethylene oxide-*b*-propylene oxide-*b*-ethylene oxide)		
	1,2-dichloroethane and water	2006LAZ
	dimethylurea and water	2008SHA
	dodecyltrimethylammonium bromide and water	2010PAT
	ethanol and water	2008SHA
	ethanol and water	2007BHA
	formic acid and water	2008SHA
	1-hexanol and water	2007BHA
	hydrochloric acid and water	2008SHA
	magnesium chloride and water	2008SHA
	methanol and water	2008SHA
	methanol and water	2007BHA
	monosodium phosphate and water	2011DEY
	nicotinamide and water	2008SHA
	octadecyl-trimethylammonium bromide and water	2010PAT
	poly(ethylene oxide-*b*-propylene oxide-*b*-ethylene oxide) and water	2009NAN
	1-pentanol and water	2007BHA
	1-propanol and water	2008SHA
	1-propanol and water	2007BHA
	sodium acetate and water	2008SHA
	sodium benzenesulfonate and water	2004VAR
	sodium bis(2-ethylhexyl)sulfosuccinate and water	2010RO2
	sodium bromide and water	2011DEY
	sodium carbonate and water	2011DEY
	sodium chloride and water	2011DEY
	sodium chloride and water	2008SHA
	sodium dodecyl sulfate and water	2008SHA
	sodium fluoride and water	2011DEY
	sodium formate and water	2008SHA
	sodium hydroxide and water	2008SHA
	sodium iodide and water	2011DEY
	sodium nitrate and water	2011DEY
	sodium oxalate and water	2008SHA
	sodium phosphate and water	2008SHA
	sodium salicylate and water	2008SHA
	sodium sulfate and water	2011DEY
	sodium sulfate and water	2008BER
	sodium sulfate and water	2008SHA
	sodium thiocyanate and water	2011DEY
	sodium thiocyanate and water	2008SHA
	sodium thiosulfate and water	2011DEY
	sodium toluenesulfonate and water	2004VAR
	sodium xylenesulfonate and water	2004VAR

Polymer (B)	Second and third component	Ref.
Poly(ethylene oxide-*b*-propylene oxide-*b*-ethylene oxide)		
	tributyl phosphate and water	2006CAU
	tetradecyltrimethylammonium bromide and water	2010PAT
	Tween surfactant and water	2004MAH
	urea and water	2008SHA
Poly(ethylene oxide sulfide)		
	potassium phosphate and water	2006HER
Poly(ethylenimine)		
	isobutyric acid and water	2008NOR
	poly(sodium acrylate) and water	2005GRA
Poly(ethylenimine) (branched, *N*-acylated)		
	sodium chloride and water	2011KIM
Poly(*N*-ethyl-*N*-methylacrylamide)		
	ammonium sulfate and water	2004PAN
	tert-butanol and water	2004PAN
	methanol and water	2004PAN
	2-methyl-1-propanol and water	2004PAN
	potassium bromide and water	2004PAN
	potassium carbonate and water	2004PAN
	potassium chloride and water	2004PAN
	potassium fluoride and water	2004PAN
	potassium hydroxide and water	2004PAN
	potassium iodide and water	2004PAN
	potassium nitrate and water	2004PAN
	1-propanol and water	2004PAN
Poly(2-ethyl-2-oxazoline)		
	lithium acetate and water	2010BLO
	lithium iodide and water	2010BLO
	lithium perchlorate and water	2010BLO
	lithium sulfate and water	2010BLO
	sodium acetate and water	2010BLO
	sodium chloride and water	2010BLO
	sodium iodide and water	2010BLO
	sodium perchlorate and water	2010BLO
	sodium sulfate and water	2010BLO
	sodium thiocyanate and water	2010BLO

Polymer (B)	Second and third component	Ref.
Poly[2-ethyl-2-oxazoline-*co*-2-(dec-9-enyl)-2-oxazoline] (sugar-substituted)	sodium chloride and water	2011KEM
Poly(2-ethyl-2-oxazoline-*b*-2-nonyl-2-oxazoline)	ethanol and water	2009LAM
Poly(2-ethyl-2-oxazoline-*co*-2-nonyl-2-oxazoline)	ethanol and water	2009LAM
Poly(2-ethyl-2-oxazoline-*co*-2-phenyl-2-oxazoline)	ethanol and water	2008HO2
Poly(*N*-ethyl-4-vinylpyridinium) bromide	poly(acrylic acid) and water	2006IZU
	poly(divinyl ether-*alt*-maleic anhydride) and water	2004VOL
	poly(divinyl ether-*alt*-maleic anhydride) and water	2005IZU
	poly(methacrylic acid) and water	2006IZU
Poly(2-hydroxyethyl acrylate-*co*-butyl acrylate)	poly(acrylic acid) and water	2007MUN
Poly(2-hydroxypropyl methacrylate)	methanol and water	2004GAR
Polyimide	*N,N*-dimethylformamide and water	2003MAT
Poly(isoprene-*b*-ethylene oxide)	oligo(ethylene oxide) monododecyl ether and water	2004KUN
Poly(*N*-isopropylacrylamide)	acetonitrile and water	2004PAN
	benzoic acid and water	2009HOF

Polymer (B)	Second and third component	Ref.
Poly(*N*-isopropylacrylamide)		
	1-benzyl-3-methylimidazolium tetrafluoroborate and water	2011RE1
	1-butanol and water	2004PAN
	tert-butanol and water	2004PAN
	calcium chloride and water	2005VA1
	calcium chloride and water	2008BUR
	cesium chloride and water	2004MAO
	deuterium oxide and water	2004MAO
	deuterium oxide and water	2008BER
	3,4-dimethoxybenzaldehyde and water	2009HOF
	N,N-dimethylformamide and water	2004PAN
	dimethylsulfoxide and water	2007YAM
	1,4-dioxane and water	2004PAN
	1,4-dioxane and water	2006DAL
	ethanol and water	2004PAN
	ethylvanillin and water	2009HOF
	glycerol and water	2004PAN
	3-hydroxybenzaldehyde and water	2009HOF
	4-hydroxybenzaldehyde and water	2009HOF
	lithium chloride and water	2004MAO
	methanol and water	2004GAR
	methanol and water	2004PAN
	methanol and water	2005TAO
	methanol and water	2009MA3
	methanol and water	2010SU2
	methanol and water	2010ZH4
	methyl 4-hydroxybenzoate and water	2009HOF
	2-methyl-1-propanol and water	2004PAN
	poly(*N,N*-diethylacrylamide) and water	2003MAO
	poly(ethylene glycol) and water	2004HUE
	poly(*N*-isopropylmethacrylamide) and water	2005STA
	poly(sodium acrylate) and water	2006MYL
	potassium bromide and water	2004MAO
	potassium chloride and water	2004MAO
	potassium fluoride and water	2004MAO
	potassium thiocyanate and water	2010SHE
	1-propanol and water	2004PAN
	2-propanol and water	2004PAN
	2-propanone and water	2011MUN
	2-propanone and water	2004PAN
	rubidium chloride and water	2004MAO
	salicylaldehyde and water	2009HOF
	sodium bromide and water	2004MAO

Polymer (B)	Second and third component	Ref.
Poly(*N*-isopropylacrylamide)		
	sodium bromide and water	2005ZH5
	sodium carbonate and water	2005ZH5
	sodium chloride and water	2004MAO
	sodium chloride and water	2005VA1
	sodium chloride and water	2005ZH5
	sodium chloride and water	2006JHO
	sodium chloride and water	2008BUR
	sodium cholate and water	2011KUM
	sodium fluoride and water	2005ZH5
	sodium iodide and water	2004MAO
	sodium iodide and water	2005ZH5
	sodium monophosphate and water	2005ZH5
	sodium nitrate and water	2005ZH5
	sodium perchlorate and water	2005ZH5
	sodium sulfate and water	2005VA1
	sodium sulfate and water	2005ZH5
	sodium thiocyanate and water	2005ZH5
	sodium thiosulfate and water	2005ZH5
	sulfolane and water	2004PAN
	tetrahydrofuran and water	2004PAN
	tetrahydrofuran and water	2010HAO
Poly(*N*-isopropylacrylamide-*co*-acrylic acid)		
	peptides and water	2001BUL
Poly(*N*-isopropylacrylamide-*co*-acrylic acid-*co*-ethyl methacrylate)		
	potassium chloride and water	2005TIE
Poly(*N*-isopropylacrylamide-*co*-*N*-(3-aminopropyl)methacrylamide hydrochloride)		
	potassium chloride and water	2009MAI
Poly(*N*-isopropylacrylamide-*co*-*N*-(3-aminopropyl)methacrylamide hydrochloride-*b*-*N*-isopropylacrylamide)		
	potassium chloride and water	2009MAI

Polymer (B)	Second and third component	Ref.
Poly(*N*-isopropylacrylamide-*co*-benzo-15-crown-5-acrylamide)		
	cesium nitrate and water	2008MIP
	lithium nitrate and water	2008MIP
	potassium nitrate and water	2008MIP
	sodium nitrate and water	2008MIP
Poly(*N*-isopropylacrylamide-*co*-butyl acrylate-*co*-chlorophyllin sodium copper salt)		
	dextran and water	2007KON
	dextran and water	2009SHA
	poly[2-(*N,N*-dimethylamino)ethyl methacrylate-*co*-acrylic acid-*co*-butyl methacrylate] and water	2009NIN
Poly(*N*-isopropylacrylamide-*co*-*N,N*-dimethylacrylamide		
	1,4-dioxane and water	2006PAG
Poly[*N*-isopropylacrylamide-*co*-(2-hydroxyisopropyl)acrylamide]		
	sodium chloride and water	2006MA1
Poly[*N*-isopropylacrylamide-*co*-(p-methacrylamido)acetophenone thiosemicarbazone]		
	sodium chloride and water	2005LIC
Poly{*N*-isopropylacrylamide-*b*-3-[*N*-(3-methacrylamidopropyl)-*N,N*-dimethylammonio]propane sulfate}		
	sodium chloride and water	2002VIR
Poly[*N*-isopropylacrylamide-*co*-oligo(ethylene glycol) monomethacrylate]		
	sodium dodecylbenzenesulfonate and water	2004ALA
	sodium dodecylbenzenesulfonate and water	2006ALA

Polymer (B)	**Second and third component**	**Ref.**
Poly(*N*-isopropylacrylamide-*co*-sodium 2-acrylamido-2-methyl-1-propanesulfonate)		
	dodecyltrimethylammonium chloride and water	2003NOW
Poly(*N*-isopropylacrylamide-*co*-sodium 2-acrylamido-2-methyl-1-propanesulfonate-*co*-*N*-*tert*-butylacrylamide)		
	sodium chloride and water	2007SH1
Poly(*N*-isopropylacrylamide-*co*-sodium styrenesulfonate)		
	dodecyltrimethylammonium chloride and water	2004NOW
Poly(*N*-isopropylacrylamide-*co*-styrenesulfonate)		
	sodium chloride and water	2011MCF
Poly(*N*-isopropylmethacrylamide)		
	deuterium oxide and ethanol	2009KOU
	deuterium oxide and poly(vinyl methyl ether)	2005SP2
	poly(*N*-isopropylacrylamide) and water	2005STA
Poly(*N*-isopropylmethacrylamide-*co*-acrylamide)		
	deuterium oxide and ethanol	2010KOU
	deuterium oxide and 2-propanone	2010KOU
Poly(2-isopropyl-2-oxazoline)		
	lithium acetate and water	2010BLO
	lithium iodide and water	2010BLO
	lithium perchlorate and water	2010BLO
	lithium sulfate and water	2010BLO
	sodium acetate and water	2010BLO
	sodium chloride and water	2010BLO
	sodium iodide and water	2010BLO
	sodium perchlorate and water	2010BLO
	sodium sulfate and water	2010BLO
	sodium thiocyanate and water	2010BLO

Polymer (B)	Second and third component	Ref.
Poly(D,L-lactic acid)		
	1,4-dioxane and water	2010CH1
Poly(L-lactic acid)		
	1,4-dioxane and water	2004TAN
	1,4-dioxane and water	2009HEL
Poly(maleic acid-*co*-acrylic acid)		
	hydroxypropylcellulose and water	2005BUM
Poly(maleic acid-*co*-styrene)		
	hydroxypropylcellulose and water	2005BUM
Poly(maleic acid-*co*-vinyl acetate)		
	hydroxypropylcellulose and water	2005BUM
Poly(methacrylic acid)		
	poly(divinyl ether-*alt*-maleic anhydride) and water	2004VOL
	poly(divinyl ether-*alt*-maleic anhydride) and water	2005IZU
	poly(*N*-ethyl-4-vinylpyridinium) bromide and water	2006IZU
Poly(2-methacryloxyethyl phosphorylcholine)		
	ethanol and water	2008MAT
	1-propanol and water	2008MAT
	2-propanol and water	2008MAT
Poly(*N*-methacryloyl-L-β-isopropylasparagine)		
	sodium chloride and water	2010LUO
Poly[2-(2-methoxyethoxy)ethyl methacrylate-*co*-oligo(ethylene glycol) methacrylate]		
	sodium chloride and water	2006LUT
Poly(methyl acrylate)		
	ethanol and water	2011CAN

Polymer (B)	Second and third component	Ref.
Poly(methyl acrylate-*b*-styrene)		
	ethanol and water	2011CAN
Poly(2-methyl-2-oxazoline-*co*-2-phenyl-2-oxazoline)		
	ethanol and water	2008HO2
Poly[oligo(ethylene glycol) diglycidyl ether-*co*-piperazine-*co*-oligo(propylene glycol) diglycidyl ether)]		
	sodium chloride and water	2009REN
Polypeptide		
	sodium dodecyl sulfate and water	2006ZH1
	sodium salts and water	2008CHO
Poly(propylene oxide-*b*-ethylene oxide-*b*-propylene oxide)		
	cetyltrimethylammonium bromide and water	2010PAT
	decyltrimethylammonium bromide and water	2010PAT
	dodecyltrimethylammonium bromide and water	2010PAT
	octadecyl-trimethylammonium bromide and water	2010PAT
	tetradecyltrimethylammonium bromide and water	2010PAT
Poly(2-propyl-2-oxazoline)		
	lithium acetate and water	2010BLO
	lithium iodide and water	2010BLO
	lithium perchlorate and water	2010BLO
	lithium sulfate and water	2010BLO
	sodium acetate and water	2010BLO
	sodium chloride and water	2010BLO
	sodium iodide and water	2010BLO
	sodium perchlorate and water	2010BLO
	sodium sulfate and water	2010BLO
	sodium thiocyanate and water	2010BLO

Polymer (B)	Second and third component	Ref.
Poly(pyrrolidinoacrylamide)		
	tert-butanol and water	2004PAN
	1,4-dioxane and water	2004PAN
	ethanol and water	2004PAN
	methanol and water	2004PAN
	2-methyl-1-propanol and water	2004PAN
	1-propanol and water	2004PAN
	2-propanol and water	2004PAN
Poly(sodium acrylate)		
	poly(allylamine) hydrochloride and water	2010CHO
	poly(ethylene glycol) and water	2008JOH
	poly(ethylenimine) and water	2005GRA
	poly(*N*-isopropylacrylamide) and water	2006MYL
Poly(sodium styrenesulfonate)		
	sodium dodecyl sulfate and water	2010KAL
Poly[(styrene-*alt*-maleic anhydride)-*g*-poly(amidoamine) dendrons]		
	sodium chloride and water	2009GAO
Poly(styrene-*b*-methyl acrylate)		
	ethanol and water	2011CAN
Poly(styrene-*alt*-sodium maleate)		
	barium chloride and water	1995JAR
	calcium chloride and water	1995JAR
	magnesium chloride and water	1995JAR
Poly(sulfobetaine methacrylate-*co*-*N*-isopropylacrylamide)		
	methanol and water	2009CHA2
Polysulfone		
	1-methyl-2-pyrrolidinone and water	2004LEE
	1-methyl-2-pyrrolidinone and water	2010ARO

Polymer (B)	Second and third component	Ref.
Poly(*N*-tetrahydrofurfurylacryl-amide)		
	methanol and water	2010MAE
	1-propanol and water	2010MAE
Poly(*N*-tetrahydrofurfurylacryl-amide-*co*-*N*-isopropylacrylamide)		
	methanol and water	2010MAE
	1-propanol and water	2010MAE
Poly(*N*-tetrahydrofurfuryl-methacrylamide)		
	methanol and water	2010MAE
	1-propanol and water	2010MAE
Poly(*N*-tetrahydrofurfuryl-methacrylamide-*co*-*N*-isopropyl-acrylamide)		
	methanol and water	2010MAE
	1-propanol and water	2010MAE
Poly(*N*-vinylacetamide-*co*-methyl acrylate)		
	1-butanol and water	2004MOR
	ethanol and water	2004MOR
	magnesium sulfate and water	2004MOR
	methanol and water	2004MOR
	potassium sulfate and water	2004MOR
	1-propanol and water	2004MOR
	2-propanol and water	2004MOR
	sodium chloride and water	2004MOR
	sodium nitrate and water	2004MOR
	sodium sulfate and water	2004MOR
	sodium sulfite and water	2004MOR
Poly(vinyl alcohol)		
	dimethylsulfoxide and water	2007TAK
	hydroxypropylated polyrotaxane and water	2007KAT

Polymer (B)	Second and third component	Ref.
Poly(vinyl alcohol-*co*-sodium acrylate)		
	sodium chloride and water	2008PAN
Poly(*N*-vinylcaprolactam)		
	acetonitril and water	1996KIR
	2-aminoethanol and water	1996KIR
	tert-butanol and water	1996KIR
	cadmium chloride and water	2011PAK
	copper chloride and water	2011PAK
	ethanol and water	1996KIR
	methanol and water	1996KIR
	phenol and water	1996KIR
	1-propanol and water	1996KIR
	2-propanol and water	1996KIR
	poly(ethylene oxide) and water	2003VE1
	sodium chloride and water	2005DUB
	sodium dodecyl sulfate and water	2005DUB
	urea and water	1996KIR
Poly(*N*-vinylcaprolactam-*co*-1-vinylimidazole)		
	potassium bromide and water	2007SH2
	sodium thiocyanate and water	2007SH2
Poly(*N*-vinylcaprolactam-*co*-1-vinyl-2-methylimidazole)		
	potassium bromide and water	2007SH2
	sodium thiocyanate and water	2007SH2
Poly(vinyl chloride)		
	tetrahydrofuran and water	2010MAG
Poly(*N*-vinyl formamide-*co*-acrylic acid)		
	sodium sulfate and water	2008CHE
Poly(*N*-vinylformamide-*co*-vinyl acetate)		
	potassium chloride and water	2003SET
	potassium iodide and water	2003SET
	potassium sulfate and water	2003SET

Polymer (B)	Second and third component	Ref.
Poly(vinylidene fluoride)		
	N,N-dimethylacetamide and water	2003YEO
	N,N-dimethylformamide and water	2003YEO
	N,N-dimethylformamide and water	2010GUO
	1-methyl-2-pyrrolidinone and water	2003YEO
	triethyl phosphate and water	2003YEO
	triethyl phosphate and water	2006LIN
Poly[vinylidene fluoride-*g*-poly(*N*-isopropylacrylamide)]		
	N,N-dimethylformamide and water	2010GUO
Poly(vinyl isobutyl ether-*b*-vinyl methyl ether-*b*-vinyl isobutyl ether)		
	n-decane and water	2003VE2
Poly(vinyl methyl ether)		
	alkyl triphenylphosphonium bromide and water	2005PRA
	calcium chloride and water	2005VA1
	deuterium oxide and poly(*N*-isopropyl-methacrylamide)	2005SP2
	ethanol and water	1010LAB
	potassium bromide and water	2003MAE
	potassium chloride and water	2003MAE
	potassium fluoride and water	2003MAE
	potassium iodide and water	2003MAE
	tetrabutylammonium bromide and water	2004MA2
	tetraethylammonium bromide and water	2004MA2
	tetrahydrofuran and water	2004BER
	tetramethylammonium bromide and water	2004MA2
	tetrapropylammonium bromide and water	2004MA2
Poly(vinyl methyl ether-*b*-vinyl isobutyl ether)		
	n-decane and water	2003VE2
Poly[4-vinylpyridine-*g*-poly(ethylene oxide)]		
	sodium chloride and water	2008REN

Polymer (B)	Second and third component	Ref.
Poly(1-vinyl-2-pyrrolidinone)		
	dextran and water	2008AHM
	isobutyric acid and water	2009PAK
	isopentanoic acid and water	2009PAK
	magnesium sulfate and water	2008DAN
	poly(ethylene glycol) and water	2004INA
	potassium bromide and water	2008DAN
	potassium carbonate and water	2007NAK
	potassium chloride and water	2008DAN
	potassium dihydrogenphosphate and water	2007NAK
	potassium fluoride and water	2007NAK
	sodium chloride and water	2007NAK
	sodium chloride and water	2008DAN
	sodium phosphate and water	2008DAN
	sodium sulfate and water	2007NAK
	sodium sulfate and water	2008DAN
	sodium trichloroacetate and water	2009PAK
	trichloroacetic acid and water	2009PAK
	zinc chloride and water	2011PAK
	zinc sulfate and water	2011PAK
Poly(vinyl sulfonic acid)		
	Tween 80 and water	2010JAD
Poly(5-vinyltetrazole-*co*-*N*-vinylcaprolactam)		
	sulfuric acid and water	2010KIZ
Sodium alginate		
	casein and water	2007AN2
	sodium caseinate and water	2009AN2
Sodium caseinate		
	sodium alginate and water	2009AN2
Sodium hyaluronate		
	2-propanol and water	2006CAL
Starch		
	xanthan and water	2004MAN

Polymer (B)	Second and third component	Ref.
Tetra(ethylene glycol) monooctyl ether		
	urea and water	2011BIA
Xanthan		
	amylose and water	2004MAN
	starch and water	2004MAN

3.5. Cloud-point and/or coexistence curves of quasiquaternary solutions containing water and at least one polymer

3.5.1. Nonelectrolyte solutions

No new data were found.

3.5.2. Electrolyte solutions

Polymer (B): **poly(ethylene glycol)** **2003HAG**
Characterization: M_n/g.mol^{-1} = 1500, Merck, Darmstadt, Germany
Solvent (A): **water** **H_2O** **7732-18-5**
Salt (C): **dipotassium phosphate** **K_2HPO_4** **7758-11-4**
Polymer (D): **bovine serum albumin**
Characterization: M_n/g.mol^{-1} = 68000, Merck, Darmstadt, Germany

Type of data: coexistence data (tie lines)

T/K = 298.15 pH = 9.1

Total system			Top phase			Bottom phase		
w_B	w_C	w_D	w_B	w_C	w_D	w_B	w_C	w_D
0.1399	0.1115	0.000500	0.2964	0.0413	0.000097	0.0390	0.1509	0.000868
0.1398	0.1026	0.000500	0.2656	0.0439	0.000093	0.0480	0.1333	0.000943
0.1598	0.1019	0.000500	0.3062	0.0379	0.000105	0.0473	0.1483	0.001448
0.1395	0.0934	0.000500	0.2249	0.0597	0.000034	0.0570	0.1205	0.000567
0.1598	0.0925	0.000500	0.2638	0.0456	0.000163	0.0507	0.1398	0.001108
0.1798	0.0916	0.000500	0.2987	0.0371	0.000122	0.0416	0.1471	0.000821
0.1797	0.0826	0.000500	0.2583	0.0455	0.000322	0.0447	0.1402	0.001358
0.2000	0.0816	0.000500	0.2952	0.0391	0.000097	0.0380	0.1518	0.001341

Polymer (B): **poly(ethylene glycol)** **2009KHE**
Characterization: M_n/g.mol^{-1} = 4000, Merck, Darmstadt, Germany
Solvent (A): **water** **H_2O** **7732-18-5**
Salt (C): **dipotassium phosphate** **K_2HPO_4** **7758-11-4**
Component (D): **cephalexin** **$C_{16}H_{17}N_3O_4S$** **15686-71-2**

Type of data: coexistence data (tie lines)

continued

continued

Total system			Top phase			Bottom phase		
w_B	w_C	w_D	w_B	w_C	w_D	w_B	w_C	w_D
T/K = 301.35								
0.2647	0.1140	0.00017	0.4839	0.0171	0.00019	0.0709	0.2485	0.00021
0.2647	0.1139	0.00009	0.4742	0.0167	0.00014	0.0710	0.2511	0.00018
0.2706	0.0942	0.00018	0.3551	0.0901	0.00011	0.0178	0.2216	0.00027
0.1661	0.1292	0.00020	0.4066	0.0185	0.00017	0.0708	0.1815	0.00030
T/K = 307.35								
0.2647	0.1140	0.00017	0.4174	0.0778	0.00033	0.0035	0.2949	0.00033
0.2647	0.1140	0.00009	0.4231	0.0635	0.00021	0.0000	0.2920	0.00027
0.2706	0.0942	0.00018	0.4437	0.0091	0.00024	0.0750	0.2096	0.00038
0.1661	0.1292	0.00020	0.3391	0.0802	0.00032	0.0000	0.2599	0.00041
T/K = 310.35								
0.2647	0.1140	0.00017	0.4474	0.0732	0.00040	0.0752	0.2569	0.00041
0.2647	0.1140	0.00009	0.4297	0.0630	0.00026	0.0205	0.2585	0.00031
0.2705	0.0942	0.00018	0.4860	0.0655	0.00024	0.0348	0.2632	0.00042
0.1661	0.1292	0.00020	0.3881	0.0691	0.00038	0.0000	0.2574	0.00045

Polymer (B):	**poly(ethylene glycol)**		**2009KHE**
Characterization:	M_n/g.mol^{-1} = 10000, Merck, Darmstadt, Germany		
Solvent (A):	**water**	**H_2O**	**7732-18-5**
Salt (C):	**dipotassium phosphate**	**K_2HPO_4**	**7758-11-4**
Component (D):	**cephalexin**	**$C_{16}H_{17}N_3O_4S$**	**15686-71-2**

Type of data: coexistence data (tie lines)

Total system			Top phase			Bottom phase		
w_B	w_C	w_D	w_B	w_C	w_D	w_B	w_C	w_D
T/K = 301.35								
0.2647	0.1140	0.00017	0.4851	0.0064	0.00019	0.0737	0.2114	0.00026
0.2647	0.1018	0.00007	0.4801	0.0109	0.00015	0.0768	0.2079	0.00022
0.2706	0.0929	0.00018	0.4452	0.0191	0.00012	0.0722	0.2022	0.00033
0.1661	0.0889	0.00020	0.4030	0.0176	0.00016	0.0491	0.1973	0.00036
T/K = 307.35								
0.2647	0.1140	0.00017	0.4330	0.0535	0.00027	0.0175	0.2945	0.00031
0.2647	0.1140	0.00007	0.4289	0.0572	0.00019	0.0000	0.2717	0.00026
0.2706	0.0942	0.00018	0.4010	0.0591	0.00017	0.0780	0.3050	0.00035
0.1661	0.0889	0.00020	0.3456	0.0694	0.00026	0.0000	0.2937	0.00037

continued

continued

T/K = 310.35

0.2647	0.1140	0.00017	0.4231	0.0686	0.00033	0.0000	0.2808	0.00034
0.2647	0.1140	0.00008	0.4141	0.0675	0.00024	0.0050	0.2659	0.00030
0.2706	0.0942	0.00018	0.4245	0.0649	0.00020	0.0000	0.2783	0.00037
0.1661	0.0889	0.00020	0.3750	0.0741	0.00033	0.0000	0.2801	0.00042

Polymer (B): **poly(ethylene glycol)** **2008YAN**
Characterization: M_w/g.mol^{-1} = 10000, M_w/M_n = 1.05
Fluka AG, Buchs, Switzerland
Solvent (A): **water** **H_2O** **7732-18-5**
Salt (C): **dipotassium phosphate** **K_2HPO_4** **7758-11-4**
Component (D): **lactic acid** **$C_3H_6O_3$** **50-21-5**

Type of data: cloud points

T/K = 298.15

Comments: The concentration of lactic acid is 53.0 g/L.

w_B	0.3934	0.3659	0.3236	0.3151	0.2953	0.2709	0.2571	0.2339	0.2266
w_C	0.04142	0.04589	0.05141	0.05276	0.05602	0.06073	0.06220	0.06573	0.06748
w_B	0.2155	0.2020	0.1901	0.1774	0.1609	0.1520	0.1374	0.1217	0.1087
w_C	0.07094	0.07417	0.07649	0.07954	0.08305	0.08584	0.08859	0.09336	0.09718
w_B	0.09676	0.08106	0.06858	0.05955	0.04887	0.03522	0.02646	0.01672	0.00862
w_C	0.09985	0.1034	0.1074	0.1109	0.1147	0.1194	0.1233	0.1259	0.1308
w_B	0.00439	0.00244	0.00183						
w_C	0.1358	0.1466	0.1596						

Polymer (B): **poly(ethylene glycol)** **2008YAN**
Characterization: M_w/g.mol^{-1} = 20000, M_w/M_n = 1.05
Fluka AG, Buchs, Switzerland
Solvent (A): **water** **H_2O** **7732-18-5**
Salt (C): **dipotassium phosphate** **K_2HPO_4** **7758-11-4**
Component (D): **lactic acid** **$C_3H_6O_3$** **50-21-5**

Type of data: cloud points

T/K = 298.15

Comments: The concentration of lactic acid is 53.0 g/L.

w_B	0.3787	0.3675	0.3562	0.3448	0.3337	0.3227	0.3119	0.3014	0.2910
w_C	0.03907	0.04128	0.04350	0.04572	0.04793	0.05015	0.05236	0.05458	0.05680
w_B	0.2807	0.2705	0.2604	0.2503	0.2402	0.2302	0.2003	0.1905	0.1807
w_C	0.05901	0.06123	0.06344	0.06566	0.06788	0.07009	0.07674	0.07896	0.08117

continued

continued

w_B	0.1517	0.1422	0.1235	0.1052	0.09625	0.06267	0.04779	0.03469	0.02366
w_C	0.08782	0.09004	0.09447	0.09890	0.1011	0.1100	0.1144	0.1188	0.1233
w_B	0.01481	0.00804	0.00129						
w_C	0.1277	0.1321	0.1388						

Polymer (B): **poly(ethylene glycol)** **2010MIR**
Characterization: M_w/g.mol^{-1} = 4000, Merck, Darmstadt, Germany
Solvent (A): **water** H_2O **7732-18-5**
Salt (C): **dipotassium phosphate** K_2HPO_4 **7758-11-4**
Component (D): **L-lysine monohydrochloride** $C_6H_{15}ClN_2O_2$ **657-27-2**

Type of data: coexistence data (tie lines)

Total system			Top phase			Bottom phase		
w_B	w_C	w_D	w_B	w_C	w_D	w_B	w_C	w_D
T/K = 293.15	pH around 9.7							
0.264685	0.113960	0.000182	0.463093	0.015945	0.000010	0.092671	0.167435	0.000055
0.264710	0.113971	0.000090	0.463825	0.014663	0.000008	0.127625	0.138552	0.000068
0.270589	0.094194	0.000187	0.428687	0.017336	0.000015	0.123197	0.118059	0.000051
0.167012	0.129098	0.000207	0.392427	0.021033	0.000016	0.085056	0.128385	0.000046
T/K = 298.15	pH around 9.7							
0.264685	0.113960	0.000182	0.467425	0.011986	0.000009	0.079787	0.175532	0.000057
0.264709	0.113971	0.000089	0.462139	0.013750	0.000014	0.122438	0.132250	0.000046
0.270589	0.094194	0.000187	0.430288	0.016485	0.000016	0.003242	0.132558	0.000070
0.167012	0.129098	0.000207	0.402965	0.018762	0.000002	0.042498	0.162380	0.000007
T/K = 303.15	pH around 9.7							
0.264685	0.113960	0.000182	0.494365	0.011900	0.000008	0.105876	0.158432	0.000081
0.264709	0.113971	0.000089	0.512261	0.010878	0.000007	0.091469	0.185098	0.000195
0.270589	0.094194	0.000187	0.458604	0.014528	0.000008	0.092494	0.144284	0.000025
0.167012	0.129098	0.000207	0.427853	0.017484	0.000006	0.084769	0.131714	0.000015

Polymer (B): **poly(ethylene glycol)** **2010MIR**
Characterization: M_w/g.mol^{-1} = 10000, Merck, Darmstadt, Germany
Solvent (A): **water** H_2O **7732-18-5**
Salt (C): **dipotassium phosphate** K_2HPO_4 **7758-11-4**
Component (D): **L-lysine monohydrochloride** $C_6H_{15}ClN_2O_2$ **657-27-2**

Type of data: coexistence data (tie lines)

continued

continued

Total system			Top phase			Bottom phase		
w_B	w_C	w_D	w_B	w_C	w_D	w_B	w_C	w_D
T/K = 293.15	pH around 9.3							
0.264685	0.113960	0.000182	0.459646	0.014124	0.000026	0.094637	0.160374	0.000032
0.264710	0.113971	0.000090	0.460005	0.013799	0.000030	0.093204	0.161057	0.000031
0.270589	0.094196	0.000187	0.440534	0.016081	0.000003	0.052863	0.179771	0.000025
0.167013	0.129098	0.000207	0.402031	0.019634	0.000027	0.049593	0.155123	0.000029
T/K = 298.15	pH around 9.3							
0.264685	0.113960	0.000182	0.486066	0.012302	0.000029	0.049829	0.208284	0.000036
0.264709	0.113971	0.000089	0.481061	0.013150	0.000040	0.077266	0.179775	0.000049
0.270590	0.094194	0.000187	0.448068	0.015399	0.000012	0.052988	0.178431	0.000018
0.167012	0.129098	0.000208	0.414081	0.018546	0.000040	0.046508	0.156074	0.000046
T/K = 303.15	pH around 9.3							
0.264685	0.113960	0.000182	0.507610	0.011212	0.000023	0.117119	0.149233	0.000027
0.264709	0.113972	0.000090	0.500893	0.011769	0.000030	0.109422	0.150677	0.000034
0.270589	0.094195	0.000187	0.479632	0.014443	0.000014	0.082583	0.149809	0.000020
0.167012	0.129098	0.000206	0.435461	0.015150	0.000038	0.042793	0.170478	0.000039

Polymer (B):	**poly(ethylene glycol)**	**2003HAG**
Characterization:	M_n/g.mol^{-1} = 1500, Merck, Darmstadt, Germany	
Solvent (A):	**water** H_2O	**7732-18-5**
Salt (C):	**dipotassium phosphate** K_2HPO_4	**7758-11-4**
Polymer (D):	**lysozyme**	
Characterization:	M_n/g.mol^{-1} = 14300, Merck, Darmstadt, Germany	

Type of data: coexistence data (tie lines)

T/K = 298.15 pH = 9.1

Total system			Top phase			Bottom phase		
w_B	w_C	w_D	w_B	w_C	w_D	w_B	w_C	w_D
0.1401	0.1119	0.000500	0.3041	0.0372	0.000566	0.0140	0.1747	0.000497
0.1400	0.1026	0.000500	0.2715	0.0436	0.000521	0.0453	0.1403	0.000530
0.1601	0.1018	0.000500	0.3076	0.0379	0.000550	0.0443	0.1492	0.000488
0.1404	0.0936	0.000500	0.2189	0.0563	0.000512	0.0682	0.1205	0.000532
0.1599	0.0928	0.000500	0.2646	0.0448	0.000515	0.0547	0.1339	0.000528
0.1798	0.0917	0.000500	0.2969	0.0374	0.000533	0.0437	0.1484	0.000526
0.1800	0.0825	0.000500	0.2662	0.0433	0.000516	0.0426	0.1381	0.000505
0.2002	0.0817	0.000500	0.3033	0.0380	0.000533	0.0429	0.1472	0.000509

Polymer (B): **poly(ethylene glycol)** **2003HAG**
Characterization: M_n/g.mol^{-1} = 4000, Merck, Darmstadt, Germany
Solvent (A): **water** H_2O **7732-18-5**
Salt (C): **dipotassium phosphate** K_2HPO_4 **7758-11-4**
Polymer (D): **lysozyme**
Characterization: M_n/g.mol^{-1} = 14300, Merck, Darmstadt, Germany

Type of data: coexistence data (tie lines)

T/K = 298.15 pH = 9.1

Total system			Top phase			Bottom phase		
w_B	w_C	w_D	w_B	w_C	w_D	w_B	w_C	w_D
0.1125	0.1082	0.000194	0.2910	0.0336	0.000085	0.0065	0.1627	0.000295
0.1120	0.1033	0.000194	0.2442	0.0408	0.000092	0.0022	0.1579	0.000295
0.1130	0.0993	0.000194	0.2530	0.0423	0.000097	0.0058	0.1446	0.000296
0.1138	0.0938	0.000196	0.2341	0.0462	0.000111	0.0090	0.1382	0.000256
0.1136	0.0903	0.000196	0.2211	0.0470	0.000136	0.0145	0.1332	0.000264

Polymer (B): **poly(ethylene glycol)** **2003HAG**
Characterization: M_n/g.mol^{-1} = 4000, Merck, Darmstadt, Germany
Solvent (A): **water** H_2O **7732-18-5**
Salt (C): **dipotassium phosphate/monopotassium phosphate**
K_2HPO_4/H_2KPO_4 **7758-11-4/7778-77-0**
Polymer (D): **α-amylase**
Characterization: M_n/g.mol^{-1} = 54000, from *Bacillus subtilis*,
Merck, Darmstadt, Germany

Type of data: coexistence data (tie lines)

T/K = 298.15 pH = 7.1 (prepared by a mixture of dibasic and monobasic potassium phosphate)

Total system			Top phase			Bottom phase		
w_B	w_C	w_D	w_B	w_C	w_D	w_B	w_C	w_D
0.0888	0.1074	0.000890	0.2322	0.0468	0.001017	0.0312	0.1299	0.000702
0.0952	0.1034	0.000890	0.2260	0.0487	0.001008	0.0207	0.1355	0.000699
0.0956	0.0947	0.000890	0.1778	0.0620	0.000859	0.0376	0.1148	0.000722
0.1089	0.0942	0.000890	0.2070	0.0542	0.000873	0.0144	0.1397	0.000682
0.1093	0.0898	0.000890	0.1759	0.0618	0.000816	0.0443	0.1153	0.000749
0.1229	0.0849	0.000890	0.1864	0.0591	0.000807	0.0397	0.1185	0.000700

Polymer (B): **poly(ethylene glycol)** **2003HAG**
Characterization: M_n/g.mol^{-1} = 4000, Merck, Darmstadt, Germany
Solvent (A): **water** $\mathbf{H_2O}$ **7732-18-5**
Salt (C): **dipotassium phosphate/monopotassium phosphate**
$\mathbf{K_2HPO_4/H_2KPO_4}$ **7758-11-4/7778-77-0**
Polymer (D): **lysozyme**
Characterization: M_n/g.mol^{-1} = 14300, Merck, Darmstadt, Germany

Type of data: coexistence data (tie lines)

T/K = 298.15 pH = 7.2 (prepared by a mixture of dibasic and monobasic potassium phosphate)

Total system			Top phase			Bottom phase		
w_B	w_C	w_D	w_B	w_C	w_D	w_B	w_C	w_D
0.1125	0.1081	0.000185	0.2410	0.0523	0.000102	0.0047	0.1646	0.000300
0.1139	0.1018	0.000185	0.2283	0.0538	0.000108	0.1617		0.000296
0.1149	0.0929	0.000185	0.1773	0.0699	0.000144	0.0177	0.1406	0.000281
0.1156	0.0889	0.000185	0.1471	0.0786	0.000181	0.0409	0.1267	0.000258

Polymer (B): **poly(ethylene glycol)** **2010SH2**
Characterization: M_n/g.mol^{-1} = 6000, Merck, Darmstadt, Germany
Solvent (A): **water** $\mathbf{H_2O}$ **7732-18-5**
Salt (C): **monopotassium phosphate** $\mathbf{H_2KPO_4}$ **7778-77-0**
Polymer (D): **β-amylase**
Characterization: M_n/g.mol^{-1} = 60000, from barley

Type of data: coexistence data (tie lines)

Total system			Top phase			Bottom phase		
w_B	w_C	w_D	w_B	w_C	w_D	w_B	w_C	w_D
T/K = 301.65	feed pH is about 4.5							
0.1852	0.10009	0.000279	0.25592	0.08358	0.00014	0.02209	0.17940	0.00060
0.1852	0.10009	0.000136	0.25079	0.08999	0.00008	0.02772	0.18000	0.00028
0.1804	0.12358	0.000272	0.32583	0.07270	0.00020	0.02396	0.19657	0.00035
0.1471	0.10477	0.000292	0.24885	0.07967	0.00013	0.02454	0.17973	0.00048
T/K = 304.65	feed pH is about 4.5							
0.1852	0.10009	0.000279	0.25926	0.06669	0.00015	0.01996	0.18544	0.00056
0.1852	0.10009	0.000136	0.25400	0.06054	0.00008	0.01913	0.18648	0.00026
0.1804	0.12358	0.000272	0.31877	0.04419	0.00019	0.01477	0.20464	0.00037
0.1471	0.10477	0.000292	0.27991	0.07910	0.00013	0.01133	0.19367	0.00046

Polymer (B): **poly(ethylene glycol)** **2010SH2**
Characterization: M_n/g.mol^{-1} = 10000, Merck, Darmstadt, Germany
Solvent (A): **water** **H_2O** **7732-18-5**
Salt (C): **monopotassium phosphate** **H_2KPO_4** **7778-77-0**
Polymer (D): **β-amylase**
Characterization: M_n/g.mol^{-1} = 60000, from barley

Type of data: coexistence data (tie lines)

Total system			Top phase			Bottom phase		
w_B	w_C	w_D	w_B	w_C	w_D	w_B	w_C	w_D
T/K = 301.65	feed pH is about 4.4							
0.1852	0.10009	0.000279	0.24558	0.07680	0.00015	0.01428	0.18050	0.00063
0.1852	0.10009	0.000136	0.26957	0.06782	0.00007	0.01977	0.17952	0.00027
0.1804	0.12358	0.000272	0.31694	0.06248	0.00015	0.02349	0.19350	0.00042
0.1471	0.10477	0.000292	0.19194	0.08326	0.00013	0.01882	0.16546	0.00075
T/K = 304.65	feed pH is about 4.4							
0.1852	0.10009	0.000279	0.26942	0.06632	0.00015	0.01419	0.18146	0.00055
0.1852	0.10009	0.000136	0.26220	0.06946	0.00008	0.01595	0.18094	0.00027
0.1804	0.12358	0.000272	0.32292	0.06004	0.00016	0.01624	0.19839	0.00040
0.1471	0.10477	0.000292	0.20279	0.07724	0.00013	0.01417	0.16622	0.00067

Polymer (B): **poly(ethylene glycol)** **2009KHE**
Characterization: M_n/g.mol^{-1} = 4000, Merck, Darmstadt, Germany
Solvent (A): **water** **H_2O** **7732-18-5**
Salt (C): **sodium citrate** **$Na_3C_6H_5O_7$** **68-04-2**
Component (D): **cephalexin** **$C_{16}H_{17}N_3O_4S$** **15686-71-2**

Type of data: coexistence data (tie lines)

Total system			Top phase			Bottom phase		
w_B	w_C	w_D	w_B	w_C	w_D	w_B	w_C	w_D
T/K = 301.35								
0.2647	0.0852	0.00017	0.3938	0.0605	0.00013	0.1604	0.1734	0.00010
0.2647	0.0852	0.00008	0.3674	0.0614	0.00009	0.1611	0.1614	0.00007
0.2706	0.0704	0.00018	0.3531	0.0596	0.00018	0.1422	0.1679	0.00008
0.1661	0.0966	0.00020	0.2868	0.0651	0.00013	0.1165	0.1506	0.00009

continued

continued

T/K = 307.35								
0.2647	0.0852	0.00017	0.4148	0.0385	0.00021	0.1464	0.1861	0.00014
0.2647	0.0852	0.00008	0.3837	0.0444	0.00015	0.1376	0.1839	0.00011
0.2706	0.0704	0.00018	0.3637	0.0485	0.00034	0.1401	0.1688	0.00014
0.1661	0.0966	0.00020	0.2863	0.0656	0.00020	0.1189	0.1467	0.00014
T/K = 310.35								
0.2647	0.0852	0.00017	0.4098	0.0403	0.00027	0.1454	0.1730	0.00018
0.2647	0.0852	0.00009	0.4005	0.0487	0.00021	0.1479	0.1707	0.00015
0.2705	0.0704	0.00018	0.3563	0.0458	0.00038	0.1406	0.1488	0.00015
0.1661	0.0966	0.00020	0.3078	0.0544	0.00026	0.1258	0.1334	0.00017

Polymer (B):	**poly(ethylene glycol)**		**2009KHE**
Characterization:	M_n/g.mol^{-1} = 10000, Merck, Darmstadt, Germany		
Solvent (A):	**water**	H_2O	**7732-18-5**
Salt (C):	**sodium citrate**	$Na_3C_6H_5O_7$	**68-04-2**
Component (D):	**cephalexin**	$C_{16}H_{17}N_3O_4S$	**15686-71-2**

Type of data: coexistence data (tie lines)

Total system			Top phase			Bottom phase		
w_B	w_C	w_D	w_B	w_C	w_D	w_B	w_C	w_D
T/K = 301.35								
0.2646	0.0852	0.00017	0.3866	0.0531	0.00009	0.1467	0.1675	0.00007
0.2647	0.0852	0.00008	0.3979	0.0484	0.00006	0.1507	0.1606	0.00005
0.2706	0.0704	0.00018	0.3727	0.0577	0.00011	0.1559	0.1518	0.00006
0.1661	0.0966	0.00020	0.3108	0.0643	0.00009	0.1401	0.1162	0.00007
T/K = 307.35								
0.2647	0.0852	0.00017	0.1779	0.2693	0.00015	0.1467	0.1675	0.00011
0.2647	0.0852	0.00008	0.4244	0.0221	0.00011	0.1503	0.1606	0.00008
0.2706	0.0704	0.00018	0.4025	0.0268	0.00021	0.1560	0.1518	0.00009
0.1661	0.0889	0.00020	0.3392	0.0347	0.00014	0.1401	0.1163	0.00010
T/K = 310.35								
0.2647	0.0852	0.00017	0.4235	0.0352	0.00022	0.1453	0.1738	0.00016
0.2647	0.0852	0.00008	0.4183	0.0413	0.00017	0.1639	0.1637	0.00012
0.2706	0.0704	0.00018	0.3958	0.0289	0.00033	0.1278	0.1428	0.00014
0.1661	0.0966	0.00020	0.3332	0.0453	0.00022	0.1288	0.1341	0.00015

Polymer (B): **poly(ethylene glycol)** **2010MIR**
Characterization: M_w/g.mol^{-1} = 4000, Merck, Darmstadt, Germany
Solvent (A): **water** H_2O **7732-18-5**
Salt (C): **sodium citrate** $Na_3C_6H_5O_7$ **68-04-2**
Component (D): **L-lysine monohydrochloride** $C_6H_{15}ClN_2O_2$ **657-27-2**

Type of data: coexistence data (tie lines)

Total system			Top phase			Bottom phase		
w_B	w_C	w_D	w_B	w_C	w_D	w_B	w_C	w_D
T/K = 293.15	pH around 8							
0.264685	0.113960	0.000183	0.375181	0.043526	0.000098	0.057937	0.237833	0.000098
0.264709	0.113971	0.000089	0.369121	0.047057	0.000076	0.106404	0.184310	0.000411
0.270589	0.094194	0.000187	0.348153	0.057059	0.000037	0.046838	0.232592	0.000066
0.167012	0.129098	0.000208	0.249419	0.085228	0.000118	0.062100	0.171680	0.000189
T/K = 298.15	pH around 8							
0.264685	0.113960	0.000182	0.404617	0.045730	0.000077	0.075720	0.228355	0.000247
0.264709	0.113971	0.000089	0.391369	0.046250	0.000063	0.069865	0.227459	0.000457
0.270589	0.094194	0.000187	0.362321	0.053470	0.000007	0.047089	0.231628	0.000107
0.167013	0.129098	0.000207	0.283236	0.067401	0.000147	0.042767	0.198229	0.000973
T/K = 303.15	pH around 8							
0.264685	0.113960	0.000182	0.456423	0.042788	0.000159	0.147199	0.195698	0.000867
0.264709	0.113971	0.000089	0.411488	0.039945	0.000101	0.045768	0.254779	0.000271
0.270589	0.094194	0.000187	0.373604	0.048687	0.000173	0.057758	0.220464	0.000325
0.167013	0.129098	0.000207	0.344175	0.051390	0.000277	0.075453	0.178775	0.000735

Polymer (B): **poly(ethylene glycol)** **2010MIR**
Characterization: M_w/g.mol^{-1} = 10000, Merck, Darmstadt, Germany
Solvent (A): **water** H_2O **7732-18-5**
Salt (C): **sodium citrate** $Na_3C_6H_5O_7$ **68-04-2**
Component (D): **L-lysine monohydrochloride** $C_6H_{15}ClN_2O_2$ **657-27-2**

Type of data: coexistence data (tie lines)

Total system			Top phase			Bottom phase		
w_B	w_C	w_D	w_B	w_C	w_D	w_B	w_C	w_D
T/K = 293.15	pH around 8							
0.264685	0.113960	0.000182	0.390883	0.029106	0.000053	0.047722	0.239192	0.000593
0.264709	0.113972	0.000089	0.392084	0.036992	0.000030	0.060574	0.227987	0.000705
0.270589	0.094194	0.000187	0.356686	0.035034	0.000152	0.043845	0.218630	0.000247
0.167012	0.129098	0.000207	0.301350	0.045301	0.000020	0.018457	0.208410	0.001863

continued

continued

T/K = 298.15 pH around 8

0.264685	0.113960	0.000182	0.414090	0.031754	0.000031	0.067688	0.218511	0.000597
0.264710	0.113971	0.000089	0.495093	0.00000	0.000190	0.262727	0.002442	0.001807
0.270589	0.094194	0.000187	0.388065	0.037644	0.000072	0.098908	0.171426	0.000256
0.167012	0.129098	0.000207	0.402215	0.033520	0.000116	0.095991	0.171638	0.002580

T/K = 303.15 pH around 8

0.264684	0.113960	0.000182	0.413861	0.031991	0.000046	0.058559	0.225159	0.000463
0.264709	0.113971	0.000089	0.424519	0.031485	0.000142	0.076533	0.213563	0.000822
0.270590	0.094195	0.000187	0.394109	0.028574	0.000038	0.095815	0.170416	0.000334
0.167013	0.129097	0.000207	0.363683	0.031298	0.000120	0.041806	0.198972	0.001172

Polymer (B): **poly(ethylene glycol)** **2003HAG**
Characterization: M_n/g.mol^{-1} = 1500, Merck, Darmstadt, Germany
Solvent (A): **water** H_2O **7732-18-5**
Salt (C): **sodium sulfate** Na_2SO_4 **7757-82-6**
Polymer (D): **α-amylase**
Characterization: M_n/g.mol^{-1} = 54000, from *Bacillus subtilis*, Merck, Darmstadt, Germany

Type of data: coexistence data (tie lines)

T/K = 298.15 pH = 5.4

Total system			Top phase			Bottom phase		
w_B	w_C	w_D	w_B	w_C	w_D	w_B	w_C	w_D
0.0886	0.0810	0.000900	0.2229	0.0380	0.000602	0.0194	0.1024	0.000913
0.0888	0.0781	0.000900	0.1964	0.0430	0.000625	0.0062	0.1186	0.000880
0.0957	0.0775	0.000900	0.2136	0.0402	0.000635	0.0215	0.1004	0.000857
0.0886	0.0748	0.000900	0.1712	0.0502	0.000655	0.0406	0.0887	0.000856
0.0954	0.0746	0.000900	0.1939	0.0454	0.000642	0.0317	0.0939	0.000873
0.1092	0.0706	0.000900	0.1917	0.0448	0.000631	0.0130	0.1100	0.000906

Polymer (B): **poly(ethylene glycol)** **2010SH2**
Characterization: M_n/g.mol^{-1} = 4000, Merck, Darmstadt, Germany
Solvent (A): **water** H_2O **7732-18-5**
Salt (C): **sodium sulfate** Na_2SO_4 **7757-82-6**
Polymer (D): **amyloglucosidase**
Characterization: M_n/g.mol^{-1} = 97000, from *Aspergillus niger*, Fluka AG, Buchs, Switzerland

Type of data: coexistence data (tie lines)

continued

continued

Total system			Top phase			Bottom phase		
w_B	w_C	w_D	w_B	w_C	w_D	w_B	w_C	w_D
T/K = 301.65	feed pH is about 5.5							
0.1852	0.10009	0.000279	0.43185	0.01643	0.00028	0.02125	0.16044	0.00028
0.1852	0.10009	0.000136	0.42339	0.01700	0.00020	0.01934	0.16022	0.00009
0.1804	0.12358	0.000272	0.47343	0.01226	0.00051	0.03916	0.16712	0.00016
0.1471	0.10477	0.000292	0.39772	0.02168	0.00051	0.01113	0.15584	0.00018
T/K = 305.65	feed pH is about 5.5							
0.1852	0.10009	0.000279	0.44366	0.01608	0.00030	0.01987	0.16106	0.00026
0.1852	0.10009	0.000136	0.43427	0.01685	0.00022	0.01865	0.16154	0.00008
0.1804	0.12358	0.000272	0.49268	0.01026	0.00052	0.02983	0.17322	0.00015
0.1471	0.10477	0.000292	0.40738	0.01930	0.00052	0.00758	0.15649	0.00017

Polymer (B): **poly(ethylene glycol)** **2010SH2**
Characterization: M_n/g.mol^{-1} = 6000, Merck, Darmstadt, Germany
Solvent (A): **water** H_2O **7732-18-5**
Salt (C): **sodium sulfate** Na_2SO_4 **7757-82-6**
Polymer (D): **amyloglucosidase**
Characterization: M_n/g.mol^{-1} = 97000, from *Aspergillus niger*, Fluka AG, Buchs, Switzerland

Type of data: coexistence data (tie lines)

Total system			Top phase			Bottom phase		
w_B	w_C	w_D	w_B	w_C	w_D	w_B	w_C	w_D
T/K = 301.65	feed pH is about 5.3							
0.1852	0.10009	0.000279	0.40354	0.01575	0.00028	0.02243	0.15825	0.00028
0.1852	0.10009	0.000136	0.40892	0.01581	0.00018	0.01923	0.16007	0.00011
0.1804	0.12358	0.000272	0.44575	0.01232	0.00045	0.03670	0.16701	0.00018
0.1471	0.10477	0.000292	0.38476	0.01954	0.00035	0.01212	0.15163	0.00026
T/K = 305.65	feed pH is about 5.3							
0.1852	0.10009	0.000279	0.42291	0.01542	0.00038	0.01621	0.15903	0.00020
0.1852	0.10009	0.000136	0.41001	0.01541	0.00020	0.01188	0.16264	0.00008
0.1804	0.12358	0.000272	0.46484	0.01092	0.00048	0.03173	0.17126	0.00017
0.1471	0.10477	0.000292	0.38984	0.01923	0.00036	0.00966	0.15205	0.00025

Polymer (B): **poly(ethylene glycol)** **2010SH2**
Characterization: M_n/g.mol^{-1} = 10000, Merck, Darmstadt, Germany
Solvent (A): **water** H_2O **7732-18-5**
Salt (C): **sodium sulfate** Na_2SO_4 **7757-82-6**
Polymer (D): **amyloglucosidase**
Characterization: M_n/g.mol^{-1} = 97000, from *Aspergillus niger*

Type of data: coexistence data (tie lines)

Total system			Top phase			Bottom phase		
w_B	w_C	w_D	w_B	w_C	w_D	w_B	w_C	w_D
T/K = 301.65		feed pH is about 5.0						
0.1852	0.10009	0.000279	0.39931	0.01510	0.00029	0.02109	0.15333	0.00027
0.1852	0.10009	0.000136	0.40462	0.01519	0.00020	0.01931	0.15310	0.00009
0.1804	0.12358	0.000272	0.44258	0.01159	0.00027	0.03515	0.16353	0.00027
0.1471	0.10477	0.000292	0.37737	0.01884	0.00029	0.00946	0.15081	0.00029
T/K = 305.65		feed pH is about 4.7						
0.1852	0.10009	0.000279	0.43877	0.01135	0.00033	0.01982	0.15460	0.00024
0.1852	0.10009	0.000136	0.43603	0.01073	0.00021	0.01773	0.15400	0.00008
0.1804	0.12358	0.000272	0.47124	0.00921	0.00032	0.03522	0.16413	0.00025
0.1471	0.10477	0.000292	0.40602	0.01580	0.00036	0.00740	0.15355	0.00025

Polymer (B): **poly(ethylene glycol)** **2008MOK**
Characterization: M_n/g.mol^{-1} = 1500, Merck, Darmstadt, Germany
Solvent (A): **water** H_2O **7732-18-5**
Salt (C): **sodium sulfate** Na_2SO_4 **7757-82-6**
Component (D): **ciprofloxacin** $C_{17}H_{18}FN_3O_3$ **85721-33-1**

Type of data: coexistence data (tie lines)

Total system			Top phase			Bottom phase		
w_B	w_C	w_D	w_B	w_C	w_D	w_B	w_C	w_D
T/K = 286.85								
0.2544	0.1060	0.00018	0.6621	0.0065	0.000185	0.0570	0.1974	0.000125
0.2544	0.1060	0.00007	0.6617	0.0059	0.000076	0.0552	0.1965	0.000048
0.2544	0.0890	0.00018	0.4686	0.0136	0.000195	0.0634	0.1817	0.000076
0.2544	0.0890	0.00007	0.4675	0.0133	0.000086	0.0617	0.1808	0.000033
0.1424	0.1060	0.00018	0.4805	0.0059	0.000165	0.0490	0.1879	0.000109
0.1424	0.1060	0.00007	0.4797	0.0052	0.000066	0.0486	0.1870	0.000028
0.1424	0.0890	0.00018	0.3712	0.0244	0.000173	0.0767	0.1478	0.000050
0.1424	0.0890	0.00007	0.3705	0.0238	0.000073	0.0756	0.1468	0.000023

continued

continued

T/K = 295.65								
0.1960	0.0980	0.00013	0.4793	0.0117	0.000140	0.0523	0.1752	0.000077
0.1275	0.0980	0.00013	0.4130	0.0120	0.000160	0.0559	0.1697	0.000065
0.2750	0.0980	0.00013	0.5537	0.0074	0.000165	0.0458	0.1897	0.000088
0.1960	0.0850	0.00013	0.3916	0.0197	0.000145	0.0862	0.1512	0.000047
0.1960	0.1100	0.00013	0.5568	0.0043	0.000140	0.0618	0.1974	0.000098
0.1960	0.0980	0.00005	0.4786	0.0117	0.000060	0.0530	0.1752	0.000035
0.1960	0.0980	0.00020	0.4793	0.0123	0.000190	0.0534	0.1755	0.000113
T/K = 304.45								
0.2544	0.1060	0.00018	0.6610	0.0056	0.000190	0.0549	0.1968	0.000126
0.2544	0.1060	0.00007	0.6586	0.0046	0.000080	0.0538	0.1959	0.000053
0.2544	0.0890	0.00018	0.4631	0.0130	0.000200	0.0620	0.1805	0.000078
0.2544	0.0890	0.00007	0.4729	0.0126	0.000090	0.0608	0.1789	0.000034
0.1424	0.1060	0.00018	0.4793	0.0056	0.000170	0.0482	0.1866	0.000111
0.1424	0.1060	0.00007	0.4780	0.0049	0.000070	0.0471	0.1851	0.000028
0.1424	0.0890	0.00018	0.3702	0.0241	0.000178	0.0753	0.1465	0.000054
0.1424	0.0890	0.00007	0.3674	0.0234	0.000079	0.0748	0.1450	0.000019

Polymer (B):	**poly(ethylene glycol)**		**2008MOK**
Characterization:	M_n/g.mol^{-1} = 2000, Merck, Darmstadt, Germany		
Solvent (A):	**water**	H_2O	**7732-18-5**
Salt (C):	**sodium sulfate**	Na_2SO_4	**7757-82-6**
Component (D):	**ciprofloxacin**	$C_{17}H_{18}FN_3O_3$	**85721-33-1**

Type of data: coexistence data (tie lines)

Total system			Top phase			Bottom phase		
w_B	w_C	w_D	w_B	w_C	w_D	w_B	w_C	w_D
T/K = 286.85								
0.2544	0.1060	0.00018	0.6492	0.0059	0.000188	0.0429	0.1913	0.000127
0.2544	0.1060	0.00007	0.6488	0.0052	0.000079	0.0411	0.1903	0.000050
0.2544	0.0890	0.00018	0.4556	0.0130	0.000198	0.0493	0.1755	0.000078
0.2544	0.0890	0.00007	0.4545	0.0126	0.000089	0.0475	0.1746	0.000035
0.1424	0.1060	0.00018	0.4675	0.0052	0.000168	0.0349	0.1817	0.000111
0.1424	0.1060	0.00007	0.4667	0.0046	0.000069	0.0344	0.1808	0.000030
0.1424	0.0890	0.00018	0.3582	0.0238	0.000176	0.0626	0.1416	0.000052
0.1424	0.0890	0.00007	0.3575	0.0231	0.000076	0.0615	0.1407	0.000025
T/K = 295.65								

continued

continued

0.1960	0.0980	0.00013	0.4663	0.0111	0.000143	0.0382	0.1691	0.000077
0.1275	0.0980	0.00013	0.4000	0.0114	0.000163	0.0418	0.1635	0.000067
0.2750	0.0980	0.00013	0.5407	0.0068	0.000168	0.0316	0.1836	0.000090
0.1960	0.0850	0.00013	0.3786	0.0191	0.000148	0.0721	0.1450	0.000049
0.1960	0.1100	0.00013	0.5439	0.0037	0.000143	0.0476	0.1913	0.000100
0.1960	0.0980	0.00005	0.4656	0.0111	0.000063	0.0388	0.1691	0.000037
0.1960	0.0980	0.00020	0.4663	0.0117	0.000193	0.0392	0.1694	0.000115

T/K = 304.45

0.2544	0.1060	0.00018	0.6481	0.0049	0.000193	0.0408	0.1907	0.000128
0.2544	0.1060	0.00007	0.6456	0.0040	0.000082	0.0397	0.1897	0.000055
0.2544	0.0890	0.00018	0.4501	0.0123	0.000203	0.0478	0.1743	0.000080
0.2544	0.0890	0.00007	0.4599	0.0120	0.000093	0.0467	0.1728	0.000036
0.1424	0.1060	0.00018	0.4664	0.0049	0.000173	0.0341	0.1805	0.000113
0.1424	0.1060	0.00007	0.4650	0.0043	0.000073	0.0329	0.1789	0.000030
0.1424	0.0890	0.00018	0.3572	0.0234	0.000181	0.0611	0.1404	0.000056
0.1424	0.0890	0.00007	0.3544	0.0228	0.000082	0.0606	0.1388	0.000022

Polymer (B): **poly(ethylene glycol)** **2008MOK**
Characterization: M_n/g.mol^{-1} = 4000, Merck, Darmstadt, Germany
Solvent (A): **water** **H_2O** **7732-18-5**
Salt (C): **sodium sulfate** **Na_2SO_4** **7757-82-6**
Component (D): **ciprofloxacin** **$C_{17}H_{18}FN_3O_3$** **85721-33-1**

Type of data: coexistence data (tie lines)

Total system			Top phase			Bottom phase		
w_B	w_C	w_D	w_B	w_C	w_D	w_B	w_C	w_D
T/K = 286.85								
0.2544	0.1060	0.00018	0.5966	0.0040	0.000191	0.0135	0.1666	0.000128
0.2544	0.1060	0.00007	0.5962	0.0034	0.000080	0.0117	0.1657	0.000055
0.2544	0.0890	0.00018	0.4030	0.0111	0.000200	0.0199	0.1509	0.000080
0.2544	0.0890	0.00007	0.4020	0.0108	0.000091	0.0181	0.1499	0.000036
0.1424	0.1060	0.00018	0.4149	0.0034	0.000171	0.0055	0.1570	0.000113
0.1424	0.1060	0.00007	0.4142	0.0028	0.000071	0.0050	0.1561	0.000030
0.1424	0.0890	0.00018	0.3057	0.0219	0.000176	0.0332	0.1169	0.000051
0.1424	0.0890	0.00007	0.3050	0.0213	0.000080	0.0321	0.1160	0.000021

T/K = 295.65

continued

continued

0.1960	0.0980	0.00013	0.4137	0.0093	0.000146	0.0088	0.1444	0.000082
0.1275	0.0980	0.00013	0.3474	0.0096	0.000145	0.0123	0.1388	0.000070
0.2750	0.0980	0.00013	0.4882	0.0049	0.000172	0.0022	0.1589	0.000094
0.1960	0.0850	0.00013	0.3260	0.0173	0.000152	0.0427	0.1203	0.000051
0.1960	0.1100	0.00013	0.4913	0.0019	0.000146	0.0182	0.1666	0.000104
0.1960	0.0980	0.00005	0.4131	0.0093	0.000066	0.0094	0.1444	0.000041
0.1960	0.0980	0.00020	0.4138	0.0099	0.000197	0.0098	0.1447	0.000119

T/K = 304.45

0.2544	0.1060	0.00018	0.5955	0.0031	0.000196	0.0114	0.1660	0.000132
0.2544	0.1060	0.00007	0.5930	0.0022	0.000085	0.0103	0.1650	0.000059
0.2544	0.0890	0.00018	0.3975	0.0105	0.000206	0.0184	0.1496	0.000084
0.2544	0.0890	0.00007	0.4074	0.0102	0.000096	0.0173	0.1481	0.000040
0.1424	0.1060	0.00018	0.4139	0.0031	0.000176	0.0047	0.1558	0.000117
0.1424	0.1060	0.00007	0.4125	0.0025	0.000076	0.0035	0.1543	0.000034
0.1424	0.0890	0.00018	0.3046	0.0216	0.000184	0.0317	0.1157	0.000060
0.1424	0.0890	0.00007	0.3019	0.0210	0.000085	0.0312	0.1141	0.000025

Polymer (B): **poly(ethylene glycol)** **2003HAG**
Characterization: M_n/g.mol^{-1} = 1500, Merck, Darmstadt, Germany
Solvent (A): **water** H_2O **7732-18-5**
Salt (C): **sodium sulfate** Na_2SO_4 **7757-82-6**
Polymer (D): **lysozyme**
Characterization: M_n/g.mol^{-1} = 14300, Merck, Darmstadt, Germany

Type of data: coexistence data (tie lines)

T/K = 298.15 pH = 4.6

Total system			Top phase			Bottom phase		
w_B	w_C	w_D	w_B	w_C	w_D	w_B	w_C	w_D
0.1001	0.0990	0.000400	0.2341	0.0470	0.000400	0.0289	0.1307	0.000450
0.1199	0.0914	0.000400	0.2357	0.0474	0.000410	0.0461	0.1186	0.000460
0.1401	0.0840	0.000400	0.2406	0.0447	0.000390	0.0403	0.1222	0.000470
0.1597	0.0764	0.000400	0.2392	0.0454	0.000410	0.0392	0.1240	0.000470

Polymer (B): **poly(ethylene glycol)** **2003HAG**
Characterization: M_n/g.mol^{-1} = 4000, Merck, Darmstadt, Germany
Solvent (A): **water** H_2O **7732-18-5**
Salt (C): **sodium sulfate** Na_2SO_4 **7757-82-6**
Polymer (D): **lysozyme**
Characterization: M_n/g.mol^{-1} = 14300, Merck, Darmstadt, Germany

continued

continued

Type of data: coexistence data (tie lines)

T/K = 298.15 pH = 9.1

Total system			Top phase			Bottom phase		
w_B	w_C	w_D	w_B	w_C	w_D	w_B	w_C	w_D
0.1362	0.0920	0.000190	0.3471	0.0201	0.000089	0.0000	0.1487	0.000228
0.1362	0.0873	0.000190	0.3256	0.0271	0.000146	0.0118	0.1326	0.000216
0.1364	0.0836	0.000190	0.3133	0.0252	0.000204	0.0086	0.1315	0.000221
0.1360	0.0786	0.000190	0.2870	0.0334	0.000153	0.0000	0.1378	0.000211
0.1362	0.0755	0.000190	0.2785	0.0312	0.000168	0.0149	0.1192	0.000293
0.1360	0.0696	0.000190	0.2535	0.0319	0.000177	0.0170	0.1117	0.000231

Polymer (B): **poly(ethylene glycol)** **2011FE1**
Characterization: M_w/g.mol^{-1} = 8000, Sigma Chemical Co., Inc., St. Louis, MO
Solvent (A): **water** **H_2O** **7732-18-5**
Salt (C): **sodium sulfate** **Na_2SO_4** **7757-82-6**
Salt (D): **potassium chloride** **KCl** **7447-40-7**

Type of data: coexistence data (tie lines)

T/K = 296.15

Total system		Top phase		Bottom phase	
w_B	w_C	w_B	w_C	w_B	w_C
c_D/(mol/L) = 0.10					
0.1104	0.0600	0.1813	0.0375	0.0134	0.0883
0.1152	0.0630	0.2113	0.0315	0.0071	0.0959
0.1204	0.0660	0.2330	0.0277	0.0044	0.1009
0.1248	0.0689	0.2585	0.0235	0.0029	0.1051
c_D/(mol/L) = 0.25					
0.1103	0.0571	0.1871	0.0324	0.0134	0.0883
0.1149	0.0591	0.2008	0.0297	0.0071	0.0959
0.1204	0.0611	0.2240	0.0255	0.0044	0.1009
0.1248	0.0630	0.2422	0.0224	0.0029	0.1051
c_D/(mol/L) = 0.50					
0.1001	0.0552	0.1714	0.0320	0.0135	0.0832
0.1047	0.0571	0.1945	0.0274	0.0083	0.0889
0.1103	0.0590	0.2163	0.0235	0.0049	0.0944
0.1152	0.0611	0.2383	0.0198	0.0031	0.0987

continued

continued

c_D/(mol/L) = 0.75					
0.1000	0.0530	0.1758	0.0285	0.0104	0.0819
0.1050	0.0551	0.2010	0.0234	0.0058	0.0879
0.1107	0.0571	0.2192	0.0201	0.0030	0.0939
0.1151	0.0592	0.2375	0.0170	0.0018	0.0982
c_D/(mol/L) = 1.00					
0.0952	0.0530	0.1882	0.0239	0.0090	0.0801
0.1000	0.0552	0.2064	0.0206	0.0048	0.0861
0.1051	0.0570	0.2290	0.0169	0.0030	0.0901
0.1105	0.0591	0.2472	0.0142	0.0017	0.0948

Polymer (B):	**poly(ethylene glycol)**		**2011FE1**
Characterization:	M_w/g.mol^{-1} = 8000, Sigma Chemical Co., Inc., St. Louis, MO		
Solvent (A):	**water**	**H_2O**	**7732-18-5**
Salt (C):	**sodium sulfate**	**Na_2SO_4**	**7757-82-6**
Salt (D):	**sodium chloride**	**NaCl**	**7647-14-5**

Type of data: coexistence data (tie lines)

T/K = 296.15

Total system		Top phase		Bottom phase	
w_B	w_C	w_B	w_C	w_B	w_C
c_D/(mol/L) = 0.10					
0.1097	0.0600	0.1805	0.0356	0.0077	0.0950
0.1155	0.0630	0.2061	0.0306	0.0035	0.1032
0.1206	0.0661	0.2353	0.0257	0.0021	0.1078
0.1254	0.0690	0.2550	0.0226	0.0011	0.1136
c_D/(mol/L) = 0.25					
0.1097	0.0569	0.1830	0.0316	0.0070	0.0924
0.1156	0.0589	0.2050	0.0275	0.0040	0.0981
0.1195	0.0610	0.2298	0.0232	0.0030	0.1010
0.1253	0.0630	0.2605	0.0185	0.0022	0.1036
c_D/(mol/L) = 0.50					
0.1009	0.0549	0.1793	0.0284	0.0088	0.0861
0.1050	0.0569	0.1963	0.0253	0.0053	0.0914
0.1103	0.0589	0.2159	0.0216	0.0030	0.0968
0.1149	0.0610	0.2331	0.0186	0.0017	0.1017
c_D/(mol/L) = 0.75					
0.1065	0.0529	0.1892	0.0249	0.0062	0.0869
0.1050	0.0550	0.1986	0.0232	0.0050	0.0889
0.1099	0.0570	0.2188	0.0200	0.0032	0.0931
0.1157	0.0589	0.2388	0.0169	0.0019	0.0977

continued

continued

c_D/(mol/L) = 1.00

0.0950	0.0529	0.1785	0.0253	0.0089	0.0815
0.1003	0.0550	0.2014	0.0214	0.0055	0.0865
0.1052	0.0569	0.2250	0.0181	0.0040	0.0897

Polymer (B): **poly(ethylene oxide-*b*-propylene oxide-*b*-ethylene oxide)** **2010MAH**

Characterization: M_w/g.mol^{-1} = 2036, (EO)2.5(PO)31(EO)2.5, Aldrich Chem. Co., Inc., Milwaukee, WI

Solvent (A): **water** **H_2O** **7732-18-5**

Component (C): **L-alanine** **$C_3H_7NO_2$** **56-41-7**

Component (D): **hexadecyltrimethylammonium bromide** **$C_{19}H_{42}BrN$** **57-09-0**

Type of data: cloud points (LCST-behavior)

w_B = 0.01 in the binary aqueous solution was kept constant
c_D/(mol/l) = 0.02 was kept constant

c_C/(mol/l)	0.00	0.05	0.10	0.25	0.50
T/K	326.95	326.15	325.15	323.75	321.65

Polymer (B): **poly(ethylene oxide-*b*-propylene oxide-*b*-ethylene oxide)** **2010MAH**

Characterization: M_w/g.mol^{-1} = 2900, 40.0 wt% ethylene oxide, L64, (EO)13(PO)30(EO)13, Aldrich Chem. Co., Inc., Milwaukee, WI

Solvent (A): **water** **H_2O** **7732-18-5**

Component (C): **L-alanine** **$C_3H_7NO_2$** **56-41-7**

Component (D): **hexadecyltrimethylammonium bromide** **$C_{19}H_{42}BrN$** **57-09-0**

Type of data: cloud points (LCST-behavior)

w_B = 0.01 in the binary aqueous solution was kept constant
c_D/(mol/l) = 0.005 was kept constant

c_C/(mol/l)	0.00	0.05	0.10	0.25	0.50
T/K	350.65	346.25	344.25	342.15	340.25

Polymer (B): **poly(ethylene oxide-*b*-propylene oxide-*b*-ethylene oxide)** **2010MAH**

Characterization: M_w/g.mol^{-1} = 2036, (EO)2.5(PO)31(EO)2.5, Aldrich Chem. Co., Inc., Milwaukee, WI

Solvent (A): **water** **H_2O** **7732-18-5**

Component (C): **L-alanine** **$C_3H_7NO_2$** **56-41-7**

Component (D): **sodium dodecyl sulfate** **$C_{12}H_{25}NaO_4S$** **151-21-3**

Type of data: cloud points (LCST-behavior)

w_B = 0.01 in the binary aqueous solution was kept constant
c_D/(mol/l) = 0.02 was kept constant

c_C/(mol/l)	0.00	0.05	0.10	0.25	0.50
T/K	329.15	327.95	326.75	325.65	324.55

Polymer (B): **poly(ethylene oxide-*b*-propylene oxide-*b*-ethylene oxide)** **2010MAH**
Characterization: M_w/g.mol^{-1} = 2900, 40.0 wt% ethylene oxide, L64, (EO)13(PO)30(EO)13, Aldrich Chem. Co., Inc., Milwaukee, WI
Solvent (A): **water** **H_2O** **7732-18-5**
Component (C): **L-alanine** **$C_3H_7NO_2$** **56-41-7**
Component (D): **sodium dodecyl sulfate** **$C_{12}H_{25}NaO_4S$** **151-21-3**

Type of data: cloud points (LCST-behavior)

w_B = 0.01 in the binary aqueous solution was kept constant
c_D/(mol/l) = 0.005 was kept constant

c_C/(mol/l)	0.00	0.05	0.10	0.25	0.50
T/K	345.65	343.65	342.25	341.45	340.45

Polymer (B): **poly(ethylene oxide-*b*-propylene oxide-*b*-ethylene oxide)** **2005SIL**
Characterization: M_n/g.mol^{-1} = 1706, M_w/g.mol^{-1} = 1945, 50.0 wt% ethylene oxide, (EO)11-(PO)16-(EO)11, L35, Aldrich Chem. Co., Inc., Milwaukee, WI
Solvent (A): **water** **H_2O** **7732-18-5**
Salt (C): **dipotassium phosphate** **K_2HPO_4** **7758-11-4**
Salt (D): **potassium hydroxide** **KOH** **1310-58-3**

Comments: The mixture of K_2HPO_4/KOH was chosen in a ratio that provides a pH = 12.

Type of data: coexistence data (tie lines)

Total system			Top phase			Bottom phase		
w_A	w_B	w_C	w_A	w_B	w_C	w_A	w_B	w_C
T/K = 283.15								
0.4617	0.2869	0.2514	0.4246	0.5699	0.0055	0.5113	0.0596	0.4291
0.3736	0.3402	0.2862	0.3386	0.6601	0.0013	0.4037	0.0365	0.5598
0.2845	0.3895	0.3260	0.2913	0.7060	0.0027	0.3205	0.0286	0.6509
0.2059	0.4265	0.3676	0.2036	0.7930	0.0034	0.1961	0.0188	0.7851
T/K = 298.15								
0.5965	0.2404	0.1634	0.3654	0.6322	0.0024	0.7113	0.0443	0.2444
0.5091	0.2881	0.2028	0.2859	0.7115	0.0026	0.6305	0.0417	0.3278
0.4300	0.3414	0.2286	0.1981	0.8010	0.0009	0.5908	0.0239	0.3853
0.3509	0.3897	0.2594	0.1411	0.8586	0.0003	0.5242	0.0088	0.4670
T/K = 313.15								
0.5708	0.2889	0.1403	0.3103	0.6862	0.0035	0.7423	0.0252	0.2325
0.5103	0.3205	0.1692	0.2706	0.7269	0.0025	0.6897	0.0110	0.2993
0.4383	0.3738	0.1879	0.2123	0.7860	0.0017	0.6192	0.0255	0.3553
0.3643	0.4265	0.2092	0.1324	0.8672	0.0004	0.5846	0.0131	0.4023

Polymer (B):	**poly(ethylene oxide-*b*-propylene oxide-*b*-ethylene oxide)**		**2010MAH**
Characterization:	M_w/g.mol^{-1} = 2036, (EO)2.5(PO)31(EO)2.5, Aldrich Chem. Co., Inc., Milwaukee, WI		
Solvent (A):	**water**	**H_2O**	**7732-18-5**
Component (C):	**glycine**	**$C_2H_5NO_2$**	**56-40-6**
Component (D):	**hexadecyltrimethylammonium bromide**	**$C_{19}H_{42}BrN$**	**57-09-0**

Type of data: cloud points (LCST-behavior)

w_B = 0.01 in the binary aqueous solution was kept constant
c_D/(mol/l) = 0.02 was kept constant

c_C/(mol/l)	0.00	0.05	0.10	0.25	0.50
T/K	326.95	324.85	323.65	322.55	321.25

Polymer (B):	**poly(ethylene oxide-*b*-propylene oxide-*b*-ethylene oxide)**		**2010MAH**
Characterization:	M_w/g.mol^{-1} = 2900, 40.0 wt% ethylene oxide, L64, (EO)13(PO)30(EO)13, Aldrich Chem. Co., Inc., Milwaukee, WI		
Solvent (A):	**water**	**H_2O**	**7732-18-5**
Component (C):	**glycine**	**$C_2H_5NO_2$**	**56-40-6**
Component (D):	**hexadecyltrimethylammonium bromide**	**$C_{19}H_{42}BrN$**	**57-09-0**

Type of data: cloud points (LCST-behavior)

w_B = 0.01 in the binary aqueous solution was kept constant
c_D/(mol/l) = 0.005 was kept constant

c_C/(mol/l)	0.00	0.05	0.10	0.25	0.50
T/K	350.65	346.45	342.75	340.35	338.15

Polymer (B):	**poly(ethylene oxide-*b*-propylene oxide-*b*-ethylene oxide)**		**2010MAH**
Characterization:	M_w/g.mol^{-1} = 2036, (EO)2.5(PO)31(EO)2.5, Aldrich Chem. Co., Inc., Milwaukee, WI		
Solvent (A):	**water**	**H_2O**	**7732-18-5**
Component (C):	**glycine**	**$C_2H_5NO_2$**	**56-40-6**
Component (D):	**sodium dodecyl sulfate**	**$C_{12}H_{25}NaO_4S$**	**151-21-3**

Type of data: cloud points (LCST-behavior)

w_B = 0.01 in the binary aqueous solution was kept constant
c_D/(mol/l) = 0.02 was kept constant

c_C/(mol/l)	0.00	0.05	0.10	0.25	0.50
T/K	329.15	327.75	326.95	325.25	323.45

Polymer (B):	**poly(ethylene oxide-*b*-propylene oxide-*b*-ethylene oxide)**		**2010MAH**
Characterization:	M_w/g.mol^{-1} = 2900, 40.0 wt% ethylene oxide, L64, (EO)13(PO)30(EO)13, Aldrich Chem. Co., Inc., Milwaukee, WI		
Solvent (A):	**water**	**H_2O**	**7732-18-5**
Component (C):	**glycine**	**$C_2H_5NO_2$**	**56-40-6**
Component (D):	**sodium dodecyl sulfate**	**$C_{12}H_{25}NaO_4S$**	**151-21-3**

Type of data: cloud points (LCST-behavior)

w_B = 0.01 in the binary aqueous solution was kept constant
c_D/(mol/l) = 0.005 was kept constant

c_C/(mol/l)	0.00	0.05	0.10	0.25	0.50
T/K	345.65	343.95	342.05	340.25	339.15

Polymer (B):	**poly(ethylene oxide-*b*-propylene oxide-*b*-ethylene oxide)**		**2010MAH**
Characterization:	M_w/g.mol^{-1} = 2036, (EO)2.5(PO)31(EO)2.5, Aldrich Chem. Co., Inc., Milwaukee, WI		
Solvent (A):	**water**	**H_2O**	**7732-18-5**
Component (C):	**glycylglycine**	**$C_4H_8N_2O_3$**	**556-50-3**
Component (D):	**hexadecyltrimethylammonium bromide**	**$C_{19}H_{42}BrN$**	**57-09-0**

Type of data: cloud points (LCST-behavior)

w_B = 0.01 in the binary aqueous solution was kept constant
c_D/(mol/l) = 0.02 was kept constant

c_C/(mol/l)	0.00	0.05	0.10	0.25	0.50
T/K	326.95	323.35	322.15	321.05	318.75

Polymer (B):	**poly(ethylene oxide-*b*-propylene oxide-*b*-ethylene oxide)**		**2010MAH**
Characterization:	M_w/g.mol^{-1} = 2900, 40.0 wt% ethylene oxide, L64, (EO)13(PO)30(EO)13, Aldrich Chem. Co., Inc., Milwaukee, WI		
Solvent (A):	**water**	**H_2O**	**7732-18-5**
Component (C):	**glycylglycine**	**$C_4H_8N_2O_3$**	**556-50-3**
Component (D):	**hexadecyltrimethylammonium bromide**	**$C_{19}H_{42}BrN$**	**57-09-0**

Type of data: cloud points (LCST-behavior)

w_B = 0.01 in the binary aqueous solution was kept constant
c_D/(mol/l) = 0.005 was kept constant

c_C/(mol/l)	0.00	0.05	0.10	0.25	0.50
T/K	350.65	340.95	337.35	330.95	322.15

Polymer (B):	**poly(ethylene oxide-*b*-propylene oxide-*b*-ethylene oxide)**		**2010MAH**
Characterization:	M_w/g.mol^{-1} = 2036, (EO)2.5(PO)31(EO)2.5, Aldrich Chem. Co., Inc., Milwaukee, WI		
Solvent (A):	**water**	**H_2O**	**7732-18-5**
Component (C):	**glycylglycine**	**$C_4H_8N_2O_3$**	**556-50-3**
Component (D):	**sodium dodecyl sulfate**	**$C_{12}H_{25}NaO_4S$**	**151-21-3**

Type of data: cloud points (LCST-behavior)

w_B = 0.01 in the binary aqueous solution was kept constant
c_D/(mol/l) = 0.02 was kept constant

c_C/(mol/l)	0.00	0.05	0.10	0.25	0.50
T/K	329.15	326.25	324.55	322.45	320.75

Polymer (B):	**poly(ethylene oxide-*b*-propylene oxide-*b*-ethylene oxide)**		**2010MAH**
Characterization:	M_w/g.mol^{-1} = 2900, 40.0 wt% ethylene oxide, L64, (EO)13(PO)30(EO)13, Aldrich Chem. Co., Inc., Milwaukee, WI		
Solvent (A):	**water**	**H_2O**	**7732-18-5**
Component (C):	**glycylglycine**	**$C_4H_8N_2O_3$**	**556-50-3**
Component (D):	**sodium dodecyl sulfate**	**$C_{12}H_{25}NaO_4S$**	**151-21-3**

Type of data: cloud points (LCST-behavior)

w_B = 0.01 in the binary aqueous solution was kept constant
c_D/(mol/l) = 0.005 was kept constant

c_C/(mol/l)	0.00	0.05	0.10	0.25	0.50
T/K	345.65	342.95	340.75	338.95	337.05

Polymer (B):	**poly(ethylene oxide-*b*-propylene oxide-*b*-ethylene oxide)**		**2010MAH**
Characterization:	M_w/g.mol^{-1} = 2036, (EO)2.5(PO)31(EO)2.5, Aldrich Chem. Co., Inc., Milwaukee, WI		
Solvent (A):	**water**	**H_2O**	**7732-18-5**
Component (C):	**glycyl-DL-valine**	**$C_7H_{14}N_2O_3$**	**2325-17-9**
Component (D):	**hexadecyltrimethylammonium bromide**	**$C_{19}H_{42}BrN$**	**57-09-0**

Type of data: cloud points (LCST-behavior)

w_B = 0.01 in the binary aqueous solution was kept constant
c_D/(mol/l) = 0.02 was kept constant

c_C/(mol/l)	0.00	0.05	0.10	0.25	0.50
T/K	326.95	321.65	320.65	319.55	316.95

Polymer (B):	**poly(ethylene oxide-*b*-propylene oxide-*b*-ethylene oxide)**		**2010MAH**
Characterization:	M_w/g.mol^{-1} = 2900, 40.0 wt% ethylene oxide, L64, (EO)13(PO)30(EO)13, Aldrich Chem. Co., Inc., Milwaukee, WI		
Solvent (A):	**water**	H_2O	**7732-18-5**
Component (C):	**glycyl-DL-valine**	$C_7H_{14}N_2O_3$	**2325-17-9**
Component (D):	**hexadecyltrimethylammonium bromide**	$C_{19}H_{42}BrN$	**57-09-0**

Type of data: cloud points (LCST-behavior)

w_B = 0.01 in the binary aqueous solution was kept constant
c_D/(mol/l) = 0.005 was kept constant

c_C/(mol/l)	0.00	0.05	0.10	0.25	0.50
T/K	350.65	339.15	334.85	328.55	320.45

Polymer (B):	**poly(ethylene oxide-*b*-propylene oxide-*b*-ethylene oxide)**		**2010MAH**
Characterization:	M_w/g.mol^{-1} = 2036, (EO)2.5(PO)31(EO)2.5, Aldrich Chem. Co., Inc., Milwaukee, WI		
Solvent (A):	**water**	H_2O	**7732-18-5**
Component (C):	**glycyl-DL-valine**	$C_7H_{14}N_2O_3$	**2325-17-9**
Component (D):	**sodium dodecyl sulfate**	$C_{12}H_{25}NaO_4S$	**151-21-3**

Type of data: cloud points (LCST-behavior)

w_B = 0.01 in the binary aqueous solution was kept constant
c_D/(mol/l) = 0.02 was kept constant

c_C/(mol/l)	0.00	0.05	0.10	0.25	0.50
T/K	329.15	325.25	323.55	321.25	319.55

Polymer (B):	**poly(ethylene oxide-*b*-propylene oxide-*b*-ethylene oxide)**		**2010MAH**
Characterization:	M_w/g.mol^{-1} = 2900, 40.0 wt% ethylene oxide, L64, (EO)13(PO)30(EO)13, Aldrich Chem. Co., Inc., Milwaukee, WI		
Solvent (A):	**water**	H_2O	**7732-18-5**
Component (C):	**glycyl-DL-valine**	$C_7H_{14}N_2O_3$	**2325-17-9**
Component (D):	**sodium dodecyl sulfate**	$C_{12}H_{25}NaO_4S$	**151-21-3**

Type of data: cloud points (LCST-behavior)

w_B = 0.01 in the binary aqueous solution was kept constant
c_D/(mol/l) = 0.005 was kept constant

c_C/(mol/l)	0.00	0.05	0.10	0.25	0.50
T/K	345.65	342.65	339.35	337.95	335.65

Polymer (B): **poly(ethylene oxide-*b*-propylene oxide-*b*-ethylene oxide)** **2010MAH**
Characterization: M_w/g.mol^{-1} = 2036, (EO)2.5(PO)31(EO)2.5, Aldrich Chem. Co., Inc., Milwaukee, WI
Solvent (A): **water** **H_2O** **7732-18-5**
Salt (C): **magnesium chloride** **$MgCl_2$** **7786-30-3**
Component (D): **hexadecyltrimethylammonium bromide** **$C_{19}H_{42}BrN$** **57-09-0**

Type of data: cloud points (LCST-behavior)

w_B = 0.01 in the binary aqueous solution was kept constant
c_D/(mol/l) = 0.02 was kept constant

c_C/(mol/l)	0.00	0.10	0.20	0.30	0.40
T/K	326.95	326.05	323.95	323.05	321.15

Polymer (B): **poly(ethylene oxide-*b*-propylene oxide-*b*-ethylene oxide)** **2010MAH**
Characterization: M_w/g.mol^{-1} = 2900, 40.0 wt% ethylene oxide, L64, (EO)13(PO)30(EO)13, Aldrich Chem. Co., Inc., Milwaukee, WI
Solvent (A): **water** **H_2O** **7732-18-5**
Salt (C): **magnesium chloride** **$MgCl_2$** **7786-30-3**
Component (D): **hexadecyltrimethylammonium bromide** **$C_{19}H_{42}BrN$** **57-09-0**

Type of data: cloud points (LCST-behavior)

w_B = 0.01 in the binary aqueous solution was kept constant
c_D/(mol/l) = 0.005 was kept constant

c_C/(mol/l)	0.00	0.10	0.20	0.30	0.40	0.50
T/K	350.65	347.05	343.55	341.45	340.05	338.65

Polymer (B): **poly(ethylene oxide-*b*-propylene oxide-*b*-ethylene oxide)** **2010MAH**
Characterization: M_w/g.mol^{-1} = 2036, (EO)2.5(PO)31(EO)2.5, Aldrich Chem. Co., Inc., Milwaukee, WI
Solvent (A): **water** **H_2O** **7732-18-5**
Salt (C): **magnesium chloride** **$MgCl_2$** **7786-30-3**
Component (D): **sodium dodecyl sulfate** **$C_{12}H_{25}NaO_4S$** **151-21-3**

Type of data: cloud points (LCST-behavior)

w_B = 0.01 in the binary aqueous solution was kept constant
c_D/(mol/l) = 0.02 was kept constant

c_C/(mol/l)	0.00	0.10	0.20	0.30	0.40
T/K	329.15	328.35	326.15	325.05	323.25

Polymer (B):	**poly(ethylene oxide-*b*-propylene oxide-*b*-ethylene oxide)**		**2010MAH**
Characterization:	M_w/g.mol^{-1} = 2900, 40.0 wt% ethylene oxide, L64, (EO)13(PO)30(EO)13, Aldrich Chem. Co., Inc., Milwaukee, WI		
Solvent (A):	**water**	**H_2O**	**7732-18-5**
Salt (C):	**magnesium chloride**	**$MgCl_2$**	**7786-30-3**
Component (D):	**sodium dodecyl sulfate**	**$C_{12}H_{25}NaO_4S$**	**151-21-3**

Type of data: cloud points (LCST-behavior)

w_B = 0.01 in the binary aqueous solution was kept constant
c_D/(mol/l) = 0.005 was kept constant

c_C/(mol/l)	0.00	0.10	0.20	0.30	0.40	0.50
T/K	344.65	342.15	339.25	336.85	335.05	333.85

Polymer (B):	**poly(ethylene oxide-*b*-propylene oxide-*b*-ethylene oxide)**		**2010MAH**
Characterization:	M_w/g.mol^{-1} = 2036, (EO)2.5(PO)31(EO)2.5, Aldrich Chem. Co., Inc., Milwaukee, WI		
Solvent (A):	**water**	**H_2O**	**7732-18-5**
Component (C):	**L-proline**	**$C_5H_9NO_2$**	**147-85-3**
Component (D):	**hexadecyltrimethylammonium bromide**	**$C_{19}H_{42}BrN$**	**57-09-0**

Type of data: cloud points (LCST-behavior)

w_B = 0.01 in the binary aqueous solution was kept constant
c_D/(mol/l) = 0.02 was kept constant

c_C/(mol/l)	0.00	0.05	0.10	0.25	0.50
T/K	326.95	325.35	323.95	322.95	319.25

Polymer (B):	**poly(ethylene oxide-*b*-propylene oxide-*b*-ethylene oxide)**		**2010MAH**
Characterization:	M_w/g.mol^{-1} = 2900, 40.0 wt% ethylene oxide, L64, (EO)13(PO)30(EO)13, Aldrich Chem. Co., Inc., Milwaukee, WI		
Solvent (A):	**water**	**H_2O**	**7732-18-5**
Component (C):	**L-proline**	**$C_5H_9NO_2$**	**147-85-3**
Component (D):	**hexadecyltrimethylammonium bromide**	**$C_{19}H_{42}BrN$**	**57-09-0**

Type of data: cloud points (LCST-behavior)

w_B = 0.01 in the binary aqueous solution was kept constant
c_D/(mol/l) = 0.005 was kept constant

c_C/(mol/l)	0.00	0.05	0.10	0.25	0.50
T/K	350.65	342.05	337.75	336.45	329.95

Polymer (B):	**poly(ethylene oxide-*b*-propylene oxide-*b*-ethylene oxide)**		**2010MAH**
Characterization:	M_w/g.mol^{-1} = 2036, (EO)2.5(PO)31(EO)2.5, Aldrich Chem. Co., Inc., Milwaukee, WI		
Solvent (A):	**water**	**H_2O**	**7732-18-5**
Component (C):	**L-proline**	**$C_5H_9NO_2$**	**147-85-3**
Component (D):	**sodium dodecyl sulfate**	**$C_{12}H_{25}NaO_4S$**	**151-21-3**

Type of data: cloud points (LCST-behavior)

w_B = 0.01 in the binary aqueous solution was kept constant
c_D/(mol/l) = 0.02 was kept constant

c_C/(mol/l)	0.00	0.05	0.10	0.25	0.50
T/K	329.15	326.05	324.65	322.55	321.15

Polymer (B):	**poly(ethylene oxide-*b*-propylene oxide-*b*-ethylene oxide)**		**2010MAH**
Characterization:	M_w/g.mol^{-1} = 2900, 40.0 wt% ethylene oxide, L64, (EO)13(PO)30(EO)13, Aldrich Chem. Co., Inc., Milwaukee, WI		
Solvent (A):	**water**	**H_2O**	**7732-18-5**
Component (C):	**L-proline**	**$C_5H_9NO_2$**	**147-85-3**
Component (D):	**sodium dodecyl sulfate**	**$C_{12}H_{25}NaO_4S$**	**151-21-3**

Type of data: cloud points (LCST-behavior)

w_B = 0.01 in the binary aqueous solution was kept constant
c_D/(mol/l) = 0.005 was kept constant

c_C/(mol/l)	0.00	0.05	0.10	0.25	0.50
T/K	345.65	342.85	339.35	338.25	337.35

Polymer (B):	**poly(ethylene oxide-*b*-propylene oxide-*b*-ethylene oxide)**		**2010MAH**
Characterization:	M_w/g.mol^{-1} = 2036, (EO)2.5(PO)31(EO)2.5, Aldrich Chem. Co., Inc., Milwaukee, WI		
Solvent (A):	**water**	**H_2O**	**7732-18-5**
Salt (C):	**sodium chloride**	**NaCl**	**7647-14-5**
Component (D):	**hexadecyltrimethylammonium bromide**	**$C_{19}H_{42}BrN$**	**57-09-0**

Type of data: cloud points (LCST-behavior)

w_B = 0.01 in the binary aqueous solution was kept constant
c_D/(mol/l) = 0.02 was kept constant

c_C/(mol/l)	0.00	0.10	0.20	0.30	0.40
T/K	326.95	325.15	323.35	320.65	317.65

Polymer (B):	**poly(ethylene oxide-*b*-propylene oxide-*b*-ethylene oxide)**		**2010MAH**
Characterization:	M_w/g.mol^{-1} = 2900, 40.0 wt% ethylene oxide, L64, (EO)13(PO)30(EO)13, Aldrich Chem. Co., Inc., Milwaukee, WI		
Solvent (A):	**water**	H_2O	**7732-18-5**
Salt (C):	**sodium chloride**	**NaCl**	**7647-14-5**
Component (D):	**hexadecyltrimethylammonium bromide**	$C_{19}H_{42}BrN$	**57-09-0**

Type of data: cloud points (LCST-behavior)

w_B = 0.01 in the binary aqueous solution was kept constant
c_D/(mol/l) = 0.005 was kept constant

c_C/(mol/l)	0.00	0.10	0.20	0.30	0.40	0.50
T/K	350.65	346.65	343.35	339.75	336.45	333.35

Polymer (B):	**poly(ethylene oxide-*b*-propylene oxide-*b*-ethylene oxide)**		**2010MAH**
Characterization:	M_w/g.mol^{-1} = 2036, (EO)2.5(PO)31(EO)2.5, Aldrich Chem. Co., Inc., Milwaukee, WI		
Solvent (A):	**water**	H_2O	**7732-18-5**
Salt (C):	**sodium chloride**	**NaCl**	**7647-14-5**
Component (D):	**sodium dodecyl sulfate**	$C_{12}H_{25}NaO_4S$	**151-21-3**

Type of data: cloud points (LCST-behavior)

w_B = 0.01 in the binary aqueous solution was kept constant
c_D/(mol/l) = 0.02 was kept constant

c_C/(mol/l)	0.00	0.10	0.20	0.30	0.40
T/K	329.15	327.75	324.95	322.65	319.75

Polymer (B):	**poly(ethylene oxide-*b*-propylene oxide-*b*-ethylene oxide)**		**2010MAH**
Characterization:	M_w/g.mol^{-1} = 2900, 40.0 wt% ethylene oxide, L64, (EO)13(PO)30(EO)13, Aldrich Chem. Co., Inc., Milwaukee, WI		
Solvent (A):	**water**	H_2O	**7732-18-5**
Salt (C):	**sodium chloride**	**NaCl**	**7647-14-5**
Component (D):	**sodium dodecyl sulfate**	$C_{12}H_{25}NaO_4S$	**151-21-3**

Type of data: cloud points (LCST-behavior)

w_B = 0.01 in the binary aqueous solution was kept constant
c_D/(mol/l) = 0.005 was kept constant

c_C/(mol/l)	0.00	0.10	0.20	0.30	0.40	0.50
T/K	344.65	341.75	338.95	334.55	331.25	328.35

Polymer (B):	**poly(ethylene oxide-*b*-propylene oxide-*b*-ethylene oxide)**		**2010MAH**
Characterization:	M_w/g.mol^{-1} = 2036, (EO)2.5(PO)31(EO)2.5, Aldrich Chem. Co., Inc., Milwaukee, WI		
Solvent (A):	**water**	H_2O	**7732-18-5**
Salt (C):	**sodium phosphate**	Na_3PO_4	**7601-54-9**
Component (D):	**hexadecyltrimethylammonium bromide**	$C_{19}H_{42}BrN$	**57-09-0**

Type of data: cloud points (LCST-behavior)

w_B = 0.01 in the binary aqueous solution was kept constant
c_D/(mol/l) = 0.02 was kept constant

c_C/(mol/l)	0.00	0.10	0.20	0.30	0.40
T/K	326.95	323.05	320.15	318.05	314.95

Polymer (B):	**poly(ethylene oxide-*b*-propylene oxide-*b*-ethylene oxide)**		**2010MAH**
Characterization:	M_w/g.mol^{-1} = 2900, 40.0 wt% ethylene oxide, L64, (EO)13(PO)30(EO)13, Aldrich Chem. Co., Inc., Milwaukee, WI		
Solvent (A):	**water**	H_2O	**7732-18-5**
Salt (C):	**sodium phosphate**	Na_3PO_4	**7601-54-9**
Component (D):	**hexadecyltrimethylammonium bromide**	$C_{19}H_{42}BrN$	**57-09-0**

Type of data: cloud points (LCST-behavior)

w_B = 0.01 in the binary aqueous solution was kept constant
c_D/(mol/l) = 0.005 was kept constant

c_C/(mol/l)	0.00	0.10	0.20	0.30	0.40	0.50
T/K	350.65	341.75	337.45	333.55	328.85	325.35

Polymer (B):	**poly(ethylene oxide-*b*-propylene oxide-*b*-ethylene oxide)**		**2010MAH**
Characterization:	M_w/g.mol^{-1} = 2036, (EO)2.5(PO)31(EO)2.5, Aldrich Chem. Co., Inc., Milwaukee, WI		
Solvent (A):	**water**	H_2O	**7732-18-5**
Salt (C):	**sodium phosphate**	Na_3PO_4	**7601-54-9**
Component (D):	**sodium dodecyl sulfate**	$C_{12}H_{25}NaO_4S$	**151-21-3**

Type of data: cloud points (LCST-behavior)

w_B = 0.01 in the binary aqueous solution was kept constant
c_D/(mol/l) = 0.02 was kept constant

c_C/(mol/l)	0.00	0.10	0.20	0.30	0.40
T/K	329.15	326.05	323.25	321.15	318.05

Polymer (B):	**poly(ethylene oxide-*b*-propylene oxide-*b*-ethylene oxide)**		**2010MAH**
Characterization:	M_w/g.mol^{-1} = 2900, 40.0 wt% ethylene oxide, L64, (EO)13(PO)30(EO)13, Aldrich Chem. Co., Inc., Milwaukee, WI		
Solvent (A):	**water**	H_2O	**7732-18-5**
Salt (C):	**sodium phosphate**	Na_3PO_4	**7601-54-9**
Component (D):	**sodium dodecyl sulfate**	$C_{12}H_{25}NaO_4S$	**151-21-3**

Type of data: cloud points (LCST-behavior)

w_B = 0.01 in the binary aqueous solution was kept constant
c_D/(mol/l) = 0.005 was kept constant

c_C/(mol/l)	0.00	0.10	0.20	0.30	0.40	0.50
T/K	344.65	335.85	330.45	326.05	324.05	319.55

Polymer (B):	**poly(ethylene oxide-*b*-propylene oxide-*b*-ethylene oxide)**		**2010MAH**
Characterization:	M_w/g.mol^{-1} = 2036, (EO)2.5(PO)31(EO)2.5, Aldrich Chem. Co., Inc., Milwaukee, WI		
Solvent (A):	**water**	H_2O	**7732-18-5**
Salt (C):	**sodium sulfate**	Na_2SO_4	**7757-82-6**
Component (D):	**hexadecyltrimethylammonium bromide**	$C_{19}H_{42}BrN$	**57-09-0**

Type of data: cloud points (LCST-behavior)

w_B = 0.01 in the binary aqueous solution was kept constant
c_D/(mol/l) = 0.02 was kept constant

c_C/(mol/l)	0.00	0.10	0.20	0.30	0.40
T/K	326.95	324.75	322.25	319.95	317.45

Polymer (B):	**poly(ethylene oxide-*b*-propylene oxide-*b*-ethylene oxide)**		**2010MAH**
Characterization:	M_w/g.mol^{-1} = 2900, 40.0 wt% ethylene oxide, L64, (EO)13(PO)30(EO)13, Aldrich Chem. Co., Inc., Milwaukee, WI		
Solvent (A):	**water**	H_2O	**7732-18-5**
Salt (C):	**sodium sulfate**	Na_2SO_4	**7757-82-6**
Component (D):	**hexadecyltrimethylammonium bromide**	$C_{19}H_{42}BrN$	**57-09-0**

Type of data: cloud points (LCST-behavior)

w_B = 0.01 in the binary aqueous solution was kept constant
c_D/(mol/l) = 0.005 was kept constant

c_C/(mol/l)	0.00	0.10	0.20	0.30	0.40	0.50
T/K	350.65	343.45	340.75	337.45	332.95	329.35

Polymer (B):	**poly(ethylene oxide-*b*-propylene oxide-*b*-ethylene oxide)**		**2010MAH**
Characterization:	M_w/g.mol^{-1} = 2036, (EO)2.5(PO)31(EO)2.5, Aldrich Chem. Co., Inc., Milwaukee, WI		
Solvent (A):	**water**	**H_2O**	**7732-18-5**
Salt (C):	**sodium sulfate**	**Na_2SO_4**	**7757-82-6**
Component (D):	**sodium dodecyl sulfate**	**$C_{12}H_{25}NaO_4S$**	**151-21-3**

Type of data: cloud points (LCST-behavior)

w_B = 0.01 in the binary aqueous solution was kept constant
c_D/(mol/l) = 0.02 was kept constant

c_C/(mol/l)	0.00	0.10	0.20	0.30	0.40
T/K	329.15	327.85	325.35	323.15	321.65

Polymer (B):	**poly(ethylene oxide-*b*-propylene oxide-*b*-ethylene oxide)**		**2010MAH**
Characterization:	M_w/g.mol^{-1} = 2900, 40.0 wt% ethylene oxide, L64, (EO)13(PO)30(EO)13, Aldrich Chem. Co., Inc., Milwaukee, WI		
Solvent (A):	**water**	**H_2O**	**7732-18-5**
Salt (C):	**sodium sulfate**	**Na_2SO_4**	**7757-82-6**
Component (D):	**sodium dodecyl sulfate**	**$C_{12}H_{25}NaO_4S$**	**151-21-3**

Type of data: cloud points (LCST-behavior)

w_B = 0.01 in the binary aqueous solution was kept constant
c_D/(mol/l) = 0.005 was kept constant

c_C/(mol/l)	0.00	0.10	0.20	0.30	0.40	0.50
T/K	344.65	337.65	333.85	329.35	326.25	323.75

Polymer (B):	**poly(ethylene oxide-*b*-propylene oxide-*b*-ethylene oxide)**		**2010MAH**
Characterization:	M_w/g.mol^{-1} = 2036, (EO)2.5(PO)31(EO)2.5, Aldrich Chem. Co., Inc., Milwaukee, WI		
Solvent (A):	**water**	**H_2O**	**7732-18-5**
Component (C):	**L-threonine**	**$C_4H_9NO_3$**	**72-19-5**
Component (D):	**hexadecyltrimethylammonium bromide**	**$C_{19}H_{42}BrN$**	**57-09-0**

Type of data: cloud points (LCST-behavior)

w_B = 0.01 in the binary aqueous solution was kept constant
c_D/(mol/l) = 0.02 was kept constant

c_C/(mol/l)	0.00	0.05	0.10	0.25	0.50
T/K	326.95	325.65	324.55	322.75	319.55

Polymer (B):	**poly(ethylene oxide-*b*-propylene oxide-*b*-ethylene oxide)**		**2010MAH**
Characterization:	M_w/g.mol^{-1} = 2900, 40.0 wt% ethylene oxide, L64, (EO)13(PO)30(EO)13, Aldrich Chem. Co., Inc., Milwaukee, WI		
Solvent (A):	**water**	**H_2O**	**7732-18-5**
Component (C):	**L-threonine**	**$C_4H_9NO_3$**	**72-19-5**
Component (D):	**hexadecyltrimethylammonium bromide**	**$C_{19}H_{42}BrN$**	**57-09-0**

Type of data: cloud points (LCST-behavior)

w_B = 0.01 in the binary aqueous solution was kept constant
c_D/(mol/l) = 0.005 was kept constant

c_C/(mol/l)	0.00	0.05	0.10	0.25	0.50
T/K	350.65	343.25	341.15	339.05	337.05

Polymer (B):	**poly(ethylene oxide-*b*-propylene oxide-*b*-ethylene oxide)**		**2010MAH**
Characterization:	M_w/g.mol^{-1} = 2036, (EO)2.5(PO)31(EO)2.5, Aldrich Chem. Co., Inc., Milwaukee, WI		
Solvent (A):	**water**	**H_2O**	**7732-18-5**
Component (C):	**L-threonine**	**$C_4H_9NO_3$**	**72-19-5**
Component (D):	**sodium dodecyl sulfate**	**$C_{12}H_{25}NaO_4S$**	**151-21-3**

Type of data: cloud points (LCST-behavior)

w_B = 0.01 in the binary aqueous solution was kept constant
c_D/(mol/l) = 0.02 was kept constant

c_C/(mol/l)	0.00	0.05	0.10	0.25	0.50
T/K	329.15	327.45	325.45	323.15	321.75

Polymer (B):	**poly(ethylene oxide-*b*-propylene oxide-*b*-ethylene oxide)**		**2010MAH**
Characterization:	M_w/g.mol^{-1} = 2900, 40.0 wt% ethylene oxide, L64, (EO)13(PO)30(EO)13, Aldrich Chem. Co., Inc., Milwaukee, WI		
Solvent (A):	**water**	**H_2O**	**7732-18-5**
Component (C):	**L-threonine**	**$C_4H_9NO_3$**	**72-19-5**
Component (D):	**sodium dodecyl sulfate**	**$C_{12}H_{25}NaO_4S$**	**151-21-3**

Type of data: cloud points (LCST-behavior)

w_B = 0.01 in the binary aqueous solution was kept constant
c_D/(mol/l) = 0.005 was kept constant

c_C/(mol/l)	0.00	0.05	0.10	0.25	0.50
T/K	345.65	343.45	342.15	340.25	338.45

Polymer (B):	**poly(ethylene oxide-*b*-propylene oxide-*b*-ethylene oxide)**		**2010MAH**
Characterization:	M_w/g.mol^{-1} = 2036, (EO)2.5(PO)31(EO)2.5, Aldrich Chem. Co., Inc., Milwaukee, WI		
Solvent (A):	**water**	**H_2O**	**7732-18-5**
Component (C):	**L-valine**	**$C_5H_{11}NO_2$**	**72-18-4**
Component (D):	**hexadecyltrimethylammonium bromide**	**$C_{19}H_{42}BrN$**	**57-09-0**

Type of data: cloud points (LCST-behavior)

w_B = 0.01 in the binary aqueous solution was kept constant
c_D/(mol/l) = 0.02 was kept constant

c_C/(mol/l)	0.00	0.05	0.10	0.25	0.50
T/K	326.95	325.45	342.35	323.15	321.05

Polymer (B):	**poly(ethylene oxide-*b*-propylene oxide-*b*-ethylene oxide)**		**2010MAH**
Characterization:	M_w/g.mol^{-1} = 2900, 40.0 wt% ethylene oxide, L64, (EO)13(PO)30(EO)13, Aldrich Chem. Co., Inc., Milwaukee, WI		
Solvent (A):	**water**	**H_2O**	**7732-18-5**
Component (C):	**L-valine**	**$C_5H_{11}NO_2$**	**72-18-4**
Component (D):	**hexadecyltrimethylammonium bromide**	**$C_{19}H_{42}BrN$**	**57-09-0**

Type of data: cloud points (LCST-behavior)

w_B = 0.01 in the binary aqueous solution was kept constant
c_D/(mol/l) = 0.005 was kept constant

c_C/(mol/l)	0.00	0.05	0.10	0.25	0.50
T/K	350.65	345.95	343.75	340.45	338.25

Polymer (B):	**poly(ethylene oxide-*b*-propylene oxide-*b*-ethylene oxide)**		**2010MAH**
Characterization:	M_w/g.mol^{-1} = 2036, (EO)2.5(PO)31(EO)2.5, Aldrich Chem. Co., Inc., Milwaukee, WI		
Solvent (A):	**water**	**H_2O**	**7732-18-5**
Component (C):	**L-valine**	**$C_5H_{11}NO_2$**	**72-18-4**
Component (D):	**sodium dodecyl sulfate**	**$C_{12}H_{25}NaO_4S$**	**151-21-3**

Type of data: cloud points (LCST-behavior)

w_B = 0.01 in the binary aqueous solution was kept constant
c_D/(mol/l) = 0.02 was kept constant

c_C/(mol/l)	0.00	0.05	0.10	0.25	0.50
T/K	329.15	327.35	325.45	323.75	328.15

Polymer (B):	**poly(ethylene oxide-*b*-propylene oxide-*b*-ethylene oxide)**		**2010MAH**
Characterization:	M_w/g.mol^{-1} = 2900, 40.0 wt% ethylene oxide, L64, (EO)13(PO)30(EO)13, Aldrich Chem. Co., Inc., Milwaukee, WI		
Solvent (A):	**water**	**H_2O**	**7732-18-5**
Component (C):	**L-valine**	**$C_5H_{11}NO_2$**	**72-18-4**
Component (D):	**sodium dodecyl sulfate**	**$C_{12}H_{25}NaO_4S$**	**151-21-3**

Type of data: cloud points (LCST-behavior)

w_B = 0.01 in the binary aqueous solution was kept constant
c_D/(mol/l) = 0.005 was kept constant

c_C/(mol/l)	0.00	0.05	0.10	0.25	0.50
T/K	345.65	343.65	341.85	340.55	338.95

Polymer (B):	**poly(1-vinyl-2-pyrrolidinone)**		**2010FED**
Characterization:	M_n/g.mol^{-1} = 3880, PVPK17, BASF AG, Ludwigshafen, Germany		
Solvent (A):	**water**	**H_2O**	**7732-18-5**
Solvent (C):	**1-vinyl-2-pyrrolidinone**	**C_6H_9NO**	**88-12-0**
Salt (D):	**sodium sulfate**	**Na_2SO_4**	**7757-82-6**

Type of data: cloud points

T/K = 298.15

w_B	0.0423	0.0916	0.1853	0.3347	0.0667	0.1469	0.2516	0.0828	0.1202
w_C	0.0440	0.0484	0.0878	0.0849	0.0576	0.0654	0.0741	0.0494	0.0568
w_D	0.0983	0.0897	0.0533	0.0121	0.0911	0.0658	0.0371	0.0864	0.0740
w_B	0.2124	0.0226	0.0229	0.0233	0.0168	0.0088	0.0087	0.0054	0.0066
w_C	0.0770	0.0463	0.0353	0.0287	0.0201	0.0155	0.0147	0.0112	0.0175
w_D	0.0476	0.0888	0.0995	0.0982	0.1124	0.1137	0.1086	0.1032	0.1100
w_B	0.0069	0.0025	0.0019	0.0014	0.0006				
w_C	0.0096	0.0024	0.0019	0.0014	0.0006				
w_D	0.1160	0.1295	0.1505	0.1824	0.2003				

Type of data: coexistence data (tie lines)

T/K = 298.15

Total system			Top phase			Bottom phase		
w_B	w_C	w_D	w_B	w_C	w_D	w_B	w_C	w_D
0.0491	0.0539	0.1795	0.2889	0.2193	0.0253	0.0087	0.0168	0.2104
0.0592	0.0681	0.1178	0.3381	0.1883	0.0227	0.0047	0.0399	0.1495
0.0630	0.0820	0.1086	0.2560	0.2429	0.0143	0.0119	0.0439	0.1312

Polymer (B): **poly(1-vinyl-2-pyrrolidinone)** **2010FED**
Characterization: M_n/g.mol^{-1} = 17750, PVPK30,
BASF AG, Ludwigshafen, Germany
Solvent (A): **water** **H_2O** **7732-18-5**
Solvent (C): **1-vinyl-2-pyrrolidinone** **C_6H_9NO** **88-12-0**
Salt (D): **sodium sulfate** **Na_2SO_4** **7757-82-6**

Type of data: cloud points

T/K = 298.15

w_B	0.0517	0.0843	0.1135	0.1498	0.2050	0.2598	0.1288	0.3075	0.0402
w_C	0.0521	0.0554	0.0778	0.0612	0.0771	0.0781	0.0630	0.0774	0.0535
w_D	0.0794	0.0711	0.0646	0.0586	0.0462	0.0341	0.0614	0.0291	0.0824
w_B	0.0634	0.2277	0.0389	0.0305	0.0284	0.0212	0.0122	0.0061	0.0074
w_C	0.0543	0.0713	0.0636	0.0452	0.0429	0.0386	0.0320	0.0328	0.0243
w_D	0.0761	0.0407	0.0681	0.0774	0.0807	0.0869	0.0942	0.0899	0.0901
w_B	0.0067	0.0082	0.0010	0.0007	0.0006	0.0006			
w_C	0.0233	0.0147	0.0010	0.0008	0.0006	0.0006			
w_D	0.0937	0.0903	0.1239	0.1527	0.1851	0.2067			

Type of data: coexistence data (tie lines)

T/K = 298.15

Total system			Top phase			Bottom phase		
w_B	w_C	w_D	w_B	w_C	w_D	w_B	w_C	w_D
0.0429	0.0426	0.1242	0.2856	0.2415	0.0316	0.0239	0.0141	0.1651
0.0354	0.0369	0.1563	0.2116	0.1585	0.0571	0.0045	0.0208	0.1749
0.0614	0.0473	0.1191	0.3512	0.1472	0.0148	0.0076	0.0320	0.1278
0.0442	0.0391	0.1157	0.3158	0.0990	0.0234	0.0123	0.0274	0.1221
0.0354	0.0356	0.1606	0.3479	0.1950	0.0159	0.0040	0.0221	0.1765
0.0303	0.0319	0.1482	0.3890	0.1256	0.0215	0.0047	0.0210	0.1653

Polymer (B): **poly(1-vinyl-2-pyrrolidinone)** **2010FED**
Characterization: M_n/g.mol^{-1} = 138600, PVPK90,
BASF AG, Ludwigshafen, Germany
Solvent (A): **water** **H_2O** **7732-18-5**
Solvent (C): **1-vinyl-2-pyrrolidinone** **C_6H_9NO** **88-12-0**
Salt (D): **sodium sulfate** **Na_2SO_4** **7757-82-6**

continued

continued

Type of data: cloud points

T/K = 298.15

w_B	0.0085	0.0157	0.0271	0.0332	0.0328	0.0397	0.0505	0.0617	0.0713
w_C	0.0555	0.0630	0.0553	0.0562	0.0465	0.0571	0.0576	0.0745	0.0632
w_D	0.0795	0.0788	0.0760	0.0736	0.0744	0.0724	0.0703	0.0656	0.0654
w_B	0.0735	0.0816	0.0424	0.0097	0.0088	0.0091	0.0097	0.0050	0.0042
w_C	0.0594	0.0601	0.0549	0.0561	0.0479	0.0484	0.0381	0.0203	0.0315
w_D	0.0677	0.0680	0.0754	0.0814	0.0820	0.0821	0.0835	0.0958	0.0763
w_B	0.0045	0.0042	0.0027	0.0014	0.0010	0.0008	0.0008		
w_C	0.0175	0.0155	0.0106	0.0013	0.0010	0.0008	0.0008		
w_D	0.0759	0.0804	0.0799	0.1250	0.1480	0.1812	0.2103		

Type of data: coexistence data (tie lines)

T/K = 298.15

Total system			Top phase			Bottom phase		
w_B	w_C	w_D	w_B	w_C	w_D	w_B	w_C	w_D
0.0184	0.0186	0.1207	0.3021	0.1160	0.0285	0.0033	0.0295	0.1362
0.0354	0.0402	0.1218	0.3287	0.1688	0.0265	0.0018	0.0237	0.1696

3.6. Table of systems where quaternary LLE data were published only in graphical form as phase diagrams or related figures

Polymer (B)	Second/third and fourth component	Ref.
Dextran		
	bovine serum albumin/guar-*g*-poly(1-vinyl-2-pyrrolidinone] and water	2009KAU
	poly(ethylene oxide)/polystyrene and water	2005OL1
	poly(ethylene oxide)/silica and water	2005OL1
	poly(ethylene oxide-*co*-propylene oxide)/poly(ethylene oxide-*co*-propylene oxide)/silica and water	2005OL1
	polystyrene and water	2005OL1
Methyl(hydroxypropyl)cellulose		
	sodium carboxymethylcellulose/sodium dodecyl sulfate and water	2010KA1
Poly(acrylic acid)		
	poly(*N*,*N*-diallyl-*N*,*N*-dimethylammonium chloride)/HCl and water	2010LIT
	poly(ethylene glycol)/NaCl and water	2008SAR
Poly(ε-caproamide)		
	N,*N*-dimethylacetamide/lithium chloride and water	2004FEN
Poly(*N*,*N*-diallyl-*N*,*N*-dimethyl-ammonium chloride)		
	1-pentanol/sulfobetaine/toluene and water	2003KOT
	poly(acrylic acid)/HCl and water	2010LIT
Poly(*N*,*N*-dimethylacrylamide-*co*-*tert*-butylacrylamide)		
	poly(ethylene glycol)/monopotassium phosphate and water	2009FO2

Polymer (B)	Second/third and fourth component	Ref.
Poly(ethylene glycol)		
	ammonium carbamate/proteins and water	2007DAL
	β-amylase/KH_2PO_4 and water	2010SH1
	β-amylase/Na_2SO_4 and water	2010SH1
	amyloglucosidase/KH_2PO_4 and water	2010SH1
	amyloglucosidase/Na_2SO_4 and water	2010SH1
	bovine serum albumin/dipotassium phosphate and water	2007FAR
	bovine serum albumin/potassium citrate and water	2010LUY
	bovine serum albumin/sodium citrate and water	2011PER
	cetyltrimethylammonium bromide/cobalt sulfate and water	2010KAL
	cetyltrimethylammonium bromide/copper sulfate and water	2010KAL
	cetyltrimethylammonium bromide/nickel nitrate and water	2010KAL
	cetyltrimethylammonium bromide/sodium chloride and water	2010KAL
	cetyltrimethylammonium bromide/sodium dodecyl sulfate and water	2011WA1
	cetyltrimethylammonium bromide/sodium nitrate and water	2010KAL
	diammonium phosphate/Brij 58 and water	2009FO1
	dipotassium phosphate/sodium dihydrogen phosphate and water	2008RAH
	dipotassium phosphate/sodium dihydrogen phosphate/guanidine hydrochloride and water	2008RAH
	glycomacropeptide/sodium citrate and water	2009SIL
	hemoglobin/poly(sodium acrylate) and water	2008JOH
	hexadecylpyridinium bromide/sodium chloride and water	2010KAL
	lysozyme/poly(sodium acrylate) and water	2008JOH
	poly(acrylic acid)/sodium chloride and water	2008SAR
	poly(*N,N*-dimethylacrylamide-*co*-*tert*-butyl-acrylamide)/monopotassium phosphate and water	2009FO2
	potassium phosphate/sodium chloride and water	2006WON
	proteins/salts and water	2011ASE
	sodium adipate/poly(sodium acrylate) and water	2011JOH
	sodium azelate/poly(sodium acrylate) and water	2011JOH
	sodium chloride/poly(sodium acrylate) and water	2011JOH
	sodium citrate/citric acid and water	2007POR
	sodium citrate/xylanase and water	2011LIX
	sodium dodecyl sulfate/gemini surfactant (12-3-12) and water	2009ALH

Polymer (B)	Second/third and fourth component	Ref.
Poly(ethylene glycol)		
	sodium dodecyl sulfate/potassium chloride/ gemini surfactant (12-3-12) and water	2009ALH
	sodium dodecyl sulfate/sodium bromide/ gemini surfactant (12-3-12) and water	2009ALH
	sodium dodecyl sulfate/sodium chloride/ gemini surfactant (12-3-12) and water	2009ALH
	sodium sulfate/amino acids and water	2011FE2
	sodium sulfate/pectinase and water	2009AN1
	sodium sulfate/poly(sodium acrylate) and water	2011JOH
	Triton-X/surfactants and water	2011NAQ
Poly(ethylene glycol) p-octyl-phenyl ether		
	amino acids/magnesium sulfate and water	2011SAL1
	amino acids/sodium citrate and water	2011SAL1
	poly(ethylene glycol)/surfactants and water	2011NAQ
Poly(ethylene oxide)		
	dextran/polystyrene and water	2005OL1
	dextran/silica and water	2005OL1
Poly(ethylene oxide-*co*-propylene oxide)		
	dextran/polystyrene and water	2005OL1
	dextran/silica and water	2005OL1
Poly(ethylene oxide-*b*-propylene oxide-*b*-ethylene oxide)		
	ammonium carbamate/bovine serum albumin and water	2007OLI
	ammonium carbamate/γ-globulin and water	2007OLI
	ammonium carbamate/β-lactoglobulin and water	2007OLI
	ammonium carbamate/lysozyme and water	2007OLI
	1-butyl-3-methylimidazolium hexafluorophosphate/ 1-butanol and water	2009WA1
	1-butyl-3-methylimidazolium hexafluorophosphate/ ethanol and water	2009WA1
	1-butyl-3-methylimidazolium hexafluorophosphate/ 1-propanol and water	2009WA1
	cetyltrimethylammonium bromide/sodium bromide and water	2010PAT
	surfactants/sodium bromide and water	2006BAK

Polymer (B)	Second/third and fourth component	Ref.
Poly[*N*-isopropylacrylamide-*co*-(p-methacrylamido)acetophenone thiosemicarbazone]		
	monopotassium phosphate/disodium phosphate and water	2005LIC
Polyoxyethylene(20) sorbitan monooleate (Tween-80)		
	1-butanol/n-hexane and water	2011MUK
Poly(propylene glycol)		
	amino acids/magnesium sulfate and water	2007SAL
	L-methionine/disodium phosphate and water	2011SAL2
	L-methionine/monosodium phosphate and water	2011SAL2
	L-methionine/sodium phosphate and water	2011SAL2
Poly(propylene oxide-*b*-ethylene oxide-*b*-propylene oxide)		
	cetyltrimethylammonium bromide/sodium bromide and water	2010PAT
Poly(sodium acrylate)		
	hemoglobin/poly(ethylene glycol) and water	2008JOH
	lysozyme/poly(ethylene glycol) and water	2008JOH
	1-pentanol/sulfobetaine/toluene and water	2003KOT
	sodium adipate/poly(ethylene glycol) and water	2011JOH
	sodium azelate/poly(ethylene glycol) and water	2011JOH
	sodium chloride/poly(ethylene glycol) and water	2011JOH
	sodium sulfate/poly(ethylene glycol) and water	2011JOH
Poly(sodium styrenesulfonate)		
	sodium chloride/sodium dodecyl sulfate and water	2010KAL
Polysulfone		
	1-methyl-2-pyrrolidinone/ethanol and water	2010ARO
	1-methyl-2-pyrrolidinone/glycerol and water	2010ARO
	1-methyl-2-pyrrolidinone/poly(ethylene glycol) and water	2010ARO
	1-methyl-2-pyrrolidinone/poly(1-vinyl-2-pyrrolidinone) and water	2010ARO

Polymer (B)	Second/third and fourth component	Ref.
Poly(vinyl alcohol)		
	hexadecylpyridinium bromide/cobalt sulfate and water	2010KAL
	hexadecylpyridinium bromide/nickel nitrate and water	2010KAL
	hexadecylpyridinium bromide/sodium chloride and water	2010KAL
Poly(*N*-vinylcaprolactam)		
	cadmium chloride/HCl and water	2011PAK
	copper chloride/sodium chloride and water	2011PAK
Poly(vinylidene fluoride)		
	N,N-dimethylacetamide/ethanol and water	2003YEO
	N,N-dimethylacetamide/lithium perchlorate and water	2003YEO
	N,N-dimethylacetamide/poly(1-vinyl-2-pyrrolidinone) and water	2003YEO
Poly(1-vinyl-2-pyrrolidinone)		
	ammonium sulfate/phenol and water	2011CHU
	ammonium sulfate/substituted phenols and water	2011CHU
	cadmium chloride/HCl and water	2011PAK
	cadmium chloride/potassium chloride and water	2011PAK
	copper chloride/potassium chloride and water	2011PAK
	copper chloride/sodium chloride and water	2011PAK
	copper chloride/sodium chloride/HCl and water	2011PAK
	N,N-dimethylacetamide/poly(vinylidene fluoride) and water	2003YEO
	potassium chloride/zinc chloride and water	2011PAK
	potassium phosphate/proteins and water	2011MOK
	trichloroacetic acid/HCl and water	2009PAK
	zinc chloride/HCl and water	2011PAK
Sodium alginate		
	sodium caseinate/sodium dextransulfate and water	2009AN2
Sodium caseinate		
	sodium alginate/sodium dextransulfate and water	2009AN2
Sodium dextransulfate		
	sodium alginate/sodium caseinate and water	2009AN2

3.7. Lower critical (LCST) and/or upper critical (UCST) solution temperatures of aqueous polymer solutions

Polymer (B)	M_n g/mol	M_w g/mol	**Solvent (A)**	**UCST/ K**	**LCST/ K**	**Ref.**
Hexa(ethylene glycol) monodecyl ether						
	424		water		335.25	2005IMA
Penta(ethylene glycol) monodecyl ether						
	378		water		317.45	2005IMA
Poly(acrylamide-*co*-hydroxypropyl acrylate)						
(50 mol% hydroxypropyl acrylate)			water		348.15	1975TAY
(60 mol% hydroxypropyl acrylate)			water		324.15	1975TAY
(70 mol% hydroxypropyl acrylate)			water		305.15	1975TAY
Poly(*N*-cyclopropylacrylamide)						
			water		330.15	1975TAY
Poly(diacetone acrylamide-*co*-hydroxyethyl acrylate)						
(70 mol% hydroxyethyl acrylate)			water		284.15	1975TAY
(80 mol% hydroxyethyl acrylate)			water		298.15	1975TAY
(90 mol% hydroxyethyl acrylate)			water		326.15	1975TAY
Poly(*N,N*-diethylacrylamide)						
	9600	13300	water		306.05	2003LES
			water		305.15	1999LIU
			water		305.15	2007CAO
		<5000	water		305	2004PAN
	19200	40300	water		304.15	2001LES
	19200	40300	water		304.15	2003LES
	32500	58500	water		304.05	2001LES
	32500	58500	water		304.05	2003LES
	81600	165700	water		303.25	2001LES
	81600	165700	water		303.25	2003LES
	90400	132000	water		302.85	2001LES
	96900	252000	water		302.65	2001LES
	96900	252000	water		302.65	2003LES

Polymer (B)	M_n g/mol	M_w g/mol	Solvent (A)	UCST/ K	LCST/ K	Ref.
Poly(*N,N*-diethylacrylamide)						
	180900	376300	water		302.45	2001LES
	180900	376300	water		302.45	2003LES
	218000	311700	water		302.25	2001LES
	363600	709000	water		301.55	2001LES
	363600	709000	water		301.55	2003LES
	593600	890400	water		301.75	2001LES
	1295000	1547000	water		301.75	2003LES
	1300000	1547000	water		301.75	2001LES
Poly(*N,N*-diethylacrylamide-*co*-acrylic acid)						
(5.98 mol% acrylic acid)		319000	water		305.05	2001CAI
(13.22 mol% acrylic acid)		306000	water		304.15	2001CAI
(20.64 mol% acrylic acid)		308000	water		300.15	2001CAI
Poly(*N,N*-diethylacrylamide-*co*-*N*-ethylacrylamide)						
(40 mol% *N*-ethylacrylamide)			water		305.15	1999LIU
Poly[di(ethylene glycol) methacrylate]						
	13000		water		299.15	2010LUZ
Poly(3,3-dimethyl-1-vinyl-2-pyrrolidinone)						
	29000		water		314.15	2010CH4
Poly[3-(1,3-dioxolan-2-yl)ethyl-1-vinyl-2-pyrrolidinone]						
	11000		water		320.65	2010CH4
Poly[3-(2-ethoxyethyl)-1-vinyl-2-pyrrolidinone]						
	6100		water		309.15	2010CH4
Poly(*N*-ethylacrylamide)						
	3300	5400	water		358.65	2003XUE
	5800	12200	water		355.15	2003XUE
			water		355.15	1999LIU
	5900	12400	water		354.35	2003XUE
	6400	18300	water		351.95	2003XUE
			water		347.15	1975TAY
	7400	33600	water		345.95	2003XUE

Polymer (B)	M_n g/mol	M_w g/mol	Solvent (A)	UCST/ K	LCST/ K	Ref.
Poly(*N*-ethylacrylamide-*co*-*N*-isopropylacrylamide)						
(50 mol% *N*-isopropylacrylamide)			water		321.15	1975TAY
(60 mol% *N*-isopropylacrylamide)			water		325.15	1975TAY
(70 mol% *N*-isopropylacrylamide)			water		329.15	1975TAY
(80 mol% *N*-isopropylacrylamide)			water		335.15	1975TAY
(90 mol% *N*-isopropylacrylamide)			water		340.15	1975TAY
Poly[(ethylene glycol) monomethacrylate-*co*-methyl methacrylate]						
(60 mol% methyl methacrylate)			water		328.95	2004ALI
(70 mol% methyl methacrylate)			water		322.95	2004ALI
(76 mol% methyl methacrylate)			water		315.85	2004ALI
Poly(*N*-ethylmethacrylamide)						
			water		331.15	1975TAY
Poly(*N*-ethyl-*N*-methylacrylamide)						
			water		343.15	2007CAO
		<5000	water		345	2004PAN
Poly(2-ethyl-2-oxazoline)						
	6700		water		363.75	2008HO1
	9000		water		358.45	2008HO1
	13300		water		351.45	2008HO1
	21000		water		346.65	2008HO1
	37300		water		342.45	2008HO1
		170000	water		335.15	2003CHR
Poly(2-ethyl-2-oxazoline-*stat*-2-propyl-2-oxazoline)						
(EO25/PO75)	11000		water		309.45	2007PAR
(EO50/PO50)	10000		water		314.95	2007PAR
(EO75/PO25)	14000		water		323.75	2007PAR
(EO95/PO05)	10100		water		348.25	2007PAR
(EO40/PO10)	3500		water		370.45	2008HO1
(EO35/PO15)	3700		water		355.15	2008HO1
(EO30/PO20)	3700		water		345.35	2008HO1
(EO25/PO25)	4000		water		332.95	2008HO1
(EO20/PO30)	5400		water		324.45	2008HO1
(EO15/PO35)	3900		water		318.95	2008HO1
(EO10/PO40)	3800		water		313.15	2008HO1
(EO05/PO45)	4000		water		307.35	2008HO1

Polymer (B)	**M_n g/mol**	**M_w g/mol**	**Solvent (A)**	**UCST/ K**	**LCST/ K**	**Ref.**
Poly(2-ethyl-2-oxazoline-*stat*-2-propyl-2-oxazoline)						
(EO90/PO10)	15200		water		354.75	2008HO1
(EO80/PO20)	13600		water		348.65	2008HO1
(EO70/PO30)	12600		water		337.95	2008HO1
(EO60/PO40)	13000		water		329.05	2008HO1
(EO50/PO50)	10700		water		324.25	2008HO1
(EO40/PO60)	10200		water		317.35	2008HO1
(EO30/PO70)	9700		water		313.15	2008HO1
(EO20/PO80)	9600		water		307.95	2008HO1
(EO10/PO90)	7800		water		304.35	2008HO1
(EO135/PO15)	17000		water		344.55	2008HO1
(EO120/PO30)	17700		water		336.25	2008HO1
(EO105/PO45)	17200		water		326.85	2008HO1
(EO90/PO60)	17900		water		321.35	2008HO1
(EO75/PO75)	17600		water		315.75	2008HO1
(EO60/PO90)	18600		water		310.45	2008HO1
(EO45/PO105)	17900		water		307.25	2008HO1
(EO30/PO120)	18500		water		302.25	2008HO1
(EO15/PO135)	18600		water		297.65	2008HO1
Poly(2-ethyl-*N*-vinylimidazole)						
			water		311.15	1975TAY
Poly(3-ethyl-1-vinyl-2-pyrrolidinone)						
			water		300	2009TRE
	6700		water		298.15	2010CH4
Poly[3-ethyl-1-vinyl-2-pyrrolidinone-*co*-3-(1,3-dioxolan-2-yl)ethyl-1-vinyl-2-pyrrolidinone]						
(5.7 mol% DOE)	15000		water		300.85	2010CH4
(26.0 mol% DOE)	16000		water		303.65	2010CH4
(50.5 mol% DOE)	16000		water		305.55	2010CH4
Poly[3-ethyl-1-vinyl-2-pyrrolidinone-*co*-3-(2,3-epoxypropyl)-1-vinyl-2-pyrrolidinone]						
(10.7 mol% EOP)	12000		water		305.15	2010CH4
(23.3 mol% EOP)	14000		water		312.85	2010CH4
(46.1 mol% EOP)	15000		water		333.75	2010CH4
Poly[3-ethyl-1-vinyl-2-pyrrolidinone-*co*-3-(2-methoxyethyl)-1-vinyl-2-pyrrolidinone]						
(25.4 mol% MOE)	11000		water		307.65	2010CH4
(53.4 mol% MOE)	16000		water		314.45	2010CH4
(72.9 mol% MOE)	13000		water		324.15	2010CH4

Polymer (B)	M_n g/mol	M_w g/mol	Solvent (A)	UCST/ K	LCST/ K	Ref.
Poly(3-ethyl-1-vinyl-2-pyrrolidinone-*co*-1-vinyl-2-pyrrolidinone)						
(12 mol% 1-vinyl-2-pyrrolidinone)			water		303	2009TRE
(24 mol% 1-vinyl-2-pyrrolidinone)			water		305	2009TRE
Poly(2-hydroxyethyl acrylate-*co*-2-hydroxypropyl acrylate)						
(20 mol% hydroxypropyl acrylate)			water		333.15	1975TAY
(30 mol% hydroxypropyl acrylate)			water		317.15	1975TAY
(40 mol% hydroxypropyl acrylate)			water		307.15	1975TAY
(50 mol% hydroxypropyl acrylate)			water		305.15	1975TAY
(60 mol% hydroxypropyl acrylate)			water		298.15	1975TAY
(70 mol% hydroxypropyl acrylate)			water		296.15	1975TAY
Poly(2-hydroxypropyl acrylate)						
			water		289.15	1975TAY
Poly[*N*-(2-hydroxypropyl)methacrylamide dilactate]						
	6300	10700	water		286.15	2004SOG
Poly[*N*-(2-hydroxypropyl)methacrylamide monolactate]						
	11400	24400	water		338.15	2004SOG
Poly[*N*-(2-hydroxypropyl)methacrylamide monolactate-*co*-*N*-(2-hydroxypropyl)methacrylamide dilactate]						
(25 mol% dilactate)	7500	17600	water		323.65	2004SOG
(49 mol% dilactate)	8100	16900	water		309.65	2004SOG
(74 mol% dilactate)	6800	14000	water		298.65	2004SOG
Poly(isobutyl vinyl ether-*co*-2-hydroxyethyl vinyl ether)						
(12 mol% isobutyl vinylether)			water		338.15	2004SUG
(20 mol% isobutyl vinylether)			water		314.15	2004SUG
(33 mol% isobutyl vinylether)			water		289.15	2004SUG
Poly(*N*-isopropylacrylamide)						
		<5000	water		307	2004PAN
			water		309.15	2007CAO

Polymer (B)	M_n g/mol	M_w g/mol	Solvent (A)	UCST/ K	LCST/ K	Ref.
Poly(*N*-isopropylacrylamide-*co*-*N,N*-dimethylacrylamide)						
(10.7 mol% *N,N*-dimethylacrylamide)			water		310.05	2006SHE
(13.0 mol% *N,N*-dimethylacrylamide)			water		309.15	2003BAR
(14.7 mol% *N,N*-dimethylacrylamide)			water		311.25	2006SHE
(17.3 mol% *N,N*-dimethylacrylamide)			water		312.95	2006SHE
(21.2 mol% *N,N*-dimethylacrylamide)			water		314.25	2006SHE
(27.0 mol% *N,N*-dimethylacrylamide)			water		312.15	2003BAR
(30.0 mol% *N,N*-dimethylacrylamide)			water		315.15	2003BAR
(31.4 mol% *N,N*-dimethylacrylamide)			water		319.15	2006SHE
(50.0 mol% *N,N*-dimethylacrylamide)			water		323.15	2003BAR
(60.0 mol% *N,N*-dimethylacrylamide)			water		336.15	2003BAR
(66.0 mol% *N,N*-dimethylacrylamide)			water		345.15	2003BAR
Poly(*N*-isopropylacrylamide-*co*-2-hydroxyethyl methacrylate-*co*-acrylic acid)						
(92.0 mol% NIPAM, 5.0 mol% HEMA, 3.0 mol% acrylic acid)		160000	water		308.15	2005LE2
(92.0 mol% NIPAM, 4.3 mol% HEMA, 3.7 mol% acrylic acid)		120000	water		309.75	2005LE2
(89.8 mol% NIPAM, 4.2 mol% HEMA, 6.0 mol% acrylic acid)		140000	water		316.25	2005LE2
Poly(*N*-isopropylacrylamide-*co*-2-hydroxyethyl methacrylate lactate-*co*-acrylic acid)						
(84.4 mol% NIPAM, 9.4 mol% HEMA lactate, 6.2 mol% acrylic acid)		230000	water		308.05	2005LE2
(81.7 mol% NIPAM, 12.3 mol% HEMA lactate, 6.0 mol% acrylic acid)		240000	water		301.55	2005LE2
(75.8 mol% NIPAM, 18.5 mol% HEMA lactate, 5.7 mol% acrylic acid)		160000	water		296.65	2005LE2
Poly[*N*-isopropylacrylamide-*co*-(2-hydroxyisopropyl)acrylamide]						
(10 mol% HIPAAM)			water		309.85	2006MA2
(30 mol% HIPAAM)			water		314.95	2006MA2
(50 mol% HIPAAM)			water		328.15	2006MA2
(80 mol% HIPAAM)			water		353.15	2006MA2
Poly[*N*-isopropylacrylamide-*co*-oligo(ethylene glycol) monomethacrylate]						
			deuterium oxide		309.15	2003KOH
			water		312.65	2004ALA
Poly(*N*-isopropylacrylamide-*b*-propylene glycol-*b*-*N*-isopropylacrylamide)						
	6600	12700	water		305.15	2005CHE

Polymer (B)	M_n g/mol	M_w g/mol	Solvent (A)	UCST/ K	LCST/ K	Ref.
Poly(2-isopropyl-2-oxazoline)						
	1700	1785	water		345.65	2004DIA
	2400	2500	water		335.95	2004DIA
	3900	4250	water		320.15	2008HU1
	4300	4430	water		324.45	2004DIA
	5700	5870	water		321.25	2004DIA
	10200		water		311.85	2007PAR
Poly(2-isopropyl-2-oxazoline-*co*-2-butyl-2-oxazoline)						
(5.8 mol% 2-but)	2750	3470	water		311.15	2008HU1
(8.0 mol% 2-but)	3110	3920	water		308.15	2008HU1
(12 mol% 2-but)	3220	3960	water		306.15	2008HU1
(20 mol% 2-but)	3120	4060	water		294.15	2008HU1
Poly(2-isopropyl-2-oxazoline-*co*-2-nonyl-2-oxazoline)						
(5.9 mol% 2-non)	3510	4140	water		288.15	2008HU1
(7.7 mol% 2-non)	3140	3670	water		284.15	2008HU1
(12 mol% 2-non)	3350	4020	water		282.15	2008HU1
Poly(2-isopropyl-2-oxazoline-*co*-2-propyl-2-oxazoline)						
(11.5 mol% 2-prop)	2660	3000	water		319.15	2008HU1
(20.8 mol% 2-prop)	2860	3670	water		310.15	2008HU1
Poly[3-(2-methoxyethyl)-1-vinyl-2-pyrrolidinone]						
	15000		water		337.15	2010CH4
Poly[*N*-(3-methoxypropyl)acrylamide]						
			water		283.15	1975TAY
Poly(3-methyl-1-vinyl-2-pyrrolidinone)						
	6300		water		337.15	2010CH4
Poly(*N*-propylacrylamide)						
			water		298.15	2004MAO
			water		298.15	2007CAO

Polymer (B)	M_n g/mol	M_w g/mol	Solvent (A)	UCST/ K	LCST/ K	Ref.
Poly(*N*-propylacrylamide), syndiotactic						
(51.8 rac. diads)	34400	96300	deuterium oxide		296.35	2009MOR
(56.8 rac. diads)	27100	73400	deuterium oxide		297.45	2009MOR
(65.6 rac. diads)	38600	96500	deuterium oxide		298.45	2009MOR
(68.2 rac. diads)	55700	145000	deuterium oxide		298.85	2009MOR
(70.9 rac. diads)	55000	165000	deuterium oxide		299.05	2009MOR
(51.8 rac. diads)	34400	96300	water		295.45	2009MOR
(56.8 rac. diads)	27100	73400	water		297.05	2009MOR
(65.6 rac. diads)	38600	96500	water		297.85	2009MOR
(68.2 rac. diads)	55700	145000	water		298.25	2009MOR
(70.9 rac. diads)	55000	165000	water		298.35	2009MOR
Poly(propylene oxide-*b*-ethylene oxide-*b*-propylene oxide)						
PPO_{14}-PEO_{24}-PPO_{14}	2670	2830	deuterium oxide		316.78	2011HUF
PPO_{14}-PEO_{24}-PPO_{14}	2670	2830	water		317.85	2011HUF
Poly(2-propyl-2-oxazoline)						
	3070	3470	water		298.15	2008HU1
	3100		water		316.05	2008HO1
	3700		water		312.15	2008HO1
	4300		water		310.65	2008HO1
	6200		water		303.45	2008HO1
	8140		water		302.75	2008HO1
	10800		water		296.95	2007PAR
	12300		water		298.65	2008HO1
	15500		water		297.25	2008HO1
	18000		water		295.65	2008HO1
Poly(pyrrolidinoacrylamide)						
		<5000	water		345	2004PAN
Poly(styrene-*b*-*N*-isopropylacrylamide)						
	13910	29100	water		306.65	2011SHI
Poly(*N*-tetrahydrofurfurylacrylamide)						
		31000	water		311.15	2010MAE
Poly(*N*-tetrahydrofurfurylmethacrylamide)						
		20000	water		316.15	2010MAE

Polymer (B)	M_n **g/mol**	M_w **g/mol**	**Solvent (A)**	**UCST/ K**	**LCST/ K**	**Ref.**
Poly(*N*-vinylcaprolactam)						
		330000	deuterium oxide		304.35	2004LAU
	5600		water		303.15	2003OKH
		21000	water		310.65	2004LAU
		30000	water		308.85	2004LAU
		330000	water		304.95	2004LAU
		1300000	water		303.95	2004LAU
		1500000	water		304.15	2004LAU
Poly(*N*-vinylcaprolactam-*co*-methacrylic acid)						
(9 mol% methacrylic acid)			water		303.15	2003OKH
(12 mol% methacrylic acid)			water		310.65	2003OKH
(18 mol% methacrylic acid)			water		313.85	2003OKH
(37 mol% methacrylic acid)			water		304.15	2003OKH

3.8. References

1975TAY Taylor, L.D. and Cerankowski, L.D., Preparation of films exhibiting a balanced temperature dependence to permeation by aqueous solutions. A study of lower consolute behavior, *J. Polym. Sci.: Polym. Chem. Ed.*, 13, 2551, 1975.

1991SAM Samii, A.A., Karlström, G., and Lindman, B., Phase behavior of nonionic block copolymer in a mixed-solvent system, *J. Phys. Chem.*, 95, 7887, 1991.

1993OTA Otake, K., Karaki, R., Ebina, T., Yokoyama, C., and Takahashi, S., Pressure effects on the aggregation of poly(*N*-isopropylacrylamide) and poly(*N*-isopropylacrylamide-*co*-acrylic acid) in aqueous solutions, *Macromolecules*, 26, 2194, 1993.

1994HOG Ho-Gutierrez, I.V., Cheluget, E.L., Vera, J.H., and Weber, M.E., Liquid-liquid equilibrium of aqueous mixtures of poly(ethylene glycol) with Na_2SO_4 or NaCl, *J. Chem. Eng. Data*, 39, 245, 1994.

1995CHE Chen, G. and Hoffman, A.S., A new temperature- and pH-responsive copolymer for possible use in protein conjugation, *Macromol. Chem. Phys.*, 196, 1251, 1995.

1995JAR Jarm, V., Sertic, S., and Segudovic, N., Stability of aqueous solutions of poly[(maleic acid)-*alt*-styrene] sodium salts in the presence of divalent cations, *J. Appl. Polym. Sci.*, 58, 1973, 1995.

1995THU Thuresson, K., Karlström, G., and Lindman, B., Phase diagrams of mixtures of a nonionic polymer, hexanol, and water. An experimental and theoretical study of the effect of hydrophobic modification, *J. Phys. Chem.*, 99, 3823, 1995.

1996KIR Kirsh, Yu.E., Krylov, A.V., Belova, T.A., Abdelsadek, G.G., and Pashkin, I.I., Phase transition of poly(*N*-vinylcaprolactam) in water-organic solvent systems (Russ.), *Zh. Fiz. Khim.*, 70, 1403, 1996.

1996THU Thuresson, K., Nilsson, S., and Lindman, B., Influence of cosolutes on phase behavior and viscosity of a nonionic cellulose ether. The effect of hydrophobic modification, *Langmuir*, 12, 2412, 1996.

1997YOO Yoo, M.K., Sung, Y.K., Cho, C.S., and Lee, Y.M., Effect of polymer complex formation on the cloud-point of poly(*N*-isopropylacrylamide) (PNIPAAm) in the poly(NIPAAm-*co*-acrylic acid): Polyelectrolyte complex between poly(acrylic acid) and poly(allylamine), *Polymer*, 38, 2759, 1997

1998GOS Gosselet, N.M., Borie, C., Amiel, C., and Sebille, B., Aqueous two-phase systems from cyclodextrin polymers and hydrophobically modified acrylic polymers, *J. Dispersion Sci.Technol.*, 19, 805, 1998.

1999JON Jones, M.S., Effect of pH on the lower critical solution temperature of random copolymers of *N*-isopropylacrylamide and acrylic acid, *Eur. Polym. J.*, 35, 795, 1999.

1999KUN Kunugi, S., Yamazaki, Y., Takano, K., and Tanaka, N., Effects of ionic additives and ionic comonomers on the temperature and pressure responsive behavior of thermoresponsive polymers in aqueous solutions, *Langmuir*, 15, 4056, 1999.

1999LIU Liu, H.Y. and Zhu, X.X., Lower critical solution temperatures of *N*-substituted acrylamide copolymers in aqueous solutions, *Polymer*, 40, 6985, 1999.

1999SVE Svensson, M., Berggren, K., Veide, A., and Tjerneld, F., Aqueous two-phase systems containing self-associating block copolymers. Partitioning of hydrophilic and hydrophobic biomolecules, *J. Chromatogr. A*, 839, 71, 1999.

2000BIG Bignotti, F., Penco, M., Sartore, L., Peroni, I., Mendichi, R., and Casolaro, M., Synthesis, characterisation and solution behaviour of thermo- and pH-responsive polymers bearing L-leucine residues in the side chains, *Polymer*, 41, 8247, 2000.

2000KUC Kuckling, D., Adler, H.-J.P., Arndt, K.F., Ling, L., and Habicher, W.D., Temperature and pH dependent solubility of novel poly(*N*-isopropylacrylamide) copolymers, *Macromol. Chem. Phys.*, 201, 273, 2000.

2000PRI Principi, T., Goh, C.C.E., Liu, R.C.W., and Winnik, F.M., Solution properties of hydrophobically modified copolymers of *N*-isopropylacrylamide and *N*-glycineacrylamide: A study by microcalorimetry and fluorescence spectroscopy, *Macromolecules*, 33, 2958, 2000.

2000SHI Shi, X., Li, J., Sun, C., and Wu, S., The aggregation and phase separation behavior of a hydrophobically modified poly(*N*-isopropylacrylamide), *Coll. Surfaces A*, 175, 41, 2000.

2001BUL Bulmus, V., Patir, S., Tuncel, S.A., and Piskin, E., Stimuli-responsive properties of conjugates of *N*-isopropylacrylamide-*co*-acrylic acid oligomers with alanine, glycine and serine mono-, di- and tri-peptides, *J. Control. Release*, 76, 265, 2001.

2001CAI Cai, W.S., Gan, L.H., and Tam, K.C., Phase transition of aqueous solutions of poly(*N,N*-diethylacrylamide-*co*-acrylic acid) by differential scanning calorimetric and spectrophotometric methods, *Colloid Polym. Sci.*, 279, 793, 2001.

2001LES Lessard, D.G., Ousalem, M., and Zhu, X.X., Effect of the molecular weight on the lower critical solution temperature of poly(*N,N*-diethylacrylamide) in aqueous solutions, *Can. J. Chem.*, 79, 1870, 2001.

2001TOK To, K., Coexistence curve exponent of a binary mixture with a high molecular weight polymer, *Phys. Rev. E*, 63, 026108/1-4, 2001.

2002DIN Dincer, S., Tuncel, A., and Piskin, E., A potential gene delivery vector: *N*-isopropylacrylamide-ethylenimine block copolymers, *Macromol. Chem. Phys.*, 203, 1460, 2002.

2002VIR Virtanen, J., Arotcarena, M., Heise, B., Ishaya, S., Laschewsky, A., and Tenhu, H., Dissolution and aggregation of a poly(NIPA-*block*-sulfobetaine) copolymer in water and saline aqueous solutions, *Langmuir*, 18, 5360, 2002.

2002YIN Yin, X. and Stoever, H.D.H., Thermosensitive and pH-sensitive polymers based on maleic anhydride copolymers, *Macromolecules*, 35, 100178, 2002.

2003ABU Abu-Sharkh, B.F., Hamad, E.Z., and Ali, S.A., Influence of hydrophobe content on phase coexistence curves of aqueous two-phase solutions of associative polyacrylamide copolymers and poly(ethylene glycol), *J. Appl. Polym. Sci.*, 89, 1351, 2003.

2003BAR Barker, I.C., Cowie, J.M.G., Huckeby, T.N., Shaw, D.A., Soutar, I., and Swanson, L., Studies of the 'smart' thermoresponsive behavior of copolymers of *N*-isopropylacrylamide and *N,N*-dimethylacrylamide in dilute aqueous solution, *Macromolecules*, 36, 7765, 2003.

2003CAM Campese, G.M., Rodriguez, E.M.G., Tambourgi, E.B., and Pessoa, Jr, A., Determination of cloud-point temperatures for different copolymers, *Brazil. J. Chem. Eng.*, 20, 335, 2003.

2003CHR Christova, D., Velichkova, R., Loos, W., Goethals, E.J., and DuPrez, F., New thermoresponsive polymer materials based on poly(2-ethyl-2-oxazoline) segments, *Polymer*, 44, 2255, 2003.

2003DAV David, G., Alupei, V., Simionescu, B.C., Dincer, S., and Piskin, E., Poly(*N*-isopropylacrylamide)/poly[(*N*-acetylimino)ethylene] thermosensitive block and graft copolymers, *Eur. Polym. J.*, 39, 1209, 2003.

2003EDE Edelman, M.W., van der Linden, E., and Tromp, R.H., Phase separation of aqueous mixtures of poly(ethylene oxide) and dextran, *Macromolecules*, 36, 7783, 2003.

2003ELI Eliassi, A., Modarress, H., and Mansoori, G.A., The effect of electrolytes on the cloud point of polyethylene glycol aqueous solutions, *Iran. J. Sci. Technol., Trans. B*, 26, 319, 2003.

2003HAG Haghtalab, A., Mokhtarani, B., and Maurer, G., Experimental results and thermodynamic modeling of the partitioning of lysozyme, bovine serum albumin, and α-amylase in aqueous two-phase systems of PEG and (K_2HPO_4 or Na_2SO_4), *J. Chem. Eng. Data*, 48, 1170, 2003.

2003HEL Hellebust, S., Nilsson, S., and Blokhus, A.M., Phase behavior of anionic polyelectrolyte mixtures in aqueous solution. Effects of molecular weights, polymer charge density, and ionic strength of solution, *Macromolecules*, 36, 5372, 2003.

2003HUD Huddleston, J.G., Willauer, H.D., and Rogers, R.D., Phase diagram data for several PEG + salt aqueous biphasic systems at 25°C, *J. Chem. Eng. Data*, 48, 1230, 2003.

2003KAL Kalaycioglu, E., Patir, S., and Pisukin, E., Poly(*N*-isopropylacrylamide-*co*-2-methacryloamidohistidine) copolymers and their interactions with human immunoglobulin-G, *Langmuir*, 19, 9538, 2003.

2003KOH Koh, A.Y.C., Heenan, R.K., and Saunders, B.R., A study of temperature-induced aggregation of responsive comb copolymers in aqueous solution, *Phys. Chem. Chem. Phys.*, 5, 2417, 2003.

2003KOT Koetz, J., Guenther, C., Kosmella, S., Kleinpeter, E., and Wolf, G., Polyelectrolyte-induced structural changes in the isotropic phase of sulfobetaine/pentanol/toluene/water system, *Progr. Colloid Polym. Sci.*, 122, 27, 2003.

2003LES Lessard, D.G., Ousalem, M., Zhu, X.X., Eisenberg, A., and Carreau, P.J., Study of the phase transition of poly(*N,N*-diethylacrylamide) in water by rheology and dynamic light scattering, *J. Polym. Sci.: Part B: Polym. Phys.*, 41, 1627, 2003.

2003LI1 Liu, S. and Liu, M., Synthesis and characterization of temperature- and pH-sensitive poly(*N,N*-diethylacrylamide-*co*-methacrylic acid), *J. Appl. Polym. Sci.*, 90, 3563, 2003.

2003LI2 Liu, S.-Q., Yang, Y.-Y., Liu, X.-M., and Tong, Y.-W., Preparation and characterization of temperature-sensitive poly(*N*-isopropylacrylamide)-*b*-poly(DL-lactide) microspheres for protein delivery, *Biomacromolecules*, 4, 1784, 2003.

2003LUT Lutz, A., Vergleich des Assoziationsverhaltens von Ethyl-Hydroxyethyl-Cellulose mit dem seines hydrophob modifizierten Analogons, *Dissertation*, Johannes-Gutenberg-Universität Mainz, 2003.

2003MAE Maeda, Y., Mochiduki, H., Yamamoto, H., Nishimura, Y., and Ikeda, I., Effects of ions on two-step phase separation of poly(vinyl methyl ether) in water as studied by IR and Raman spectroscopy, *Langmuir*, 19, 10357, 2003.

2003MAO Mao, H., Li, C., Zhang, Y., Bergbreiter, D.E., and Cremer, P.S., Measuring LCSTs by novel temperature gradient methods: Evidence for intermolecular interactions in mixed polymer solutions, *J. Amer. Chem. Soc.*, 125, 2850, 2003.

2003MAT Matsuyama, H., Nakagawa, K., Maki, T., and Teramoto, M., Studies on phase separation rate in porous polyimide membrane formation by immersion precipitation, *J. Appl. Polym. Sci.*, 90, 292, 2003.

2003NOW Nowakowska, M., Szczubialka, K., and Grebosz, M., Interactions of temperature-responsive anionic polyelectrolytes with a cationic surfactant, *J. Colloid Interface Sci.*, 265, 214, 2003.

2003NOZ Nozary, S., Modarress, H., and Eliassi, A., Cloud-point measurements for salt + poly(ethylene glycol) + water systems by viscometry and laser beam scattering methods, *J. Appl. Polym. Sci.*, 89, 1983, 2003.

2003OKH Okhapkin, I.M., Nasimova, I.R., Makhaeva, E.E., and Khokhlov, A.R., Effect of complexation of monomer units on pH- and temperature-sensitive properties of poly(*N*-vinylcaprolactam-*co*-methacrylic acid), *Macromolecules*, 36, 8130, 2003.

2003SEI Seiler, M., Köhler, D., and Arlt, W., Hyperbranched polymers: New selective solvents for extractive distillation and solvent extraction, *Separation Purification Technol.*, 30, 179, 2003.

2003SET Seto, Y., Kameyama, K., Tanaka, N., Kunugi, S., Yamamoto, K., and Akashi, M., High-pressure studies on the coacervation of copoly(*N*-vinylformamide-vinylacetate) and copoly(*N*-vinylacetylamide-vinylacetate), *Colloid Polym. Sci.*, 281, 690, 2003.

2003SHT Shtanko, N.I., Lequieu, W., Goethals, E.J., and DuPrez, F.E., pH- and thermo-responsive properties of poly(*N*-vinylcaprolactam-*co*-acrylic acid) copolymers, *Polym. Int.* 52, 1605, 2003.

2003STI Stieger, M. and Richtering, W., Shear-induced phase separation in aqueous polymer solutions: Temperature-sensitive microgels and linear polymer chains, *Macromolecules*, 36, 8811, 2003.

2003TAC Tachibana, Y., Kurisawa, M., Uyama, H., and Kobayashi, S., Thermo- and pH-responsive biodegradable poly(α-*N*-substituted γ-glutamine), *Biomacromolecules*, 4, 1132, 2003.

2003TSY Tsypina, N.A., Kizhnyaev, V.N., and Adamova, L.V., Triazole-containing polymers: Solubility and thermodynamic behavior in solutions (Russ.), *Vysokomol. Soedin., Ser. A*, 45, 1718, 2003.

2003VE1 Verbrugghe, S., Bernaerts, K., and DuPrez, F.E., Thermo-responsive and emulsifying properties of poly(*N*-vinylcaprolactam) based graft copolymers, *Macromol. Chem. Phys.*, 204, 1217, 2003.

2003VE2 Verdonck, B., Goethals, E.J., and DuPrez, F.E., Block copolymers of methyl vinyl ether and isobutyl vinyl ether with thermo-adjustable amphiphilic properties, *Macromol. Chem. Phys.*, 204, 2090, 2003.

2003XUE Xue, W., Huglin, M.B., and Jones, T.G.J., Parameters affecting the lower critical solution temperature of linear and crosslinked poly(*N*-ethylacrylamide) in aqueous media, *Macromol. Chem. Phys.*, 204, 1956, 2003.

2003YAM Yamaoka, T. Tamura, T., Seto, Y., Tada, T., Kunugi, S., and Tirrell, D.A., Mechanism for the phase transition of a genetically engineered elastin model peptide (VPGIG)40 in aqueous solution, *Biomacromolecules*, 4, 1680, 2003.

2003YAN Yang, Z., Crothers, M., Attwood, D., Collett, J.H., Ricardo, N.M.P., Martini, L.G.A., and Booth, C., Association properties of ethylene oxide/styrene oxide diblock copolymer E17S8 in aqueous solution, *J. Colloid Interface Sci.*, 263, 312, 2003.

2003YEO Yeow, M.L., Liu, Y.T., and Li, K., Isothermal phase diagrams and phase-inversion behavior of poly(vinylidene fluoride)/solvents/additives/water systems, *J. Appl. Polym. Sci.*, 90, 2150, 2003.

2003YIN Yin, X. and Stoever, H.D.H., Hydrogel microspheres by thermally induced coacervation of poly(*N,N*-dimethylacrylamide-*co*-glycidyl methacrylate) aqueous solutions, *Macromolecules*, 36, 9817, 2003.

2004ALA Alava, C. and Saunders, B.R., Effect of added surfactant on temperature-induced gelation of emulsions, *Langmuir*, 20, 3107, 2004.

2004ALI Ali, M.M. and Stoever, H.D.H., Well-defined amphiphilic thermosensitive copolymers based on poly(ethylene glycol monomethacrylate) and methyl methacrylate prepared by atom transfer radical polymerization, *Macromolecules*, 37, 5219, 2004.

2004ANT Antonov, Y.A., Van Puyvelde, P., Moldenaers, P., and Leuven, K.U., Effect of shear flow on the phase behavior of an aqueous gelatin-dextran emulsion, *Biomacromolecules*, 5, 276, 2004.

2004BER Berge, B., Koningsveld, R., and Berghmans, H., Influence of added components on the miscibility behavior of the (quasi-) binary system water/poly(vinyl methyl ether) and on the swelling behavior of the corresponding hydrogels. 1. Tetrahydrofuran, *Macromolecules*, 37, 8082, 2004.

2004CHE Chen, J., Spear, S.K., Huddleston, J.G., Holbrey, J.D., Swatloski, R.P., and Rogers, R.D., Application of poly(ethylene glycol)-based aqueous biphasic systems as reaction and reactive extraction media, *Ind. Eng. Chem. Res.*, 43, 5358, 2004.

2004DER D'Errico, G., Paduano, L., and Khan, A., Temperature and concentration effects on supramolecular aggregation and phase behavior for poly(propylene oxide)-*b*-poly(ethylene oxide)-*b*-poly(propylene oxide) copolymers of different composition in aqueous mixtures, *J. Coll. Interface Sci.*, 279, 379, 2004.

2004DIA Diab, C., Akiyama, Y., Kataoka, K., and Winnik, F.M., Microcalorimetric study of the temperature-induced phase separation in aqueous solutions of poly(2-isopropyl-2-oxazolines), *Macromolecules*, 37, 2556, 2004.

2004DOR Dormidontova, E.E., Influence of end groups on phase behavior and properties of PEO in aqueous solutions, *Macromolecules*, 37, 7747, 2004.

2004DU1 Durme, K. van, Verbrugghe, S., DuPrez, F.E., and Mele, B. van, Influence of poly(ethylene oxide) grafts on kinetics of LCST behavior in aqueous poly(*N*-vinylcaprolactam) solutions and networks studied by modulated temperature DSC, *Macromolecules*, 37, 1054, 2004.

2004DU2 Durme, K., van, Assche, G. van, and Mele, B. van, Kinetics of demixing and remixing in poly(*N*-isopropylacrylamide)/water studied by modulated temperature DSC, *Macromolecules*, 37, 9596, 2004.

2004ECK Eckelt, J., Fraktionierung und Membranbildung von Polysacchariden, *Dissertation*, Johannes-Gutenberg-Universität Mainz, 2004.

2004EEC Eeckman, F., Moës, A.J., and Amighi, K., Synthesis and characterization of thermosensitive copolymers for oral controlled drug delivery, *Eur. Polym. J.*, 40, 873, 2004.

2004FEN Fen'ko, L.A., Bil'dyukevich, A.V., and Soldatov, V.S., Phase diagrams of a poly(ε-caproamide)-dimethylacetamide-lithium chloride system (Russ.), *Vysokomol. Soedin., Ser. A*, 46, 706, 2004.

2004GAR Garay, M.T., Rodriguez, M., Vilas, J.L., and Leon, L.M., Study of polymer-polymer complexes of poly(*N*-isopropylacrylamide) with hydroxyl-containing polymers, *J. Macromol. Sci. Part B-Phys.*, 43, 437, 2004.

2004GRA Graber, T.A., Galvez, M.E., Galleguillos, H.R., and Alvarez-Benedi, J., Liquid-liquid equilibrium of the aqueous two-phase system water + PEG 4000 + lithium sulfate at different temperatures. Experimental determination and correlation, *J. Chem. Eng. Data*, 49, 1661, 2004.

2004HAG Haghtalab, A. and Mokhtarani, B., The new experimental data and a new thermodynamic model based on group contribution for correlation liquid–liquid equilibria in aqueous two-phase systems of PEG and (K_2HPO_4 or Na_2SO_4), *Fluid Phase Equil.*, 215, 151, 2004.

2004HAR Haraguchi, L.H., Mohamed, R.S., Loh, W., and Pessoa Filho, P.A., Phase equilibrium and insulin partitioning in aqueous two-phase systems containing block copolymers and potassium phosphate, *Fluid Phase Equil.*, 215, 1, 2004.

2004HAS Hasan, E., Zhang, M., Müller, A.H.E., and Tsvetanov, Ch.B., Thermoassociative block copolymers of poly(*N*-isopropylacrylamide) and poly(propylene oxide), *J. Macromol. Sci. Part A-Pure Appl. Chem.*, 41, 467, 2004.

2004HOP Hopkinson, I., Myatt, M., and Tajbakhsh, A., Static and dynamic studies of phase composition in a polydisperse system (exp. data by I. Hopkinson), *Polymer*, 45, 4307, 2004.

2004HUE Hüther, A. and Maurer, G., Swelling of *N*-isopropyl acrylamide hydrogels in aqueous solutions of poly(ethylene glycol), *Fluid Phase Equil.*, 226, 321, 2004.

2004HUM Hu, M., Zhai, Q., Jiang, Y., Jin, J., and Liu, Z., Liquid-liquid and liquid-liquid-solid equilibrium in PEG + Cs_2SO_4 + H_2O, *J. Chem. Eng. Data*, 49, 1440, 2004.

2004INA Inamura, I., Jinbo, Y., Kittaka, M., and Asano, A., Solution properties of water-poly(ethylene glycol)-poly(*N*-vinylpyrrolidone) ternary system, *Polym. J.*, 36, 108, 2004.

2004KAV Kavlak, S., Can, H.K., and Guener, A., Miscibility studies on poly(ethylene glycol)/dextran blends in aqueous solutions by dilute solution viscometry, *J. Appl. Polym. Sci.*, 94, 453, 2004.

2004KEL Kelarakis, A., Ming, X.-T., Yuan, X.-F., and Booth, C., Aqueous micellar solutions of mixed triblock and diblock copolymers studied using oscillatory shear, *Langmuir*, 20, 2036, 2004.

2004KIT Kitano, H., Hirabayashi, T., Gemmei-Ide, M., and Kyogoku, M., Effect of macrocycles on the temperature-responsiveness of poly[(methoxy diethylene glycol methacrylate)-*graft*-PEG], *Macromol. Chem. Phys.*, 205, 1651, 2004.

2004KUN Kunieda, H., Kaneko, M., Lopez-Quintela, M.A., and Tsukahara, M., Phase behavior of a mixture of poly(isoprene)-poly(oxyethylene) diblock copolymer and poly(oxyethylene) surfactant in water, *Langmuir*, 20, 2164, 2004.

2004KUR Kurata, K. and Dobashi, A., Novel temperature- and pH-responsive linear polymers and crosslinked hydrogels comprised of acidic L-α-amino acid derivatives, *J. Macromol.-Sci., Phys. B*, 43, 143, 2004.

2004LAU Laukkanen, A., Valtola, L., Winnik, F.M., and Tenhu, H., Formation of colloidally stable phase separated poly(*N*-vinylcaprolactam) in water: A study by dynamic light scattering, microcalorimetry, and pressure perturbation calorimetry, *Macromolecules*, 37, 2268, 2004.

2004LEE Lee, H.J., Jung, B., Kang, Y.S., and Lee, H., Phase separation of polymer casting solution by nonsolvent vapor, *J. Membrane Sci.*, 245, 103, 2004.

2004MA1 Maeda, Y., Yamamoto, H., and Ikeda, I., Effects of ionization on the phase behavior of poly(*N*-isopropylacrylamide-*co*-acrylic acid) and poly(*N, N*-diethylacrylamide-*co*-acrylic acid) in water, *Coll. Polym. Sci.*, 282, 1268, 2004.

2004MA2 Maeda, Y., Yamamoto, H., and Ikeda, I., The association of tetraalkylammonium ions with poly(vinyl methyl ether) in water and its effect on phase-separation behavior as studied by micro-Raman spectroscopy, *Macromol. Rapid Commun.*, 25, 720, 2004.

2004MAH Mahajan, R.K., Chawla, J., and Bakshi, M.S., Depression in the cloud point of Tween in the presence of glycol additives and triblock polymers, *Colloid Polym. Sci.*, 282, 1165, 2004.

2004MAN Mandala, I., Michon, C., and Launay, B., Phase and rheological behavior of xanthan/amylose and xanthan/starch mixed systems, *Carbohydrate Polym.*, 58, 285, 2004.

2004MAO Mao, H., Li, C., Zhang, Y. Furyk, S., Cremer, P.S., and Bergbreiter, D.E., High-throughput studies of the effects of polymer structure and solution components on the phase separation of thermoresponsive polymers, *Macromolecules*, 37, 1031, 2004.

2004MEY Meyer, D.E. and Chilkoti, A., Quantification of the effects of chain length and concentration on the thermal behavior of elastin-like polypeptides, *Biomacromolecules*, 5, 846, 2004.

2004MOR Mori, T., Fukuda, Y., Okamura, H., Minagawa, K., Masuda, S., and Tanaka, M., Thermosensitive copolymers having soluble and insoluble monomer units, poly(*N*-vinylacetamide-*co*-methyl acrylate)s: Effect of additives on their lower critical solution temperatures, *J. Polym. Sci.: Part A: Polym. Chem.*, 42, 2651, 2004.

2004NED Nedelcheva, A.N., Vladimirov, N.G., Novakov, C.P., and Berlinova, I.V., Associative block copolymers comprising poly(*N*-isopropylacrylamide) and poly(ethylene oxide) end-functionalized with a fluorophilic or hydrophilic group. Synthesis and aqueous solution properties, *J. Polym. Sci.: Part B: Polym. Phys.*, 42, 5736, 2004.

2004NON Nonaka, T., Hanada, Y., Watanabe, T., Ogata, T., and Kurihara, S., Formation of thermosensitive water-soluble copolymers with phosphinic acid groups and the thermosensitivity of the copolymers and copolymer/metal complexes, *J. Appl. Polym. Sci.*, 92, 116, 2004.

2004NOW Nowakowska, M., Szczubialka, K., and Grebosz, M., Modifying the thermosensitivity of copolymers of sodium styrenesulfonate and *N*-isopropyl-acrylamide with dodecyltrimethylammonium chloride, *Coll. Polym. Sci.*, 283, 291, 2004.

2004OLS Olsson, M., Joabsson, F., and Piculell, L., Particle-induced phase separation in quasi-binary polymer solutions, *Langmuir*, 20, 1605, 2004.

2004PAG Pagonis, K. and Bokias, G., Upper critical solution temperature-type cononsolvency of poly(*N,N*-dimethylacrylamide) in water-organic solvent mixtures, *Polymer*, 45, 2149, 2004.

2004PAN Panayiotou, M., Garret-Flaudy, F., and Freitag, R., Co-nonsolvency effects in the thermoprecipitation of oligomeric polyacrylamides from hydroorganic solutions, *Polymer*, 45, 3055, 2004.

2004PA1 Park, M.J. and Char, K., Gelation of PEO-PLGA-PEO triblock copolymers induced by macroscopic phase separation, *Langmuir*, 20, 2456, 2004.

2004PA2 Park, J.-S., Akiyama, Y., Winnik, F.M., and Kataoka, K., Versatile synthesis of end-functionalized thermosensitive poly(2-isopropyl-2-oxazolines), *Macromolecules*, 37, 6786, 2004.

2004PER Pereira, M., Wu, Y.T., Madeira, P., Venancio, A., Macedo, E., and Teixeira, J., Liquid-liquid equilibrium phase diagrams of new aqueous two-phase systems: Ucon 50-HB5100 + ammonium sulfate + water, Ucon 50-HB5100 + poly(vinyl alcohol) + water, Ucon 50-HB5100 + hydroxypropylstarch + water, and poly(ethylene glycol) 8000 + poly(vinyl alcohol) + water, *J. Chem. Eng. Data* 49, 43, 2004.

2004RAN Rangel-Yagui, C.O., Pessoa Jr, A., and Blankschtein, D., Two-phase aqueous micellar systems: An alternative method for protein purification, *Brazil. J. Chem. Eng.*, 21, 531, 2004.

2004SA1 Salgado-Rodriguez, R., Licea-Claverie, A., and Arndt, K.F., Random copolymers of *N*-isopropylacrylamide and methacrylic acid monomers with hydrophobic spacers: pH-tunable temperature sensitive materials, *Eur. Polym. J.*, 40, 1931, 2004.

2004SA2 Salabat, A. and Dashti, H., Phase compositions, viscosities and densities of systems PPG425 + Na_2SO_4 + H_2O and PPG425 + $(NH_4)_2SO_4$ + H_2O at 298.15 K, *Fluid Phase Equil.*, 216, 153, 2004.

2004SAN Sanchez, M.S., Hanykova, L., Ilavsky, M., and Pradas, M.M., Thermal transitions of poly(*N*-isopropylmethacrylamide) in aqueous solutions, *Polymer*, 45, 4087, 2004.

2004SHA Shang, Q.K., Li, W., Jia, Q., and Li, D.Q., Partitioning behavior of amino acids in aqueous two-phase systems containing polyethylene glycol and phosphate buffer, *Fluid Phase Equil.*, 219, 195, 2004.

2004SHI Shibayama, M., Isono, K., Okabe, S., Karino, T., and Nagao, M., SANS study on pressure-induced phase separation of poly(*N*-isopropylacrylamide) aqueous solutions and gels, *Macromolecules*, 37, 2909, 2004.

2004SOG Soga, O., Nostrum, C.F. van, and Hennik, W.E., Poly[*N*-(2-hydroxypropyl)methacrylamide mono/dilactate]: A new class of biodegradable polymers with tuneable thermosensitivity, *Biomacromolecules*, 5, 818, 2004.

2004SUG Sugihara, S., Kanaoka, S., and Aoshima, S., Thermosensitive random copolymers of hydrophilic and hydrophobic monomers obtained by living cationic copolymerization, *Macromolecules*, 37, 1711, 2004.

2004TAD Tada, E. dos S., Loh, W., and Pessoa-Filho, P. de A., Phase equilibrium in aqueous two-phase systems containing ethylene oxide-propylene oxide block copolymers and dextran, *Fluid Phase Equil.*, 218, 221, 2004.

2004TAK Takeda, N., Nakamura, E., Yokoyama, M., and Okano, T., Temperature-responsive polymeric carriers incorporating hydrophobic monomers for effective transfection in small doses, *J. Control. Release*, 95, 343, 2004.

2004TAN Tanaka, T. and Lloyd, D.R., Formation of poly(L-lactic acid) membranes via thermally induced phase separation, *J. Membrane Sci.*, 238, 65, 2004.

2004VAR Varade, D., Sharma, R., Aswal, V.K., Goyal, P.S., and Bahadur, P., Effect of hydrotopes on the solution behavior of PEO/PPO/PEO block copolymer L62 in aqueous solutions, *Eur. Polym. J.*, 40, 2457, 2004.

2004VIV Vivares, D. and Bonnete, F., Liquid-liquid phase separations in urate oxidase/PEG mixtures: Characterization and implications for protein crystallization, *J. Phys. Chem. B*, 108, 6498, 2004.

2004VOL Volkova, I.F., Gorshkova, M.Yu., Izumrudov, V.A., and Stotskaya, L.L., Interaction of a polycation with divinyl ether-maleic anhydride copolymer in aqueous solutions (Russ.), *Vysokomol. Soedin., Ser. A*, 46, 1388, 2004.

2004WAZ Waziri, S.M., Abu-Sharkh, B.F., and Ali, S.A., Protein partitioning in aqueous two-phase systems composed of a pH-responsive copolymer and poly(ethylene glycol), *Biotechnol. Progr.*, 20, 526, 2004.

2004WEI Weiss-Malik, R.A., Solis, F.J., and Vernon, B.L., Independent control of lower critical solution temperature and swelling behavior with pH for poly(*N*-isopropylacrylamide-*co*-maleic acid), *J. Appl. Polym. Sci.*, 94, 2110, 2004.

2004YOS Yoshimura, S., Shirai, S., and Einaga, Y., Light-scattering characterization of the wormlike micelles of hexaoxyethylene dodecyl $C_{12}E_6$ and hexaoxyethylene tetradecyl $C_{14}E_6$ ethers in dilute aqueous solution, *J. Phys. Chem. B*, 108, 15477, 2004.

2004ZA1 Zafarani-Moattar, M.T. and Sadeghi, R., Phase behavior of aqueous two-phase PEG + NaOH system at different temperatures, *J. Chem. Eng. Data*, 49, 297, 2004.

2004ZA2 Zafarani-Moattar, M.T., Sadeghi, R., and Hamidi, A.A., Liquid–liquid equilibria of an aqueous two-phase system containing polyethylene glycol and sodium citrate: Experiment and correlation, *Fluid Phase Equil.*, 219, 149, 2004.

2004ZHO Zhou, S., Xu, C., Wang, J., Golas, P., and Batteas, J., Phase behavior of cationic hydroxyethyl cellulose-sodium dodecyl sulfate mixtures: Effects of molecular weight and ethylene oxide side chain length of polymers, *Langmuir*, 20, 8482, 2004.

2005AUB Aubrecht, K.B. and Grubbs, R.B., Synthesis and characterization of thermo-responsive amphiphilic block copolymers incorporating a poly(ethylene oxide-*stat*-propylene oxide) block, *J. Polym. Sci.: Part A: Polym. Chem.*, 43, 5156, 2005.

2005BAE Bae, S.J., Suh, J.M., Sohn, Y.S., Bae, Y.H., Kim, S.W., and Jeong, B., Thermogelling of poly(caprolactone-*b*-ethylene glycol-*b*-caprolactone) aqueous solutions, *Macromolecules*, 38, 5260, 2005.

2005BIS Bisht, H.S., Wan, L., Mao, G., and Oupicky, D., pH-Controlled association of PEG-containing terpolymers of *N*-isopropylacrylamide and 1-vinylimidazole, *Polymer*, 46, 7945, 2005.

2005BOL Bolognese, B., Nerli, B., and Pico, G., Application of the aqueous two-phase systems of ethylene and propylene oxide copolymer-maltodextrin for protein purification, *J. Chromatogr. B*, 814, 347, 2005.

2005BUM Bumbu, G.-G., Vasile, C., Chitanu, G.C., and Staikos, G., Interpolymer complexes between hydroxypropylcellulose and copolymers of maleic acid: A comparative study, *Macromol. Chem. Phys.*, 206, 540, 2005.

2005CAO Cao, Z., Liu, W., Gao, P., Yao, K., Li, H., and Wang, G., Toward an understanding of thermoresponsive transition behavior of hydrophobically modified *N*-isopropyl-acrylamide copolymer solution, *Polymer*, 46, 5268, 2005.

2005CAR Carter, S., Hunt, B., and Rimmer, S., Highly branched poly(*N*-isopropylacrylamide)s with imidazole end groups prepared by radical polymerization in the presence of a styryl monomer containing a dithioester group, *Macromolecules*, 38, 4595, 2005.

2005CAS Castro, B.D. and Aznar, M., Liquid-liquid equilibrium of water + PEG8000 + magnesium sulfate or sodium sulfate aqueous two-phase systems at 35°C: Experimental determination and thermodynamic modeling, *Braz. J. Chem. Eng.*, 22, 463, 2005.

2005CHA Chaibundit, C., Sumanatrakool, P., Chinchew, S., Kanatharana, P., Tattershall, C.E., Booth, C., and Yuan, X.-F., Association properties of diblock copolymer of ethylene oxide and 1,2-butylene oxide: $E_{17}B_{12}$ in aqueous solution, *J. Coll. Interface Sci.*, 283, 544, 2005.

2005CHE Chen, X., Ding, X., Zheng, Z., and Peng, Y., Thermosensitive polymeric vesicles self-assembled by PNIPAAm-*b*-PPG-*b*-PNIPAAm triblock copolymers, *Coll. Polym. Sci.*, 283, 452, 2005.

2005CIM Cimen, E.K., Rzaev, Z.M.O., and Piskin, E., Bioengineering functional copolymers. V. Synthesis, LCST, and thermal behavior of poly(*N*-isopropylacrylamide-*co*-p-vinyl-phenylboronic acid), *J. Appl. Polym. Sci.*, 95, 573, 2005.

2005DEN Deng, G., Yao, S.-L., and Lin, D.-Q., Partitioning of proteins using a hydrophobically modified ethylene oxide/SDS aqueous two-phase system, *World J. Microbiol. Biotechnol.*, 21, 1209, 2005.

2005DUB Dubovik, A.S., Makhaeva, E.E., Grinberg, V.Ya., and Khokhlov, A.R., Energetics of cooperative transitions of *N*-vinylcaprolactam polymers in aqueous solutions, *Macromol. Chem. Phys.*, 206, 915, 2005.

2005DUR Durme, K. van, Delellio, L., Kudryashov, E., Buckin, V., and Mele, B. van, Exploration of high-resolution ultrasonic spectroscopy as an analytical tool to study demixing and remixing in poly(*N*-isopropyl acrylamide)/water solutions, *J. Polym. Sci.: Part B: Polym. Phys.*, 43, 1283, 2005.

2005EIN Einaga, Y., Kusumoto, A., and Noda, A., Effects of hydrophobic chain length on the micelles of heptaoxyethylene hexadecyl $C_{16}E_7$ and octadecyl $C_{18}E_7$ ethers, *Polym. J.*, 37, 368, 2005.

2005GAO Gaoa, C., Möhwald, H., and Shen, J., Thermosensitive poly(allylamine)-*g*-poly(*N*-isopropylacrylamide): Synthesis, phase separation and particle formation, *Polymer*, 46, 4088, 2005.

2005GRA Gratson, G.M. and Lewis, J.A., Phase behavior and rheological properties of polyelectrolyte inks for direct-write assembly, *Langmuir*, 21, 457, 2005.

2005GUP Gupta, A., Mohanty, B., and Bohidar, H.B., Flory temperature and upper critical solution temperature of gelatin solutions, *Biomacromolecules*, 6, 1623, 2005.

2005HAM Hamada, N. and Einaga, Y., Effects of hydrophilic chain length on the characteristics of the micelles of octaoxyethylene tetradecyl $C_{14}E_8$, hexadecyl $C_{16}E_8$, and octadecyl $C_{18}E_8$ ethers, *J. Phys. Chem. B*, 109, 6990, 2005.

2005HEI Heijkants, R.G.J.C., Calck, R.V.van, DeGroot, J.H, Pennings, A.J., and Schouten, A.J., Phase transitions in segmented polyesterurethane-DMSO-water systems, *J. Polym. Sci.: Part B: Polym. Phys.*, 43, 716, 2005.

2005HEY Hey, M.J., Jackson. D.P., and Yan, H., The salting-out effect and phase separation in aqueous solutions of electrolytes and poly(ethylene glycol) (exp. data by M.J. Hey), *Polymer*, 46, 2567, 2005.

2005HO1 Holyst, R., Staniszewski, K., and Demyanchuk, I., Ordering in surfactant mixtures induced by polymers, *J. Phys. Chem. B*, 109, 4881, 2005

2005HO2 Holyst, R., Staniszewski, K., Patkowski, A., and Gapinski, J., Hidden minima of the Gibbs free energy revealed in a phase separation in polymer/surfactant/water mixture, *J. Phys. Chem. B*,109, 8533, 2005.

2005IHA Ihata, O., Kayaki, Y., and Ikariya, T., Aliphatic poly(urethane-amine)s synthesized by copolymerization of aziridines and supercritical carbon dioxide, *Macromolecules*, 38, 6429, 2005.

2005IMA Imanishi, K. and Einaga, Y., Effects of hydrophilic chain length on the characteristics of the micelles of pentaoxyethylene n-decyl $C_{10}E_5$ and hexaoxyethylene n-decyl $C_{10}E_6$ ethers, *J. Phys. Chem. B*, 109, 7574, 2005.

2005IZU Izumrudov, V.A., Gorshkova, M.Yu., and Volkova, I.F., Controlled phase separations in solutions of soluble polyelectrolyte complex of DIVEMA (copolymer of divinyl ether and maleic anhydride), *Eur. Polym. J.*, 41, 1251, 2005.

2005JON Jones, J.A., Novo, N., Flagler, K., Pagnucco, C.D., Carew, S., Cheong, C., Kong, Z., Burke, N.A.D., and Stoever, H.D.H., Thermoresponsive copolymers of methacrylic acid and poly(ethylene glycol) methyl ether methacrylate, *J. Polym. Sci.: Part A: Polym. Chem.*, 43, 6095, 2005.

2005KJ1 Kjøniksen, A.-L., Laukkanen, A., Galant, C., Knudsen, K.D., Tenhu, H., and Nyström, B., Association in aqueous solutions of a thermoresponsive PVCL-*g*-$C_{11}EO_{42}$ copolymer, *Macromolecules*, 38, 948, 2005.

2005KJ2 Kjøniksen, A.-L., Knudsen, K.D., and Nyström, B., Phase separation and structural properties of semidilute aqueous mixtures of ethyl(hydroxyethyl)cellulose and an ionic surfactant, *Eur. Polym. J.*, 41, 1954, 2005.

2005KOH Koh, A.Y.C. and Saunders, B.R., Small-angle neutron scattering study of temperature-induced emulsion gelation: The role of sticky microgel particles, *Langmuir*, 21, 6734, 2005.

2005LAU Laukkanen, A., Valtola, L., Winnik, F.M., and Tenhu, H., Thermosensitive graft copolymers of an amphiphilic macromonomer and *N*-vinylcaprolactam: Synthesis and solution properties in dilute aqueous solutions below and above the LCST, *Polymer*, 46, 7055, 2005.

2005LE2 Lee, B.H. and Vernon, B., Copolymers of *N*-isopropylacrylamide, HEMA-lactate and acrylic acid with time-dependent lower critical solution temperature as a bioresorbable carrier, *Polym. Int.*, 54, 418, 2005.

2005LIC Li, C., Meng, L.-Z., Lu, X.-J., Wu, Z.-Q., Zhang, L.-F., and He, Y.-B., Thermo- and pH-sensitivities of thiosemicarbazone-incorporated, fluorescent and amphiphilic poly(*N*-isopropylacrylamide), *Macromol. Chem. Phys.*, 206, 1870, 2005.

2005LIU Liu, S.Q., Tong, Y.W., and Yang, Y.-Y., Incorporation and in vitro release of doxorubicin in thermally sensitive micelles made from poly(*N*-isopropylacrylamide-*co*-*N,N*-dimethylacrylamide)-*b*-poly(D,L-lactide-*co*-glycolide) with varying compositions, *Biomaterials*, 26, 5064, 2005.

2005LUN Lundell, C., de Hoog, E.H.A., Tromp, R.H., and Hermansson, A.-M., Effects of confined geometry on phase-separated dextran/gelatine mixtures exposed to shear, *J. Coll. Interface Sci.*, 288, 222, 2005.

2005MAB Ma, B., Hu, M., Li, S., Jiang, Y., and Liu, Z., Liquid-liquid phase equilibrium in the system poly(ethylene glycol) + Cs_2CO_3 + H_2O, *J. Chem. Eng. Data*, 50, 792, 2005.

2005MAT Matsuda, Y., Miyazaki, Y., Sugihara, S., Aoshima, S., Saito, K., and Sato, T., Phase separation behavior of aqueous solutions of a thermoresponsive polymer (exp. data by Y. Matsuda), *J. Polym. Sci.: Part B: Polym. Phys.*, 43, 2937, 2005.

2005MOO Moody, M.L., Willauer, H.D., Griffin, S.T., Huddleston, J.G., and Rogers, R.D., Solvent property characterization of poly(ethylene glycol)/dextran aqueous biphasic systems using the free energy of transfer of a methylene group and a linear solvation energy relationship, *Ind. Eng. Chem. Res.*, 44, 3749, 2008.

2005MUR Murugesan, T. and Perumalsamy, M., Liquid-liquid equilibria of poly(ethylene glycol) 2000 + sodium citrate + water at (25, 30, 35, 40, and 45)°C, *J. Chem. Eng. Data*, 50, 1392, 2005.

2005NED Nedelcheva, A.N., Novakov, C.P., Miloshev, S.M., and Berlinova, I.V., Electrostatic self-assembly of thermally responsive zwitterionic poly(*N*-isopropylacrylamide) and poly(ethylene oxide) modified with ionic groups, *Polymer*, 46, 2059, 2005.

2005NIE Nies, E., Ramzi, A., Berghmans, H., Li, T., Heenan, R.K., and King, S.M., Composition fluctuations, phase behavior, and complex formation in poly(vinyl methyl ether)/D_2O investigated by small-angle neutron scattering, *Macromolecules*, 38, 915, 2005.

2005OL1 Olsson, M., Joabsson, F., and Piculell, L., Particle-induced phase separation in mixed polymer solutions, *Langmuir*, 21, 1560, 2005.

2005OL2 Olsson, M., Boström, G., Karlson, L., and Piculell, L., Added surfactant can change the phase behavior of aqueous polymer-particle mixtures, *Langmuir*, 21, 2743, 2005.

2005PAU Paul, A., Griffiths, P.C., James, R., Willock, D.J., and Rogueda, P.G., Explaining the phase behaviour of the pharmaceutically relevant polymers poly(ethylene glycol) and poly(vinyl pyrrolidone) in semi-fluorinated liquids, *J. Pharm. Pharmacol.*, 57, 973, 2005.

2005PRA Prasad, M., Moulik, S.P., Wardian, A.A., Moore, S., Bommel, A. van, and Palepu, R., Alkyl (C_{10}, C_{12}, C_{14} and C_{16}) triphenyl phosphonium bromide influenced cloud points of nonionic surfactants (Triton X 100, Brij 56 and Brij 97) and the polymer poly(vinyl methyl ether), *Coll. Polym. Sci.*, 283, 887, 2005.

2005RAY Ray, B., Okamoto, Y., Kamigito, M., Sawamoto, M., Seno, K., Kanaoka, S., and Aoshima, S., Effect of tacticity of poly(*N*-isopropylacrylamide) on the phase separation temperature of its aqueous solutions, *Polym. J.*, 37, 234, 2005.

2005SAD Sadeghi, R., Measurement and correlation of phase equilibria for several PVP + salt aqueous two-phase systems at 303.15K, *Fluid Phase Equil.*, 237, 40, 2005.

2005SAL Salabat, A., Shamshiri, L., and Sardrodi, J.J., Liquid-liquid equilibrium data, viscosities, and densities of aqueous mixtures of poly(propylene glycol) with tri-sodium citrate at 298.15 K, *J. Chem. Eng. Data*, 50, 154, 2005.

2005SET Seto, Y., Aoki, T., and Kunugi, S., Temperature- and pressure-responsive properties of L- and DL-forms of poly(*N*-(1-hydroxymethyl)propylmethacrylamide) in aqueous solutions, *Colloid Polym. Sci.*, 283, 1137, 2005.

2005SHI Shirai, S. and Einaga, Y., Wormlike micelles of polyoxyethylene dodecyl $C_{12}E_j$ and heptaoxyethylene alkyl C_iE_7 ethers. Hydrophobic and hydrophilic chain length dependence of the micellar characteristics, *Polym. J.*, 37, 913, 2005.

2005SIL Silva, L.H.M. da, Silva, M. do C.H. da, Mesquita, A.F., Nascimento, K.S. do, Coimbra, J.S.R., and Minim, L.A., Equilibrium phase behavior of triblock copolymer + salt + water two-phase systems at different temperatures and pH, *J. Chem. Eng. Data*, 50, 1457, 2005.

2005SP1 Spelzini, D., Rigatusso, R., Farruggia B., and Pico, G., Thermal aggregation of methyl cellulose in aqueous solution: A thermodynamic study and protein partitioning behaviour, *Cellulose*, 12, 293, 2005.

2005SP2 Spevacek, J., Phase separation in aqueous polymer solutions as studied by NMR methods, *Macromol. Symp.*, 222, 1, 2005.

2005STA Starovoytova, L., Spevacek, J., and Ilavsky, M., ^{1}H-NMR study of temperature-induced phase transitions in D_2O solutions of poly(*N*-isopropylmethacrylamide)/poly(*N*-isopropylacrylamide) mixtures and copolymers, *Polymer*, 46, 677, 2005.

2005SUG Sugaya, R., Thermodynamics of aqueous biopolymer solutions and fractionation of dextran, *Dissertation*, Johannes-Gutenberg-Universität Mainz, 2005.

2005TAB Taboada, M.E., Galleguillos, H.R., Graber, T.A., and Bolado, S., Compositions, densities, conductivities, and refractive indices of potassium chloride or/and sodium chloride + PEG4000 + water at 298.15 and liquid-liquid equilibrium of potassium chloride or sodium chloride + PEG4000 + water at 333.15 K, *J. Chem. Eng. Data*, 50, 264, 2005.

2005TAD Tada, E. dos S., Loh, W., and Pessoa-Filho, P. de A., Erratum to [*Fluid Phase Equilibr.* 218 (2004) 221–228], *Fluid Phase Equil.*, 231, 250, 2005.

2005TAO Tao, C.-T. and Young, T.-H., Phase behavior of poly(*N*-isopropylacrylamide) in water-methanol cononsolvent mixtures and its relevance to membrane formation, *Polymer*, 46, 10077, 2005.

2005TIE Tiera, M.J., Santos, G.R. dos, Oliveira Tiera, V.A. de, Vieira, N.A.B., Frolini, E., Silva, R.C. da, and Loh, W., Aqueous solution behavior of thermosensitive (*N*-isopropylacrylamide-acrylic acid-ethyl methacrylate) terpolymers, *Coll. Polym. Sci.*, 283, 662, 2005.

2005UGU Uguzdogan, E., Camh, T., Kabasakal, O.S., Patir, S., Öztürk, E., Denkbas, E.B., and Tuncel, A., A new temperature-sensitive polymer: Poly(ethoxypropylacrylamide), *Eur. Polym. J.*, 41, 2142, 2005.

2005VA1 Van Durme, K., Rahier, H., and Van Mele, B., Influence of additives on the thermo-responsive behavior of polymers in aqueous solution, *Macromolecules*, 38, 10155, 2005.

2005VA2 Van Durme, K., Loozen, E., Nies, E., and Van Mele, B., Phase behavior of poly(vinyl methyl ether) in deuterium oxide, *Macromolecules*, 38, 10234, 2005.

2005VER Verdonck, B., Gohy, J.-F., Khousakoun, E., Jerome, R., and DuPrez, F., Association behavior of thermo-responsive block copolymers based on poly(vinyl ethers), *Polymer*, 46, 9899, 2005.

2005VIE Vieira, N.A.B., Neto, J.R., and Tiera, M.J., Synthesis, characterization and solution properties of amphiphilic *N*-isopropylacrylamide–poly(ethylene glycol)–dodecyl methacrylate thermosensitive polymers, *Colloids Surfaces A*, 262, 251, 2005.

2005XIA Xia, Y., Yin, X., Burke, N.A.D., and Stoever, H.D., Thermal response of narrow-disperse poly(*N*-isopropylacrylamide) prepared by atom transfer radical polymerization, *Macromolecules*, 38, 5937, 2005.

2005XUY Xu, Y. and Li, L., Thermoreversible and salt-sensitive turbidity of methylcellulose in aqueous solution, *Polymer*, 46, 7410, 2005.

2005YAM Yamazaki, R., Iizuka, K., Hiraoka, K., and Nose, T., Phase behavior and mechanical properties of mixed poly(ethylene glycol) monododecyl ether aqueous solutions, *Macromol. Chem. Phys.*, 206, 439, 2005.

2005YI1 Yin, X. and Stoever, H.D.H., Probing the influence of polymer architecture on liquid-liquid phase transitions of aqueous poly(*N,N*-dimethylacrylamide) copolymer solutions, *Macromolecules*, 38, 2109, 2005.

2005YI2 Yin, X. and Stoever, H.D.H., Temperature-sensitive hydrogel microspheres formed by liquid-liquid phase transitions of aqueous solutions of poly(*N,N*-dimethylacrylamide-*co*-allyl methacrylate), *J. Polym. Sci.: Part A: Polym. Chem.*, 43, 1641, 2005.

2005YUE Yue, Z., Eccleston, M.E., and Slater, N.K.H., PEGylation and aqueous solution behaviour of pH responsive poly(L-lysine isophthalamide), *Polymer*, 46, 2497, 2005.

2005ZA1 Zafarani-Moattar, M.T. and Sadeghi, R., Phase diagram data for several PPG + salt aqueous biphasic systems at 25°C, *J. Chem. Eng. Data*, 50, 947, 2005.

2005ZA2 Zafarani-Moattar, M.T. and Sadeghi, R., Effect of temperature on the phase equilibrium of aqueous two-phase systems containing polyvinylpyrrolidone and disodium hydrogen phosphate or trisodium phosphate, *Fluid Phase Equil.*, 238, 129, 2005.

2005ZA3 Zafarani-Moattar, M.T. and Hamzehzadeh, Sh., Liquid-liquid equilibria of aqueous two-phase systems containing polyethylene glycol and sodium succinate or sodium formate, *CALPHAD*, 29, 1, 2005.

2005ZH1 Zhang, D., Macias, C., and Ortiz, C., Synthesis and solubility of (mono-)end-functionalized poly(2-hydroxyethyl ethacrylate-*g*-ethylene glycol) graft copolymers with varying macromolecular architectures, *Macromolecules*, 38, 2530, 2005.

2005ZH2 Zhang, R., Liu, J., Han, B., Wang, B., Sun, D., and He, J., Effect of PEO–PPO–PEO structure on the compressed ethylene-induced reverse micelle formation and water solubilization, *Polymer*, 46, 3936, 2005.

2005ZH3 Zhang, W., Shi, L., Wu, K., and An, Y., Thermoresponsive micellization of poly(ethylene glycol)-*b*-poly(*N*-isopropylacrylamide) in water, *Macromolecules*, 38, 5743, 2005.

2005ZH4 Zhang, W., Shi, L., Ma, R., An, Y., Xu, Y., and Wu, K., Micellization of thermo- and pH-responsive triblock copolymer of poly(ethylene glycol)-*b*-poly(4-vinylpyridine)-*b*-poly(*N*-isopropylacrylamide), *Macromolecules*, 38, 8850, 2005.

2005ZH5 Zhang, Y., Furyk, S., Bergbreiter, D.E., and Cremer, P.S., Specific ion effects on the water solubility of macromolecules: PNIPAM and the Hofmeister series, *J. Amer. Chem. Soc.*, 127, 14505, 2005.

2006ALA Alava, C. and Saunders, B.R., Polymer stabilisers for temperature-induced dispersion gelation: Versatility and control, *J. Colloid Interface Sci.*, 293, 93, 2006.

2006ANT Antonov, Yu.A. and Wolf, B.A., Phase behavior of aqueous solutions of bovine serum albumin in the presence of dextran, at rest, and under shear, *Biomacromolecules*, 7, 1562, 2006.

2006BAK Bakshi, M.S., Kaur, N., Mahajan, R.K., Singh, J., and Sing, N., Estimation of degree of counterion binding and related parameters of monomeric and dimeric cationic surfactants from cloud point measurements by using triblock polymer as probe, *Coll. Polym. Sci.*, 284, 879, 2006.

2006CAL Calciu, D., Eckelt, J., Haase, T., and Wolf, B.A., Inverse spin fractionation: A tool to fractionate sodium hyaluronate, *Biomacromolecules*, 7, 3544, 2006.

2006CAO Cao, Z., Liu, W., Ye, G., Zhao, X., Lin, X., Gao, P., and Yao, K., *N*-Isopropylacrylamide/2-hydroxyethyl methacrylate star diblock copolymers: Synthesis and thermoresponsive behavior, *Macromol. Chem. Phys.*, 207, 2329, 2006.

2006CAU Causse, J., Lagerge, S., Menorval, L.C. de, and Faure, S., Micellar solubilization of tributylphosphate in aqueous solutions of Pluronic block copolymers Part I. Effect of the copolymer structure and temperature on the phase behavior, *J. Colloid Interface Sci.*, 300, 713, 2006.

2006CHE Chethana, S., Rastogi, N.K., and Raghavarao, K.S.M.S., New aqueous two phase system comprising polyethylene glycol and xanthan, *Biotechnol. Lett.*, 28, 25, 2006.

2006CHO Choi, C., Chae, S.Y., and Nah, J.-W., Thermosensitive poly(*N*-isopropylacrylamide)-*b*-poly(ε-caprolactone) nanoparticles for efficient drug delivery system, *Polymer*, 47, 4571, 2006.

2006DAL Dalkas, G., Pagonis, K., and Bokias, G., Control of the lower critical solution temperature-type cononsolvency properties of poly(*N*-isopropylacrylamide) in water-dioxane mixtures through copolymerisation with acrylamide, *Polymer*, 47, 243, 2006.

2006DI1 Dincer, S., Rzaev, Z.M.O., and Piskin, E., Synthesis and characterization of stimuli-responsive poly(*N*-isopropylacrylamide-*co*-*N*-vinyl-2-pyrrolidone), *J. Polym. Res.*, 13, 121, 2006.

2006DI2 Ding, H., Wu, F., Huang, Y., Zhang, Z., and Nie, Y., Synthesis and characterization of temperature-responsive copolymer of PELGA modified poly(*N*-isopropyl-acrylamide), *Polymer*, 47, 1575, 2006.

2006DUA Duan, Q., Narumi, A., Miura, Y., Shen, X., Sato, S.-I., Satoh, T., and Kakuchi, T., Thermoresponsive property controlled by end-functionalization of poly(*N*-isopropyl-acrylamide) with phenyl, biphenyl, triphenyl groups, *Polym. J.*, 38, 306, 2006.

2006EDA Edahiro, J., Sumaru, K., Takagi, T., Shinbo, T., and Kanamori, T., Photoresponse of an aqueous two-phase system composed of photochromic dextran, *Langmuir*, 22, 5224, 2006.

2006EIN Einaga, Y., Inaba, Y., and Syakado, M., Wormlike micelles of pentaoxyethylene tetradecyl ether $C_{14}E_5$ and hexaoxyethylene hexadecyl ether $C_{16}E_6$, *Polym. J.*, 38, 64, 2006.

2006GEE Geever, L.M., Devine, D.M., Nugent, M.J.D., Kennedy, J.E., Lyons, J.G., Hanley, A., and Higginbotham, C.L., Lower critical solution temperature control and swelling behaviour of physically crosslinked thermosensitive copolymers based on *N*-isopropylacrylamide, *Eur. Polym. J.*, 42, 2540, 2006.

2006HAB Haba, Y., Kojima, C., Harada, A., and Kono, K., Control of temperature-sensitive properties of poly(amidoamine) dendrimers using peripheral modification with various alkylamide groups, *Macromolecules*, 39, 7451, 2006.

2006HAL Halacheva, S., Rangelov, S., and Tsvetanov, C., Poly(glycidol)-based analogues to pluronic block copolymers. Synthesis and aqueous solution properties, *Macromolecules*, 39, 6845, 2006.

2006HER Hernandez-Mireles, T. and Rito-Palomares, M., New aqueous two-phase systems based on poly(ethylene oxide sulfide) (PEOS) and potassium phosphate for the potential recovery of proteins, *J. Chem. Technol. Biotechnol.*, 81, 997, 2006.

2006HU1 Hua, F., Jiang, X., and Zhao, B., Well-defined thermosensitive, water-soluble polyacrylates and polystyrenics with short pendant oligo(ethylene glycol) groups synthesized by nitroxide-mediated radical polymerization, *J. Polym. Sci.: Part A: Polym. Chem.*, 44, 2454, 2006.

2006HU2 Hua, F., Jiang, X., and Zhao, B., Temperature-induced self-association of doubly thermosensitive diblock copolymers with pendant methoxytris(oxyethylene) groups in dilute aqueous solutions, *Macromolecules*, 39, 3476, 2006.

2006INO Inoue, T. and Yamashita, K., Aggregation behavior of polypropylene oxide with electric charges at both ends in aqueous solution, *J. Colloid Interface Sci.*, 300, 774, 2006.

2006IZU Izumrudov, V.A. and Sybachin, A.V., Phase separation in solutions of polyelectrolyte complexes: The decisive effect of a host polyion, *Polym. Sci., Ser. A*, 48, 1849, 2006.

2006JHO Jhon, Y.K., Bhat, R.R., Jeong, C., Rojas, O.J., Szleifer, I., and Genzer, J., Salt-induced depression of lower critical solution temperature in a surface-grafted neutral thermoresponsive polymer, *Macromol. Rapid Commun.*, 27, 697, 2006.

2006KI1 Kim, Y.-C., Kil, D.-S., and Kim, J.C., Synthesis and phase separation of poly(*N*-isopropylacrylamide-*co*-methoxy polyethyleneglycol monomethacrylate), *J. Appl. Polym. Sci.*, 101, 1833, 2006.

2006KI2 Kim, Y.C., Bang, M.-S., and Kim, J.-C., Synthesis and characterization of poly(*N*-isopropyl acrylamide) copolymer with methoxy polyethyleneglycol monomethacrylate, *J. Ind. Eng. Chem.*, 12, 446, 2006.

2006KIR Kirpach, A. and Adolf, D., High pressure induced coil-globule transitions of smart polymers, *Macromol. Symp.*, 237, 7, 2006.

2006KUJ Kujawa, P., Segui, F., Shaban, S., Diab, C., Okada, Y., Tanaka, F., and Winnik, F.M., Impact of end-group association and main-chain hydration on the thermosensitive properties of hydrophobically modified telechelic poly(*N*-isopropylacrylamides) in water, *Macromolecules*, 39, 341, 2006.

2006LAZ Lazzara, G., Milioto, S., and Gradzielski, M., The solubilisation behaviour of some dichloroalkanes in aqueous solutions of PEO–PPO–PEO triblock copolymers: A dynamic light scattering, fluorescence spectroscopy, and SANS study, *Phys. Chem. Chem. Phys.*, 8, 2299, 2006.

2006LIC Li, C., Lee, D.H., Kim, J.K., Ryu, D.Y., and Russel, T.P., Closed-loop phase behavior for weakly interacting block copolymers, *Macromolecules*, 39, 5926, 2006.

2006LIN Lin, D.-J., Chang, H.-H., Chen, T.-C., Lee, Y.-C., and Cheng, L.-P., Formation of porous poly(vinylidene fluoride) membranes with symmetric or asymmetric morphology by immersion precipitation in the water/TEP/PVDF system, *Eur. Polym. J.*, 42, 1581, 2006.

2006LOO Loozen, E., Nies, E., Heremans, K., and Berghmans, H., The influence of pressure on the lower critical solution temperature miscibility behavior of aqueous solutions of poly(vinyl methyl ether) and the relation to the compositional curvature of the volume of mixing, *J. Phys. Chem. B*, 110, 7793, 2006.

2006LOZ Lozinskii, V.I., Simenel, I.A., Semenova, M.G., Belyakova, L.E., Ilin, M.M., Grinberg, V. Ya., Dubovik, A.S., and Khokhlov, A.R., Behavior of protein-like *N*-vinylcaprolactam and *N*-vinylimidazole copolymers in aqueous solutions, *Polym. Sci., Ser. A*, 48, 435, 2006.

2006LUC Lu, C., Guo, S.-R., Zhang, Y., and Yin, M., Synthesis and aggregation behavior of four different shaped PCL-PEG block copolymers, *Polym. Int.*, 55, 694, 2006.

2006LUT Lutz, J.-F., Akdemir, O., and Hoth, A., Point by point comparison of two thermosensitive polymers exhibiting a similar LCST: Is the age of poly(NIPAM) over?, *J. Amer. Chem. Soc.*, 128, 13046, 2006.

2006MA1 Maeda, T., Kanda, T., Yonekura, Y., Yamamoto, Y., and Aoyagi, T., Hydroxylated poly(*N*-isopropylacrylamide) as functional thermoresponsive materials, *Biomacromolecules*, 7, 545, 2006.

2006MA2 Maeda, T., Yamamoto, K., and Aoyagi, T., Importance of bound water in hydration–dehydration behavior of hydroxylated poly(*N*-isopropylacrylamide), *J. Colloid Interface Sci.*, 302, 467, 2006.

2006MOH Mohsen-Nia, M., Rasa, H., and Modarress, H., Cloud-point measurements for {water + poly(ethylene glycol) + salt} ternary mixtures by refractometry method, *J. Chem. Eng. Data*, 51, 1316, 2006.

2006MOR Mori, T., Nakashima, M., Fukuda, Y., Minagawa, K., Tanaka, M., and Maeda, Y., Soluble-insoluble-soluble transitions of aqueous poly(*N*-vinylacetamide-*co*-acrylic acid) solutions, *Langmuir*, 22, 4336, 2006.

2006MUN Mun, G.A., Nurkeeva, Z.S., Akhmetkalieva, G.T., Shmakov, S.N., Khutoryanskiy, V.V., Lee, S.C., and Park, K., Novel temperature-responsive water-soluble copolymers based on 2-hydroxyethylacrylate and vinyl butyl ether and their interactions with poly(carboxylic acids), *J. Polym. Sci.: Part B: Polym. Phys.*, 44, 195, 2006.

2006MYL Mylonas, Y., Bokias, G., Iliopoulos, I., and Staikos, G., Interpolymer association between hydrophobically modified poly(sodium acrylate) and poly(*N*-isopropylacrylamide) in water: The role of hydrophobic interactions and polymer structure, *Eur. Polym. J.*, 42, 849, 2006.

2006NIE Nies, E., Li, T., Berghmans, H., Heenan, R.K., and King, S.M., Upper critical solution temperature phase behavior, composition fluctuations, and complex formation in poly(vinyl methyl ether)/D_2O solutions: Small-angle neutron-scattering experiments and Wertheim lattice thermodynamic perturbation theory predictions, *J. Phys. Chem. B*, 110, 5321, 2006.

2006OKA Okada, Y., Tanaka, F., Kujawa, P., and Winnik, F.M., Unified model of association-induced lower critical solution temperature phase separation and its application to solutions of telechelic poly(ethylene oxide) and of telechelic poly(N-isopropylacrylamide) in water, *J. Chem. Phys.*, 125, 244902, 2006.

2006OSA Osaka, N., Okabe, S., Karino, T., Hirabaru, Y., Aoshima, S., and Shibayama, M., Micro- and macrophase separations of hydrophobically solvated block copolymer aqueous solutions induced by pressure and temperature, *Macromolecules*, 39, 5875, 2006.

2006PAG Pagonis, K. and Bokias, G., Simultaneous lower and upper critical solution temperature-type co-nonsolvency behaviour exhibited in water-dioxane mixtures by linear copolymers and hydrogels containing *N*-isopropylacrylamide and *N,N*-dimethylacrylamide, *Polym. Int.*, 55, 1254, 2006.

2006PAR Parrott, M.C., Valliant, J.F., and Adronov, A., Thermally induced phase separation of carborane functionalized aliphatic polyester dendrimers in aqueous media, *Langmuir*, 22, 5251, 2006.

2006PER Perumalsamy, M. and Murugesan, T., Prediction of liquid-liquid equilibria for PEG2000-sodium citrate based aqueous two-phase systems, *Fluid Phase Equil.*, 244, 52, 2006.

2006PLU Plummer, R., Hill, D.J.T., and Whittaker, A.K., Solution properties of star and linear poly(*N*-isopropylacrylamide), *Macromolecules*, 39, 8379, 2006.

2006QIN Qin, S., Geng, Y., Discher, D.E., and Yang, S., Temperature-controlled assembly and release from polymer vesicles of poly(ethylene oxide)-*block*-poly(*N*-isopropyl acrylamide), *Adv. Mater.*, 18, 2905, 2006.

2006QIU Qiu, G.-M., Zhu, B.-K., Xu, Y.-Y., and Geckeler, K.E., Synthesis of ultrahigh molecular weight poly(styrene-*alt*-maleic anhydride) in supercritical carbon dioxide, *Macromolecules*, 39, 3231, 2006.

2006SAL Salabat, A., Moghadasi, M.A., Zalaghi, P., and Sadeghi, R., (Liquid + liquid) equilibria for ternary mixtures of (polyvinylpyrrolidone + $MgSO_4$ + water) at different temperatures, *J. Chem. Thermodyn.*, 38, 1479, 2006.

2006SA1 Sadeghi, R., Aqueous two-phase systems of poly(vinylpyrrolidone) and potassium citrate at different temperatures. Experimental results and modeling of liquid-liquid equilibrium data, *Fluid Phase Equil.*, 246, 89, 2006.

2006SA2 Sadeghi, R., Rafiei, H.R., and Motamedi, M., Phase equilibrium in aqueous two-phase systems containing poly(vinylpyrrolidone) and sodium citrate at different temperatures. Experimental and modeling, *Thermochim. Acta*, 451, 163, 2006.

2006SAR Saravanan, S., Reena, J.A., Rao, J.R., Murugesan, T., and Nair, B.U., Phase equilibrium compositions, densities, and viscosities of aqueous two-phase poly(ethylene glycol) + poly(acrylic acid) system at various temperatures, *J. Chem. Eng. Data*, 51, 1246, 2006.

2006SCH Schagerlöf, H., Johansson, M., Richardson, S., Brinkmalm, G., Wittgren, B., and Tjerneld, F., Substituent distribution and clouding behavior of hydroxypropyl methyl cellulose analyzed using enzymatic degradation, *Biomacromolecules*, 7, 3474, 2006.

2006SEE Seetapan, N., Mai-ngam, K., Plucktaveesak, N., and Sirivat, A., Linear viscoelasticity of thermoassociative chitosan-*g*-poly(*N*-isopropylacrylamide) copolymer, *Rheol. Acta*, 45, 1011, 2006.

2006SHA Sharma, S.C., Acharya, D.P., Garcia-Roman, M., Itami, Y., and Kunieda, H., Phase behavior and surface tensions of amphiphilic fluorinated random copolymer aqueous solutions, *Coll. Surfaces A*, 280, 140, 2006.

2006SHE Shen, Z., Terao, K., Maki, Y., Dobashi, T., Ma, G., and Yamamoto, T., Synthesis and phase behavior of aqueous poly(*N*-isopropylacrylamide-*co*-acrylamide), poly(*N*-iso-propylacrylamide-*co*-*N,N*-dimethylacrylamide) and poly(*N*-isopropylacrylamide-*co*-2-hydroxyethyl methacrylate), *Colloid Polym. Sci.*, 284, 1001, 2006.

2006SIL Silva, M. do C.H. da, Silva, L.H.M. da, Amim, J., Guimaraes, R.O., and Martins, J.P., Liquid-liquid equilibrium of aqueous mixture of triblock copolymers L35 and F68 with Na_2SO_4, Li_2SO_4, or $MgSO_4$, *J. Chem. Eng. Data*, 51, 2260, 2006.

2006SUG Sugi, R., Ohishi, T., Yokoyama, A., and Yokozawa, T., Novel water-soluble poly(m-benzamide)s: Precision synthesis and thermosensitivity in aqueous solution, *Macromol. Rapid Commun.*, 27, 716, 2006.

2006THO Tho, I., Kjøniksen, A.-L., Knudsen, K.-D., and Nyström, B., Effect of solvent composition on the association behavior of pectin in methanol–water mixtures, *Eur. Polym. J.*, 42, 1164, 2006.

2006TON Tono, Y., Kojima, C., Haba, Y., Takahashi, T., Harada, A., Yagi, S., and Kono, K., Thermosensitive properties of poly(amidoamine) dendrimers with peripheral phenylalanine residues, *Langmuir*, 22, 4920, 2006.

2006TUB Tubio, G., Pellegrini, L., Nerli, B.B., and Pico, G.A., Liquid-liquid equilibria of aqueous two-phase systems containing poly(ethylene glycols) of different molecular weight and sodium citrate, *J. Chem. Eng. Data*, 51, 209, 2006.

2006VAN Van Durme, K., Van Mele, B., Bernaerts, K.V., Verdonck, B., and DuPrez, F.E., End-group modified poly(methyl vinyl ether): Characterization and LCST demixing behavior in water, *J. Polym. Sci.: Part B: Polym. Phys.*, 44, 461, 2006

2006VER Verezhnikov, V.N., Plaksitskaya, T.V., and Poyarkova, T.N., pH-Thermosensitive behavior of *N,N*-dimethylaminoethyl methacrylate (co)polymers with *N*-vinylcaprolactam, *Polym. Sci., Ser. A*, 48, 870, 2006.

2006WEI Wei, H., Zhang, X.-Z., Zhou, Y., Cheng, S.-X., and Zhuo, R.-X., Self-assembled thermoresponsive micelles of poly(*N*-isopropylacrylamide-*b*-methyl methacrylate), *Biomaterials*, 27, 2028, 2006.

2006WEN Weng, Y., Ding, Y., and Zhang, G., Microcalorimetric investigation on the lower critical solution temperature behavior of *N*-isopropylacrylamide-*co*-acrylic acid copolymer in aqueous solution, *J. Phys. Chem. B*, 110, 11813, 2006.

2006WON Wongmongkol, N. and Prichanont, S., Partition of alkaline protease in aqueous two-phase systems of poly(ethylene glycol) 1000 and potassium phosphate, *Korean J. Chem. Eng.*, 23, 71, 2006.

2006XIA Xia, Y., Burke, N.A.D., and Stoever, H.D.H., End group effect on the thermal response of narrow-disperse poly(*N*-isopropylacrylamide) prepared by atom transfer radical polymerization, *Macromolecules*, 39, 2275, 2006.

2006YAN Yankov, D.S., Stateva, R.P., Trusler, J.P.M., and Cholakov, G.St., Liquid-liquid equilibria in aqueous two-phase systems of poly(ethylene glycol) and poly(ethylenimine): Experimental measurements and correlation, *J. Chem. Eng. Data*, 51, 1056, 2006.

2006YIN Yin, X., Hoffman, A.S., and Stayton, P.S., Poly(*N*-isopropylacrylamide-*co*-propylacrylic acid) copolymers that respond sharply to temperature and pH, *Biomacromolecules*, 7, 1381, 2006.

2006ZH1 Zhang, Y., Trabbic-Carlson, K., Albertorio, F., Chilkoti, A, and Cremer, P.S., Aqueous two-phase system formation kinetics for elastin-like polypeptides of varying chain length, *Biomacromolecules*, 7, 2192, 2006.

2006ZH2 Zhao, G. and Chen, S.B., Clouding and phase behavior of nonionic surfactants in hydrophobically modified hydroxyethyl cellulose solutions, *Langmuir*, 22, 9129, 2006.

2007AN1 Antonov, Y., Lefebvre, J., and Doublier, J.-L., Phase separation in aqueous casein-guar gum systems, *Polym. Bull.*, 58, 723, 2007.

2007AN2 Antonov, Y. and Friedrich, C., Aqueous phase-separated biopolymer mixture compatibilized by physical interactions of the constituents, *Polym. Bull.*, 58, 969, 2007.

2007BAR Barzin, J. and Sadatnia, B., Theoretical phase diagram calculation and membrane morphology evaluation for water/solvent/polyethersulfone systems, *Polymer*, 48, 1620, 2007.

2007BHA Bharatiya, B., Guo, C., Ma, J.H., Hassan, P.A., and Bahadur, P., Aggregation and clouding behavior of aqueous solution of EO–PO block copolymer in presence of n-alkanols, *Eur. Polym. J.*, 43, 1883, 2007.

2007CAO Cao, Y., Zhu, X.X., Luo, J., and Liu, H., Effects of substitution groups on the RAFT polymerization of *N*-alkylacrylamides in the preparation of thermosensitive block copolymers, *Macromolecules*, 40, 6481, 2007.

2007CAR Carvalho, C.P., Coimbra, J.S.R., Costa, I.A.F., Minim, L.A., Silva, L.H.M., and Maffia, M.C., Equilibrium data for PEG 4000 + salt + water systems from (278.15 to 318.15) K, *J. Chem. Eng. Data*, 52, 351, 2007.

2007DAL Dallora, N.L.P., Klemz, J.G.D., and Filho, P. de A.P., Partitioning of model proteins in aqueous two-phase systems containing polyethylene glycol and ammonium carbamate, *Biochem. Eng. J.*, 34, 92, 2007.

2007DE1 Derkaoui, N., Said, S., Grohens, Y., Olier, R., and Privat, M., PEG400 novel phase description in water, *J. Colloid Interface Sci.*, 305, 330, 2007.

2007DE2 DeRosa, M.E., DeRosa, R.L., Noni, L.M., and Hendrick, E.S., Phase separation of poly(*N*-isopropylacrylamide) solutions and gels using a near infrared fiber laser, *J. Appl. Polym. Sci.*, 105, 2083, 2007.

2007DIM Dimitrov, P., Jamroz-Piegza, M., Trzebicka, B., and Dworak, A., The influence of hydrophobic substitution on self-association of poly(ethylene oxide)-*b*-poly(n-alkyl glycidyl carbamate)s-*b*-poly(ethylene oxide) triblock copolymers in aqueous media, *Polymer*, 48, 1866, 2007.

2007ERB Erbil, C., Gökcören, A.T., and Polat, Y.A., *N*-isopropylacrylamide-acrylamide copolymers initiated by ceric ammonium nitrate in water, *Polym. Int.*, 56, 547, 2007.

2007FAR Faravash, R.S., Modarress, H., and Nasernejad, B., Structural and partitioning studies of bovine serum albumin in mixture of {poly(ethylene glycol) + K_2HPO_4 + H_2O}, *J. Chem. Eng. Data*, 52, 71, 2007.

2007FED Fedicheva, N., Ninni, L., and Maurer, G., Aqueous two-phase systems of poly(vinyl pyrrolidone) and sodium sulfate: Experimental results and correlation/prediction, *J. Chem. Eng. Data*, 52, 1858, 2007.

2007FOR Foroutan, M., Liquid-liquid equilibria of aqueous two-phase poly(vinylpyrrolidone) and K_2HPO_4/KH_2PO_4 buffer: Effects of pH and temperature, *J. Chem. Eng. Data*, 52, 859, 2007.

2007FOU Fournier, D., Hoogenboom, R., Thijs, H.M.L., Paulus, R.M., and Schubert, U.S., Tunable pH- and temperature-sensitive copolymer libraries by reversible addition-fragmentation chain transfer copolymerizations of methacrylates, *Macromolecules*, 40, 915, 2007.

2007FRA Francis, R., Jijil, C.P., Prabhu, C.A., and Suresh, C.H., Synthesis of poly(*N*-isopropylacrylamide) copolymer containing anhydride and imide comonomers: A theoretical study on reversal of LCST, *Polymer*, 48, 6707, 2007.

2007GRA Graber, T. A., Medina, H., Galleguillos, H.R., and Taboada, M.E., Phase equilibrium and partition of iodide in an aqueous biphasic system formed by $(NH_4)_2SO_4$ + PEG + H_2O at 25°C, *J. Chem. Eng. Data*, 52, 1262, 2007.

2007HOU Housni, A. and Narain, R., Aqueous solution behavior of poly(*N*-isopropyl acrylamide) in the presence of water-soluble macromolecular species, *Eur. Polym. J.*, 43, 4344, 2007.

2007IMA Imanishi, K. and Einaga, Y., Wormlike micelles of polyoxyethylene alkyl ether mixtures $C_{10}E_5$ + $C_{14}E_5$ and $C_{14}E_5$ + $C_{14}E_7$: Hydrophobic and hydrophilic chain length dependence of the micellar characteristics, *J. Phys. Chem. B*, 111, 62, 2007.

2007JAY Jayapal, M., Regupathi, I., and Murugesan, T., Liquid-liquid equilibrium of poly(ethylene glycol) 2000 + potassium citrate + water at (25, 35, and 45)°C, *J. Chem. Eng. Data*, 52, 56, 2007.

2007KAT Kataoka, T., Nagaoa, Y., Kidowaki, M., Araki, J., and Ito, K., Liquid-liquid equilibria of polyrotaxane and poly(vinyl alcohol), *Colloids Surfaces B: Biointerfaces*, 56, 270, 2007.

2007KON Kong, F.Q., Cao, X., Xia, J., and Hur, B.-K., Synthesis and application of a light-sensitive polymer forming aqueous two-phase systems, *J. Ind. Eng. Chem.*, 13, 424, 2007.

2007LIU Liu, H., Chen, Y., and Shen, Z., Thermoresponsive hyperbranched polyethylenimines with isobutyramide functional groups, *J. Polym. Sci.: Part A: Polym. Chem.*, 45, 1177, 2007.

2007LUT Lutz, J.-F., Weichenhan, K., Akdemir, O., and Hoth, A., About the phase transitions in aqueous solutions of thermoresponsive copolymers and hydrogels based on 2-(2-methoxyethoxy)ethyl methacrylate and oligo(ethylene glycol) methacrylate, *Macromolecules*, 40, 2503, 2007.

2007MAE Maeda, Y., Yamauchi, H., Fujisawa, M., Sugihara, S., and Ikeda, I., Infrared spectroscopic investigation of poly(2-methoxyethyl vinyl ether) during thermo-sensitive phase separation in water, *Langmuir*, 23, 6561, 2007.

2007MAN Mansur, C., Spinelli, L., Gonzalez, G., and Lucas, E.F., Evaluation of the physical-chemical properties of poly(ethylene oxide)-*b*-poly(propylene oxide) by different characterization techniques, *Macromol. Symp.*, 258, 5, 2007.

2007MAO Mao, Z., Ma, L., Yan, J., Yan, M., Gao, C., and Shen, J., The gene transfection efficiency of thermoresponsive *N,N,N*-trimethylchitosan chloride-*g*-poly(*N*-isopropyl acrylamide) copolymer, *Biomaterials*, 28, 4488, 2007.

2007MIY Miyake, M. and Einaga, Y., Characteristics of wormlike pentaoxyethylene decyl ether $C_{10}E_5$ micelles containing n-dodecanol, *J. Phys. Chem. B*, 111, 535, 2007.

2007MUN Mun, G.A., Nurkeeva, Z.S., Beissegul, A.B., Dubolazov, A.V., Urkimbaeva, P.I., Park, K., and Khutoryanskiy, V.V., Temperature-responsive water-soluble copolymers based on 2-hydroxyethyl acrylate and butyl acrylate, *Macromol. Chem. Phys.*, 208, 979, 2007.

2007NAK Nakhmanovich, B.I., Pakuro, N.I., Akhmet'eva, E.I., Litvinenko, G.I., and Arest-Yakubovich, A.A., Thermal sensitivity of poly(*N*-vinylpyrrolidone) solutions in water-saline media, *Polym. Sci., Ser. B*, 49, 136, 2007.

2007NOR Norman, A.I., Ho, D.L., and Greer, S.C., Partitioning, fractionation, and conformations of star poly(ethylene glycol) in isobutyric acid and water, *Macromolecules*, 40, 9628, 2007.

2007OEZ Oezyuerek, Z., Komber, H., Gramm, S., Schmaljohann, D., Mueller, A.H.E., and Voit, B., Thermoresponsive glycopolymers via controlled radical polymerization, *Macromol. Chem. Phys.*, 208, 1035, 2007.

2007OLI Oliveira, M.C. de, Filho, M.A.N. de A., and Filho, P. de A.P., Phase equilibrium and protein partitioning in aqueous two-phase systems containing ammonium carbamate and block copolymers PEO–PPO–PEO, *Biochem. Eng. J.*, 37, 311, 2007.

2007PAR Park, J.-S. and Kataoka, K., Comprehensive and accurate control of thermosensitivity of poly(2-alkyl-2-oxazoline)s via well-defined gradient or random copolymerization, *Macromolecules*, 40, 3599, 2007.

2007PER Perumalsamy, M., Bathmalakshmi, A., and Murugesan, T., Experiment and correlation of liquid-liquid equilibria of an aqueous salt polymer system containing PEG6000 + sodium citrate, *J. Chem. Eng. Data*, 52, 1186, 2007.

2007PL1 Plashchina, I.G., Zhuravleva, I.L., and Antonov, Y.A., Phase behavior of gelatin in the presence of pectin in water-acid medium, *Polym. Bull.*, 58, 587, 2007.

2007PL2 Plaksitskaya, T. V., Verezhnikov, V. N., Poyarkova, T. N., Panarina, T. V., and Ivanov, V. N., Spontaneous change in thermosensitive properties of aqueous solutions of poly(*N,N*-dimethylaminoethyl methacrylate) and its mixtures with sodium dodecyl sulfate, *Polym. Sci., Ser. A*, 49, 1538, 2007.

2007PL3 Plamper, F.A., Ruppel, M., Schmalz, A., Borisov, O., Ballauff, M., and Müller, A.H.E., Tuning the thermoresponsive properties of weak polyelectrolytes: Aqueous solutions of star-shaped and linear poly(*N,N*-dimethylaminoethyl methacrylate), *Macromolecules*, 40, 8361, 2007.

2007POR Porto, T.S., Pessoa-Filho, P.A., Neto, B.B., Filho, J.L.L., Converti, A., Porto, A.L.F., and Pessoa, A., Removal of proteases from Clostridium perfringens fermented broth by aqueous two-phase systems, *J. Ind. Microbiol. Biotechnol.*, 34, 547, 2007.

2007QIU Qiu, X.-P., Tanaka, F., and Winnik, F.M., Temperature-induced phase transition of well-defined cyclic poly(*N*-isopropylacrylamide)s in aqueous solution, *Macromolecules*, 40, 7069, 2007.

2007REN Ren, L. and Agarwal, S., Synthesis, characterization, and properties evaluation of poly[(*N*-isopropylacrylamide)-*co*-ester]s, *Macromol. Chem. Phys.*, 208, 245, 2007.

2007ROD Rodriguez, O., Silverio, S.C., Madeira, P.P., Teixeira, J.A., and Macedo, E.A., Physicochemical characterization of the PEG8000-Na2SO4 aqueous two-phase system, *Ind. Eng. Chem. Res.*, 46, 8199, 2007.

2007SAL Salabat, A., Abnosi, M.H., and Bahar, A.R., Partitioning of some amino acids in aqueous two-phase system of poly(propylene glycol) and magnesium sulfate, *J. Chromatogr. B*, 858, 234, 2007.

2007SCH Schaink, H.M. and Smit, J.A.M., Protein-polysaccharide interactions: The determination of the osmotic second virial coefficients in aqueous solutions of β-lactoglobulin and dextran, *Food Hydrocoll.*, 21, 1389, 2007.

2007SH1 Shao, D. and Ni, C., Preparations and properties of thermosensitive terpolymers of *N*-isopropylacrylamide, sodium 2-acrylamido-2-methyl-propanesuphonate, and *N-tert*-butylacrylamide, *J. Appl. Polym. Sci.*, 105, 2299, 2007.

2007SH2 Shatalov, G.V., Churilina, E.V., Kuznetsov, V.A., and Verezhnikov, V.N., Copolymerization of *N*-vinylcaprolactam with *N*-vinyl(benz)imidazoles and the properties of aqueous solutions of the copolymers, *Polym. Sci., Ser. B*, 49, 57, 2007.

2007SHI Shibayama, M., Studies on pressure induced phase separation and hydrophobic interaction of polymer solutions and gels by neutron scattering and light scattering (Jap.), *Koatsuryoko Kagaku Gijutsu*, 17, 131, 2007.

2007SIL Silva, T.M., Minim, L.A., Maffia, M.C., Coimbra, J.S.R., Minim, V.P.R., and Mendes da Silva, L.H., Equilibrium data for poly(propylene glycol) + sucrose + water and poly(propylene glycol) + fructose + water systems from (15 to 45)°C, *J. Chem. Eng. Data*, 52, 1649, 2007.

2007SKR Skrabania, K., Kristen, J., Laschewsky, A., Akdemir, O., Hoth, A., and Lutz, J.-F., Design, synthesis, and aqueous aggregation behavior of nonionic single and multiple thermoresponsive polymers, *Langmuir*, 23, 84, 2007.

2007TAK Takahashi, N., Kanaya, T., Nishida, K., and Kaji, K., Gelation-induced phase separation of poly(vinyl alcohol) in mixed solvents of dimethyl sulfoxide and water, *Macromolecules*, 40, 8750, 2007.

2007VA1 Van Durme, K., Van Assche, G., Nies, E., and Van Mele, B., Phase transformations in aqueous low molar mass poly(vinyl methyl ether) solutions: Theoretical prediction and experimental validation of the peculiar solvent melting line, bimodal LCST, and (adjacent) UCST miscibility gaps, *J. Phys. Chem. B*, 111, 1288, 2007.

2007VA2 Van Durme, K., Van Assche, G., Aseyev, V., Raula, J., Tenhu, H., and Van Mele, B., Influence of macromolecular architecture on the thermal response rate of amphiphilic copolymers, based on poly(*N*-isopropylacrylamide) and poly(oxyethylene), in water, *Macromolecules*, 40, 3765, 2007.

2007VS1 Vshivkov, S.A., Adamova, L.V., Rusinova, E.V., Safronov, A.P., Dreval, V.E., and Galyas, A.G., Thermodynamics of liquid-crystalline solutions of hydroxypropyl cellulose in water and ethanol, *Polym. Sci., Ser. A*, 49, 578, 2007.

2007VS2 Vshivkov, S.A. and Rusinova, E.V., Phase diagrams of a hydroxypropyl cellulose–water system under static conditions and in the shear field, *Polym. Sci., Ser. B*, 49, 209, 2007.

2007WAN Wang, Y. and Annunziata, O., Comparison between protein-polyethylene glycol (PEG) interactions and the effect of PEG on protein-protein interactions using the liquid-liquid phase transition, *J. Phys. Chem. B*, 111, 1222, 2007.

2007WEN Wentzel, N. and Gunton, J.D., Liquid-liquid coexistence surface for lysozyme: Role of salt type and salt concentration, *J. Phys. Chem. B*, 111, 1478, 2007.

2007XUJ Xu, J., Ye, J., and Liu, S., Synthesis of well-defined cyclic poly(*N*-isopropyl acrylamide) via click chemistry and its unique thermal phase transition behavior, *Macromolecules*, 40, 9103, 2007.

2007XUY Xu, Y., Shi, L., Ma, R., Zhang, W., An, Y., and Zhu, X.X., Synthesis and micellization of thermo- and pH-responsive block copolymer of poly(*N*-isopropylacrylamide)-*block*-poly(4-vinylpyridine), *Polymer*, 48, 1711, 2007.

2007YAM Yamauchi, H. and Maeda, Y., LCST and UCST behavior of poly(*N*-isopropyl-acrylamide) in DMSO/water mixed solvents studied by IR and micro-Raman spectroscopy, *J. Phys. Chem. B*, 111, 12964, 2007.

2007ZHA Zhao, G. and Chen, S.B., Three-phase separation in nonionic surfactant/hydrophobically modified polymer aqueous mixtures, *Langmuir*, 23, 9967, 2007.

2008AHM Ahmed, A.A., Jayaraju, J., Sherigara, B.S., Bhojya, H.S.N., and Keshavayya, J., Miscibility studies of dextran/poly(vinyl pyrrolidone) blend in solution, *J. Macromol. Sci.: Part A: Pure Appl. Chem.*, 45, 1055, 2008.

2008ALV Alves, J.G.L.F., Brenneisen, J., Ninni, L., Meirelles, A.J.A., and Maurer, G., Aqueous two-phase systems of poly(ethylene glycol) and sodium citrate: Experimental results and modeling, *J. Chem. Eng. Data*, 53, 1587, 2008.

2008AMA Amaresh, S.P., Murugesan, S., Regupathi, I., and Murugesan, T., Liquid-liquid equilibrium of poly(ethylene glycol) 4000 + diammonium hydrogen phosphate + water at different temperatures, *J. Chem. Eng. Data*, 53, 1574, 2008.

2008BEC Becer, C.R., Hahn, S., Fijten, M.W.M., Thijs, H.M.L., Hoogenboom, R., and Schubert, U.S., Libraries of methacrylic acid and oligo(ethylene glycol) methacrylate copolymers with LCST behavior, *J. Polym. Sci.: Part A: Polym. Chem.*, 46, 7138, 2008.

2008BER Bergbreiter, D.E. and Fu, H., Thermodynamic cloud point assays, *J. Polym. Sci.: Part A: Polym. Chem.*, 46, 186, 2008.

2008BUR Burba, C.M., Carter, S.M., Meyer, K.J., and Rice, C.V., Salt effects on poly(*N*-isopropylacrylamide) phase transition thermodynamics from NMR spectroscopy, *J. Phys. Chem. B*, 112, 10399, 2008.

2008CAR Carvalho, C.P., Coimbra, J.S.R., Costa, I.A.F., Minim, L.A., Maffia, M.C., and Silva, L.H.M., Influence of the temperature and type of salt on the phase equilibrium of PEG1500 + potassium phosphate and PEG1500 + sodium citrate aqueous two-phase systems, *Quim. Nova*, 31, 209, 2008.

2008CHA Chang, C., Wei, H., Quan, C.-Y., Li, Y.-Y., Liu, J., Wang, Z.-C., Cheng, S.-X., Zhang, X.-Z., and Zhuo, R.-X., Fabrication of thermosensitive PCL-PNIPAAm-PCL triblock copolymeric micelles for drug delivery, *J. Polym. Sci.: Part A: Polym. Chem.*, 46, 3048, 2008.

2008CHE Chen, Q., Liu, X., Xu, K., Song, C., Zhang, W., and Wang, P., Phase behavior and self-assembly of poly[*N*-vinylformamide-*co*-(acrylic acid)] copolymers under highly acidic conditions, *J. Appl. Polym. Sci.*, 109, 2802, 2008.

2008CHO Cho, Y., Zhang, Y., Christensen, T., Sagle, L.B., Chilkoti, A., and Cremer, P.S., Effects of Hofmeister anions on the phase transition temperature of elastin-like polypeptides, *J. Phys. Chem. B*, 112, 13765, 2008.

2008COT Cote, M., Rogueda, P.G.A., and Griffiths, P.C., Effect of molecular weight and end-group nature on the solubility of ethylene oxide oligomers in 2H,3H-decafluoropentane and its fully fluorinated analogue, perfluoropentane, *J. Pharm. Pharmacol.*, 60, 593, 2008.

2008CRI Cristobal, G., Berret, J.-F., Chevallier, C., Talingting-Pabalan, R., Joanicot, M., and Grillo, I., Phase behavior of polyelectrolyte block copolymers in mixed solvents, *Macromolecules*, 41, 1872, 2008.

2008DAN Dan, A., Ghosh, S., and Moulik, S.P., The solution behavior of poly(vinylpyrrolidone): Its clouding in salt solution, solvation by water and isopropanol, and interaction with sodium dodecyl sulfate, *J. Phys. Chem. B*, 112, 3617, 2008.

2008EGG Eggenhuisen, T.M., Becer, C.R., Fijten, M.W.M., Eckardt, R., Hoogenboom, R., and Schubert, U.S., Libraries of statistical hydroxypropyl acrylate containing copolymers with LCST properties prepared by NMP, *Macromolecules*, 41, 5132, 2008.

2008FAN Fang, J., Bian, F., and Shen, W., A study on solution properties of poly(*N,N*-diethylacrylamide-*co*-acrylic acid), *J. Appl. Polym. Sci.*, 110, 3373, 2008.

2008FAR Fares, M.M. and Othman, A.A., Lower critical solution temperature determination of smart, thermosensitive *N*-isopropylacrylamide-*alt*-2-hydroxyethyl methacrylate copolymers: Kinetics and physical properties, *J. Appl. Polym. Sci.*, 110, 2815, 2008.

2008FO1 Foroutan, M. and Khomami, M.H., Activities of polymer, salt and water in liquid-liquid equilibria of polyvinylpyrrolidone and K_2HPO_4/KH_2PO_4 buffer using the Flory-Huggins model with Debye-Hueckel equation and the osmotic virial model: Effects of pH and temperature, Fluid Phase Equil., 265, 17, 2008.

2008FO2 Foroutan, M. and Zarrabi, M., Activities of water, polymer and salt in liquid-liquid equilibria of polyvinylpyrrolidone and $(NH_4)_2HPO_4/(NH_4)H_2PO_4$ buffer using the Flory-Huggins model with Debye-Hueckel and Pitzer-Debye-Hueckel equations and the osmotic virial model: Effects of pH and temperature, Fluid Phase Equil., 266, 164, 2008.

2008FO3 Foroutan, M. and Zarrabi, M., Quaternary (liquid + liquid) equilibria of aqueous two-phase polyethylene glycol, poly(*N*-vinylcaprolactam), and KH_2PO_4: Experimental and the generalized Flory-Huggins theory, *J. Chem. Thermodyn.*, 40, 935, 2008.

2008HE1 He, C., Zhao, C., Chen, X., Guo, Z., Zhuang, X., and Jing, X., Novel pH- and temperature-responsive block copolymers with tunable pH-responsive range, *Macromol. Rapid Commun.*, 29, 490, 2008.

2008HE2 He, C., Zhao, C., Guo, X., Guo, Z., Chen, X., Zhuang, X., Liu, S., and Jing, X., Novel temperature- and pH-responsive graft copolymers composed of poly(L-glutamic acid) and poly(*N*-isopropylacrylamide), *J. Polym. Sci.: Part A: Polym. Chem.*, 46, 4140, 2008.

2008HO1 Hoogenboom, R., Thijs, H.M.L., Jochems, M.J.H.C., Lankvelt, B.M. van, Fijten, M.W.M., and Schubert, U.S., Tuning the LCST of poly(2-oxazoline)s by varying composition and molecular weight: Alternatives to poly(*N*-isopropylacrylamide), *Chem. Commun.*, 44, 5758, 2008.

2008HO2 Hoogenboom, R., Thijs, H.M.L., Wouters, D., Hoeppener, S., and Schubert, U.S., Tuning solution polymer properties by binary water–ethanol solvent mixtures, *Soft Matter*, 4, 103, 2008.

2008HU1 Huber, S. and Jordan, R., Modulation of the lower critical solution temperature of 2-alkyl-2-oxazoline copolymers, *Coll. Polym. Sci.*, 286, 395, 2008.

2008HU2 Huber, S., Hutter, N., and Jordan, R., Effect of end group polarity upon the lower critical solution temperature of poly(2-isopropyl-2-oxazoline), *Colloid Polym. Sci.*, 286, 1653, 2008.1

2008HUH Hu, H.-Y., Du, J., Meng, Q.-B., Li, Z.-Y., and Zhu, X.X., Thermosensitivity of narrow-dispersed poly(*N*-n-propylacrylamide) prepared by atom transfer radical polymerization, *Chin. J. Polym. Sci.*, 26, 187, 2008.

2008ISH Ishizone, T., Seki, A., Hagiwara, M., Han, S., Yokoyama, H., Oyane, A., Deffieux, A., and Carlotti, S., Anionic polymerizations of oligo(ethylene glycol) alkyl ether methacrylates: Effect of side chain length and ω-alkyl group of side chain on cloud point in water, *Macromolecules*, 41, 2963, 2008.

2008JOH Johansson, H.-O., Magaldi, F.M., Feitosa, E., and Pessoa Jr., A., Protein partitioning in poly(ethylene glycol)/sodium polyacrylate aqueous two-phase systems, *J. Chromatogr. A*, 1178, 145, 2008.

2008KAT Katsumoto, Y. and Kubosaki, N., Tacticity effects on the phase diagram for poly(*N*-isopropylacrylamide) in water, *Macromolecules*, 41, 5955, 2008.

2008KAW Kawaguchi, T., Kojima, Y., Osa, M., and Yoshizaki, T., Cloud points in aqueous poly(*N*-isopropylacrylamide) solutions, *Polym. J.*, 40, 455, 2008.

2008KHU Khutoryanskaya, O.V., Mayeva, Z.A., Mun, G.A., and Khutoryanskiy, V.V., Designing temperature-responsive biocompatible copolymers and hydrogels based on 2-hydroxyethyl(meth)acrylates, *Biomacromolecules*, 9, 3353, 2008.

2008KOT Kotsuchinashi, Y., Kuboshima, Y., Yamamoto, K., and Aoyagi, T., Synthesis and characterization of double thermo-responsive block copolymer consisting *N*-isopropylacrylamide by atom transfer radical polymerization, *J. Polym. Sci.: Part A: Polym. Chem.*, 46, 6142, 2008.

2008LI1 Liu, S., Liu, X., Li, F., Fang, Y., Wang, Y., and Yu, J., Phase behavior of temperature- and pH-sensitive poly(acrylic acid-*g*-*N*-isopropylacrylamide) in dilute aqueous solution, *J. Appl. Polym. Sci.*, 109, 4036, 2008.

2008LI2 Liu, R., DeLeonardis, P., Cellesi, F., Tirelli, N., and Saunders, B.R., Cationic temperature-responsive poly(*N*-isopropyl acrylamide) graft copolymers: From triggered association to gelation, *Langmuir*, 24, 7099, 2008.

2008LI3 Liu, L., Wu, C., Zhang, J., Zhang, M., Liu, Y., Wang, X., and Fu, G., Controlled polymerization of 2-(diethylamino)ethyl methacrylate and its block copolymer with *N*-isopropylacrylamide by RAFT polymerization, *J. Polym. Sci.: Part A: Polym. Chem.*, 46, 3294, 2008.

2008LIX Li, X., Deng, F., Hua, Y., Qui, A., Yang, C., and Cui, S., Effect of molecular weight of dextran on the phase behavior and microstructure of preheated soy protein/dextran mixtures, *Carbohydrate Polym.*, 72, 160, 2008.

2008LIY Li, Y., Liu, R., Liu, W., Kang, H., Wu, M., and Huang, Y., Synthesis, self-assembly, and thermosensitive properties of ethyl cellulose-*g*-p(PEGMA) amphiphilic copolymers, *J. Polym. Sci.: Part A: Polym. Chem.*, 46, 6907, 2008.

2008LOH Loh, X.J., Wu, Y.-L., Seow, W.T.J., Norimzan, M.N.I., Zhang, Z.-X., Xu, F.-J., Kang, E.-T., Neoh, K.-G., and Li, J., Micellization and phase transition behavior of thermosensitive poly(*N*-isopropylacrylamide)-poly(ε-caprolactone)-poly(*N*-isopropyl acrylamide) triblock copolymers, *Polymer*, 49, 5084, 2008.

2008MAD Madeira, P.P., Teixeira, J.A., Macedo, E.A., Mikheeva, L.M., and Zaslavsky, B.Y., $\Delta G(CH_2)$ as solvent descriptor in polymer/polymer aqueous two-phase systems, *J. Chromatogr. A*, 1185, 85, 2008.

2008MAL Malpiedi, L.P., Fernandez, C., Pico, G., and Nerli, B., Liquid–liquid equilibrium phase diagrams of polyethyleneglycol + sodium tartrate + water two-phase systems, *J. Chem. Eng. Data*, 53, 1175, 2008.

2008MA1 Martins, J.P., Carvalho, C. de P., Silva, L.H.M. da, Coimbra, J.S. dos R., Silva, M. do C.H. da, Rodrigues, G.D., and Minim, L.A., Liquid-liquid equilibria of an aqueous two-phase system containing poly(ethylene glycol) 1500 and sulfate salts at different temperatures, *J. Chem. Eng. Data*, 53, 238, 2008.

2008MA2 Martins, J.P., Oliveira, F.C. de, Coimbra, J.S. dos R., Silva, L.H.M. da, Silva, M.C.H. da, and Nascimento, I.S.B. do, Equilibrium phase behavior for ternary mixtures of poly(ethylene glycol) 6000 + water + sulfate salts at different temperatures, *J. Chem. Eng. Data*, 53, 2441, 2008.

2008MAS Masci, G., Diociaiuti, M., and Crescenzi, V., ATRP synthesis and association properties of thermoresponsive anionic block copolymers, *J. Polym. Sci.: Part A: Polym. Chem.*, 46, 4830, 2008.

2008MAT Matsuda, Y., Kobayashi, M., Annaka, M., Ishihara, K., and Takahara, A., UCST-type cononsolvency behavior of poly(2-methacryloxyethyl phosphorylcholine) in the mixture of water and ethanol, *Polym. J.*, 40, 479, 2008.

2008MIK Mikhailov, Yu. M., Ganina, L. V., Shapaeva, N. V., and Makarova, E. N., Interdiffusion and phase equilibrium in systems based on cellulose nitrates and low-volatile solvents, *Polym. Sci., Ser. B*, 50, 268, 2008.

2008MIP Mi, P., Chu, L.-Y., Ju, X.-J., and Niu, C.H., A smart polymer with ion-induced negative shift of the lower critical solution temperature for phase transition, *Macromol. Rapid Commun.*, 29, 27, 2008.

2008MIY Miyake, M., Asano, A., and Einaga, Y., Size change of the wormlike micelles of pentaoxyethylene, hexaoxyethylene, and heptaoxyethylene dodecyl ethers with uptake of n-dodecane, *J. Phys. Chem. B*, 112, 4648, 2008.

2008MOH Mohsen-Nia, M., Rasa, H., and Modarress, H., Liquid-liquid equilibria for the poly(ethylene glycol) + water + copper sulfate system at different temperatures, *J. Chem. Eng. Data*, 53, 946, 2008.

2008MOK Mokhtarani, B., Karimzadeh, R., Amini, M.H., and Manesh, S.D., Partitioning of Ciprofloxacin in aqueous two-phase system of poly(ethylene glycol) and sodium sulphate, *Biochem. Eng. J.*, 38, 241, 2008.

2008NOR Norman, A.I., Manvilla, B.A., Frank, E.L., Niamke, J.N., Smith, G.D., and Greer, S.C., Partitioning of poly(ethylene oxide), poly(ethylenimine), and bovine serum albumin in isobutyric acid + water, *Macromolecules*, 41, 997, 2008.

2008NU1 Nuopponen, M., Kalliomaeki, K., Laukkanen, A., Hietala, S., and Tenhu, H., A-B-A stereoblock copolymers of *N*-isopropylacrylamide, *J. Polym. Sci.: Part A: Polym. Chem.*, 46, 38, 2008.

2008NU2 Nuopponen, M., Kalliomaeki, K., Aseyev, V., and Tenhu, H., Spontaneous and thermally induced self-organization of A-B-A stereoblock polymers of *N*-isopropyl-acrylamide in aqueous solutions, *Macromolecules*, 41, 4881, 2008.

2008OL1 Oliveira, R.M. de, Reis Coimbra, J.S. dos, Minim, L.A., Silva, L.H.M. da, and Fontes, M.P.F., Liquid-liquid equilibria of biphasic systems composed of sodium citrate + poly(ethylene glycol) 1500 or 4000 at different temperatures, *J. Chem. Eng. Data*, 53, 895, 2008.

2008OL2 Oliveira, R.M. de, Reis Coimbra, J.S. dos, Francisco, K.R., Minim, L.A., Silva, L.H.M. da, and Pereira, J.A.M., Liquid-liquid equilibrium of aqueous two-phase systems containing poly(ethylene glycol) 4000 and zinc sulfate at different temperatures, *J. Chem. Eng. Data*, 53, 919, 2008.

2008OL3 Oliveira, R.M. de, Reis Coimbra, J.S. dos, Francisco, K.R., Minim, L.A., Silva, L.H.M. da, and Rojas, E.A.G., Equilibrium data of the biphasic system poly(ethylene oxide) 4000 + copper sulfate + water at (5, 10, 35, and 45)°C, *J. Chem. Eng. Data*, 53, 1571, 2008.

2008PAN Pan, Y., Xiao, H., Zhao, G., and He, B., Synthesis and characterization of temperature-responsive poly(vinyl alcohol)-based copolymers, *J. Appl. Polym. Sci.*, 110, 2698, 2008.

2008QIN Qin, W. and Cao, X.-J., Synthesis of a novel pH-sensitive methacrylate amphiphilic polymer and its primary application in aqueous two-phase systems, *Appl. Biochem. Biotechnol.*, 150, 171, 2008.

2008QUI Quiroga, C.C. and Bergenstahl, B., Phase segregation of amylopectin and β-lactoglobulin in aqueous system, *Carbohydrate Polym.*, 72, 151, 2008.

2008RAH Rahimpour, F. and Pirdashti, M., The effect of guanidine hydrochloride on phase diagram of PEG-phosphate aqueous two-phase system, *Int. J. Chem. Biomol. Eng.*, 1, 35, 2008.

2008RAS Rasa, H., Mohsen-Nia, M., and Modarress, H., Phase separation in aqueous two-phase systems containing poly(ethylene glycol) and magnesium sulphate at different temperatures, *J. Chem. Thermodyn.*, 40, 573, 2008.

2008REN Ren, Y., Jiang, X., and Yin, J., Copolymer of poly(4-vinylpyridine)-*g*-poly(ethylene oxide) respond sharply to temperature, pH and ionic strength, *Eur. Polym. J.*, 44, 4108, 2008.

2008SA1 Sadeghi, R., Hosseini, R., and Jamehbozorg, B., Effect of sodium phosphate salts on the thermodynamic properties of aqueous solutions of poly(ethylene oxide) 6000 at different temperatures, *J. Chem. Thermodyn.*, 40, 1364, 2008.

2008SA2 Sadeghi, R. and Jamehbozorg, B., Effect of temperature on the salting-out effect and phase separation in aqueous solutions of sodium dihydrogen phosphate and poly(propylene glycol), *Fluid Phase Equil.*, 271, 13, 2008.

2008SAR Saravanan, S., Rao, J.R., Nair, B.U., and Ramasami, T., Aqueous two-phase poly(ethylene glycol)-poly(acrylic acid) system for protein partitioning: Influence of molecular weight, pH and temperature, *Process Biochemistry*, 43, 905, 2008.

2008SHA Shaheen, A., Kaur, N., and Mahajan, R.K., Influence of various series of additives on the clouding behavior of aqueous solutions of triblock copolymers, *Coll. Polym. Sci.*, 286, 319, 2008.

2008SI1 Silva, S.M.C., Pinto, F.V., Antunes, F.E., Miguel, M.G., Sousa, J.J.S., and Pais, A.A.C.C., Aggregation and gelation in hydroxypropylmethyl cellulose aqueous solutions, J. Colloid Interface Sci., 327, 333, 2008.

2008SI2 Silverio, S.C., Madeira, P.P., Rodriguez, O., Teixeira, J.A., and Macedo, E.A., $\Delta G(CH_2)$ in PEG-salt and Ucon-salt aqueous two-phase systems, *J. Chem. Eng. Data*, 53, 1622, 2008.

2008TAN Tan, L., Pan, D., and Pan, N., Thermodynamic study of a water-dimethylformamide-polyacrylonitrile ternary system, *J. Appl. Polym. Sci.*, 110, 3439, 2008.

2008TRO Troll, K., Kulkarni, A., Wang, W. Darko, C., Koumba, A.M.B., Laschewsky, A., Müller-Buschbaum, P., and Papadakis, C.M., The collapse transition of poly(styrene-*b*-(*N*-isopropylacrylamide)) diblock copolymers in aqueous solution and in thin films, *Colloid Polym. Sci.*, 286, 1079, 2008.

2008VSH Vshivkov, S.A. and Rusinova, E.V., Effect of magnetic field on phase transitions in solutions of cellulose derivatives, *Polym. Sci., Ser. A*, 50, 725, 2008.

2008WIN Wintgens, V. and Amiel, C., Physical gelation of amphiphilic poly(*N*-isopropylacrylamide): Influence of the hydrophobic groups, *Macromol. Chem. Phys.*, 209, 1553, 2008.

2008XUJ Xu, J., Jiang, X., and Liu, S., Synthesis of low-polydispersity poly(*N*-ethyl-*N*-methylacrylamide) by controlled radical polymerizations and their thermal phase transition behavior, *J. Polym. Sci.: Part A: Polym. Chem.*, 46, 60, 2008.

2008YAM Yamamoto, S.-I., Pietrasik, J., and Matyjaszewski, K., The effect of structure on the thermoresponsive nature of well-defined poly(oligo(ethylene oxide) methacrylates) synthesized by ATRP, *J. Polym. Sci.: Part A: Polym. Chem.*, 46, 194, 2008.

2008YAN Yankov, D.S., Trusler, J.P.M., Yordanov, B.Y., and Stateva, R.P., Influence of lactic acid on the formation of aqueous two-phase systems containing poly(ethylene glycol) and phosphates, *J. Chem. Eng. Data*, 53, 1309, 2008.

2008ZA1 Zafarani-Moattar, M.T., Emamian, S., and Hamzehzadeh, S., Effect of temperature on the phase equilibrium of the aqueous two-phase poly(propylene glycol) + tripotassium citrate system, *J. Chem. Eng. Data*, 53, 456, 2008.

2008ZA2 Zafarani-Moattar, M.T., Hamzehzadeh, S., and Hosseinzadeh, S., Phase diagrams for liquid-liquid equilibrium of ternary poly(ethylene glycol) + disodium tartrate aqueous system and vapor-liquid equilibrium of constituting binary aqueous systems at T = (298.15, 308.15, and 318.15)K. Experiment and correlation, *Fluid Phase Equil.*, 268, 142, 2008.

2008ZA3 Zafarani-Moattar, M.T. and Nikjoo, D., Liquid-liquid and liquid-liquid-solid equilibrium of the poly(ethylene glycol) dimethyl ether 2000 + sodium sulfate + water system, *J. Chem. Eng. Data*, 53, 2666, 2008.

2008ZA4 Zafarani-Moattar, M.T. and Tolouei, S., Liquid-liquid equilibria of aqueous two-phase systems containing poly(ethylene glycol) 4000 and dipotassium tartrate, potassium sodium tartrate, or dipotassium oxalate: Experiment and correlation, *CALPHAD*, 32, 655, 2008.

2008ZHA Zhao, C., Zhuang, X., He, C., Chen, X., and Jing, X., Synthesis of novel thermo- and pH-responsive poly(L-lysine)-based copolymer and its micellization in water, *Macromol. Rapid Commun.*, 29, 1810, 2008.

2008ZHO Zhou, X., Li, J., Wu, C., and Zheng, B., Constructing the phase diagram of an aqueous solution of poly(*N*-isopropyl acrylamide) by controlled microevaporation in a nanoliter microchamber, *Macromol. Rapid Commun.*, 29, 1363, 2008.

2008ZOU Zou, Y., Brooks, D.E., and Kizhakkedathu, J.N., A novel functional polymer with tunable LCST, *Macromolecules*, 41, 5393, 2008.

2009ALH Al-Hakimi, A.M., Shang, Y., Liu, H., and Hu, Y., Salt effects on the aqueous two-phase system of gemini(12-3-12, 2Br-)/SDS/PEG, *J. Solution Chem.*, 38, 1307, 2009.

2009ALV Alvarez-Ramírez, J.G., Fernández, V.V.A., Macíasa, E.R., Rharbi, Y., Taboada, P., Gámez-Corrales, R., Puig, J.E., and Soltero, J.F.A., Phase behavior of the Pluronic P103/water system in the dilute and semi-dilute regimes, *J. Colloid Interface Sci.*, 333, 655, 2009.

2009AN1 Antov, M. and Omorjan, R., Pectinase partitioning in poly(ethylene glycol)1000/ Na_2SO_4 aqueous two-phase system: Statistical modeling of the experimental results, *Bioprocess Biosyst. Eng.*, 32, 235, 2009.

2009AN2 Antonov, Y.A. and Moldenaers, P., Inducing demixing of semidilute and highly compatible biopolymer mixtures in the presence of a strong polyelectrolyte, *Biomacromolecules*, 10, 3235, 2009.

2009CAO Cao, Y., Zhao, N., Wu, K., and Zhu, X.X., Solution properties of a thermosensitive triblock copolymer of *N*-alkyl substituted acrylamides, *Langmuir*, 25, 1699, 2009.

2009CHA1 Chang, C., Wei, H., Feng, J., Wang, Z.-C., Wu, X.-J., Wu, D.-Q., Cheng, S.-X., Zhang, X.-Z., and Zhuo, R.-X., Temperature and pH double responsive hybrid cross-linked micelles based on P(NIPAAm-*co*-MPMA)-*b*-P(DEA): RAFT synthesis and schizophrenic micellization, *Macromolecules*, 42, 4838, 2009.

2009CHA2 Chang, Y., Chen, W.-Y., Yandi, W., Shih, Y.-J., Chu, W.-L., Liu, Y.-L., Chu, C.-W., Ruaan, R.-C., and Higuchi, A., Dual-thermoresponsive phase behavior of blood compatible zwitterionic copolymers containing nonionic poly(*N*-isopropyl-acrylamide), *Biomacromolecules*, 10, 2092, 2009.

2009CHE1 Chen, W., Pelton, R., and Leung, V., Solution properties of polyvinylamine derivatized with phenylboronic acid, *Macromolecules*, 42, 1300, 2009.

2009CHE2 Chernysheva, M.G., Tyasto, Z.A., and Badun, G.A., Anomalous behavior of poly(ethylene glycol) p-*tert*-octylphenyl ether (Triton X-100) in the water-cyclohexane system, *Russ. J. Phys. Chem. A*, 83, 285, 2009.

2009CRI Cristiano, C.M.Z., Soldi, V., Li, C., Armes, S.P., Rochas, C., Pignot-Paintrand, I., and Borsali, R., Thermo-responsive copolymers based on poly(*N*-isopropyl acrylamide) and poly[2-(methacryloyloxy)ethyl phosphorylcholine]: Light scattering and microscopy experiments, Macromol. Chem. Phys., 210, 1726, 2009.

2009CUN Cunha, E.V.S. and Aznar, M., Liquid-liquid equilibrium in aqueous two-phase (water + PEG 8000 + salt): Experimental determination and thermodynamic modeling, *J. Chem. Eng. Data*, 54, 3242, 2009.

2009DEN Deng, K., Tian, H., Zhang, P., Zhong, H., Ren, X., and Wang, H., pH–temperature responsive poly(HPA-*co*-AMHS) hydrogel as a potential drug-release carrier, *J. Appl. Polym. Sci.*, 114, 176, 2009.

2009DON Dong, R., Zhao, J., Zhang, Y., and Pan, D., Morphology control of polyacrylonitrile (PAN) fibers by phase separation technique, *J. Polym. Sci.: Part B: Polym. Phys.*, 47, 261, 2009.

2009DUR Durme, K. van, Assche, G. van, Rahier, H., and Mele, B. van, LCST demixing in poly(vinyl methyl ether)/water studied by means of a high-resolution ultrasonic resonator, *J. Therm. Anal. Calorim.*, 98, 495, 2009.

2009FO1 Foroutan, M., Heidari, N., Mohammadlou, M., and Sojahrood, A.J., (Surfactant + polymer) interaction parameter studied by (liquid + liquid) equilibrium data of quaternary aqueous solution containing surfactant, polymer, and salt, *J. Chem. Thermodyn.*, 41, 227, 2009.

2009FO2 Foroutan, M. and Khomami, M.H., Quaternary (liquid + liquid) equilibria of aqueous two-phase poly(ethylene glycol), poly(DMAM-TBAM), and KH_2PO_4: Experimental and generalized Flory-Huggins theory, *J. Chem. Thermodyn.*, 41, 604, 2009.

2009FUC Fuchise, K., Kakuchi, R., Lin, S.-T., Sakai, R., Sato, S.-I., Satoh, T., Chen, W.-C., and Kakuchi, T., Control of thermoresponsive property of urea end-functionalized poly(*N*-isopropylacrylamide) based on the hydrogen bond-assisted self-assembly in water, *J. Polym. Sci.: Part A: Polym. Chem.*, 47, 6259, 2009.

2009GAO Gao, M., Jia, X., Kuang, G., Li, Y., Liang, D., and Wei, Y., Thermo- and pH-responsive dendronized copolymers of styrene and maleic anhydride pendant with poly(amidoamine) dendrons as side groups, *Macromolecules*, 42, 4273, 2009.

2009GRU Grünfelder, T., Pessoa Filho, P.A., and Maurer, G., Liquid-liquid equilibrium of aqueous two-phase systems containing some synthetic polyelectrolytes and polyethylene glycol, *J. Chem. Eng. Data*, 54, 198, 2009.

2009HEL He, L., Zhan, Y., Zeng, X., Quan, D., Liao, S., Zeng, Y., Lu, J., and Ramakrishna, S., Fabrication and characterization of poly(L-lactic acid) 3D nanofibrous scaffolds with controlled architecture by liquid–liquid phase separation from a ternary polymer-solvent system, *Polymer*, 50, 4128, 2009.

2009HIR Hirano, T., Kamikubo, T., Okumura, Y., Bando, Y., Yamaoka, R., Mori, T., and Ute, K., Heterotactic-specific radical polymerization of *N*-isopropylacrylamide and phase transition behavior of aqueous solution of heterotactic poly(*N*-isopropylacrylamide), *J. Polym. Sci.: Part A: Polym. Chem.*, 47, 2539, 2009.

2009HOF Hofmann, C. and Schoenhoff, M., Do additives shift the LCST of poly(*N*-isopropyl-acrylamide) by solvent quality changes or by direct interactions?, *Colloid Polym. Sci.*, 287, 1369, 2009.

2009IMA Imani, A., Modarress, H., Eliassi, A., and Abdous, M., Cloud-point measurement for (sulphate salts + polyethylene glycol 15000 + water) systems by the particle counting method, *J. Chem. Thermodyn.*, 41, 893, 2009.

2009JOC Jochum, F.D. and Theato, P., Temperature and light sensitive copolymers containing azobenzene moieties prepared via a polymer analogous reaction, *Polymer*, 50, 3079, 2009.

2009KAU Kautharapu, K., Pujari, N.S., Golegaonkar, S.B., Ponrathnam, S., Nene, S.N., and Bhatnagar, D., Vinyl-2-pyrrolidone derivatized guar gum based aqueous two-phase system, *Separation Purification Technol.*, 65, 9, 2009.

2009KHE Khederlou, K., Pazuki, G.R., Taghikhani, V., Vossoughi, M., and Ghotbi, C., Measurement and modeling process partitioning of cephalexin antibiotic in aqueous two-phase systems containing poly(ethylene glycol) 4000, 10000 and K_2HPO_4, Na_3citrate, *J. Chem. Eng. Data*, 54, 2239, 2009.

2009KIZ Kizhnyaev, V.N., Adamova, L.V., Pokatilov, F.A., Krakhotkina, E.A., Petrova, T.L., and Smirnov, A.I., Thermodynamics of water interaction with 5-vinyltetrazole copolymers with different hydrophilic–hydrophobic balances, *Polym. Sci., Ser. A*, 51, 168, 2009.

2009KOB Kobayashi, K., Yamada, S., Nagaoka, K., Kawaguchi, T., Osa, M., and Yoshizaki, T., Characterization of linear poly(*N*-isopropylacrylamide) and cloud points in its aqueous solutions (exp. data by T. Yoshizaki), *Polym. J.*, 41, 416, 2009.

2009KOS Kostko, A.F., Harden, J.L., and McHugh, M.A., Dynamic light scattering study of concentrated triblock copolymer micellar solutions under pressure, *Macromolecules*, 42, 5328, 2009.

2009KOT Kotsuchibashi, Y., Yamamoto, K., and Aoyagi, T., Assembly behavior of double thermo-responsive block copolymers with controlled response temperature in aqueous solution, *J. Colloid Interface Sci.*, 336, 67, 2009.

2009KOU Kourilova, H., Hanykova, L., and Spevacek, J., NMR Study of phase separation in D_2O/ethanol solutions of poly(*N*-isopropylmethacrylamide) induced by solvent composition and temperature, *Eur. Polym. J.*, 45, 2935, 2009.

2009LAM Lambermont-Thijs, H.M.L., Hoogenboom, R., Fustin, C.A., Bomal-D'Haese, C., Gohy, J.-F., and Schubert, U.S., Solubility behavior of amphiphilic block and random copolymers based on 2-ethyl-2-oxazoline and 2-nonyl-2-oxazoline in binary water–ethanol mixtures, *J. Polym. Sci.: Part A: Polym. Chem.*, 47, 515, 2009.

2009LIN Lin, J.-J. and Hsu, Y.-C., Temperature and pH-responsive properties of poly(styrene-*co*-maleic anhydride)-grafting-poly(oxypropylene)amines, *J. Colloid Interface Sci.*, 336, 82, 2009.

2009LIP Li, P.-F., Wang, W., Xie, R., Yang, M., Ju, X.-J., and Chu, L.-Y., Lower critical solution temperatures of thermo-responsive poly(*N*-isopropylacrylamide) copolymers with racemate or single enantiomer groups, *Polym. Int.*, 58, 202, 2009.

2009LIT Litmanovich, E.A., Zakharchenko, S.O., Stoychev, G.V., and Zezin, A.B., Phase separation in a poly(acrylic acid)-polycation system in acidic solutions, *Polym. Sci., Ser. A*, 51, 616, 2009.

2009LIX Li, X., Yin, M. Zhang, G., and Zhang, F., Synthesis and characterization of novel temperature and pH responsive hydroxylpropyl cellulose-based graft copolymers, *Chin. J. Chem. Eng.*, 17, 145, 2009.

2009LI1 Liu, C., He, J., Zhao, Q., Zhang, M., and Ni, P., Well-defined poly[(dimethylamimo)ethyl methacrylate]-*b*-poly(fluoroalkyl methacrylate) diblock copolymers: Effects of different fluoroalkyl groups on the solution properties, *J. Polym. Sci.: Part A: Polym. Chem.*, 47, 2702, 2009.

2009LI2 Liu, R., Cellesi, F., Tirelli, N., Saunders, B.R., A study of thermoassociative gelation of aqueous cationic poly(*N*-isopropylacrylamide) graft copolymer solutions, *Polymer*, 50, 1456, 2009.

2009MA1 Maeda, Y., Yamauchi, H., and Kubota, T., Confocal micro-Raman and infrared spectroscopic study on the phase separation of aqueous poly(2-(2-methoxyethoxy) ethyl (meth)acrylate) solutions, *Langmuir*, 25, 479, 2009.

2009MA2 Maeda, Y. and Yamabe, M., A unique phase behavior of random copolymer of *N*-isopropylacrylamide and *N,N*-diethylacrylamide in water, *Polymer*, 50, 519, 2009.

2009MA3 Maeda, Y., Sakamoto, J., Wang, S.-Y., and Mizuno, Y., Lower critical solution temperature behavior of poly(*N*-(2-ethoxyethyl)acrylamide) as compared with poly(*N*-isopropylacrylamide), *J. Phys. Chem. B*, 113, 12456, 2009.

2009MAH Mahdavi, H., Sadeghzadeh, M., and Qazvini, N.T., Phase behavior study of poly(*N*-*tert*-butylacrylamide-*co*-acrylamide) in the mixture of water–methanol: The role of polymer–nonsolvent second-order interactions, *J. Polym. Sci.: Part B: Polym. Phys.*, 47, 455, 2009.

2009MAI Mai-ngam, K., Boonkitpattarakul, K., Sakulsombat, M., Chumningan, P., and Mai-ngam, B., Synthesis and phase separation of amine-functional temperature responsive copolymers based on poly(*N*-isopropylacrylamide), *Eur. Polym. J.*, 45, 1260, 2009.

2009MAR Martins, J.P., Mageste, A.B., Silva, M.C.H. da, Silva, L.H.M. da, Patricio, P. da R., Coimbra, J.S. dos R., and Minim, L.A., Liquid-liquid equilibria of an aqueous two-phase system formed by a triblock copolymer and sodium salts at different temperatures, *J. Chem. Eng. Data*, 54, 2891, 2009.

2009MOO Moon, J.R., Park, Y.H., and Kim, J.-H., Synthesis and characterization of novel thermo- and pH-responsive copolymers based on amphiphilic polyaspartamides, *J. Appl. Polym. Sci.*, 111, 998, 2009.

2009MOR Mori, T., Hirano, T., Maruyama, A., Katayama, Y., Niidome, T., Bando, Y., Ute, K., Takaku, S., and Maeda, Y., Syndiotactic poly(*N*-n-propylacrylamide) shows highly cooperative phase transition, *Langmuir*, 25, 48, 2009.

2009NAN Nandni, D., Vohra, K.K., and Mahajan, R.K., Study of micellar and phase separation behavior of mixed systems of triblock polymers, *J. Colloid Interface Sci.*, 338, 420, 2009.

2009NIN Ning, B., Wan J., and Cao X., Preparation and recycling of aqueous two-phase systems with pH-sensitive amphiphilic terpolymer PADB, *Biotechnol. Progr.*, 25, 820, 2009.

2009OBE Obeid, R., Tanaka, F., and Winnik, F.M., Heat-induced phase transition and crystallization of hydrophobically end-capped poly(2-isopropyl-2-oxazoline)s in water, *Macromolecules*, 42, 5818, 2009.

2009OSA Osa, M., Aqueous solution properties of poly(*N*-isopropylacrylamide), *Kobunshi Ronbunshu*, 66, 273, 2009.

2009PAK Pakuro, N., Yakimansky, A., Chibirova, F., and Arest-Yakubovich, A., Thermo- and pH-sensitivity of aqueous poly(*N*-vinylpyrrolidone) solutions in the presence of organic acids, *Polymer*, 50, 148, 2009.

2009PAM Pamies, R., Zhu, K., Kjøniksen, A.-L., and Nyström, B., Thermal response of low molecular weight poly(*N*-isopropylacrylamide) polymers in aqueous solution, *Polym. Bull.*, 62, 487, 2009.

2009PAT Patrizi, M.L., Piantanida, G., Coluzza, C., and Masci, G., ATRP synthesis and association properties of temperature responsive dextran copolymers grafted with poly(*N*-isopropylacrylamide), *Eur. Polym. J.*, 45, 2779, 2009.

2009PER Perumalsamy, M. and Murugesan, T., Phase compositions, molar mass, and temperature effect on densities, viscosities, and liquid-liquid equilibrium of polyethylene glycol and salt-based aqueous two-phase systems, *J. Chem. Eng. Data*, 54, 1359, 2009.

2009REC Recillas, M., Silva, L.L., Peniche, C., Goycoolea, F.M., Rinaudo, M., and Argüelles-Monal, W.M., Thermoresponsive behavior of chitosan-*g*-*N*-isopropylacrylamide copolymer solutions, *Biomacromolecules*, 10, 1633, 2009.

2009REG Regupathi, I., Murugesan, S., Govindarajan, R., Amaresh, S.P., and Thanapalan, M., Liquid-liquid equilibrium of poly(ethylene glycol) 6000 + triammonium citrate + water systems at different temperatures, *J. Chem. Eng. Data*, 54, 1094, 2009.

2009REN Ren, Y., Jiang, X., and Yin, J., Poly(ether *tert*-amine): A novel family of multiresponsive polymer, *J. Polym. Sci.: Part A: Polym. Chem.*, 47, 1292, 2009.

2009ROD Rodrigues, G.D., Silva, M. do C.H. da, Silva, L.H.M. da, Teixeira, L. da S., and Andrade, V.M. de, Liquid-liquid phase equilibrium of triblock copolymer L64, poly(ethylene oxide-*b*-propylene oxide-*b*-ethylene oxide), with sulfate salts from (278.15 to 298.15) K, *J. Chem. Eng. Data*, 54, 1894, 2009.

2009SAD Sadeghi, R. and Jamehbozorg, B., The salting-out effect and phase separation in aqueous solutions of sodium phosphate salts and poly(propylene glycol), *Fluid Phase Equil.*, 280, 68, 2009

2009SHA Shao, Z., Kong, F., and Cao, X., Phase diagram prediction of recycling aqueous two-phase systems formed by a light-sensitive copolymer and dextran, *Korean J. Chem. Eng.*, 26, 147, 2009.

2009SHI Shinde, V.S. and Pawar, V.U., Synthesis of thermosensitive glycopolymers containing D-glucose residue: Copolymers with *N*-isopropylacrylamide, *J. Appl. Polym. Sci.*, 111, 2607, 2009.

2009SIL Silva, C.A.S. da, Coimbra, J.S.R., Rojas, E.E.G., and Teixeira, J.A.C., Partitioning of glycomacropeptide in aqueous two-phase systems, *Process Biochem.*, 44, 1213, 2009.

2009TAI Tai, H., Wang, W., Vermonden, T., Heath, F., Hennink, W.E., Alexander, C., Shakesheff, K.M., and Howdle, S.M., Thermoresponsive and photocrosslinkable PEGMEMA-PPGMA-EGDMA copolymers from a one-step ATRP synthesis, *Biomacromolecules*, 10, 822, 2009.

2009TAN Tan, L., Liu, S., and Pan, D., Water effect on the gelation behavior of polyacrylonitrile/dimethyl sulfoxide solution, *Colloids Surfaces A*, 340, 168, 2009.

2009TAU Tauer, K., Gau, D., Schulze, S., Völkel, A., and Dimova, R., Thermal property changes of poly(*N*-isopropylacrylamide) microgel particles and block copolymers, *Colloid Polym. Sci.*, 287, 299, 2009.

2009TRE Trellenkamp, T. and Ritter, H., 3-Ethylated *N*-vinyl-2-pyrrolidone with LCST properties in water, *Macromol. Rapid Commun.*, 30, 1736, 2009.

2009TUB Tubio, G., Nerli, B.B., Pico, G.A., Venancio, A., and Teixeira, J., Liquid-liquid equilibrium of the Ucon 50-HB5100/sodium citrate aqueous two-phase systems, *Separation Purification Technol.*, 65, 3, 2009.

2009VIR Viriden, A., Wittgren, B., Andersson, T., Abrahmsen-Alami, S., and Larsson, A., Influence of substitution pattern on solution behavior of hydroxypropyl methylcellulose, *Biomacromolecules*, 10, 522, 2009.

2009WA1 Wang, T., Peng, C., Liu, H., and Hu, Y., Phase behavior and microstructure of the system consisting of 1-butyl-3-methylimidazolium hexafluorophosphate, water, triblock copolymer F127 and short-chain alcohols, *J. Mol. Liq.*, 146, 89, 2009.

2009WA2 Wang, W., Liang, H., Al Ghanami, R.C., Hamilton, L., Fraylich, M., Shakesheff, K.M., Saunders, B., and Alexander, C., Biodegradable thermoresponsive microparticle dispersions for injectable cell delivery prepared using a single-step process, *Adv. Mater*, 21, 1809, 2009.

2009WA3 Wang, Y.-C., Xia, H., Yang, X.-Z., and Wang, J., Synthesis and thermoresponsive behaviors of biodegradable Pluronic analogs, *J. Polym. Sci.: Part A: Polym. Chem.*, 47, 6168, 2009.

2009WEB Weber, C., Becer, C.R., Hoogenboom, R., and Schubert, U.S., Lower critical solution temperature behavior of comb and graft shaped poly[oligo(2-ethyl-2-oxazoline) methacrylate]s, *Macromolecules*, 42, 2965, 2009.

2009XUJ Xu, J. and Liu, S., Synthesis of well-defined 7-arm and 21-arm poly(*N*-isopropyl-acrylamide) star polymers with beta-cyclodextrin cores via click chemistry and their thermal phase transition behavior in aqueous solution, *J. Polym. Sci.: Part A: Polym. Chem.*, 47, 404, 2009.

2009YOU You, X., Qin, W., and Dai, Y., Phase separation behavior of cocamidopropyl betaine/water/polyethylene glycol system, *Chin. J. Chem. Eng.*, 17, 746, 2009.

2009YUB Yu, B. and Lowe, A.B., Synthesis of di- and tritertiary amine containing methacrylic monomers and their (co)polymerization via RAFT, *J. Polym. Sci.: Part A: Polym. Chem.*, 47, 1877, 2009.

2009ZA1 Zafarani-Moattar, M.T. and Zaferanloo, A., Measurement and correlation of phase equilibria in aqueous two-phase systems containing polyvinylpyrrolidone and dipotassium tartrate or dipotassium oxalate at different temperatures, *J. Chem. Thermodyn.*, 41, 864, 2009.

2009ZA2 Zafarani-Moattar, M.T. and Nikjoo, D., Phase diagrams for liquid-liquid and liquid-solid equilibrium of the ternary poly(ethylene glycol) dimethyl ether 2000 + sodium carbonate + water system, *J. Chem. Eng. Data*, 54, 2918, 2009.

2009ZHA Zhao, C., Zhuang, X., He, P., Xiao, C., He, C., Sun, J., Chen, X., and Jing, X., Synthesis of biodegradable thermo- and pH-responsive hydrogels for controlled drug release, *Polymer*, 50, 4308, 2009.

2010ARO Aroon, M.A., Ismail, A.F., Montazer-Rahmati, M.M., and Matsuura, T., Morphology and permeation properties of polysulfone membranes for gas separation: Effects of non-solvent additives and co-solvent, *Separation Purification Technol.*, 72, 194, 2010.

2010AZI Azimaie, R., Pazuki, G.R., Taghikhani, V., Vossoughi, M., and Ghotbi, C., Liquid–liquid phase equilibrium of $MgSO_4$ and PEG1500 aqueous two-phase system, *Phys. Chem. Liq.*, 48, 764, 2010.

2010BAO Bao, H., Li, L., Leong, W.C., and Gan, L.H., Thermo-responsive association of chitosan-*graft*-poly(*N*-isopropylacrylamide) in aqueous solutions, *J. Phys. Chem. B*, 114, 10666, 2010.

2010BER Berber, M.R., Mori, H., Hafez, I.H., Minagawa, K., Tanaka, M., Niidome, T., Katayama, Y., Maruyama, A., Hirano, T., Maeda, Y., and Mori, T., Unusually large hysteresis of temperature-responsive poly(*N*-ethyl-2-propionamidoacrylamide) studied by microcalorimetry and FT-IR, *J. Phys. Chem. B*, 114, 7784, 2010.

2010BIV Bivigou-Koumba, A.M., Görnitz, E., Laschewsky, A., Müller-Buschbaum, P., and Papadakis, C.M., Thermoresponsive amphiphilic symmetrical triblock copolymers with a hydrophilic middle block made of poly(*N*-isopropylacrylamide): Synthesis, self-organization, and hydrogel formation, *Colloid Polym. Sci.*, 288, 499, 2010.

2010BLO Bloksma, M.M., Bakker, D.J., Weber, C., Hoogenboom, R., and Schubert, U.S., The effect of Hofmeister salts on the LCST transition of poly(2-oxazoline)s with varying hydrophilicity, *Macromol. Rapid Commun.*, 31, 724, 2010.

2010CH1 Chen, J.-S., Tu, S.-L., and Tsay, R.-Y., A morphological study of porous polylactide scaffolds prepared by thermally induced phase separation, *J. Taiwan Inst. Chem. Eng.*, 41, 229, 2010.

2010CH2 Cheng, H., Xie, S., Zhou, Y., Huang, W., Yan, D., Yang, J., and Ji, B., Effect of degree of branching on the thermoresponsive phase transition behaviors of hyperbranched multiarm copolymers: Comparison of systems with LCST transition based on coil-to-globule transition or hydrophilic-hydrophobic balance, *J. Phys. Chem. B*, 114, 6291, 2010.

2010CH3 Chen, B.-K., Lo, S.-H., and Lee, S.-F., Temperature responsive methacrylamide polymers with antibacterial activity, *Chin. J. Polym. Sci.*, 28, 607, 2010.

2010CH4 Chen, G.-T., Wang, C.-H., Zhang, J.-G., Wang, Y., Zhang, R., Du, F.-S., Yan, N., Kou, Y., and Li, Z.-C., Toward functionalization of thermoresponsive poly(*N*-vinyl-2-pyrrolidone), *Macromolecules*, 43, 9972, 2010.

2010CH5 Chen, W. and Shan, G., Phase separation of ammonium sulfate-water-polyacrylamide and cationic polyacrylamide aqueous two-phase systems, *Huagong Xuebao*, 61, 1560, 2010.

2010CHO Chollakup, R., Smitthipong, W., Eisenbach, C.D., and Tirrell, M., Phase behavior and coacervation of aqueous poly(acrylic acid)-poly(allylamine) solutions, *Macromolecules*, 43, 2518, 2010.

2010DIP Dilip, M., Griffin, S.T., Spear, S.K., Rodriguez, H., Rijksen, C., and Rogers, R.D., Comparison of temperature effects on the salting out of poly(ethylene glycol) versus poly(ethylene oxide)-poly(propylene oxide) random copolymer, *Ind. Eng. Chem. Res.*, 49, 2371, 2010.

2010DO1 Dong, A., Zhai, Y., Xiao, L., Qi, H., Tian, Q., Deng, L., and Guo, R., Thermo-sensitive behavior of poly(ethylene glycol)/poly(2-(*N,N*-dimethylamino)ethyl methacrylate) double hydrophilic block copolymers, *J. Polym. Sci.: Part B: Polym. Phys.*, 48, 503, 2010.

2010DO2 Dong, H. and Matyjaszewski, K., Thermally responsive P(M(EO)2MA-*co*-OEOMA) copolymers via AGET ATRP in miniemulsion, *Macromolecules*, 43, 4623, 2010.

2010DUB Dubovika, A.S., Grinberg, N.V., and Grinberg, V.Ya., Energetics of phase separation in aqueous solutions of poly(*N*-isopropylacrylamide), *Polym. Sci., Ser. A*, 52, 565, 2010.

2010ELI Eliassi, A. and Parach, A., Cloud points of poly(propylene glycol) aqueous mixtures at various concentrations, *J. Chem. Eng. Data*, 55, 4010, 2010.

2010FAR Fares, M.M. and Othman, A.A., Smart pH-sensitive alternating copolymers of (methylacrylamide-hydroxyethyl methacrylate): Kinetic and physical properties, *J. Macromol. Sci.: Part A: Pure Appl. Chem.*, 47, 61, 2010.

2010FED Fedicheva, N., Ninni, L., and Maurer, G., Aqueous two-phase systems containing *N*-vinylpyrrolidone: Experimental results and correlation/prediction, *Fluid Phase Equil.*, 299, 127, 2010.

2010FRI Frith, W.J., Mixed biopolymer aqueous solutions – phase behaviour and rheology, *Adv. Colloid Interface Sci.*, 161, 48, 2010.

2010GAO Gao, M., Jia, X., Li, Y., Liang, D., and Wei, Y., Synthesis and thermo-/pH- dual responsive properties of poly(amidoamine) dendronized poly(2-hydroxyethyl) methacrylate, *Macromolecules*, 43, 4314, 2010.

2010GOL Golova, L.K., Makarov, I.S., Matukhina, E.V., and Kulichikhin, V.G., Solutions of cellulose and its blends with synthetic polymers in *N*-methylmorpholine-*N*-oxide: Preparation, phase state, structure, and properties, *Polym. Sci., Ser. A*, 52, 1209, 2010.

2010GON Gonsior, N., Schmitz, S., and Ritter, H., Thermal sensitivity of *tert*-butyloxy-carbonylmethyl-modified polyquats in condensed phase and solubility properties of copolymers with *N*-isopropylacrylamide, *Macromol. Chem. Phys.*, 211, 1695, 2010.

2010GRI Grinberg, V.Ya., Grinberg, N.V., Burova, T.V., Dubovik, A.S., Tur, D.R., and Papkov, V.S., Phase separation in aqueous solutions of polyethylaminophosphazene hydrochloride during heating, *Polym. Sci., Ser. A*, 52, 1220, 2010.

2010GUO Guo, Y., Feng, X., Chen, L., Zhao, Y., and Bai, J., Influence of the coagulation-bath temperature on the phase-separation process of poly(vinylidene fluoride)-*graft*-poly(*N*-isopropylacrylamide) solutions and membrane structures, *J. Appl. Polym. Sci.*, 116, 1005, 2010.

2010HAO Hao, J., Cheng, H., Butler, P., Zhang, L., and Han, C.C., Origin of cononsolvency, based on the structure of tetrahydrofuran-water mixture, *J. Chem. Phys.*, 132, 154902, 2010.

2010JAD Jadhav, V.B. and Patil, T.J., Influence of polyvinyl sulphonic acid (PVSA) on the thermodynamics of clouding behaviour of non ionic surfactant Tween 80, *Oriental J. Chem.*, 26, 623, 2010.

2010JIM	Jimenez, Y.P. and Galleguillos, H.R., (Liquid + liquid) equilibrium of ($NaClO_4$ + PEG 4000 + H_2O) ternary system at different temperatures, *J. Chem. Thermodyn.*, 42, 419, 2010.
2010JOC	Jochum, F.D., Roth, P.J., Kessler, D., and Theato, P., Double thermoresponsive block copolymers featuring a biotin end group, *Biomacromolecules*, 11, 2432, 2010.
2010KAL	Kalwarczyk, E., Golos, M., Holyst, R., and Fialkowski, M., Polymer-induced ordering and phase separation in ionic surfactants, *J. Colloid Interface Sci.*, 342, 93, 2010.
2010KA1	Katona, J.M., Sovilj, V.J., Petrovic, L.B., and Mucic, N.Z., Tensiometric investigation of the interaction and phase separation in a polymer mixture–ionic surfactant ternary system, *J. Serb. Chem. Soc.*, 75, 823, 2010.
2010KA2	Katsumoto, Y., Etoh, Y., and Shimoda, N., Phase diagrams of stereocontrolled poly(*N,N*-diethylacrylamide) in water, *Macromolecules*, 43, 3120, 2010.
2010KIM	Kim, H.C. and Kim, J.-D., The polydispersity effect of distributed oxyethylene chains on the cloud points of nonionic surfactants, *J. Colloid Interface Sci.*, 352, 444, 2010.
2010KIZ	Kizhnyaev, V.N., Krakhotkina, E.A., Petrova, T.L., Ratovskii, G.V., Tyukalova, O.V., Pokatilov, F.A., and Smirnov, A.I., Copolymerization of 5-vinyltetrazole with *N*-vinyllactams and properties of the copolymers, *Polym. Sci., Ser. B*, 52, 480, 2010.
2010KLO	Kloxin, C.J. and Zanten, J.H. van, High pressure phase diagram of an aqueous PEO-PPO-PEO triblock copolymer system via probe diffusion measurements, *Macromolecules*, 43, 2084, 2010.
2010KOJ	Kojima, C., Yoshimura, K., Harada, A., Sakanishi, Y., and Kono, K., Temperature-sensitive hyperbranched poly(glycidol)s with oligo(ethylene glycol) monoethers, *J. Polym. Sci.: Part A: Polym. Chem.*, 48, 4047, 2010.
2010KOU	Kourilova, H., Stastna, J., Hanykova, L., Sedlakova, Z., and Spevacek, J., ^{1}H NMR study of temperature-induced phase separation in solutions of poly(*N*-isopropyl-methacrylamide-*co*-acrylamide) copolymers, *Eur. Polym. J.*, 46, 1299, 2010.
2010KUL	Kulichikhin, V.G., Makarova, V.V., Tolstykh, M.Yu., and Vasil'ev, G.B., Phase equilibria in solutions of cellulose derivatives and the rheological properties of solutions in various phase states, *Polym. Sci., Ser. A*, 52, 1196, 2010.
2010LAB	Labuta, J., Hill, J.P., Hanykova, L., Ishihara, S., and Ariga, K., Probing the micro-phase separation of thermo-responsive amphiphilic polymer in water/ethanol solution, *J. Nanosci. Nanotechnol.*, 10, 8408, 2010.
2010LE1	Lee, R.-S., Huang, Y.-T., and Chen, W.-H., Synthesis and characterization of temperature-sensitive block copolymers from poly(*N*-isopropylacrylamide) and 4-methyl-ε-caprolactone or 4-phenyl-ε-caprolactone, *J. Appl. Polym. Sci.*, 118, 1634, 2010.
2010LE2	Lee, R.-S., Chen, W.-H., and Huang, Y.-T., Synthesis and characterization of dual stimuli-responsive block copolymers based on poly(*N*-isopropylacrylamide)-*b*-poly(pseudoamino acid), *Polymer*, 51, 5942, 2010.
2010LEM	Lemos, L.R. de, Santos, I.J.B., Rodrigues, G.D., Ferreira, G.M.D., Silva, L.H.M. da, Silva, M. do C.H. da, and Carvalho, R.M.M. de, Phase compositions of aqueous two-phase systems formed by L35 and salts at different temperatures, *J. Chem. Eng. Data*, 55, 1193, 2010.
2010LIT	Litmanovich, E.A., Chernikova, E.V., Stoychev, G.V., and Zakharchenko, S.O., Unusual phase behavior of the mixture of poly(acrylic acid) and poly(diallyldimethyl ammonium chloride) in acidic media, *Macromolecules*, 43, 6871, 2010.

2010LI1 Liu, X., Ni, P., He, J., and Zhang, M., Synthesis and micellization of pH/temperature-responsive double-hydrophilic diblock copolymers polyphosphoester-*block*-poly[2-(dimethylamino)ethyl methacrylate] prepared via ROP and ATRP, *Macromolecules*, 43, 4771, 2010.

2010LI2 Liu, Z., Hu, J., Sun, J., He, G., Li, Y., and Zhang, G., Preparation of thermoresponsive polymers bearing amino acid diamide derivatives via RAFT polymerization, *J. Polym. Sci.: Part A: Polym. Chem.*, 48, 3573, 2010.

2010LI3 Liu, Y., Kemmer, A., Keim, K., Curdy, C., Petersen, H., Kissel, T., Poly(ethylene carbonate) as a surface-eroding biomaterial for in situ forming parenteral drug delivery systems: A feasibility study, *Eur. J. Pharmaceut. Biopharmaceut.*, 76, 222, 2010.

2010LIY Li, Y. and Cao, X., Prediction of phase diagrams for new pH-thermo sensitive recycling aqueous two-phase systems, *Fluid Phase Equil.*, 298, 206, 2010.

2010LUE Lue, T., Shan, G., and Shang, S., Intermolecular interaction in aqueous solution of binary blends of poly(acrylamide) and poly(ethylene glycol), *J. Appl. Polym. Sci.*, 118, 2572, 2010.

2010LUO Luo, C., Liu, Y., and Li, Z., Thermo- and pH-responsive polymer derived from methacrylamide and aspartic acid, *Macromolecules*, 43, 8101, 2010.

2010LUY Lu, Y.-M., Yang, Y.-Z., Zhao, X.-D., and Xia, C.-B., Bovine serum albumin partitioning in polyethylene glycol (PEG)/potassium citrate aqueous two-phase systems, *Food Bioproducts Processing*, 88, 40, 2010.

2010LUZ Luzon, M., Boyer, C., Peinado, C., Corrales, T., Whittaker, M., Tao, L., and Davis, T.P., Water-soluble, thermoresponsive, hyperbranched copolymers based on PEG-methacrylates: Synthesis, characterization, and LCST behavior, *J. Polym. Sci.: Part A: Polym. Chem.*, 48, 2783, 2010.

2010MAD Madeira, P.P., Reis, C.A., Rodrigues, A.E., Mikheeva, L.M., and Zaslavsky, B.Y., Solvent properties governing solute partitioning in polymer/polymer aqueous two-phase systems: Nonionic compounds, *J. Phys. Chem. B*, 114, 457, 2010.

2010MAE Maeda, Y. and Takaku, S., Lower critical solution temperature behavior of poly(*N*-tetrahydrofurfuryl(meth)acrylamide) in water and alcohol-water mixtures, *J. Phys. Chem. B*, 114, 13110, 2010.

2010MAG Maghsoud, Z., Famili, M.H.M., and Madaeni, S.S., Phase diagram calculations of water/tetrahydrofuran/poly(vinyl chloride) ternary system based on a compressible regular solution model, *Iran. Polym. J.*, 19, 581, 2010.

2010MAH Mahajan, S., Shaheen, A., Banipal, T.S., and Mahajan, R.K., Cloud point and surface tension studies of triblock copolymer-ionic surfactant mixed systems in the presence of amino acids or dipeptides and electrolytes, *J. Chem. Eng. Data*, 55, 3995, 2010.

2010MAL Ma, L., Liu, R., Tan, J., Wang, D., Jin, X., Kang, H., Wu, M., and Huang, Y., Self-assembly and dual-stimuli sensitivities of hydroxypropylcellulose-*graft*-poly(*N,N*-dimethyl aminoethyl methacrylate) copolymers in aqueous solution, *Langmuir*, 26, 8697, 2010.

2010MA1 Martins, J.P., Coimbra, J.S. dos R., Oliveira, F.C. de, Sanaiotti, G., Silva, C.A.S. da, Silva, L.H.M. da, and Silva, M. do C.D. da, Liquid-liquid equilibrium of aqueous two-phase system composed of poly(ethylene glycol) 400 and sulfate salts, *J. Chem. Eng. Data*, 55, 1247, 2010.

2010MA2 Martins, J.P., Silva, M. do C.H. da, Silva, L.H.M. da, Senra, T.D.A., Ferreira, G.M.D., Coimbra, J.S. dos R., and Minim, L.A., Liquid-liquid phase equilibrium of triblock copolymer F68, poly(ethylene oxide)-*b*-poly(propylene oxide)-*b*-poly(ethylene oxide), with sulfate salts, *J. Chem. Eng. Data*, 55, 1618, 2010.

2010MEN Mendrek, S., Mendrek, A., Adler, H.-J., Dworak, A., and Kuckling, D., Temperature-sensitive behaviour of poly(glycidol)-*b*-poly(*N*-isopropylacrylamide) block copoly-mers, *Colloid Polym. Sci.*, 288, 777, 2010.

2010MIR Mirsiaghi, M., Pazuki, G., Vossoughi, M., and Alemzadeh, I., Partitioning of L-lysine monohydrochloride in aqueous two-phase systems of poly(ethylene glycol) and dipotassium hydrogen phosphate or trisodium citrate 5-hydrate, *J. Chem. Eng. Data*, 55, 3005, 2010.

2010MO1 Mori, H., Kato, I., Saito, S., and Endo, T., Proline-based block copolymers displaying upper and lower critical solution temperatures, *Macromolecules*, 43, 1289, 2010.

2010MO2 Mori, T., Beppu, S., Berber, M.R., Mori, H., Makimura, T., Tsukamoto, A., Minagawa, K., Hirano, T., Tanaka, M., Niidome, T., Katayama, Y., Hirano, T., and Maeda, Y., Design of temperature-responsive polymers with enhanced hysteresis: α,α-Disubstituted vinyl polymers, *Langmuir*, 26, 9224, 2010.

2010PAT Patel, T., Bahadur, P., and Mata, J., The clouding behaviour of PEO–PPO based triblock copolymers in aqueous ionic surfactant solutions: A new approach for cloud point measurements, *J. Colloid Interface Sci.*, 345, 346, 2010.

2010PAZ Pazuki, G., Vosoughi, M., and Taghikhani, V., Partitioning of Penicillin G acylase in aqueous two-phase systems of poly(ethylene glycol) 20000 or 35000 and potassium dihydrogen phosphate or sodium citrate, *J. Chem. Eng. Data*, 55, 243, 2010.

2010PEN Peng, B., Hao, Y., Kang, H., Han, X., Peng, C., and Liu, H., Aggregation of *N*-carboxyethylchitosan in aqueous solution: Effects of pH, polymer concentration, and presence of a gemini surfactant, *Carbohydrate Res.*, 345, 101, 2010.

2010PER Perumalsamy, M. and Murugesan, T., Extended NRTL model for PEG + salt based aqueous two-phase system, *Asia-Pacific J. Chem. Eng.*, 5, 355, 2010.

2010QIA Qiao, Z.-Y., Du, F.-S., Zhang, R., Liang, D.-H., and Li, Z.-C., Biocompatible thermoresponsive polymers with pendent oligo(ethylene glycol) chains and cyclic ortho ester groups, *Macromolecules*, 43, 6485, 2010.

2010RO1 Rodrigues, G.D., Teixeira, L. da S., Ferreira, G.M.D., Silva, M. do C.H. da, Silva, L.H.M. da, and Carvalho, R.M.M. de, Phase diagrams of aqueous two-phase systems with organic salts and F68 triblock copolymer at different temperatures, *J. Chem. Eng. Data*, 55, 1158, 2010.

2010RO2 Rodríguez-Abreu, C., Sanchez-Domínguez, M., Sarac, B., Rogac, M.B., Shrestha, R.G., Shrestha, L.K., Varade, D., Ghosh, G., and Aswal, V.K., Solution behavior of aqueous mixtures of low and high molecular weight hydrophobic amphiphiles, *Colloid Polym. Sci.*, 288, 739, 2010.

2010ROT Roth, P.J., Jochum, F.D., Forst, F.R., Zentel, R., and Theato, P., Influence of end groups on the stimulus-responsive behavior of poly[oligo(ethylene glycol) methacrylate] in water, *Macromolecules*, 43, 4638, 2010.

2010RUE Rueda, J.C., Zschoche, S., Komber, H., Krahl, F., Arndt, K.-F., and Voit, B., New thermo-sensitive graft copolymers based on a poly(*N*-isopropylacrylamide) backbone and functional polyoxazoline grafts with random and diblock structure, *Macromol. Chem. Phys.*, 211, 706, 2010.

2010SAD Sadeghi, R. and Golabiazar, R., Thermodynamics of phase equilibria of aqueous poly(ethylene glycol) + sodium tungstate two-phase systems, *J. Chem. Eng. Data*, 55, 74, 2005.

2010SA1 Salabat, A.R., Abnosi, M.H., and Motahari, A., Application of aqueous mixtures of polypropylene glycol or polyethylene glycol with salts in proteomic analysis, *J. Iran. Chem. Soc.*, 7, 142, 2010.

2010SA2 Salabat, A., Tiani, S.M., and Rahmati, M.F., Liquid-liquid equilibria of aqueous two-phase systems composed of Triton X-100 and sodium citrate or magnesium sulfate salts, *CALPHAD*, 34, 81, 2010.

2010SAR Sardar, N., Ali, M.S., Kamil, M., and Kabir-ud-Din, Phase behavior of nonionic polymer hydroxypropylmethyl cellulose: Effect of gemini and single-chain surfactants on the energetics at the cloud point, *J. Chem. Eng. Data*, 55, 4990, 2010.

2010SCH Schmalz, A., Hanisch, M., Schmalz, H., and Müller, A.H.E., Double stimuli-responsive behavior of linear and star-shaped poly(*N,N*-diethylaminoethyl methacrylate) in aqueous solution, *Polymer*, 51, 1213, 2010.

2010SH1 Shahriari, Sh., Taghikhani, V., Vossoughi, M., Safekordi, A.A., Alemzadeh, I., and Pazuki, G.R., Measurement of partition coefficients of β-amylase and amyloglucosidase enzymes in aqueous two-phase systems containing poly(ethylene glycol) and Na_2SO_4/KH_2PO_4 at different temperatures, *Fluid Phase Equil.*, 292, 80, 2010.

2010SH2 Shahriari, S., Vossoughi, M., Taghikhani, V., Safekordi, A.A., and Alemzadeh, I., Experimental study and mathematical modeling of partitioning of β-amylase and amyloglucosidase in PEG-salt aqueous two-phase systems, *J. Chem. Eng. Data*, 55, 4968, 2010.

2010SHE Shechter, I., Ramon, O., Portnaya, I., Paz, Y., and Livney, Y.D., Microcalorimetric study of the effects of a chaotropic salt, KSCN, on the lower critical solution temperature (LCST) of aqueous poly(*N*-isopropylacrylamide) (PNIPA) solutions, *Macromolecules*, 43, 480, 2010.

2010SHI Shibayama, M. and Osaka, N., Pressure- and temperature-induced phase separation transition in homopolymer, block copolymer, and protein in water, *Macromol. Symp.*, 291-292, 115, 2010.

2010SIB Siband, E., Tran, Y., and Hourdet, D., pH- and thermo-responsive polymer assemblies in aqueous solution, *Progr. Colloid Polym. Sci.*, 137, 19, 2010.

2010SIL Silverio, S.C., Rodriguez, O., Teixeira, J.A., and Macedo, E.A., Liquid-liquid equilibria of UCON + (sodium or potassium) phosphate salt aqueous two-phase systems at 23 °C, *J. Chem. Eng. Data*, 55, 1285, 2010.

2010SU1 Sun, J., Peng, Y., Chen, Y., Liu, Y., Deng, J., Lu, L., and Cai, Y., Effect of molecular structure on thermoresponsive behaviors of pyrrolidone-based water-soluble polymers, *Macromolecules*, 43, 4041, 2010.

2010SU2 Sun, S. and Wu, P., Role of water/methanol clustering dynamics on thermosensitivity of poly(*N*-isopropylacrylamide) from spectral and calorimetric insights, *Macromolecules*, 43, 9501, 2010.

2010TAN Tang, X., Liang, X., Gao, L., Fand, X., and Zhou, Q., Water-soluble triply-responsive homopolymers of *N,N*-dimethylaminoethyl methacrylate with a terminal azobenzene moiety, *J. Polym. Sci.: Part B: Polym. Chem.*, 48, 2564, 2010.

2010TSU Tsui, H.-W., Wang, J.-H., Hsu, Y.-H., and Chen, L.-J., Study of heat of micellization and phase separation for Pluronic aqueous solutions by using a high sensitivity differential scanning calorimetry, *Colloid Polym. Sci.*, 288, 1687, 2010.

2010VIR Virtuoso, L.S., Silva, L.M. de S., Malaquias, B.S., Vello, K.A.S.F., Cindra, C.P.R., Silva, L.H.M. da, Mesquita, A.F., Silva, M.C.H. da, and Carvalho, R.M.M. de, Equilibrium phase behavior of triblock copolymer + sodium or + potassium hydroxides + water two-phase systems at different temperatures, *J. Chem. Eng. Data*, 55, 3847, 2010.

2010VSH Vshivkov, S.A., Adamova, L.V., and Galyas, G.A., Thermodynamics of mixtures and solutions of hydroxypropylcellulose with poly(ethylene glycol), *Russ. J. Appl. Chem.*, 83, 1196, 2010.

2010WAN Wang, Z. and Feng, H., Double cloud point of ethylene oxide–propylene oxide triblock copolymer in an aqueous solution, *Colloids Surfaces A*, 362, 110, 2010.

2010WAR Ward, M.A. and Georgiou, T.K., Thermoresponsive terpolymers based on methacrylate monomers: Effect of architecture and composition, *J. Polym. Sci.: Part A: Polym. Chem.*, 48, 775, 2010.

2010WEB Weber, C., Becer, C.R., Guenther, W., Hoogenboom, R., and Schubert, U.S., Dual responsive methacrylic acid and oligo(2-ethyl-2-oxazoline) containing graft copolymers, *Macromolecules*, 43, 160, 2010.

2010WUC Wu, C., Wang, J., Pei, Y., Wang, H., and Li, Z., Salting-out effect of ionic liquids on poly(propylene glycol) (PPG): Formation of PPG + ionic liquid aqueous two-phase systems, *J. Chem. Eng. Data*, 55, 5004, 2010.

2010XI1 Xie, X., Yan, Y., Han, J., Wang, Y., and Guan, W., Liquid-liquid equilibrium of aqueous two-phase systems of PPG400 and biodegradable salts at temperatures of (298.15, 308.15, and 318.15) K, *J. Chem. Eng. Data*, 55, 2857, 2010.

2010XI2 Xie, X., Han, J., Wang, Y., Yan, Y., Yin, G., and Guan, W., Measurement and correlation of the phase diagram data for PPG400 + (K_3PO_4, K_2CO_3, and K_2HPO_4) + H_2O aqueous two-phase systems at T = 298.15 K, *J. Chem. Eng. Data*, 55, 4741, 2010.

2010XIN Xing, D.Y., Peng, N., and Chung, T.-S., Formation of cellulose acetate membranes via phase inversion using ionic liquid, [BMIM]SCN, as the solvent, *Ind. Eng. Chem. Res.*, 49, 8761, 2010.

2010YEX Ye, X., Ding, Y., and Li, J., Scaling of the molecular weight-dependent thermal volume transition of poly(*N*-isopropylacrylamide), *J. Polym. Sci.: Part B: Polym. Phys.*, 48, 1388, 2010.

2010YUB Yu, B., Jiang, X., Yin, G., and Yin, J., Multistimuli-responsive hyperbranched poly(ether amine)s, *J. Polym. Sci.: Part A: Polym. Chem.*, 48, 4252, 2010.

2010ZA1 Zafarani-Moattar, M.T. and Nasiri, S., Phase diagrams for liquid-liquid and liquid-solid equilibrium of the ternary polyethylene glycol dimethyl ether 2000 + trisodium phosphate + water system at different temperatures and ambient pressure, *CALPHAD*, 34, 222, 2010.

2010ZA2 Zafarani-Moattar, M.T. and Nasiri, S., (Liquid + liquid) and (liquid + solid) equilibrium of aqueous two-phase systems containing poly(ethylene glycol) dimethyl ether 2000 and disodium hydrogen phosphate, *J. Chem. Thermodyn.*, 42, 1071, 2010.

2010ZA3 Zafarani-Moattar, M.T. and Nemati-Kande, E., Study of liquid-liquid and liquid-solid equilibria of the ternary aqueous system containing poly(ethylene glycol) dimethyl ether 2000 and tripotassium phosphate at different temperatures: Experiment and correlation, *CALPHAD*, 34, 478, 2010.

2010ZH1 Zhang, X., Zhou, L., Zhang, X., and Dai, H., Synthesis and solution properties of temperature-sensitive copolymers based on NIPAM, *J. Appl. Polym. Sci.*, 116, 1099, 2010.

2010ZH2 Zhang, S., Liu, S.-X., Han, X.-Y., Dang, L., Qi, X.-J., and Yang, X., Synthesis of P(DEAM-*co*-NAS) and the environmental effects of its temperature sensitivity, *Acta Phys.-Chim. Sin.*, 26, 2189, 2010.

2010ZH3 Zhao, Y., Guo, K., Wang, C., and Wang, L., Effect of inclusion complexation with cyclodextrin on the cloud point of poly(2-(dimethylamino)ethyl methacrylate) solution, *Langmuir*, 26, 8966, 2010.

2010ZH4 Zhao, R.-L., Lin, K., Zhou, X.-G., and Liu, S.-L., Solubility of poly(*N*-isopropylacrylamide) in aqueous methanol from Raman spectroscopy, *Acta Phys.-Chim. Sin.*, 26, 1915, 2010.

2010ZH5 Zhao, J., Hoogenboom, R., Van Assche, G., and Van Mele, B., Demixing and remixing kinetics of poly(2-isopropyl-2-oxazoline) (PIPOZ) aqueous solutions studied by modulated temperature differential scanning calorimetry, *Macromolecules*, 43, 6853, 2010.

2010ZHE Zheng, S., Shi, S., Xia, Y., Wu, Q., Su, Z., and Chen, X., Study on micellization of poly(*N*-isopropylacrylamide-butyl acrylate) macromonomers in aqueous solution, *J. Appl. Polym. Sci.*, 118, 671, 2010.

2010ZHO Zhou, L., Wan, J., and Cao, X, Synthesis of thermo-sensitive copolymer with affinity butyl ligand and its application in lipase purification, *J. Chromatogr. B*, 878, 1025, 2010.

2011AGR Agrawal, A.K., Jassal, M., Sahoo, A., and Garapati, S.K., Phase behavior and mechanism of formation of protofiber morphology of solution spun poly(acrylonitrile) copolymers in DMF-water system, *J. Appl. Polym. Sci.*, 119, 837, 2011.

2011AND Andrade, V.M. de, Rodrigues, G.D., Carvalho, R.M.M. de, Silva, L.H.M. da, and Silva, M.C.H. da, Aqueous two-phase systems of copolymer L64 + organic salt + water: Enthalpic L64-salt interaction and Othmer-Tobias, NRTL and UNIFAC thermodynamic modeling, *Chem. Eng. J.*, 171, 9, 2011.

2011ASE Asenjo, J.A. and Andrews, B.A., Aqueous two-phase systems for protein separation: A perspective, *J. Chromatography A*, 1218, 8826, 2011.

2011ASS Assche, G.V., Mele, B.V., Li, T., and Nies, E., Adjacent UCST phase behavior in aqueous solutions of poly(vinyl methyl ether): Detection of a narrow low temperature UCST in the lower concentration range, *Macromolecules*, 44, 993, 2011.

2011BAT Batigoec, C. and Akbas, H., Spectrophotometric determination of cloud point of Brij 35 nonionic surfactant, *Fluid Phase Equil.*, 303, 91, 2011.

2011BIA Bianco, C.L., Schneider, C.S., Santonicola, M., Lenhoff, A.M., and Kaler, E.W., Effects of urea on the microstructure and phase behavior of aqueous solutions of poly(oxyethylene) surfactants, *Ind. Eng. Chem. Res.*, 50, 85, 2011.

2011CAN Can, A., Höppener, S., Guillet, P., Gohy, J.-F., Hoogenboom, R., and Schubert, U.S., Upper critical solution temperature switchable micelles based on polystyrene-*block*-poly(methyl acrylate) block copolymers, *J. Polym. Sci.: Part A: Polym. Chem.*, 49, 3681, 2011.

2011CAO Cao, Y. and He, W., Functionalized biocompatible poly(*N*-vinyl-2-caprolactam) with pH-dependent lower critical solution temperature behaviors, *Macromol. Chem. Phys.*, 212, 2503, 2011.

2011CHE1 Chen, C., Wang, Z., and Li, Z., Thermoresponsive polypeptides from pegylated poly(L-glutamate)s, *Biomacromolecules*, 12, 2859, 2011.

2011CHE2 Chen, J., Liu, M., Gong, H., Huang, Y., and Chen, C., Synthesis and self-assembly of thermoresponsive PEG-*b*-PNIPAM-*b*-PCL ABC triblock copolymer through the combination of atom transfer radical polymerization, ring-opening polymerization, and click chemistry, *J. Phys. Chem. B*, 115, 14947, 2011.

2011CHU Churilina, E.V., Sukhanov, P.T., Korenman, Ya.I., Il'in, A.N., Shatalov, G.V., and Bolotov, V.M., The distribution coefficients of phenol and substituted phenols in the ammonium sulfate–poly(*N*-vinylpyrrolidone)–water system, *Russ. J. Phys. Chem. A*, 85, 568, 2011.

2011CUI Cui, Q., Wu, F., and Wang, E., Thermosensitive behavior of poly(ethylene glycol)-based block copolymer (PEG-*b*-PADMO) controlled via self-assembled microstructure, *J. Phys. Chem. B*, 115, 5913, 2011.

2011DEY Deyerle, B.A. and Zhang, Y., Effects of Hofmeister anions on the aggregation behavior of PEO-PPO-PEO triblock copolymers, *Langmuir*, 27, 9203, 2011.

2011EIS Eissa, A.M. and Khosravi, E., Synthesis of a new smart temperature responsive glycopolymer via click-polymerisation, *Eur. Polym. J.*, 47, 61, 2011.

2011FEN Feng, L., Liu, Y., Hao, J., Li, X., Xiong, C., and Deng, X., Synthesis and properties of novel thermoresponsive polyesters with oligo(ethylene glycol) pendent chains, *Macromol. Chem. Phys.*, 212, 2626, 2011.

2011FE1 Ferreira, L.A. and Teixeira, J.A., Salt effect on the aqueous two-phase system PEG 8000-sodium sulfate, *J. Chem. Eng. Data*, 53, 133, 2011.

2011FE2 Ferreira, L.A. and Teixeira, J.A., Salt effect on the (polyethylene glycol 8000 + sodium sulfate) aqueous two-phase system: Relative hydrophobicity of the equilibrium phases, *J. Chem. Thermodyn.*, 43, 1299, 2011.

2011FET Fettaka, M., Issaadi, R., Moulai-Mostefa, N., Dez, I., LeCerf, D., and Picton, L., Thermo sensitive behavior of cellulose derivatives in dilute aqueous solutions: From macroscopic to mesoscopic scale, *J. Colloid Interface Sci.*, 357, 372, 2011.

2011FIS Fischer, F., Zufferey, D., and Tahoces, R., Lower critical solution temperature in superheated water: the highest in the poly(*N,N*-dialkylacrylamide) series, *Polym. Int.*, 60, 1259, 2011.

2011GAL Gallow, K.C., Jhon, Y.K., Tang, W., Genzer, J., and Loo, Y.-L., Cloud point suppression in dilute solutions of model gradient copolymers with prespecified composition profiles, *J. Polym. Sci.: Part B: Polym. Phys.*, 49, 629, 2011.

2011GLA Glatzel, S., Laschewsky, A., and Lutz, J.-F., Well-defined uncharged polymers with a sharp UCST in water and in physiological milieu, *Macromolecules*, 44, 413, 2011.

2011HIR Hirao, A., Inushima, R., Nakayama, T., Watanabe, T., Yoo, H.-S., Ishizone, T., Sugiyama, K., Kakuchi, T., Carlotti, S., and Deffieux, A., Precise synthesis of thermo-responsive and water-soluble star-branched polymers and star block copolymers by living anionic polymerization, *Eur. Polym. J.*, 47, 713, 2011.

2011HUF Huff, A., Patton, K., Odhner, H., Jacobs, D.T., Clover, B.C., and Greer, S.C., Micellization and phase separation for triblock copolymer 17R4 in H_2O and in D_2O, *Langmuir*, 27, 1707, 2011.

2011IEO Ieong, N.S., Redhead, M., Bosquillon, C., Alexander, C., Kelland, M., and O'Reilly, R.K., The missing lactam − thermoresponsive and biocompatible poly(*N*-vinylpiperidone) polymers by xanthate-mediated RAFT polymerization, *Macromolecules*, 44, 886, 2011.

2011ISE Ise, T., Nagaoka, K., Osa, M., and Yoshizaki, T., Cloud points in aqueous solutions of poly(*N*-isopropylacrylamide) synthesized by aqueous redox polymerization (exp. data by T. Yoshizaki), *Polym. J.*, 43, 164, 2011.

2011JI1 Jimenez, Y.P., Taboada, M.E., and Galleguillos, H.R., Cloud-point measurements of the {H_2O + poly(ethylene glycol) + $NaNO_3$} system, *J. Chem. Thermodyn.*, 43, 1204, 2011.

2011JI2 Jimenez, Y.P. and Galleguillos, H.R., (Liquid + liquid) equilibrium of ($NaNO_3$ + PEG 4000 + H_2O) ternary system at different temperatures, *J. Chem. Thermodyn.*, 43, 1573, 2011.

2011JOH Johansson, H.-O., Feitosa, E., and Pessoa, A., Phase diagrams of the aqueous two-phase systems of poly(ethylene glycol)/sodium polyacrylate/salts, *Polymers*, 3, 587, 2011.

2011JUN Jung, S.-H., Song, H.-Y., Lee, Y., Jeong, H.M., and Lee, H.-I., Novel thermoresponsive polymers tunable by pH, *Macromolecules*, 44, 1628, 2011.

2011KEL Kelland, M.A., Tuning the thermoresponsive properties of hyperbranched poly(ester amide)s based on diisopropanolamine and cyclic dicarboxylic anhydrides, *J. Appl. Polym. Sci.*, 121, 2282, 2011.

2011KEM Kempe, K., Weber, C., Babiuch, K., Gottschaldt, M., Hoogenboom, R., and Schubert, U.S., Responsive glyco-poly(2-oxazoline)s: Synthesis, cloud point tuning, and lectin binding, *Biomacromolecules*, 12, 2591, 2011.

2011KHA Khayati, G., Daghbandan, A., Gilvari, H., and Pheyz-Sani, N., Liquid-liquid equilibria of aqueous two-phase systems containing polyethylene glycol 4000 and two different salts of ammonium, *Res. J. Appl. Sci., Eng. Technol.*, 3, 96, 2011.

2011KIM Kim, H., Lee, S., Noh, M., Lee, S.H., Moka, Y., Jin, G.-W., Seo, J.-H., and Lee, Y., Thermosensitivity control of polyethlyenimine by simple acylation, *Polymer*, 52, 1367, 2011.

2011KOW Kowalczuk,A., Kronek, J., Bosowska, K., Trzebicka, B., and Dworak, A., Star poly(2-ethyl-2-oxazoline)s: Synthesis and thermosensitivity, *Polym. Int.*, 60, 1001, 2011.

2011KUM Kumar, A.C., Erothu, H., Bohidar, H.B., and Mishra, A.K., Bile-salt-induced aggregation of poly(*N*-isopropylacrylamide) and lowering of the lower critical solution temperature in aqueous solutions, *J. Phys. Chem. B*, 115, 433, 2011.

2011LEM Lemos, L.R. de, Patrício, P. da R., Rodrigues, G.D., Carvalho, R.M.M. de, Silva, M.C.H. da, and Silva, L.H.M. da, Liquid–liquid equilibrium of aqueous two-phase systems composed of poly(ethylene oxide) 1500 and different electrolytes ($(NH_4)_2SO_4$, $ZnSO_4$ and K_2HPO_4): Experimental and correlation, *Fluid Phase Equil.*, 305, 19, 2011.

2011LIB Libera, M., Trzebicka, B., Kowalczuk, A., Walach, W., and Dworak, A., Synthesis and thermoresponsive properties of four arm, amphiphilic poly(*tert*-butyl glycidylether)-*block*-polyglycidol stars, *Polymer*, 52, 250, 2011.

2011LIX Li, X., Lian, Z., Dong, B., Xu, Y., Yong, Q., and Yu, S., Extractive bioconversion of xylan for production of xylobiose and xylotriose using a PEG 6000/sodium citrate aqueous two-phase system, *Korean J. Chem. Eng.*, 28, 1897, 2011.

2011LON Longenecker, R., Mu, T., Hanna, M., Burke, N.A.D., and Stöver, H.D.H., Thermally responsive 2-hydroxyethyl methacrylate polymers: Soluble-insoluble and soluble-insoluble-soluble transitions, *Macromolecules*, 44, 8962, 2011.

2011MAL Maleki, A., Zhu, K., Pamies, R., Schmidt, R.R., Kjøniksen, A.-L., Karlsson, G., Cifre, J.G.H., Garcia de la Torre, J., and Nyström, B., Effect of polyethylene glycol (PEG) length on the association properties of temperature-sensitive amphiphilic triblock copolymers (PNIPAAMm-*b*-PEGn-*b*-PNIPAAMm) in aqueous solution, *Soft Matter*, 7, 8111, 2011.

2011MAN Mangold, C., Obermeier, B., Wurm, F., and Frey, H., From an epoxide monomer toolkit to functional PEG copolymers with adjustable LCST behavior, *Macromol. Rapid Commun.*, 32, 1930, 2011.

2011MCF McFaul, C.A., Alb, A.M., Drenski, M.F., and Reed, W.F., Simultaneous multiple sample light scattering detection of LCST during copolymer synthesis, *Polymer*, 52, 4825, 2011.

2011MIZ Mizuntani, M., Satoh, K., and Kamigaito, M., Degradable poly(*N*-isopropylacrylamide) with tunable thermosensitivity by simultaneous chain- and step-growth radical polymerization, *Macromolecules*, 44, 2382, 2011.

2011MOK Mokhtarani, B., Mortaheb, H.R., Mafi, M., and Amini, M.H., Partitioning of α-lactalbumin and β-lactoglobulin in aqueous two-phase systems of poly(*N*-vinylpyrrolidone) and potassium phosphate, *J. Chromatogr. B*, 879, 721, 2011.

2011MUK Mukherjee, P., Padhan, S.K., Dash, S., Patel, S., Mohapatra, P.K., and Mishra, B.K., Effect of temperature on pseudoternary system Tween-80–butanol–hexane–water, *J. Colloid Interface Sci.*, 355, 157, 2011.

2011MUN Munk, T., Hietala, S., Kalliomaeki, K., Nuopponen, M., Tenhu, H., Tian, F., Rantanen, J., and Baldursdottir, S., Behaviour of stereoblock poly(*N*-isopropyl acrylamide) in acetone–water mixtures, *Polym. Bull.*, 67, 677, 2011.

2011NAK Nakano, S., Ogiso, T., Kita, R., Shinyashiki, N., Yagihara, S., Yoneyama, M., and Katsumoto, Y., Thermoreversible gelation of isotactic-rich poly(*N*-isopropylacrylamide) in water, *J. Chem. Phys.*, 135, 114903, 2011.

2011NAQ Naqvi, A.Z., Khatoon, S., and Kabir-ud-Din, Phase separation phenomenon in non-ionic surfactant TX-114 micellar solutions: Effect of added surfactants and polymers, *J. Solution Chem.*, 40, 643, 2011.

2011NAS Nascimento, K.S., Yelo, S., Cavada, B.S., Azevedo, A.M., and Aires-Barros, M.R., Liquid-liquid equilibrium data for aqueous two-phase systems composed of ethylene oxide propylene oxide copolymers, *J. Chem. Eng. Data*, 56, 190, 2011.

2011PAK Pakuro, N.I., Nakhmanovich, B.I., Pergushov, D.V., and Chibirova, F.Kh., Thermosensitivity of poly(*N*-vinylpyrrolidone): Effect of transition-metal halides, *Polym. Sci., Ser. A*, 53, 6, 2011.

2011PAR Paris, R., Liras, M., and Quijada-Garrido, I., Thermoresponsive behavior of mixtures of epoxy functionalized oligo(ethylene glycol) methacrylate copolymers, *Macromol. Chem. Phys.*, 212, 1859, 2011.

2011PAT Patrício, P. da R., Mageste, A.B., Lemos, L.R. de, Carvalho, R.M.M. de, Silva, L.H.M. da, and Silva, M.C.H. da, Phase diagram and thermodynamic modeling of PEO + organic salts + H_2O and PPO + organic salts + H_2O aqueous two-phase systems, *Fluid Phase Equil.*, 305, 1, 2011.

2011PER Perumalsamy, M. and Batcha, M.I., Synergistic extraction of bovine serum albumin using poly(ethylene glycol) based aqueous biphasic system, *Process Biochem.*, 46, 494, 2011.

2011PHA Phan, H.T.T., Zhu, K., Kjoniksen, A.L., and Nyström, B., Temperature-responsive self-assembly of charged and uncharged hydroxyethylcellulose-*graft*-poly(*N*-isopropylacrylamide) copolymer in aqueous solution, *Colloid Polym. Sci.*, 289, 993, 2011.

2011POS Poschlad, K. and Enders, S., Thermodynamics of aqueous solutions containing poly(*N*-isopropylacrylamide), *J. Chem. Thermodyn.*, 43, 262, 2011.

2011RE1 Reddy, P.M. and Venkatesu, P., Ionic liquid modifies the lower critical solution temperature (LCST) of poly(*N*-isopropylacrylamide) in aqueous solution, *J. Phys. Chem. B*, 115, 4752, 2011.

2011RE2 Reddy, P.M., Venkatesu, P., and Bohidar, H.B., Influence of polymer molecular weight and concentration on coexistence curve of isobutyric acid + water, *J. Phys. Chem. B*, 115, 12065, 2011.

2011REG Regupathi, I., Srikanth, C.K., and Sindhu, N., Liquid-liquid equilibrium of poly(ethylene glycol) 2000 + diammonium hydrogen citrate + water system at different temperatures, *J. Chem. Eng. Data*, 56, 3643, 2011.

2011ROT Roth, P.J., Jochum, F.D., and Theato, P., UCST-type behavior of poly[oligo(ethylene glycol) methyl ether methacrylate] (POEGMA) in aliphatic alcohols: Solvent, co-solvent, molecular weight, and end group dependences, *Soft Matter*, 7, 2484, 2011.

2011RUG Ru, G. and Feng, J., Effects of end groups on phase transition and segmental mobility of poly(*N*-isopropylacrylamide) chains in D_2O, *J. Polym. Sci.: Part B: Polym. Phys.*, 49, 749, 2011.

2011SAD Sadeghi, R. and Kahaki, H.B., Thermodynamics of aqueous solutions of poly(ethylene glycol) dimethyl ethers in the presence or absence of ammonium phosphate salts, *Fluid Phase Equil.*, 306, 219, 2011.

2011SAL1 Salabat, A., Far, M.R., and Moghadam, S.T., Partitioning of amino acids in surfactant based aqueous two-phase systems containing the nonionic surfactant (Triton X-100) and salts, *J. Solution Chem.*, 40, 61, 2011.

2011SAL2 Salabat, A., Sadeghi, R., Moghadam, S.T., and Jamehbozorg, B., Partitioning of L-methionine in aqueous two-phase systems containing poly(propylene glycol) and sodium phosphate salts, *J. Chem. Thermodyn.*, 43, 1525, 2011.

2011SAR Sardar, N., Kamil, M., Kabir-ud-Din, and Ali, M.S., Solution behavior of nonionic polymer hydroxypropylmethyl cellulose: Effect of salts on the energetics at the cloud point, *J. Chem. Eng. Data*, 56, 984, 2011.

2011SHI Shinde, V.S., Girme, M.R., and Pawar, V.U., Thermoresponsive polystyrene-*b*-poly(*N*-isopropylacrylamide) copolymers by atom transfer radical polymerization, *Indian J. Chem.*, 50A, 781, 2011.

2011SON Song, X., Zhang, Y., Yang, D., Yuan, L., Hu, J., LU, G., and Huang, X., Convenient synthesis of thermo-responsive PtBA-*g*-PPEGMEMA well-defined amphiphilic graft copolymer without polymeric functional group transformation, *J. Polym. Sci.: Part A: Polym. Chem.*, 49, 3328, 2011.

2011TIA Tian, H.-Y., Yan, J.-J., Wang, D., Gu, C., You, Y.-Z., and Chen, X.-S., Synthesis of thermo-responsive polymers with both tunable UCST and LCST, *Macromol. Rapid Commun.*, 32, 660, 2011.

2011VIL Villetti, M.A., Soldi, V., Rochas, C., and Borsali, R., Phase-separation kinetics and mechanism in a methylcellulose/salt aqueous solution studied by time-resolved small-angle light scattering (SALS), *Macromol. Chem. Phys.*, 212, 1063, 2011.

2011WA1 Wang, F., Chen, T., Shang, Y., and Liu, H., Two-phase aqueous systems of cetyltrimethylammonium bromide/sodium dodecyl sulfate with and without polyethylene glycol, *Korean J. Chem. Eng.*, 28, 923, 2011.

2011WA2 Wang, H., Sun, S., and Wu, P., Thermodynamics of hyperbranched poly(ethylenimine) with isobutyramide residues during phase transition: An insight into the molecular mechanism, *J. Phys. Chem. B*, 115, 8832, 2011.

2011WA3 Wang, J., Dan, Y., Yang, Y., Wang, Y., Hu, Y., and Xie, Y., The measurements of coexistence curves and critical behavior of a binary mixture with a high molecular weight polymer, *J. Mol. Liq.*, 161, 115, 2011.

2011WEB Weber, C., Krieg, A., Paulus, R.M., Lambermont-Thijs, H.M.L., Becer, C.R., Hoogenboom, R., and Schubert, U.S., Thermal properties of oligo(2-ethyl-2-oxazoline) containing comb and graft copolymers and their aqueous solutions, *Macromol. Symp.*, 308, 17, 2011.

2011WUG Wu, G., Chen, S.-C., Zhan, Q., and Wang, Y.-Z., Well-defined amphiphilic biodegradable comb-like graft copolymers: Their unique architecture-determined LCST and UCST thermoresponsivity, *Macromolecules*, 44, 999, 2011.

2011XIA Xia, Y., Wang, Y., Wang, Y., Wang, D., Deng, H., Zhuang, Y., Yan, D., Zhu, B., and Zhu, X., Backbone-thermoresponsive hyperbranched polyglycerol by random copolymerization of glycidol and 3-methyl-3-(hydroxymethyl)oxetane, *Macromol. Chem. Phys.*, 212, 1056, 20.11.

2011XIO Xiong, Z., Peng, B., Han, X., Peng, C., Liu, H., and Hu, Y., Dual-stimuli responsive behaviors of diblock polyampholyte PDMAEMA-*b*-PAA in aqueous solution, *J. Colloid Interface Sci.*, 356, 557, 2011.

2011ZAF Zafarani-Moattar, M.T. and Nemati-Kande, E., Thermodynamic studies on the complete phase diagram of aqueous two phase system containing poly(ethylene glycol) dimethyl ether 2000 and dipotassium hydrogen phosphate at different temperatures, *CALPHAD*, 35, 165, 2011.

2011ZEI1 Zeiner, T., Schrader, P., Enders, S., and Browarzik, D., Phase- and interfacial behavior of hyperbranched polymer solutions, *Fluid Phase Equil.*, 302, 321, 2011.

2011ZEI2 Zeiner, T., Phase- and interfacial behavior of hyperbranched polymer solutions, *Dissertation*, TU Berlin, 2011.

2011ZEI3 Zeiner, T. and Enders, S., Phase behavior of hyperbranched polymer solutions in mixed solvents, *Chem. Eng. Sci.*, 66, 5244, 2011.

2011ZH1 Zhao, X., Xie, X., and Yan, Y., Liquid–liquid equilibrium of aqueous two-phase systems containing poly(propylene glycol) and salt ($(NH_4)_2SO_4$, $MgSO_4$, KCl, and KAc): Experiment and correlation, *Thermochim. Acta*, 516, 46, 2011.

2011ZH2 Zhao, J., Wang, Z.N., Wei, X.L., Liu, F., Zhou, W., Tang, X.L., and Wu, T.H., Phase behavior and rheological properties of the lamellar liquid crystals formed in dodecyl polyoxyethylene polyoxypropylene ether/water system, *Indian J. Chem.*, 50A, 641, 2011.

2011ZH3 Zhao, Y., Tremblay, L., and Zhao, Y., Phototunable LCST of water-soluble polymers: Exploring a topological effect, *Macromolecules*, 44, 4007, 2011.

2011ZHO Zhou, W., An, X., Gong, J., Shen, W., Chen, Z., and Wang, X., Synthesis, characteristics, and phase behavior of a thermosensitive and pH-sensitive polyelectrolyte, *J. Appl. Polym. Sci.*, 121, 2089, 2011.

4. HIGH-PRESSURE PHASE EQUILIBRIUM (HPPE) DATA OF AQUEOUS POLYMER SOLUTIONS

4.1. Experimental data of quasibinary polymer solutions

Polymer (B): **poly(*N*-isopropylacrylamide)** **2004SHI**
Characterization: synthesized in the laboratory
Solvent (A): **deuterium oxide** **D_2O** **7789-20-0**

Type of data: cloud points

c_B 0.690 mol/l was kept constant

$T/\text{K} = 308.45 - 5.99\ 10^{-4}\ (P/\text{MPa} - 48.2)^2$

Polymer (B): **poly(*N*-isopropylacrylamide)** **2004SHI**
Characterization: synthesized in the laboratory
Solvent (A): **water** **H_2O** **7732-18-5**

Type of data: cloud points

c_B 0.690 mol/l was kept constant

$T/\text{K} = 306.75 - 5.59\ 10^{-4}\ (P/\text{MPa} - 51.7)^2$

4.2. Table of systems where binary HPPE data were published only in graphical form as phase diagrams or related figures

Polymer (B)	Solvent (A)	Ref.
Poly[2-(2-ethoxy)ethoxyethyl vinyl ether-*b*-(2-methoxyethyl vinyl ether)]		
	deuterium oxide	2006OSA
	deuterium oxide	2010SHI
	water	2006OSA
Poly(ethylene oxide-*b*-propylene oxide-*b*-ethylene oxide)		
	deuterium oxide	2010KLO
	water	2009KOS
Poly(*N*-(1-hydroxymethyl)propylmethacrylamide)		
	water	2005SET
Poly(*N*-isopropylacrylamide)		
	deuterium oxide	2004SHI
	deuterium oxide	2010SHI
	water	2004SHI
	water	2010SHI
Poly(*N*-isopropylmethacrylamide)		
	water	2006KIR
Polypeptide		
	water	2003YAM

4.3. Experimental data of quasiternary or quasiquaternary solutions containing water and at least one polymer

Polymer (B): **dextran** **2005PER**
Characterization: M_w/g.mol^{-1} = 68800, Sigma-Aldrich, Inc., St. Louis, MO
Solvent (A): **carbon dioxide** CO_2 **124-38-9**
Solvent (C): **dimethylsulfoxide** C_2H_6OS **67-68-5**
Solvent (D): **water** H_2O **7732-18-5**

Type of data: cloud points

w_A	0.1997	0.1997	0.1997	0.1997	0.1997	0.1997	0.1997	0.1997	0.1997
w_B	0.0140	0.0140	0.0140	0.0140	0.0140	0.0140	0.0140	0.0140	0.0140
w_C	0.7763	0.7763	0.7763	0.7763	0.7763	0.7763	0.7763	0.7763	0.7763
w_D	0.0100	0.0100	0.0100	0.0100	0.0100	0.0100	0.0100	0.0100	0.0100
T/K	283.65	293.17	303.15	313.33	323.54	333.34	343.34	363.00	303.86
P/MPa	2.169	2.674	3.224	3.872	4.574	5.314	5.909	7.724	3.349

w_A	0.1997	0.1997	0.1997	0.1997	0.1997	0.1997	0.1997	0.1997
w_B	0.0140	0.0140	0.0140	0.0140	0.0140	0.0140	0.0140	0.0140
w_C	0.7763	0.7763	0.7763	0.7763	0.7763	0.7763	0.7763	0.7763
w_D	0.0100	0.0100	0.0100	0.0100	0.0100	0.0100	0.0100	0.0100
T/K	313.29	323.51	333.31	343.29	362.80	298.76	299.86	301.38
P/MPa	4.049	4.759	5.504	6.304	8.064	4.049	7.044	12.049

Polymer (B): **poly(ethylene glycol)** **2009MAR**
Characterization: M_w/g.mol^{-1} = 6000, Clariant, Burghausen, Germany
Solvent (A): **carbon dioxide** CO_2 **124-38-9**
Solvent (C): **water** H_2O **7732-18-5**

Type of data: coexistence data (VLE)

T/K = 353.15

P/MPa	10	10	10	10	10	10	10
w_A(liquid phase)	0.144	0.116	0.080	0.059	0.044	0.041	0.041
w_B(liquid phase)	0.8570	0.7990	0.6510	0.4750	0.2853	0.0969	0.0000
w_C(liquid phase)	0.0000	0.0852	0.2697	0.4660	0.6710	0.8617	0.9589
w_A(gas phase)	0.9995	0.9983	0.9982	0.9980	0.9967	0.9964	0.9968
w_B(gas phase)	0.00048	0.00029	0.00026	0.00025	0.00010	0.00005	0.00000
w_C(gas phase)	0.00000	0.00140	0.00160	0.00210	0.00320	0.00360	0.00320

continued

continued

P/MPa	20	20	20	20	20	20	20
w_A(liquid phase)	0.220	0.180	0.102	0.069	0.055	0.053	0.051
w_B(liquid phase)	0.7800	0.7460	0.6416	0.4700	0.2840	0.0950	0.0000
w_C(liquid phase)	0.0000	0.0710	0.2564	0.4610	0.6610	0.8520	0.9486
w_A(gas phase)	0.9996	0.9981	0.9979	0.9970	0.9965	0.9962	0.9964
w_B(gas phase)	0.00040	0.00033	0.00035	0.00034	0.00025	0.00011	0.00000
w_C(gas phase)	0.00000	0.00160	0.00180	0.00270	0.00320	0.00370	0.00360
P/MPa	30	30	30	30	30	30	30
w_A(liquid phase)	0.240	0.215	0.125	0.083	0.069	0.067	0.055
w_B(liquid phase)	0.7600	0.7160	0.6290	0.4649	0.2810	0.0980	0.0000
w_C(liquid phase)	0.0000	0.0692	0.2460	0.4520	0.6500	0.8350	0.9450
w_A(gas phase)	0.9997	0.9977	0.9973	0.9955	0.9955	0.9956	0.9960
w_B(gas phase)	0.00032	0.00053	0.00054	0.00041	0.00028	0.00013	0.00000
w_C(gas phase)	0.00000	0.00170	0.00220	0.00410	0.00420	0.00430	0.00400

T/K = 393.15

P/MPa	10	10	10	10	10	10	
w_A(liquid phase)	0.119	0.060	0.051	0.043	0.034	0.032	
w_B(liquid phase)	0.881	0.673	0.487	0.289	0.107	0.000	
w_C(liquid phase)	0.000	0.267	0.463	0.668	0.859	0.968	
w_A(gas phase)	0.9995	0.9977	0.9975	0.9965	0.9963	0.9870	
w_B(gas phase)	0.00050	0.00028	0.00022	0.00009	0.00003	0.00000	
w_C(gas phase)	0.00000	0.00206	0.00226	0.00342	0.00366	0.01300	
P/MPa	30	30	30	30	30	30	30
w_A(liquid phase)	0.220	0.190	0.130	0.097	0.079	0.080	0.058
w_B(liquid phase)	0.790	0.744	0.632	0.453	0.368	0.341	0.000
w_C(liquid phase)	0.000	0.068	0.235	0.444	0.553	0.580	0.942
w_A(gas phase)	0.9996	0.9919	0.9894	0.9890	0.9870	0.9868	0.9873
w_B(gas phase)	0.00036	0.00028	0.00026	0.00022	.00020	0.00007	0.00000
w_C(gas phase)	0.00000	0.00778	0.01034	0.01090	0.01280	0.01312	0.01264

Polymer (B):	**starch**		**1996SIN**
Characterization:	degermed yellow cornmeal, Lauhoff Grain Co., Danville, IL		
Solvent (A):	**carbon dioxide**	**CO_2**	**124-38-9**
Solvent (C):	**water**	**H_2O**	**7732-18-5**

Type of data: gas solubility

T/K = 343.15

w_A	0.0037	0.0086	0.0121	0.0162	0.0037	0.0080	0.0131	0.0171	0.0035
w_B	0.6057	0.6014	0.6006	0.5982	0.6057	0.6031	0.6000	0.5975	0.6378
w_C	0.3906	0.3900	0.3873	0.3856	0.3906	0.3889	0.3869	0.3854	0.3587
P/bar	26.39	58.41	92.11	118.31	25.41	58.76	92.82	117.58	22.92

continued

continued

w_A	0.0074	0.0117	0.0152	0.0037	0.0080	0.0127	0.0171	0.0035	0.0075
w_B	0.6353	0.6325	0.6303	0.6376	0.6349	0.6319	0.6291	0.6527	0.6501
w_C	0.3573	0.3558	0.3545	0.3587	0.3571	0.3554	0.3538	0.3438	0.3424
P/bar	55.78	90.56	114.24	22.00	56.62	92.67	117.99	24.09	57.61
w_A	0.0118	0.0157	0.0035	0.0076	0.0121	0.0167			
w_B	0.6473	0.6447	0.6527	0.6500	0.6471	0.6441			
w_C	0.3409	0.3396	0.3438	0.3424	0.3408	0.3392			
P/bar	91.14	116.85	23.07	58.27	93.88	116.82			

4.4. Table of systems where ternary or quaternary HPPE data were published only in graphical form as phase diagrams or related figures

Polymer (B)	Second and third component	Ref.
Poly(ethylene glycol) mono-2,6,8-trimethyl-4-nonyl ether		
	carbon dioxide + water	2010HAR
Poly(ethylene oxide)		
	acetonitrile/carbon dioxide and water	2003STR
Poly(ethylene oxide-*b*-propylene oxide-*b*-ethylene oxide)		
	ethene/1,4-dimethylbenzene and water	2005ZHA
Polypeptide		
	potassium chloride and water	2003YAM
	potassium iodide and water	2003YAM
	potassium sulfate and water	2003YAM
	sodium chloride and water	2003YAM
Starch		
	carbon dioxide and water	2006CHE

4.5. References

1996SIN Singh, B., Rizvi, S.S.H., and Harriott, P., Measurement of diffusivity and solubility of carbon dioxide in gelatinized starch at elevated pressures, *Ind. Eng. Chem. Res.*, 35, 4457, 1996.

2003STR Striolo, A., Elvassore, N., Parton, T., and Bertucco, A., Relationship between volume expansion, solvent-power, and precipitation in GAS processes, *AIChE-J.*, 49, 2671, 2003.

2003YAM Yamaoka, T. Tamura, T., Seto, Y., Tada, T., Kunugi, S., and Tirrell, D.A., Mechanism for the phase transition of a genetically engineered elastin model peptide (VPGIG)40 in aqueous solution, *Biomacromolecules*, 4, 1680, 2003.

2004SHI Shibayama, M., Isono, K., Okabe, S., Karino, T., and Nagao, M., SANS study on pressure-induced phase separation of poly(*N*-isopropylacrylamide) aqueous solutions and gels, *Macromolecules*, 37, 2909, 2004.

2005PER Perez de Diego, Y., Wubolts, F.E., Witkamp, G.J., Loos, Th.W. de, and Jansens, P.J., Measurements of the phase behaviour of the system dextran/DMSO/CO_2 at high pressures, *J. Supercrit. Fluids*, 35, 1, 2005.

2005SET Seto, Y., Aoki, T., and Kunugi, S., Temperature- and pressure-responsive properties of L- and DL-forms of poly(*N*-(1-hydroxymethyl)propylmethacrylamide) in aqueous solutions, *Colloid Polym. Sci.*, 283, 1137, 2005.

2005ZHA Zhang, R., Liu, J., Han, B., Wang, B., Sun, D., and He, J., Effect of PEO–PPO–PEO structure on the compressed ethylene-induced reverse micelle formation and water solubilization, *Polymer*, 46, 3936, 2005.

2006CHE Chen, K.-H.J. and Rizvi, S.S.H., Measurement and prediction of solubilities and diffusion coefficients of carbon dioxide in starch–water mixtures at elevated pressures, *J. Polym. Sci.: Part B: Polym. Phys.*, 44, 607, 2006.

2006KIR Kirpach, A. and Adolf, D., High pressure induced coil-globule transitions of smart polymers, *Macromol. Symp.*, 237, 7, 2006.

2006OSA Osaka, N., Okabe, S., Karino, T., Hirabaru, Y., Aoshima, S., and Shibayama, M., Micro- and macrophase separations of hydrophobically solvated block copolymer aqueous solutions induced by pressure and temperature, *Macromolecules*, 39, 5875, 2006.

2009KOS Kostko, A.F., Harden, J.L., and McHugh, M.A., Dynamic light scattering study of concentrated triblock copolymer micellar solutions under pressure, *Macromolecules*, 42, 5328, 2009.

2009MAR Martin, A., Pham, H.M., Kilzer, A., Kareth, S., and Weidner, E., Phase equilibria of carbon dioxide + poly(ethylene glycol) + water mixtures at high pressure: Measurements and modelling, *Fluid Phase Equil.*, 286, 162, 2009.

2010HAR Haruki, M., Matsuura, K., Kaida, Y., Kihara, S.-I., and Takishima, S., Microscopic phase behavior of supercritical carbon dioxide + non-ionic surfactant + water systems at elevated pressures, *Fluid Phase Equil.*, 289, 1, 2010.

2010KLO Kloxin, C.J. and Zanten, J.H. van, High pressure phase diagram of an aqueous PEO-PPO-PEO triblock copolymer system via probe diffusion measurements, *Macromolecules*, 43, 2084, 2010.

2010SHI Shibayama, M. and Osaka, N., Pressure- and temperature-induced phase separation transition in homopolymer, block copolymer, and protein in water, *Macromol. Symp.*, 291-292, 115, 2010.

5. ENTHALPY CHANGES FOR AQUEOUS POLYMER SOLUTIONS

5.1. Enthalpies of mixing or intermediary enthalpies of dilution, and partial enthalpies of mixing (at infinite dilution), or polymer (first) integral enthalpies of solution

Polymer (B): **cellobiose** **2008RAD, 2011RA2**
Characterization: M_w/g.mol^{-1} = 342.3, Merck, Darmstadt, Germany
Solvent (A): **water** **H_2O** **7732-18-5**

T/K = 298.15 $\Delta_{sol}H_B$ = –28.45 J/(g polymer)

Comments: $\Delta_{sol}H_B$ is independent on concentration in the range of 2-10%.

Polymer (B): **hydroxypropylcellulose** **2007VSH**
Characterization: M_w/g.mol^{-1} = 95000, D.S. = 3.4, 10-15% crystallinity, Klucel-JF, Hercules, Wilmington, DE
Solvent (A): **water** **H_2O** **7732-18-5**

T/K = 298.15

w_B	0.10	0.22	0.33	0.42	0.51	0.52	0.60
$\Delta_{dil}H^{12}$/(J/g polymer)	−3.6	−6.8	−10.7	−11.4	−17.7	−20.7	−32.7
w_B	0.61	0.71	0.80	0.83	0.89	0.90	0.98
$\Delta_{dil}H^{12}$/(J/g polymer)	−34.2	−47.0	−53.1	−59.0	−77.6	−71.3	−95.0
w_B	1.00						
$\Delta_{dil}H^{12}$/(J/g polymer)	−105.2						

Comments: The table gives the mass fraction of the polymer in the starting solution before diluting with pure water. $\Delta_{dil}H^{12}$ is measured as the heat released or absorbed upon the interaction of the mixtures with a large excess of water under stirring. The relative error in measuring the thermal effects did not exceed 2%.

Polymer (B): **maltodextrin** **2005NIN**
Characterization: M_n/g.mol^{-1} = 1475, M_w/g.mol^{-1} = 13710, Aldrich Chem. Co., Inc., Milwaukee, WI
Solvent (A): **water** **H_2O** **7732-18-5**

T/K = 313.15

$m_{(A+B)}'$/g	28.850	27.346	28.801	29.752	30.587	29.850	39.874	39.484
w_B'	0.290	0.290	0.300	0.398	0.404	0.398	0.508	0.481
m_A''/g	15.938	15.857	15.902	15.916	15.855	15.905	14.805	14.976
$\Delta_{dil}H^{12}$/J	−1.19	−1.23	−1.22	−1.68	−2.05	−1.82	−3.16	−2.80

continued

continued

$m_{(A+B)}'/g$	40.436	40.564	41.342	41.518	40.569	40.309	42.170	43.049
w_B'	0.548	0.548	0.579	0.579	0.638	0.638	0.694	0.694
m_A''/g	14.783	14.822	14.979	15.000	14.866	14.772	14.854	14.855
$\Delta_{dil}H^{12}/J$	−3.64	−3.66	−4.49	−4.49	−5.89	−5.46	−6.69	−6.77

Comments: The table gives the mass of the polymer solution, the mass fraction of the polymer in this solution before diluting with pure water, and the mass of water added to this solution as used in the experiment of a mixing calorimeter. The superscripts ' and " designate the polymer solution or water as placed in the two parts of a mixing cell, respectively. $\Delta_{dil}H^{12}$ is therefore an extensive quantity obtained for the given masses in the table in the result of the mixing process.

Polymer (B): **maltodextrin** **2005NIN**
Characterization: M_n/g.mol^{-1} = 1140, M_w/g.mol^{-1} = 8283, Aldrich Chem. Co., Inc., Milwaukee, WI
Solvent (A): **water** **H_2O** **7732-18-5**

T/K = 313.15

$m_{(A+B)}'/g$	38.488	37.672	39.794	38.575	40.396	40.276	41.918	41.314
w_B'	0.404	0.393	0.501	0.511	0.548	0.543	0.608	0.608
m_A''/g	15.014	14.953	14.779	15.472	15.057	15.030	15.038	15.039
$\Delta_{dil}H^{12}/J$	−3.50	−2.62	−4.49	−3.90	−6.43	−6.03	−8,94	−8.48
$m_{(A+B)}'/g$	42.519	42.182	41.810	42.727				
w_B'	0.649	0.637	0.700	0.700				
m_A''/g	14.871	14.953	14.801	15.021				
$\Delta_{dil}H^{12}/J$	−10.34	−9.61	−12.21	−12.19				

Comments: The table gives the mass of the polymer solution, the mass fraction of the polymer in this solution before diluting with pure water, and the mass of water added to this solution as used in the experiment of a mixing calorimeter. The superscripts ' and " designate the polymer solution or water as placed in the two parts of a mixing cell, respectively. $\Delta_{dil}H^{12}$ is therefore an extensive quantity obtained for the given masses in the table in the result of the mixing process.

Polymer (B): **poly(propylene glycol) bis(2-aminopropyl ether)** **2006INO**
Characterization: M_n/g.mol^{-1} = 2000, Aldrich Chem. Co., Inc., Milwaukee, WI
Solvent (A): **water** **H_2O** **7732-18-5**

T/K = at solid-liquid equilibrium

c_A = 1 mol/dm^3 was kept constant

w_B	0.049	0.102	0.152	0.202	0.252	0.301	0.347	0.402
$\Delta_M H/(J/g)$	−31.3	−61.5	−86.5	−108	−126	−141	−152	−161

Polymer (B): **poly(1-vinyl-2-pyrrolidinone)** **2009SUG**
Characterization: M_w/g.mol^{-1} = 10000, Aldrich Chem. Co., Inc., Milwaukee, WI
Solvent (A): **water** **H_2O** **7732-18-5**

T/K = 298.15 $\Delta_{sol}H_B$ = –119 J/(g polymer)

Comments: The final concentration is about 0.75 g polymer/100 g solvent.

Polymer (B): **poly(1-vinyl-2-pyrrolidinone)** **2011RA1**
Characterization: M_w/g.mol^{-1} = 10000, Aldrich Chem. Co., Inc., Milwaukee, WI
Solvent (A): **water** **H_2O** **7732-18-5**

T/K = 298.15 $\Delta_{sol}H_B$ = –174.1 J/(g polymer)
T/K = 303.15 $\Delta_{sol}H_B$ = –173.5 J/(g polymer)
T/K = 308.15 $\Delta_{sol}H_B$ = –157.9 J/(g polymer)
T/K = 318.15 $\Delta_{sol}H_B$ = –146.4 J/(g polymer)

Comments: The final polymer concentration is about 1.6 g/l.

Polymer (B): **poly(1-vinyl-2-pyrrolidinone)** **2011RA1**
Characterization: M_w/g.mol^{-1} = 29000, Aldrich Chem. Co., Inc., Milwaukee, WI
Solvent (A): **water** **H_2O** **7732-18-5**

T/K = 298.15 $\Delta_{sol}H_B$ = –185.3 J/(g polymer)
T/K = 303.15 $\Delta_{sol}H_B$ = –180.5 J/(g polymer)
T/K = 308.15 $\Delta_{sol}H_B$ = –171.0 J/(g polymer)
T/K = 318.15 $\Delta_{sol}H_B$ = –154.0 J/(g polymer)

Comments: The final polymer concentration is about 1.6 g/l.

Polymer (B): **poly(1-vinyl-2-pyrrolidinone)** **2011RA1**
Characterization: M_w/g.mol^{-1} = 55000, Aldrich Chem. Co., Inc., Milwaukee, WI
Solvent (A): **water** **H_2O** **7732-18-5**

T/K = 298.15 $\Delta_{sol}H_B$ = –192.7 J/(g polymer)
T/K = 303.15 $\Delta_{sol}H_B$ = –191.6 J/(g polymer)
T/K = 308.15 $\Delta_{sol}H_B$ = –177.0 J/(g polymer)
T/K = 318.15 $\Delta_{sol}H_B$ = –165.9 J/(g polymer)

Comments: The final polymer concentration is about 1.6 g/l.

Polymer (B): **sodium carboxymethylcellulose** **2011RA2**
Characterization: M_w/g.mol^{-1} = 23000, D.S. = 0.77, Hercules, Wilmington, DE
Solvent (A): **water** **H_2O** **7732-18-5**

T/K = 298.15 $\Delta_{sol}H_B$ = –115.5 J/(g polymer)

5.2. Partial molar enthalpies of mixing at infinite dilution of water and enthalpies of solution of water vapor in molten polymers from inverse gas-liquid chromatography (IGC)

Polymer (B): **polyester (hyperbranched)** **2009DOM**
Characterization: M_n/g.mol^{-1} = 6500, M_w/g.mol^{-1} = 9750, fatty acid modified, Boltorn U3000, Perstorp Specialty Chemicals AB, Perstorp, Sweden

Solvent (A)	T-range/ K	$\Delta_M H_A^\infty$/ kJ/mol	$\Delta_{sol} H_{A(vap)}^\infty$/ kJ/mol
water	308.15-348.15	13.1	−30.2

Polymer (B): **polyester (hyperbranched)** **2010DO1**
Characterization: M_n/g.mol^{-1} = 3200, fatty acid modified, Boltorn H2004, density (298.15) = 1.0765 cm^3/g, T_g/K = 215.0, Perstorp Specialty Chemicals AB, Perstorp, Sweden

Solvent (A)	T-range/ K	$\Delta_M H_A^\infty$/ kJ/mol	$\Delta_{sol} H_{A(vap)}^\infty$/ kJ/mol
water	308.15-348.15	9.2	−34.8

Polymer (B): **polyester (hyperbranched)** **2010DO2**
Characterization: M_n/g.mol^{-1} = 9000, fatty acid modified, Boltorn W3000, density (308.15) = 1.0408 cm^3/g, T_g/K = 205.2, Perstorp Specialty Chemicals AB, Perstorp, Sweden

Solvent (A)	T-range/ K	$\Delta_M H_A^\infty$/ kJ/mol	$\Delta_{sol} H_{A(vap)}^\infty$/ kJ/mol
water	308.15-348.15	3.3	−41.4

5.3. Table of systems where additional information on enthalpy effects in aqueous polymer solutions can be found

Polymer (B)	Solvent (A)	Enthalpy	*T*-range	Ref.
Cellobiose				
	water	$\Delta_{sol}H$	298 K	2008RAD
	water/dimethylsulfoxide	$\Delta_{sol}H$	298 K	2008RAD
Gelatine				
	water	$\Delta_{dil}H^{12}$	318.15	1999CES
Maltodextrin				
	water	$\Delta_{dil}H^{12}$	318.15	1999CES
Penta(ethylene glycol) monooctyl ether				
	water	$\Delta_{sol}H$	298 K	2004DER
Poly(acrylic acid)				
	water	$\Delta_{sol}H$	298 K	2004DER
	water	$\Delta_{sol}H_{A(vap)}^{\infty}$	303-333 K	2006SMI
Poly(ether urethane)				
	water	$\Delta_{sol}H_{A(vap)}^{\infty}$	298 K	2004SMI
Poly(ethylene glycol)				
	water	$\Delta_{M}H$	298.15	2002JAB
	water	$\Delta_{dil}H^{12}$	298.15	2002JAB
	water	$\Delta_{M}H$	298 K	2003JAB
	water	ΔH		2006ATT
	water	$\Delta_{M}H$	298 K	2009MOH

Polymer (B)	Solvent (A)	Enthalpy	*T*-range	Ref.
Poly(ethylene oxide)				
	water	$\Delta_{sol}H_{A(vap)}^{\infty}$	303-333 K	2006SMI
	water/sodium dodecyl sulfate	ΔH	298 K	2004SIL
Poly(ethylene oxide-*b*-propylene oxide-*b*-ethylene oxide)				
	water	$\Delta_M H$	298 K	2002THU
	water/surfactant	ΔH		2004DEL
	water	ΔH		2004JAN
	water	ΔH		2010TSU
Poly(L-glutamic acid)				
	water	$\Delta_{dil}H^{12}$		2005GOD
Poly(*N*-isopropylacrylamide)				
	water	ΔH		2003CHO
	water/surfactant	ΔH		2004LOH
	water	ΔH		2006DIN
	water	ΔH		2010DUB
	water	ΔH		2010SHE
Poly(*N*-isopropylacrylamide-*co*-acrylic acid)				
	water	ΔH		2006WEN
Poly(styrene oxide-*b*-ethylene oxide)				
	water/sodium dodecyl sulfate	ΔH	298 K	2005CAS
Poly(vinyl methyl ether)				
	water	ΔH		2007VAN

Polymer (B)	Solvent (A)	Enthalpy	*T*-range	Ref.
Poly(2-vinyl-5-methyl-1,3,4-oxadiazole)	water	$\Delta_M H$	298.15	2006KIZ
Poly(1-vinyl-2-pyrrolidinone)	water/ethanol	$\Delta_{sol} H_B$	298.15	2011RA1
Sodium carboxymethyl-cellulose	water	$\Delta_{dil} H^{12}$	298.15	2009SAF
Sodium carboxymethyl-starch	water	$\Delta_{dil} H^{12}$	298.15	2009SAF
Starch	water	$\Delta_{dil} H^{12}$	298.15	2009SAF

5.4. References

1999CES Cesaro, A., Cuppo, F., Fabri, D., and Sussich, F., Thermodynamic behavior of mixed biopolymers in solution and in gel phase, *Thermochim. Acta*, 328, 143, 1999.

2002JAB Jablonski, P., Kalorimetrische Untersuchungen des Systems Polyethylenglykol/Wasser, *Dissertation*, Gerhard-Mercator-Universität Gesamthochschule Duisburg, 2002.

2002THU Thurn, T., Couderc, S., Sidhu, J., Bloor, D.M., Penfold, J., Holzwarth, J.F., and Wyn Jones, E., Study of mixed micelles and interaction parameters for triblock copolymers of the type EO_m-PO_n-EO_m and ionic surfactants: Equilibrium and structure, *Langmuir*, 18, 9267, 2002.

2003CHO Cho, E.C., Lee, J., and Cho, K., Role of bound water and hydrophobic interaction in phase transition of poly(*N*-isopropylacrylamide) aqueous solution, *Macromolecules*, 36, 9929, 2003.

2003JAB Jablonski, P., Müller-Blecking, A., and Borchard, W., A method to determine mixing enthalpies by DSC, *J. Therm. Anal. Calorim.*, 74, 779, 2003.

2004DEL DeLisi, R., Lazzara, G., Milioto, S., and Muratore, N., Thermodynamics of aqueous poly(ethylene oxide)-poly(propylene oxide)-poly(ethylene oxide)/surfactant mixtures. Effect of the copolymer molecular weight and the surfactant alkyl chain length, *J. Phys. Chem. B*, 108, 18214, 2004.

2004DER D'Errico, G., Ciccarelli, D., Ortona, O., Paduano, L., and Sartorio, R., Interaction between pentaethylene glycol n-octyl ether and low-molecular-weight poly(acrylic acid), *Coll. Interface Sci.*, 270, 490, 2004.

2004JAN Jansson, J., Schillen, K., Olofsson, G., Cardoso da Silva, R., and Loh, W., The interaction between PEO-PPO-PEO triblock copolymers and ionic surfactants in aqueous solution studied using light scattering and calorimetry, *J. Phys. Chem. B*, 108, 82, 2004.

2004LOH Loh, W., Teixeira, L.A.C., and Lee, L.-T., Isothermal calorimetric investigation of the interaction of poly(*N*-isopropylacrylamide) and ionic surfactants, *J.Phys.Chem. B*, 108, 3196, 2004.

2004SIL Silva, R.C. da, Loh, W., and Olofsson, G., Calorimetric investigation of temperature effect on the interaction between poly(ethylene oxide) and sodium dodecylsulfate in water, *Thermochimica Acta*, 417, 295, 2004.

2004SMI Smith, A.L., Mulligan, R.B., and Shirazi, H.M., Determining the effects of vapor sorption in polymers with the quartz crystal microbalance/heat conduction calorimeter, *J. Polym. Sci.: Part B: Polym. Phys.*, 42, 3893, 2004.

2005CAS Castro, E., Taboada, P., and Mosquera, V., Behavior of a styrene oxide-ethylene oxide diblock copolymer/surfactant system: A thermodynamic and spectroscopy study, *J. Phys. Chem. B*, 109, 5592, 2005.

2005GOD Godec, A. and Skerjanc, J., Enthalpy changes upon dilution and ionization of poly(L-glutamic acid) in aqueous solutions, *J. Phys. Chem. B*, 109, 13363, 2005.

2005NIN Ninni, L., Meirelles, A.J.A., and Maurer, G., Thermodynamic properties of aqueous solutions of maltodextrins from laser-light scattering, calorimetry and isopiestic investigations, *Carbohydrate Polym.*, 59, 289, 2005.

2006ATT	Attanasio, F., Rialdi, G., Swierzewski, R., and Zielenkiewicz, W., Pressure perturbation calorimetry of poly(ethylene glycol) solutions in water, *J. Therm. Anal. Calorim.*, 83, 637, 2006.
2006DIN	Ding, Y. and Zhang, G., Microcalorimetric investigation on association and dissolution of poly(*N*-isopropylacrylamide) chains in semidilute solutions, *Macromolecules*, 39, 9654, 2006.
2006INO	Inoue, T. and Yamashita, K., Aggregation behavior of polypropylene oxide with electric charges at both ends in aqueous solution, *J. Colloid Interface Sci.*, 300, 774, 2006.
2006KIZ	Kizhnyaev, V.N., Pokatilov, A.F., Vereshchagin, L.I., Adamova, L.V., Safronov, A.P., and Smirnov, A.I., Synthesis and properties of carbon-chain polymers with pendant 1,3,4-oxadiazole rings, *Russ. J. Appl. Chem.*, 79, 1167, 2006.
2006SMI	Smith, A.L., Ashcraft, J.N., and Hammond, P.T., Sorption isotherms, sorption enthalpies, diffusion coefficients and permeabilities of water in a multilayer PEO/PAA polymer film using the quartz crystal microbalance/heat conduction calorimeter, *Thermochim. Acta*, 450, 118, 2006.
2006WEN	Weng, Y., Ding, Y., and Zhang, G., Microcalorimetric investigation on the lower critical solution temperature behavior of *N*-isopropylacrylamide-*co*-acrylic acid copolymer in aqueous solution, *J. Phys. Chem. B*, 110, 11813, 2006.
2007VAN	Van Durme, K., Van Assche, G., Nies, E., and Van Mele, B., Phase transformations in aqueous low molar mass poly(vinyl methyl ether) solutions: Theoretical prediction and experimental validation of the peculiar solvent melting line, bimodal LCST, and (adjacent) UCST miscibility gaps, *J. Phys. Chem. B*, 111, 1288, 2007.
2007VSH	Vshivkov, S.A., Adamova, L.V., Rusinova, E.V., Safronov, A.P., Dreval, V.E., and Galyas, A.G., Thermodynamics of liquid-crystalline solutions of hydroxypropyl cellulose in water and ethanol, *Polym. Sci., Ser. A*, 49, 578, 2007.
2008RAD	Radugin, M.V., Prusov, A.N., Lebedeva, T.N., and Zakharov, A.G., Heat of dissolution of cellulose in water and water-dimethyl sulfoxide mixtures, *Fibre Chem.*, 40, 533, 2008.
2009DOM	Domanska, U. and Zolek-Tryznowska, Z., Thermodynamic properties of hyperbranched polymer, Boltorn U3000, using inverse gas chromatography, *J. Phys. Chem. B*, 113, 15312, 2009.
2009MOH	Mohite, L.V. and Juvekar, V.A., Quantification of thermodynamics of aqueous solutions of poly(ethylene glycols): Role of calorimetry, *Fluid Phase Equil.*, 278, 41, 2009.
2009SAF	Safronov, A.P., Tyukova, I.S., and Suvorova, A.I., The nature of thermodynamic compatibility of components of aqueous solutions of starch and cellulose carboxymethyl derivatives, *Polym. Sci., Ser. A*, 51, 174, 2009.
2009SUG	Sugiura, T., Kimura, F., and Ogawa, H., Thermodynamic study on the polyvinylpyrrolidone + water, + alkanol (n = 1−4), + 1-methyl-2-pyrrolidone at 298.15 K, *Netsu Sokutei*, 36, 130, 2009.
2010DO1	Domanska, U. and Zolek-Tryznowska, Z., Measurements of mass-fraction activity coefficient at infinite dilution of aliphatic and aromatic hydrocarbons, thiophene, alcohols, water, ethers, and ketones in hyperbranched polymer, Boltorn H2004, using inverse gas chromatography, *J. Chem. Thermodyn.*, 42, 363, 2010.

2010DO2 Domanska, U. and Zolek-Tryznowska, Z., Mass-fraction activity coefficients at infinite dilution measurements for organic solutes and water in the hyperbranched polymer Boltorn W3000 using inverse gas chromatography, *J. Chem. Eng. Data*, 55, 1258, 2010.

2010DUB Dubovika, A.S., Grinberg, N.V., and Grinberg, V.Ya., Energetics of phase separation in aqueous solutions of poly(*N*-isopropylacrylamide), *Polym. Sci., Ser. A*, 52, 565, 2010.

2010SHE Shechter, I., Ramon, O., Portnaya, I., Paz, Y., and Livney, Y.D., Microcalorimetric study of the effects of a chaotropic salt, KSCN, on the lower critical solution temperature (LCST) of aqueous poly(*N*-isopropylacrylamide) (PNIPA) solutions, *Macromolecules*, 43, 480, 2010.

2010TSU Tsui, H.-W., Wang, J.-H., Hsu, Y.-H., and Chen, L.-J., Study of heat of micellization and phase separation for Pluronic aqueous solutions by using a high sensitivity differential scanning calorimetry, *Colloid Polym. Sci.*, 288, 1687, 2010.

2011RA1 Radugin, M.V., Lebedeva, T.N., Prusov, A.N., and Zakharov, A.G., Dependence of the enthalpy of polyvinylpyrrolidone dissolution on the composition of water–ethanol solvent at 298 K, *Russ. J. Phys. Chem. A*, 85, 897, 2011.

2011RA2 Radugin, M.V., Lebedeva, T.N., Prusov, A.N., and Zakharov, A.G., Enthalpies of interaction between cellobiose and the derivatives of cellulose in aqueous solutions, *Russ. J. Phys. Chem. A*, 85, 1863, 2011.

6. PVT DATA OF POLYMERS AND AQUEOUS POLYMER SOLUTIONS

6.1. PVT data of some polymers

Polymer (B): **poly(ε-caprolactone)** **2011SCH**
Characterization: M_w/g.mol^{-1} = 80000, CAPA® FB100, Solvay Warrington, Cheshire, United Kingdom

P/MPa	*T*/K 400.93	421.04	431.10	441.31	451.42	471.87
	ρ/(g/cm^3)					
110	1.0691	1.0591	1.0538	1.0486	1.0437	1.0336
120	1.0732	1.0631	1.0580	1.0529	1.0480	1.0382
130	1.0769	1.0671	1.0622	1.0571	1.0522	1.0425
140	1.0807	1.0710	1.0660	1.0610	1.0563	1.0469
150	1.0843	1.0747	1.0700	1.0649	1.0601	1.0510
160	1.0878	1.0783	1.0736	1.0687	1.0641	1.0549
170	1.0910	1.0820	1.0774	1.0725	1.0679	1.0590
180	1.0944	1.0854	1.0809	1.0761	1.0715	1.0628
190	1.0977	1.0889	1.0844	1.0797	1.0751	1.0666
200	1.1009	1.0921	1.0878	1.0832	1.0786	1.0702

Polymer (B): **polyester (hyperbranched, aliphatic)** **2006SEI**
Characterization: M_n/g.mol^{-1} = 1620, M_w/g.mol^{-1} = 2100, 16 OH groups per macromolecule, hydroxyl no. = 490-520 mg KOH/g, acid no. = 5-9 mg KOH/g, Boltorn H20, Perstorp Specialty Chemicals AB, Sweden

P/MPa	*T*/K 343.15	383.15	423.15
	ρ/kg m^{-3}		
5	1241.6	1214.7	1188.3
10	1245.1	1217.6	1191.5
20	1249.2	1222.3	1197.0
35	1255.6	1229.4	1205.4
60	1265.2	1240.2	1216.2

Polymer (B): **poly(ethylene-*co*-vinyl alcohol)** **2007FUN**
Characterization: 15 mol% ethene, ρ(293.15 K) = 1.2522 g/cm^3, T_g/K = 341, T_m/K = 484, Kuraray Specialities, Japan

P/MPa	*T*/K 483.55	493.80	503.38	514.05	523.80	534.11
	V_{spec}/cm^{3}g^{-1}					
0.1	0.8783	0.8852	0.8908	0.8964	0.9034	0.9133
10	0.8744	0.8810	0.8863	0.8918	0.8987	0.9083
20	0.8706	0.8769	0.8818	0.8874	0.8940	0.9036
30	0.8671	0.8734	0.8783	0.8837	0.8901	0.8994
40	0.8640	0.8701	0.8749	0.8801	0.8865	0.8958
50	0.8609	0.8670	0.8718	0.8768	0.8831	0.8921
60	0.8581	0.8642	0.8686	0.8736	0.8799	0.8887
70	0.8551	0.8614	0.8659	0.8707	0.8768	0.8854
80	0.8523	0.8587	0.8629	0.8679	0.8738	0.8823
90	0.8492	0.8562	0.8604	0.8651	0.8709	0.8795
100	0.8463	0.8536	0.8577	0.8624	0.8681	0.8765
110	0.8433	0.8512	0.8552	0.8600	0.8655	0.8740
120	0.8404	0.8488	0.8528	0.8574	0.8628	0.8712
130	0.8374	0.8464	0.8505	0.8550	0.8604	0.8687
140	0.8341	0.8443	0.8482	0.8526	0.8581	0.8661
150	0.8313	0.8421	0.8461	0.8505	0.8556	0.8638
160	0.8284	0.8400	0.8439	0.8483	0.8534	0.8615
170	0.8258	0.8379	0.8418	0.8461	0.8513	0.8593
180	0.8234	0.8358	0.8398	0.8440	0.8491	0.8572
190	0.8208	0.8337	0.8379	0.8419	0.8470	0.8550
200	0.8185	0.8316	0.8360	0.8400	0.8451	0.8531

Polymer (B): **poly(ethylene-*co*-vinyl alcohol)** **2007FUN**
Characterization: 27 mol% ethene, ρ(293.15 K) = 1.1959 g/cm^3, T_g/K = 339, T_m/K = 466, Kuraray Specialities, Japan

P/MPa	*T*/K 466.52	477.39	487.28	497.68	508.00	518.11	528.90	539.09
	V_{spec}/cm^{3}g^{-1}							
0.1	0.9234	0.9290	0.9340	0.9392	0.9453	0.9576	0.9702	0.9821
10	0.9189	0.9243	0.9292	0.9341	0.9402	0.9522	0.9641	0.9754
20	0.9146	0.9196	0.9244	0.9292	0.9352	0.9469	0.9584	0.9691
30	0.9109	0.9158	0.9204	0.9250	0.9309	0.9425	0.9534	0.9638

continued

continued

40	0.9075	0.9121	0.9167	0.9212	0.9271	0.9385	0.9488	0.9589
50	0.9042	0.9088	0.9132	0.9175	0.9235	0.9346	0.9445	0.9543
60	0.9010	0.9055	0.9098	0.9141	0.9200	0.9309	0.9405	0.9500
70	0.8981	0.9024	0.9066	0.9107	0.9167	0.9273	0.9366	0.9458
80	0.8951	0.8994	0.9035	0.9075	0.9136	0.9239	0.9330	0.9419
90	0.8924	0.8965	0.9006	0.9045	0.9105	0.9208	0.9295	0.9382
100	0.8897	0.8938	0.8977	0.9016	0.9076	0.9175	0.9260	0.9346
110	0.8872	0.8910	0.8949	0.8987	0.9048	0.9146	0.9228	0.9311
120	0.8846	0.8885	0.8923	0.8960	0.9020	0.9117	0.9197	0.9278
130	0.8821	0.8860	0.8897	0.8934	0.8994	0.9088	0.9166	0.9246
140	0.8798	0.8835	0.8871	0.8909	0.8968	0.9060	0.9137	0.9215
150	0.8772	0.8812	0.8847	0.8883	0.8944	0.9033	0.9109	0.9186
160	0.8746	0.8789	0.8824	0.8858	0.8919	0.9008	0.9081	0.9156
170	0.8717	0.8766	0.8801	0.8835	0.8895	0.8982	0.9055	0.9129
180	0.8685	0.8745	0.8779	0.8813	0.8873	0.8958	0.9028	0.9101
190	0.8650	0.8724	0.8757	0.8790	0.8850	0.8933	0.9003	0.9075
200	0.8616	0.8703	0.8736	0.8769	0.8828	0.8910	0.8978	0.9049

Polymer (B): **poly(ethylene-*co*-vinyl alcohol)** **2007FUN**
Characterization: 32 mol% ethene, ρ(293.15 K) = 1.1810 g/cm^3, T_g/K = 335, T_m/K = 457, Kuraray Specialities, Japan

P/MPa	T/K							
	467.46	477.59	488.37	498.75	508.97	519.72	529.57	539.30
	V_{spec}/cm^{3}g^{-1}							
0.1	0.9363	0.9420	0.9479	0.9536	0.9599	0.9673	0.9753	0.9841
10	0.9318	0.9372	0.9428	0.9483	0.9543	0.9613	0.9691	0.9775
20	0.9273	0.9325	0.9378	0.9432	0.9488	0.9555	0.9632	0.9712
30	0.9235	0.9286	0.9335	0.9388	0.9444	0.9507	0.9582	0.9660
40	0.9200	0.9249	0.9297	0.9348	0.9401	0.9464	0.9536	0.9612
50	0.9164	0.9213	0.9260	0.9308	0.9361	0.9423	0.9493	0.9566
60	0.9132	0.9179	0.9225	0.9272	0.9323	0.9383	0.9451	0.9524
70	0.9100	0.9147	0.9190	0.9238	0.9287	0.9347	0.9413	0.9483
80	0.9070	0.9115	0.9159	0.9204	0.9253	0.9311	0.9375	0.9445
90	0.9042	0.9086	0.9128	0.9173	0.9220	0.9277	0.9340	0.9407
100	0.9014	0.9057	0.9098	0.9142	0.9190	0.9245	0.9306	0.9371
110	0.8988	0.9028	0.9069	0.9112	0.9158	0.9213	0.9274	0.9337
120	0.8962	0.9002	0.9042	0.9084	0.9129	0.9183	0.9242	0.9304
130	0.8936	0.8976	0.9015	0.9057	0.9100	0.9153	0.9212	0.9273
140	0.8911	0.8951	0.8989	0.9031	0.9073	0.9125	0.9182	0.9243
150	0.8888	0.8926	0.8964	0.9005	0.9047	0.9097	0.9154	0.9213

continued

continued

160	0.8866	0.8903	0.8940	0.8980	0.9021	0.9071	0.9126	0.9184
170	0.8842	0.8880	0.8916	0.8955	0.8996	0.9045	0.9099	0.9157
180	0.8820	0.8856	0.8893	0.8931	0.8972	0.9020	0.9073	0.9130
190	0.8799	0.8835	0.8871	0.8907	0.8947	0.8996	0.9047	0.9103
200	0.8778	0.8813	0.8849	0.8885	0.8925	0.8972	0.9023	0.9079

Polymer (B): **poly(ethylene-*co*-vinyl alcohol)** **2007FUN**

Characterization: 38 mol% ethene, ρ(293.15 K) = 1.1690 g/cm^3, T_g/K = 332, T_m/K = 450, Kuraray Specialities, Japan

P/MPa	T/K							
	467.82	477.75	488.37	499.04	509.02	519.75	529.66	539.20
	V_{spec}/cm^{3}g^{-1}							
0.1	0.9530	0.9585	0.9643	0.9705	0.9769	0.9858	0.9963	1.0071
10	0.9482	0.9535	0.9590	0.9649	0.9711	0.9798	0.9899	1.0002
20	0.9434	0.9486	0.9539	0.9595	0.9656	0.9740	0.9837	0.9936
30	0.9394	0.9444	0.9495	0.9549	0.9608	0.9690	0.9784	0.9880
40	0.9355	0.9403	0.9452	0.9505	0.9563	0.9642	0.9733	0.9828
50	0.9318	0.9365	0.9413	0.9464	0.9521	0.9598	0.9686	0.9777
60	0.9283	0.9329	0.9376	0.9425	0.9481	0.9556	0.9642	0.9730
70	0.9250	0.9294	0.9339	0.9388	0.9443	0.9515	0.9600	0.9686
80	0.9218	0.9261	0.9305	0.9353	0.9406	0.9477	0.9560	0.9644
90	0.9187	0.9229	0.9272	0.9318	0.9372	0.9441	0.9521	0.9603
100	0.9156	0.9199	0.9241	0.9286	0.9337	0.9407	0.9485	0.9565
110	0.9128	0.9170	0.9211	0.9253	0.9306	0.9373	0.9449	0.9528
120	0.9100	0.9139	0.9181	0.9224	0.9275	0.9341	0.9416	0.9492
130	0.9073	0.9112	0.9152	0.9194	0.9244	0.9310	0.9383	0.9458
140	0.9047	0.9086	0.9125	0.9167	0.9216	0.9280	0.9353	0.9425
150	0.9021	0.9060	0.9099	0.9138	0.9188	0.9251	0.9322	0.9394
160	0.8998	0.9034	0.9073	0.9112	0.9160	0.9223	0.9292	0.9363
170	0.8974	0.9010	0.9047	0.9086	0.9133	0.9196	0.9263	0.9333
180	0.8951	0.8986	0.9023	0.9062	0.9108	0.9170	0.9236	0.9303
190	0.8927	0.8963	0.8999	0.9037	0.9083	0.9144	0.9209	0.9276
200	0.8906	0.8941	0.8977	0.9013	0.9058	0.9118	0.9183	0.9248

Polymer (B): **poly(ethylene-*co*-vinyl alcohol)** **2007FUN**
Characterization: 44 mol% ethene, ρ(293.15 K) = 1.1359 g/cm^3, T_g/K = 327, T_m/K = 439, Kuraray Specialities, Japan

P/MPa	T/K							
	450.10	460.43	471.16	481.64	492.41	502.47	512.67	523.20
	V_{spec}/cm^{3}g^{-1}							
0.1	0.9751	0.9813	0.9879	0.9945	1.0012	1.0075	1.0141	1.0214
10	0.9702	0.9762	0.9825	0.9888	0.9952	1.0013	1.0077	1.0145
20	0.9655	0.9713	0.9774	0.9834	0.9895	0.9954	1.0015	1.0081
30	0.9613	0.9669	0.9727	0.9787	0.9845	0.9901	0.9961	1.0024
40	0.9574	0.9627	0.9684	0.9742	0.9798	0.9853	0.9910	0.9971
50	0.9535	0.9587	0.9643	0.9700	0.9752	0.9807	0.9863	0.9921
60	0.9499	0.9550	0.9603	0.9659	0.9711	0.9764	0.9817	0.9875
70	0.9465	0.9515	0.9566	0.9620	0.9671	0.9722	0.9774	0.9830
80	0.9431	0.9479	0.9530	0.9582	0.9632	0.9682	0.9733	0.9788
90	0.9400	0.9446	0.9497	0.9546	0.9596	0.9644	0.9694	0.9748
100	0.9369	0.9415	0.9464	0.9513	0.9560	0.9608	0.9656	0.9709
110	0.9339	0.9385	0.9432	0.9480	0.9526	0.9573	0.9621	0.9671
120	0.9310	0.9354	0.9401	0.9449	0.9493	0.9539	0.9585	0.9636
130	0.9282	0.9327	0.9373	0.9418	0.9462	0.9508	0.9553	0.9601
140	0.9256	0.9300	0.9344	0.9389	0.9433	0.9477	0.9521	0.9569
150	0.9231	0.9272	0.9316	0.9361	0.9403	0.9447	0.9489	0.9537
160	0.9204	0.9247	0.9290	0.9334	0.9375	0.9418	0.9459	0.9507
170	0.9180	0.9221	0.9264	0.9307	0.9347	0.9389	0.9432	0.9476
180	0.9155	0.9197	0.9238	0.9281	0.9321	0.9362	0.9403	0.9449
190	0.9128	0.9173	0.9215	0.9257	0.9295	0.9335	0.9376	0.9420
200	0.9096	0.9150	0.9191	0.9231	0.9269	0.9309	0.9349	0.9392

continued

P/MPa	T/K	
	533.89	543.95
	V_{spec}/cm^{3}g^{-1}	
0.1	1.0293	1.0378
10	1.0220	1.0301
20	1.0153	1.0230
30	1.0093	1.0166
40	1.0037	1.0108
50	0.9985	1.0054
60	0.9937	1.0003
70	0.9891	0.9954

continued

continued

80	0.9846	0.9909
90	0.9804	0.9866
100	0.9764	0.9825
110	0.9726	0.9785
120	0.9689	0.9747
130	0.9654	0.9710
140	0.9620	0.9675
150	0.9588	0.9641
160	0.9556	0.9608
170	0.9525	0.9578
180	0.9495	0.9547
190	0.9467	0.9517
200	0.9438	0.9488

Polymer (B): **poly(ethylene-*co*-vinyl alcohol)** **2007FUN**
Characterization: 48 mol% ethene, ρ(293.15 K) = 1.1243 g/cm^3, T_g/K = 321, T_m/K = 436, Kuraray Specialities, Japan

P/MPa	T/K 447.09	456.99	467.74	478.12	488.02	493.16	503.16	519.30
	V_{spec}/cm^{3}g^{-1}							
0.1	0.9886	0.9947	1.0010	1.0072	1.0135	1.0199	1.0264	1.0336
10	0.9833	0.9891	0.9951	1.0011	1.0071	1.0132	1.0193	1.0260
20	0.9781	0.9838	0.9895	0.9954	1.0010	1.0067	1.0127	1.0188
30	0.9736	0.9791	0.9845	0.9902	0.9957	1.0013	1.0069	1.0128
40	0.9694	0.9748	0.9800	0.9854	0.9906	0.9961	1.0014	1.0070
50	0.9654	0.9704	0.9757	0.9809	0.9860	0.9912	0.9964	1.0018
60	0.9616	0.9665	0.9715	0.9767	0.9816	0.9866	0.9916	0.9969
70	0.9579	0.9628	0.9676	0.9726	0.9773	0.9824	0.9873	0.9922
80	0.9545	0.9592	0.9638	0.9687	0.9733	0.9781	0.9829	0.9877
90	0.9511	0.9556	0.9602	0.9649	0.9695	0.9742	0.9788	0.9836
100	0.9479	0.9522	0.9568	0.9615	0.9658	0.9705	0.9749	0.9796
110	0.9447	0.9491	0.9535	0.9580	0.9623	0.9667	0.9711	0.9757
120	0.9417	0.9460	0.9504	0.9547	0.9588	0.9632	0.9676	0.9721
130	0.9388	0.9429	0.9473	0.9516	0.9556	0.9599	0.9641	0.9685
140	0.9361	0.9402	0.9443	0.9485	0.9524	0.9567	0.9608	0.9651
150	0.9333	0.9373	0.9414	0.9456	0.9494	0.9535	0.9575	0.9618
160	0.9307	0.9346	0.9386	0.9427	0.9466	0.9505	0.9545	0.9586
170	0.9281	0.9321	0.9359	0.9399	0.9437	0.9476	0.9516	0.9556
180	0.9257	0.9295	0.9334	0.9373	0.9408	0.9448	0.9486	0.9525
190	0.9232	0.9270	0.9308	0.9346	0.9382	0.9421	0.9458	0.9498
200	0.9209	0.9246	0.9282	0.9322	0.9355	0.9394	0.9431	0.9468

continued

continued

P/MPa	*T*/K 528.86	539.38
	V_{spec}/cm^{3}g^{-1}	
0.1	1.0406	1.0491
10	1.0328	1.0408
20	1.0254	1.0330
30	1.0191	1.0264
40	1.0134	1.0202
50	1.0078	1.0146
60	1.0029	1.0093
70	0.9979	1.0042
80	0.9935	0.9996
90	0.9891	0.9951
100	0.9849	0.9909
110	0.9810	0.9867
120	0.9772	0.9827
130	0.9735	0.9790
140	0.9700	0.9754
150	0.9666	0.9719
160	0.9633	0.9685
170	0.9601	0.9653
180	0.9571	0.9621
190	0.9541	0.9590
200	0.9512	0.9562

Polymer (B): **poly(ethylene glycol)** **2006LEE**
Characterization: M_n/g.mol^{-1} = 260, M_w/g.mol^{-1} = 280

P/MPa	*T*/K 298.15	318.15	348.15
	V_{spec}/cm^{3}g^{-1}		
0.1	1.1200	1.1038	1.0802
10	1.1243	1.1087	1.0854
15	1.1264	1.1110	1.0880
20	1.1284	1.1134	1.0905
25	1.1303	1.1157	1.0929
30	1.1323	1.1179	1.0952
35	1.1342	1.1201	1.0975
40	1.1361	1.1222	1.0998
45	1.1379	1.1243	1.1019
50	1.1397	1.1263	1.1041

Polymer (B): **poly(ethylene glycol)** **2009AFZ**

Characterization: M_n/g.mol^{-1} = 414.5, Pluriol-E400, BASF AG, Ludwigshafen, Germany

P/MPa = 0.1

T/K	283.16	285.65	288.15	290.65	293.15	295.65	298.14	300.65	303.15
ρ/(g/cm^3)	1.13465	1.13254	1.13045	1.12840	1.12635	1.12430	1.12224	1.12018	1.11813
T/K	305.65	308.15	310.65	313.15	315.65	318.15	320.64	323.15	325.65
ρ/(g/cm^3)	1.11607	1.11402	1.11197	1.10992	1.10788	1.10583	1.10379	1.10175	1.09971
T/K	328.15	330.65	333.15	335.65	338.15	340.65	343.15	345.64	348.15
ρ/(g/cm^3)	1.09767	1.09564	1.09360	1.09157	1.08954	1.08751	1.08548	1.08345	1.08142
T/K	350.65	353.14	355.65	358.15	360.64	363.14			
ρ/(g/cm^3)	1.07939	1.07737	1.07534	1.07332	1.07129	1.06927			

Polymer (B): **poly(ethylene glycol) monomethyl ether** **2003LEE**

Characterization: M_n/g.mol^{-1} = 366, M_w/g.mol^{-1} = 373, PEGME-350, Aldrich Chem. Co., Inc., Milwaukee, WI

P/MPa	T/K		
	298.15	318.15	348.15
	V_{spec}/cm^{3}g^{-1}		
0.1	0.9223	0.9371	0.9596
10	0.9181	0.9321	0.9540
15	0.9159	0.9298	0.9512
20	0.9140	0.9276	0.9485
25	0.9120	0.9253	0.9459
30	0.9101	0.9233	0.9434
35	0.9083	0.9212	0.9410
40	0.9065	0.9192	0.9386
45	0.9046	0.9173	0.9363
50	0.9029	0.9153	0.9342

Polymer (B): **poly(glycerol), hyperbranched** **2006SEI**
Characterization: M_n/g.mol^{-1} = 2000, M_w/g.mol^{-1} = 3000,
synthesized in the laboratory

P/MPa	*T*/K 343.15	383.15	423.15
	ρ/kg m^{-3}		
5	1263.5	1234.9	1211.8
10	1266.7	1238.8	1216.8
20	1270.5	1244.7	1222.4
35	1274.7	1250.4	1226.2
60	1282.4	1258.8	1235.9

Polymer (B): **poly(propylene glycol)** **2004BOL**
Characterization: M_n/g.mol^{-1} = 425, M_w/M_n = 1.086,
Merck, Darmstadt, Germany

P/MPa	*T*/K 293.15	303.15	313.15	323.15	333.15	343.15	353.15	363.15	373.15
	V_{spez}/cm^{3}g^{-1}								
0.1	0.9804	0.9881	0.9950	1.0040	1.0121	1.0215	1.0309	1.0395	1.0493
10	0.9747	0.9823	0.9891	0.9970	1.0050	1.0142	1.0225	1.0320	1.0406
20	0.9699	0.9766	0.9843	0.9911	0.9990	1.0070	1.0152	1.0235	1.0331
30	0.9653	0.9718	0.9785	0.9852	0.9921	1.0000	1.0081	1.0163	1.0246
40	0.9606	0.9671	0.9737	0.9794	0.9872	0.9940	1.0020	1.0101	1.0183
50	0.9560	0.9625	0.9690	0.9747	0.9814	0.9881	0.9960	1.0040	1.0121
60	0.9524	0.9588	0.9643	0.9699	0.9766	0.9833	0.9911	0.9990	1.0060
70	0.9488	0.9551	0.9606	0.9662	0.9728	0.9794	0.9862	0.9940	1.0010
80	0.9452	0.9515	0.9569	0.9625	0.9690	0.9756	0.9823	0.9891	0.9960
90	0.9425	0.9479	0.9533	0.9588	0.9653	0.9718	0.9785	0.9852	0.9921
100	0.9398	0.9452	0.9506	0.9560	0.9615	0.9681	0.9747	0.9814	0.9881

Polymer (B): **poly(propylene glycol)** **2003LEE**
Characterization: M_n/g.mol^{-1} = 485, M_w/g.mol^{-1} = 500,
PPG-425, Aldrich Chem. Co., Inc., Milwaukee, WI

P/MPa	T/K 298.15	318.15	348.15
	V_{spec}/cm^{3}g^{-1}		
0.1	0.9969	1.0130	1.0384
10	0.9910	1.0062	1.0302
15	0.9881	1.0030	1.0264
20	0.9853	0.9999	1.0227
25	0.9827	0.9969	1.0193
30	0.9801	0.9941	1.0158
35	0.9775	0.9913	1.0127
40	0.9752	0.9885	1.0095
45	0.9728	0.9861	1.0064
50	0.9705	0.9836	1.0036

Polymer (B): **poly(vinyl alcohol)** **2007FUN**
Characterization: M_w/g.mol^{-1} = 195000, ρ(293.15 K) = 1.2906 g/cm^3, T_g/K = 350,
T_m/K = 493, Mowiol 56-98, Kuraray Specialities, Japan

P/MPa	T/K 499.16	504.17	508.69	513.70	518.52	523.47
	V_{spec}/cm^{3}g^{-1}					
0.1	0.8581	0.8611	0.8646	0.8679	0.8726	0.8782
10	0.8544	0.8573	0.8606	0.8640	0.8684	0.8738
20	0.8507	0.8536	0.8566	0.8601	0.8644	0.8697
30	0.8477	0.8504	0.8535	0.8568	0.8608	0.8659
40	0.8448	0.8475	0.8504	0.8535	0.8576	0.8628
50	0.8420	0.8446	0.8476	0.8505	0.8546	0.8596
60	0.8392	0.8419	0.8447	0.8476	0.8516	0.8567
70	0.8369	0.8392	0.8421	0.8450	0.8488	0.8539
80	0.8342	0.8367	0.8395	0.8424	0.8461	0.8512
90	0.8318	0.8343	0.8371	0.8399	0.8435	0.8486
100	0.8296	0.8319	0.8346	0.8375	0.8411	0.8460
110	0.8272	0.8296	0.8322	0.8350	0.8387	0.8436
120	0.8252	0.8274	0.8301	0.8327	0.8364	0.8410

continued

continued

130	0.8231	0.8253	0.8279	0.8306	0.8344	0.8388
140	0.8210	0.8232	0.8259	0.8284	0.8321	0.8366
150	0.8190	0.8212	0.8237	0.8265	0.8300	0.8345
160	0.8171	0.8194	0.8218	0.8245	0.8280	0.8324
170	0.8153	0.8175	0.8200	0.8226	0.8260	0.8304
180	0.8135	0.8156	0.8182	0.8207	0.8242	0.8285
190	0.8118	0.8139	0.8162	0.8188	0.8222	0.8265
200	0.8102	0.8123	0.8146	0.8171	0.8205	0.8248

Polymer (B): **poly(1-vinyl-2-pyrrolidinone)** **2008SHI**
Characterization: M_w/g.mol^{-1} = 55000, Aldrich Chem. Co., Inc., St. Louis, MO

P/MPa	*T*/K				
	445.15	455.15	465.15	475.15	485.15
	V_{spec}/cm^{3}g^{-1}				
10	0.7418	0.7445	0.7474	0.7503	0.7533
20	0.7397	0.7423	0.7450	0.7476	0.7506
30	0.7377	0.7400	0.7426	0.7452	0.7479
40	0.7358	0.7382	0.7405	0.7431	0.7455
50	0.7340	0.7364	0.7387	0.7410	0.7434
60	0.7325	0.7346	0.7369	0.7390	0.7413
70	0.7310	0.7330	0.7351	0.7371	0.7395
80	0.7295	0.7315	0.7335	0.7356	0.7377
90	0.7282	0.7300	0.7320	0.7340	0.7361
100	0.7269	0.7287	0.7305	0.7325	0.7343
110	0.7257	0.7274	0.7292	0.7310	0.7330
120	0.7244	0.7262	0.7279	0.7297	0.7315
130	0.7234	0.7249	0.7267	0.7284	0.7302
140	0.7224	0.7239	0.7257	0.7272	0.7290
150	0.7214	0.7229	0.7244	0.7262	0.7277
160	0.7205	0.7219	0.7234	0.7249	0.7267
170	0.7197	0.7209	0.7224	0.7239	0.7254
180	0.7187	0.7202	0.7217	0.7229	0.7244
190	0.7180	0.7195	0.7207	0.7222	0.7234
200	0.7173	0.7185	0.7200	0.7212	0.7227

Polymer (B): **poly(1-vinyl-2-pyrrolidinone)** **2003KIM**
Characterization: –

P/MPa	T/K								
	439.85	449.45	458.95	467.95	476.35	485.95	495.05	504.55	513.35
	V_{spec}/cm^{3}g^{-1}								
0.1	1.0035	1.0088	1.0137	1.0191	1.0239	1.0291	1.0343	1.0396	1.0453
10	0.9989	1.0039	1.0087	1.0137	1.0183	1.0234	1.0283	1.0333	1.0385
20	0.9945	0.9993	1.0038	1.0086	1.0130	1.0178	1.0225	1.0271	1.0319
30	0.9905	0.9952	0.9996	1.0041	1.0083	1.0131	1.0176	1.0220	1.0265
40	0.9868	0.9913	0.9955	1.0000	1.0041	1.0087	1.0130	1.0172	1.0216
50	0.9832	0.9875	0.9918	0.9960	1.0001	1.0046	1.0088	1.0127	1.0169
60	0.9799	0.9840	0.9881	0.9923	0.9963	1.0005	1.0046	1.0085	1.0127
70	0.9768	0.9806	0.9845	1.0000	0.9927	0.9969	1.0009	1.0044	1.0084
80	0.9737	0.9774	0.9812	0.9852	1.0000	0.9931	0.9970	1.0006	1.0050
90	0.9709	0.9743	0.9781	0.9819	0.9857	0.9898	0.9936	0.9970	1.0008
100	0.9682	0.9713	0.9750	0.9788	0.9826	0.9864	0.9901	0.9935	0.9972

6.2. Excess volumes and/or densities of aqueous polymer solutions

Polymer (B): **chitosan** **2005WAL**
Characterization: medium molecular weight sample, Aldrich Chem. Co., Inc., Milwaukee, WI
Solvent (A): **water** **H_2O** **7732-18-5**

T/K = 308.15

w_B	0.0005	0.0010	0.0015	0.0020	0.0025
ρ/(g/cm^3)	0.99698	0.99707	0.99240	0.99733	0.99753

Polymer (B): **chitosan** **2006MUT**
Characterization: Aldrich Chem. Co., Inc., Milwaukee, WI
Solvent (A): **water** **H_2O** **7732-18-5**

T/K = 298.15

w_B	0.1	0.2	0.3	0.4	0.5	0.6	0.7	0.8
ρ/(g/cm^3)	0.99816	0.99881	0.99947	1.00012	1.00078	1.00143	1.00209	1.00274

w_B	0.9	1.0
ρ/(g/cm^3)	1.00339	1.00405

T/K = 308.15

w_B	0.1	0.2	0.3	0.4	0.5	0.6	0.7	0.8
ρ/(g/cm^3)	0.99514	0.99579	0.99643	0.99708	0.99772	0.99836	0.99901	0.99965

w_B	0.9	1.0
ρ/(g/cm^3)	1.00030	1.00094

T/K = 318.15

w_B	0.1	0.2	0.3	0.4	0.5	0.6	0.7	0.8
ρ/(g/cm^3)	0.99072	0.99142	0.99212	0.99281	0.99352	0.99421	0.99491	0.99561

w_B	0.9	1.0
ρ/(g/cm^3)	0.99631	0.99701

T/K = 328.15

w_B	0.1	0.2	0.3	0.4	0.5	0.6	0.7	0.8
ρ/(g/cm^3)	0.98399	0.98488	0.98576	0.98664	0.98752	0.98841	0.98929	0.99017

w_B	0.9	1.0
ρ/(g/cm^3)	0.99106	0.99194

continued

continued

T/K = 338.15

w_B	0.1	0.2	0.3	0.4	0.5	0.6	0.7	0.8
ρ/(g/cm^3)	0.97401	0.97512	0.97622	0.97732	0.97843	0.97953	0.98064	0.98174

w_B	0.9	1.0
ρ/(g/cm^3)	0.98284	0.98395

Polymer (B): **guar gum** **2006MUT**
Characterization: S.D. Fine Chemicals, Mumbai, India
Solvent (A): **water** **H_2O** **7732-18-5**

T/K = 298.15

w_B	0.1	0.2	0.3	0.4	0.5	0.6	0.7	0.8
ρ/(g/cm^3)	0.99788	0.99811	0.99842	0.99887	0.99925	0.99973	1.00014	1.00057

w_B	0.9	1.0
ρ/(g/cm^3)	1.00122	1.00141

T/K = 308.15

w_B	0.1	0.2	0.3	0.4	0.5	0.6	0.7	0.8
ρ/(g/cm^3)	0.99484	0.99508	0.99541	0.99587	0.99623	0.99671	0.99713	0.99754

w_B	0.9	1.0
ρ/(g/cm^3)	0.99820	0.99837

T/K = 318.15

w_B	0.1	0.2	0.3	0.4	0.5	0.6	0.7	0.8
ρ/(g/cm^3)	0.99093	0.99098	0.99140	0.99199	0.99240	0.99285	0.99327	0.99369

w_B	0.9	1.0
ρ/(g/cm^3)	0.99436	0.99452

T/K = 328.15

w_B	0.1	0.2	0.3	0.4	0.5	0.6	0.7	0.8
ρ/(g/cm^3)	0.98203	0.98224	0.98359	0.98691	0.98784	0.98812	0.98854	0.98914

w_B	0.9	1.0
ρ/(g/cm^3)	0.98980	0.98996

T/K = 338.15

w_B	0.1	0.2	0.3	0.4	0.5	0.6	0.7	0.8
ρ/(g/cm^3)	0.97160	0.97191	0.97303	0.97603	0.97785	0.97830	0.98176	0.98302

w_B	0.9	1.0
ρ/(g/cm^3)	0.98447	0.98560

Polymer (B): **hydroxyethylcellulose** **2006MUT**
Characterization: Loba Chemicals, Mumbai, India
Solvent (A): **water** **H_2O** **7732-18-5**

T/K = 298.15

w_B	0.1	0.2	0.3	0.4	0.5	0.6	0.7	0.8
ρ/(g/cm^3)	0.99789	0.99829	0.99869	0.99909	0.99949	0.99989	1.00029	1.00068

w_B	0.9	1.0
ρ/(g/cm^3)	1.00108	1.00148

T/K = 308.15

w_B	0.1	0.2	0.3	0.4	0.5	0.6	0.7	0.8
ρ/(g/cm^3)	0.99489	0.99529	0.99568	0.99607	0.99646	0.99686	0.99725	0.99764

w_B	0.9	1.0
ρ/(g/cm^3)	0.99804	0.99843

T/K = 318.15

w_B	0.1	0.2	0.3	0.4	0.5	0.6	0.7	0.8
ρ/(g/cm^3)	0.99046	0.99091	0.99135	0.99180	0.99223	0.99268	0.99312	0.99356

w_B	0.9	1.0
ρ/(g/cm^3)	0.99401	0.99445

T/K = 328.15

w_B	0.1	0.2	0.3	0.4	0.5	0.6	0.7	0.8
ρ/(g/cm^3)	0.98373	0.98435	0.98497	0.98559	0.98621	0.98683	0.98745	0.98807

w_B	0.9	1.0
ρ/(g/cm^3)	0.98869	0.98931

T/K = 338.15

w_B	0.1	0.2	0.3	0.4	0.5	0.6	0.7	0.8
ρ/(g/cm^3)	0.97412	0.97533	0.97654	0.97775	0.97897	0.98018	0.98139	0.98261

w_B	0.9	1.0
ρ/(g/cm^3)	0.98381	0.98502

Polymer (B): **hydroxyethylcellulose** **2005WAL**
Characterization: low-viscosity-grade sample, Polysciences, Mumbai, India
Solvent (A): **water** **H_2O** **7732-18-5**

T/K = 308.15

w_B	0.0005	0.0010	0.0015	0.0020	0.0025
ρ/(g/cm^3)	0.99692	0.99701	0.99713	0.99725	0.99738

Polymer (B):	**hydroxyethylcellulose**		**2005WAL**
Characterization:	low-viscosity-grade sample, Polysciences, Mumbai, India		
Solvent (A):	**water**	**H_2O**	**7732-18-5**
Polymer (C):	**chitosan**		
Characterization:	medium molecular weight sample, Aldrich Chem. Co., Inc., Milwaukee, WI		

T/K = 308.15

w_C/w_B = 20/80 was kept constant

w_{B+C}	0.0005	0.0010	0.0015	0.0020	0.0025
ρ/(g/cm^3)	0.99693	0.99702	0.99715	0.99727	0.99740

w_C/w_B = 40/60 was kept constant

w_{B+C}	0.0005	0.0010	0.0015	0.0020	0.0025
ρ/(g/cm^3)	0.99694	0.99703	0.99716	0.99729	0.99743

w_C/w_B = 50/50 was kept constant

w_{B+C}	0.0005	0.0010	0.0015	0.0020	0.0025
ρ/(g/cm^3)	0.99694	0.99704	0.99717	0.99729	0.99744

w_C/w_B = 60/40 was kept constant

w_{B+C}	0.0005	0.0010	0.0015	0.0020	0.0025
ρ/(g/cm^3)	0.99696	0.99705	0.99719	0.99730	0.99747

w_C/w_B = 80/20 was kept constant

w_{B+C}	0.0005	0.0010	0.0015	0.0020	0.0025
ρ/(g/cm^3)	0.99697	0.99706	0.99721	0.99732	0.99750

Polymer (B):	**hydroxypropyl(methyl)cellulose**		**2006MUT**
Characterization:	Aldrich Chem. Co., Inc., Milwaukee, WI		
Solvent (A):	**water**	**H_2O**	**7732-18-5**

T/K = 298.15

w_B	0.1	0.2	0.3	0.4	0.5	0.6	0.7	0.8
ρ/(g/cm^3)	0.99770	0.99803	0.99831	0.99854	0.99878	0.99905	0.99934	0.99957

w_B	0.9	1.0
ρ/(g/cm^3)	0.99999	1.00018

T/K = 308.15

w_B	0.1	0.2	0.3	0.4	0.5	0.6	0.7	0.8
ρ/(g/cm^3)	0.99470	0.99502	0.99529	0.99551	0.99576	0.99603	0.99632	0.99657

w_B	0.9	1.0
ρ/(g/cm^3)	0.99696	0.99715

continued

continued

T/K = 318.15

w_B	0.1	0.2	0.3	0.4	0.5	0.6	0.7	0.8
ρ/(g/cm^3)	0.99088	0.99177	0.99146	0.99167	0.99193	0.99220	0.99247	0.99271
w_B	0.9	1.0						
ρ/(g/cm^3)	0.99310	0.99326						

T/K = 328.15

w_B	0.1	0.2	0.3	0.4	0.5	0.6	0.7	0.8
ρ/(g/cm^3)	0.98633	0.98662	0.98691	0.98713	0.98738	0.98761	0.98793	0.98812
w_B	0.9	1.0						
ρ/(g/cm^3)	0.98852	0.98870						

T/K = 338.15

w_B	0.1	0.2	0.3	0.4	0.5	0.6	0.7	0.8
ρ/(g/cm^3)	0.98060	0.98082	0.98160	0.98179	0.98196	0.98211	0.98262	0.98299
w_B	0.9	1.0						
ρ/(g/cm^3)	0.98322	0.98343						

Polymer (B): **maltodextrin** **2005NIN**
Characterization: M_n/g.mol^{-1} = 1140, M_w/g.mol^{-1} = 8285,
Aldrich Chem. Co., Inc., Milwaukee, WI
Solvent (A): **water** **H_2O** **7732-18-5**

T/K = 298.15

w_B	0.0500	0.0995	0.1416	0.1988	0.2506	0.2878	0.3489	0.3980
ρ/(g/cm^3)	1.01591	1.03525	1.05224	1.07614	1.09869	1.11537	1.14380	1.16776
w_B	0.4507	0.4938	0.5392	0.5941				
ρ/(g/cm^3)	1.19390	1.21650	1.24056	1.27196				

Polymer (B): **maltodextrin** **2005NIN**
Characterization: M_n/g.mol^{-1} = 1475, M_w/g.mol^{-1} = 13710,
Aldrich Chem. Co., Inc., Milwaukee, WI
Solvent (A): **water** **H_2O** **7732-18-5**

T/K = 298.15

w_B	0.0499	0.0999	0.1487	0.2004	0.2497	0.2996	0.3512	0.4025
ρ/(g/cm^3)	1.01573	1.03516	1.05472	1.07628	1.09761	1.11995	1.14387	1.16867
w_B	0.4992	0.5491						
ρ/(g/cm^3)	1.21786	1.24457						

Polymer (B):	**poly(acrylic acid)**	**2006SAR**
Characterization:	M_n/g.mol^{-1} = 2100, Aldrich Chem. Co., Inc., Milwaukee, WI	
Solvent (A):	**water** H_2O	**7732-18-5**

T/K = 293.15

w_B	0.00	0.05	0.10	0.15	0.20	0.25	0.30	0.35	0.40
ρ/(g/cm^3)	0.9982	1.0148	1.0309	1.0492	1.0669	1.0856	1.1020	1.1201	1.1412
w_B	0.45	0.50							
ρ/(g/cm^3)	1.1572	1.1751							

T/K = 298.15

w_B	0.00	0.05	0.10	0.15	0.20	0.25	0.30	0.35	0.40
ρ/(g/cm^3)	0.9970	1.0136	1.0297	1.0478	1.0659	1.0827	1.0984	1.1178	1.1382
w_B	0.45	0.50							
ρ/(g/cm^3)	1.1545	1.1717							

T/K = 303.15

w_B	0.00	0.05	0.10	0.15	0.20	0.25	0.30	0.35	0.40
ρ/(g/cm^3)	0.9956	1.0113	1.0289	1.0464	1.0642	1.0802	1.0987	1.1168	1.1362
w_B	0.45	0.50							
ρ/(g/cm^3)	1.1526	1.1690							

T/K = 308.15

w_B	0.00	0.05	0.10	0.15	0.20	0.25	0.30	0.35	0.40
ρ/(g/cm^3)	0.9940	1.0099	1.0273	1.0453	1.0619	1.0784	1.0981	1.1140	1.1319
w_B	0.45	0.50							
ρ/(g/cm^3)	1.1492	1.1667							

T/K = 313.15

w_B	0.00	0.05	0.10	0.15	0.20	0.25	0.30	0.35	0.40
ρ/(g/cm^3)	0.9921	1.0104	1.0268	1.0450	1.0607	1.0774	1.0966	1.1116	1.1293
w_B	0.45	0.50							
ρ/(g/cm^3)	1.1468	1.1632							

T/K = 318.15

w_B	0.00	0.05	0.10	0.15	0.20	0.25	0.30	0.35	0.40
ρ/(g/cm^3)	0.9901	1.0084	1.0249	1.0422	1.0580	1.0755	1.0915	1.1082	1.1265
w_B	0.45	0.50							
ρ/(g/cm^3)	1.1447	1.1611							

T/K = 323.15

w_B	0.00	0.05	0.10	0.15	0.20	0.25	0.30	0.35	0.40
ρ/(g/cm^3)	0.9880	1.0056	1.0209	1.0386	1.0548	1.0702	1.0844	1.1035	1.1199
w_B	0.45	0.50							
ρ/(g/cm^3)	1.1359	1.1524							

Polymer (B): **poly(acrylic acid)** **2010MOL**
Characterization: M_η/g.mol^{-1} = 450000, Aldrich Chem. Co., Inc., Milwaukee, WI
Solvent (A): **water** **H_2O** **7732-18-5**

T/K = 298.15

c_B/(mol/kg)	0.0000	0.0101	0.0179	0.0300	0.0399	0.0490	0.0601	0.0699
ρ/(g/cm^3)	0.99704	0.99737	0.99752	0.99808	0.99807	0.99831	0.99859	0.99884
c_B/(mol/kg)	0.0800	0.0901	0.0998	0.1093	0.1200	0.1293	0.1398	0.1502
ρ/(g/cm^3)	0.99909	0.99934	0.99958	0.99980	1.00010	1.00032	1.00059	1.00086
c_B/(mol/kg)	0.1586	0.1699	0.1808	0.1901	0.1998			
ρ/(g/cm^3)	1.00108	1.00133	1.00162	1.00184	1.00208			

Comments: Molality is calculated using 72.064 g/mol for the repetitive unit.

Polymer (B): **poly(ethylene glycol)** **2010TRI**
Characterization: M_w/g.mol^{-1} = 200, Central Drug House, New Delhi, India
Solvent (A): **water** **H_2O** **7732-18-5**

T/K = 283.15

w_B	0.00	0.10	0.20	0.30	0.40	0.50	0.60	0.70	0.80
ρ/(g/cm^3)	0.9997	1.0158	1.0331	1.0510	1.0691	1.0865	1.1031	1.1147	1.1242
w_B	0.90	1.00							
ρ/(g/cm^3)	1.1298	1.1327							

T/K = 293.15

w_B	0.00	0.10	0.20	0.30	0.40	0.50	0.60	0.70	0.80
ρ/(g/cm^3)	0.9982	1.0136	1.0299	1.0468	1.0638	1.0802	1.0961	1.1070	1.1163
w_B	0.90	1.00							
ρ/(g/cm^3)	1.1218	1.1246							

T/K = 303.15

w_B	0.00	0.10	0.20	0.30	0.40	0.50	0.60	0.70	0.80
ρ/(g/cm^3)	0.9957	1.0106	1.0262	1.0422	1.0583	1.0737	1.0888	1.0995	1.1084
w_B	0.90	1.00							
ρ/(g/cm^3)	1.1138	1.1166							

T/K = 313.15

w_B	0.00	0.10	0.20	0.30	0.40	0.50	0.60	0.70	0.80
ρ/(g/cm^3)	0.9923	1.0066	1.0217	1.0370	1.0522	1.0667	1.0814	1.0918	1.1005
w_B	0.90	1.00							
ρ/(g/cm^3)	1.1059	1.1086							

continued

continued

T/K = 323.15

w_B	0.00	0.10	0.20	0.30	0.40	0.50	0.60	0.70	0.80
ρ/(g/cm^3)	0.9880	1.0018	1.0164	1.0311	1.0457	1.0599	1.0738	1.0838	1.0925

w_B	0.90	1.00
ρ/(g/cm^3)	1.0979	1.1006

T/K = 333.15

w_B	0.00	0.10	0.20	0.30	0.40	0.50	0.60	0.70	0.80
ρ/(g/cm^3)	0.9831	0.9989	1.0114	1.0261	1.0406	1.0546	1.0686	1.0785	1.0872

w_B	0.90	1.00
ρ/(g/cm^3)	1.0925	1.0953

T/K = 343.15

w_B	0.00	0.10	0.20	0.30	0.40	0.50	0.60	0.70	0.80
ρ/(g/cm^3)	0.9776	0.9914	1.0058	1.0204	1.0348	1.0489	1.0627	1.0725	1.0811

w_B	0.90	1.00
ρ/(g/cm^3)	1.0865	1.0892

T/K = 353.15

w_B	0.00	0.10	0.20	0.30	0.40	0.50	0.60	0.70	0.80
ρ/(g/cm^3)	0.9716	0.9852	0.9995	1.0140	1.0284	1.0424	1.0561	1.0658	1.0744

w_B	0.90	1.00
ρ/(g/cm^3)	1.0797	1.0824

T/K = 363.15

w_B	0.00	0.10	0.20	0.30	0.40	0.50	0.60	0.70	0.80
ρ/(g/cm^3)	0.9651	0.9784	0.9926	1.0070	1.0213	1.0351	1.0488	1.0585	1.0669

w_B	0.90	1.00
ρ/(g/cm^3)	1.0722	1.0749

Polymer (B): **poly(ethylene glycol)** **2008DHO**
Characterization: M_n/g.mol^{-1} = 200, Merck, Darmstadt, Germany
Solvent (A): **water** **H_2O** **7732-18-5**

T/K = 275.15

x_B	0.0000	0.0018	0.0037	0.0057	0.0078	0.0099	0.0181	0.0299	0.0492
ρ/(g/cm^3)	0.9999	1.0035	1.0066	1.0099	1.0130	1.0166	1.0289	1.0453	1.0662

x_B	0.0782	0.0956
ρ/(g/cm^3)	1.0888	1.0974

continued

continued

T/K = 277.15

x_B	0.0000	0.0018	0.0037	0.0057	0.0078	0.0099	0.0181	0.0299	0.0492
ρ/(g/cm^3)	1.0000	1.0036	1.0065	1.0098	1.0129	1.0164	1.0286	1.0448	1.0653
x_B	0.0782	0.0956							
ρ/(g/cm^3)	1.0873	1.0963							

T/K = 279.15

x_B	0.0000	0.0018	0.0037	0.0057	0.0078	0.0099	0.0181	0.0299	0.0492
ρ/(g/cm^3)	0.9999	1.0035	1.0065	1.0096	1.0127	1.0161	1.0282	1.0444	1.0645
x_B	0.0782	0.0956							
ρ/(g/cm^3)	1.0869	1.0949							

T/K = 281.15

x_B	0.0000	0.0018	0.0037	0.0057	0.0078	0.0099	0.0181	0.0299	0.0492
ρ/(g/cm^3)	0.9999	1.0033	1.0063	1.0094	1.0125	1.0158	1.0279	1.0436	1.0637
x_B	0.0782	0.0956							
ρ/(g/cm^3)	1.0853	1.0937							

T/K = 283.15

x_B	0.0000	0.0018	0.0037	0.0057	0.0078	0.0099	0.0181	0.0299	0.0492
ρ/(g/cm^3)	0.9997	1.0031	1.0062	1.0092	1.0123	1.0155	1.0275	1.0428	1.0628
x_B	0.0782	0.0956							
ρ/(g/cm^3)	1.0842	1.0927							

Polymer (B): **poly(ethylene glycol)** **2008AYR**
Characterization: M_w/g.mol^{-1} = 200, Sigma Chemical Co., St. Louis, MO
Solvent (A): **water** **H_2O** **7732-18-5**

T/K = 288.15

c_B/(mol/kg)	0.0082	0.0519	0.1034	0.1506	0.2157	0.3055	0.3493	0.3970
ρ/(g/cm^3)	0.999343	1.000640	1.002146	1.003581	1.005493	1.008044	1.009225	1.010576

T/K = 298.15

c_B/(mol/kg)	0.0082	0.0519	0.1034	0.1506	0.2157	0.3055	0.3493	0.3970
ρ/(g/cm^3)	0.997279	0.998531	0.999977	1.001357	1.003189	1.005639	1.006765	1.008606

T/K = 308.15

c_B/(mol/kg)	0.0082	0.0519	0.1034	0.1506	0.2157	0.3055	0.3493	0.3970
ρ/(g/cm^3)	0.994258	0.995472	0.996870	0.998203	0.999975	1.002331	1.003422	1.004669

Polymer (B): **poly(ethylene glycol)** **2005ELI**
Characterization: M_n/g.mol^{-1} = 200, Merck, Darmstadt, Germany
Solvent (A): **water** **H_2O** **7732-18-5**

T/K = 298.15

w_B	0.0999	0.2003	0.2996	0.3998	0.4998	0.6000	0.7002	0.8000	0.8997
ρ/(g/cm^3)	1.0186	1.0347	1.0508	1.0659	1.0805	1.0984	1.1069	1.1190	1.1231
w_B	1.0000								
ρ/(g/cm^3)	1.1256								

T/K = 308.15

w_B	0.0999	0.2003	0.2996	0.3998	0.4998	0.6000	0.7002	0.8000	0.8997
ρ/(g/cm^3)	1.0152	1.0325	1.0466	1.0624	1.0769	1.0910	1.1039	1.1122	1.1179
w_B	1.0000								
ρ/(g/cm^3)	1.1196								

T/K = 318.15

w_B	0.0999	0.2003	0.2996	0.3998	0.4998	0.6000	0.7002	0.8000	0.8997
ρ/(g/cm^3)	1.0111	1.0280	1.0411	1.0547	1.0707	1.0843	1.0964	1.1054	1.1102
w_B	1.0000								
ρ/(g/cm^3)	1.1122								

T/K = 328.15

w_B	0.0999	0.2003	0.2996	0.3998	0.4998	0.6000	0.7002	0.8000	0.8997
ρ/(g/cm^3)	1.0037	1.0220	1.0356	1.0486	1.0640	1.0766	1.0888	1.0966	1.1022
w_B	1.0000								
ρ/(g/cm^3)	1.1044								

T/K = 338.15

w_B	0.0999	0.2003	0.2996	0.3998	0.4998	0.6000	0.7002	0.8000	0.8997
ρ/(g/cm^3)	1.0005	1.0155	1.0288	1.0378	1.0560	1.0682	1.0800	1.0889	1.0934
w_B	1.0000								
ρ/(g/cm^3)	1.0964								

Polymer (B): **poly(ethylene glycol)** **2003RAH**
Characterization: M_n/g.mol^{-1} = 200, Merck, Darmstadt, Germany
Solvent (A): **water** **H_2O** **7732-18-5**

T/K = 313.15

w_B	0.8991	0.8011	0.7120	0.6411	0.5197	0.4481	0.2220	0.1473
ρ/(g/cm^3)	1.1049	1.1018	1.0925	1.0811	1.0707	1.0613	1.0260	1.0032

continued

continued

T/K = 333.15

w_B	0.8375	0.7328	0.5586	0.4670	0.3828	0.2329	0.1493
ρ/(g/cm^3)	1.0852	1.0800	1.0572	1.0499	1.0437	1.0167	1.0063

T/K = 338.15

w_B	0.3552	0.2931	0.1974	0.1913	0.1076	0.0982	0.0793	0.0601	0.0399
ρ/(g/cm^3)	1.0364	1.0260	1.0094	1.0073	0.9938	0.9917	0.9845	0.9814	0.9782

Polymer (B): **poly(ethylene glycol)** **2008AYR**
Characterization: M_w/g.mol^{-1} = 300, Sigma Chemical Co., St. Louis, MO
Solvent (A): **water** **H_2O** **7732-18-5**

T/K = 288.15

c_B/(mol/kg)	0.0500	0.0993	0.1490	0.1975	0.2498	0.2973	0.3411	0.3983
ρ/(g/cm^3)	1.001442	1.003698	1.005935	1.008055	1.010339	1.012351	1.014186	1.016524

T/K = 298.15

c_B/(mol/kg)	0.0500	0.0993	0.1490	0.1975	0.2498	0.2973	0.3411	0.3983
ρ/(g/cm^3)	0.999297	1.001464	1.003606	1.005637	1.007824	1.009746	1.001150	1.013731

T/K = 308.15

c_B/(mol/kg)	0.0500	0.0993	0.1490	0.1975	0.2498	0.2973	0.3411	0.3983
ρ/(g/cm^3)	1.001442	1.003698	1.005935	1.008055	1.010339	1.012351	1.014186	1.016524

Polymer (B): **poly(ethylene glycol)** **2005ELI**
Characterization: M_n/g.mol^{-1} = 300, Merck, Darmstadt, Germany
Solvent (A): **water** **H_2O** **7732-18-5**

T/K = 298.15

w_B	0.0998	0.2002	0.3002	0.4001	0.4997	0.6000	0.7000	0.8007	0.9000
ρ/(g/cm^3)	1.0231	1.0363	1.0553	1.0682	1.0821	1.1019	1.1096	1.1226	1.1269
w_B	1.0000								
ρ/(g/cm^3)	1.1304								

T/K = 308.15

w_B	0.0998	0.2002	0.3002	0.4001	0.4997	0.6000	0.7000	0.8007	0.9000
ρ/(g/cm^3)	1.0208	1.0332	1.0534	1.0653	1.0804	1.0960	1.1066	1.1152	1.1205
w_B	1.0000								
ρ/(g/cm^3)	1.1224								

continued

continued

T/K = 318.15

w_B	0.0998	0.2002	0.3002	0.4001	0.4997	0.6000	0.7000	0.8007	0.9000
ρ/(g/cm^3)	1.0169	1.0275	1.0471	1.0572	1.0704	1.0870	1.0984	1.1065	1.1120

w_B	1.0000
ρ/(g/cm^3)	1.1156

T/K = 328.15

w_B	0.0998	0.2002	0.3002	0.4001	0.4997	0.6000	0.7000	0.8007	0.9000
ρ/(g/cm^3)	1.0112	1.0202	1.0424	1.0531	1.0612	1.0776	1.0882	1.0988	1.1043

w_B	1.0000
ρ/(g/cm^3)	1.1080

T/K = 338.15

w_B	0.0998	0.2002	0.3002	0.4001	0.4997	0.6000	0.7000	0.8007	0.9000
ρ/(g/cm^3)	1.0067	1.0150	1.0348	1.0447	1.0563	1.0701	1.0786	1.0895	1.0962

w_B	1.0000
ρ/(g/cm^3)	1.1000

Polymer (B): **poly(ethylene glycol)** **2003RAH**
Characterization: M_n/g.mol^{-1} = 300, Merck, Darmstadt, Germany
Solvent (A): **water** **H_2O** **7732-18-5**

T/K = 308.15

w_B	0.3926	0.2987	0.2500	0.1998	0.1080
ρ/(g/cm^3)	1.0561	1.0416	1.0317	1.0219	1.0104

T/K = 318.15

w_B	0.3926	0.2987	0.2500	0.1998	0.1080
ρ/(g/cm^3)	1.0447	1.0323	1.0251	1.0177	1.0053

T/K = 328.15

w_B	0.3926	0.2987	0.2500	0.1998	0.1080
ρ/(g/cm^3)	1.0385	1.0271	1.0199	1.0125	1.0021

T/K = 338.15

w_B	0.3926	0.2987	0.2500	0.1998	0.1080
ρ/(g/cm^3)	1.0333	1.0198	1.0146	1.0084	0.9897

Polymer (B):	**poly(ethylene glycol)**	**2008HAN**
Characterization:	M_n/g.mol^{-1} = 380-420, Beijing Reagent Company, China	
Solvent (A):	**water** H_2O	**7732-18-5**

T/K = 298.15

x_B	0.0000	0.0050	0.0190	0.0291	0.0431	0.0634	0.0951	0.1504
ρ/(g/cm^3)	0.9970	1.0126	1.0468	1.0645	1.0814	1.0965	1.1082	1.1158
V^E/(cm^3/g)	0.000	−0.085	−0.357	−0.553	−0.774	−1.013	−1.223	−1.345
x_B	0.2719	0.3403	0.4330	0.5199	0.6873	0.7996	0.8923	1.0000
ρ/(g/cm^3)	1.1200	1.1207	1.1213	1.1215	1.1217	1.1217	1.1218	1.1218
V^E/(cm^3/g)	−1.288	−1.196	−1.066	−0.913	−0.606	−0.377	−0.216	0.000

T/K = 303.15

x_B	0.0000	0.0050	0.0190	0.0291	0.0431	0.0634	0.0951	0.1504
ρ/(g/cm^3)	0.9957	1.0114	1.0448	1.0616	1.0781	1.0931	1.1041	1.1118
V^E/(cm^3/g)	0.000	−0.093	−0.359	−0.538	−0.753	−0.994	−1.179	−1.305
x_B	0.2719	0.3403	0.4330	0.5199	0.6873	0.7996	0.8923	1.0000
ρ/(g/cm^3)	1.1160	1.1168	1.1173	1.1176	1.1178	1.1179	1.1180	1.1180
V^E/(cm^3/g)	−1.246	−1.164	−1.020	−0.881	−0.574	−0.371	−0.213	0.000

T/K = 308.15

x_B	0.0000	0.0050	0.0190	0.0291	0.0431	0.0634	0.0951	0.1504
ρ/(g/cm^3)	0.9941	1.0095	1.0421	1.0586	1.0746	1.0892	1.1007	1.1081
V^E/(cm^3/g)	0.000	−0.089	−0.346	−0.522	−0.728	−0.958	−1.168	−1.285
x_B	0.2719	0.3403	0.4330	0.5199	0.6873	0.7996	0.8923	1.0000
ρ/(g/cm^3)	1.1119	1.1127	1.1132	1.1134	1.1136	1.1141	1.1142	1.1142
V^E/(cm^3/g)	−1.230	−1.152	−1.015	−0.864	−0.563	−0.361	−0.210	0.000

T/K = 313.15

x_B	0.0000	0.0050	0.0190	0.0291	0.0431	0.0634	0.0951	0.1504
ρ/(g/cm^3)	0.9922	1.0075	1.0394	1.0557	1.0710	1.0850	1.0964	1.1035
V^E/(cm^3/g)	0.000	−0.093	−0.346	−0.524	−0.717	−0.935	−1.147	−1.255
x_B	0.2719	0.3403	0.4330	0.5199	0.6873	0.7996	0.8923	1.0000
ρ/(g/cm^3)	1.1077	1.1085	1.1090	1.1092	1.1095	1.1096	1.1097	1.1097
V^E/(cm^3/g)	−1.209	−1.134	−0.999	−0.851	−0.577	−0.359	−0.207	0.000

T/K = 318.15

x_B	0.0000	0.0050	0.0190	0.0291	0.0431	0.0634	0.0951	0.1504
ρ/(g/cm^3)	0.9902	1.0049	1.0361	1.0518	1.0673	1.0809	1.0918	1.0993
V^E/(cm^3/g)	0.000	−0.084	−0.330	−0.497	−0.700	0.909	−1.102	−1.231
x_B	0.2719	0.3403	0.4330	0.5199	0.6873	0.7996	0.8923	1.0000
ρ/(g/cm^3)	1.1036	1.1043	1.1049	1.1050	1.1052	1.1056	1.1057	1.1057
V^E/(cm^3/g)	−1.200	−1.116	−0.999	−0.839	−0.547	−0.354	−0.205	0.000

continued

continued

T/K = 323.15

x_B	0.0000	0.0050	0.0190	0.0291	0.0431	0.0634	0.0951	0.1504
ρ/(g/cm^3)	0.9880	1.0025	1.0332	1.0486	1.0635	1.0768	1.0877	1.0954
V^E/(cm^3/g)	0.000	−0.083	−0.326	−0.489	−0.678	−0.879	−1.073	−1.214
x_B	0.2719	0.3403	0.4330	0.5199	0.6873	0.7996	0.8923	1.0000
ρ/(g/cm^3)	1.0995	1.1004	1.1009	1.1011	1.1013	1.1015	1.1016	1.1017
V^E/(cm^3/g)	−1.166	−1.104	−0.974	−0.830	−0.541	−0.350	−0.202	0.000

Polymer (B): **poly(ethylene glycol)** **2010TRI**
Characterization: M_w/g.mol^{-1} = 400, Central Drug House, New Delhi, India
Solvent (A): **water** **H_2O** **7732-18-5**

T/K = 283.15

w_B	0.00	0.10	0.20	0.30	0.40	0.50	0.60	0.70	0.80
ρ/(g/cm^3)	0.9997	1.0170	1.0353	1.0540	1.0733	1.0919	1.1087	1.1210	1.1272
w_B	0.90	1.00							
ρ/(g/cm^3)	1.1335	1.1351							

T/K = 293.15

w_B	0.00	0.10	0.20	0.30	0.40	0.50	0.60	0.70	0.80
ρ/(g/cm^3)	0.9982	1.0147	1.0321	1.0497	1.0678	1.0853	1.1012	1.1129	1.1191
w_B	0.90	1.00							
ρ/(g/cm^3)	1.1251	1.1265							

T/K = 303.15

w_B	0.00	0.10	0.20	0.30	0.40	0.50	0.60	0.70	0.80
ρ/(g/cm^3)	0.9957	1.0114	1.0280	1.0447	1.0618	1.0784	1.0934	1.1047	1.1108
w_B	0.90	1.00							
ρ/(g/cm^3)	1.1167	1.1181							

T/K = 313.15

w_B	0.00	0.10	0.20	0.30	0.40	0.50	0.60	0.70	0.80
ρ/(g/cm^3)	0.9923	1.0075	1.0234	1.0393	1.0555	1.0712	1.0856	1.0964	1.1025
w_B	0.90	1.00							
ρ/(g/cm^3)	1.1083	1.1098							

T/K = 323.15

w_B	0.00	0.10	0.20	0.30	0.40	0.50	0.60	0.70	0.80
ρ/(g/cm^3)	0.9880	1.0025	1.0181	1.0333	1.0489	1.0638	1.0776	1.0881	1.0941
w_B	0.90	1.00							
ρ/(g/cm^3)	1.1001	1.1015							

continued

continued

T/K = 333.15

w_B	0.00	0.10	0.20	0.30	0.40	0.50	0.60	0.70	0.80
ρ/(g/cm^3)	0.9831	0.9977	1.0132	1.0279	1.0437	1.0585	1.0719	1.0814	1.0888
w_B	0.90	1.00							
ρ/(g/cm^3)	1.0947	1.0962							

T/K = 343.15

w_B	0.00	0.10	0.20	0.30	0.40	0.50	0.60	0.70	0.80
ρ/(g/cm^3)	0.9776	0.9922	1.0075	1.0223	1.0379	1.0526	1.0659	1.0754	1.0828
w_B	0.90	1.00							
ρ/(g/cm^3)	1.0887	1.0901							

T/K = 353.15

w_B	0.00	0.10	0.20	0.30	0.40	0.50	0.60	0.70	0.80
ρ/(g/cm^3)	0.9716	0.9860	1.0013	1.0159	1.0315	1.0461	1.0593	1.0686	1.0760
w_B	0.90	1.00							
ρ/(g/cm^3)	1.0819	1.0833							

T/K = 363.15

w_B	0.00	0.10	0.20	0.30	0.40	0.50	0.60	0.70	0.80
ρ/(g/cm^3)	0.9651	0.9791	0.9943	1.0088	1.0243	1.0388	1.0519	1.0612	1.0686
w_B	0.90	1.00							
ρ/(g/cm^3)	1.0744	1.0758							

Polymer (B): **poly(ethylene glycol)** **2008AYR**
Characterization: M_w/g.mol^{-1} = 400, Sigma Chemical Co., St. Louis, MO
Solvent (A): **water** H_2O **7732-18-5**

T/K = 288.15

c_B/(mol/kg)	0.0500	0.0985	0.1360	0.2161	0.3006	0.3500	0.4001
ρ/(g/cm^3)	1.00228	1.00528	1.00754	1.01219	1.01687	1.01961	1.02221

T/K = 298.15

c_B/(mol/kg)	0.0500	0.0985	0.1360	0.2161	0.3006	0.3500	0.4001
ρ/(g/cm^3)	1.00010	1.00297	1.00514	1.00958	1.01405	1.01666	1.01914

T/K = 308.15

c_B/(mol/kg)	0.0500	0.0985	0.1360	0.2161	0.3006	0.3500	0.4001
ρ/(g/cm^3)	0.99698	0.99975	1.00183	1.00612	1.01041	1.01291	1.01526

Polymer (B): **poly(ethylene glycol)** **2008DHO**
Characterization: M_n/g.mol^{-1} = 400, Qualigens Fine Chemicals, Mumbai, India
Solvent (A): **water** **H_2O** **7732-18-5**

T/K = 275.15

x_B	0.0000	0.0023	0.0047	0.0073	0.0108	0.0143	0.0186	0.0267	0.0346
ρ/(g/cm^3)	0.9999	1.0087	1.0168	1.0254	1.0361	1.0489	1.0614	1.0740	1.0856
x_B	0.0616	0.0867							
ρ/(g/cm^3)	1.1119	1.1243							

T/K = 277.15

x_B	0.0000	0.0023	0.0047	0.0073	0.0108	0.0143	0.0186	0.0267	0.0346
ρ/(g/cm^3)	1.0000	1.0086	1.0166	1.0251	1.0357	1.0484	1.0604	1.0740	1.0844
x_B	0.0616	0.0867							
ρ/(g/cm^3)	1.1102	1.1228							

T/K = 279.15

x_B	0.0000	0.0023	0.0047	0.0073	0.0108	0.0143	0.0186	0.0267	0.0346
ρ/(g/cm^3)	0.9999	1.0085	1.0163	1.0247	1.0350	1.0478	1.0601	1.0719	1.0840
x_B	0.0616	0.0867							
ρ/(g/cm^3)	1.1073	1.1190							

T/K = 281.15

x_B	0.0000	0.0023	0.0047	0.0073	0.0108	0.0143	0.0186	0.0267	0.0346
ρ/(g/cm^3)	0.9999	1.0083	1.0160	1.0243	1.0346	1.0472	1.0590	1.0710	1.0824
x_B	0.0616	0.0867							
ρ/(g/cm^3)	1.1075	1.1194							

T/K = 283.15

x_B	0.0000	0.0023	0.0047	0.0073	0.0108	0.0143	0.0186	0.0267	0.0346
ρ/(g/cm^3)	0.9997	1.0080	1.0157	1.0238	1.0341	1.0466	1.0580	1.0700	1.0811
x_B	0.0616	0.0867							
ρ/(g/cm^3)	1.1073	1.1811							

Polymer (B): **poly(ethylene glycol)** **2007KUS**
Characterization: M_n/g.mol^{-1} = 400, Merck, Darmstadt, Germany
Solvent (A): **water** **H_2O** **7732-18-5**

T/K = 298.15

c_B/(mol/kg)	0.00000	0.04106	0.08692	0.13264	0.18218	0.22888	0.28080	0.33438
ρ/(g/cm^3)	0.99705	0.99958	1.00235	1.00502	1.00785	1.01084	1.01318	1.01696
c_B/(mol/kg)	0.38898	0.44096	0.49970					
ρ/(g/cm^3)	1.01893	1.02148	1.02412					

Polymer (B): **poly(ethylene glycol)** **2009AFZ**
Characterization: M_n/g.mol^{-1} = 414.5, Pluriol-E400, BASF AG, Ludwigshafen, Germany
Solvent (A): **water** **H_2O** **7732-18-5**

T/K = 278.15

x_A	0.9963	0.9928	0.9876	0.9822	0.9768	0.9627	0.9571	0.9472
ρ/(g/cm^3)	1.01322	1.02461	1.03967	1.05342	1.06513	1.08817	1.09491	1.10446
V^E/(cm^3/mol)	−0.070	−0.142	−0.255	−0.379	−0.500	−0.788	−0.886	−1.038
x_A	0.9007	0.8536	0.6867	0.3411	0.1852			
ρ/(g/cm^3)	1.12581	1.13238	1.13729	1.13876	1.13859			
V^E/(cm^3/mol)	−1.436	−1.553	−1.465	−0.716	−0.963			

T/K = 288.15

x_A	0.9963	0.9928	0.9876	0.9822	0.9768	0.9627	0.9571	0.9472
ρ/(g/cm^3)	1.01167	1.02242	1.03659	1.04947	1.06044	1.08206	1.08841	1.09743
V^E/(cm^3/mol)	−0.064	−0.130	−0.234	−0.346	−0.456	−0.719	−0.809	−0.948
x_A	0.9007	0.8536	0.6867	0.3411	0.1852			
ρ/(g/cm^3)	1.11786	1.12431	1.12899	1.13020	1.13034			
V^E/(cm^3/mol)	−1.315	−1.421	−1.276	−0.659	−0.358			

T/K = 298.15

x_A	0.9963	0.9928	0.9876	0.9822	0.9768	0.9627	0.9571	0.9472
ρ/(g/cm^3)	1.00907	1.01933	1.03276	1.04494	1.05529	1.07566	1.08166	1.09020
V^E/(cm^3/mol)	−0.062	−0.124	−0.222	−0.326	−0.429	−0.674	−0.759	−0.890
x_A	0.9007	0.8536	0.6867	0.3411	0.1852			
ρ/(g/cm^3)	1.10979	1.11613	1.12086	1.12202	1.12215			
V^E/(cm^3/mol)	−1.242	−1.350	−1.228	−0.640	−0.350			

T/K = 308.15

x_A	0.9963	0.9928	0.9876	0.9822	0.9768	0.9627	0.9571	0.9472
ρ/(g/cm^3)	1.00561	1.01545	1.02829	1.03987	1.04968	1.06898	1.07466	1.08278
V^E/(cm^3/mol)	−0.060	−0.120	−0.213	−0.312	−0.408	−0.638	−0.717	−0.840
x_A	0.9007	0.8536	0.6867	0.3411	0.1852			
ρ/(g/cm^3)	1.10161	1.10786	1.11267	1.11383	1.11395			
V^E/(cm^3/mol)	−1.178	−1.287	−1.185	−0.624	−0.342			

T/K = 318.15

x_A	0.9963	0.9928	0.9876	0.9822	0.9768	0.9627	0.9571	0.9472
ρ/(g/cm^3)	1.00141	1.01090	1.02321	1.03429	1.04364	1.06199	1.06740	1.07514
V^E/(cm^3/mol)	−0.058	−0.117	−0.205	−0.299	−0.390	−0.606	−0.681	−0.797
x_A	0.9007	0.8536	0.6867	0.3411	0.1852			
ρ/(g/cm^3)	1.09333	1.09953	1.10445	1.10565	1.10577			
V^E/(cm^3/mol)	−1.120	−1.229	−1.145	−0.607	−0.335			

continued

continued

T/K = 328.15

x_A	0.9963	0.9928	0.9876	0.9822	0.9768	0.9627	0.9571	0.9472
ρ/(g/cm^3)	0.99656	1.00573	1.01760	1.02822	1.03718	1.05471	1.05989	1.06730
V^E/(cm^3/mol)	−0.057	−0.114	−0.199	−0.288	−0.374	−0.577	−0.647	−0.757
x_A	0.9007	0.8536	0.6867	0.3411	0.1852			
ρ/(g/cm^3)	1.08494	1.09111	1.09619	1.09747	1.09760			
V^E/(cm^3/mol)	−1.066	−1.175	−1.105	−0.589	−0.326			

T/K = 338.15

x_A	0.9963	0.9928	0.9876	0.9822	0.9768	0.9627	0.9571	0.9472
ρ/(g/cm^3)	0.99112	1.00001	1.01148	1.02172	1.03032	1.04714	1.05211	1.05925
V^E/(cm^3/mol)	−0.056	−0.110	−0.192	−0.277	−0.358	−0.550	−0.616	−0.719
x_A	0.9007	0.8536	0.6867	0.3411	0.1852			
ρ/(g/cm^3)	1.07644	1.08262	1.08789	1.08929	1.08945			
V^E/(cm^3/mol)	−1.014	−1.122	−1.064	−0.569	−0.314			

T/K = 348.15

x_A	0.9963	0.9928	0.9876	0.9822	0.9768	0.9627	0.9571	0.9472
ρ/(g/cm^3)	0.98514	0.99378	1.00489	1.01478	1.02307	1.03928	1.04407	1.05098
V^E/(cm^3/mol)	−0.054	−0.107	−0.185	−0.266	−0.343	−0.523	−0.585	−0.682
x_A	0.9007	0.8536	0.6867	0.3411	0.1852			
ρ/(g/cm^3)	1.06781	1.07403	1.07954	1.08113	1.08131			
V^E/(cm^3/mol)	−0.964	−1.070	−1.022	−0.552	−0.304			

T/K = 358.15

x_A	0.9963	0.9928	0.9876	0.9822	0.9768	0.9627	0.9571	0.9472
ρ/(g/cm^3)	0.97866	0.98707	0.99786	1.00744	1.01546	1.03114	1.03578	1.04249
V^E/(cm^3/mol)	−0.052	−0.103	−0.178	−0.254	−0.327	−0.496	−0.554	−0.646
x_A	0.9007	0.8536	0.6867	0.3411	0.1852			
ρ/(g/cm^3)	1.05904	1.06533	1.07115	1.07295	1.07317			
V^E/(cm^3/mol)	−0.914	−1.018	−0.978	−0.530	−0.292			

T/K = 363.15

x_A	0.9963	0.9928	0.9876	0.9822	0.9768	0.9627	0.9571	0.9472
ρ/(g/cm^3)	–	0.98354	0.99418	1.00362	1.01152	1.02696	1.03153	1.03816
V^E/(cm^3/mol)	–	−0.101	−0.173	−0.248	−0.318	−0.482	−0.539	−0.627
x_A	0.9007	0.8536	0.6867	0.3411	0.1852			
ρ/(g/cm^3)	1.05461	1.06095	1.06692	–	–			
V^E/(cm^3/mol)	−0.888	−0.991	−0.954	–	–			

Polymer (B): **poly(ethylene glycol)** **2008AYR**
Characterization: M_w/g.mol^{-1} = 550, Sigma Chemical Co., St. Louis, MO
Solvent (A): **water** H_2O **7732-18-5**

T/K = 288.15

c_B/(mol/kg)	0.0500	0.1058	0.1498	0.2001	0.2492	0.2996	0.3492	0.4032
ρ/(g/cm^3)	1.00358	1.00834	1.01197	1.01597	1.01973	1.02338	1.02683	1.03054

T/K = 298.15

c_B/(mol/kg)	0.0500	0.1058	0.1498	0.2001	0.2492	0.2996	0.3492	0.4032
ρ/(g/cm^3)	1.00134	1.00590	1.00937	1.01319	1.01617	1.02025	1.02352	1.02705

T/K = 308.15

c_B/(mol/kg)	0.0500	0.1058	0.1498	0.2001	0.2492	0.2996	0.3492	0.4032
ρ/(g/cm^3)	0.99818	1.00256	1.00590	1.00957	1.01300	1.01640	1.01950	1.02290

Polymer (B): **poly(ethylene glycol)** **2011ZHA**
Characterization: M_n/g.mol^{-1} = 600, Beijing Reagent Company, China
Solvent (A): **water** H_2O **7732-18-5**

T/K = 298.15

x_B	0.0000	0.0075	0.0196	0.0431	0.1001	0.1998	0.3001	0.3994
ρ/(g/cm^3)	0.9970	1.0309	1.0662	1.0995	1.1156	1.1217	1.1225	1.1226
V^E/(cm^3/mol)	0.000	−0.230	−0.585	1.107	−1.349	−1.436	−1.269	−1.047
x_B	0.4983	0.6977	0.8849	1.0000				
ρ/(g/cm^3)	1.1228	1.1230	1.1232	1.1233				
V^E/(cm^3/mol)	−0.907	−0.490	−0.149	0.000				

T/K = 303.15

x_B	0.0000	0.0075	0.0196	0.0431	0.1001	0.1998	0.3001	0.3994
ρ/(g/cm^3)	0.9957	1.0287	1.0629	1.0951	1.1120	1.1179	1.1187	1.1189
V^E/(cm^3/mol)	0.000	−0.223	−0.562	−1.056	−1.340	−1.422	−1.274	−1.072
x_B	0.4983	0.6977	0.8849	1.0000				
ρ/(g/cm^3)	1.1190	1.1192	1.1194	1.1195				
V^E/(cm^3/mol)	−1.874	−0.503	−0.183	0.000				

T/K = 308.15

x_B	0.0000	0.0075	0.0196	0.0431	0.1001	0.1998	0.3001	0.3994
ρ/(g/cm^3)	0.9940	1.0259	1.0598	1.0906	1.1070	1.1133	1.1139	1.1140
V^E/(cm^3/mol)	0.000	−0.210	−0.555	−1.024	−1.291	−1.412	−1.259	−1.046
x_B	0.4983	0.6977	0.8849	1.0000				
ρ/(g/cm^3)	1.1141	1.1144	1.1145	1.1147				
V^E/(cm^3/mol)	−0.852	−0.497	−0.161	0.000				

Polymer (B):	**poly(ethylene glycol)**	**2010TRI**
Characterization:	M_w/g.mol^{-1} = 600, Central Drug House, New Delhi, India	
Solvent (A):	**water** H_2O	**7732-18-5**

T/K = 283.15

w_B	0.00	0.10	0.20	0.30	0.40	0.50	0.60	0.70	0.80
ρ/(g/cm^3)	0.9997	1.0170	1.0354	1.0547	1.0743	1.0933	1.1100	1.1226	1.1306

w_B	0.90
ρ/(g/cm^3)	1.1344

T/K = 293.15

w_B	0.00	0.10	0.20	0.30	0.40	0.50	0.60	0.70	0.80
ρ/(g/cm^3)	0.9982	1.0148	1.0322	1.0504	1.0688	1.0866	1.1023	1.1143	1.1221

w_B	0.90	1.00
ρ/(g/cm^3)	1.1259	1.1271

T/K = 303.15

w_B	0.00	0.10	0.20	0.30	0.40	0.50	0.60	0.70	0.80
ρ/(g/cm^3)	0.9957	1.0115	1.0281	1.0454	1.0627	1.0796	1.0945	1.1060	1.1136

w_B	0.90	1.00
ρ/(g/cm^3)	1.1176	1.1186

T/K = 313.15

w_B	0.00	0.10	0.20	0.30	0.40	0.50	0.60	0.70	0.80
ρ/(g/cm^3)	0.9923	1.0076	1.0234	1.0399	1.0564	1.0724	1.0865	1.0977	1.1052

w_B	0.90	1.00
ρ/(g/cm^3)	1.1090	1.1102

T/K = 323.15

w_B	0.00	0.10	0.20	0.30	0.40	0.50	0.60	0.70	0.80
ρ/(g/cm^3)	0.9880	1.0029	1.0182	1.0340	1.0487	1.0649	1.0784	1.0892	1.0966

w_B	0.90	1.00
ρ/(g/cm^3)	1.1006	1.1019

T/K = 333.15

w_B	0.00	0.10	0.20	0.30	0.40	0.50	0.60	0.70	0.80
ρ/(g/cm^3)	0.9831	0.9980	1.0133	1.0291	1.0447	1.0598	1.0734	1.0840	1.0913

w_B	0.90	1.00
ρ/(g/cm^3)	1.0952	1.0965

continued

continued

T/K = 343.15

w_B	0.00	0.10	0.20	0.30	0.40	0.50	0.60	0.70	0.80
ρ/(g/cm^3)	0.9776	0.9924	1.0077	1.0234	1.0390	1.0539	1.0674	1.0780	1.0852
w_B	0.90	1.00							
ρ/(g/cm^3)	1.0892	1.0904							

T/K = 353.15

w_B	0.00	0.10	0.20	0.30	0.40	0.50	0.60	0.70	0.80
ρ/(g/cm^3)	0.9716	0.9863	1.0014	1.0170	1.0324	1.0474	1.0608	1.0712	1.0785
w_B	0.90	1.00							
ρ/(g/cm^3)	1.0824	1.0836							

T/K = 363.15

w_B	0.00	0.10	0.20	0.30	0.40	0.50	0.60	0.70	0.80
ρ/(g/cm^3)	0.9651	0.9794	0.9945	1.0099	1.0253	1.0401	1.0534	1.0638	1.0710
w_B	0.90	1.00							
ρ/(g/cm^3)	1.0749	1.0761							

Polymer (B): **poly(ethylene glycol)** **2008AYR**
Characterization: M_w/g.mol^{-1} = 600, Sigma Chemical Co., St. Louis, MO
Solvent (A): **water** H_2O **7732-18-5**

T/K = 288.15

c_B/(mol/kg)	0.0499	0.1499	0.1984	0.2348	0.2925	0.3500	0.4013
ρ/(g/cm^3)	1.00420	1.01330	1.01728	1.02013	1.02464	1.02900	1.03253

T/K = 298.15

c_B/(mol/kg)	0.0499	0.1499	0.1984	0.2348	0.2925	0.3500	0.4013
ρ/(g/cm^3)	1.00199	1.01061	1.01448	1.01717	1.02144	1.02563	1.02890

T/K = 308.15

c_B/(mol/kg)	0.0499	0.1499	0.1984	0.2348	0.2925	0.3500	0.4013
ρ/(g/cm^3)	0.99881	1.00708	1.01076	1.01339	1.01747	1.02143	1.02462

Polymer (B): **poly(ethylene glycol)** **2005ELI**
Characterization: M_n/g.mol^{-1} = 600, Merck, Darmstadt, Germany
Solvent (A): **water** H_2O **7732-18-5**

T/K = 298.15

w_B	0.1117	0.2008	0.3006	0.4027	0.5015	0.6026	0.7007	0.8007	0.9004
ρ/(g/cm^3)	1.0144	1.0304	1.0473	1.0642	1.0809	1.0953	1.1068	1.1169	1.1212
w_B	1.0000								
ρ/(g/cm^3)	1.1225								

continued

continued

T/K = 308.15

w_B	0.1117	0.2008	0.3006	0.4027	0.5015	0.6026	0.7007	0.8007	0.9004
ρ/(g/cm^3)	1.0094	1.0249	1.0417	1.0580	1.0743	1.0883	1.0995	1.1080	1.1137
w_B	1.0000								
ρ/(g/cm^3)	1.1151								

T/K = 318.15

w_B	0.1117	0.2008	0.3006	0.4027	0.5015	0.6026	0.7007	0.8007	0.9004
ρ/(g/cm^3)	1.0058	1.0205	1.0371	1.0527	1.0683	1.0817	1.0932	1.1018	1.1057
w_B	1.0000								
ρ/(g/cm^3)	1.1065								

T/K = 328.15

w_B	0.1117	0.2008	0.3006	0.4027	0.5015	0.6026	0.7007	0.8007	0.9004
ρ/(g/cm^3)	1.0022	1.0167	1.0317	1.0476	1.0621	1.0746	1.0851	1.0935	1.0984
w_B	1.0000								
ρ/(g/cm^3)	1.0992								

T/K = 338.15

w_B	0.1117	0.2008	0.3006	0.4027	0.5015	0.6026	0.7007	0.8007	0.9004
ρ/(g/cm^3)	0.9975	1.0119	1.0272	1.0427	1.0581	1.0709	1.0814	1.0904	1.0949
w_B	1.0000								
ρ/(g/cm^3)	1.0965								

Polymer (B): **poly(ethylene glycol)** **2005MOH**
Characterization: M_n/g.mol^{-1} = 950-1050, Merck, Darmstadt, Germany
Solvent (A): **water** H_2O **7732-18-5**

T/K = 298.15

w_B	0.10	0.15	0.20	0.25	0.30	0.35	0.40	0.45	0.50
ρ/(g/cm^3)	1.0169	1.0250	1.0329	1.0428	1.0510	1.0585	1.0680	1.0785	1.0868

T/K = 303.15

w_B	0.10	0.15	0.20	0.25	0.30	0.35	0.40	0.45	0.50
ρ/(g/cm^3)	1.0149	1.0237	1.0321	1.0407	1.0505	1.0576	1.0672	1.0778	1.0864

T/K = 308.15

w_B	0.10	0.15	0.20	0.25	0.30	0.35	0.40	0.45	0.50
ρ/(g/cm^3)	1.0140	1.0221	1.0306	1.0400	1.0484	1.0557	1.0650	1.0742	1.0843

continued

continued

T/K = 313.15

w_B	0.10	0.15	0.20	0.25	0.30	0.35	0.40	0.45	0.50
ρ/(g/cm^3)	1.0110	1.0192	1.0294	1.0380	1.0454	1.0529	1.0610	1.0702	1.0800

T/K = 318.15

w_B	0.10	0.15	0.20	0.25	0.30	0.35	0.40	0.45	0.50
ρ/(g/cm^3)	1.0092	1.0175	1.0270	1.0345	1.0421	1.0500	1.0591	1.0662	1.0749

T/K = 323.15

w_B	0.10	0.15	0.20	0.25	0.30	0.35	0.40	0.45	0.50
ρ/(g/cm^3)	1.0080	1.0149	1.0225	1.0304	1.0395	1.0473	1.0550	1.0645	1.0719

T/K = 328.15

w_B	0.10	0.15	0.20	0.25	0.30	0.35	0.40	0.45	0.50
ρ/(g/cm^3)	1.0047	1.0128	1.0211	1.0281	1.0351	1.0442	1.0521	1.0606	1.0681

Polymer (B): **poly(ethylene glycol)** **2008AYR**
Characterization: M_w/g.mol^{-1} = 1000, Sigma Chemical Co., St. Louis, MO
Solvent (A): **water** H_2O **7732-18-5**

T/K = 288.15

c_B/(mol/kg)	0.0498	0.0990	0.1501	0.1996	0.2493	0.2995	0.3489	0.3998
ρ/(g/cm^3)	1.00730	1.01479	1.02204	1.02863	1.03468	1.04039	1.04591	1.05104

T/K = 298.15

c_B/(mol/kg)	0.0498	0.0990	0.1501	0.1996	0.2493	0.2995	0.3489	0.3998
ρ/(g/cm^3)	1.00489	1.01206	1.01896	1.02523	1.03097	1.03649	1.04162	1.04647

T/K = 308.15

c_B/(mol/kg)	0.0498	0.0990	0.1501	0.1996	0.2493	0.2995	0.3489	0.3998
ρ/(g/cm^3)	1.00150	1.00841	1.01500	1.02101	1.02659	1.03179	1.03674	1.04135

Polymer (B): **poly(ethylene glycol)** **2007KUS**
Characterization: M_n/g.mol^{-1} = 1000, Merck, Darmstadt, Germany
Solvent (A): **water** H_2O **7732-18-5**

T/K = 298.15

c_B/(mol/kg)	0.00000	0.03667	0.07503	0.11528	0.15830	0.20642	0.25416	0.31263
ρ/(g/cm^3)	0.99705	1.00275	1.00836	1.01401	1.01967	1.02557	1.03106	1.03745
c_B/(mol/kg)	0.37010	0.43100	0.50037					
ρ/(g/cm^3)	1.04310	1.04871	1.05468					

Polymer (B): **poly(ethylene glycol)** **2008AYR**
Characterization: M_w/g.mol^{-1} = 1450, Sigma Chemical Co., St. Louis, MO
Solvent (A): **water** **H_2O** **7732-18-5**

T/K = 288.15

c_B/(mol/kg)	0.0500	0.0996	0.1487	0.1994	0.2501	0.2983	0.3509	0.3988
ρ/(g/cm^3)	1.01149	1.02235	1.03219	1.04123	1.04926	1.05649	1.06319	1.06868

T/K = 298.15

c_B/(mol/kg)	0.0500	0.0996	0.1487	0.1994	0.2501	0.2983	0.3509	0.3988
ρ/(g/cm^3)	1.00890	1.01926	1.02863	1.03722	1.04484	1.05171	1.05805	1.06327

T/K = 308.15

c_B/(mol/kg)	0.0500	0.0996	0.1487	0.1994	0.2501	0.2983	0.3509	0.3988
ρ/(g/cm^3)	1.00543	1.01538	1.02436	1.03257	1.03985	1.04640	1.05245	1.05742

Polymer (B): **poly(ethylene glycol)** **2010TRI**
Characterization: M_w/g.mol^{-1} = 1500, Central Drug House, New Delhi, India
Solvent (A): **water** **H_2O** **7732-18-5**

T/K = 283.15

w_B	0.00	0.10	0.20	0.30	0.40	0.50	0.60	0.70
ρ/(g/cm^3)	0.9997	1.0173	1.0359	1.0557	1.0752	1.0952	1.1112	1.1249

T/K = 293.15

w_B	0.00	0.10	0.20	0.30	0.40	0.50	0.60	0.70	0.80
ρ/(g/cm^3)	0.9982	1.0149	1.0326	1.0511	1.0694	1.0882	1.1032	1.1161	1.1200

T/K = 303.15

w_B	0.00	0.10	0.20	0.30	0.40	0.50	0.60	0.70	0.80
ρ/(g/cm^3)	0.9957	1.0118	1.0284	1.0462	1.0635	1.0811	1.0953	1.1077	1.1130

T/K = 313.15

w_B	0.00	0.10	0.20	0.30	0.40	0.50	0.60	0.70	0.80
ρ/(g/cm^3)	0.9923	1.0078	1.0237	1.0407	1.0571	1.0738	1.0873	1.0991	1.1064

w_B	0.90
ρ/(g/cm^3)	1.1094

T/K = 323.15

w_B	0.00	0.10	0.20	0.30	0.40	0.50	0.60	0.70	0.80
ρ/(g/cm^3)	0.9880	1.0030	1.0183	1.0346	1.0501	1.0660	1.0789	1.0902	1.0976

w_B	0.90	1.00
ρ/(g/cm^3)	1.1008	1.1067

continued

continued

T/K = 333.15

w_B	0.00	0.10	0.20	0.30	0.40	0.50	0.60	0.70	0.80
ρ/(g/cm^3)	0.9831	0.9981	1.0134	1.0295	1.0450	1.0608	1.0736	1.0849	1.0923
w_B	0.90	1.00							
ρ/(g/cm^3)	1.0954	1.1040							

T/K = 343.15

w_B	0.00	0.10	0.20	0.30	0.40	0.50	0.60	0.70	0.80
ρ/(g/cm^3)	0.9776	0.9926	1.0078	1.0238	1.0392	1.0549	1.0677	1.0789	1.0862
w_B	0.90	1.00							
ρ/(g/cm^3)	1.0894	1.0920							

T/K = 353.15

w_B	0.00	0.10	0.20	0.30	0.40	0.50	0.60	0.70	0.80
ρ/(g/cm^3)	0.9716	0.9864	1.0015	1.0174	1.0328	1.0483	1.0610	1.0722	1.0795
w_B	0.90	1.00							
ρ/(g/cm^3)	1.0826	1.0852							

T/K = 363.15

w_B	0.00	0.10	0.20	0.30	0.40	0.50	0.60	0.70	0.80
ρ/(g/cm^3)	0.9651	0.9795	0.9945	1.0104	1.0256	1.0412	1.0537	1.0648	1.0720
w_B	0.90	1.00							
ρ/(g/cm^3)	1.0751	1.0774							

Polymer (B): **poly(ethylene glycol)** **2005ELI**
Characterization: M_n/g.mol^{-1} = 2000, Merck, Darmstadt, Germany
Solvent (A): **water** H_2O **7732-18-5**

T/K = 298.15

w_B	0.0547	0.1002	0.1515	0.2055	0.2513	0.3071	0.3512	0.4039	0.4522
ρ/(g/cm^3)	1.0131	1.0216	1.0306	1.0389	1.0478	1.0563	1.0651	1.0736	1.0820
w_B	0.5004								
ρ/(g/cm^3)	1.0906								

T/K = 308.15

w_B	0.0547	0.1002	0.1515	0.2055	0.2513	0.3071	0.3512	0.4039	0.4522
ρ/(g/cm^3)	1.0054	1.0162	1.0251	1.0343	1.0429	1.0516	1.0599	1.0689	1.0781
w_B	0.5004								
ρ/(g/cm^3)	1.0861								

continued

continued

T/K = 318.15

w_B	0.0547	0.1002	0.1515	0.2055	0.2513	0.3071	0.3512	0.4039	0.4522
ρ/(g/cm^3)	0.9942	1.0078	1.0187	1.0279	1.0372	1.0460	1.0546	1.0634	1.0719
w_B	0.5004								
ρ/(g/cm^3)	1.0803								

T/K = 328.15

w_B	0.0547	0.1002	0.1515	0.2055	0.2513	0.3071	0.3512	0.4039	0.4522
ρ/(g/cm^3)	0.9889	1.0026	1.0134	1.0228	1.0323	1.0407	1.0493	1.0580	1.0671
w_B	0.5004								
ρ/(g/cm^3)	1.0751								

T/K = 338.15

w_B	0.0547	0.1002	0.1515	0.2055	0.2513	0.3071	0.3512	0.4039	0.4522
ρ/(g/cm^3)	0.9865	1.0003	1.0107	1.0205	1.0301	1.0391	1.0468	1.0556	1.0649
w_B	0.5004								
ρ/(g/cm^3)	1.0726								

Polymer (B):	**poly(ethylene glycol)**		**2004GRA**
Characterization:	M_n/g.mol^{-1} = 2000, M_w/g.mol^{-1} = 2080, Merck, Darmstadt, Germany		
Solvent (A):	**water**	**H_2O**	**7732-18-5**
Salt (C):	**sodium carbonate**	**Na_2CO_3**	**497-19-8**

T/K = 293.15

w_B	0.00000	0.00000	0.00000	0.00000	0.00000	0.00000	0.00000	0.10000	0.10001
w_C	0.00999	0.02000	0.03000	0.04000	0.05000	0.06000	0.07000	0.01000	0.02000
ρ/(g/cm^3)	1.00866	1.01919	1.02943	1.03991	1.05028	1.06085	1.07135	1.02665	1.03703
w_B	0.10000	0.10000	0.09999	0.09997	0.19999	0.20000	0.20000	0.30000	0.29999
w_C	0.03001	0.04003	0.05007	0.06001	0.01001	0.02000	0.03000	0.01001	0.02002
ρ/(g/cm^3)	1.04756	1.05755	1.06788	1.07833	1.04386	1.05401	1.06521	1.06195	1.07295
w_B	0.40000	0.10000	0.19999	0.29999	0.40003				
w_C	0.01001	0.00000	0.00000	0.00000	0.00000				
ρ/(g/cm^3)	1.08124	1.01565	1.03342	1.05126	1.07076				

T/K = 298.15

w_B	0.00000	0.00000	0.00000	0.00000	0.00000	0.00000	0.00000	0.10000	0.10001
w_C	0.00999	0.02000	0.03000	0.04000	0.05000	0.06000	0.07000	0.01000	0.02000
ρ/(g/cm^3)	1.00738	1.01803	1.02794	1.03848	1.04862	1.05929	1.06952	1.02413	1.03456
w_B	0.10000	0.10000	0.09999	0.09997	0.19999	0.20000	0.20000	0.30000	0.29999
w_C	0.03001	0.04003	0.05007	0.06001	0.01001	0.02000	0.03000	0.01001	0.02002
ρ/(g/cm^3)	1.04526	1.05539	1.06594	1.07631	1.04177	1.05185	1.06309	1.05949	1.07031

continued

continued

w_B	0.40000	0.10000	0.19999	0.29999	0.40003
w_C	0.01001	0.00000	0.00000	0.00000	0.00000
ρ/(g/cm^3)	1.07813	1.01362	1.03117	1.04889	1.06771

T/K = 303.15

w_B	0.00000	0.00000	0.00000	0.00000	0.00000	0.00000	0.00000	0.10000	0.10001
w_C	0.00999	0.02000	0.03000	0.04000	0.05000	0.06000	0.07000	0.01000	0.02000
ρ/(g/cm^3)	1.00589	1.01629	1.02628	1.03665	1.04679	1.05718	1.06757	1.02229	1.03263
w_B	0.10000	0.10000	0.09999	0.09997	0.19999	0.20000	0.20000	0.30000	0.29999
w_C	0.03001	0.04003	0.05007	0.06001	0.01001	0.02000	0.03000	0.01001	0.02002
ρ/(g/cm^3)	1.04303	1.05346	1.06376	1.07407	1.03986	1.04957	1.06022	1.05678	1.06756
w_B	0.40000	0.10000	0.19999	0.29999	0.40003				
w_C	0.01001	0.00000	0.00000	0.00000	0.00000				
ρ/(g/cm^3)	1.07495	1.01189	1.02887	1.04627	1.06460				

T/K = 308.15

w_B	0.00000	0.00000	0.00000	0.00000	0.00000	0.00000	0.00000	0.10000	0.10001
w_C	0.00999	0.02000	0.03000	0.04000	0.05000	0.06000	0.07000	0.01000	0.02000
ρ/(g/cm^3)	1.00421	1.01453	1.02446	1.03477	1.04483	1.05523	1.06549	1.02030	1.03052
w_B	0.10000	0.10000	0.09999	0.09997	0.19999	0.20000	0.20000	0.30000	0.29999
w_C	0.03001	0.04003	0.05007	0.06001	0.01001	0.02000	0.03000	0.01001	0.02002
ρ/(g/cm^3)	1.04090	1.05107	1.06146	1.07173	1.03702	1.04735	1.05774	1.05398	1.06474
w_B	0.40000	0.10000	0.19999	0.29999	0.40003				
w_C	0.01001	0.00000	0.00000	0.00000	0.00000				
ρ/(g/cm^3)	1.07162	1.00998	1.02647	1.04356	1.06138				

Polymer (B): **poly(ethylene glycol)** **2005MUR**
Characterization: M_n/g.mol^{-1} = 2000, Merck, Darmstadt, Germany
Solvent (A): **water** **H_2O** **7732-18-5**

T/K = 298.15

w_B	0.00	0.05	0.10	0.15	0.20	0.25	0.30	0.35	0.40
ρ/(g/cm^3)	0.99700	1.00487	1.01327	1.02219	1.03160	1.04037	1.04924	1.05854	1.06756
w_B	0.45	0.50							
ρ/(g/cm^3)	1.07612	1.08394							

T/K = 303.15

w_B	0.00	0.05	0.10	0.15	0.20	0.25	0.30	0.35	0.40
ρ/(g/cm^3)	0.99560	1.00314	1.01168	1.02082	1.02976	1.03830	1.04754	1.05638	1.06562
w_B	0.45	0.50							
ρ/(g/cm^3)	1.07446	1.08250							

continued

continued

T/K = 308.15

w_B	0.00	0.05	0.10	0.15	0.20	0.25	0.30	0.35	0.40
ρ/(g/cm^3)	0.99400	1.00135	1.00939	1.01787	1.02673	1.03501	1.04412	1.05284	1.06209
w_B	0.45	0.50							
ρ/(g/cm^3)	1.07097	1.07789							

T/K = 313.15

w_B	0.00	0.05	0.10	0.15	0.20	0.25	0.30	0.35	0.40
ρ/(g/cm^3)	0.99220	0.99926	1.00732	1.01488	1.02344	1.03250	1.04056	1.04962	1.05868
w_B	0.45	0.50							
ρ/(g/cm^3)	1.06674	1.07380							

T/K = 318.15

w_B	0.00	0.05	0.10	0.15	0.20	0.25	0.30	0.35	0.40
ρ/(g/cm^3)	0.99020	0.99728	1.00466	1.01190	1.02025	1.02882	1.03766	1.04640	1.05431
w_B	0.45	0.50							
ρ/(g/cm^3)	1.06248	1.07048							

Polymer (B):	**poly(ethylene glycol)**		**2005MUR**
Characterization:	M_n/g.mol^{-1} = 2000, Merck, Darmstadt, Germany		
Solvent (A):	**water**	H_2O	**7732-18-5**
Salt (C):	**sodium citrate**	$Na_3C_6H_5O_7$	**68-04-2**

T/K = 298.15

w_B	0.40	0.35	0.30	0.25	0.20	0.15	0.10	0.05	0.00
w_C	0.02	0.02	0.02	0.02	0.02	0.02	0.02	0.02	0.02
ρ/(g/cm^3)	1.08221	1.07330	1.06371	1.05506	1.04665	1.03650	1.02712	1.01810	1.01060
w_B	0.35	0.30	0.25	0.20	0.15	0.10	0.05	0.00	0.25
w_C	0.04	0.04	0.04	0.04	0.04	0.04	0.04	0.04	0.06
ρ/(g/cm^3)	1.08400	1.07445	1.06524	1.05644	1.04720	1.03800	1.02903	1.02141	1.07904
w_B	0.20	0.15	0.10	0.05	0.00	0.20	0.15	0.10	0.05
w_C	0.06	0.06	0.06	0.06	0.06	0.08	0.08	0.08	0.08
ρ/(g/cm^3)	1.06960	1.06004	1.05148	1.04186	1.03314	1.08219	1.07365	1.06436	1.05549
w_B	0.00	0.15	0.10	0.05	0.00				
w_C	0.08	0.10	0.10	0.10	0.10				
ρ/(g/cm^3)	1.04682	1.08621	1.07755	1.06912	1.06070				

T/K = 308.15

w_B	0.40	0.35	0.30	0.25	0.20	0.15	0.10	0.05	0.00
w_C	0.02	0.02	0.02	0.02	0.02	0.02	0.02	0.02	0.02
ρ/(g/cm^3)	1.07840	1.06950	1.06135	1.05190	1.04200	1.03330	1.02370	1.01435	1.00479

continued

continued

w_B	0.35	0.30	0.25	0.20	0.15	0.10	0.05	0.00	0.25
w_C	0.04	0.04	0.04	0.04	0.04	0.04	0.04	0.04	0.06
ρ/(g/cm^3)	1.08110	1.07160	1.06300	1.05304	1.04418	1.03498	1.02650	1.01657	1.07570
w_B	0.20	0.15	0.10	0.05	0.00	0.20	0.15	0.10	0.05
w_C	0.06	0.06	0.06	0.06	0.06	0.08	0.08	0.08	0.08
ρ/(g/cm^3)	1.06650	1.05801	1.04800	1.03940	1.02970	1.07910	1.06950	1.06060	1.05196
w_B	0.00	0.15	0.10	0.05	0.00				
w_C	0.08	0.10	0.10	0.10	0.10				
ρ/(g/cm^3)	1.04201	1.08266	1.07364	1.06548	1.05769				

T/K = 318.15

w_B	0.40	0.35	0.30	0.25	0.20	0.15	0.10	0.05	0.00
w_C	0.02	0.02	0.02	0.02	0.02	0.02	0.02	0.02	0.02
ρ/(g/cm^3)	1.06857	1.06134	1.05358	1.04600	1.03800	1.02890	1.01944	1.01083	1.00089
w_B	0.35	0.30	0.25	0.20	0.15	0.10	0.05	0.00	0.25
w_C	0.04	0.04	0.04	0.04	0.04	0.04	0.04	0.04	0.06
ρ/(g/cm^3)	1.07202	1.06560	1.05803	1.04930	1.04110	1.03189	1.02292	1.01330	1.07100
w_B	0.20	0.15	0.10	0.05	0.00	0.20	0.15	0.10	0.05
w_C	0.06	0.06	0.06	0.06	0.06	0.08	0.08	0.08	0.08
ρ/(g/cm^3)	1.06222	1.05363	1.04365	1.03436	1.02523	1.07510	1.06647	1.05660	1.04720
w_B	0.00	0.15	0.10	0.05	0.00				
w_C	0.08	0.10	0.10	0.10	0.10				
ρ/(g/cm^3)	1.03840	1.07923	1.07059	1.06216	1.05356				

Polymer (B): **poly(ethylene glycol)** **2004TEN**
Characterization: M_n/g.mol^{-1} = 2000, Merck, Darmstadt, Germany
Solvent (A): **water** **H_2O** **7732-18-5**

T/K = 293.15

c_B/(mol/kg)	0.05054	0.10009	0.15051	0.20054	0.25050	0.30131	0.40157	0.50198
ρ/(g/cm^3)	1.01322	1.02642	1.03817	1.04875	1.05730	1.06531	1.07766	1.08846

T/K = 298.15

c_B/(mol/kg)	0.05054	0.10009	0.15051	0.20054	0.25050	0.30131	0.40157	0.50198
ρ/(g/cm^3)	1.01195	1.02491	1.03634	1.04610	1.05497	1.06305	1.07542	1.08531

T/K = 303.15

c_B/(mol/kg)	0.05054	0.10009	0.15051	0.20054	0.25050	0.30131	0.40157	0.50198
ρ/(g/cm^3)	1.01028	1.02320	1.03425	1.04381	1.05225	1.05990	1.07254	1.08147

Polymer (B): **poly(ethylene glycol)** **2004TEN**
Characterization: M_n/g.mol^{-1} = 2000, Merck, Darmstadt, Germany
Solvent (A): **water** H_2O **7732-18-5**
Salt (C): **sodium nitrate** $NaNO_3$ **7631-99-4**

T/K = 293.15

c_B/(mol/kg)	0.015	0.015	0.066	0.070	0.150	0.151	0.251	0.251
c_C/(mol/kg)	0.493	0.993	0.375	0.797	0.297	0.793	0.197	0.698
ρ/(g/cm^3)	1.02961	1.05529	1.03612	1.05670	1.0S063	1.07053	1.06432	1.08135

c_B/(mol/kg)	0.351
c_C/(mol/kg)	0.199
ρ/(g/cm^3)	1.07789

T/K = 298.15

c_B/(mol/kg)	0.015	0.015	0.066	0.070	0.150	0.151	0.251	0.251
c_C/(mol/kg)	0.493	0.993	0.375	0.797	0.297	0.793	0.197	0.698
ρ/(g/cm^3)	1.02769	1.0S327	1.03429	1.05484	1.04865	1.06807	1.06141	1.07814

c_B/(mol/kg)	0.351
c_C/(mol/kg)	0.199
ρ/(g/cm^3)	1.07513

T/K = 303.15

c_B/(mol/kg)	0.015	0.015	0.066	0.070	0.150	0.151	0.251	0.251
c_C/(mol/kg)	0.493	0.993	0.375	0.797	0.297	0.793	0.197	0.698
ρ/(g/cm^3)	1.02679	1.05148	1.03253	1.05269	1.04631	1.06592	1.05912	1.07604

c_B/(mol/kg)	0.351
c_C/(mol/kg)	0.199
ρ/(g/cm^3)	1.07234

Polymer (B): **poly(ethylene glycol)** **2004TEN**
Characterization: M_n/g.mol^{-1} = 2000, Merck, Darmstadt, Germany
Solvent (A): **water** H_2O **7732-18-5**
Component (C): **sucrose** $C_{12}H_{22}O_{11}$ **57-50-1**

T/K = 293.15

c_B/(mol/kg)	0.015	0.015	0.070	0.070	0.151	0.151	0.252	0.351
c_C/(mol/kg)	0.201	0.200	0.201	0.201	0.201	0.352	0.201	0.050
ρ/(g/cm^3)	1.02809	1.06098	1.04130	1.07089	1.05760	1.07100	1.07396	1.07632

c_B/(mol/kg)	0.352
c_C/(mol/kg)	0.126
ρ/(g/cm^3)	1.08154

continued

continued

T/K = 298.15

c_B/(mol/kg)	0.015	0.015	0.070	0.070	0.151	0.151	0.252	0.351
c_C/(mol/kg)	0.201	0.200	0.201	0.201	0.201	0.352	0.201	0.050
ρ/(g/cm^3)	1.02647	1.05903	1.03931	1.05530	1.05530	1.06866	1.07118	1.07305
c_B/(mol/kg)	0.352							
c_C/(mol/kg)	0.126							
ρ/(g/cm^3)	1.07832							

T/K = 303.15

c_B/(mol/kg)	0.015	0.015	0.070	0.070	0.151	0.151	0.252	0.351
c_C/(mol/kg)	0.201	0.200	0.201	0.201	0.201	0.352	0.201	0.050
ρ/(g/cm^3)	1.02447	1.05717	1.03739	1.06678	1.05282	1.06607	1.06823	1.06971
c_B/(mol/kg)	0.352							
c_C/(mol/kg)	0.126							
ρ/(g/cm^3)	1.07486							

Polymer (B): **poly(ethylene glycol)** **2004TEN**
Characterization: M_n/g.mol^{-1} = 2000, Merck, Darmstadt, Germany
Solvent (A): **water** **H_2O** **7732-18-5**
Component (C): **sucrose** **$C_{12}H_{22}O_{11}$** **57-50-1**
Salt (D): **sodium chloride** **NaCl** **7647-14-5**

T/K = 293.15

c_B/(mol/kg)	0.015	0.015	0.050	0.050	0.100	0.100	0.201	0.201
c_C/(mol/kg)	0.200	0.201	0.200	0.200	0.301	0.301	0.201	0.201
c_D/(mol/kg)	0.497	0.998	0.499	1.005	0.509	1.005	0.203	0.503
ρ/(g/cm^3)	1.04605	1.06389	1.05361	1.07051	1.07303	1.08773	1.07170	1.07964
c_B/(mol/kg)	0.251	0.250	0.301					
c_C/(mol/kg)	0.100	0.100	0.100					
c_D/(mol/kg)	0.103	0.300	0.100					
ρ/(g/cm^3)	1.06862	1.07360	1.07559					

T/K = 298.15

c_B/(mol/kg)	0.015	0.015	0.050	0.050	0.100	0.100	0.201	0.201
c_C/(mol/kg)	0.200	0.201	0.200	0.200	0.301	0.301	0.201	0.201
c_D/(mol/kg)	0.497	0.998	0.499	1.005	0.509	1.005	0.203	0.503
ρ/(g/cm^3)	1.04444	1.06198	1.05186	1.06829	1.07076	1.08530	1.06896	1.07693
c_B/(mol/kg)	0.251	0.250	0.301					
c_C/(mol/kg)	0.100	0.100	0.100					
c_D/(mol/kg)	0.103	0.300	0.100					
ρ/(g/cm^3)	1.06573	1.07047	1.07245					

continued

continued

T/K = 303.15

c_B/(mol/kg)	0.015	0.015	0.050	0.050	0.100	0.100	0.201	0.201
c_C/(mol/kg)	0.200	0.201	0.200	0.200	0.301	0.301	0.201	0.201
c_D/(mol/kg)	0.497	0.998	0.499	1.005	0.509	1.005	0.203	0.503
ρ/(g/cm^3)	1.04248	1.05994	1.04981	1.06617	1.06836	1.08290	1.06631	1.07420

c_B/(mol/kg)	0.251	0.250	0.301
c_C/(mol/kg)	0.100	0.100	0.100
c_D/(mol/kg)	0.103	0.300	0.100
ρ/(g/cm^3)	1.06296	1.06781	1.06935

Polymer (B): **poly(ethylene glycol)** **2010SAL**
Characterization: M_n/g.mol^{-1} = 3000, Merck, Darmstadt, Germany
Solvent (A): **water** H_2O **7732-18-5**

T/K = 298.15

w_B	0.000	0.005	0.010	0.015	0.020	0.025	0.030	0.035	0.040
ρ/(g/cm^3)	0.99704	0.99782	0.99859	0.99937	1.00016	1.00093	1.00174	1.00253	1.00334

w_B	0.045	0.050
ρ/(g/cm^3)	1.00417	1.00497

T/K = 308.15

w_B	0.000	0.005	0.010	0.015	0.020	0.025	0.030	0.035	0.040
ρ/(g/cm^3)	0.99406	0.99478	0.99553	0.99633	0.99714	0.99789	0.99871	0.99951	1.00033

w_B	0.045	0.050
ρ/(g/cm^3)	1.00113	1.00199

T/K = 318.15

w_B	0.000	0.005	0.010	0.015	0.020	0.025	0.030	0.035	0.040
ρ/(g/cm^3)	0.99024	0.99077	0.99137	0.99206	0.99278	0.99359	0.99426	0.99497	0.99573

w_B	0.045	0.050
ρ/(g/cm^3)	0.99655	0.99731

Polymer (B): **poly(ethylene glycol)** **2009CRU**
Characterization: M_n/g.mol^{-1} = 3000, Merck, Darmstadt, Germany
Solvent (A): **water** H_2O **7732-18-5**

T/K = 283.15

m_B/m_A	0.0546	0.1119	0.1784	0.2544	0.3418	0.4322	0.5516	0.6657	0.8200
ρ/(g/cm^3)	1.00903	1.01824	1.02737	1.03769	1.04789	1.05714	1.06792	1.07721	1.08724

m_B/m_A	0.8988	0.9982	1.1035	1.2201	1.3519	1.4932
ρ/(g/cm^3)	1.09117	1.09689	1.10060	1.10551	1.10914	1.11409

continued

continued

T/K = 288.15

m_B/m_A	0.0546	0.1119	0.1784	0.2544	0.3418	0.4322	0.5516	0.6657	0.8200
$\rho/(g/cm^3)$	1.00819	1.01716	1.02604	1.03606	1.04592	1.05492	1.06536	1.07436	1.08407

m_B/m_A	0.8988	0.9982	1.1035	1.2201	1.3519	1.4932
$\rho/(g/cm^3)$	1.08788	1.09344	1.09707	1.10180	1.10524	1.11008

T/K = 293.15

m_B/m_A	0.0546	0.1119	0.1784	0.2544	0.3418	0.4322	0.5516	0.6657	0.8200
$\rho/(g/cm^3)$	1.00708	1.01584	1.02449	1.03425	1.04382	1.05256	1.06269	1.07141	1.08084

m_B/m_A	0.8988	0.9982	1.1035	1.2201	1.3519	1.4932
$\rho/(g/cm^3)$	1.08452	1.08992	1.09343	1.09803	1.10141	1.10602

T/K = 298.15

m_B/m_A	0.0546	0.1119	0.1784	0.2544	0.3418	0.4322	0.5516	0.6657	0.8200
$\rho/(g/cm^3)$	1.00573	1.01429	1.02275	1.03226	1.04158	1.05007	1.05990	1.06836	1.07752

m_B/m_A	0.8988	0.9982	1.1035	1.2201	1.3519	1.4932
$\rho/(g/cm^3)$	1.08110	1.08633	1.08974	1.09423	1.09752	1.10203

T/K = 303.15

m_B/m_A	0.0546	0.1119	0.1784	0.2544	0.3418	0.4322	0.5516	0.6657	0.8200
$\rho/(g/cm^3)$	1.00417	1.01254	1.02081	1.03009	1.03919	1.04745	1.05700	1.06523	1.07412

m_B/m_A	0.8988	0.9982	1.1035	1.2201	1.3519	1.4932
$\rho/(g/cm^3)$	1.07759	1.08268	1.08601	1.09037	1.09358	1.09800

T/K = 308.15

m_B/m_A	0.0546	0.1119	0.1784	0.2544	0.3418	0.4322	0.5516	0.6657	0.8200
$\rho/(g/cm^3)$	1.00240	1.01060	1.01870	1.02778	1.03665	1.04470	1.05400	1.06200	1.07065

m_B/m_A	0.8988	0.9982	1.1035	1.2201	1.3519	1.4932
$\rho/(g/cm^3)$	1.07402	1.07898	1.08221	1.08647	1.08960	1.09392

T/K = 313.15

m_B/m_A	0.0546	0.1119	0.1784	0.2544	0.3418	0.4322	0.5516	0.6657	0.8200
$\rho/(g/cm^3)$	1.00044	1.00846	1.01643	1.02531	1.03396	1.04182	1.05089	1.05868	1.06710

m_B/m_A	0.8988	0.9982	1.1035	1.2201	1.3519	1.4932
$\rho/(g/cm^3)$	1.07038	1.07520	1.07837	1.08252	1.08557	1.08981

Polymer (B): **poly(ethylene glycol)** **2008AYR**
Characterization: M_w/g.mol^{-1} = 3350, Sigma Chemical Co., St. Louis, MO
Solvent (A): **water** H_2O **7732-18-5**

T/K = 288.15

c_B/(mol/kg)	0.0100	0.0500	0.1000	0.1436	0.1946	0.2502	0.2992	0.3500
ρ/(g/cm^3)	1.00480	1.02487	1.04510	1.05938	1.07303	1.08424	1.09242	1.09926
c_B/(mol/kg)	0.4004							
ρ/(g/cm^3)	1.10487							

T/K = 298.15

c_B/(mol/kg)	0.0100	0.0500	0.1000	0.1436	0.1946	0.2502	0.2992	0.3500
ρ/(g/cm^3)	1.00248	1.02165	1.04095	1.05500	1.06711	1.07767	1.08538	1.09185
c_B/(mol/kg)	0.4004							
ρ/(g/cm^3)	1.09715							

T/K = 308.15

c_B/(mol/kg)	0.0100	0.0500	0.1000	0.1436	0.1946	0.2502	0.2992	0.3500
ρ/(g/cm^3)	0.99925	1.01766	1.03604	1.04903	1.06112	1.07079	1.07807	1.08421
c_B/(mol/kg)	0.4004							
ρ/(g/cm^3)	1.08924							

Polymer (B): **poly(ethylene glycol)** **2009RE1**
Characterization: M_n/g.mol^{-1} = 3500-4500, PEG 4000,
Merck-Schuchardt, Hohenbrunn, Germany
Solvent (A): **water** H_2O **7732-18-5**

T/K = 298.15

w_B	0.0000	0.0500	0.1000	0.1500	0.2000	0.2500	0.3000	0.3500	0.4000
ρ/(g/cm^3)	0.9970	1.0049	1.0136	1.0223	1.0313	1.0407	1.0492	1.0591	1.0684
w_B	0.4500	0.5000							
ρ/(g/cm^3)	1.0778	1.0856							

T/K = 303.15

w_B	0.0000	0.0500	0.1000	0.1500	0.2000	0.2500	0.3000	0.3500	0.4000
ρ/(g/cm^3)	0.9956	1.0035	1.0123	1.0212	1.0297	1.0387	1.0474	1.0565	1.0656
w_B	0.4500	0.5000							
ρ/(g/cm^3)	1.0745	1.0833							

T/K = 308.15

w_B	0.0000	0.0500	0.1000	0.1500	0.2000	0.2500	0.3000	0.3500	0.4000
ρ/(g/cm^3)	0.9940	1.0019	1.0103	1.0184	1.0270	1.0357	1.0448	1.0533	1.0627
w_B	0.4500	0.5000							
ρ/(g/cm^3)	1.0709	1.0791							

continued

continued

T/K = 313.15

w_B	0.0000	0.0500	0.1000	0.1500	0.2000	0.2500	0.3000	0.3500	0.4000
ρ/(g/cm^3)	0.9922	0.9996	1.0081	1.0164	1.0247	1.0333	1.0419	1.0511	1.0597
w_B	0.4500	0.5000							
ρ/(g/cm^3)	1.0684	1.0760							

T/K = 318.15

w_B	0.0000	0.0500	0.1000	0.1500	0.2000	0.2500	0.3000	0.3500	0.4000
ρ/(g/cm^3)	0.9902	0.9977	1.0057	1.0138	1.0225	1.0305	1.0389	1.0471	1.0557
w_B	0.4500	0.5000							
ρ/(g/cm^3)	1.0638	1.0711							

Polymer (B): **poly(ethylene glycol)** **2009RE1**
Characterization: M_n/g.mol^{-1} = 3500-4500, PEG 4000, Merck-Schuchardt, Hohenbrunn, Germany
Solvent (A): **water** **H_2O** **7732-18-5**
Salt (C): **diammonium phosphate** **$(NH_4)_2HPO_4$** **7783-28-0**

T/K = 298.15

w_B	0.0500	0.1000	0.1500	0.2000	0.2500	0.3000	0.3500	0.4000	0.0500
w_C	0.0100	0.0100	0.0100	0.0100	0.0100	0.0100	0.0100	0.0100	0.0200
ρ/(g/cm^3)	1.0111	1.0202	1.0282	1.0373	1.0463	1.0555	1.0647	1.0738	1.0180
w_B	0.1000	0.1500	0.2000	0.2500	0.3000	0.0500	0.1000	0.1500	0.2000
w_C	0.0200	0.0200	0.0200	0.0200	0.0200	0.0300	0.0300	0.0300	0.0300
ρ/(g/cm^3)	1.0258	1.0348	1.0435	1.0492	1.0619	1.0239	1.0322	1.0415	1.0498
w_B	0.2500	0.0500	0.1000	0.1500	0.2000	0.0500	0.1000	0.1500	0.0500
w_C	0.0300	0.0400	0.0400	0.0400	0.0400	0.0500	0.0500	0.0500	0.0600
ρ/(g/cm^3)	1.0571	1.0310	1.0385	1.0467	1.0562	1.0371	1.0447	1.0541	1.0414
w_B	0.1000	0.0500	0.0500						
w_C	0.0600	0.0700	0.0800						
ρ/(g/cm^3)	1.0506	1.0493	1.0545						

T/K = 303.15

w_B	0.0500	0.1000	0.1500	0.2000	0.2500	0.3000	0.3500	0.4000	0.0500
w_C	0.0100	0.0100	0.0100	0.0100	0.0100	0.0100	0.0100	0.0100	0.0200
ρ/(g/cm^3)	1.0105	1.0186	1.0272	1.0358	1.0442	1.0537	1.0619	1.0707	1.0166
w_B	0.1000	0.1500	0.2000	0.2500	0.3000	0.0500	0.1000	0.1500	0.2000
w_C	0.0200	0.0200	0.0200	0.0200	0.0200	0.0300	0.0300	0.0300	0.0300
ρ/(g/cm^3)	1.0252	1.0340	1.0428	1.0516	1.0599	1.0227	1.0307	1.0399	1.0480
w_B	0.2500	0.0500	0.1000	0.1500	0.2000	0.0500	0.1000	0.1500	0.0500
w_C	0.0300	0.0400	0.0400	0.0400	0.0400	0.0500	0.0500	0.0500	0.0600
ρ/(g/cm^3)	1.0564	1.0292	1.0375	1.0570	1.0549	1.0351	1.0429	1.0519	1.0411

continued

continued

w_B	0.1000	0.0500	0.0500
w_C	0.0600	0.0700	0.0800
ρ/(g/cm^3)	1.0498	1.0469	1.0533

T/K = 308.15

w_B	0.0500	0.1000	0.1500	0.2000	0.2500	0.3000	0.3500	0.4000	0.0500
w_C	0.0100	0.0100	0.0100	0.0100	0.0100	0.0100	0.0100	0.0100	0.0200
ρ/(g/cm^3)	1.0080	1.0163	1.0245	1.0334	1.0415	1.0502	1.0587	1.0674	1.0145

w_B	0.1000	0.1500	0.2000	0.2500	0.3000	0.0500	0.1000	0.1500	0.2000
w_C	0.0200	0.0200	0.0200	0.0200	0.0200	0.0300	0.0300	0.0300	0.0300
ρ/(g/cm^3)	1.0226	1.0312	1.0397	1.0482	1.0570	1.0202	1.0283	1.0370	1.0449

w_B	0.2500	0.0500	0.1000	0.1500	0.2000	0.0500	0.1000	0.1500	0.0500
w_C	0.0300	0.0400	0.0400	0.0400	0.0400	0.0500	0.0500	0.0500	0.0600
ρ/(g/cm^3)	1.0538	1.0266	1.0348	1.0539	1.0518	1.0322	1.0404	1.0490	1.0390

w_B	0.1000	0.0500	0.0500
w_C	0.0600	0.0700	0.0800
ρ/(g/cm^3)	1.0470	1.0443	1.0509

T/K = 313.15

w_B	0.0500	0.1000	0.1500	0.2000	0.2500	0.3000	0.3500	0.4000	0.0500
w_C	0.0100	0.0100	0.0100	0.0100	0.0100	0.0100	0.0100	0.0100	0.0200
ρ/(g/cm^3)	1.0063	1.0138	1.0222	1.0308	1.0390	1.0479	1.0564	1.0663	1.0126

w_B	0.1000	0.1500	0.2000	0.2500	0.3000	0.0500	0.1000	0.1500	0.2000
w_C	0.0200	0.0200	0.0200	0.0200	0.0200	0.0300	0.0300	0.0300	0.0300
ρ/(g/cm^3)	1.0206	1.0286	1.0367	1.0449	1.0536	1.0190	1.0263	1.0352	1.0429

w_B	0.2500	0.0500	0.1000	0.1500	0.2000	0.0500	0.1000	0.1500	0.0500
w_C	0.0300	0.0400	0.0400	0.0400	0.0400	0.0500	0.0500	0.0500	0.0600
ρ/(g/cm^3)	1.0498	1.0257	1.0325	1.0402	1.0491	1.0314	1.0386	1.0476	1.0359

w_B	0.1000	0.0500	0.0500
w_C	0.0600	0.0700	0.0800
ρ/(g/cm^3)	1.0446	1.0437	1.0491

T/K = 318.15

w_B	0.0500	0.1000	0.1500	0.2000	0.2500	0.3000	0.3500	0.4000	0.0500
w_C	0.0100	0.0100	0.0100	0.0100	0.0100	0.0100	0.0100	0.0100	0.0200
ρ/(g/cm^3)	1.0036	1.0119	1.0204	1.0283	1.0364	1.0441	1.0528	1.0609	1.0100

w_B	0.1000	0.1500	0.2000	0.2500	0.3000	0.0500	0.1000	0.1500	0.2000
w_C	0.0200	0.0200	0.0200	0.0200	0.0200	0.0300	0.0300	0.0300	0.0300
ρ/(g/cm^3)	1.0177	1.0267	1.0339	1.0419	1.0502	1.0167	1.0241	1.0322	1.0398

continued

continued

w_B	0.2500	0.0500	0.1000	0.1500	0.2000	0.0500	0.1000	0.1500	0.0500
w_C	0.0300	0.0400	0.0400	0.0400	0.0400	0.0500	0.0500	0.0500	0.0600
$\rho/(g/cm^3)$	1.0461	1.0232	1.0304	1.0383	1.0470	1.0282	1.0356	1.0454	1.0330

w_B	0.1000	0.0500	0.0500
w_C	0.0600	0.0700	0.0800
$\rho/(g/cm^3)$	1.0425	1.0405	1.0458

Polymer (B): **poly(ethylene glycol)** **2009PER**
Characterization: M_n/g.mol^{-1} = 4000, Merck-Schuchardt, Hohenbrunn, Germany
Solvent (A): **water** **H_2O** **7732-18-5**

T/K = 298.15

w_B	0.00	0.05	0.10	0.15	0.20	0.25	0.30	0.35	0.40
$\rho/(g/cm^3)$	0.99700	1.00487	1.01327	1.02219	1.03160	1.04037	1.04924	1.05854	1.06756

w_B	0.45	0.50
$\rho/(g/cm^3)$	1.07612	1.08394

T/K = 308.15

w_B	0.00	0.05	0.10	0.15	0.20	0.25	0.30	0.35	0.40
$\rho/(g/cm^3)$	0.99560	1.00314	1.01168	1.02082	1.02976	1.03830	1.04754	1.05638	1.06562

w_B	0.45	0.50
$\rho/(g/cm^3)$	1.07446	1.08250

T/K = 318.15

w_B	0.00	0.05	0.10	0.15	0.20	0.25	0.30	0.35	0.40
$\rho/(g/cm^3)$	0.99400	1.00135	1.00939	1.01787	1.02673	1.03501	1.04412	1.05284	1.06209

w_B	0.45	0.50
$\rho/(g/cm^3)$	1.07097	1.07789

Polymer (B): **poly(ethylene glycol)** **2009PER**
Characterization: M_n/g.mol^{-1} = 4000, Merck-Schuchardt, Hohenbrunn, Germany
Solvent (A): **water** **H_2O** **7732-18-5**
Salt (C): **sodium citrate** **$Na_3C_6H_5O_7$** **68-04-2**

T/K = 298.15

w_B	0.00	0.05	0.10	0.15	0.20	0.25	0.30	0.35	0.40
w_C	0.02	0.02	0.02	0.02	0.02	0.02	0.02	0.02	0.02
$\rho/(g/cm^3)$	1.01117	1.01933	1.02835	1.03751	1.04588	1.05629	1.06494	1.07236	1.08344

w_B	0.00	0.05	0.10	0.15	0.20	0.25	0.30	0.35	0.00
w_C	0.04	0.04	0.04	0.04	0.04	0.04	0.04	0.04	0.06
$\rho/(g/cm^3)$	1.02264	1.03026	1.04071	1.04843	1.05767	1.06647	1.07568	1.08422	1.03437

continued

continued

w_B	0.05	0.10	0.15	0.20	0.25	0.00	0.05	0.10	0.15
w_C	0.06	0.06	0.06	0.06	0.06	0.08	0.08	0.08	0.08
ρ/(g/cm^3)	1.04309	1.05271	1.06127	1.07083	1.08027	1.04805	1.05672	1.06559	1.07488

w_B	0.20	0.25	0.00	0.05	0.10	0.15
w_C	0.08	0.08	0.10	0.10	0.10	0.10
ρ/(g/cm^3)	1.08342	1.08985	1.06193	1.07035	1.07878	1.08744

T/K = 308.15

w_B	0.00	0.05	0.10	0.15	0.20	0.25	0.30	0.35	0.40
w_C	0.02	0.02	0.02	0.02	0.02	0.02	0.02	0.02	0.02
ρ/(g/cm^3)	1.00682	1.01558	1.02493	1.03484	1.04364	1.05213	1.06158	1.06970	1.07802

w_B	0.00	0.05	0.10	0.15	0.20	0.25	0.30	0.35	0.00
w_C	0.04	0.04	0.04	0.04	0.04	0.04	0.04	0.04	0.06
ρ/(g/cm^3)	1.01818	1.02882	1.03621	1.04541	1.05427	1.06423	1.07152	1.08148	1.03093

w_B	0.05	0.10	0.15	0.20	0.25	0.00	0.05	0.10	0.15
w_C	0.06	0.06	0.06	0.06	0.06	0.08	0.08	0.08	0.08
ρ/(g/cm^3)	1.03958	1.04923	1.05924	1.06773	1.07612	1.04324	1.05319	1.06321	1.07239

w_B	0.20	0.25	0.00	0.05	0.10	0.15
w_C	0.08	0.08	0.10	0.10	0.10	0.10
ρ/(g/cm^3)	1.08033	1.08705	1.05892	1.06671	1.07487	1.08389

T/K = 318.15

w_B	0.00	0.05	0.10	0.15	0.20	0.25	0.30	0.35	0.40
w_C	0.02	0.02	0.02	0.02	0.02	0.02	0.02	0.02	0.02
ρ/(g/cm^3)	1.00282	1.01206	1.02067	1.02911	1.03775	1.04621	1.05481	1.06257	1.06980

w_B	0.00	0.05	0.10	0.15	0.20	0.25	0.30	0.35	0.00
w_C	0.04	0.04	0.04	0.04	0.04	0.04	0.04	0.04	0.06
ρ/(g/cm^3)	1.01415	1.02415	1.03312	1.04073	1.04863	1.05926	1.06576	1.07325	1.02646

w_B	0.05	0.10	0.15	0.20	0.25	0.00	0.05	0.10	0.15
w_C	0.06	0.06	0.06	0.06	0.06	0.08	0.08	0.08	0.08
ρ/(g/cm^3)	1.03559	1.04488	1.05486	1.06345	1.07078	1.03968	1.04726	1.05783	1.06770

w_B	0.20	0.25	0.00	0.05	0.10	0.15
w_C	0.08	0.08	0.10	0.10	0.10	0.10
ρ/(g/cm^3)	1.07543	1.08146	1.05467	1.06339	1.07182	1.08046

Polymer (B): **poly(ethylene glycol)** **2007KUS**
Characterization: M_n/g.mol^{-1} = 4000, Merck, Darmstadt, Germany
Solvent (A): **water** H_2O **7732-18-5**

T/K = 298.15

c_B/(mol/kg)	0.00000	0.02749	0.05906	0.09871	0.14892	0.21606
ρ/(g/cm^3)	0.997047	1.013559	1.029182	1.045862	1.062210	1.078940

Polymer (B): **poly(ethylene glycol)** **2009SIN**
Characterization: M_w/g.mol^{-1} = 4000, Sigma Chemical Co., St. Louis, MO
Solvent (A): **water** **H_2O** **7732-18-5**

w_B = 0.10 was kept constant

T/K	308	313	318	323	324	325	326	327	328
ρ/(kg/m^3)	1015.76	1013.94	1010.31	1005.81	1004.92	1003.13	1002.24	1001.36	1000.47

T/K	329	333	338
ρ/(kg/m^3)	998.70	995.18	991.69

Polymer (B): **poly(ethylene glycol)** **2007SIL**
Characterization: M_w/g.mol^{-1} = 4000, Isofar, Rio de Janeiro, Brazil
Solvent (A): **water** **H_2O** **7732-18-5**

T/K = 278.15

w_B	0.05	0.10	0.20	0.30
ρ/(g/cm^3)	1.00738	1.01633	1.03372	1.05228

T/K = 298.15

w_B	0.05	0.10	0.20	0.30
ρ/(g/cm^3)	1.00689	1.01466	1.03820	1.05272

T/K = 308.15

w_B	0.05	0.10	0.20	0.30
ρ/(g/cm^3)	1.00377	1.01097	1.02861	1.04423

T/K = 318.15

w_B	0.05	0.10	0.20	0.30
ρ/(g/cm^3)	0.99955	1.00681	1.02654	1.04162

Polymer (B): **poly(ethylene glycol)** **2005TAB**
Characterization: M_w/g.mol^{-1} = 4000, M_w/M_n = 1.1,
Merck, Darmstadt, Germany
Solvent (A): **water** **H_2O** **7732-18-5**
Salt (C): **potassium chloride** **KCl** **7447-40-7**

T/K = 333.15

w_B	0.0000	0.0446	0.0955	0.1273	0.1880	0.2120	0.2211	0.3005	0.3129
w_C	0.2667	0.2443	0.2227	0.2067	0.1852	0.1803	0.1774	0.1514	0.1474
ρ/(g/cm^3)	1.17720	1.17064	1.16816	1.16606	1.16052	1.16046	1.15798	1.15475	1.15422

w_B	0.4018	0.4528	0.4932	0.5271
w_C	0.1226	0.1080	0.0995	0.0885
ρ/(g/cm^3)	1.15277	1.15262	1.15280	1.15211

Polymer (B):	**poly(ethylene glycol)**		**2005TAB**
Characterization:	M_w/g.mol^{-1} = 4000, M_w/M_n = 1.1, Merck, Darmstadt, Germany		
Solvent (A):	**water**	**H_2O**	**7732-18-5**
Salt (C):	**sodium chloride**	**NaCl**	**7647-14-5**

T/K = 333.15

w_B	0.0000	0.0446	0.0713	0.1272	0.1856	0.2120	0.2747	0.3597	0.3852
w_C	0.2628	0.2530	0.2402	0.2202	0.1957	0.1919	0.1646	0.1407	0.1337
ρ/(g/cm^3)	1.19778	1.19276	1.19000	1.18484	1.17990	1.17811	1.17328	1.16735	1.16789

w_B	0.4430	0.4735	0.4866
w_C	0.1172	0.1086	0.1047
ρ/(g/cm^3)	1.16570	1.16448	1.16369

Polymer (B):	**poly(ethylene glycol)**		**2005ZAF**
Characterization:	M_n/g.mol^{-1} = 5890, Merck, Darmstadt, Germany		
Solvent (A):	**water**	**H_2O**	**7732-18-5**

T/K = 298.15

c_B/(mol/kg)	0.0036	0.0186	0.0408	0.0557	0.0881
ρ/(g/cm^3)	1.000477	1.013575	1.030120	1.039721	1.057243

T/K = 308.15

c_B/(mol/kg)	0.0036	0.0186	0.0408	0.0557	0.0881
ρ/(g/cm^3)	0.997353	1.009922	1.025751	1.034919	1.051547

T/K = 318.15

c_B/(mol/kg)	0.0036	0.0186	0.0408	0.0557	0.0881
ρ/(g/cm^3)	0.993436	1.005571	1.020756	1.029526	1.045141

Polymer (B):	**poly(ethylene glycol)**		**2005ZAF**
Characterization:	M_n/g.mol^{-1} = 5890, Merck, Darmstadt, Germany		
Solvent (A):	**water**	**H_2O**	**7732-18-5**
Salt (C):	**disodium succinate**	**$C_4H_4Na_2O_4$**	**150-90-3**

T/K = 298.15

c_B/(mol/kg)	0.0022	0.0064	0.1250	0.0173	0.0332	0.0409	0.0463	0.0829
c_C/(mol/kg)	0.7921	0.7288	0.6916	0.7052	0.5402	0.3293	0.4886	0.3909
ρ/(g/cm^3)	1.07407	1.07086	1.07089	1.07447	1.06928	1.05692	1.07168	1.08040

c_B/(mol/kg)	0.1015
c_C/(mol/kg)	0.3399
ρ/(g/cm^3)	1.08389

continued

continued

T/K = 308.15

c_B/(mol/kg)	0.0022	0.0064	0.1250	0.0173	0.0332	0.0409	0.0463	0.0829
c_C/(mol/kg)	0.7921	0.7288	0.6916	0.7052	0.5402	0.3293	0.4886	0.3909
ρ/(g/cm^3)	1.07003	1.06676	1.06668	1.07011	1.06461	1.05222	1.06673	1.07460
c_B/(mol/kg)	0.1015							
c_C/(mol/kg)	0.3399							
ρ/(g/cm^3)	1.07771							

T/K = 318.15

c_B/(mol/kg)	0.0022	0.0064	0.1250	0.0173	0.0332	0.0409	0.0463	0.0829
c_C/(mol/kg)	0.7921	0.7288	0.6916	0.7052	0.5402	0.3293	0.4886	0.3909
ρ/(g/cm^3)	1.06552	1.06216	1.06196	1.06528	1.05948	1.04702	1.06127	1.06839
c_B/(mol/kg)	0.1015							
c_C/(mol/kg)	0.3399							
ρ/(g/cm^3)	1.07123							

Polymer (B): **poly(ethylene glycol)** **2005ZAF**
Characterization: M_n/g.mol^{-1} = 5890, Merck, Darmstadt, Germany
Solvent (A): **water** **H_2O** **7732-18-5**
Salt (C): **sodium formate** **$CHNaO_2$** **141-53-7**

T/K = 298.15

c_B/(mol/kg)	0.0021	0.0067	0.0128	0.0169	0.0218	0.0331	0.0471	0.0575
c_C/(mol/kg)	3.0622	2.9567	2.6354	2.5645	2.3062	2.2243	2.1152	1.1330
ρ/(g/cm^3)	1.10872	1.10690	1.09910	1.09845	1.09247	1.09422	1.09612	1.07354
c_B/(mol/kg)	0.0642	0.0841						
c_C/(mol/kg)	1.9323	2.1382						
ρ/(g/cm^3)	1.09694	1.10733						

T/K = 308.15

c_B/(mol/kg)	0.0021	0.0067	0.0128	0.0169	0.0218	0.0331	0.0471	0.0575
c_C/(mol/kg)	3.0622	2.9567	2.6354	2.5645	2.3062	2.2243	2.1152	1.1330
ρ/(g/cm^3)	1.10398	1.10220	1.09434	1.09358	1.08764	1.08916	1.09076	1.06819
c_B/(mol/kg)	0.0642	0.0841						
c_C/(mol/kg)	1.9323	2.1382						
ρ/(g/cm^3)	1.09130	1.10118						

T/K = 318.15

c_B/(mol/kg)	0.0021	0.0067	0.0128	0.0169	0.0218	0.0331	0.0471	0.0575
c_C/(mol/kg)	3.0622	2.9567	2.6354	2.5645	2.3062	2.2243	2.1152	1.1330
ρ/(g/cm^3)	1.09889	1.09700	1.08917	1.08831	1.08237	1.08369	1.08503	1.06254
c_B/(mol/kg)	0.0642	0.0841						
c_C/(mol/kg)	1.9323	2.1382						
ρ/(g/cm^3)	1.08524	1.09481						

Polymer (B): **poly(ethylene glycol)** **2003RAH**
Characterization: M_n/g.mol^{-1} = 6000, Merck, Darmstadt, Germany
Solvent (A): **water** **H_2O** **7732-18-5**

T/K = 308.15

w_B	0.3552	0.2931	0.1974	0.1913	0.1076	0.0982	0.0793	0.0601	0.0399
ρ/(g/cm^3)	1.0530	1.0447	1.0281	1.0260	1.0115	1.0084	1.0042	0.9990	0.9960

T/K = 318.15

w_B	0.3552	0.2931	0.1974	0.1913	0.1076	0.0982	0.0793	0.0601	0.0399
ρ/(g/cm^3)	1.0478	1.0343	1.0230	1.0208	1.0053	1.0021	0.9980	0.9938	0.9897

T/K = 328.15

w_B	0.3552	0.2931	0.1974	0.1913	0.1076	0.0982	0.0793	0.0601	0.0399
ρ/(g/cm^3)	1.0416	1.0314	1.0150	1.0146	1.0000	0.9959	0.9928	0.9897	0.9866

Polymer (B): **poly(ethylene glycol)** **2009RE2**
Characterization: M_n/g.mol^{-1} = 6000, Merck-Schuchardt, Hohenbrunn, Germany
Solvent (A): **water** **H_2O** **7732-18-5**

T/K = 298.15

w_B	0.0000	0.0500	0.1000	0.1500	0.2000	0.2500	0.3000	0.3500	0.4000
ρ/(g/cm^3)	0.9970	1.0051	1.0137	1.0217	1.0311	1.0397	1.0487	1.0584	1.0679
w_B	0.4500	0.5000							
ρ/(g/cm^3)	1.0766	1.0856							

T/K = 303.15

w_B	0.0000	0.0500	0.1000	0.1500	0.2000	0.2500	0.3000	0.3500	0.4000
ρ/(g/cm^3)	0.9956	1.0047	1.0125	1.0213	1.0297	1.0386	1.0474	1.0566	1.0656
w_B	0.4500	0.5000							
ρ/(g/cm^3)	1.0742	1.0833							

T/K = 308.15

w_B	0.0000	0.0500	0.1000	0.1500	0.2000	0.2500	0.3000	0.3500	0.4000
ρ/(g/cm^3)	0.9940	1.0022	1.0105	1.0190	1.0273	1.0358	1.0448	1.0539	1.0628
w_B	0.4500	0.5000							
ρ/(g/cm^3)	1.0714	1.0799							

T/K = 313.15

w_B	0.0000	0.0500	0.1000	0.1500	0.2000	0.2500	0.3000	0.3500	0.4000
ρ/(g/cm^3)	0.9922	1.0006	1.0088	1.0167	1.0248	1.0339	1.0428	1.0512	1.0597
w_B	0.4500	0.5000							
ρ/(g/cm^3)	1.0694	1.0779							

continued

continued

T/K = 318.15

w_B	0.0000	0.0500	0.1000	0.1500	0.2000	0.2500	0.3000	0.3500	0.4000
ρ/(g/cm^3)	0.9902	0.9977	1.0060	1.0135	1.0222	1.0305	1.0390	1.0470	1.0559
w_B	0.4500	0.5000							
ρ/(g/cm^3)	1.0638	1.0716							

Polymer (B): **poly(ethylene glycol)** **2009RE2**
Characterization: M_n/g.mol^{-1} = 6000, Merck-Schuchardt, Hohenbrunn, Germany
Solvent (A): **water** H_2O **7732-18-5**
Salt (C): **ammonium citrate** $(NH_4)_3C_6H_5O_7$ **3458-72-8**

T/K = 298.15

w_B	0.5000	0.4500	0.4000	0.3500	0.3000	0.2500	0.2000	0.1500	0.1000
w_C	0.0200	0.0200	0.0200	0.0200	0.0200	0.0200	0.0200	0.0200	0.0200
ρ/(g/cm^3)	1.0948	1.0862	1.0767	1.0680	1.0590	1.0498	1.0412	1.0319	1.0230
w_B	0.0500	0.4000	0.3500	0.3000	0.2500	0.2000	0.1500	0.1000	0.0500
w_C	0.0200	0.0400	0.0400	0.0400	0.0400	0.0400	0.0400	0.0400	0.0400
ρ/(g/cm^3)	1.0148	1.0859	1.0771	1.0683	1.0588	1.0502	1.0409	1.0322	1.0238
w_B	0.2500	0.2000	0.1500	0.1000	0.0500	0.2000	0.1500	0.1000	0.0500
w_C	0.0600	0.0600	0.0600	0.0600	0.0600	0.0800	0.0800	0.0800	0.0800
ρ/(g/cm^3)	1.0676	1.0589	1.0498	1.0412	1.0323	1.0678	1.0589	1.0502	1.0410
w_B	0.1000	0.0500							
w_C	0.1000	0.1000							
ρ/(g/cm^3)	1.0592	1.0502							

T/K = 303.15

w_B	0.5000	0.4500	0.4000	0.3500	0.3000	0.2500	0.2000	0.1500	0.1000
w_C	0.0200	0.0200	0.0200	0.0200	0.0200	0.0200	0.0200	0.0200	0.0200
ρ/(g/cm^3)	1.0929	1.0846	1.0750	1.0662	1.0575	1.0483	1.0386	1.0307	1.0213
w_B	0.0500	0.4000	0.3500	0.3000	0.2500	0.2000	0.1500	0.1000	0.0500
w_C	0.0200	0.0400	0.0400	0.0400	0.0400	0.0400	0.0400	0.0400	0.0400
ρ/(g/cm^3)	1.0129	1.0838	1.0755	1.0668	1.0577	1.0478	1.0382	1.0299	1.0213
w_B	0.2500	0.2000	0.1500	0.1000	0.0500	0.2000	0.1500	0.1000	0.0500
w_C	0.0600	0.0600	0.0600	0.0600	0.0600	0.0800	0.0800	0.0800	0.0800
ρ/(g/cm^3)	1.0666	1.0570	1.0487	1.0402	1.0320	1.0668	1.0579	1.0491	1.0402
w_B	0.1000	0.0500							
w_C	0.1000	0.1000							
ρ/(g/cm^3)	1.0584	1.0492							

continued

continued

T/K = 308.15

w_B	0.5000	0.4500	0.4000	0.3500	0.3000	0.2500	0.2000	0.1500	0.1000
w_C	0.0200	0.0200	0.0200	0.0200	0.0200	0.0200	0.0200	0.0200	0.0200
ρ/(g/cm^3)	1.0875	1.0798	1.0710	1.0624	1.0532	1.0447	1.0354	1.0278	1.0180
w_B	0.0500	0.4000	0.3500	0.3000	0.2500	0.2000	0.1500	0.1000	0.0500
w_C	0.0200	0.0400	0.0400	0.0400	0.0400	0.0400	0.0400	0.0400	0.0400
ρ/(g/cm^3)	1.0097	1.0795	1.0710	1.0629	1.0535	1.0449	1.0351	1.0282	1.0189
w_B	0.2500	0.2000	0.1500	0.1000	0.0500	0.2000	0.1500	0.1000	0.0500
w_C	0.0600	0.0600	0.0600	0.0600	0.0600	0.0800	0.0800	0.0800	0.0800
ρ/(g/cm^3)	1.0632	1.0540	1.0456	1.0367	1.0290	1.0632	1.0545	1.0461	1.0370
w_B	0.1000	0.0500							
w_C	0.1000	0.1000							
ρ/(g/cm^3)	1.0550	1.0467							

T/K = 313.15

w_B	0.5000	0.4500	0.4000	0.3500	0.3000	0.2500	0.2000	0.1500	0.1000
w_C	0.0200	0.0200	0.0200	0.0200	0.0200	0.0200	0.0200	0.0200	0.0200
ρ/(g/cm^3)	1.0835	1.0759	1.0669	1.0591	1.0497	1.0419	1.0329	1.0254	1.0165
w_B	0.0500	0.4000	0.3500	0.3000	0.2500	0.2000	0.1500	0.1000	0.0500
w_C	0.0200	0.0400	0.0400	0.0400	0.0400	0.0400	0.0400	0.0400	0.0400
ρ/(g/cm^3)	1.0094	1.0757	1.0679	1.0585	1.0508	1.0408	1.0339	1.0252	1.0179
w_B	0.2500	0.2000	0.1500	0.1000	0.0500	0.2000	0.1500	0.1000	0.0500
w_C	0.0600	0.0600	0.0600	0.0600	0.0600	0.0800	0.0800	0.0800	0.0800
ρ/(g/cm^3)	1.0597	1.0501	1.0432	1.0336	1.0263	1.0592	1.0515	1.0420	1.0347
w_B	0.1000	0.0500							
w_C	0.1000	0.1000							
ρ/(g/cm^3)	1.0514	1.0436							

T/K = 318.15

w_B	0.5000	0.4500	0.4000	0.3500	0.3000	0.2500	0.2000	0.1500	0.1000
w_C	0.0200	0.0200	0.0200	0.0200	0.0200	0.0200	0.0200	0.0200	0.0200
ρ/(g/cm^3)	1.0805	1.0718	1.0643	1.0567	1.0483	1.0398	1.0306	1.0231	1.0149
w_B	0.0500	0.4000	0.3500	0.3000	0.2500	0.2000	0.1500	0.1000	0.0500
w_C	0.0200	0.0400	0.0400	0.0400	0.0400	0.0400	0.0400	0.0400	0.0400
ρ/(g/cm^3)	1.0064	1.0732	1.0642	1.0562	1.0472	1.0398	1.0323	1.0244	1.0160
w_B	0.2500	0.2000	0.1500	0.1000	0.0500	0.2000	0.1500	0.1000	0.0500
w_C	0.0600	0.0600	0.0600	0.0600	0.0600	0.0800	0.0800	0.0800	0.0800
ρ/(g/cm^3)	1.0562	1.0492	1.0408	1.0326	1.0235	1.0566	1.0491	1.0409	1.0333
w_B	0.1000	0.0500							
w_C	0.1000	0.1000							
ρ/(g/cm^3)	1.0491	1.0412							

Polymer (B):	**poly(ethylene glycol)**	**2004CRU**
Characterization:	M_w/g.mol^{-1} = 6000, Isofar, Rio de Janeiro, Brazil	
Solvent (A):	**water** H_2O	**7732-18-5**

T/K = 283.15

w_B	0.0483	0.1009	0.1494	0.1976	0.2500	0.3018	0.3412	0.3981	0.4525
ρ/(g/cm^3)	1.00794	1.01633	1.02352	1.03010	1.03700	1.04312	1.04766	1.05381	1.05921
w_B	0.5031	0.5493							
ρ/(g/cm^3)	1.06387	1.06799							

T/K = 288.15

w_B	0.0483	0.1009	0.1494	0.1976	0.2500	0.3018	0.3412	0.3981	0.4525
ρ/(g/cm^3)	1.00714	1.01532	1.02231	1.02872	1.03542	1.04135	1.04576	1.05173	1.05696
w_B	0.5031	0.5493							
ρ/(g/cm^3)	1.06150	1.06548							

T/K = 293.15

w_B	0.0483	0.1009	0.1494	0.1976	0.2500	0.3018	0.3412	0.3981	0.4525
ρ/(g/cm^3)	1.00606	1.01405	1.02087	1.02710	1.03363	1.03940	1.04368	1.04948	1.05455
w_B	0.5031	0.5493							
ρ/(g/cm^3)	1.05897	1.06281							

T/K = 298.15

w_B	0.0483	0.1009	0.1494	0.1976	0.2500	0.3018	0.3412	0.3981	0.4525
ρ/(g/cm^3)	1.00474	1.01252	1.01921	1.02530	1.03166	1.03727	1.04144	1.04708	1.05201
w_B	0.5031	0.5493							
ρ/(g/cm^3)	1.05631	1.06006							

T/K = 303.15

w_B	0.0483	0.1009	0.1494	0.1976	0.2500	0.3018	0.3412	0.3981	0.4525
ρ/(g/cm^3)	1.00319	1.01084	1.01736	1.02330	1.02952	1.03500	1.03905	1.04455	1.04936
w_B	0.5031	0.5493							
ρ/(g/cm^3)	1.05353	1.05718							

T/K = 308.15

w_B	0.0483	0.1009	0.1494	0.1976	0.2500	0.3018	0.3412	0.3981	0.4525
ρ/(g/cm^3)	1.00144	1.00894	1.01533	1.02115	1.02723	1.03257	1.03653	1.04189	1.04657
w_B	0.5031	0.5493							
ρ/(g/cm^3)	1.05064	1.05419							

T/K = 313.15

w_B	0.0483	0.1009	0.1494	0.1976	0.2500	0.3018	0.3412	0.3981	0.4525
ρ/(g/cm^3)	0.99949	1.00679	1.01312	1.01882	1.02477	1.02999	1.03387	1.03911	1.04368
w_B	0.5031	0.5493							
ρ/(g/cm^3)	1.04764	1.05108							

Polymer (B):	**poly(ethylene glycol)**		**2009CRU**
Characterization:	M_n/g.mol^{-1} = 6000, Merck, Darmstadt, Germany		
Solvent (A):	**water**	**H_2O**	**7732-18-5**

T/K = 283.15

m_B/m_A	0.0553	0.1128	0.1786	0.2525	0.3344	0.4299	0.5407	0.6689	0.7412
ρ/(g/cm^3)	1.00872	1.01832	1.02812	1.03744	1.04792	1.05722	1.06849	1.07761	1.08222
m_B/m_A	0.8201	0.9026	1.0014	1.1056	1.2190	1.3521	1.4986		
ρ/(g/cm^3)	1.08736	1.09181	1.09696	1.10129	1.10608	1.10980	1.11390		

T/K = 288.15

m_B/m_A	0.0553	0.1128	0.1786	0.2525	0.3344	0.4299	0.5407	0.6689	0.7412
ρ/(g/cm^3)	1.00789	1.01724	1.02677	1.03582	1.04600	1.05500	1.06594	1.07473	1.07923
m_B/m_A	0.8201	0.9026	1.0014	1.1056	1.2190	1.3521	1.4986		
ρ/(g/cm^3)	1.08418	1.08843	1.09344	1.09763	1.10229	1.10590	1.10990		

T/K = 293.15

m_B/m_A	0.0553	0.1128	0.1786	0.2525	0.3344	0.4299	0.5407	0.6689	0.7412
ρ/(g/cm^3)	1.00679	1.01591	1.02520	1.03401	1.04391	1.05264	1.06324	1.07178	1.07614
m_B/m_A	0.8201	0.9026	1.0014	1.1056	1.2190	1.3521	1.4986		
ρ/(g/cm^3)	1.08096	1.08507	1.08986	1.09393	1.09846	1.10198	1.10588		

T/K = 298.15

m_B/m_A	0.0553	0.1128	0.1786	0.2525	0.3344	0.4299	0.5407	0.6689	0.7412
ρ/(g/cm^3)	1.00545	1.01437	1.02344	1.03202	1.04167	1.05014	1.06045	1.06873	1.07296
m_B/m_A	0.8201	0.9026	1.0014	1.1056	1.2190	1.3521	1.4986		
ρ/(g/cm^3)	1.07766	1.08164	1.08624	1.09018	1.09460	1.09801	1.10181		

T/K = 303.15

m_B/m_A	0.0553	0.1128	0.1786	0.2525	0.3344	0.4299	0.5407	0.6689	0.7412
ρ/(g/cm^3)	1.00388	1.01261	1.02148	1.02987	1.03928	1.04752	1.05754	1.06559	1.06969
m_B/m_A	0.8201	0.9026	1.0014	1.1056	1.2190	1.3521	1.4986		
ρ/(g/cm^3)	1.07425	1.07813	1.08262	1.08641	1.09066	1.09400	1.09772		

T/K = 308.15

m_B/m_A	0.0553	0.1128	0.1786	0.2525	0.3344	0.4299	0.5407	0.6689	0.7412
ρ/(g/cm^3)	1.00212	1.01068	1.01935	1.02756	1.03675	1.04476	1.05454	1.06237	1.06636
m_B/m_A	0.8201	0.9026	1.0014	1.1056	1.2190	1.3521	1.4986		
ρ/(g/cm^3)	1.07079	1.07456	1.07893	1.08262	1.08676	1.08994	1.09358		

T/K = 313.15

m_B/m_A	0.0553	0.1128	0.1786	0.2525	0.3344	0.4299	0.5407	0.6689	0.7412
ρ/(g/cm^3)	1.00016	1.00855	1.01706	1.02509	1.03409	1.04190	1.05141	1.05903	1.06294
m_B/m_A	0.8201	0.9026	1.0014	1.1056	1.2190	1.3521	1.4986		
ρ/(g/cm^3)	1.06725	1.07092	1.07518	1.07878	1.08281	1.08589	1.08940		

Polymer (B): **poly(ethylene glycol)** **2010SA2**
Characterization: M_n/g.mol^{-1} = 6000, Merck, Darmstadt, Germany
Solvent (A): **water** H_2O **7732-18-5**
Salt (C): **sodium 1-pentanesulfonate** $C_5H_{11}NaO_3S$ **22767-49-3**

w_B = 0.02 was kept constant

T/K = 288.15

c_C/(mol/kg)	0.0297	0.1318	0.2930	0.4322	0.6394	0.8485	1.1688	1.4643
ρ/(g/cm^3)	1.00427	1.01011	1.01898	1.02629	1.03668	1.04649	1.06055	1.07181
c_C/(mol/kg)	1.9705	2.4025	3.0987					
ρ/(g/cm^3)	1.08828	1.10046	1.11787					

T/K = 293.15

c_C/(mol/kg)	0.0297	0.1318	0.2930	0.4322	0.6394	0.8485	1.1688	1.4643
ρ/(g/cm^3)	1.00327	1.00901	1.01772	1.02491	1.03511	1.04473	1.05849	1.06955
c_C/(mol/kg)	1.9705	2.4025	3.0987					
ρ/(g/cm^3)	1.08581	1.09784	1.11506					

T/K = 298.15

c_C/(mol/kg)	0.0297	0.1318	0.2930	0.4322	0.6394	0.8485	1.1688	1.4643
ρ/(g/cm^3)	1.00201	1.00766	1.01624	1.02331	1.03334	1.04278	1.05629	1.06716
c_C/(mol/kg)	1.9705	2.4025	3.0987					
ρ/(g/cm^3)	1.08322	1.09514	1.11217					

T/K = 303.15

c_C/(mol/kg)	0.0297	0.1318	0.2930	0.4322	0.6394	0.8485	1.1688	1.4643
ρ/(g/cm^3)	1.00052	1.00610	1.01455	1.02152	1.03139	1.04067	1.05395	1.06466
c_C/(mol/kg)	1.9705	2.4025	3.0987					
ρ/(g/cm^3)	1.08055	1.09234	1.10922					

T/K = 308.15

c_C/(mol/kg)	0.0297	0.1318	0.2930	0.4322	0.6394	0.8485	1.1688	1.4643
ρ/(g/cm^3)	0.99883	1.00433	1.01268	1.01955	1.02927	1.03841	1.05147	1.06205
c_C/(mol/kg)	1.9705	2.4025	3.0987					
ρ/(g/cm^3)	1.07778	1.08947	1.10619					

T/K = 313.15

c_C/(mol/kg)	0.0297	0.1318	0.2930	0.4322	0.6394	0.8485	1.1688	1.4643
ρ/(g/cm^3)	0.99694	1.00238	1.01062	1.01740	1.02699	1.03600	1.04887	1.05932
c_C/(mol/kg)	1.9705	2.4025	3.0987					
ρ/(g/cm^3)	1.07492	1.08652	1.10310					

Polymer (B):	**poly(ethylene glycol)**		**2010SA1**
Characterization:	M_n/g.mol^{-1} = 6000, Merck, Darmstadt, Germany		
Solvent (A):	**water**	**H_2O**	**7732-18-5**
Salt (C):	**sodium tungstate**	**Na_2WO_4**	**13472-45-2**

w_B = 0.02 was kept constant

T/K = 288.15

c_C/(mol/kg)	0.0000	0.0118	0.0906	0.1227	0.1907	0.2575	0.3282	0.4393
ρ/(g/cm^3)	1.00245	1.00559	1.02642	1.03493	1.05271	1.06990	1.08806	1.11657
c_C/(mol/kg)	0.4811	0.5588	0.6967	0.9125	0.9733			
ρ/(g/cm^3)	1.12716	1.14637	1.18039	1.23265	1.24740			

T/K = 293.15

c_C/(mol/kg)	0.0000	0.0118	0.0906	0.1227	0.1907	0.2575	0.3282	0.4393
ρ/(g/cm^3)	1.00148	1.00460	1.02533	1.03379	1.05149	1.06860	1.08668	1.11508
c_C/(mol/kg)	0.4811	0.5588	0.6967	0.9125	0.9733			
ρ/(g/cm^3)	1.12562	1.14476	1.17864	1.23073	1.24543			

T/K = 298.15

c_C/(mol/kg)	0.0000	0.0118	0.0906	0.1227	0.1907	0.2575	0.3282	0.4393
ρ/(g/cm^3)	1.00025	1.00335	1.02400	1.03242	1.05005	1.06709	1.08510	1.11339
c_C/(mol/kg)	0.4811	0.5588	0.6967	0.9125	0.9733			
ρ/(g/cm^3)	1.12389	1.14296	1.17673	1.22865	1.24330			

T/K = 303.15

c_C/(mol/kg)	0.0000	0.0118	0.0906	0.1227	0.1907	0.2575	0.3282	0.4393
ρ/(g/cm^3)	0.99878	1.00188	1.02245	1.03084	1.04840	1.06538	1.08332	1.11152
c_C/(mol/kg)	0.4811	0.5588	0.6967	0.9125	0.9733			
ρ/(g/cm^3)	1.12199	1.14100	1.17466	1.22643	1.24104			

T/K = 308.15

c_C/(mol/kg)	0.0000	0.0118	0.0906	0.1227	0.1907	0.2575	0.3282	0.4393
ρ/(g/cm^3)	0.99711	1.00020	1.02069	1.02906	1.04656	1.06349	1.08137	1.10948
c_C/(mol/kg)	0.4811	0.5588	0.6967	0.9125	0.9733			
ρ/(g/cm^3)	1.11992	1.13887	1.17244	1.22407	1.23875			

Polymer (B):	**poly(ethylene glycol)**		**2010MIN**
Characterization:	M_w/g.mol^{-1} = 6000, Isofar, Rio de Janeiro, Brazil		
Solvent (A):	**water**	**H_2O**	**7732-18-5**

T/K = 288.15

w_B	0.04	0.06	0.08	0.10	0.12
ρ/(kg/m^3)	1007.3	1011.1	1014.4	1018.0	1021.5

continued

continued

T/K = 293.15

w_B	0.04	0.06	0.08	0.10	0.12
ρ/(kg/m^3)	1004.4	1008.4	1011.6	1015.4	1018.8

T/K = 298.15

w_B	0.04	0.06	0.08	0.10	0.12
ρ/(kg/m^3)	1004.1	1006.8	1010.5	1013.1	1017.5

T/K = 303.15

w_B	0.04	0.06	0.08	0.10	0.12
ρ/(kg/m^3)	1000.6	1005.0	1008.4	1010.9	1014.3

T/K = 308.15

w_B	0.04	0.06	0.08	0.10	0.12
ρ/(kg/m^3)	999.6	1004.3	1007.0	1010.4	1012.8

T/K = 313.15

w_B	0.04	0.06	0.08	0.10	0.12
ρ/(kg/m^3)	998.2	1001.2	1004.5	1007.6	1010.1

T/K = 318.15

w_B	0.04	0.06	0.08	0.10	0.12
ρ/(kg/m^3)	996.6	999.8	1002.6	1005.3	1007.2

Polymer (B): **poly(ethylene glycol)** **2010MIN**
Characterization: M_w/g.mol^{-1} = 6000, Isofar, Rio de Janeiro, Brazil
Solvent (A): **water** **H_2O** **7732-18-5**
Polymer (C): **poly(sodium acrylate)**
Characterization: M_w/g.mol^{-1} = 8000, Aldrich Chem. Co., Inc., Milwaukee, WI

T/K = 288.15

w_B	0.04	0.04	0.04	0.04	0.04	0.06	0.06	0.06	0.06
w_C	0.04	0.06	0.08	0.10	0.12	0.04	0.06	0.08	0.10
ρ/(kg/m^3)	1021.8	1031.1	1039.4	1046.8	1056.1	1023.6	1032.3	1040.2	1049.6
w_B	0.06	0.08	0.08	0.08	0.08	0.08	0.10	0.10	0.10
w_C	0.12	0.04	0.06	0.08	0.10	0.12	0.04	0.06	0.08
ρ/(kg/m^3)	1056.5	1024.5	1034.7	1042.8	1050.3	1059.2	1027.5	1035.3	1043.6
w_B	0.10	0.10	0.12	0.12	0.12	0.12	0.12		
w_C	0.10	0.12	0.04	0.06	0.08	0.10	0.12		
ρ/(kg/m^3)	1052.6	1061.4	1029.0	1037.8	1045.9	1054.2	1062.7		

continued

continued

T/K = 293.15

w_B	0.04	0.04	0.04	0.04	0.04	0.06	0.06	0.06	0.06
w_C	0.04	0.06	0.08	0.10	0.12	0.04	0.06	0.08	0.10
ρ/(kg/m^3)	1019.0	1027.8	1036.5	1044.2	1052.6	1020.5	1029.1	1037.8	1046.6
w_B	0.06	0.08	0.08	0.08	0.08	0.08	0.10	0.10	0.10
w_C	0.12	0.04	0.06	0.08	0.10	0.12	0.04	0.06	0.08
ρ/(kg/m^3)	1053.7	1022.6	1031.0	1039.2	1048.1	1056.1	1024.8	1033.3	1041.6
w_B	0.10	0.10	0.12	0.12	0.12	0.12	0.12		
w_C	0.10	0.12	0.04	0.06	0.08	0.10	0.12		
ρ/(kg/m^3)	1049.9	1058.6	1026.4	1034.5	1043.9	1051.2	1059.8		

T/K = 298.15

w_B	0.04	0.04	0.04	0.04	0.04	0.06	0.06	0.06	0.06
w_C	0.04	0.06	0.08	0.10	0.12	0.04	0.06	0.08	0.10
ρ/(kg/m^3)	1017.0	1026.7	1034.4	1042.3	1050.9	1018.6	1026.5	1035.5	1044.8
w_B	0.06	0.08	0.08	0.08	0.08	0.08	0.10	0.10	0.10
w_C	0.12	0.04	0.06	0.08	0.10	0.12	0.04	0.06	0.08
ρ/(kg/m^3)	1052.2	1020.8	1029.8	1037.1	1045.1	1053.3	1022.7	1030.6	1038.8
w_B	0.10	0.10	0.12	0.12	0.12	0.12	0.12		
w_C	0.10	0.12	0.04	0.06	0.08	0.10	0.12		
ρ/(kg/m^3)	1047.7	1055.5	1023.7	1032.1	1040.9	1048.8	1057.3		

T/K = 303.15

w_B	0.04	0.04	0.04	0.04	0.04	0.06	0.06	0.06	0.06
w_C	0.04	0.06	0.08	0.10	0.12	0.04	0.06	0.08	0.10
ρ/(kg/m^3)	1015.0	1024.1	1031.6	1040.0	1048.3	1017.0	1024.9	1033.4	1042.3
w_B	0.06	0.08	0.08	0.08	0.08	0.08	0.10	0.10	0.10
w_C	0.12	0.04	0.06	0.08	0.10	0.12	0.04	0.06	0.08
ρ/(kg/m^3)	1049.5	1019.0	1027.0	1034.6	1043.0	1051.2	1020.3	1028.1	1036.1
w_B	0.10	0.10	0.12	0.12	0.12	0.12	0.12		
w_C	0.10	0.12	0.04	0.06	0.08	0.10	0.12		
ρ/(kg/m^3)	1044.1	1053.6	1022.6	1029.7	1038.2	1046.3	1054.8		

T/K = 308.15

w_B	0.04	0.04	0.04	0.04	0.04	0.06	0.06	0.06	0.06
w_C	0.04	0.06	0.08	0.10	0.12	0.04	0.06	0.08	0.10
ρ/(kg/m^3)	1013.6	1022.3	1030.4	1038.6	1046.5	1017.3	1023.2	1031.7	1040.0
w_B	0.06	0.08	0.08	0.08	0.08	0.08	0.10	0.10	0.10
w_C	0.12	0.04	0.06	0.08	0.10	0.12	0.04	0.06	0.08
ρ/(kg/m^3)	1048.2	1018.7	1025.3	1032.6	1041.2	1049.4	1021.0	1026.5	1034.6

continued

continued

w_B	0.10	0.10	0.12	0.12	0.12	0.12	0.12		
w_C	0.10	0.12	0.04	0.06	0.08	0.10	0.12		
ρ/(kg/m^3)	1042.9	1051.4	1022.2	1028.2	1036.0	1044.7	1052.9		

T/K = 313.15

w_B	0.04	0.04	0.04	0.04	0.04	0.06	0.06	0.06	0.06
w_C	0.04	0.06	0.08	0.10	0.12	0.04	0.06	0.08	0.10
ρ/(kg/m^3)	1010.4	1019.8	1027.8	1036.2	1044.6	1014.1	1021.4	1029.6	1037.8
w_B	0.06	0.08	0.08	0.08	0.08	0.08	0.10	0.10	0.10
w_C	0.12	0.04	0.06	0.08	0.10	0.12	0.04	0.06	0.08
ρ/(kg/m^3)	1045.1	1016.3	1023.5	1030.7	1038.2	1047.1	1018.0	1024.0	1032.0
w_B	0.10	0.10	0.12	0.12	0.12	0.12	0.12		
w_C	0.10	0.12	0.04	0.06	0.08	0.10	0.12		
ρ/(kg/m^3)	1040.6	1048.7	1018.9	1025.6	1034.2	1042.7	1050.7		

T/K = 318.15

w_B	0.04	0.04	0.04	0.04	0.04	0.06	0.06	0.06	0.06
w_C	0.04	0.06	0.08	0.10	0.12	0.04	0.06	0.08	0.10
ρ/(kg/m^3)	1008.7	1017.0	1024.7	1032.6	1040.3	1012.6	1017.8	1026.2	1033.0
w_B	0.06	0.08	0.08	0.08	0.08	0.08	0.10	0.10	0.10
w_C	0.12	0.04	0.06	0.08	0.10	0.12	0.04	0.06	0.08
ρ/(kg/m^3)	1040.5	1013.1	1019.3	1026.3	1034.0	1041.4	1016.3	1021.3	1028.9
w_B	0.10	0.10	0.12	0.12	0.12	0.12	0.12		
w_C	0.10	0.12	0.04	0.06	0.08	0.10	0.12		
ρ/(kg/m^3)	1036.7	1044.5	1017.8	1027.3	1029.8	1037.9	1045.1		

Polymer (B): **poly(ethylene glycol)** **2008AYR**
Characterization: M_w/g.mol^{-1} = 8000, Sigma Chemical Co., St. Louis, MO
Solvent (A): **water** **H_2O** **7732-18-5**

T/K = 288.15

c_B/(mol/kg)	0.0010	0.0050	0.0100	0.0250	0.0493	0.0740	0.0998
ρ/(g/cm^3)	1.00047	1.00575	1.01200	1.02870	1.05090	1.06838	1.08260

T/K = 298.15

c_B/(mol/kg)	0.0010	0.0050	0.0100	0.0250	0.0493	0.0740	0.0998
ρ/(g/cm^3)	0.99835	1.00339	1.00931	1.02532	1.04640	1.06278	1.07600

T/K = 308.15

c_B/(mol/kg)	0.0010	0.0050	0.0100	0.0250	0.0493	0.0740	0.0998
ρ/(g/cm^3)	0.99530	1.00017	1.00592	1.02118	1.04117	1.05671	1.06930

Polymer (B): **poly(ethylene glycol)** **2009CRU**
Characterization: M_n/g.mol^{-1} = 10000, Merck, Darmstadt, Germany
Solvent (A): **water** **H_2O** **7732-18-5**

T/K = 283.15

m_B/m_A	0.0552	0.1134	0.1865	0.2444	0.3385	0.4361	0.5454	0.6646	0.8184
ρ/(g/cm^3)	1.00918	1.01829	1.02864	1.03722	1.04758	1.05722	1.06771	1.07745	1.08749

m_B/m_A	0.9997	1.2241
ρ/(g/cm^3)	1.09756	1.10602

T/K = 288.15

m_B/m_A	0.0552	0.1134	0.1865	0.2444	0.3385	0.4361	0.5454	0.6646	0.8184
ρ/(g/cm^3)	1.00836	1.01721	1.02729	1.03561	1.04567	1.05501	1.06521	1.07454	1.08429

m_B/m_A	0.9997	1.2241
ρ/(g/cm^3)	1.09405	1.10219

T/K = 293.15

m_B/m_A	0.0552	0.1134	0.1865	0.2444	0.3385	0.4361	0.5454	0.6646	0.8184
ρ/(g/cm^3)	1.00725	1.01589	1.02570	1.03381	1.04359	1.05266	1.06257	1.07161	1.08101

m_B/m_A	0.9997	1.2241
ρ/(g/cm^3)	1.09048	1.09839

T/K = 298.15

m_B/m_A	0.0552	0.1134	0.1865	0.2444	0.3385	0.4361	0.5454	0.6646	0.8184
ρ/(g/cm^3)	1.00590	1.01434	1.02393	1.03182	1.04136	1.05017	1.05980	1.06858	1.07765

m_B/m_A	0.9997	1.2241
ρ/(g/cm^3)	1.08683	1.09452

T/K = 303.15

m_B/m_A	0.0552	0.1134	0.1865	0.2444	0.3385	0.4361	0.5454	0.6646	0.8184
ρ/(g/cm^3)	1.00433	1.01260	1.02393	1.02968	1.03898	1.04755	1.05692	1.06547	1.07421

m_B/m_A	0.9997	1.2241
ρ/(g/cm^3)	1.08314	1.09063

T/K = 308.15

m_B/m_A	0.0552	0.1134	0.1865	0.2444	0.3385	0.4361	0.5454	0.6646	0.8184
ρ/(g/cm^3)	1.00254	1.01066	1.01982	1.02737	1.03645	1.04481	1.05394	1.06226	1.07075

m_B/m_A	0.9997	1.2241
ρ/(g/cm^3)	1.07939	1.08668

T/K = 313.15

m_B/m_A	0.0552	0.1134	0.1865	0.2444	0.3385	0.4361	0.5454	0.6646	0.8184
ρ/(g/cm^3)	1.00056	1.00854	1.01752	1.02491	1.03379	1.04195	1.05083	1.05894	1.06724

m_B/m_A	0.9997	1.2241
ρ/(g/cm^3)	1.07556	1.08269

Polymer (B):	**poly(ethylene glycol)**	**2008AYR**
Characterization:	M_w/g.mol^{-1} = 10000, Sigma Chemical Co., St. Louis, MO	
Solvent (A):	**water** H_2O	**7732-18-5**

T/K = 288.15

c_B/(mol/kg)	0.0010	0.0050	0.0100	0.0200	0.0299	0.0399	0.0500
ρ/(g/cm^3)	1.00077	1.00733	1.01497	1.02870	1.04080	1.05142	1.06078

T/K = 298.15

c_B/(mol/kg)	0.0010	0.0050	0.0100	0.0200	0.0299	0.0399	0.0500
ρ/(g/cm^3)	0.99864	1.00492	1.01221	1.02529	1.03676	1.04679	1.05565

T/K = 308.15

c_B/(mol/kg)	0.0010	0.0050	0.0100	0.0200	0.0299	0.0399	0.0500
ρ/(g/cm^3)	0.99557	1.00160	1.00860	1.02112	1.03207	1.04161	1.05001

Polymer (B):	**poly(ethylene glycol)**	**2005MOH**
Characterization:	M_n/g.mol^{-1} = 9000-11250, Merck, Darmstadt, Germany	
Solvent (A):	**water** H_2O	**7732-18-5**

T/K = 298.15

w_B	0.10	0.15	0.20	0.25	0.30	0.35	0.40	0.45	0.50
ρ/(g/cm^3)	1.0186	1.0278	1.0384	1.0466	1.0549	1.0655	1.0737	1.0839	1.0935

T/K = 303.15

w_B	0.10	0.15	0.20	0.25	0.30	0.35	0.40	0.45	0.50
ρ/(g/cm^3)	1.0159	1.0271	1.0369	1.0447	1.0538	1.0627	1.0732	1.0811	1.0895

T/K = 308.15

w_B	0.10	0.15	0.20	0.25	0.30	0.35	0.40	0.45	0.50
ρ/(g/cm^3)	1.0150	1.0254	1.0350	1.0438	1.0505	1.0606	1.0698	1.0780	1.0860

T/K = 313.15

w_B	0.10	0.15	0.20	0.25	0.30	0.35	0.40	0.45	0.50
ρ/(g/cm^3)	1.0143	1.0244	1.0317	1.0407	1.0490	1.0570	1.0677	1.0761	1.0826

T/K = 318.15

w_B	0.10	0.15	0.20	0.25	0.30	0.35	0.40	0.45	0.50
ρ/(g/cm^3)	1.0120	1.0200	1.0291	1.0373	1.0460	1.0552	1.0639	1.0720	1.0804

T/K = 323.15

w_B	0.10	0.15	0.20	0.25	0.30	0.35	0.40	0.45	0.50
ρ/(g/cm^3)	1.0102	1.0197	1.0268	1.0344	1.0425	1.0510	1.0612	1.0683	1.0746

T/K = 328.15

w_B	0.10	0.15	0.20	0.25	0.30	0.35	0.40	0.45	0.50
ρ/(g/cm^3)	1.0069	1.0168	1.0232	1.0313	1.0399	1.0484	1.0570	1.0647	1.0711

Polymer (B):	**poly(ethylene glycol)**		**2009CRU**
Characterization:	M_n/g.mol^{-1} = 20000, Merck, Darmstadt, Germany		
Solvent (A):	**water**	**H_2O**	**7732-18-5**

T/K = 283.15

m_B/m_A	0.0530	0.1138	0.1860	0.2554	0.3345	0.4292	0.5405	0.6684	0.7440
ρ/(g/cm^3)	1.00913	1.01884	1.02951	1.03836	1.04810	1.05712	1.06740	1.07807	1.08319
m_B/m_A	0.8219	0.9078	1.0034	1.1013	1.2264				
ρ/(g/cm^3)	1.08962	1.09305	1.09877	1.10208	1.10712				

T/K = 288.15

m_B/m_A	0.0530	0.1138	0.1860	0.2554	0.3345	0.4292	0.5405	0.6684	0.7440
ρ/(g/cm^3)	1.00828	1.01775	1.02812	1.03673	1.04619	1.05487	1.06485	1.07516	1.08013
m_B/m_A	0.8219	0.9078	1.0034	1.1013	1.2264				
ρ/(g/cm^3)	1.08636	1.08951	1.09504	1.09832	1.10329				

T/K = 293.15

m_B/m_A	0.0530	0.1138	0.1860	0.2554	0.3345	0.4292	0.5405	0.6684	0.7440
ρ/(g/cm^3)	1.00718	1.01641	1.02652	1.03491	1.04410	1.05254	1.06217	1.07218	1.07699
m_B/m_A	0.8219	0.9078	1.0034	1.1013	1.2264				
ρ/(g/cm^3)	1.08303	1.08594	1.09132	1.09444	1.09937				

T/K = 298.15

m_B/m_A	0.0530	0.1138	0.1860	0.2554	0.3345	0.4292	0.5405	0.6684	0.7440
ρ/(g/cm^3)	1.00583	1.01485	1.02472	1.03290	1.04185	1.05008	1.05937	1.06909	1.07377
m_B/m_A	0.8219	0.9078	1.0034	1.1013	1.2264				
ρ/(g/cm^3)	1.07963	1.08245	1.08767	1.09053	1.09538				

T/K = 303.15

m_B/m_A	0.0530	0.1138	0.1860	0.2554	0.3345	0.4292	0.5405	0.6684	0.7440
ρ/(g/cm^3)	1.00426	1.01309	1.02274	1.03073	1.03946	1.04749	1.05908	1.06592	1.07047
m_B/m_A	0.8219	0.9078	1.0034	1.1013	1.2264				
ρ/(g/cm^3)	1.07617	1.07890	1.08398	1.08675	1.09133				

T/K = 308.15

m_B/m_A	0.0530	0.1138	0.1860	0.2554	0.3345	0.4292	0.5405	0.6684	0.7440
ρ/(g/cm^3)	1.00248	1.01114	1.02059	1.02841	1.03693	1.04476	1.05345	1.06266	1.06708
m_B/m_A	0.8219	0.9078	1.0034	1.1013	1.2264				
ρ/(g/cm^3)	1.07263	1.07528	1.08022	1.08291	1.08723				

T/K = 313.15

m_B/m_A	0.0530	0.1138	0.1860	0.2554	0.3345	0.4292	0.5405	0.6684	0.7440
ρ/(g/cm^3)	1.00052	1.00902	1.01827	1.02592	1.03425	1.04190	1.05038	1.05929	1.06361
m_B/m_A	0.8219	0.9078	1.0034	1.1013	1.2264				
ρ/(g/cm^3)	1.06901	1.07160	1.07641	1.07903	1.08313				

Polymer (B): **poly(ethylene glycol) dimethyl ether** **2011SA2**
Characterization: M_n/g.mol^{-1} = 366, Merck, Darmstadt, Germany
Solvent (A): **water** H_2O **7732-18-5**

T/K = 283.15

w_B	0.0000	0.0050	0.0101	0.0200	0.0400	0.0599	0.0999	0.1198	0.1594
ρ/(g/cm^3)	0.99970	1.00024	1.00078	1.00186	1.00408	1.00636	1.01108	1.01353	1.01849
w_B	0.2001	0.2798	0.3599	0.4298	0.5098	0.5992	0.6790	0.7505	0.8388
ρ/(g/cm^3)	1.02374	1.03436	1.04490	1.05341	1.06142	1.06714	1.06876	1.06748	1.06299
w_B	0.9093	0.9596	1.0000						
ρ/(g/cm^3)	1.05752	1.05276	1.04857						

T/K = 288.15

w_B	0.0000	0.0050	0.0101	0.0200	0.0400	0.0599	0.0999	0.1198	0.1594
ρ/(g/cm^3)	0.99910	0.99962	1.00013	1.00118	1.00331	1.00549	1.01002	1.01236	1.01708
w_B	0.2001	0.2798	0.3599	0.4298	0.5098	0.5992	0.6790	0.7505	0.8388
ρ/(g/cm^3)	1.02206	1.03212	1.04207	1.05009	1.05762	1.06294	1.06437	1.06301	1.05849
w_B	0.9093	0.9596	1.0000						
ρ/(g/cm^3)	1.05300	1.04822	1.04402						

T/K = 293.15

w_B	0.0000	0.0050	0.0101	0.0200	0.0400	0.0599	0.0999	0.1198	0.1594
ρ/(g/cm^3)	0.99820	0.99870	0.99920	1.00021	1.00226	1.00436	1.00870	1.01095	1.01546
w_B	0.2001	0.2798	0.3599	0.4298	0.5098	0.5992	0.6790	0.7505	0.8388
ρ/(g/cm^3)	1.02019	1.02973	1.03914	1.04669	1.05376	1.05871	1.05995	1.05853	1.05398
w_B	0.9093	0.9596	1.0000						
ρ/(g/cm^3)	1.04848	1.04368	1.03948						

T/K = 298.15

w_B	0.0000	0.0050	0.0101	0.0200	0.0400	0.0599	0.0999	0.1198	0.1594
ρ/(g/cm^3)	0.99704	0.99752	0.99801	0.99898	1.00096	1.00299	1.00717	1.00932	1.01363
w_B	0.2001	0.2798	0.3599	0.4298	0.5098	0.5992	0.6790	0.7505	0.8388
ρ/(g/cm^3)	1.01815	1.02721	1.03611	1.04322	1.04985	1.05445	1.05551	1.05403	1.04946
w_B	0.9093	0.9596	1.0000						
ρ/(g/cm^3)	1.04395	1.03914	1.03496						

T/K = 303.15

w_B	0.0000	0.0050	0.0101	0.0200	0.0400	0.0599	0.0999	0.1198	0.1594
ρ/(g/cm^3)	0.99564	0.99611	0.99658	0.99752	0.99944	1.00140	1.00542	1.00749	1.01162
w_B	0.2001	0.2798	0.3599	0.4298	0.5098	0.5992	0.6790	0.7505	0.8388
ρ/(g/cm^3)	1.01594	1.02456	1.03297	1.03967	1.04589	1.05014	1.05104	1.04950	1.04493
w_B	0.9093	0.9596	1.0000						
ρ/(g/cm^3)	1.03941	1.03461	1.03044						

continued

continued

T/K = 308.15

w_B	0.0000	0.0050	0.0101	0.0200	0.0400	0.0599	0.0999	0.1198	0.1594
ρ/(g/cm^3)	0.99403	0.99448	0.99494	0.99585	0.99771	0.99960	1.00348	1.00547	1.00944
w_B	0.2001	0.2798	0.3599	0.4298	0.5098	0.5992	0.6790	0.7505	0.8388
ρ/(g/cm^3)	1.01357	1.02177	1.02974	1.03605	1.04187	1.04581	1.04655	1.04496	1.04038
w_B	0.9093	0.9596	1.0000						
ρ/(g/cm^3)	1.03487	1.03008	1.02594						

Polymer (B): **poly(ethylene glycol) dimethyl ether** **2011SA2**
Characterization: M_n/g.mol^{-1} = 698, Merck, Darmstadt, Germany
Solvent (A): **water** **H_2O** **7732-18-5**

T/K = 283.15

w_B	0.0050	0.0100	0.0199	0.0400	0.0600	0.0999	0.1200	0.1602	0.1998
ρ/(g/cm^3)	1.00038	1.00108	1.00245	1.00527	1.00815	1.01396	1.01703	1.02321	1.02937
w_B	0.2797	0.3597	0.4290	0.5098	0.6000	0.6793	0.7510	0.8376	0.9082
ρ/(g/cm^3)	1.04235	1.05544	1.06641	1.07795	1.08770	1.09278	1.09452	1.09352	1.09075
w_B	0.9601	1.0000							
ρ/(g/cm^3)	1.08874	solid							

T/K = 288.15

w_B	0.0050	0.0100	0.0199	0.0400	0.0600	0.0999	0.1200	0.1602	0.1998
ρ/(g/cm^3)	0.99976	1.00043	1.00176	1.00450	1.00728	1.01290	1.01585	1.02181	1.02774
w_B	0.2797	0.3597	0.4290	0.5098	0.6000	0.6793	0.7510	0.8376	0.9082
ρ/(g/cm^3)	1.04018	1.05271	1.06321	1.07424	1.08359	1.08847	1.09014	1.08912	1.08636
w_B	0.9601	1.0000							
ρ/(g/cm^3)	1.08425	solid							

T/K = 293.15

w_B	0.0050	0.0100	0.0199	0.0400	0.0600	0.0999	0.1200	0.1602	0.1998
ρ/(g/cm^3)	0.99885	0.99950	1.00079	1.00344	1.00614	1.01159	1.01445	1.02020	1.02590
w_B	0.2797	0.3597	0.4290	0.5098	0.6000	0.6793	0.7510	0.8376	0.9082
ρ/(g/cm^3)	1.03787	1.04987	1.05992	1.07048	1.07944	1.08413	1.08573	1.08470	1.08196
w_B	0.9601	1.0000							
ρ/(g/cm^3)	1.07953	1.07660							

T/K = 298.15

w_B	0.0050	0.0100	0.0199	0.0400	0.0600	0.0999	0.1200	0.1602	0.1998
ρ/(g/cm^3)	0.99767	0.99830	0.99956	1.00214	1.00477	1.01005	1.01282	1.01839	1.02390

continued

continued

w_B	0.2797	0.3597	0.4290	0.5098	0.6000	0.6793	0.7510	0.8376	0.9082
ρ/(g/cm^3)	1.03541	1.04693	1.05655	1.06667	1.07525	1.07977	1.08131	1.08027	1.07755
w_B	0.9601	1.0000							
ρ/(g/cm^3)	1.07480	1.07209							

T/K = 303.15

w_B	0.0050	0.0100	0.0199	0.0400	0.0600	0.0999	0.1200	0.1602	0.1998
ρ/(g/cm^3)	0.99625	0.99687	0.99810	1.00062	1.00318	1.00831	1.01100	1.01639	1.02172
w_B	0.2797	0.3597	0.4290	0.5098	0.6000	0.6793	0.7510	0.8376	0.9082
ρ/(g/cm^3)	1.03282	1.04389	1.05311	1.06280	1.07103	1.07547	1.07687	1.07584	1.07313
w_B	0.9601	1.0000							
ρ/(g/cm^3)	1.07016	1.06771							

T/K = 308.15

w_B	0.0050	0.0100	0.0199	0.0400	0.0600	0.0999	0.1200	0.1602	0.1998
ρ/(g/cm^3)	0.99462	0.99523	0.99643	0.99888	1.00138	1.00637	1.00899	1.01422	1.01938
w_B	0.2797	0.3597	0.4290	0.5098	0.6000	0.6793	0.7510	0.8376	0.9082
ρ/(g/cm^3)	1.03009	1.04074	1.04959	1.05888	1.06677	1.07097	1.07241	1.07140	1.06872
w_B	0.9601	1.0000							
ρ/(g/cm^3)	1.06576	1.06337							

Polymer (B): **poly(ethylene glycol) dimethyl ether** **2011SA2**
Characterization: M_n/g.mol^{-1} = 698, Merck, Darmstadt, Germany
Solvent (A): **water** H_2O **7732-18-5**
Salt (C): **diammonium phosphate** $(NH_4)_2HPO_4$ **7783-28-0**

Comments: The concentration of the salt is 0.5 mol/kg $(NH_4)_2HPO_4$ in water.

T/K = 283.15

w_B	0.0000	0.0050	0.0100	0.0200	0.0400	0.0599	0.0998	0.1600	0.1999
ρ/(g/cm^3)	1.03394	1.03456	1.03504	1.03606	1.03812	1.04024	1.04464	1.05155	1.05642
w_B	0.2801								
ρ/(g/cm^3)	1.06607								

T/K = 288.15

w_B	0.0000	0.0050	0.0100	0.0200	0.0400	0.0599	0.0998	0.1600	0.1999
ρ/(g/cm^3)	1.03288	1.03349	1.03396	1.03495	1.03693	1.03898	1.04322	1.04985	1.05450
w_B	0.2801								
ρ/(g/cm^3)	1.06369								

continued

continued

T/K = 293.15

w_B	0.0000	0.0050	0.0100	0.0200	0.0400	0.0599	0.0998	0.1600	0.1999
ρ/(g/cm^3)	1.03161	1.03219	1.03265	1.03361	1.03553	1.03751	1.04160	1.04797	1.05243
w_B	0.2801								
ρ/(g/cm^3)	1.06119								

T/K = 298.15

w_B	0.0000	0.0050	0.0100	0.0200	0.0400	0.0599	0.0998	0.1600	0.1999
ρ/(g/cm^3)	1.03012	1.03069	1.03114	1.03207	1.03393	1.03585	1.03980	1.04593	1.05020
w_B	0.2801								
ρ/(g/cm^3)	1.05855								

T/K = 303.15

w_B	0.0000	0.0050	0.0100	0.0200	0.0400	0.0599	0.0998	0.1600	0.1999
ρ/(g/cm^3)	1.02844	1.02900	1.02943	1.03034	1.03214	1.03400	1.03782	1.04373	1.04783
w_B	0.2801								
ρ/(g/cm^3)	1.05580								

T/K = 308.15

w_B	0.0000	0.0050	0.0100	0.0200	0.0400	0.0599	0.0998	0.1600	0.1999
ρ/(g/cm^3)	1.02658	1.02713	1.02755	1.02843	1.03018	1.03199	1.03568	1.04137	1.04531
w_B	0.2801								
ρ/(g/cm^3)	1.05293								

Polymer (B):	**poly(ethylene glycol) dimethyl ether**		**2011SA2**
Characterization:	M_n/g.mol^{-1} = 698, Merck, Darmstadt, Germany		
Solvent (A):	**water**	H_2O	**7732-18-5**
Salt (C):	**monoammonium phosphate**	$(NH_4)H_2PO_4$	**7722-76-1**

Comments: The concentration of the salt is 0.5 mol/kg $(NH_4)H_2PO_4$ in water.

T/K = 283.15

w_B	0.0000	0.0050	0.0100	0.0200	0.0400	0.0600	0.1001	0.1599	0.2001
ρ/(g/cm^3)	1.03096	1.03153	1.03211	1.03323	1.03546	1.03779	1.04251	1.04982	1.05496
w_B	0.2799	0.3599	0.4305						
ρ/(g/cm^3)	1.06532	1.07557	1.08418						

T/K = 288.15

w_B	0.0000	0.0050	0.0100	0.0200	0.0400	0.0600	0.1001	0.1599	0.2001
ρ/(g/cm^3)	1.03004	1.03059	1.03116	1.03225	1.03440	1.03664	1.04119	1.04821	1.05312
w_B	0.2799	0.3599	0.4305						
ρ/(g/cm^3)	1.06300	1.07274	1.08090						

continued

continued

T/K = 293.15

w_B	0.0000	0.0050	0.0100	0.0200	0.0400	0.0600	0.1001	0.1599	0.2001
$\rho/(g/cm^3)$	1.02889	1.02942	1.02997	1.03103	1.03311	1.03528	1.03966	1.04641	1.05112
w_B	0.2799	0.3599	0.4305						
$\rho/(g/cm^3)$	1.06055	1.06982	1.07755						

T/K = 298.15

w_B	0.0000	0.0050	0.0100	0.0200	0.0400	0.0600	0.1001	0.1599	0.2001
$\rho/(g/cm^3)$	1.02752	1.02804	1.02857	1.02960	1.03162	1.03372	1.03796	1.04445	1.04897
w_B	0.2799	0.3599	0.4305						
$\rho/(g/cm^3)$	1.05799	1.06680	1.07413						

T/K = 303.15

w_B	0.0000	0.0050	0.0100	0.0200	0.0400	0.0600	0.1001	0.1599	0.2001
$\rho/(g/cm^3)$	1.02596	1.02646	1.02698	1.02798	1.02994	1.03197	1.03607	1.04234	1.04668
w_B	0.2799	0.3599	0.4305						
$\rho/(g/cm^3)$	1.05530	1.06369	1.07064						

T/K = 308.15

w_B	0.0000	0.0050	0.0100	0.0200	0.0400	0.0600	0.1001	0.1599	0.2001
$\rho/(g/cm^3)$	1.02421	1.02470	1.02520	1.02618	1.02808	1.03006	1.03403	1.04007	1.04424
w_B	0.2799	0.3599	0.4305						
$\rho/(g/cm^3)$	1.05247	1.06049	1.06708						

Polymer (B): **poly(ethylene glycol) dimethyl ether** **2011SA2**
Characterization: M_n/g.mol^{-1} = 2145, Merck, Darmstadt, Germany
Solvent (A): **water** H_2O **7732-18-5**

T/K = 283.15

w_B	0.0040	0.0080	0.0120	0.0200	0.0299	0.0350	0.0400	0.0500	0.0600
$\rho/(g/cm^3)$	1.00043	1.00114	1.00184	1.00326	1.00504	1.00596	1.00684	1.00856	1.01030
w_B	0.0801	0.0999	0.1200	0.1494	0.1797	0.1957			
$\rho/(g/cm^3)$	1.01387	1.01756	1.02144	1.02708	1.03262	1.03571			

T/K = 288.15

w_B	0.0040	0.0080	0.0120	0.0200	0.0299	0.0350	0.0400	0.0500	0.0600
$\rho/(g/cm^3)$	0.99980	1.00049	1.00118	1.00256	1.00430	1.00519	1.00605	1.00773	1.00942
w_B	0.0801	0.0999	0.1200	0.1494	0.1797	0.1957			
$\rho/(g/cm^3)$	1.01289	1.01647	1.02025	1.02574	1.03110	1.03410			

continued

continued

T/K = 293.15

w_B	0.0040	0.0080	0.0120	0.0200	0.0299	0.0350	0.0400	0.0500	0.0600
ρ/(g/cm^3)	0.99889	0.99956	1.00023	1.00158	1.00328	1.00415	1.00499	1.00662	1.00827
w_B	0.0801	0.0999	0.1200	0.1494	0.1797	0.1957			
ρ/(g/cm^3)	1.01166	1.01515	1.01883	1.02417	1.02939	1.03230			

T/K = 298.15

w_B	0.0040	0.0080	0.0120	0.0200	0.0299	0.0350	0.0400	0.0500	0.0600
ρ/(g/cm^3)	0.99771	0.99837	0.99902	1.00035	1.00201	1.00286	1.00368	1.00528	1.00689
w_B	0.0801	0.0999	0.1200	0.1494	0.1797	0.1957			
ρ/(g/cm^3)	1.01019	1.01361	1.01720	1.02241	1.02748	1.03032			

T/K = 303.15

w_B	0.0040	0.0080	0.0120	0.0200	0.0299	0.0350	0.0400	0.0500	0.0600
ρ/(g/cm^3)	0.99630	0.99695	0.99759	0.99888	1.00051	1.00134	1.00214	1.00371	1.00528
w_B	0.0801	0.0999	0.1200	0.1494	0.1797	0.1957			
ρ/(g/cm^3)	1.00851	1.01185	1.01537	1.02046	1.02540	1.02817			

T/K = 308.15

w_B	0.0040	0.0080	0.0120	0.0200	0.0299	0.0350	0.0400	0.0500	0.0600
ρ/(g/cm^3)	0.99468	0.99531	0.99593	0.99721	0.99880	0.99961	1.00040	1.00194	1.00348
w_B	0.0801	0.0999	0.1200	0.1494	0.1797	0.1957			
ρ/(g/cm^3)	1.00665	1.00991	1.01335	1.01833	1.02316	1.02586			

Polymer (B): **poly(ethylene glycol) monomethyl ether** **2010ZAF**
Characterization: M_n/g.mol^{-1} = 350, Fluka AG, Buchs, Switzerland
Solvent (A): **water** **H_2O** **7732-18-5**

T/K = 283.15

x_B	0.0000	0.0028	0.0059	0.0130	0.0224	0.0342	0.0466	0.0732
ρ/(g/cm^3)	0.99970	1.00677	1.01383	1.02935	1.04634	1.06256	1.07477	1.08954
V^E/(cm^3/g)	0.000	−0.050	−0.102	−0.252	−0.457	−0.696	−0.913	−1.238
x_B	0.0974	0.1696	0.2025	0.3025	0.4077	0.5000	0.6016	0.6984
ρ/(g/cm^3)	1.09577	1.10034	1.10086	1.09975	1.09893	1.09824	1.09769	1.09736
V^E/(cm^3/g)	−1.412	−1.561	−1.580	−1.437	−1.266	−1.083	−0.878	−0.694
x_B	0.8021	0.8922	1.0000					
ρ/(g/cm^3)	1.09691	1.09662	1.09628					
V^E/(cm^3/g)	−0.457	−0.255	0.000					

continued

continued

T/K = 293.15

x_B	0.0000	0.0028	0.0059	0.0130	0.0224	0.0342	0.0466	0.0732
ρ/(g/cm^3)	0.99809	1.00439	1.01075	1.02572	1.04176	1.05673	1.06801	1.08173
V^E/(cm^3/g)	0.000	−0.041	−0.086	−0.236	−0.434	−0.657	−0.861	−1.169
x_B	0.0974	0.1696	0.2025	0.3025	0.4077	0.5000	0.6016	0.6984
ρ/(g/cm^3)	1.08724	1.09195	1.09235	1.09156	1.09042	1.08973	1.08913	1.08879
V^E/(cm^3/g)	−1.325	−1.495	−1.512	−1.406	−1.225	−1.053	−0.853	−0.679
x_B	0.8021	0.8922	1.0000					
ρ/(g/cm^3)	1.08831	1.08803	1.08766					
V^E/(cm^3/g)	−0.446	−0.257	0.000					

T/K = 303.15

x_B	0.0000	0.0028	0.0059	0.0130	0.0224	0.0342	0.0466	0.0732
ρ/(g/cm^3)	0.99564	1.00202	1.00814	1.02136	1.03664	1.05053	1.06097	1.07375
V^E/(cm^3/g)	0.000	−0.048	−0.093	−0.220	−0.414	−0.623	−0.814	−1.107
x_B	0.0974	0.1696	0.2025	0.3025	0.4077	0.5000	0.6016	0.6984
ρ/(g/cm^3)	1.07864	1.08345	1.08379	1.08252	1.08186	1.08118	1.08057	1.08023
V^E/(cm^3/g)	−1.245	−1.433	−1.450	−1.307	−1.184	−1.022	−0.830	−0.664
x_B	0.8021	0.8922	1.0000					
ρ/(g/cm^3)	1.07974	1.07945	1.07906					
V^E/(cm^3/g)	−0.437	−0.252	0.000					

T/K = 313.15

x_B	0.0000	0.0028	0.0059	0.0130	0.0224	0.0342	0.0466	0.0732
ρ/(g/cm^3)	0.99207	0.99832	1.00417	1.01668	1.03101	1.04397	1.05367	1.06560
V^E/(cm^3/g)	0.000	−0.050	−0.095	−0.217	−0.401	−0.598	−0.778	−1.055
x_B	0.0974	0.1696	0.2025	0.3025	0.4077	0.5000	0.6016	0.6984
ρ/(g/cm^3)	1.06993	1.07487	1.07515	1.07406	1.07330	1.07262	1.07201	1.07167
V^E/(cm^3/g)	−1.177	−1.379	−1.396	−1.284	−1.149	−0.995	−0.808	−0.650
x_B	0.8021	0.8922	1.0000					
ρ/(g/cm^3)	1.07117	1.07089	1.07049					
V^E/(cm^3/g)	−0.426	−0.249	0.000					

T/K = 323.15

x_B	0.0000	0.0028	0.0059	0.0130	0.0224	0.0342	0.0466	0.0732
ρ/(g/cm^3)	0.98531	0.99115	0.99775	1.01188	1.02491	1.03706	1.04609	1.05728
V^E/(cm^3/g)	0.000	−0.044	−0.106	−0.269	−0.434	−0.620	−0.788	−1.051
x_B	0.0974	0.1696	0.2025	0.3025	0.4077	0.5000	0.6016	0.6984
ρ/(g/cm^3)	1.06144	1.06621	1.06647	1.06548	1.06470	1.06405	1.06348	1.06313
V^E/(cm^3/g)	−1.169	−1.367	−1.381	−1.277	−1.141	−0.989	−0.809	−0.647
x_B	0.8021	0.8922	1.0000					
ρ/(g/cm^3)	1.06266	1.06236	1.06196					
V^E/(cm^3/g)	−0.431	−0.249	0.000					

Polymer (B):	**poly(ethylene oxide)**		**2008SAD**
Characterization:	M_n/g.mol^{-1} = 4000, Merck, Darmstadt, Germany		
Solvent (A):	**water**	**H_2O**	**7732-18-5**

T/K = 293.15

c_B/(mol/kg)	0.0040	0.0082	0.0123	0.0204	0.0253	0.0310	0.0363	0.0417
ρ/(g/cm^3)	0.99887	0.99956	1.00023	1.00155	1.00233	1.00326	1.00409	1.00496
c_B/(mol/kg)	0.0472	0.0527	0.0636	0.0752	0.0869	0.1109	0.1361	0.1761
ρ/(g/cm^3)	1.00580	1.00666	1.00833	1.01007	1.01181	1.01527	1.01880	1.02413
c_B/(mol/kg)	0.2195	0.2480						
ρ/(g/cm^3)	1.02961	1.03301						

T/K = 298.15

c_B/(mol/kg)	0.0040	0.0082	0.0123	0.0204	0.0253	0.0310	0.0363	0.0417
ρ/(g/cm^3)	0.99770	0.99837	0.99902	1.00032	1.00108	1.00200	1.00280	1.00365
c_B/(mol/kg)	0.0472	0.0527	0.0636	0.0752	0.0869	0.1109	0.1361	0.1761
ρ/(g/cm^3)	1.00447	1.00532	1.00695	1.00865	1.01035	1.01374	1.01718	1.02239
c_B/(mol/kg)	0.2195	0.2480						
ρ/(g/cm^3)	1.02773	1.03105						

T/K = 303.15

c_B/(mol/kg)	0.0040	0.0082	0.0123	0.0204	0.0253	0.0310	0.0363	0.0417
ρ/(g/cm^3)	0.99628	0.99694	0.99759	0.99885	0.99961	1.00050	1.00129	1.00212
c_B/(mol/kg)	0.0472	0.0527	0.0636	0.0752	0.0869	0.1109	0.1361	0.1761
ρ/(g/cm^3)	1.00293	1.00375	1.00535	1.00702	1.00869	1.01200	1.01537	1.02046
c_B/(mol/kg)	0.2195	0.2480						
ρ/(g/cm^3)	1.02568	1.02891						

Polymer (B):	**poly(ethylene oxide)**		**2008SAD**
Characterization:	M_n/g.mol^{-1} = 4000, Merck, Darmstadt, Germany		
Solvent (A):	**water**	**H_2O**	**7732-18-5**
Salt (C):	**disodium phosphate**	**Na_2HPO_4**	**7558-79-4**

c_C/(mol/kg) = 0.1440 was kept constant

T/K = 293.15

c_B/(mol/kg)	0.0079	0.0306	0.0359	0.0462	0.0528	0.0642	0.0747	0.0862
ρ/(g/cm^3)	1.01916	1.02249	1.02324	1.02472	1.02563	1.02723	1.02864	1.03017
c_B/(mol/kg)	0.1106	0.1361	0.1720	0.2168	0.2506			
ρ/(g/cm^3)	1.03334	1.03655	1.04087	1.04597	1.04961			

continued

continued

T/K = 298.15

c_B/(mol/kg)	0.0079	0.0306	0.0359	0.0462	0.0528	0.0642	0.0747	0.0862
ρ/(g/cm^3)	1.01778	1.02103	1.02177	1.02322	1.02411	1.02566	1.02705	1.02854
c_B/(mol/kg)	0.1106	0.1361	0.1720	0.2168	0.2506			
ρ/(g/cm^3)	1.03164	1.03477	1.03898	1.04395	1.04750			

T/K = 303.15

c_B/(mol/kg)	0.0079	0.0306	0.0359	0.0462	0.0528	0.0642	0.0747	0.0862
ρ/(g/cm^3)	1.01619	1.01938	1.02010	1.02152	1.02239	1.02391	1.02526	1.02672
c_B/(mol/kg)	0.1106	0.1361	0.1720	0.2168	0.2506			
ρ/(g/cm^3)	1.02975	1.03282	1.03692	1.04177	1.04523			

Polymer (B): **poly(ethylene oxide)** **2008SAD**
Characterization: M_n/g.mol^{-1} = 4000, Merck, Darmstadt, Germany
Solvent (A): **water** H_2O **7732-18-5**
Salt (C): **monosodium phosphate** NaH_2PO_4 **7558-80-7**

c_C/(mol/kg) = 0.1440 was kept constant

T/K = 293.15

c_B/(mol/kg)	0.0100	0.0122	0.0150	0.0204	0.0257	0.0307	0.0362	0.0415
ρ/(g/cm^3)	1.01342	1.01375	1.01419	1.01500	1.01581	1.01656	1.01736	1.01815
c_B/(mol/kg)	0.0472	0.0525	0.0637	0.0752	0.0873	0.1109	0.1363	0.1760
ρ/(g/cm^3)	1.01897	1.01975	1.02135	1.02298	1.02466	1.02784	1.03115	1.03611
c_B/(mol/kg)	0.2193	0.2499						
ρ/(g/cm^3)	1.04121	1.04467						

T/K = 298.15

c_B/(mol/kg)	0.0100	0.0122	0.0150	0.0204	0.0257	0.0307	0.0362	0.0415
ρ/(g/cm^3)	1.01210	1.01243	1.01286	1.01365	1.01444	1.01518	1.01596	1.01673
c_B/(mol/kg)	0.0472	0.0525	0.0637	0.0752	0.0873	0.1109	0.1363	0.1760
ρ/(g/cm^3)	1.01753	1.01830	1.01986	1.02146	1.02310	1.02620	1.02943	1.03427
c_B/(mol/kg)	0.2193	0.2499						
ρ/(g/cm^3)	1.03924	1.04261						

T/K = 303.15

c_B/(mol/kg)	0.0100	0.0122	0.0150	0.0204	0.0257	0.0307	0.0362	0.0415
ρ/(g/cm^3)	1.01058	1.01089	1.01131	1.01209	1.01286	1.01359	1.01436	1.01511
c_B/(mol/kg)	0.0472	0.0525	0.0637	0.0752	0.0873	0.1109	0.1363	0.1760
ρ/(g/cm^3)	1.01589	1.01664	1.01817	1.01973	1.02134	1.02437	1.02752	1.03225
c_B/(mol/kg)	0.2193	0.2499						
ρ/(g/cm^3)	1.03711	1.04039						

Polymer (B):	**poly(ethylene oxide)**		**2008SAD**
Characterization:	M_n/g.mol^{-1} = 4000, Merck, Darmstadt, Germany		
Solvent (A):	**water**	H_2O	**7732-18-5**
Salt (C):	**sodium phosphate**	Na_3PO_4	**7601-54-9**

c_C/(mol/kg) = 0.1440 was kept constant

T/K = 293.15

c_B/(mol/kg)	0.0081	0.0100	0.0151	0.0201	0.0315	0.0362	0.0419	0.0478
ρ/(g/cm^3)	1.02781	1.02810	1.02882	1.02951	1.03109	1.03175	1.03251	1.03331
c_B/(mol/kg)	0.0525	0.0638	0.0739	0.0865	0.1105	0.1331	0.1752	0.2194
ρ/(g/cm^3)	1.03394	1.03542	1.03670	1.03833	1.04130	1.04399	1.04880	1.05353
c_B/(mol/kg)	0.2491							
ρ/(g/cm^3)	1.05658							

T/K = 298.15

c_B/(mol/kg)	0.0081	0.0100	0.0151	0.0201	0.0315	0.0362	0.0419	0.0478
ρ/(g/cm^3)	1.02634	1.02662	1.02732	1.02800	1.02955	1.03018	1.03093	1.03171
c_B/(mol/kg)	0.0525	0.0638	0.0739	0.0865	0.1105	0.1331	0.1752	0.2194
ρ/(g/cm^3)	1.03233	1.03377	1.03502	1.03661	1.03952	1.04214	1.04683	1.05144
c_B/(mol/kg)	0.2491							
ρ/(g/cm^3)	1.05441							

T/K = 303.15

c_B/(mol/kg)	0.0081	0.0100	0.0151	0.0201	0.0315	0.0362	0.0419	0.0478
ρ/(g/cm^3)	1.02467	1.02493	1.02562	1.02630	1.02781	1.02843	1.02916	1.02992
c_B/(mol/kg)	0.0525	0.0638	0.0739	0.0865	0.1105	0.1331	0.1752	0.2194
ρ/(g/cm^3)	1.03053	1.03194	1.03317	1.03472	1.03756	1.04012	1.04470	1.04919
c_B/(mol/kg)	0.2491							
ρ/(g/cm^3)	1.05209							

Polymer (B):	**poly(ethylene oxide-*b*-propylene oxide)**		**2004TAB**
Characterization:	M_n/g.mol^{-1} = 20700, $EO_{316}PO_{94}$, 77.1 mol% or 71.8 wt% ethylene oxide, synthesized in the laboratory		
Solvent (A):	**water**	H_2O	**7732-18-5**

T/K = 278.15

c_B/(g/L)	1.003	1.505	2.046	2.529	3.024	3.921	5.038	7.458	9.991
ρ/(g/cm^3)	1.00013	1.00021	1.00030	1.00038	1.00046	1.00060	1.00078	1.00118	1.00158
c_B/(g/L)	12.092	14.991	17.409	19.838					
ρ/(g/cm^3)	1.00192	1.00238	1.00277	1.00316					

continued

continued

T/K = 283.15

c_B/(g/L)	1.003	1.505	2.046	2.529	3.024	3.921	5.038	7.458	9.991
ρ/(g/cm^3)	0.99986	0.99994	1.00002	1.00010	1.00018	1.00032	1.00049	1.00087	1.00127

c_B/(g/L)	12.092	14.991	17.409	19.838
ρ/(g/cm^3)	1.00159	1.00204	1.00242	1.00279

T/K = 288.15

c_B/(g/L)	1.003	1.505	2.046	2.529	3.024	3.921	5.038	7.458	9.991
ρ/(g/cm^3)	0.99925	0.99933	0.99941	0.99948	0.99956	0.99969	0.99986	1.00023	1.00061

c_B/(g/L)	12.092	14.991	17.409	19.838
ρ/(g/cm^3)	1.00092	1.00135	1.00171	1.00207

T/K = 293.15

c_B/(g/L)	0.989	1.322	1.846	2.453	2.995	4.024	5.305	7.581	10.162
ρ/(g/cm^3)	0.99835	0.99840	0.99848	0.99856	0.99864	0.99879	0.99897	0.99929	0.99965

c_B/(g/L)	12.340	15.132	17.467	19.935
ρ/(g/cm^3)	0.99995	1.00034	1.00067	1.00101

T/K = 298.15

c_B/(g/L)	0.989	1.322	1.846	2.453	2.995	4.024	5.305	7.581	10.162
ρ/(g/cm^3)	0.99718	0.99723	0.99729	0.99737	0.99744	0.99758	0.99774	0.99804	0.99837

c_B/(g/L)	12.340	15.132	17.467	19.935
ρ/(g/cm^3)	0.99864	0.99900	0.99929	0.99960

T/K = 303.15

c_B/(g/L)	0.989	1.322	1.846	2.453	2.995	4.024	5.305	7.581	10.162
ρ/(g/cm^3)	0.99577	0.99581	0.99588	0.99596	0.99602	0.99615	0.99631	0.99660	0.99692

c_B/(g/L)	12.340	15.132	17.467	19.935
ρ/(g/cm^3)	0.99719	0.99753	0.99782	0.99812

T/K = 308.15

c_B/(g/L)	0.989	1.322	1.846	2.453	2.995	4.024	5.305	7.581	10.162
ρ/(g/cm^3)	0.99415	0.99419	0.99426	0.99433	0.99440	0.99452	0.99468	0.99495	0.99527

c_B/(g/L)	12.340	15.132	17.467	19.935
ρ/(g/cm^3)	0.99553	0.99586	0.99614	0.99644

T/K = 313.15

c_B/(g/L)	0.989	1.322	1.846	2.453	2.995	4.024	5.305	7.581	10.162
ρ/(g/cm^3)	0.99233	0.99237	0.99244	0.99251	0.99257	0.99269	0.99285	0.99312	0.99342

c_B/(g/L)	12.340	15.132	17.467	19.935
ρ/(g/cm^3)	0.99368	0.99400	0.99428	0.99456

continued

continued

T/K = 318.15

c_B/(g/L)	0.989	1.322	1.846	2.453	2.995	4.024	5.305	7.581	10.162
ρ/(g/cm^3)	0.99032	0.99037	0.99043	0.99050	0.99056	0.99068	0.99083	0.99109	0.99139
c_B/(g/L)	12.340	15.132	17.467	19.935					
ρ/(g/cm^3)	0.99164	0.99196	0.99223	0.99251					

T/K = 323.15

c_B/(g/L)	0.989	1.322	1.846	2.453	2.995	4.024	5.305	7.581	10.162
ρ/(g/cm^3)	0.98814	0.98818	0.98824	0.98831	0.98837	0.98849	0.98864	0.98890	0.98919
c_B/(g/L)	12.340	15.132	17.467	19.935					
ρ/(g/cm^3)	0.98944	0.98975	0.99001	0.99029					

Polymer (B): **poly(ethylene oxide-*b*-propylene oxide-*b*-ethylene oxide)** **2006DE1**

Characterization: M_n/g.mol^{-1} = 11400, $EO_{103}PO_{39}EO_{103}$, 15.9 mol% propylene oxide, F88, BASF AG, Ludwigshafen, Germany

Solvent (A): **water** **H_2O** **7732-18-5**

T/K = 298.15

c_B/(mol/kg)	0.00000	0.00126	0.00208	0.00382	0.00393	0.00489	0.00491	0.00559
ρ/(g/cm^3)	0.99705	0.99919	1.00057	1.00339	1.00361	1.00516	1.00515	1.00622
c_B/(mol/kg)	0.00668	0.00739	0.00749	0.00831	0.01021	0.01245	0.01376	0.01450
ρ/(g/cm^3)	1.00789	1.00900	1.00913	1.01036	1.01314	1.01634	1.01813	1.01921
c_B/(mol/kg)	0.01526	0.01667	0.01939	0.02007	0.02126	0.02350		
ρ/(g/cm^3)	1.02018	1.02201	1.02551	1.02627	1.02757	1.02998		

T/K = 308.15

c_B/(mol/kg)	0.00000	0.00126	0.00201	0.00208	0.00292	0.00302	0.00382	0.00393
ρ/(g/cm^3)	0.99404	0.99609	0.99728	0.99739	0.99870	0.99886	1.00009	1.00027
c_B/(mol/kg)	0.00489	0.00559	0.00611	0.00668	0.00713	0.00749	0.00831	0.01021
ρ/(g/cm^3)	1.00175	1.00274	1.00353	1.00428	1.00499	1.00541	1.00653	1.00905
c_B/(mol/kg)	0.01140	0.01245	0.01376	0.01450	0.01667	0.01701	0.01939	0.02126
ρ/(g/cm^3)	1.01058	1.01183	1.01339	1.01441	1.01687	1.01732	1.02010	1.02212
c_B/(mol/kg)	0.02140	0.02342	0.02350					
ρ/(g/cm^3)	1.02226	1.02443	1.02451					

Polymer (B): **poly(ethylene oxide-*b*-propylene oxide-*b*-ethylene oxide)** **2006DE1**
Characterization: M_n/g.mol^{-1} = 14600, $EO_{132}PO_{50}EO_{132}$, 15.9 mol% propylene oxide, F108, BASF AG, Ludwigshafen, Germany
Solvent (A): **water** **H_2O** **7732-18-5**

T/K = 298.15

c_B/(mol/kg)	0.00000	0.00050	0.00067	0.00113	0.00197	0.00264	0.00335	0.00371
ρ/(g/cm^3)	0.99705	0.99816	0.99853	0.99955	1.00136	1.00276	1.00425	1.00498
c_B/(mol/kg)	0.00482	0.00545	0.00606	0.00672	0.00769	0.00828	0.00835	0.00941
ρ/(g/cm^3)	1.00723	1.00849	1.00965	1.01091	1.01265	1.01370	1.01379	1.01571
c_B/(mol/kg)	0.00982	0.01058	0.01127	0.01274	0.01482	0.01636	0.01670	0.01882
ρ/(g/cm^3)	1.01630	1.01761	1.01868	1.02105	1.02432	1.02667	1.02721	1.03023

T/K = 308.15

c_B/(mol/kg)	0.00067	0.00197	0.00264	0.00335	0.00371	0.00482	0.00606	0.00672
ρ/(g/cm^3)	0.99543	0.99803	0.99930	1.00064	1.00133	1.00333	1.00555	1.00675
c_B/(mol/kg)	0.00769	0.00828	0.00835	0.00982	0.01090	0.01127	0.01274	0.01482
ρ/(g/cm^3)	1.00845	1.00947	1.00958	1.01210	1.01388	1.01446	1.01683	1.02012

Polymer (B): **poly(ethylene oxide-*b*-propylene oxide-*b*-ethylene oxide)** **2006DE1**
Characterization: M_n/g.mol^{-1} = 8350, $EO_{76}PO_{29}EO_{76}$, 16.0 mol% propylene oxide, F68, BASF AG, Ludwigshafen, Germany
Solvent (A): **water** **H_2O** **7732-18-5**

T/K = 298.15

c_B/(mol/kg)	0.00000	0.00301	0.00524	0.00712	0.00813	0.00994	0.01197	0.01443
ρ/(g/cm^3)	0.99705	1.00074	1.00339	1.00559	1.00671	1.00881	1.01096	1.01354
c_B/(mol/kg)	0.01647	0.01687	0.01900	0.02007	0.02668	0.03065	0.03905	0.04439
ρ/(g/cm^3)	1.01565	1.01607	1.01821	1.01927	1.02547	1.02895	1.03571	1.03944
c_B/(mol/kg)	0.04950	0.05235	0.05735	0.06981	0.071515			
ρ/(g/cm^3)	1.04272	1.04444	1.04732	1.05356	1.05438			

T/K = 308.15

c_B/(mol/kg)	0.00000	0.00301	0.00524	0.00712	0.00813	0.00994	0.01197	0.01443
ρ/(g/cm^3)	0.99404	0.99755	1.00007	1.00215	1.00322	1.00519	1.00725	1.00967
c_B/(mol/kg)	0.01647	0.01687	0.01900	0.02007	0.02129	0.02523	0.02668	0.03065
ρ/(g/cm^3)	1.01162	1.01204	1.01403	1.01509	1.01603	1.01930	1.02046	1.02352
c_B/(mol/kg)	0.03572	0.03905	0.04439	0.04950	0.05089	0.05235		
ρ/(g/cm^3)	1.02728	1.02939	1.03289	1.03607	1.03708	1.03772		

Polymer (B): **poly(ethylene oxide-*b*-propylene oxide-*b*-ethylene oxide)** **2006DE1**
Characterization: M_n/g.mol^{-1} = 1900, $EO_{11}PO_{16}EO_{11}$, 42.1 mol% propylene oxide, L35, BASF AG, Ludwigshafen, Germany
Solvent (A): **water** **H_2O** **7732-18-5**

T/K = 298.15

c_B/(mol/kg)	0.01529	0.04002	0.06275	0.07064	0.07566	0.1076	0.1272	0.1417
ρ/(g/cm^3)	1.00068	1.00619	1.01098	1.01252	1.01350	1.01942	1.02278	1.02518
c_B/(mol/kg)	0.1840	0.2167	0.2503	0.2987	0.3641	0.4400	0.4749	0.5812
ρ/(g/cm^3)	1.03143	1.03573	1.03969	1.04440	1.04944	1.05357	1.05507	1.05866
c_B/(mol/kg)	0.6848	0.7383	0.7981	0.9084	0.9586	1.3439	1.4572	
ρ/(g/cm^3)	1.06117	1.06241	1.06316	1.06455	1.06478	1.06722	1.06707	

T/K = 305.15

c_B/(mol/kg)	0.00000	0.00541	0.00717	0.0153	0.03872	0.06275	0.07409	0.1082
ρ/(g/cm^3)	0.99503	0.99629	0.99668	0.99852	1.00357	1.00833	1.01048	1.01651
c_B/(mol/kg)	0.1417	0.1943	0.2167	0.2503	0.3641	0.4749	0.5812	0.6167
ρ/(g/cm^3)	1.02167	1.02875	1.03134	1.03475	1.04343	0.04878	1.05238	1.05328
c_B/(mol/kg)	0.7383	0.9084	1.3439					
ρ/(g/cm^3)	1.05666	1.05864	1.06144					

T/K = 308.15

c_B/(mol/kg)	0.00000	0.00517	0.01533	0.04002	0.07064	0.07566	0.1076	0.1272
ρ/(g/cm^3)	0.99404	0.99522	0.99748	1.00270	1.00863	1.00954	1.01507	1.01810
c_B/(mol/kg)	0.1840	0.2242	0.2355	0.2503	0.2987	0.3683	0.4400	0.4731
ρ/(g/cm^3)	1.02589	1.03039	1.03152	1.03349	1.03691	1.04161	1.04500	1.04640
c_B/(mol/kg)	0.5801	0.6848	0.7981	0.9586	1.4572			
ρ/(g/cm^3)	1.04995	1.05256	1.05426	1.05644	1.05900			

Polymer (B): **poly(propylene glycol)** **2009SAD**
Characterization: M_n/g.mol^{-1} = 400, Fluka AG, Buchs, Switzerland
Solvent (A): **water** **H_2O** **7732-18-5**

T/K = 288.15

c_B/(mol/kg)	0.0000	0.0101	0.0213	0.0309	0.0413	0.0638	0.0868	0.1071
ρ/(g/cm^3)	0.99910	0.99999	1.00097	1.00180	1.00270	1.00462	1.00652	1.00818
c_B/(mol/kg)	0.1447	0.1859	0.2334	0.2969				
ρ/(g/cm^3)	1.01113	1.01420	1.01752	1.02158				

continued

continued

T/K = 293.15

c_B/(mol/kg)	0.0000	0.0101	0.0213	0.0309	0.0413	0.0638	0.0868	0.1071
ρ/(g/cm^3)	0.99820	0.99906	1.00001	1.00081	1.00167	1.00351	1.00532	1.00691
c_B/(mol/kg)	0.1447	0.1859	0.2334	0.2969				
ρ/(g/cm^3)	1.00971	1.01260	1.01571	1.01947				

T/K = 298.15

c_B/(mol/kg)	0.0000	0.0101	0.0213	0.0309	0.0413	0.0638	0.0868	0.1071
ρ/(g/cm^3)	0.99704	0.99787	0.99878	0.99955	1.00039	1.00215	1.00389	1.00539
c_B/(mol/kg)	0.1447	0.1859	0.2334	0.2969				
ρ/(g/cm^3)	1.00805	1.01078	1.01368	1.01716				

T/K = 303.15

c_B/(mol/kg)	0.0000	0.0101	0.0213	0.0309	0.0413	0.0638	0.0868	0.1071
ρ/(g/cm^3)	0.99565	0.99645	0.99733	0.99807	0.99887	1.00056	1.00222	1.00366
c_B/(mol/kg)	0.1447	0.1859	0.2334	0.2969				
ρ/(g/cm^3)	1.00618	1.00874	1.01145	1.01465				

T/K = 308.15

c_B/(mol/kg)	0.0000	0.0101	0.0213	0.0309	0.0413	0.0638	0.0868	0.1071
ρ/(g/cm^3)	0.99403	0.99480	0.99565	0.99637	0.99714	0.99876	1.00035	1.00172
c_B/(mol/kg)	0.1447	0.1859	0.2334	0.2969				
ρ/(g/cm^3)	1.00410	1.00650	1.00901	1.01195				

T/K = 313.15

c_B/(mol/kg)	0.0000	0.0101	0.0213	0.0309	0.0413	0.0638	0.0868	0.1071
ρ/(g/cm^3)	0.99221	0.99296	0.99378	0.99447	0.99521	0.99677	0.99829	0.99958
c_B/(mol/kg)	0.1447	0.1859	0.2334	0.2969				
ρ/(g/cm^3)	1.00183	1.00408	1.00639	1.00905				

Polymer (B): **poly(propylene glycol)** **2004ZAF**
Characterization: M_n/g.mol^{-1} = 400, Fluka AG, Buchs, Switzerland
Solvent (A): **water** **H_2O** **7732-18-5**

T/K = 283.15

w_B	0.0000	0.0050	0.0100	0.0199	0.0398	0.0605	0.0991	0.1199	0.1598
ρ/(g/cm^3)	0.99970	1.00018	1.00062	1.00156	1.00347	1.00551	1.00946	1.01166	1.01595
w_B	0.1992	0.2803	0.3602	0.4328	0.5152	0.6011	0.6781	0.7506	0.8398
ρ/(g/cm^3)	1.02028	1.02909	1.03684	1.04200	1.04541	1.04638	1.04544	1.04291	1.03730
w_B	0.9128	0.9615	0.9783	1.0000					
ρ/(g/cm^3)	1.02970	1.02279	1.01990	1.01579					

continued

continued

T/K = 288.15

w_B	0.0000	0.0050	0.0100	0.0199	0.0398	0.0605	0.0991	0.1199	0.1598
ρ/(g/cm^3)	0.99910	0.99956	0.99999	1.00089	1.00272	1.00468	1.00844	1.01052	1.01458
w_B	0.1992	0.2803	0.3602	0.4328	0.5152	0.6011	0.6781	0.7506	0.8398
ρ/(g/cm^3)	1.01862	1.02674	1.03374	1.03834	1.04136	1.04219	1.04128	1.03884	1.03333
w_B	0.9128	0.9615	0.9783	1.0000					
ρ/(g/cm^3)	1.02565	1.01881	1.01593	1.01186					

T/K = 293.15

w_B	0.0000	0.0050	0.0100	0.0199	0.0398	0.0605	0.0991	0.1199	0.1598
ρ/(g/cm^3)	0.99820	0.99865	0.99906	0.99993	1.00170	1.00357	1.00716	1.00913	1.01296
w_B	0.1992	0.2803	0.3602	0.4328	0.5152	0.6011	0.6781	0.7506	0.8398
ρ/(g/cm^3)	1.01673	1.02420	1.03050	1.03457	1.03724	1.03795	1.03708	1.03472	1.02932
w_B	0.9128	0.9615	0.9783	1.0000					
ρ/(g/cm^3)	1.02158	1.01488	1.01200	1.00791					

T/K = 298.15

w_B	0.0000	0.0050	0.0100	0.0199	0.0398	0.0605	0.0991	0.1199	0.1598
ρ/(g/cm^3)	0.99704	0.99747	0.99787	0.99871	1.00041	1.00221	1.00564	1.00751	1.01111
w_B	0.1992	0.2803	0.3602	0.4328	0.5152	0.6011	0.6781	0.7506	0.8398
ρ/(g/cm^3)	1.01463	1.02147	1.02710	1.03068	1.03304	1.03366	1.03283	1.03056	1.02525
w_B	0.9128	0.9615	0.9783	1.0000					
ρ/(g/cm^3)	1.01750	1.01092	1.00803	1.00393					

T/K = 303.15

w_B	0.0000	0.0050	0.0100	0.0199	0.0398	0.0605	0.0991	0.1199	0.1598
ρ/(g/cm^3)	0.99565	0.99606	0.99645	0.99725	0.99889	1.00062	1.00389	1.00566	1.00905
w_B	0.1992	0.2803	0.3602	0.4328	0.5152	0.6011	0.6781	0.7506	0.8398
ρ/(g/cm^3)	1.01233	1.01856	1.02354	1.02668	1.02876	1.02931	1.02854	1.02635	1.02119
w_B	0.9128	0.9615	0.9783	1.0000					
ρ/(g/cm^3)	1.01377	1.00692	1.00404	0.99994					

T/K = 308.15

w_B	0.0000	0.0050	0.0100	0.0199	0.0398	0.0605	0.0991	0.1199	0.1598
ρ/(g/cm^3)	0.99403	0.99443	0.99480	0.99555	0.99716	0.99882	1.00193	1.00361	1.00679
w_B	0.1992	0.2803	0.3602	0.4328	0.5152	0.6011	0.6781	0.7506	0.8398
ρ/(g/cm^3)	1.00982	1.01546	1.01982	1.02254	1.02438	1.02488	1.02420	1.02213	1.01707
w_B	0.9128	0.9615	0.9783	1.0000					
ρ/(g/cm^3)	1.00978	1.00290	1.00003	0.99594					

continued

continued

T/K = 313.15

w_B	0.0000	0.0050	0.0100	0.0199	0.0398	0.0605	0.0991	0.1199	0.1598
ρ/(g/cm^3)	0.99221	0.99261	0.99297	0.99368	0.99524	0.99683	0.99978	1.00139	1.00436
w_B	0.1992	0.2803	0.3602	0.4328	0.5152	0.6011	0.6781	0.7506	0.8398
ρ/(g/cm^3)	1.00714	1.01219	1.01592	1.01826	1.01990	1.02038	1.01983	1.01787	1.01293
w_B	0.9128	0.9615	0.9783	1.0000					
ρ/(g/cm^3)	1.00578	0.99888	0.99603	0.99194					

Polymer (B): **poly(propylene glycol)** **2009SAD**
Characterization: M_n/g.mol^{-1} = 400, Fluka AG, Buchs, Switzerland
Solvent (A): **water** H_2O **7732-18-5**
Salt (C): **disodium phosphate** Na_2HPO_4 **7558-79-4**

T/K = 288.15

w_C = 0.010 was kept constant

c_B/(mol/kg)	0.0000	0.0060	0.0102	0.0199	0.0306	0.0404	0.0611	0.0854
ρ/(g/cm^3)	1.00895	1.00943	1.00976	1.01052	1.01135	1.01212	1.01369	1.01551
c_B/(mol/kg)	0.1119	0.1482	0.1887	0.2338	0.2873			
ρ/(g/cm^3)	1.01745	1.02000	1.02272	1.02549	1.02854			

w_C = 0.020 was kept constant

c_B/(mol/kg)	0.0000	0.0060	0.0088	0.0200	0.0308	0.0420	0.0640	0.0845
ρ/(g/cm^3)	1.01829	1.01871	1.01892	1.01969	1.02043	1.02120	1.02269	1.02407
c_B/(mol/kg)	0.1253	0.1471	0.1626					
ρ/(g/cm^3)	1.02671	1.02807	1.02901					

T/K = 293.15

w_C = 0.010 was kept constant

c_B/(mol/kg)	0.0000	0.0060	0.0102	0.0199	0.0306	0.0404	0.0611	0.0854
ρ/(g/cm^3)	1.00794	1.00840	1.00871	1.00945	1.01025	1.01098	1.01248	1.01422
c_B/(mol/kg)	0.1119	0.1482	0.1887	0.2338	0.2873			
ρ/(g/cm^3)	1.01605	1.01846	1.02101	1.02353	1.02637			

w_C = 0.020 was kept constant

c_B/(mol/kg)	0.0000	0.0060	0.0088	0.0200	0.0308	0.0420	0.0640	0.0845
ρ/(g/cm^3)	1.01718	1.01758	1.01778	1.01852	1.01923	1.01996	1.02137	1.02268
c_B/(mol/kg)	0.1253	0.1471	0.1626					
ρ/(g/cm^3)	1.02517	1.02643	1.02731					

continued

continued

T/K = 298.15

w_C = 0.010 was kept constant

c_B/(mol/kg)	0.0000	0.0060	0.0102	0.0199	0.0306	0.0404	0.0611	0.0854
ρ/(g/cm^3)	1.00668	1.00712	1.00743	1.00813	1.00890	1.00960	1.01104	1.01269
c_B/(mol/kg)	0.1119	0.1482	0.1887	0.2338	0.2873			
ρ/(g/cm^3)	1.01443	1.01670	1.01908	1.02142	1.02400			

w_C = 0.020 was kept constant

c_B/(mol/kg)	0.0000	0.0060	0.0088	0.0200	0.0308	0.0420	0.0640	0.0845
ρ/(g/cm^3)	1.01583	1.01621	1.01640	1.01711	1.01779	1.01849	1.01984	1.02108
c_B/(mol/kg)	0.1253	0.1471	0.1626					
ρ/(g/cm^3)	1.02341	1.02459	1.02541					

T/K = 303.15

w_C = 0.010 was kept constant

c_B/(mol/kg)	0.0000	0.0060	0.0102	0.0199	0.0306	0.0404	0.0611	0.0854
ρ/(g/cm^3)	1.00520	1.00563	1.00592	1.00659	1.00733	1.00801	1.00938	1.01095
c_B/(mol/kg)	0.1119	0.1482	0.1887	0.2338	0.2873			
ρ/(g/cm^3)	1.01260	1.01473	1.01694	1.01913	1.02143			

w_C = 0.020 was kept constant

c_B/(mol/kg)	0.0000	0.0060	0.0088	0.0200	0.0308	0.0420	0.0640	0.0845
ρ/(g/cm^3)	1.01427	1.01464	1.01482	1.01550	1.01615	1.01682	1.01810	1.01927
c_B/(mol/kg)	0.1253	0.1471	0.1626					
ρ/(g/cm^3)	1.02146	1.02255	1.02330					

T/K = 308.15

w_C = 0.010 was kept constant

c_B/(mol/kg)	0.0000	0.0060	0.0102	0.0199	0.0306	0.0404	0.0611	0.0854
ρ/(g/cm^3)	1.00351	1.00392	1.00420	1.00485	1.00556	1.00621	1.00752	1.00902
c_B/(mol/kg)	0.1119	0.1482	0.1887	0.2338	0.2873			
ρ/(g/cm^3)	1.01057	1.01256	1.01460	1.01660	1.01867			

w_C = 0.020 was kept constant

c_B/(mol/kg)	0.0000	0.0060	0.0088	0.0200	0.0308	0.0420	0.0640	0.0845
ρ/(g/cm^3)	1.01251	1.01287	1.01304	1.01370	1.01432	1.01495	1.01617	1.01727
c_B/(mol/kg)	0.1253	0.1471	0.1626					
ρ/(g/cm^3)	1.01931	1.02032	1.02100					

continued

continued

T/K = 313.15

w_C = 0.010	was kept constant							
c_B/(mol/kg)	0.0000	0.0060	0.0102	0.0199	0.0306	0.0404	0.0611	0.0854
ρ/(g/cm^3)	1.00163	1.00203	1.00230	1.00293	1.00361	1.00423	1.00548	1.00690
c_B/(mol/kg)	0.1119	0.1482	0.1887	0.2338	0.2873			
ρ/(g/cm^3)	1.00836	1.01021	1.01209	1.01388	1.01568			
w_C = 0.020	was kept constant							
c_B/(mol/kg)	0.0000	0.0060	0.0088	0.0200	0.0308	0.0420	0.0640	0.0845
ρ/(g/cm^3)	1.01058	1.01092	1.01109	1.01171	1.01230	1.01291	1.01406	1.01510
c_B/(mol/kg)	0.1253	0.1471	0.1626					
ρ/(g/cm^3)	1.01699	1.01790	1.01851					

Polymer (B): **poly(propylene glycol)** **2009SAD**
Characterization: M_n/g.mol^{-1} = 400, Fluka AG, Buchs, Switzerland
Solvent (A): **water** **H_2O** **7732-18-5**
Salt (C): **monosodium phosphate** **NaH_2PO_4** **7558-80-7**

T/K = 288.15

w_C = 0.010	was kept constant							
c_B/(mol/kg)	0.0000	0.0079	0.0204	0.0309	0.0417	0.0869	0.1027	0.1225
ρ/(g/cm^3)	1.00676	1.00741	1.00841	1.00926	1.01012	1.01365	1.01484	1.01636
c_B/(mol/kg)	0.1477	0.1706	0.1821	0.2492	0.2844	0.3362	0.4156	0.5256
ρ/(g/cm^3)	1.01817	1.01977	1.02055	1.02484	1.02689	1.02965	1.03323	1.03714
w_C = 0.020	was kept constant							
c_B/(mol/kg)	0.0000	0.0050	0.0088	0.0198	0.0308	0.0419	0.0629	0.0846
ρ/(g/cm^3)	1.01431	1.01468	1.01497	1.01578	1.01659	1.01740	1.01891	1.02047
c_B/(mol/kg)	0.1248	0.1411	0.1624	0.1879	0.2856			
ρ/(g/cm^3)	1.02321	1.02429	1.02568	1.02729	1.03272			

T/K = 293.15

w_C = 0.010	was kept constant							
c_B/(mol/kg)	0.0000	0.0079	0.0204	0.0309	0.0417	0.0869	0.1027	0.1225
ρ/(g/cm^3)	1.00579	1.00641	1.00738	1.00818	1.00901	1.01239	1.01352	1.01496
c_B/(mol/kg)	0.1477	0.1706	0.1821	0.2492	0.2844	0.3362	0.4156	0.5256
ρ/(g/cm^3)	1.01666	1.01817	1.01889	1.02288	1.02476	1.02729	1.03051	1.03397

continued

continued

w_C = 0.020 was kept constant

c_B/(mol/kg)	0.0000	0.0050	0.0088	0.0198	0.0308	0.0419	0.0629	0.0846
ρ/(g/cm^3)	1.01326	1.01362	1.01389	1.01467	1.01544	1.01622	1.01766	1.01914
c_B/(mol/kg)	0.1248	0.1411	0.1624	0.1879	0.2856			
ρ/(g/cm^3)	1.02173	1.02274	1.02404	1.02553	1.03052			

T/K = 298.15

w_C = 0.010 was kept constant

c_B/(mol/kg)	0.0000	0.0079	0.0204	0.0309	0.0417	0.0869	0.1027	0.1225
ρ/(g/cm^3)	1.00456	1.00516	1.00609	1.00687	1.00766	1.01088	1.01196	1.01332
c_B/(mol/kg)	0.1477	0.1706	0.1821	0.2492	0.2844	0.3362	0.4156	0.5256
ρ/(g/cm^3)	1.01493	1.01634	1.01702	1.02071	1.02244	1.02472	1.02760	1.03063

w_C = 0.020 was kept constant

c_B/(mol/kg)	0.0000	0.0050	0.0088	0.0198	0.0308	0.0419	0.0629	0.0846
ρ/(g/cm^3)	1.01197	1.01231	1.01257	1.01333	1.01406	1.01481	1.01618	1.01758
c_B/(mol/kg)	0.1248	0.1411	0.1624	0.1879	0.2856			
ρ/(g/cm^3)	1.02003	1.02098	1.02218	1.02357	1.02811			

T/K = 303.15

w_C = 0.010 was kept constant

c_B/(mol/kg)	0.0000	0.0079	0.0204	0.0309	0.0417	0.0869	0.1027	0.1225
ρ/(g/cm^3)	1.00310	1.00368	1.00458	1.00532	1.00609	1.00917	1.01018	1.01147
c_B/(mol/kg)	0.1477	0.1706	0.1821	0.2492	0.2844	0.3362	0.4156	0.5256
ρ/(g/cm^3)	1.01299	1.01431	1.01494	1.01834	1.01991	1.02196	1.02450	1.02712

w_C = 0.020 was kept constant

c_B/(mol/kg)	0.0000	0.0050	0.0088	0.0198	0.0308	0.0419	0.0629	0.0846
ρ/(g/cm^3)	1.01046	1.01079	1.01104	1.01176	1.01247	1.01318	1.01449	1.01582
c_B/(mol/kg)	0.1248	0.1411	0.1624	0.1879	0.2856			
ρ/(g/cm^3)	1.01812	1.01900	1.02012	1.02141	1.02551			

T/K = 308.15

w_C = 0.010 was kept constant

c_B/(mol/kg)	0.0000	0.0079	0.0204	0.0309	0.0417	0.0869	0.1027	0.1225
ρ/(g/cm^3)	1.00144	1.00199	1.00286	1.00358	1.00431	1.00724	1.00821	1.00943
c_B/(mol/kg)	0.1477	0.1706	0.1821	0.2492	0.2844	0.3362	0.4156	0.5256
ρ/(g/cm^3)	1.01085	1.01208	1.01266	1.01577	1.01719	1.01901	1.02121	1.02342

continued

continued

w_C = 0.020 was kept constant

c_B/(mol/kg)	0.0000	0.0050	0.0088	0.0198	0.0308	0.0419	0.0629	0.0846
ρ/(g/cm^3)	1.00875	1.00906	1.00931	1.00100	1.01067	1.01135	1.01260	1.01386
c_B/(mol/kg)	0.1248	0.1411	0.1624	0.1879	0.2856			
ρ/(g/cm^3)	1.01602	1.01684	1.01787	1.01905	1.02270			

T/K = 313.15

w_C = 0.010 was kept constant

c_B/(mol/kg)	0.0000	0.0079	0.0204	0.0309	0.0417	0.0869	0.1027	0.1225
ρ/(g/cm^3)	0.99958	1.00011	1.00095	1.00164	1.00234	1.00514	1.00605	1.00720
c_B/(mol/kg)	0.1477	0.1706	0.1821	0.2492	0.2844	0.3362	0.4156	0.5256
ρ/(g/cm^3)	1.00852	1.00966	1.01019	1.01301	1.01427	1.01584	1.01768	1.01948

w_C = 0.020 was kept constant

c_B/(mol/kg)	0.0000	0.0050	0.0088	0.0198	0.0308	0.0419	0.0629	0.0846
ρ/(g/cm^3)	1.00685	1.00715	1.00739	1.00804	1.00870	1.00935	1.01053	1.01172
c_B/(mol/kg)	0.1248	0.1411	0.1624	0.1879	0.2856			
ρ/(g/cm^3)	1.01374	1.01449	1.01544	1.01650	1.01967			

Polymer (B): **poly(propylene glycol)** **2007SA1**
Characterization: M_n/g.mol^{-1} = 400, Fluka AG, Buchs, Switzerland
Solvent (A): **water** H_2O **7732-18-5**
Salt (C): **potassium citrate** $K_3C_6H_5O_7$ **866-84-2**

T/K = 288.15

w_B	0.0000	0.0015	0.0030	0.0060	0.0100	0.0199	0.0301	0.0401	0.0601
w_C	0.0206	0.0206	0.0206	0.0206	0.0206	0.0206	0.0206	0.0206	0.0206
ρ/(g/cm^3)	1.01233	1.01241	1.01253	1.01276	1.01306	1.01384	1.01465	1.01545	1.01709
w_B	0.0800	0.0998	0.1297	0.1583	0.1942	0.2468	0.2885	0.3482	0.0000
w_C	0.0206	0.0206	0.0206	0.0206	0.0206	0.0206	0.0206	0.0206	0.0397
ρ/(g/cm^3)	1.01876	1.02045	1.02305	1.02558	1.02878	1.03330	1.03661	1.04070	1.02470
w_B	0.0015	0.0030	0.0061	0.0101	0.0200	0.0299	0.0403	0.0602	0.0800
w_C	0.0397	0.0397	0.0397	0.0397	0.0397	0.0397	0.0397	0.0397	0.0397
ρ/(g/cm^3)	1.02477	1.02487	1.02508	1.02532	1.02599	1.02665	1.02737	1.02876	1.03019
w_B	0.1010	0.1300	0.1588	0.2006	0.2522	0.2983	0.3466		
w_C	0.0397	0.0397	0.0397	0.0397	0.0397	0.0397	0.0397		
ρ/(g/cm^3)	1.03173	1.03390	1.03606	1.03917	1.04280	1.04566	1.04803		

continued

continued

T/K = 293.15

w_B	0.0000	0.0015	0.0030	0.0060	0.0100	0.0199	0.0301	0.0401	0.0601
w_C	0.0206	0.0206	0.0206	0.0206	0.0206	0.0206	0.0206	0.0206	0.0206
ρ/(g/cm^3)	1.01132	1.01140	1.01151	1.01174	1.01202	1.01277	1.01355	1.01432	1.01587
w_B	0.0800	0.0998	0.1297	0.1583	0.1942	0.2468	0.2885	0.3482	0.0000
w_C	0.0206	0.0206	0.0206	0.0206	0.0206	0.0206	0.0206	0.0206	0.0397
ρ/(g/cm^3)	1.01747	1.01906	1.02151	1.02387	1.02683	1.03093	1.03389	1.03744	1.02359
w_B	0.0015	0.0030	0.0061	0.0101	0.0200	0.0299	0.0403	0.0602	0.0800
w_C	0.0397	0.0397	0.0397	0.0397	0.0397	0.0397	0.0397	0.0397	0.0397
ρ/(g/cm^3)	1.02366	1.02376	1.02396	1.02419	1.02483	1.02546	1.02614	1.02746	1.02880
w_B	0.1010	0.1300	0.1588	0.2006	0.2522	0.2983	0.3466		
w_C	0.0397	0.0397	0.0397	0.0397	0.0397	0.0397	0.0397		
ρ/(g/cm^3)	1.03024	1.03226	1.03425	1.03707	1.04027	1.04270	1.04463		

T/K = 298.15

w_B	0.0000	0.0015	0.0030	0.0060	0.0100	0.0199	0.0301	0.0401	0.0601
w_C	0.0206	0.0206	0.0206	0.0206	0.0206	0.0206	0.0206	0.0206	0.0206
ρ/(g/cm^3)	1.01007	1.01014	1.01025	1.01047	1.01074	1.01146	1.01221	1.01294	1.01443
w_B	0.0800	0.0998	0.1297	0.1583	0.1942	0.2468	0.2885	0.3482	0.0000
w_C	0.0206	0.0206	0.0206	0.0206	0.0206	0.0206	0.0206	0.0206	0.0397
ρ/(g/cm^3)	1.01594	1.01745	1.01974	1.02194	1.02466	1.02836	1.03096	1.03401	1.02226
w_B	0.0015	0.0030	0.0061	0.0101	0.0200	0.0299	0.0403	0.0602	0.0800
w_C	0.0397	0.0397	0.0397	0.0397	0.0397	0.0397	0.0397	0.0397	0.0397
ρ/(g/cm^3)	1.02233	1.02242	1.02261	1.02283	1.02344	1.02404	1.02469	1.02593	1.02720
w_B	0.1010	0.1300	0.1588	0.2006	0.2522	0.2983	0.3466		
w_C	0.0397	0.0397	0.0397	0.0397	0.0397	0.0397	0.0397		
ρ/(g/cm^3)	1.02854	1.03041	1.03223	1.03475	1.03752	1.03953	1.04103		

T/K = 303.15

w_B	0.0000	0.0015	0.0030	0.0060	0.0100	0.0199	0.0301	0.0401	0.0601
w_C	0.0206	0.0206	0.0206	0.0206	0.0206	0.0206	0.0206	0.0206	0.0206
ρ/(g/cm^3)	1.00859	1.00867	1.00877	1.00898	1.00924	1.00993	1.01065	1.01135	1.01277
w_B	0.0800	0.0998	0.1297	0.1583	0.1942	0.2468	0.2885	0.3482	0.0000
w_C	0.0206	0.0206	0.0206	0.0206	0.0206	0.0206	0.0206	0.0206	0.0397
ρ/(g/cm^3)	1.01420	1.01562	1.01777	1.01981	1.02229	1.02559	1.02784	1.03038	1.02071
w_B	0.0015	0.0030	0.0061	0.0101	0.0200	0.0299	0.0403	0.0602	0.0800
w_C	0.0397	0.0397	0.0397	0.0397	0.0397	0.0397	0.0397	0.0397	0.0397
ρ/(g/cm^3)	1.02078	1.02087	1.02105	1.02126	1.02184	1.02242	1.02303	1.02420	1.02539
w_B	0.1010	0.1300	0.1588	0.2006	0.2522	0.2983	0.3466		
w_C	0.0397	0.0397	0.0397	0.0397	0.0397	0.0397	0.0397		
ρ/(g/cm^3)	1.02664	1.02836	1.03002	1.03223	1.03456	1.03613	1.03717		

continued

continued

T/K = 308.15

w_B	0.0000	0.0015	0.0030	0.0060	0.0100	0.0199	0.0301	0.0401	0.0601
w_C	0.0206	0.0206	0.0206	0.0206	0.0206	0.0206	0.0206	0.0206	0.0206
$\rho/(g/cm^3)$	1.00691	1.00698	1.00708	1.00728	1.00753	1.00820	1.00889	1.00956	1.01091
w_B	0.0800	0.0998	0.1297	0.1583	0.1942	0.2468	0.2885	0.3482	0.0000
w_C	0.0206	0.0206	0.0206	0.0206	0.0206	0.0206	0.0206	0.0206	0.0397
$\rho/(g/cm^3)$	1.01226	1.01360	1.01560	1.01748	1.01972	1.02261	1.02450	1.02654	1.01898
w_B	0.0015	0.0030	0.0061	0.0101	0.0200	0.0299	0.0403	0.0602	0.0800
w_C	0.0397	0.0397	0.0397	0.0397	0.0397	0.0397	0.0397	0.0397	0.0397
$\rho/(g/cm^3)$	1.01904	1.01912	1.01929	1.01950	1.02005	1.02060	1.02117	1.02228	1.02339
w_B	0.1010	0.1300	0.1588	0.2006	0.2522	0.2983	0.3466		
w_C	0.0397	0.0397	0.0397	0.0397	0.0397	0.0397	0.0397		
$\rho/(g/cm^3)$	1.02455	1.02612	1.02758	1.02949	1.03131	1.03229	1.03251		

T/K = 313.15

w_B	0.0000	0.0015	0.0030	0.0060	0.0100	0.0199	0.0301	0.0401	0.0601
w_C	0.0206	0.0206	0.0206	0.0206	0.0206	0.0206	0.0206	0.0206	0.0206
$\rho/(g/cm^3)$	1.00504	1.00511	1.00520	1.00540	1.00564	1.00628	1.00694	1.00758	1.00886
w_B	0.0800	0.0998	0.1297	0.1583	0.1942	0.2468	0.2885	0.3482	0.0000
w_C	0.0206	0.0206	0.0206	0.0206	0.0206	0.0206	0.0206	0.0206	0.0397
$\rho/(g/cm^3)$	1.01014	1.01140	1.01325	1.01496	1.01695	1.01938	1.02086	1.02217	1.01706
w_B	0.0015	0.0030	0.0061	0.0101	0.0200	0.0299	0.0403	0.0602	0.0800
w_C	0.0397	0.0397	0.0397	0.0397	0.0397	0.0397	0.0397	0.0397	0.0397
$\rho/(g/cm^3)$	1.01711	1.01720	1.01736	1.01755	1.01808	1.01860	1.01915	1.02019	1.02122
w_B	0.1010	0.1300	0.1588	0.2006	0.2522				
w_C	0.0397	0.0397	0.0397	0.0397	0.0397				
$\rho/(g/cm^3)$	1.02228	1.02369	1.02494	1.02596	1.02652				

Polymer (B):	**poly(propylene glycol)**		**2007SA3**
Characterization:	M_n/g.mol^{-1} = 400, Fluka AG, Buchs, Switzerland		
Solvent (A):	**water**	H_2O	**7732-18-5**
Salt (C):	**potassium citrate**	$K_3C_6H_5O_7$	**866-84-2**

T/K = 288.15

c_B/(mol/kg)	0.000	0.000	0.000	0.000	0.000	0.000	0.000	0.000
c_C/(mol/kg)	0.0185	0.0304	0.0310	0.0558	0.0899	0.1258	0.1931	0.2653
$\rho/(g/cm^3)$	1.00299	1.00544	1.00555	1.01050	1.01735	1.02433	1.03725	1.05077
c_B/(mol/kg)	0.000	0.000	0.000	0.000	0.051	0.051	0.051	0.051
c_C/(mol/kg)	0.3412	0.4565	0.5796	0.7446	0.0308	0.0681	0.0957	0.1295
$\rho/(g/cm^3)$	1.06451	1.08471	1.10519	1.13173	1.00722	1.01458	1.01990	1.02638

continued

continued

c_B/(mol/kg)	0.051	0.051	0.051	0.051	0.051	0.104	0.104	0.104
c_C/(mol/kg)	0.1981	0.2651	0.3467	0.4719	0.5408	0.0261	0.0681	0.0991
ρ/(g/cm^3)	1.03922	1.05142	1.06583	1.08730	1.09868	1.00786	1.01597	1.02188
c_B/(mol/kg)	0.104	0.104	0.104	0.104	0.104	0.104		
c_C/(mol/kg)	0.1330	0.1968	0.2814	0.3545	0.4733	0.5805		
ρ/(g/cm^3)	1.02820	1.03991	1.05494	1.06767	1.08745	1.10464		
T/K = 293.15								
c_B/(mol/kg)	0.000	0.000	0.000	0.000	0.000	0.000	0.000	0.000
c_C/(mol/kg)	0.0185	0.0304	0.0310	0.0558	0.0899	0.1258	0.1931	0.2653
ρ/(g/cm^3)	1.00206	1.00450	1.00461	1.00951	1.01631	1.02323	1.03606	1.04948
c_B/(mol/kg)	0.000	0.000	0.000	0.000	0.051	0.051	0.051	0.051
c_C/(mol/kg)	0.3412	0.4565	0.5796	0.7446	0.0308	0.0681	0.0957	0.1295
ρ/(g/cm^3)	1.06313	1.08321	1.10358	1.12991	1.00621	1.01350	1.01879	1.02522
c_B/(mol/kg)	0.051	0.051	0.051	0.051	0.051	0.104	0.104	0.104
c_C/(mol/kg)	0.1981	0.2651	0.3467	0.4719	0.5408	0.0261	0.0681	0.0991
ρ/(g/cm^3)	1.03797	1.05008	1.06440	1.08575	1.09707	1.00679	1.01484	1.02071
c_B/(mol/kg)	0.104	0.104	0.104	0.104	0.104	0.104		
c_C/(mol/kg)	0.1330	0.1968	0.2814	0.3545	0.4733	0.5805		
ρ/(g/cm^3)	1.02698	1.03860	1.05353	1.06619	1.08585	1.10295		
T/K = 298.15								
c_B/(mol/kg)	0.000	0.000	0.000	0.000	0.000	0.000	0.000	0.000
c_C/(mol/kg)	0.0185	0.0304	0.0310	0.0558	0.0899	0.1258	0.1931	0.2653
ρ/(g/cm^3)	1.00087	1.00329	1.00340	1.00827	1.01502	1.02190	1.03465	1.04799
c_B/(mol/kg)	0.000	0.000	0.000	0.000	0.051	0.051	0.051	0.051
c_C/(mol/kg)	0.3412	0.4565	0.5796	0.7446	0.0308	0.0681	0.0957	0.1295
ρ/(g/cm^3)	1.06156	1.08154	1.10181	1.12811	1.00495	1.01219	1.01744	1.02383
c_B/(mol/kg)	0.051	0.051	0.051	0.051	0.051	0.104	0.104	0.104
c_C/(mol/kg)	0.1981	0.2651	0.3467	0.4719	0.5408	0.0261	0.0681	0.0991
ρ/(g/cm^3)	1.03650	1.04854	1.06278	1.08402	1.09529	1.00547	1.01347	1.01929
c_B/(mol/kg)	0.104	0.104	0.104	0.104	0.104	0.104		
c_C/(mol/kg)	0.1330	0.1968	0.2814	0.3545	0.4733	0.5805		
ρ/(g/cm^3)	1.02553	1.03708	1.05192	1.06451	1.08408	1.10109		
T/K = 303.15								
c_B/(mol/kg)	0.000	0.000	0.000	0.000	0.000	0.000	0.000	0.000
c_C/(mol/kg)	0.0185	0.0304	0.0310	0.0558	0.0899	0.1258	0.1931	0.2653
ρ/(g/cm^3)	0.99945	1.00186	1.00197	1.00681	1.01352	1.02035	1.03303	1.04631

continued

continued

c_B/(mol/kg)	0.000	0.000	0.000	0.000	0.051	0.051	0.051	0.051
c_C/(mol/kg)	0.3412	0.4565	0.5796	0.7446	0.0308	0.0681	0.0957	0.1295
ρ/(g/cm^3)	1.05982	1.07971	1.09990	1.12610	1.00346	1.01066	1.01588	1.02223
c_B/(mol/kg)	0.051	0.051	0.051	0.051	0.051	0.104	0.104	0.104
c_C/(mol/kg)	0.1981	0.2651	0.3467	0.4719	0.5408	0.0261	0.0681	0.0991
ρ/(g/cm^3)	1.03483	1.04682	1.06099	1.08214	1.09336	1.00392	1.01188	1.01767
c_B/(mol/kg)	0.104	0.104	0.104	0.104	0.104	0.104		
c_C/(mol/kg)	0.1330	0.1968	0.2814	0.3545	0.4733	0.5805		
ρ/(g/cm^3)	1.02387	1.03536	1.05014	1.06266	1.08215	1.09898		

T/K = 308.15

c_B/(mol/kg)	0.000	0.000	0.000	0.000	0.000	0.000	0.000	0.000
c_C/(mol/kg)	0.0185	0.0304	0.0310	0.0558	0.0899	0.1258	0.1931	0.2653
ρ/(g/cm^3)	0.99781	1.00021	1.00032	1.00514	1.01181	1.01862	1.03124	1.04446
c_B/(mol/kg)	0.000	0.000	0.000	0.000	0.051	0.051	0.051	0.051
c_C/(mol/kg)	0.3412	0.4565	0.5796	0.7446	0.0308	0.0681	0.0957	0.1295
ρ/(g/cm^3)	1.05791	1.07772	1.09785	1.12397	1.00176	1.00892	1.01411	1.02044
c_B/(mol/kg)	0.051	0.051	0.051	0.051	0.051	0.104	0.104	0.104
c_C/(mol/kg)	0.1981	0.2651	0.3467	0.4719	0.5408	0.0261	0.0681	0.0991
ρ/(g/cm^3)	1.03298	1.04492	1.05904	1.08011	1.09129	1.00217	1.01009	1.01585
c_B/(mol/kg)	0.104	0.104	0.104	0.104	0.104	0.104		
c_C/(mol/kg)	0.1330	0.1968	0.2814	0.3545	0.4733	0.5805		
ρ/(g/cm^3)	1.02203	1.03347	1.04818	1.06065	1.08000	1.09654		

T/K = 313.15

c_B/(mol/kg)	0.000	0.000	0.000	0.000	0.000	0.000	0.000	0.000
c_C/(mol/kg)	0.0185	0.0304	0.0310	0.0558	0.0899	0.1258	0.1931	0.2653
ρ/(g/cm^3)	0.99598	0.99837	0.99848	1.00328	1.00992	1.01670	1.02928	1.04245
c_B/(mol/kg)	0.000	0.000	0.000	0.000	0.051	0.051	0.051	0.051
c_C/(mol/kg)	0.3412	0.4565	0.5796	0.7446	0.0308	0.0681	0.0957	0.1295
ρ/(g/cm^3)	1.05586	1.07560	1.09567	1.12172	0.99986	1.00699	1.01217	1.01847
c_B/(mol/kg)	0.051	0.051	0.051	0.051	0.051	0.104	0.104	0.104
c_C/(mol/kg)	0.1981	0.2651	0.3467	0.4719	0.5408	0.0261	0.0681	0.0991
ρ/(g/cm^3)	1.03097	1.04286	1.05693	1.07793	1.08900	1.00023	1.00811	1.01385
c_B/(mol/kg)	0.104	0.104	0.104	0.104	0.104	0.104		
c_C/(mol/kg)	0.1330	0.1968	0.2814	0.3545	0.4733	0.5805		
ρ/(g/cm^3)	1.02000	1.03140	1.04606	1.05848	1.07749	1.09409		

Polymer (B):	**poly(propylene glycol)**		**2009SAD**
Characterization:	M_n/g.mol^{-1} = 400, Fluka AG, Buchs, Switzerland		
Solvent (A):	**water**	H_2O	**7732-18-5**
Salt (C):	**sodium phosphate**	Na_3PO_4	**7601-54-9**

T/K = 288.15

w_C = 0.010 was kept constant

c_B/(mol/kg)	0.0000	0.0104	0.0204	0.0307	0.0413	0.0626	0.0861	0.1116
ρ/(g/cm^3)	1.01107	1.01190	1.01267	1.01347	1.01427	1.01587	1.01760	1.01942
c_B/(mol/kg)	0.1486	0.1870	0.3376	0.4249				
ρ/(g/cm^3)	1.02196	1.02447	1.03276	1.03637				

w_C = 0.020 was kept constant

c_B/(mol/kg)	0.0000	0.0204	0.0308	0.0413	0.0636	0.0846	0.1070	0.1430
ρ/(g/cm^3)	1.02232	1.02377	1.02450	1.02517	1.02662	1.02796	1.02935	1.03153
c_B/(mol/kg)	0.1873	0.2496	0.2795					
ρ/(g/cm^3)	1.03401	1.03718	1.03857					

T/K = 293.15

w_C = 0.010 was kept constant

c_B/(mol/kg)	0.0000	0.0104	0.0204	0.0307	0.0413	0.0626	0.0861	0.1116
ρ/(g/cm^3)	1.01002	1.01082	1.01156	1.01233	1.01310	1.01462	1.01627	1.01800
c_B/(mol/kg)	0.1486	0.1870	0.3376	0.4249				
ρ/(g/cm^3)	1.02039	1.02274	1.03032	1.03354				

w_C = 0.020 was kept constant

c_B/(mol/kg)	0.0000	0.0204	0.0308	0.0413	0.0636	0.0846	0.1070	0.1430
ρ/(g/cm^3)	1.02114	1.02254	1.02321	1.02388	1.02525	1.02652	1.02783	1.02986
c_B/(mol/kg)	0.1873	0.2496	0.2795					
ρ/(g/cm^3)	1.03216	1.03504	1.03629					

T/K = 298.15

w_C = 0.010 was kept constant

c_B/(mol/kg)	0.0000	0.0104	0.0204	0.0307	0.0413	0.0626	0.0861	0.1116
ρ/(g/cm^3)	1.00873	1.00950	1.01022	1.01095	1.01169	1.01314	1.01471	1.01635
c_B/(mol/kg)	0.1486	0.1870	0.3376	0.4249				
ρ/(g/cm^3)	1.01859	1.02078	1.02769	1.03051				

w_C = 0.020 was kept constant

c_B/(mol/kg)	0.0000	0.0204	0.0308	0.0413	0.0636	0.0846	0.1070	0.1430
ρ/(g/cm^3)	1.01974	1.02109	1.02173	1.02236	1.02367	1.02487	1.02610	1.02799

continued

continued

c_B/(mol/kg)	0.1873	0.2496	0.2795
ρ/(g/cm^3)	1.03010	1.03270	1.03380

T/K = 303.15

w_C = 0.010 was kept constant

c_B/(mol/kg)	0.0000	0.0104	0.0204	0.0307	0.0413	0.0626	0.0861	0.1116
ρ/(g/cm^3)	1.00722	1.00796	1.00865	1.00936	1.01006	1.01145	1.01295	1.01449
c_B/(mol/kg)	0.1486	0.1870	0.3376	0.4249				
ρ/(g/cm^3)	1.01660	1.01863	1.02485	1.02729				

w_C = 0.020 was kept constant

c_B/(mol/kg)	0.0000	0.0204	0.0308	0.0413	0.0636	0.0846	0.1070	0.1430
ρ/(g/cm^3)	1.01814	1.01943	1.02004	1.02065	1.02189	1.02302	1.02417	1.02592
c_B/(mol/kg)	0.1873	0.2496	0.2795					
ρ/(g/cm^3)	1.02784	1.03014	1.03110					

T/K = 308.15

w_C = 0.010 was kept constant

c_B/(mol/kg)	0.0000	0.0104	0.0204	0.0307	0.0413	0.0626	0.0861	0.1116
ρ/(g/cm^3)	1.00551	1.00622	1.00689	1.00757	1.00824	1.00956	1.01098	1.01244
c_B/(mol/kg)	0.1486	0.1870	0.3376	0.4249				
ρ/(g/cm^3)	1.01441	1.01628	1.02182	1.02385				

w_C = 0.020 was kept constant

c_B/(mol/kg)	0.0000	0.0204	0.0308	0.0413	0.0636	0.0846	0.1070	0.1430
ρ/(g/cm^3)	1.01634	1.01759	1.01817	1.01875	1.01992	1.02098	1.02205	1.02366
c_B/(mol/kg)	0.1873	0.2496	0.2795					
ρ/(g/cm^3)	1.02539	1.02738	1.02816					

T/K = 313.15

w_C = 0.010 was kept constant

c_B/(mol/kg)	0.0000	0.0104	0.0204	0.0307	0.0413	0.0626	0.0861	0.1116
ρ/(g/cm^3)	1.00361	1.00429	1.00494	1.00559	1.00623	1.00749	1.00884	1.01021
c_B/(mol/kg)	0.1486	0.1870	0.3376	0.4249				
ρ/(g/cm^3)	1.01203	1.01374	1.01851	1.02009				

w_C = 0.020 was kept constant

c_B/(mol/kg)	0.0000	0.0204	0.0308	0.0413	0.0636	0.0846	0.1070	0.1430
ρ/(g/cm^3)	1.01437	1.01558	1.01613	1.01668	1.01779	1.01877	1.01977	1.02122
c_B/(mol/kg)	0.1873	0.2496	0.2795					
ρ/(g/cm^3)	1.02272	1.02383	1.02414					

Polymer (B): **poly(propylene glycol)** **2005ELI**
Characterization: M_n/g.mol^{-1} = 425, Merck, Darmstadt, Germany
Solvent (A): **water** **H_2O** **7732-18-5**

T/K = 298.15

w_B	0.0982	0.1995	0.3002	0.4001	0.4992	0.5914	0.6975	0.8004	0.9005
ρ/(g/cm^3)	1.0170	1.0253	1.0312	1.0391	1.0380	1.0419	1.0426	1.0347	1.0311
w_B	1.0000								
ρ/(g/cm^3)	1.0193								

T/K = 308.15

w_B	0.0982	0.1995	0.3002	0.4001	0.4992	0.5914	0.6975	0.8004	0.9005
ρ/(g/cm^3)	1.0135	1.0216	1.0277	1.0341	1.0339	1.0373	1.0373	1.0262	1.0255
w_B	1.0000								
ρ/(g/cm^3)	1.0125								

T/K = 318.15

w_B	0.6975	0.8004	0.9005	1.000
ρ/(g/cm^3)	1.0279	1.0198	1.0189	1.0054

T/K = 328.15

w_B	0.8004	0.9005	1.000
ρ/(g/cm^3)	1.0102	1.0109	0.9962

T/K = 338.15

w_B	0.9005	1.000
ρ/(g/cm^3)	1.0020	0.9881

Polymer (B): **poly(propylene oxide-*b*-ethylene oxide-*b*-propylene oxide)** **2006DE1**
Characterization: M_n/g.mol^{-1} = 1950, $PO_8EO_{23}PO_8$, 41.0 mol% propylene oxide, 10R5, BASF AG, Ludwigshafen, Germany
Solvent (A): **water** **H_2O** **7732-18-5**

T/K = 298.15

c_B/(mol/kg)	0.00000	0.00517	0.01591	0.04077	0.06899	0.1007	0.1230	0.1656
ρ/(g/cm^3)	0.99705	0.99835	1.00101	1.00684	1.01297	1.01928	1.02337	1.03045
c_B/(mol/kg)	0.2315	0.2938	0.3636	0.3937	0.5136	0.5503	0.6965	0.7908
ρ/(g/cm^3)	1.03967	1.04667	1.05283	1.05499	1.06150	1.06294	1.06702	1.06862
c_B/(mol/kg)	0.8861	1.0742	1.3691	1.7991				
ρ/(g/cm^3)	1.06979	1.07110	1.07162	1.07089				

continued

continued

T/K = 308.15

c_B/(mol/kg)	0.00000	0.00517	0.01591	0.04077	0.06899	0.1230	0.1656	0.2315
ρ/(g/cm^3)	0.99404	0.99528	0.99783	1.00332	1.00903	1.01886	1.02543	1.03378
c_B/(mol/kg)	0.2938	0.3636	0.3937	0.5136	0.5503	0.6965	0.7908	0.8861
ρ/(g/cm^3)	1.03982	1.04574	1.04723	1.05308	1.05448	1.05819	1.05983	1.06092
c_B/(mol/kg)	1.0742	1.3691	1.7991					
ρ/(g/cm^3)	1.06227	1.06305	1.06296					

Polymer (B): **poly(sodium acrylate)** **2010MIN**
Characterization: M_w/g.mol^{-1} = 8000, Aldrich Chem. Co., Inc., Milwaukee, WI
Solvent (A): **water** H_2O **7732-18-5**

T/K = 288.15

w_B	0.04	0.06	0.08	0.10	0.12
ρ/(kg/m^3)	1037.1	1053.55	1073.64	1090.45	1109.17

T/K = 293.15

w_B	0.04	0.06	0.08	0.10	0.12
ρ/(kg/m^3)	1035.1	1052.69	1068.97	1086.96	1104.37

T/K = 298.15

w_B	0.04	0.06	0.08	0.10	0.12
ρ/(kg/m^3)	1033.5	1050.40	1067.65	1084.60	1100.82

T/K = 303.15

w_B	0.04	0.06	0.08	0.10	0.12
ρ/(kg/m^3)	1030.6	1042.69	1064.33	1081.73	1098.81

T/K = 308.15

w_B	0.04	0.06	0.08	0.10	0.12
ρ/(kg/m^3)	1029.1	1045.79	1063.19	1079.82	1095.94

T/K = 313.15

w_B	0.04	0.06	0.08	0.10	0.12
ρ/(kg/m^3)	1027.0	1043.82	1060.92	1077.18	1094.57

T/K = 318.15

w_B	0.04	0.06	0.08	0.10	0.12
ρ/(kg/m^3)	1026.3	1042.02	1058.25	1077.26	1092.69

Polymer (B): **poly(vinyl alcohol)** **2005ELI**
Characterization: M_n/g.mol^{-1} = 15000, Merck, Darmstadt, Germany
Solvent (A): **water** H_2O **7732-18-5**

T/K = 298.15

w_B	0.0982	0.1995	0.3002	0.4001	0.4992	0.5914	0.6975	0.8004	0.9005
ρ/(g/cm^3)	1.0170	1.0253	1.0312	1.0391	1.0380	1.0419	1.0426	1.0347	1.0311
w_B	1.0000								
ρ/(g/cm^3)	1.0193								

T/K = 308.15

w_B	0.0982	0.1995	0.3002	0.4001	0.4992	0.5914	0.6975	0.8004	0.9005
ρ/(g/cm^3)	1.0135	1.0216	1.0277	1.0341	1.0339	1.0373	1.0373	1.0262	1.0255
w_B	1.0000								
ρ/(g/cm^3)	1.0125								

T/K = 318.15

w_B	0.6975	0.8004	0.9005	1.0000
ρ/(g/cm^3)	1.0279	1.0198	1.0189	1.0054

T/K = 328.15

w_B	0.8004	0.9005	1.0000
ρ/(g/cm^3)	1.0102	1.0109	0.9962

T/K = 338.15

w_B	0.9005	1.0000
ρ/(g/cm^3)	1.0020	0.9881

Polymer (B): **poly(vinyl alcohol)** **2010SAL**
Characterization: M_n/g.mol^{-1} = 67000, Fluka AG, Buchs, Switzerland
Solvent (A): **water** H_2O **7732-18-5**

T/K = 298.15

w_B	0.000	0.005	0.010	0.015	0.020	0.025	0.030	0.035	0.040
ρ/(g/cm^3)	0.99704	0.99890	0.99931	1.00047	1.00167	1.00275	1.00338	1.00521	1.00642
w_B	0.045	0.050							
ρ/(g/cm^3)	1.00781	1.00898							

T/K = 308.15

w_B	0.000	0.005	0.010	0.015	0.020	0.025	0.030	0.035	0.040
ρ/(g/cm^3)	0.99406	0.99507	0.99633	0.99741	0.99871	0.99993	1.00109	1.00227	1.00348
w_B	0.045	0.050							
ρ/(g/cm^3)	1.00471	1.00601							

continued

continued

T/K = 318.15

w_B	0.000	0.005	0.010	0.015	0.020	0.025	0.030	0.035	0.040
$\rho/(g/cm^3)$	0.99024	0.99121	0.99232	0.99343	0.99454	0.99562	0.99681	0.99786	0.99917

w_B	0.045	0.050
$\rho/(g/cm^3)$	1.00038	1.00161

Polymer (B): **poly(vinyl alcohol)** **2010MOL**
Characterization: M_w/g.mol^{-1} = 105000, Aldrich Chem. Co., Inc., Milwaukee, WI
Solvent (A): **water** **H_2O** **7732-18-5**

T/K = 298.15

c_B/(mol/kg)	0.0000	0.0100	0.0200	0.0300	0.0400	0.0500	0.0598	0.0701
$\rho/(g/cm^3)$	0.99704	0.99719	0.99729	0.99740	0.99750	0.99761	0.99771	0.99782
c_B/(mol/kg)	0.0796	0.0901	0.1001	0.1099	0.1373	0.1598	0.1804	0.1973
$\rho/(g/cm^3)$	0.99792	0.99804	0.99814	0.99824	0.99853	0.99877	0.99899	0.99916

Comments: Molality is calculated using 44.053 g/mol for the repetitive unit.

Polymer (B): **poly(vinyl alcohol)** **2004HON**
Characterization: M_n/g.mol^{-1} = 125100, 99.8 % hydrolysis, Aldrich Chem. Co., Inc., Milwaukee, WI
Solvent (A): **water** **H_2O** **7732-18-5**
Solvent (C): **dimethylsulfoxide** **C_2H_6OS** **67-68-5**

T/K = 298.15

Comments: The polymer concentration is c_B = 0.01 g/cm^3 in the solution. The mole fraction of DMSO, i.e., x_C of solvent (C), in the table is given on the basis of the solvent mixture without the polymer.

x_C	0.0000	0.0274	0.0596	0.0980	0.1446	0.2023	0.2756	0.3202	0.3299
$\rho/(g/cm^3)$	1.0010	1.0154	1.0306	1.0458	1.0612	1.0767	1.0897	1.0943	1.0952

x_C	0.3502	0.3717	0.5035	0.6953	1.0000
$\rho/(g/cm^3)$	1.0963	1.0984	1.1028	1.1032	1.1005

Polymer (B): **poly(vinyl alcohol)** **2010SAL**
Characterization: M_n/g.mol^{-1} = 67000, Fluka AG, Buchs, Switzerland
Solvent (A): **water** **H_2O** **7732-18-5**
Polymer (C): **poly(ethylene glycol)**
Characterization: M_n/g.mol^{-1} = 3000, Merck, Darmstadt, Germany

T/K = 298.15

continued

continued

w_B/w_C = 75/25

w_{B+C}	0.005	0.010	0.015	0.020	0.025	0.030	0.035	0.040	0.045
ρ/(g/cm^3)	0.99796	0.99901	1.00001	1.00089	1.00189	1.00304	1.00398	1.00518	1.00615

w_{B+C}	0.050
ρ/(g/cm^3)	1.00736

w_B/w_C = 50/50

w_{B+C}	0.005	0.010	0.015	0.020	0.025	0.030	0.035	0.040	0.045
ρ/(g/cm^3)	0.99794	0.99890	0.99977	1.00083	1.00171	1.00281	1.00378	1.00475	1.00577

w_{B+C}	0.050
ρ/(g/cm^3)	1.00663

w_B/w_C = 25/75

w_{B+C}	0.010	0.015	0.020	0.025	0.030	0.035	0.040	0.045	0.050
ρ/(g/cm^3)	0.99788	0.99873	0.99959	1.00046	1.00138	1.00231	1.00319	1.00407	1.00505

T/K = 308.15

w_B/w_C = 75/25

w_{B+C}	0.005	0.010	0.015	0.020	0.025	0.030	0.035	0.040	0.045
ρ/(g/cm^3)	0.99514	0.99621	0.99725	0.99828	0.99927	1.00033	1.00151	1.00255	1.00354

w_{B+C}	0.050
ρ/(g/cm^3)	1.00501

w_B/w_C = 50/50

w_{B+C}	0.005	0.010	0.015	0.020	0.025	0.030	0.035	0.040	0.045
ρ/(g/cm^3)	0.99540	0.99619	0.99686	0.99814	0.99880	0.99992	1.00102	1.00192	1.00282

w_{B+C}	0.050
ρ/(g/cm^3)	1.00372

w_B/w_C = 25/75

w_{B+C}	0.010	0.015	0.020	0.025	0.030	0.035	0.040	0.045	0.050
ρ/(g/cm^3)	0.99529	0.99602	0.99687	0.99784	0.99863	0.99951	1.00036	1.00112	1.00206

T/K = 318.15

w_B/w_C = 75/25

w_{B+C}	0.005	0.010	0.015	0.020	0.025	0.030	0.035	0.040	0.045
ρ/(g/cm^3)	0.99112	0.99232	0.99327	0.99428	0.99512	0.99617	0.99717	0.99799	0.99891

w_{B+C}	0.050
ρ/(g/cm^3)	1.00003

continued

continued

w_B/w_C = 50/50

w_{B+C}	0.005	0.010	0.015	0.020	0.025	0.030	0.035	0.040	0.045
ρ/(g/cm^3)	0.99144	0.99221	0.99308	0.99410	0.99499	0.99589	0.99688	0.99771	0.99869

w_{B+C}	0.050
ρ/(g/cm^3)	0.99951

w_B/w_C = 25/75

w_{B+C}	0.010	0.015	0.020	0.025	0.030	0.035	0.040	0.045	0.050
ρ/(g/cm^3)	0.99108	0.99192	0.99287	0.99366	0.99457	0.99535	0.99624	0.99713	0.99790

Polymer (B): **poly(vinyl alcohol)** **2010PAN**
Characterization: M_η/g.mol^{-1} = 68800, Shanghai Chemical Reagent Inc., China
Solvent (A): **water** H_2O **7732-18-5**
Component (C): **nicotinic acid** $C_6H_5NO_2$ **59-67-6**

T/K = 293.15

c_B/(g/dm^3)	0.5007	0.5007	0.5007	0.5007	0.5007	0.5007	0.5007	0.5007
c_C/(mol/kg)	0.00000	0.00410	0.00816	0.01630	0.03263	0.04907	0.06552	0.08209
ρ/(g/cm^3)	0.99829	0.99845	0.99862	0.99897	0.99975	1.00038	1.00107	1.00171
c_B/(g/dm^3)	0.5007	0.5007	0.5007	1.0006	1.0006	1.0006	1.0006	1.0006
c_C/(mol/kg)	0.09840	0.1154	0.1320	0.00000	0.00411	0.00818	0.01625	0.03267
ρ/(g/cm^3)	1.00239	1.00307	1.00378	0.99843	0.99863	0.99882	0.99923	0.99987
c_B/(g/dm^3)	1.0006	1.0006	1.0006	1.0006	1.0006	2.0003	2.0003	2.0003
c_C/(mol/kg)	0.04906	0.06540	0.08203	0.09870	0.1154	0.00000	0.00408	0.008164
ρ/(g/cm^3)	1.00055	1.00121	1.00193	1.00253	1.00322	0.99867	0.99881	0.99900
c_B/(g/dm^3)	2.0003	2.0003	2.0003	2.0003	2.0003	2.0003	2.0003	4.0005
c_C/(mol/kg)	0.01632	0.03264	0.04908	0.06554	0.08211	0.09871	0.1154	0.00000
ρ/(g/cm^3)	0.99929	1.00001	1.00068	1.00130	1.00203	1.00270	1.00340	0.99917
c_B/(g/dm^3)	4.0005	4.0005	4.0005	4.0005	4.0005	4.0005	4.0005	4.0005
c_C/(mol/kg)	0.004072	0.008164	0.01624	0.03256	0.04908	0.06553	0.08208	0.09867
ρ/(g/cm^3)	0.99928	0.99943	0.99974	1.00039	1.00110	1.00178	1.00251	1.00308
c_B/(g/dm^3)	4.0005	6.0004	6.0004	6.0004	6.0004	6.0004	6.0004	6.0004
c_C/(mol/kg)	0.11536	0.00000	0.00407	0.00817	0.01630	0.03266	0.04906	0.06515
ρ/(g/cm^3)	1.00385	0.99968	0.99978	0.99995	1.00025	1.00089	1.00164	1.00225
c_B/(g/dm^3)	6.0004	6.0004	6.0004					
c_C/(mol/kg)	0.08208	0.09868	0.1154					
ρ/(g/cm^3)	1.00289	1.00358	1.00438					

continued

continued

T/K = 298.15

c_B/(g/dm^3)	0.5007	0.5007	0.5007	0.5007	0.5007	0.5007	0.5007	0.5007
c_C/(mol/kg)	0.00000	0.00410	0.00816	0.01630	0.03263	0.04907	0.06552	0.08209
ρ/(g/cm^3)	0.99714	0.99729	0.99745	0.99775	0.99851	0.99915	0.99987	1.00054
c_B/(g/dm^3)	0.5007	0.5007	0.5007	1.0006	1.0006	1.0006	1.0006	1.0006
c_C/(mol/kg)	0.09840	0.1154	0.1320	0.00000	0.00411	0.00818	0.01625	0.03267
ρ/(g/cm^3)	1.00116	1.00191	1.00249	0.99726	0.99753	0.99767	0.99801	0.99868
c_B/(g/dm^3)	1.0006	1.0006	1.0006	1.0006	1.0006	2.0003	2.0003	2.0003
c_C/(mol/kg)	0.04906	0.06540	0.08203	0.09870	0.1154	0.00000	0.00408	0.008164
ρ/(g/cm^3)	0.99940	1.00008	1.00067	1.00135	1.00205	0.99751	0.99759	0.99785
c_B/(g/dm^3)	2.0003	2.0003	2.0003	2.0003	2.0003	2.0003	2.0003	4.0005
c_C/(mol/kg)	0.01632	0.03264	0.04908	0.06554	0.08211	0.09871	0.1154	0.00000
ρ/(g/cm^3)	0.99817	0.99887	0.99944	1.00013	1.00084	1.00146	1.00225	0.99799
c_B/(g/dm^3)	4.0005	4.0005	4.0005	4.0005	4.0005	4.0005	4.0005	4.0005
c_C/(mol/kg)	0.004072	0.008164	0.01624	0.03256	0.04908	0.06553	0.08208	0.09867
ρ/(g/cm^3)	0.99809	0.99828	0.99854	0.99920	0.99991	1.00056	1.00132	1.00191
c_B/(g/dm^3)	4.0005	6.0004	6.0004	6.0004	6.0004	6.0004	6.0004	6.0004
c_C/(mol/kg)	0.11536	0.00000	0.00407	0.00817	0.01630	0.03266	0.04906	0.06515
ρ/(g/cm^3)	1.00264	0.99849	0.99862	0.99878	0.99906	0.99969	1.00047	1.00105
c_B/(g/dm^3)	6.0004	6.0004	6.0004					
c_C/(mol/kg)	0.08208	0.09868	0.1154					
ρ/(g/cm^3)	1.00171	1.00237	1.00318					

T/K = 303.15

c_B/(g/dm^3)	0.5007	0.5007	0.5007	0.5007	0.5007	0.5007	0.5007	0.5007
c_C/(mol/kg)	0.00000	0.00410	0.00816	0.01630	0.03263	0.04907	0.06552	0.08209
ρ/(g/cm^3)	0.99573	0.99591	0.99609	0.99639	0.99714	0.99778	0.99844	0.99909
c_B/(g/dm^3)	0.5007	0.5007	0.5007	1.0006	1.0006	1.0006	1.0006	1.0006
c_C/(mol/kg)	0.09840	0.1154	0.1320	0.00000	0.00411	0.00818	0.01625	0.03267
ρ/(g/cm^3)	0.99979	1.00049	1.00102	0.99587	0.99613	0.99628	0.99662	0.99731
c_B/(g/dm^3)	1.0006	1.0006	1.0006	1.0006	1.0006	2.0003	2.0003	2.0003
c_C/(mol/kg)	0.04906	0.06540	0.08203	0.09870	0.1154	0.00000	0.00408	0.008164
ρ/(g/cm^3)	0.99795	0.99865	0.99928	0.99990	1.00060	0.99608	0.99618	0.99642
c_B/(g/dm^3)	2.0003	2.0003	2.0003	2.0003	2.0003	2.0003	2.0003	4.0005
c_C/(mol/kg)	0.01632	0.03264	0.04908	0.06554	0.08211	0.09871	0.1154	0.00000
ρ/(g/cm^3)	0.99677	0.99741	0.99805	0.99868	0.99943	1.00008	1.00070	0.99656
c_B/(g/dm^3)	4.0005	4.0005	4.0005	4.0005	4.0005	4.0005	4.0005	4.0005
c_C/(mol/kg)	0.004072	0.008164	0.01624	0.03256	0.04908	0.06553	0.08208	0.09867
ρ/(g/cm^3)	0.99663	0.99690	0.99709	0.99782	0.99853	0.99917	0.99986	1.00042

continued

continued

c_B/(g/dm^3)	4.0005	6.0004	6.0004	6.0004	6.0004	6.0004	6.0004	6.0004
c_C/(mol/kg)	0.11536	0.00000	0.00407	0.00817	0.01630	0.03266	0.04906	0.06515
ρ/(g/cm^3)	1.00118	0.99704	0.99722	0.99741	0.99771	0.99832	0.99905	0.99964
c_B/(g/dm^3)	6.0004	6.0004	6.0004					
c_C/(mol/kg)	0.08208	0.09868	0.1154					
ρ/(g/cm^3)	1.00031	1.00110	1.00162					

T/K = 308.15

c_B/(g/dm^3)	0.5007	0.5007	0.5007	0.5007	0.5007	0.5007	0.5007	0.5007
c_C/(mol/kg)	0.00000	0.00410	0.00816	0.01630	0.03263	0.04907	0.06552	0.08209
ρ/(g/cm^3)	0.99412	0.99432	0.99451	0.99477	0.99545	0.99611	0.99682	0.99747
c_B/(g/dm^3)	0.5007	0.5007	0.5007	1.0006	1.0006	1.0006	1.0006	1.0006
c_C/(mol/kg)	0.09840	0.1154	0.1320	0.00000	0.00411	0.00818	0.01625	0.03267
ρ/(g/cm^3)	0.99808	0.99878	0.99940	0.99433	0.99455	0.99473	0.99503	0.99573
c_B/(g/dm^3)	1.0006	1.0006	1.0006	1.0006	1.0006	2.0003	2.0003	2.0003
c_C/(mol/kg)	0.04906	0.06540	0.08203	0.09870	0.1154	0.00000	0.00408	0.008164
ρ/(g/cm^3)	0.99640	0.99703	0.99765	0.99832	0.99898	0.99447	0.99455	0.99481
c_B/(g/dm^3)	2.0003	2.0003	2.0003	2.0003	2.0003	2.0003	2.0003	4.0005
c_C/(mol/kg)	0.01632	0.03264	0.04908	0.06554	0.08211	0.09871	0.1154	0.00000
ρ/(g/cm^3)	0.99518	0.99580	0.99644	0.99710	0.99778	0.99838	0.99904	0.99493
c_B/(g/dm^3)	4.0005	4.0005	4.0005	4.0005	4.0005	4.0005	4.0005	4.0005
c_C/(mol/kg)	0.004072	0.008164	0.01624	0.03256	0.04908	0.06553	0.08208	0.09867
ρ/(g/cm^3)	0.99502	0.99531	0.99559	0.99617	0.99692	0.99754	0.99824	0.99877
c_B/(g/dm^3)	4.0005	6.0004	6.0004	6.0004	6.0004	6.0004	6.0004	6.0004
c_C/(mol/kg)	0.11536	0.00000	0.00407	0.00817	0.01630	0.03266	0.04906	0.06515
ρ/(g/cm^3)	0.99951	0.99540	0.99563	0.99580	0.99611	0.99670	0.99741	0.99800
c_B/(g/dm^3)	6.0004	6.0004	6.0004					
c_C/(mol/kg)	0.08208	0.09868	0.1154					
ρ/(g/cm^3)	0.99868	0.99932	0.99999					

T/K = 313.15

c_B/(g/dm^3)	0.5007	0.5007	0.5007	0.5007	0.5007	0.5007	0.5007	0.5007
c_C/(mol/kg)	0.00000	0.00410	0.00816	0.01630	0.03263	0.04907	0.06552	0.08209
ρ/(g/cm^3)	0.99235	0.99255	0.99271	0.99305	0.99371	0.99434	0.99498	0.99556
c_B/(g/dm^3)	0.5007	0.5007	0.5007	1.0006	1.0006	1.0006	1.0006	1.0006
c_C/(mol/kg)	0.09840	0.1154	0.1320	0.00000	0.00411	0.00818	0.01625	0.03267
ρ/(g/cm^3)	0.99628	0.99695	0.99761	0.99244	0.99271	0.99286	0.99312	0.99381
c_B/(g/dm^3)	1.0006	1.0006	1.0006	1.0006	1.0006	2.0003	2.0003	2.0003
c_C/(mol/kg)	0.04906	0.06540	0.08203	0.09870	0.1154	0.00000	0.00408	0.008164
ρ/(g/cm^3)	0.99451	0.99513	0.99578	0.99636	0.99703	0.99266	0.99278	0.99301

continued

continued

c_B/(g/dm^3)	2.0003	2.0003	2.0003	2.0003	2.0003	2.0003	2.0003	4.0005
c_C/(mol/kg)	0.01632	0.03264	0.04908	0.06554	0.08211	0.09871	0.1154	0.00000
ρ/(g/cm^3)	0.99331	0.99403	0.99459	0.99522	0.99593	0.99655	0.99718	0.99312
c_B/(g/dm^3)	4.0005	4.0005	4.0005	4.0005	4.0005	4.0005	4.0005	4.0005
c_C/(mol/kg)	0.004072	0.008164	0.01624	0.03256	0.04908	0.06553	0.08208	0.09867
ρ/(g/cm^3)	0.99330	0.99348	0.99381	0.99440	0.99512	0.99567	0.99634	0.99703
c_B/(g/dm^3)	4.0005	6.0004	6.0004	6.0004	6.0004	6.0004	6.0004	6.0004
c_C/(mol/kg)	0.11536	0.00000	0.00407	0.00817	0.01630	0.03266	0.04906	0.06515
ρ/(g/cm^3)	0.99765	0.99358	0.99382	0.99402	0.99429	0.99487	0.99558	0.99618
c_B/(g/dm^3)	6.0004	6.0004	6.0004					
c_C/(mol/kg)	0.08208	0.09868	0.1154					
ρ/(g/cm^3)	0.99682	0.99748	0.99812					

Polymer (B): **poly(1-vinyl-2-pyrrolidinone)** **2004SAD**
Characterization: M_w/g.mol^{-1} = 10000, Aldrich Chem. Co., Inc., Milwaukee, WI
Solvent (A): **water** **H_2O** **7732-18-5**

T/K = 298.15

w_B	0.0027	0.0055	0.0094	0.0130	0.0157	0.0208	0.0262	0.0350	0.0424
ρ/(g/cm^3)	0.99758	0.99816	0.99895	0.99967	1.00024	1.00129	1.00240	1.00421	1.00574
w_B	0.0532	0.0621	0.0735	0.0940	0.1147	0.1368	0.1806	0.2255	0.2636
ρ/(g/cm^3)	1.00800	1.00987	1.01226	1.01661	1.02098	1.02572	1.03534	1.04547	1.05426
w_B	0.3168	0.3712	0.4105	0.4649					
ρ/(g/cm^3)	1.06683	1.07997	1.08991	1.10383					

T/K = 308.15

w_B	0.0027	0.0055	0.0094	0.0130	0.0157	0.0208	0.0262	0.0350	0.0424
ρ/(g/cm^3)	0.99455	0.99513	0.99589	0.99660	0.99716	0.99819	0.99928	1.00105	1.00256
w_B	0.0532	0.0621	0.0735	0.0940	0.1147	0.1368	0.1806	0.2255	0.2636
ρ/(g/cm^3)	1.00477	1.00660	1.00895	1.01321	1.01749	1.02213	1.03154	1.04144	1.05002
w_B	0.3168	0.3712	0.4105	0.4649					
ρ/(g/cm^3)	1.06227	1.07506	1.08473	1.09828					

T/K = 318.15

w_B	0.0027	0.0055	0.0094	0.0130	0.0157	0.0208	0.0262	0.0350	0.0424
ρ/(g/cm^3)	0.99072	0.99128	0.99204	0.99273	0.99328	0.99430	0.99537	0.99711	0.99860
w_B	0.0532	0.0621	0.0735	0.0940	0.1147	0.1368	0.1806	0.2255	0.2636
ρ/(g/cm^3)	1.00077	1.00257	1.00487	1.00907	1.01326	1.01783	1.02706	1.03676	1.04516
w_B	0.3168	0.3712	0.4105	0.4649					
ρ/(g/cm^3)	1.05713	1.06964	1.07908	1.09230					

continued

continued

T/K = 328.15

w_B	0.0027	0.0055	0.0094	0.0130	0.0157	0.0208	0.0262	0.0350	0.0424
ρ/(g/cm^3)	0.98619	0.98674	0.98749	0.98819	0.98872	0.98970	0.99078	0.99249	0.99396
w_B	0.0532	0.0621	0.0735	0.0940	0.1147	0.1368	0.1806	0.2255	0.2636
ρ/(g/cm^3)	0.99609	0.99787	1.00014	1.00426	1.00840	1.01289	1.02196	1.03150	1.03975
w_B	0.3168	0.3712	0.4105	0.4649					
ρ/(g/cm^3)	1.05145	1.06376	1.07300	1.08595					

Polymer (B): **poly(1-vinyl-2-pyrrolidinone)** **2007SA2**
Characterization: M_w/g.mol^{-1} = 10000, Aldrich Chem. Co., Inc., Milwaukee, WI
Solvent (A): **water** H_2O **7732-18-5**

T/K = 283.15

w_B	0.0000	0.0015	0.0027	0.0030	0.0060	0.0090	0.0110	0.0154	0.0210
ρ/(g/cm^3)	0.99970	1.00006	1.00033	1.00038	1.00106	1.00171	1.00216	1.00313	1.00438
w_B	0.0260	0.0372	0.0471	0.0514	0.0628	0.0754	0.1065	0.1222	0.1437
ρ/(g/cm^3)	1.00547	1.00799	1.01022	1.01119	1.01377	1.01663	1.02377	1.02739	1.03251
w_B	0.1786	0.2136	0.2595	0.2961	0.3501				
ρ/(g/cm^3)	1.04078	1.04939	1.06090	1.07037	1.08447				

T/K = 288.15

w_B	0.0000	0.0015	0.0027	0.0030	0.0060	0.0090	0.0110	0.0154	0.0210
ρ/(g/cm^3)	0.99910	0.99945	0.99972	0.99977	1.00044	1.00108	1.00153	1.00249	1.00372
w_B	0.0260	0.0372	0.0471	0.0514	0.0628	0.0754	0.1065	0.1222	0.1437
ρ/(g/cm^3)	1.00480	1.00729	1.00948	1.01044	1.01299	1.01581	1.02285	1.02643	1.03148
w_B	0.1786	0.2136	0.2595	0.2961	0.3501				
ρ/(g/cm^3)	1.03964	1.04813	1.05948	1.06883	1.08260				

T/K = 293.15

w_B	0.0000	0.0015	0.0027	0.0030	0.0060	0.0090	0.0110	0.0154	0.0210
ρ/(g/cm^3)	0.99820	0.99855	0.99881	0.99887	0.99953	1.00016	1.00061	1.00155	1.00277
w_B	0.0260	0.0372	0.0471	0.0514	0.0628	0.0754	0.1065	0.1222	0.1437
ρ/(g/cm^3)	1.00383	1.00629	1.00847	1.00941	1.01193	1.01472	1.02168	1.02521	1.03020
w_B	0.1786	0.2136	0.2595	0.2961	0.3501				
ρ/(g/cm^3)	1.03826	1.04664	1.05784	1.06706	1.08063				

T/K = 298.15

w_B	0.0000	0.0015	0.0027	0.0030	0.0060	0.0090	0.0110	0.0154	0.0210
ρ/(g/cm^3)	0.99704	0.99738	0.99765	0.99770	0.99835	0.99898	0.99942	1.00035	1.00156

continued

continued

w_B	0.0260	0.0372	0.0471	0.0514	0.0628	0.0754	0.1065	0.1222	0.1437
$\rho/(g/cm^3)$	1.00261	1.00505	1.00720	1.00813	1.01062	1.01339	1.02027	1.02377	1.02870
w_B	0.1786	0.2136	0.2595	0.2961	0.3501				
$\rho/(g/cm^3)$	1.03667	1.04494	1.05601	1.06512	1.07853				

T/K = 303.15

w_B	0.0000	0.0015	0.0027	0.0030	0.0060	0.0090	0.0110	0.0154	0.0210
$\rho/(g/cm^3)$	0.99403	0.99436	0.99462	0.99467	0.99531	0.99593	0.99636	0.99728	0.99846
w_B	0.0260	0.0372	0.0471	0.0514	0.0628	0.0754	0.1065	0.1222	0.1437
$\rho/(g/cm^3)$	0.99950	1.00188	1.00400	1.00491	1.00736	1.01007	1.01682	1.02025	1.02508
w_B	0.1786	0.2136	0.2595	0.2961	0.3501				
$\rho/(g/cm^3)$	1.03289	1.04100	1.05182	1.06073	1.07386				

T/K = 308.15

w_B	0.0000	0.0015	0.0027	0.0030	0.0060	0.0090	0.0110	0.0154	0.0210
$\rho/(g/cm^3)$	0.99403	0.99436	0.99462	0.99467	0.99531	0.99593	0.99636	0.99728	0.99846
w_B	0.0260	0.0372	0.0471	0.0514	0.0628	0.0754	0.1065	0.1222	0.1437
$\rho/(g/cm^3)$	0.99950	1.00188	1.00400	1.00491	1.00736	1.01007	1.01682	1.02025	1.02508
w_B	0.1786	0.2136	0.2595	0.2961	0.3501				
$\rho/(g/cm^3)$	1.03289	1.04100	1.05182	1.06073	1.07386				

Polymer (B): **poly(1-vinyl-2-pyrrolidinone)** **2011SA1**
Characterization: M_w/g.mol^{-1} = 10000, Merck, Darmstadt, Germany
Solvent (A): **water** **H_2O** **7732-18-5**

T/K = 288.15

c_B/(kg B/kg A)	0.0000	0.0060	0.0090	0.0111	0.0156	0.0541	0.0666	0.1291
$\rho/(g/cm^3)$	0.99970	1.00036	1.00098	1.00141	1.00233	1.00995	1.01234	1.02369
c_B/(kg B/kg A)	0.1393	0.1393	0.2174	0.2712	0.3505	0.4176	0.4199	
$\rho/(g/cm^3)$	1.02547	1.02545	1.03812	1.04608	1.05695	1.06543	1.06578	

T/K = 293.15

c_B/(kg B/kg A)	0.0000	0.0060	0.0090	0.0111	0.0156	0.0541	0.0666	0.1291
$\rho/(g/cm^3)$	0.99910	0.99945	1.00006	1.00049	1.00140	1.00894	1.01129	1.02251
c_B/(kg B/kg A)	0.1393	0.1393	0.2174	0.2712	0.3505	0.4176	0.4199	
$\rho/(g/cm^3)$	1.02426	1.02425	1.03676	1.04464	1.05533	1.06376	1.06407	

T/K = 298.15

c_B/(kg B/kg A)	0.0000	0.0060	0.0090	0.0111	0.0156	0.0541	0.0666	0.1291
$\rho/(g/cm^3)$	0.99820	0.99827	0.99888	0.99930	1.00021	1.00767	1.00999	1.02109
c_B/(kg B/kg A)	0.1393	0.1393	0.2174	0.2712	0.3505	0.4176	0.4199	
$\rho/(g/cm^3)$	1.02283	1.02281	1.03518	1.04301	1.05353	1.06186	1.06216	

continued

continued

T/K = 303.15

c_B/(kg B/kg A)	0.0000	0.0060	0.0090	0.0111	0.0156	0.0541	0.0666	0.1291
ρ/(g/cm^3)	0.99704	0.99686	0.99746	0.99788	0.99878	1.00617	1.00847	1.01945
c_B/(kg B/kg A)	0.1393	0.1393	0.2174	0.2712	0.3505	0.4176	0.4199	
ρ/(g/cm^3)	1.02118	1.02116	1.03340	1.04111	1.05155	1.05979	1.06009	

T/K = 308.15

c_B/(kg B/kg A)	0.0000	0.0060	0.0090	0.0111	0.0156	0.0541	0.0666	0.1291
ρ/(g/cm^3)	0.99565	0.99524	0.99583	0.99624	0.99714	1.00446	1.00674	1.01762
c_B/(kg B/kg A)	0.1393	0.1393	0.2174	0.2712	0.3505	0.4176	0.4199	
ρ/(g/cm^3)	1.01933	1.01932	1.03144	1.03905	1.04941	1.05754	1.05786	

T/K = 313.15

c_B/(kg B/kg A)	0.0000	0.0060	0.0090	0.0111	0.0156	0.0541	0.0666	0.1291
ρ/(g/cm^3)	0.99403	0.99341	0.99399	0.99441	0.99529	1.00255	1.00482	1.01561
c_B/(kg B/kg A)	0.1393	0.1393	0.2174	0.2712	0.3505	0.4176	0.4199	
ρ/(g/cm^3)	1.01730	1.01729	1.02930	1.03685	1.04710	1.05512	1.05547	

Polymer (B): **poly(1-vinyl-2-pyrrolidinone)** **2007SA2**
Characterization: M_w/g.mol^{-1} = 10000, Aldrich Chem. Co., Inc., Milwaukee, WI
Solvent (A): **water** **H_2O** **7732-18-5**
Salt (C): **sodium citrate** **$Na_3C_6H_5O_7$** **68-04-2**

Comments: The aqueous sodium citrate solution was used with a constant mass fraction of w_C = 0.0089.

T/K = 283.15

w_B	0.0000	0.0033	0.0066	0.0135	0.0375	0.0480	0.0617	0.0742	0.0912
ρ/(g/cm^3)	1.00641	1.00709	1.00781	1.00929	1.01450	1.01678	1.01981	1.02257	1.02642
w_B	0.1004	0.1239	0.1516	0.1724	0.1994	0.2501			
ρ/(g/cm^3)	1.02846	1.03377	1.03998	1.04502	1.05138	1.06354			

T/K = 288.15

w_B	0.0000	0.0033	0.0066	0.0135	0.0375	0.0480	0.0617	0.0742	0.0912
ρ/(g/cm^3)	1.00572	1.00640	1.00711	1.00857	1.01372	1.01597	1.01895	1.02168	1.02548
w_B	0.1004	0.1239	0.1516	0.1724	0.1994	0.2501			
ρ/(g/cm^3)	1.02749	1.03273	1.03895	1.04383	1.05011	1.06209			

T/K = 293.15

w_B	0.0000	0.0033	0.0066	0.0135	0.0375	0.0480	0.0617	0.0742	0.0912
ρ/(g/cm^3)	1.00475	1.00542	1.00612	1.00756	1.01265	1.01488	1.01783	1.02053	1.02428
w_B	0.1004	0.1239	0.1516	0.1724	0.1994	0.2501			
ρ/(g/cm^3)	1.02627	1.03144	1.03763	1.04241	1.04860	1.06044			

continued

continued

T/K = 298.15

w_B	0.0000	0.0033	0.0066	0.0135	0.0375	0.0480	0.0617	0.0742	0.0912
ρ/(g/cm^3)	1.00353	1.00419	1.00489	1.00631	1.01135	1.01355	1.01647	1.01914	1.02285
w_B	0.1004	0.1239	0.1516	0.1724	0.1994	0.2501			
ρ/(g/cm^3)	1.02482	1.02994	1.03607	1.04077	1.04690	1.05859			

T/K = 303.15

w_B	0.0000	0.0033	0.0066	0.0135	0.0375	0.0480	0.0617	0.0742	0.0912
ρ/(g/cm^3)	1.00208	1.00274	1.00342	1.00484	1.00982	1.01200	1.01490	1.01754	1.02121
w_B	0.1004	0.1239	0.1516	0.1724	0.1994	0.2501			
ρ/(g/cm^3)	1.02316	1.02822	1.03429	1.03895	1.04501	1.05657			

T/K = 308.15

w_B	0.0000	0.0033	0.0066	0.0135	0.0375	0.0480	0.0617	0.0742	0.0912
ρ/(g/cm^3)	1.00042	1.00107	1.00175	1.00315	1.00809	1.01025	1.01312	1.01574	1.01938
w_B	0.1004	0.1239	0.1516	0.1724	0.1994	0.2501			
ρ/(g/cm^3)	1.02131	1.02632	1.03233	1.03694	1.04294	1.05439			

Comments: The aqueous sodium citrate solution was used with a constant mass fraction of $w_C = 0.0175$.

T/K = 283.15

w_B	0.0000	0.0016	0.0031	0.0060	0.0117	0.0246	0.0272	0.0330	0.0369
ρ/(g/cm^3)	1.01287	1.01317	1.01350	1.01409	1.01531	1.01804	1.01860	1.01984	1.02066
w_B	0.0430	0.0650	0.0681	0.0769	0.0955	0.1102	0.1492	0.1977	0.2522
ρ/(g/cm^3)	1.02200	1.02678	1.02741	1.02933	1.03345	1.03664	1.04539	1.05659	1.06944

T/K = 288.15

w_B	0.0000	0.0016	0.0031	0.0060	0.0117	0.0246	0.0272	0.0330	0.0369
ρ/(g/cm^3)	1.01210	1.01240	1.01272	1.01331	1.01451	1.01721	1.01777	1.01899	1.01980
w_B	0.0430	0.0650	0.0681	0.0769	0.0955	0.1102	0.1492	0.1977	0.2522
ρ/(g/cm^3)	1.02113	1.02584	1.02647	1.02836	1.03242	1.03558	1.04421	1.05525	1.06793

T/K = 293.15

w_B	0.0000	0.0016	0.0031	0.0060	0.0117	0.0246	0.0272	0.0330	0.0369
ρ/(g/cm^3)	1.01107	1.01137	1.01168	1.01226	1.01345	1.01612	1.01667	1.01788	1.01868
w_B	0.0430	0.0650	0.0681	0.0769	0.0955	0.1102	0.1492	0.1977	0.2522
ρ/(g/cm^3)	1.01998	1.02465	1.02527	1.02714	1.03115	1.03427	1.04279	1.05370	1.06621

T/K = 298.15

w_B	0.0000	0.0016	0.0031	0.0060	0.0117	0.0246	0.0272	0.0330	0.0369
ρ/(g/cm^3)	1.00979	1.01009	1.01040	1.01097	1.01215	1.01479	1.01533	1.01653	1.01732
w_B	0.0430	0.0650	0.0681	0.0769	0.0955	0.1102	0.1492	0.1977	0.2522
ρ/(g/cm^3)	1.01861	1.02322	1.02384	1.02569	1.02966	1.03275	1.04117	1.05196	1.06431

continued

continued

T/K = 303.15

w_B	0.0000	0.0016	0.0031	0.0060	0.0117	0.0246	0.0272	0.0330	0.0369
ρ/(g/cm^3)	1.00829	1.00859	1.00890	1.00946	1.01063	1.01324	1.01378	1.01497	1.01575
w_B	0.0430	0.0650	0.0681	0.0769	0.0955	0.1102	0.1492	0.1977	0.2522
ρ/(g/cm^3)	1.01703	1.02160	1.02221	1.02404	1.02797	1.03102	1.03936	1.05003	1.06225

T/K = 308.15

w_B	0.0000	0.0016	0.0031	0.0060	0.0117	0.0246	0.0272	0.0330	0.0369
ρ/(g/cm^3)	1.00659	1.00688	1.00719	1.00775	1.00891	1.01150	1.01203	1.01321	1.01399
w_B	0.0430	0.0650	0.0681	0.0769	0.0955	0.1102	0.1492	0.1977	0.2522
ρ/(g/cm^3)	1.01525	1.01978	1.02038	1.02219	1.02609	1.02911	1.03738	1.04793	1.06004

Polymer (B): **poly(1-vinyl-2-pyrrolidinone)** **2010SA2**
Characterization: M_w/g.mol^{-1} = 10000, Aldrich Chem. Co., Inc., Milwaukee, WI
Solvent (A): **water** H_2O **7732-18-5**
Salt (C): **sodium 1-pentanesulfonate** $C_5H_{11}NaO_3S$ **22767-49-3**

w_B = 0.02 was kept constant

T/K = 288.15

c_C/(mol/kg)	0.0612	0.1852	0.4318	0.8708	1.2028	1.5019	1.7686	2.0271
ρ/(g/cm^3)	1.00698	1.01396	1.02716	1.04857	1.06288	1.07403	1.08297	1.09096
c_C/(mol/kg)	2.4263	3.1212						
ρ/(g/cm^3)	1.10207	1.11939						

T/K = 293.15

c_C/(mol/kg)	0.0612	0.1852	0.4318	0.8708	1.2028	1.5019	1.7686	2.0271
ρ/(g/cm^3)	1.00597	1.01283	1.02581	1.04682	1.06082	1.07177	1.08059	1.08849
c_C/(mol/kg)	2.4263	3.1212						
ρ/(g/cm^3)	1.09948	1.11659						

T/K = 298.15

c_C/(mol/kg)	0.0612	0.1852	0.4318	0.8708	1.2028	1.5019	1.7686	2.0271
ρ/(g/cm^3)	1.00472	1.01147	1.02424	1.04489	1.05862	1.06939	1.07810	1.08591
c_C/(mol/kg)	2.4263	3.1212						
ρ/(g/cm^3)	1.09678	1.11372						

T/K = 303.15

c_C/(mol/kg)	0.0612	0.1852	0.4318	0.8708	1.2028	1.5019	1.7686	2.0271
ρ/(g/cm^3)	1.00323	1.00989	1.02247	1.04278	1.05628	1.06689	1.07550	1.08325
c_C/(mol/kg)	2.4263	3.1212						
ρ/(g/cm^3)	1.09400	1.11077						

continued

continued

T/K = 308.15

c_C/(mol/kg)	0.0612	0.1852	0.4318	0.8708	1.2028	1.5019	1.7686	2.0271
ρ/(g/cm^3)	1.00154	1.00811	1.02052	1.04053	1.05381	1.06429	1.07281	1.08049
c_C/(mol/kg)	2.4263	3.1212						
ρ/(g/cm^3)	1.09114	1.10776						

T/K = 313.15

c_C/(mol/kg)	0.0612	0.1852	0.4318	0.8708	1.2028	1.5019	1.7686	2.0271
ρ/(g/cm^3)	0.99965	1.00614	1.01840	1.03812	1.05128	1.06157	1.07002	1.07764
c_C/(mol/kg)	2.4263	3.1212						
ρ/(g/cm^3)	1.08819	1.10468						

Polymer (B): **poly(1-vinyl-2-pyrrolidinone)** **2009SUG**
Characterization: M_w/g.mol^{-1} = 10000, Aldrich Chem. Co., Inc., Milwaukee, WI
Solvent (A): **water** **H_2O** **7732-18-5**

T/K = 298.15

w_B	0.00984	0.02002	0.02998	0.04044	0.05076	0.08021	0.12147	0.16633	0.21117
ρ/(g/cm^3)	0.99904	1.00114	1.00319	1.00535	1.00742	1.01365	1.02246	1.03228	1.04235

Polymer (B): **poly(1-vinyl-2-pyrrolidinone)** **2011BOL**
Characterization: M_w/g.mol^{-1} = 33300, M_w/M_n = 4.05, Carl Roth, Germany
Solvent (A): **water** **H_2O** **7732-18-5**

T/K = 293.15

c_B/(g/dm^3)	0.20	0.40	0.80	1.60	3.21	6.42	10.02	15.00
ρ/(g/cm^3)	0.998	0.998	0.998	0.999	0.999	0.999	1.000	1.001
c_B/(g/dm^3)	20.05	25.00	50.02					
ρ/(g/cm^3)	1.002	1.003	1.008					

T/K = 298.15

c_B/(g/dm^3)	0.20	0.40	0.80	1.60	3.21	6.42	10.02	15.00
ρ/(g/cm^3)	0.996	0.997	0.997	0.997	0.998	0.999	0.999	1.001
c_B/(g/dm^3)	20.05	25.00	50.02					
ρ/(g/cm^3)	1.002	1.003	1.008					

T/K = 303.15

c_B/(g/dm^3)	0.20	0.40	0.80	1.60	3.21	6.42	10.02	15.00
ρ/(g/cm^3)	0.996	0.996	0.996	0.997	0.997	0.998	0.998	0.999
c_B/(g/dm^3)	20.05	25.00	50.02					
ρ/(g/cm^3)	1.001	1.001	1.007					

Polymer (B): **poly(1-vinyl-2-pyrrolidinone)** **2009SUG**
Characterization: M_w/g.mol^{-1} = 40000, Aldrich Chem. Co., Inc., Milwaukee, WI
Solvent (A): **water** **H_2O** **7732-18-5**

T/K = 298.15

w_B	0.01006	0.02006	0.03001	0.04039	0.05002	0.07415	0.09680	0.12208
ρ/(g/cm^3)	0.99914	1.00122	1.00330	1.00547	1.00749	1.01262	1.01747	1.02297

Polymer (B): **poly(1-vinyl-2-pyrrolidinone)** **2009SUG**
Characterization: M_w/g.mol^{-1} = 360000, Aldrich Chem. Co., Inc., Milwaukee, WI
Solvent (A): **water** **H_2O** **7732-18-5**

T/K = 298.15

w_B	0.01011	0.01988	0.03039	0.04048	0.04987
ρ/(g/cm^3)	0.99922	1.00133	1.00363	1.00590	1.00799

Polymer (B): **poly(1-vinyl-2-pyrrolidinone)** **2011BOL**
Characterization: M_w/g.mol^{-1} = 665000, M_w/M_n = 3.08, Carl Roth, Germany
Solvent (A): **water** **H_2O** **7732-18-5**

T/K = 293.15

c_B/(g/dm^3)	0.05	0.09	0.20	0.40	0.81	1.60	2.50	3.75
ρ/(g/cm^3)	0.999	0.999	0.998	0.998	0.999	0.999	0.999	1.000
c_B/(g/dm^3)	5.01	6.27	12.51	25.01				
ρ/(g/cm^3)	0.999	1.000	1.001	1.003				

T/K = 298.15

c_B/(g/dm^3)	0.05	0.09	0.20	0.40	0.81	1.60	2.50	3.75
ρ/(g/cm^3)	0.998	0.998	0.998	0.998	0.998	0.998	0.998	0.999
c_B/(g/dm^3)	5.01	6.27	12.51	25.01				
ρ/(g/cm^3)	0.999	0.999	1.000	1.003				

T/K = 303.15

c_B/(g/dm^3)	0.05	0.09	0.20	0.40	0.81	1.60	2.50	3.75
ρ/(g/cm^3)	0.996	0.996	0.996	0.996	0.996	0.997	0.997	0.997
c_B/(g/dm^3)	5.01	6.27	12.51	25.01				
ρ/(g/cm^3)	0.997	0.998	0.999	1.001				

T/K = 310.15

c_B/(g/dm^3)	0.05	0.09	0.20	0.40	0.81	1.60	2.50	3.75
ρ/(g/cm^3)	0.994	0.994	0.994	0.994	0.994	0.995	0.995	0.995
c_B/(g/dm^3)	5.01	6.27	12.51	25.01				
ρ/(g/cm^3)	0.995	0.995	0.997	0.999				

Polymer (B): **poly(1-vinyl-2-pyrrolidinone)** **2006SYA**
Characterization: Sisco Research Laboratories Pvt. Ltd., Bombay, India
Solvent (A): **water** **H_2O** **7732-18-5**

T/K = 298.15

w_B	0.000	0.001	0.002	0.003	0.004	0.005	0.006
ρ/(g/cm^3)	0.9970	0.9971	0.9973	0.9975	0.9977	0.9979	0.9981

T/K = 308.15

w_B	0.000	0.001	0.002	0.003	0.004	0.005	0.006
ρ/(g/cm^3)	0.9941	0.9942	0.9944	0.9946	0.9948	0.9950	0.9953

T/K = 318.15

w_B	0.000	0.001	0.002	0.003	0.004	0.005	0.006
ρ/(g/cm^3)	0.9902	0.9904	0.9905	0.9907	0.9908	0.9910	0.9912

Polymer (B): **poly(1-vinyl-2-pyrrolidinone)** **2006SYA**
Characterization: Sisco Research Laboratories Pvt. Ltd., Bombay, India
Solvent (A): **water** **H_2O** **7732-18-5**
Solvent (C): **dimethylsulfoxide** **C_2H_6OS** **67-68-5**

w_B	0.000	0.001	0.002	0.003	0.004	0.005	0.006
T/K = 298.15							
x_A/x_C = 75/25							
ρ/(g/cm^3)	1.0808	1.0816	1.0818	1.0821	1.0824	1.0827	1.0829
x_A/x_C = 50/50							
ρ/(g/cm^3)	1.0986	1.0988	1.0989	1.0991	1.0993	1.0994	1.0995
x_A/x_C = 25/75							
ρ/(g/cm^3)	1.0977	1.0979	1.0981	1.0983	1.0985	1.0986	1.0988
T/K = 308.15							
x_A/x_C = 75/25							
ρ/(g/cm^3)	1.0740	1.0741	1.0742	1.0743	1.0745	1.0746	1.0747
x_A/x_C = 50/50							
ρ/(g/cm^3)	1.0894	1.0895	1.0896	1.0897	1.0898	1.0899	1.0990
x_A/x_C = 25/75							
ρ/(g/cm^3)	1.0885	1.0886	1.0887	1.0888	1.0889	1.0890	1.0891

continued

continued

T/K = 318.15

x_A/x_C = 75/25

ρ/(g/cm^3)	1.0656	1.0657	1.0658	1.0659	1.0661	1.0662	1.0663

x_A/x_C = 50/50

ρ/(g/cm^3)	1.0801	1.0802	1.0803	1.0804	1.0805	1.0806	1.0808

x_A/x_C = 25/75

ρ/(g/cm^3)	1.0782	1.0783	1.0784	1.0785	1.0786	1.0787	1.0788

Polymer (B): **sodium alginate** **2006MUT**
Characterization: S.D. Fine Chemicals, Mumbai, India
Solvent (A): **water** **H_2O** **7732-18-5**

T/K = 298.15

w_B	0.1	0.2	0.3	0.4	0.5	0.6	0.7	0.8
ρ/(g/cm^3)	0.99785	0.99827	0.99906	0.99941	1.00018	1.00061	1.00120	1.00169
w_B	0.9	1.0						
ρ/(g/cm^3)	1.00245	1.00287						

T/K = 308.15

w_B	0.1	0.2	0.3	0.4	0.5	0.6	0.7	0.8
ρ/(g/cm^3)	0.99485	0.99525	0.99602	0.99637	0.99714	0.99757	0.99816	0.99866
w_B	0.9	1.0						
ρ/(g/cm^3)	0.99938	0.99984						

T/K = 318.15

w_B	0.1	0.2	0.3	0.4	0.5	0.6	0.7	0.8
ρ/(g/cm^3)	0.99079	0.99121	0.99212	0.99247	0.99327	0.99370	0.99426	0.99481
w_B	0.9	1.0						
ρ/(g/cm^3)	0.99544	0.99599						

T/K = 328.15

w_B	0.1	0.2	0.3	0.4	0.5	0.6	0.7	0.8
ρ/(g/cm^3)	0.98563	0.98568	0.98665	0.98716	0.98848	0.98904	0.98962	0.99025
w_B	0.9	1.0						
ρ/(g/cm^3)	0.99035	0.99142						

T/K = 338.15

w_B	0.1	0.2	0.3	0.4	0.5	0.6	0.7	0.8
ρ/(g/cm^3)	0.97630	0.97800	0.98017	0.98025	0.98139	0.98318	0.98414	O.98504
w_B	0.9	1.0						
ρ/(g/cm^3)	0.98555	0.98587						

Polymer (B):	**tetra(ethylene glycol) dimethyl ether**	**2004HEN**
Characterization:	M/g.mol^{-1} = 222.28, Lancaster, Pelham, NH	
Solvent (A):	**water** **H_2O**	**7732-18-5**

T/K = 298.15

x_B	0.0000	0.0199	0.0510	0.1066	0.1950	0.3008	0.4088	0.5116
ρ/(g/cm^3)	0.99704	1.01395	1.02021	1.03890	1.03366	1.02607	1.02042	1.01656
x_B	0.6252	0.7016	0.8037	0.9044	0.9505	1.0000		
ρ/(g/cm^3)	1.01352	1.01164	1.00981	1.00844	1.00783	1.00665		

T/K = 313.15

x_B	0.0000	0.0199	0.0510	0.1066	0.1950	0.3008	0.4088	0.5116
ρ/(g/cm^3)	0.99221	1.00673	1.01189	1.02570	1.01980	1.01212	1.00646	1.00260
x_B	0.6252	0.7016	0.8037	0.9044	0.9505	1.0000		
ρ/(g/cm^3)	0.99955	0.99766	0.99586	0.99451	0.99392	0.99328		

T/K = 323.15

x_B	0.0000	0.0199	0.0510	0.1066	0.1950	0.3008	0.4088	0.5116
ρ/(g/cm^3)	0.98804	1.00125	1.00570	1.01667	1.01040	1.00275	0.99707	0.99321
x_B	0.6252	0.7016	0.8037	0.9044	0.9505	1.0000		
ρ/(g/cm^3)	0.99020	0.98833	0.98656	0.98523	0.98467	0.98405		

T/K = 333.15

x_B	0.0000	0.0199	0.0510	0.1066	0.1950	0.3008	0.4088	0.5116
ρ/(g/cm^3)	0.98312	0.99513	0.99935	1.00747	1.00092	0.99327	0.98760	0.98377
x_B	0.6252	0.7016	0.8037	0.9044	0.9505	1.0000		
ρ/(g/cm^3)	0.98079	0.97895	0.97724	0.97595	0.97540	0.97480		

T/K = 343.15

x_B	0.0000	0.0199	0.0510	0.1066	0.1950	0.3008	0.4088	0.5116
ρ/(g/cm^3)	0.97777	0.98858	0.99192	0.99809	0.99128	0.98367	0.97807	0.97427
x_B	0.6252	0.7016	0.8037	0.9044	0.9505	1.0000		
ρ/(g/cm^3)	0.97134	0.96956	0.96788	0.96666	0.96613	0.96555		

6.3. Table of systems where additional information on volume effects in aqueous polymer solutions can be found

Polymer (B)	***T*-range**	**Ref.**
Dextran		
	298 K	2011MET
	298-323 K	2011MOR
	303-323 K	2008AHM
Polyester (hyperbranched)		
	283-363 K	2007ROL
Polyether (hyperbranched)		
	283-363 K	2007ROL
Poly(ethylene glycol)		
	298-318 K	2012ZHA
	303-323 K	2007SID
Poly(ethylene glycol) monododecyl ether		
	278-343 K	2004BRI
Poly(ethylene oxide-*b*-propylene oxide-*b*-ethylene oxide)	283-323 K	2006DE2
Poly(propylene oxide-*b*-ethylene oxide-*b*-propylene oxide)	283-323 K	2006DE2
Poly(1-vinyl-2-pyrrolidinone)		
	298 K	2011MET
	303-323 K	2008AHM
Sodium alginate		
	303-323 K	2007SID

6.4. References

2003KIM Kim, J.H., Kim, Y., Kim, C.K., Lee, J.W., and Seo, S.B., Miscibility of polysulfone blends with poly(1-vinylpyrrolidone-*co*-styrene) copolymers and their interaction energies, *J. Polym. Sci.: Part B: Polym. Phys.*, 41, 1401, 2003.

2003LEE Lee, M.-J., Tuan, Y.-C., and Lin, H.-M., Pressure-volume-temperature properties for binary and ternary polymer solutions of poly(ethylene glycol), poly(propylene glycol), and poly(ethylene glycol methyl ether) with anisole, *Polymer*, 44, 3891, 2003.

2003RAH Rahbari-Sisakht, M., Taghizadeh, M., and Eliassi, A., Densities and viscosities of binary mixtures of poly(ethylene glycol) and poly(propylene glycol) in water and ethanol in the 293.15-338.15 K temperature range, *J. Chem. Eng. Data*, 48, 1221, 2003.

2004BOL Bolotnikov, M.F., Verveyko, V.N., and Verveyko, M.V., Speeds of sound, densities, and isentropic compressibilities of poly(propylene glycol)-425 at temperatures from (293.15 to 373.15) K and pressures up to 100 MPa, *J. Chem. Eng. Data*, 49, 631, 2004.

2004BRI Briganti, G., D'Arrigo, G., and Maccarini, M., Volumetric determination of the hydration state in nonionic surfactant solutions, *J. Phys. Chem. B*, 108, 4039, 2004.

2004CRU Cruz, R. da C., Martins, R.J., Cardoso, M.J.E. de M., and Barcia, O.E., Volumetric study of aqueous solutions of polyethylene glycol in the temperature range from 283.15 K to 313.15 K and 0.1 MPa, *J. Appl. Polym. Sci.*, 91, 2685, 2004.

2004GRA Graber, T.A., Galleguillos, H.R., Cespedes, C., and Taboada, M.E., Density, refractive index, viscosity, and electrical conductivity in the Na_2CO_3 + poly(ethylene glycol) + H_2O system from (293.15 to 308.15) K, *J. Chem. Eng. Data*, 49, 1254, 2004.

2004HEN Henni, A., Tontiwachwuthikul, P., and Chakma, A., Densities, viscosities, and derived functions of binary mixtures: (Tetraethylene glycol dimethyl ether + water) from 298.15 K to 343.15 K, *J. Chem. Eng. Data*, 49, 1778, 2004.

2004HON Hong, S.-J., Huang, H.-T., and Hong, T.-D., Effects of solvent adsorption on solution properties of poly(vinyl alcohol)/dimethylsulfoxide/water ternary systems, *J. Appl. Polym. Sci.*, 92, 3211, 2004.

2004SAD Sadeghi, R. and Zafarani-Moattar, M.T., Thermodynamics of aqueous solutions of polyvinylpyrrolidone, *J. Chem. Thermodyn.*, 36, 665, 2004.

2004TAB Taboada, P., Barbosa, S., and Mosquera, V., Thermodynamic properties of a diblock copolymer of poly(oxyethylene) and poly(oxypropylene) in aqueous solution, *Langmuir*, 20, 8903, 2004.

2004TEN Teng, S.P. and Teng, T.T., Measurement and prediction of the density of aqueous multicomponent solutions involving polyethylene glycol 2000, *J. Chem. Eng. Japan*, 37, 40, 2004.

2004ZAF Zafarani-Moattar, M.T., Samadi, F., and Sadeghi, R., Volumetric and ultrasonic studies of the system (water + polypropylene glycol 400) at temperatures from (283.15 to 313.15) K, *J. Chem. Thermodyn.*, 36, 871, 2004.

2005ELI Eliassi, A. and Modarress, H., Excess volume of polymer/solvent mixtures and proposed model for prediction of activity of solvents based on excess volume data, *J. Appl. Polym. Sci.*, 95, 1219, 2005.

2005MOH Mohsen-Nia, M., Modarress, H., and Rasa, H., Measurement and modeling of density, kinematic viscosity, and refractive index for poly(ethylene glycol) aqueous solution at different temperatures, *J. Chem. Eng. Data*, 50, 1662, 2005.

2005MUR Murugesan, T. and Perumalsamy, M., Densities and viscosities of polyethylene glycol 2000 + salt + water systems from (298.15 to 318.15)K, *J. Chem. Eng. Data*, 50, 1290, 2005.

2005NIN Ninni, L., Meirelles, A.J.A., and Maurer, G., Thermodynamic properties of aqueous solutions of maltodextrins from laser-light scattering, calorimetry and isopiestic investigations, *Carbohydrate Polym.*, 59, 289, 2005.

2005TAB Taboada, M.E., Galleguillos, H.R., Graber, T.A., and Bolado, S., Compositions, densities, conductivities, and refractive indices of potassium chloride or/and sodium chloride + PEG 4000 + water at 298.15 and liquid-liquid equilibrium of potassium chloride or sodium chloride + PEG 4000 + water at 333.15 K, *J. Chem. Eng. Data*, 50, 264, 2005.

2005WAL Wali, A. C. , Naidu, B.V.K., Mallikarjuna, N.N., Sainkar, S.R., Halligudi, S.M., and Aminabhavi, T.M., Miscibility of chitosan–hydroxyethylcellulose blends in aqueous acetic acid solutions at 35°C, *J. Appl. Polym. Sci.*, 96, 1996, 2005.

2005ZAF Zafarani-Moattar, M.T. and Hamzehzadeh, Sh., Measurement and correlation of densities, ultrasonic velocities, and compressibilities for binary aqueous poly(ethylene glycol), disodium succinate, or sodium formate and ternary aqueous poly(ethylene glycol) systems containing disodium succinate or sodium formate at T = (298.15, 308.15, and 318.15) K, *J. Chem. Eng. Data*, 50, 603, 2005.

2006DE1 DeLisi, R., Lazzara, G., Lombardo, R., Milioto, S.,·Muratore, N., and Liveri, M.L.T., Thermodynamic behavior of non-ionic triblock copolymers in water at three temperatures, *J. Solution Chem.*, 35, 659, 2006.

2006DE2 DeLisi, R., Lazzara, G., Milioto, S., and Muratore, N., Volumes of aqueous block copolymers based on poly(propylene oxides) and poly(ethylene oxides) in a large temperature range: A quantitative description, *J. Chem. Thermodyn.*, 38, 1344, 2006.

2006LEE Lee, M.-J., Ho, K.-L., and Lin, H.-M., Pressure-volume-temperature properties for binary oligomeric solutions of poly(ethylene glycol) and poly(ethylene glycol methyl ether) with acetophenone up to 50 MPa, *J. Chem. Eng. Data*, 51, 1151, 2006.

2006MUT Mutalik, V., Manjeshwar, L.S., Wali, A., Sairam, M., Raju, K.V.S.N., and Aminabhavi, T.M., Thermodynamics/hydrodynamics of aqueous polymer solutions and dynamic mechanical characterization of solid films of chitosan, sodium alginate, guar gum, hydroxyethylcellulose and hydroxypropylmethylcellulose at different temperatures, *Carbohydrate Polym.*, 65, 9, 2006.

2006SAR Saravanan, S., Reena, J.A., Rao, J.R., Murugesan, T., and Nair, B.U., Phase equilibrium compositions, densities, and viscosities of aqueous two-phase poly(ethylene glycol) + poly(acrylic acid) system at various temperatures, *J. Chem. Eng. Data*, 51, 1246, 2006.

2006SEI Seiler, M., Hyperbranched polymers: Phase behavior and new applications in the field of chemical engineering, *Fluid Phase Equil.*, 241, 155, 2006.

2006SYA Syal, V.K., Chauhan, A., and Chauhan, S., A study on solutions of poly(vinylpyrrolidone) in binary mixtures of DMSO + H_2O at different temperatures by ultrasonic velocity measurements, *Indian J. Phys*, 80, 379, 2006.

2007FUN Funke, Z., Hotani, Y., Ougizawa, T., Kressler, J., and Kammer, H.-W., Equation-of-state properties and surface tension of ethylene-vinyl alcohol random copolymers (experimental data by Z. Funke), *Eur. Polym. J.*, 43, 2371, 2007.

2007KUS Kushare, S.K., Terdale, S.S., Dagade, D.H., and Patil, K.J., Compressibility and volumetric studies of polyethylene-glycols in aqueous, methanolic, and benzene solutions at $T = 298.15$ K, J. *Chem. Thermodyn.*, 39, 1125, 2007.

2007ROL Rolker, J., Seiler, M., Mokrushina, L., and Arlt, W., Potential of branched polymers in the field of gas absorption: Experimental gas solubilities and modeling, *Ind. Eng. Chem. Res.*, 46, 6572, 2007.

2007SA1 Sadeghi, R. and Ziamajidi, F., Effect of aqueous solution of tripotassium citrate on the volumetric behavior of poly(propylene glycol) 400 at $T = (288.15$ to $313.15)$ K, *J. Chem. Eng. Data*, 52, 1268, 2007.

2007SA2 Sadeghi, R. and Ziamajidi, F., Volumetric and isentropic compressibility behaviour of aqueous solutions of (polyvinylpyrrolidone + sodium citrate) at $T = (283.15$ to $308.15)$ K, *J. Chem. Thermodyn.*, 39, 1118, 2007.

2007SA3 Sadeghi, R. and Ziamajidi, F., Thermodynamic properties of tripotassium citrate in water and in aqueous solutions of polypropylene oxide 400 over a range of temperatures, *J. Chem. Eng. Data*, 52, 1753, 2007.

2007SID Siddaramaiah, M. and Swamy, T.M.M., Studies on miscibility of sodium alginate/polyethylene glycol blends, *J. Macromol. Sci.: Part A: Pure Appl. Chem.*, 44, 321, 2007.

2007SIL Silva, R.M.M., Minim, L.A., Coimbra, J.S.R., Garcia Rojas, E.E., Mendes da Silva, L.H., and Minim, V.P.R., Density, electrical conductivity, kinematic viscosity, and refractive index of binary mixtures containing poly(ethylene glycol) 4000, lithium sulfate, and water at different temperatures, *J. Chem. Eng. Data*, 52, 1567, 2007.

2008AHM Ahmed, A.A., Jayaraju, J., Sherigara, B.S., Bhojya, H.S.N., and Keshavayya, J., Miscibility studies of dextran/poly(vinyl pyrrolidone) blend in solution, *J. Macromol. Sci.: Part A: Pure Appl. Chem.*, 45, 1055, 2008.

2008AYR Ayranci, E. and Sahin, M., Interactions of polyethylene glycols with water studied by measurements of density and sound velocity, *J. Chem. Thermodyn.*, 40, 1200, 2008.

2008DHO Dhondge, S.S., Pandhurnekar, C., and Ramesh, L., Thermodynamic studies of some non-electrolytes in aqueous solutions at low temperatures, *J. Chem. Thermodyn.*, 40, 1, 2008.

2008HAN Han, F., Zhang, J., Chen, G., and Wei, X., Density, viscosity, and excess properties for aqueous poly(ethylene glycol) solutions from (298.15 to 323.15) K, *J. Chem. Eng. Data*, 53, 2598, 2008.

2008SAD Sadeghi, R., Hosseini, R., and Jamehbozorg, B., Effect of sodium phosphate salts on the thermodynamic properties of aqueous solutions of poly(ethylene oxide) 6000 at different temperatures, *J. Chem. Thermodyn.*, 40, 1364, 2008.

2008SHI Shin, M.S., Lee, J.H., and Kim, H., Phase behavior of the poly(vinylpyrrolidone) + dichloromethane + supercritical carbon dioxide system, *Fluid Phase Equil.*, 272, 42, 2008.

2009AFZ Afzal, W., Mohammadi, A.H., and Richon, D., Volumetric properties of mono-, di-, tri-, and polyethylene glycol aqueous solutions from (273.15 to 363.15) K: Experimental measurements and correlations, *J. Chem. Eng. Data*, 54, 1254, 2009.

2009CRU Cruz, R. da C., Martins, R.J., Cardoso, M.J.E. de M., and Barcia, O.E., Volumetric study of aqueous solutions of polyethylene glycol as a function of the polymer molar mass in the temperature range 283.15 to 313.15 K and 0.1 MPa, *J. Solution Chem.*, 38, 957, 2009.

2009PER Perumalsamy, M. and Murugesan, T., Phase compositions, molar mass, and temperature effect on densities, viscosities, and liquid-liquid equilibrium of polyethylene glycol and salt-based aqueous two-phase systems, *J. Chem. Eng. Data*, 54, 1359, 2009.

2009RE1 Regupathi, I., Murugesan, S., Amaresh, S.P., Govindarajan, R., and Thanapalan, M., Densities and viscosities of poly(ethylene glycol) 4000 + diammonium hydrogen phosphate + water systems, *J. Chem. Eng. Data*, 54, 1100, 2009.

2009RE2 Regupathi, I., Govindarajan, R., Amaresh, S.P., and Murugesan, T., Densities and viscosities of polyethylene glycol 6000 + triammonium citrate + water systems, *J. Chem. Eng. Data*, 54, 3291, 2009.

2009SAD Sadeghi, R. and Jamehbozorg, B., Volumetric and viscosity studies of interactions between sodium phosphate salts and poly(propylene glycol) 400 in aqueous solutions at different temperatures, *Fluid Phase Equil.*, 284, 86, 2009.

2009SIN Singh, R.K., Singh, M.P., and Chaurasia, S.K., Temperature dependent ultrasonic and conductivity studies in aqueous polymeric solution, *Fluid Phase Equil.*, 284, 10, 2009.

2009SUG Sugiura, T., Kimura, F., and Ogawa, H., Thermodynamic study on the polyvinylpyrrolidone + water, + alkanol (n = 1 - 4), + 1-methyl-2-pyrrolidone at 298.15 K, *Netsu Sokutei*, 36, 130, 2009.

2010MIN Minim, L.A., Bonomo, R.C.F., Amaral, I.V., Reis, M.F.T., Oliveira, A.A.A., and Minim, V.P.R., Density and viscosity of binary and ternary mixtures of poly(ethylene glycol) and poly(acrylic acid sodium salt) at temperatures of (288.15 to 318.15) K, *J. Chem. Eng. Data*, 55, 2328, 2010.

2010MOL Molisso, A., Mangiapia, G, D'Errico, G., and·Sartorio, R, Interaction of poly(vinyl alcohol) with poly(acrylic acid) and with sodium polyacrylate in aqueous solutions: A volumetric study at 25 °C, *J. Solution Chem.*, 39, 1627, 2010.

2010PAN Pan, H.-P., Bai, T.-S., and Wang, X.-D., Density, viscosity, and electric conductance of a ternary solution of (nicotinic acid + polyethanol + water), *J. Chem. Eng. Data*, 55, 2257, 2010.

2010SA1 Sadeghi, R., Golabiazar, R., and Ziaii, M., Vapor-liquid equilibria, density, speed of sound, and refractive index of sodium tungstate in water and in aqueous solutions of poly(ethyleneglycol) 6000, *J. Chem. Eng. Data*, 55, 125, 2010.

2010SA2 Sadeghi, R. and Ziaii, M., Thermodynamic investigation of the systems poly(ethylene glycol) + sodium pentane-1-sulfonate + water and poly(vinyl pyrrolidone) + sodium pentane-1-sulfonate + water, *J. Colloid Interface Sci.*, 346, 107, 2010.

2010SAL Salabat, A. and Mehrdad, A., Viscometric and volumetric study of dilute aqueous solutions of binary and ternary poly(ethylene glycol)/poly(vinyl alcohol) systems at different temperatures, *J. Mol. Liq.*, 157, 57, 2010.

2010TRI Trivedi, S., Bhanot, C., and Pandey, S., Densities of {poly(ethylene glycol) + water} over the temperature range (283.15 to 363.15) K, *J. Chem. Thermodyn.*, 42, 1367, 2010.

2010ZAF Zafarani-Moattar, M.T. and Kheyrabi, N., Volumetric, ultrasonic, and transport properties of an aqueous solution of polyethylene glycol monomethyl ether at different temperatures, *J. Chem. Eng. Data*, 55, 3976, 2010.

2011BOL Bolten, D. and Tuerk, M., Experimental study on the surface tension, density, and viscosity of aqueous poly(vinylpyrrolidone) solutions, *J. Chem. Eng. Data*, 56, 582, 2011.

2011MET Mete, D., Göksel, C., and Güner, A., Characterization and miscibility dynamics of dextran–poly(vinylpyrrolidone)–water system, *Macromol. Symp.*, 302, 257, 2011.

2011MO1 Morariu, S., Eckelt, J., and Wolf, B.A., Dextran-based polycations: Thermodynamic interaction with water as compared with unsubstituted dextran, 1 – Volumetric properties and light scattering, *Macromol. Chem. Phys.*, 212, 1925, 2011.

2011RE2 Reddy, P.M., Venkatesu, P., and Bohidar, H.B., Influence of polymer molecular weight and concentration on coexistence curve of isobutyric acid + water, *J. Phys. Chem. B*, 115, 12065, 2011.

2011SA1 Sadeghi, R. and Azizpour, S., Volumetric, compressibility, and viscometric measurements of binary mixtures of poly(vinylpyrrolidone) + water, + methanol, + ethanol, + acetonitrile, + 1-propanol, + 2-propanol, and + 1-butanol, *J. Chem. Eng. Data*, 56, 240, 2011.

2011SA2 Sadeghi, R. and Kahaki, H.B., Thermodynamics of aqueous solutions of poly(ethylene glycol) dimethyl ethers in the presence or absence of ammonium phosphate salts, *Fluid Phase Equil.*, 306, 219, 2011.

2011SCH Scherillo, G., Sanguigno, L., Sansone, L., DiMaio, E., Galizia, M., and Mensitieri, G., Thermodynamics of water sorption in poly(ε-caprolactone): A comparative analysis of lattice fluid models including hydrogen bond contributions (exp. data by G. Scherillo and G. Mansitieri), *Fluid Phase Equil.*, 313, 127, 2011.

2011ZHA Zhang, K., Yang, J., Yu, X., Zhang, J., and Wei, X., Densities and viscosities for binary mixtures of poly(ethylene glycol) 400 + dimethyl sulfoxide and poly(ethylene glycol) 600 + water at different temperatures, *J. Chem. Eng. Data*, 56, 3083, 2011.

2012ZHA Zhang, N., Zhang, J., Zhang, Y., Bai, J., Huo, T., and Wei, X., Excess molar volumes and viscosities of poly(ethylene glycol) 300 + water at different temperatures, *Fluid Phase Equil.*, 313, 7, 2012.

7. SECOND VIRIAL COEFFICIENTS (A_2) OF AQUEOUS POLYMER SOLUTIONS

7.1. Experimental A_2 data

Polymer	M_n/ g/mol	M_w/ g/mol	Solvent	T/ K	A_2 10^4/ cm^{3}mol/g^2	Ref.
Bovine serum albumin						
pH = 4.5	68000		water (0.15 M $(NH_4)_2SO_4$)	298.15	−0.260	2009LUY
pH = 4.5	68000		water (0.50 M $(NH_4)_2SO_4$)	298.15	−0.036	2009LUY
pH = 4.5	68000		water (1.00 M $(NH_4)_2SO_4$)	298.15	−0.239	2009LUY
pH = 4.5	68000		water (1.50 M $(NH_4)_2SO_4$)	298.15	−0.765	2009LUY
pH = 4.8	68000		water (0.15 M $(NH_4)_2SO_4$)	298.15	−0.263	2009LUY
pH = 4.8	68000		water (0.50 M $(NH_4)_2SO_4$)	298.15	−0.199	2009LUY
pH = 4.8	68000		water (1.00 M $(NH_4)_2SO_4$)	298.15	−0.779	2009LUY
pH = 4.8	68000		water (1.50 M $(NH_4)_2SO_4$)	298.15	−1.159	2009LUY
pH = 5.4	68000		water (0.15 M $(NH_4)_2SO_4$)	298.15	−0.095	2009LUY
pH = 5.4	68000		water (0.50 M $(NH_4)_2SO_4$)	298.15	−0.107	2009LUY
pH = 5.4	68000		water (1.00 M $(NH_4)_2SO_4$)	298.15	−0.101	2009LUY
pH = 5.4	68000		water (1.50 M $(NH_4)_2SO_4$)	298.15	−0.347	2009LUY
pH = 7.4	68000		water (0.15 M $(NH_4)_2SO_4$)	298.15	−0.031	2009LUY
pH = 7.4	68000		water (0.50 M $(NH_4)_2SO_4$)	298.15	0.172	2009LUY
pH = 7.4	68000		water (1.00 M $(NH_4)_2SO_4$)	298.15	0.011	2009LUY
pH = 7.4	68000		water (1.50 M $(NH_4)_2SO_4$)	298.15	−0.027	2009LUY
pH = 3.6		108879	water (0.1 M KCl)	298.15	0.26	2009PAR
pH = 4.6		74541	water (0.1 M KCl)	298.15	1.13	2009PAR
pH = 5.6		63420	water (0.1 M KCl)	298.15	0.92	2009PAR
pH = 7.6		101010	water (0.1 M KCl)	298.15	1.71	2009PAR
pH = 4.6		77927	water (1 M KCl)	298.15	0.63	2009PAR
pH = 5.6		80316	water (1 M KCl)	298.15	1.26	2009PAR
pH = 7.6		135544	water (1 M KCl)	298.15	2.29	2009PAR
pH = 5.6		107814	water (3 M KCl)	298.15	1.39	2009PAR
pH = 7.6		83623	water (3 M KCl)	298.15	1.35	2009PAR
pH = 4.6		107559	water (0.01 M LiCl)	298.15	0.61	2009PAR
pH = 5.6		93821	water (0.01 M LiCl)	298.15	0.80	2009PAR
pH = 7.6		82510	water (0.01 M LiCl)	298.15	1.06	2009PAR
pH = 4.6		94415	water (0.1 M LiCl)	298.15	0.64	2009PAR
pH = 5.6		82470	water (0.1 M LiCl)	298.15	0.65	2009PAR
pH = 7.6		117730	water (0.1 M LiCl)	298.15	1.15	2009PAR

Polymer	M_n/ g/mol	M_w/ g/mol	Solvent	T/ K	A_2 10^4/ cm^3mol/g^2	Ref.
Bovine serum albumin						
pH = 4.0		69100	water (buffered, 15 mM)	298.15	2.88	2011YAD
pH = 5.0		67500	water (buffered, 15 mM)	298.15	−0.34	2011YAD
pH = 6.0		73900	water (buffered, 15 mM)	298.15	1.31	2011YAD
pH = 7.0		73300	water (buffered, 15 mM)	298.15	3.38	2011YAD
Carboxymethylcellulose						
		680000	water (0.5 M NaCl)	298.15	5.2	2008VID
Carboxymethylcellulose-*g*-poly(*N,N*-dihexylacrylamide)						
(1.1 mol% *N,N*-dihexylacrylamide)		730000	water (0.5 M NaCl)	298.15	2.4	2008VID
(5.9 mol% *N,N*-dihexylacrylamide)		590000	water (0.5 M NaCl)	298.15	0.71	2008VID
Carboxymethylstarch (from potato starch)						
(D.S. = 0.40)		1240000	water	298.15	8.26	2009STO
(D.S. = 0.40)		2320000	water (0.01 M NaCl)	298.15	2.010	2009STO
(D.S. = 0.40)		1710000	water (0.05 M NaCl)	298.15	0.298	2009STO
(D.S. = 0.40)		1920000	water (0.10 M NaCl)	298.15	0.862	2009STO
(D.S. = 0.40)		2170000	water (0.50 M NaCl)	298.15	0.455	2009STO
(D.S. = 0.81)		838000	water	298.15	11.10	2009STO
(D.S. = 0.81)		1160000	water (0.01 M NaCl)	298.15	3.33	2009STO
(D.S. = 0.81)		895000	water (0.05 M NaCl)	298.15	1.84	2009STO
(D.S. = 0.81)		857000	water (0.10 M NaCl)	298.15	1.59	2009STO
(D.S. = 0.81)		972000	water (0.50 M NaCl)	298.15	1.04	2009STO
(D.S. = 0.90)		8090000	water	298.15	1.81	2009STO
(D.S. = 0.90)		18200000	water (0.01 M NaCl)	298.15	0.374	2009STO
(D.S. = 0.90)		20900000	water (0.05 M NaCl)	298.15	0.123	2009STO
(D.S. = 0.90)		20800000	water (0.10 M NaCl)	298.15	0.084	2009STO
(D.S. = 0.90)		20500000	water (0.50 M NaCl)	298.15	0.041	2009STO
Carboxymethylstarch (from corn starch)						
(D.S. = 0.50)		3520000	water	298.15	1.93	2009STO
(D.S. = 0.50)		5550000	water (0.01 M NaCl)	298.15	0.064	2009STO
(D.S. = 0.50)		4880000	water (0.05 M NaCl)	298.15	0.007	2009STO
(D.S. = 0.50)		4990000	water (0.10 M NaCl)	298.15	−0.005	2009STO
(D.S. = 0.50)		4820000	water (0.50 M NaCl)	298.15	−0.030	2009STO
(D.S. = 0.70)		8630000	water	298.15	0.085	2009STO
(D.S. = 0.70)		13500000	water (0.01 M NaCl)	298.15	0.048	2009STO
(D.S. = 0.70)		13300000	water (0.05 M NaCl)	298.15	0.005	2009STO
(D.S. = 0.70)		19300000	water (0.10 M NaCl)	298.15	0.016	2009STO
(D.S. = 0.70)		15200000	water (0.50 M NaCl)	298.15	0.011	2009STO
(D.S. = 1.10)		4570000	water	298.15	1.51	2009STO
(D.S. = 1.10)		5860000	water (0.01 M NaCl)	298.15	0.073	2009STO

Polymer	M_n/ g/mol	M_w/ g/mol	Solvent	T/ K	A_2 10^4/ cm^{3}mol/g^2	Ref.
Carboxymethylstarch (from corn starch)						
(D.S. = 1.10)		6320000	water (0.05 M NaCl)	298.15	0.032	2009STO
(D.S. = 1.10)		6380000	water (0.10 M NaCl)	298.15	0.029	2009STO
(D.S. = 1.10)		6290000	water (0.50 M NaCl)	298.15	0.011	2009STO
Cellulose						
		27000	water (4.6 wt% LiOH/ 15 wt% urea)	298.15	33.1	2006CAI
		65000	water (4.6 wt% LiOH/ 15 wt% urea)	298.15	33.3	2006CAI
		68000	water (4.6 wt% LiOH/ 15 wt% urea)	298.15	31.0	2006CAI
		83000	water (4.6 wt% LiOH/ 15 wt% urea)	298.15	44.1	2006CAI
		84000	water (4.6 wt% LiOH/ 15 wt% urea)	298.15	39.4	2006CAI
		86000	water (4.6 wt% LiOH/ 15 wt% urea)	298.15	33.3	2006CAI
		92000	water (4.6 wt% LiOH/ 15 wt% urea)	298.15	37.7	2006CAI
		120000	water (4.6 wt% LiOH/ 15 wt% urea)	298.15	14.9	2006CAI
		148000	water (4.6 wt% LiOH/ 15 wt% urea)	298.15	20.4	2006CAI
		195000	water (4.6 wt% LiOH/ 15 wt% urea)	298.15	13.5	2006CAI
		412000	water (4.6 wt% LiOH/ 15 wt% urea)	298.15	15.8	2006CAI
Dextran						
		40600	water	298.15	5.7	2005ROT
		161000	water	298.15	2.59	2005SUG
		595000	water	298.15	2.1	2005ROT
	347000		water	302.05	3.0	2007SCH
Hydroxyethylcellulose (derivative with triethylammonium chloride)						
		260000	water (phosphate buffered saline, pH = 7.4)	308.15	9.3	2009GAO
		420000	water (phosphate buffered saline, pH = 7.4)	308.15	8.9	2009GAO
		1030000	water (phosphate buffered saline, pH = 7.4)	308.15	7.9	2009GAO

Polymer	M_n/ g/mol	M_w/ g/mol	Solvent	T/ K	A_2 10^4/ cm^3mol/g^2	Ref.
Hydroxypropylguar (D.S.) = 1.2		5000000	water	313.15	−1.0	2010RIS
Hydroxypropylstarch						
		9700000	water	293.15	0.21	2004VOR
		43600000	water	293.15	0.11	2004VOR
Inulin						
		1030000	water	298.15	2.22	2009DAN
β-Lactoglobulin						
	36500		water	302.05	−2.2	2007SCH
Maltodextrin						
	2000	10000	water	298.15	2.13	2005NIN
Maltotriose						
	504		water	298.15	5.72	2005NIN
Methylguar (D.S. = 0.3)		2600000	water	313.15	1.4	2010RIS
Methyl(hydroxypropyl)guar (D.S. = 0.4/1.2)		1000000	water	313.15	1.5	2010RIS
Poly(acrylamide)						
		74800	water	298.15	6.30	2004ALB
		463000	water	298.15	4.80	2004ALB
		1829000	water	298.15	4.46	2004ALB
Poly(acrylic acid)						
		94000	water (0.1 M HCl)	333.15	3.0	2010LIT
		94000	water (0.2 M HCl)	333.15	5.8	2010LIT
		161000	water (0.1 M HCl)	313.15	1.8	2009LIT

Polymer	M_n/ g/mol	M_w/ g/mol	Solvent	T/ K	A_2 10^4/ cm^3mol/g^2	Ref.
Poly(*N*-decylacrylamide-*b*-*N,N*-diethylacrylamide)						
(micellar)		815000	water	293.15	8.6	2008ZHO
Poly(*N,N*-diallyl-*N,N*-dimethylammonium chloride)						
		400000	water (0.1 M HCl)	293.15	15	2010LIT
		400000	water (0.1 M HCl)	318.15	18	2010LIT
		415000	water (0.1 M HCl)	313.15	17.7	2009LIT
Poly(*N,N*-diallyl-*N*-methylammonium acetate)						
		22000	water (1 M NaCl)	298.15	2.8	2011LEZ
		60000	water (1 M NaCl)	298.15	3.2	2011LEZ
		104000	water (1 M NaCl)	298.15	3.5	2011LEZ
		112000	water (1 M NaCl)	298.15	2.5	2011LEZ
		178000	water (1 M NaCl)	298.15	2.1	2011LEZ
Poly(*N,N*-diethylacrylamide)						
		17000000	water	300.15	0.2	2008ZHO
		17000000	water	301.15	0.1	2008ZHO
		17000000	water	301.65	0.0	2008ZHO
Poly(*N,N*-dimethylacrylamide)						
	15600	16700	water	298.15	8.0	2006LAM
	47300	53900	water	298.15	7.8	2006LAM
Poly(*N,N*-dimethylacrylamide–*b*-*N*-isopropylacrylamide-*b*-*N,N*-dimethylacrylamide)						
(72 mol% NIPAM)	41300	42500	water	323.15	3.78	2006CON
(78 mol% NIPAM)	53000	60400	water	323.15	2.58	2006CON
Poly[2-(2-ethoxy)ethoxyethyl vinyl ether]						
		21900	water	298.15	6.6	2005MAT
		21900	water	308.15	2.4	2005MAT
		21900	water	311.15	0.7	2005MAT
		21900	water	313.15	−1.1	2005MAT
		30400	water	298.15	6.6	2005MAT
		30400	water	308.15	2.7	2005MAT
		30400	water	311.15	0.5	2005MAT
		30400	water	313.15	−0.6	2005MAT

Polymer	**M_n/ g/mol**	**M_w/ g/mol**	**Solvent**	**T/ K**	**A_2 10^4/ cm^3mol/g^2**	**Ref.**
Poly(ethoxyethyl glycidyl ether)						
		103000	water	298.15	5.59	2007RA1
		199000	water	298.15	6.15	2007RA1
		446000	water	298.15	5.35	2007RA1
		1093000	water	298.15	4.74	2007RA1
		1806000	water	298.15	4.31	2007RA1
		2330000	water	298.15	4.40	2007RA1
Poly(ethylene glycol)						
	1290	1730	water	298.15	24.0	2006OZD
	1290	1730	water	303.15	16.9	2006OZD
	1290	1730	water	308.15	14.8	2006OZD
	1290	1730	water	313.15	11.6	2006OZD
	3540	4600	water	298.15	28.9	2006OZD
	3540	4600	water	303.15	19.5	2006OZD
	3540	4600	water	308.15	16.9	2006OZD
	3540	4600	water	313.15	13.0	2006OZD
	7020	8000	water	298.15	33.2	2006OZD
	7020	8000	water	303.15	21.8	2006OZD
	7020	8000	water	308.15	18.4	2006OZD
	7020	8000	water	313.15	13.9	2006OZD
	5370	10640	water	298.15	36.2	2006OZD
	5370	10640	water	303.15	23.4	2006OZD
	5370	10640	water	308.15	20.0	2006OZD
	5370	10640	water	313.15	14.9	2006OZD
Poly(ethylene oxide)						
		625000	water	298.15	6.57	2006KHA
		896000	water	298.15	0.30	2006KHA
		1038000	water	298.15	9.65	2006KHA
Poly(ethylene oxide-*b*-(R)-3-hydroxybutyrate-*b*-ethylene oxide)						
(81.8 wt% EO)	2000-810-2000		water	298.15	−3.2	2006LIX
(92.4 wt% eEO)	5000-780-5000		water	298.15	0.27	2006LIX
Poly(glycidol-*b*-propylene oxide-*b*-glycidol)						
(20 wt% propylene oxide)			water	298.15	0.384	2007RA2
(20 wt% propylene oxide)			water	313.15	0.226	2007RA2
(20 wt% propylene oxide)			water	323.15	0.573	2007RA2
(20 wt% propylene oxide)			water	333.15	0.076	2007RA2

Polymer	M_n/ g/mol	M_w/ g/mol	Solvent	T/ K	A_2 10^4/ cm^3mol/g^2	Ref.
Poly(glycidol-*b*-propylene oxide-*b*-glycidol)						
(30 wt% propylene oxide)			water	298.15	0.074	2007RA2
(30 wt% propylene oxide)			water	313.15	0.454	2007RA2
(30 wt% propylene oxide)			water	323.15	0.266	2007RA2
(30 wt% propylene oxide)			water	333.15	0.283	2007RA2
(70 wt% propylene oxide)			water	298.15	−0.194	2007RA2
(70 wt% propylene oxide)			water	313.15	−0.482	2007RA2
(70 wt% propylene oxide)			water	323.15	−0.873	2007RA2
(70 wt% propylene oxide)			water	333.15	−0.499	2007RA2
Poly(*N*-isopropylacrylamide)						
		3900000	water	295.35	1.4	2003STI
		3900000	water	298.95	1.2	2003STI
		3900000	water	303.45	0.50	2003STI
		3900000	water	306.75	0.18	2003STI
		3900000	water	308.95	−0.009	2003STI
		3900000	water	309.65	−0.01	2003STI
Poly(*N*-isopropylacrylamide) stereoblock polymer (isotactic-atactic-isotactic)						
i2-a28-i2	36700		water	293.15	9.7	2008NUO
i2-a40-i2	47100		water	293.15	2.2	2008NUO
Poly(*N*-isopropylacrylamide-*b*-*N,N*-dimethylacrylamide)						
(33 wt% NIPAM)	35300	39500	water	305.15	0.2	2006LAM
(42 mol% NIPAM)	17900	19700	water	323.15	3.03	2006CON
(64 mol% NIPAM)	29600	34000	water	323.15	1.11	2006CON
(71 mol% NIPAM)	38600	43200	water	323.15	1.14	2006CON
(82 mol% NIPAM)	61900	74900	water	323.15	1.10	2006CON
Poly(*N*-isopropylacrylamide-*b*-L-glutamic acid)						
(NIPAM55-*b*-LGA35)		3370000	water (pH = 3)	298.15	−2.05	2008DEN
Poly{*N*-isopropylacrylamide-*co*-3-[*N*-(3-methacrylamidopropyl)-*N,N*-dimethylammonio] propane sulfate}						
(2 mol% 3-[MAPDMA]PS)		300000	water	298.15	−11	2005NED
(3 mol% 3-[MAPDMA]PS)		240000	water	298.15	−14	2005NED
(5 mol% 3-[MAPDMA]PS)		250000	water	298.15	−11	2005NED
(10 mol% 3-[MAPDMA]PS)		380000	water	298.15	−18	2005NED

Polymer	**M_n/ g/mol**	**M_w/ g/mol**	**Solvent**	**T/ K**	**A_2 10^4/ cm^3mol/g^2**	**Ref.**
Poly(sodium acrylate)						
		73000	water (0.1 M NaCl)	298.15	24.1	2003SCH
		90000	water (0.1 M NaCl)	298.15	23.5	2003SCH
		113000	water (0.1 M NaCl)	298.15	15.6	2003SCH
		174000	water (0.1 M NaCl)	298.15	19.1	2003SCH
		491000	water (0.1 M NaCl)	298.15	14.3	2003SCH
		575000	water (0.1 M NaCl)	298.15	8.70	2003SCH
		950000	water (0.1 M NaCl)	298.15	11.2	2003SCH
		2390000	water (0.1 M NaCl)	298.15	6.39	2003SCH
		3300000	water (0.1 M NaCl)	298.15	6.56	2003SCH
		171000	water (0.1 M NaCl)	298.15	1.24	2003SCH
		473000	water (0.1 M NaCl)	298.15	0.77	2003SCH
		700000	water (0.1 M NaCl)	298.15	0.03	2003SCH
		907000	water (0.1 M NaCl)	298.15	0.92	2003SCH
		2610000	water (0.1 M NaCl)	298.15	0.56	2003SCH
		3150000	water (0.1 M NaCl)	298.15	0.053	2003SCH
Poly(styrene-*b*-ethylene oxide) (PS:PEO = 1:2)		689000	water	298.15	0.42	2010AHM
Poly(vinyl alcohol)						
		85500	water	303.15	1.8	2000HON
Poly(*N*-vinylcaprolactam-*co*-1-vinylimidazole)						
(26.0 mol% 1-vinylimidazole)		254000	water	293.15	−4.0	2006LOZ
(26.0 mol% 1-vinylimidazole)		254000	water	323.15	−12.0	2006LOZ
(27.0 mol% 1-vinylimidazole)		30000	water	293.15	−18.0	2006LOZ
(27.0 mol% 1-vinylimidazole)		30000	water	323.15	9.0	2006LOZ
(38.0 mol% 1-vinylimidazole)		66000	water	293.15	9.0	2006LOZ
(38.0 mol% 1-vinylimidazole)		66000	water	323.15	5.0	2006LOZ
(38.5 mol% 1-vinylimidazole)		28000	water	293.15	19.0	2006LOZ
(38.5 mol% 1-vinylimidazole)		28000	water	323.15	14.0	2006LOZ
Poly(1-vinylimidazole)						
		35000	water (0.2 M NaCl, pH =3)	298.15	34.6	2009MAS
		193000	water (0.2 M NaCl, pH =3)	298.15	29.2	2009MAS

Polymer	M_n/ g/mol	M_w/ g/mol	Solvent	T/ K	A_2 10^4/ $cm^3 mol/g^2$	Ref.
Poly(1-vinyl-2-pyrrolidinone)						
		360000	water	298.15	5.02	2008KAR
		360000	water	323.15	6.78	2008KAR
Sodium hyaluronate						
		453000	water (0.1 M sodium phosphate buffer)	298.15	27.8	2010POD
		453000	water (0.05 M Na_2SO_4)	298.15	34.5	2010POD

7.2. References

2000HON Hong, P.-D. and Huang, H.-T., Effect of co-solvent complex on preferential adsorption phenomenon in polyvinyl alcohol ternary solutions, Polymer, 41, 6195, 2000.

2003SCH Schweins, R., Hollmann, J., and Huber, K., Dilute solution behaviour of sodium polyacrylate chains in aqueous NaCl solutions, *Polymer*, 44, 7131, 2003.

2003STI Stieger, M. and Richtering, W., Shear-induced phase separation in aqueous polymer solutions: Temperature-sensitive microgels and linear polymer chains, *Macromolecules*, 36, 8811, 2003.

2004ALB Alb, A.M., Mignard, E., Drenski, M.F., and Reed, W.F., In situ time-dependent signatures of light scattered from solutions undergoing polymerization reactions, *Macromolecules*, 37, 2578, 2004.

2004VOR Vorwerg, W., Dijksterhuis, J., Borghuis, J., Radosta, S., and Kröger, A., Film properties of hydroxypropyl starch, *Starch/Stärke*, 56, 297, 2004.

2005MAT Matsuda, Y., Miyazaki, Y., Sugihara, S., Aoshima, S., Saito, K., and Sato, T., Phase separation behavior of aqueous solutions of a thermoresponsive polymer, *J. Polym. Sci.: Part B: Polym. Phys.*, 43, 2937, 2005.

2005NED Nedelcheva, A.N., Novakov, C.P., Miloshev, S.M., and Berlinova, I.V., Electrostatic self-assembly of thermally responsive zwitterionic poly(*N*-isopropylacrylamide) and poly(ethylene oxide) modified with ionic groups, *Polymer*, 46, 2059, 2005.

2005NIN Ninni, L., Meirelles, A.J.A., and Maurer, G., Thermodynamic properties of aqueous solutions of maltodextrins from laser-light scattering, calorimetry and isopiestic investigations, *Carbohydrate Polym.*, 59, 289, 2005.

2005ROT Rotureau, E., Chassenieux, C., Dellacherie, E., and Durand, A., Neutral polymeric surfactants derived from dextran: A study of their aqueous solution behavior, *Macromol. Chem. Phys.*, 206, 2038, 2005.

2005SUG Sugaya, R., Thermodynamics of aqueous biopolymer solutions and fractionation of dextran, *Dissertation*, Johannes-Gutenberg-Universität Mainz, 2005.

2006CAI Cai, J., Liu, Y., and Zhang, L., Dilute solution properties of cellulose in LiOH/urea aqueous system, *J. Polym. Sci.: Part B: Polym. Phys.*, 44, 3093, 2006.

2006CON Convertine, A.J., Lokitz, B.S., Vasileva, Y., Myrick, L.J., Scales, C.W., Lowe, A.B., and McCormick, C.L., Direct synthesis of thermally responsive DMA/NIPAM diblock and DMA/NIPAM/DMA triblock copolymers via aqueous, room temperature RAFT polymerization, *Macromolecules*, 39, 1724, 2006.

2006KHA Khan, M.S., Aggregate formation in poly(ethylene oxide) solutions, *J. Appl. Polym. Sci.*, 102, 2578, 2006.

2006LAM Lambeth, R.H., Ramakrishnan, S., Mueller, R., Poziemski, J.P., Miguel, G.S., Markoski, L.J., Zukoski, C.F., and Moore, J.S., Synthesis and aggregation behavior of thermally responsive star polymers, *Langmuir*, 22, 6352, 2006.

2006LIX Li, X., Mya, K.Y., Ni, X., He, C., Leong, K.W., and Li, J., Dynamic and static light scattering studies on self-aggregation behavior of biodegradable amphiphilic poly(ethylene oxide)-poly[(R)-3-hydroxybutyrate]-poly(ethylene oxide) triblock copolymers in aqueous solution, *J. Phys. Chem. B*, 110, 5920, 2006.

2006LOZ Lozinskii, V.I., Simenel, I.A., Semenova, M.G., Belyakova, L.E., Ilin, M.M., Grinberg, V. Ya., Dubovik, A.S., and Khokhlov, A.R., Behavior of protein-like *N*-vinylcaprolactam and *N*-vinylimidazole copolymers in aqueous solutions, *Polym. Sci., Ser. A*, 48, 435, 2006.

2006OZD Ozdemir, C. and Guner, A., Solution thermodynamics of poly(ethylene glycol)/water systems, *J. Appl. Polym. Sci.*, 101, 203, 2006.

2007RA1 Rangelov, S., Trzebicka, B., Jamroz-Piegza, M., and Dworak, A., Hydrodynamic behavior of high molar mass linear polyglycidol in dilute aqueous solution, *J. Phys. Chem. B*, 111, 11127, 2007.

2007RA2 Rangelov, S., Almgren, M., Halacheva, S., and Tsvetanov, C., Polyglycidol-based analogues of pluronic block copolymers. Light scattering and cryogenic transmission electron microscopy studies, *J. Phys. Chem. C*, 111, 13185, 2007.

2007SCH Schaink, H.M. and Smit, J.A.M., Protein-polysaccharide interactions: The determination of the osmotic second virial coefficients in aqueous solutions of beta-lactoglobulin and dextran, *Food Hydrocoll.*, 21, 1389, 2007.

2008DEN Deng, L., Shi, K., Zhang, Y., Wang, H., Zeng, J., Guo, X., Du, Z., and Zhang, B., Synthesis of well-defined poly(*N*-isopropylacrylamide)-*b*-poly(L-glutamic acid) by a versatile approach and micellization, *J. Colloid Interface Sci.*, 323, 169, 2008.

2008KAR Karimi, M., Albrecht, W., Heuchel, M., Weigel, Th., and Lendlein, A., Determination of solvent/polymer interaction parameters of moderately concentrated polymer solutions by vapor pressure osmometry, *Polymer*, 49, 2587, 2008.

2008NUO Nuopponen, M., Kalliomaeki, K., Aseyev, V., and Tenhu, H., Spontaneous and thermally induced self-organization of A-B-A stereoblock polymers of *N*-isopropylacrylamide in aqueous solutions, *Macromolecules*, 41, 4881, 2008.

2008VID Vidal, R.R.L., Balaban, R., and Borsali, R., Amphiphilic derivatives of carboxy-methylcellulose: Evidence for intra- and intermolecular hydrophobic associations in aqueous solutions, *Polym. Eng. Sci.*, 48, 2011, 2008.

2008ZHO Zhou, K., Lu, Y., Li, J., Shen, L., Zhang, G., Xie, Z., and Wu, C., The coil-to-globule-to-coil transition of linear polymer chains in dilute aqueous solutions: Effect of intrachain hydrogen bonding, *Macromolecules*, 41, 8927, 2008.

2009DAN Dan, A., Ghosh, S., and Moulik, S.P., Physicochemical studies on the biopolymer inulin: A critical evaluation of its self-aggregation, aggregate-morphology, interaction with water, and thermal stability, *Biopolymers*, 91, 687, 2009.

2009GAO Gao, W., Liu, X.M., and Gross, R.A., Determination of molar mass and solution properties of cationic hydroxyethyl cellulose derivatives by multi-angle laser light scattering with simultaneous refractive index detection, *Polym. Int.*, 58, 1115, 2009.

2009LIT Litmanovich, E.A., Zakharchenko, S.O., Stoychev, G.V., and Zezin, A.B., Phase separation in a poly(acrylic acid)-polycation system in acidic solutions, *Polym. Sci., Ser. A*, 51, 616, 2009.

2009LUY Lu, Y., Chen, D.-J., Wang, G.-K., and Yan, C.-L., Study of interactions of bovine serum albumin in aqueous $(NH_4)_2SO_4$ solution at 25°C by osmotic pressure measurements, *J. Chem. Eng. Data*, 54, 1975, 2009.

2009MAS Masaki, M., Ogawa, K., and Kokufuta, E., Unusual behavior in light scattering experiments of poly(*N*-vinylimidazole) prepared by precipitation polymerization, *Colloid Polym. Sci.*, 287, 1405, 2009.

2009PAR Park, Y.K. and Choi, G., Effects of pH, salt type, and ionic strength on the second virial coefficients of aqueous bovine serum albumin solutions, *Korean J. Chem. Eng.*, 26, 193, 2009.

2009STO Stojanovic, Z.P., Jeremic, K., Jovanovic, S., Nierling, W., and Lechner, M.D., Light scattering and viscosity investigation of dilute aqueous solutions of carboxymethyl starch, *Starch/Stärke*, 61, 199, 2009.

2010AHM Ahmad, F., Baloch, M.K., Jamil, M., and Jeon, Y.J., Characterization of polystyrene-*b*-poly(ethylene oxide) diblock copolymer and investigation of its micellization behavior in water, *J. Appl. Polym. Sci.*, 118, 1704, 2010.

2010LIT Litmanovich, E.A., Chernikova, E.V., Stoychev, G.V., and Zakharchenko, S.O., Unusual phase behavior of the mixture of poly(acrylic acid) and poly(diallyldimethyl ammonium chloride) in acidic media, *Macromolecules*, 43, 6871, 2010.

2010MAR Marcelo, G., Prazeres, T.J.V., Charreyre, M.-T., Martinho, J.M.G., and Farinha, J.P.S., Thermoresponsive micelles of phenanthrene-R-end-labeled poly(*N*-decyl-acrylamide-*b*-*N,N*-diethylacrylamide) in water, *Macromolecules*, 43, 501, 2010.

2010NAG Nagy, M., Thermodynamic study of aqueous solutions of polyelectrolytes of low and medium charge density without added salt by direct measurement of osmotic pressure, *J. Chem. Thermodyn.*, 42, 387, 2010.

2010POD Podzimek, S., Hermannova, M., Bilerova, H., Bezakova, Z., and Velebny, V., Solution properties of hyaluronic acid and comparison of SEC-MALS-VIS data with off-line capillary viscometry, *J. Appl. Polym. Sci.*, 116, 3013, 2010.

2010RIS Risica, D., Barbetta, A., Vischetti, L., Cametti, C., and Dentini, M., Rheological properties of guar and its methyl, hydroxypropyl and hydroxypropyl-methyl derivatives in semidilute and concentrated aqueous solutions, *Polymer*, 51, 1972, 2010.

2010YAN Yang, L., Qi, X., Liu, P., El Ghzaoui, A., and Li, S., Aggregation behavior of self-assembling polylactide/poly(ethylene glycol) micelles for sustained drug delivery, *Int. J. Pharmaceutics*, 394, 43, 2010.

2011LEZ Lezov, A.V., Vlasov, P.S., Lezov, A.A., Domnina, N.S., and Polushina, G.E., Molecular properties of poly(carboxybetaine) in solutions with different ionic strengths and pH values, *Polym. Sci., Ser. A*, 53, 1012, 2011.

2011MAI Maia, A.M.S., Villetti, M.A., Vidal, R.R.L., Borsali, R., and Balaban, R.C., Solution properties of a hydrophobically associating polyacrylamide and its polyelectrolyte derivatives determined by light scattering, small angle X-ray scattering and viscometry, *J. Braz. Chem. Soc.*, 22, 489, 2011.

2011MOR Morariu, S., Eckelt, J., and Wolf, B.A., Dextran-based polycations: Thermodynamic interaction with water as compared with unsubstituted dextran, 1 – Volumetric properties and light scattering, *Macromol. Chem. Phys.*, 212, 1925, 2011.

2011NAI Naik, S.S., Ray, J.G., and Savin, D.A., Temperature- and pH-responsive self-assembly of poly(propylene oxide)-*b*-poly(lysine) block copolymers in aqueous solution, *Langmuir*, 27, 7231, 2011.

2011NUH Nuhn, H. and Klok, H.-A., Aqueous solution self-assembly of polystyrene-*b*-poly(L-lysine) diblock oligomers, *Eur. Polym. J.*, 47, 782, 2011.

2011YAD Yadav, S., Shire, S.J., and Kalonia, D.S., Viscosity analysis of high concentration bovine serum albumin aqueous solutions, *Pharm. Res.*, 28, 1973, 2011.

APPENDICES

Appendix 1 List of polymers in alphabetical order

Polymer	Page(s)
Agarose	445
α-Amylase	479, 484
β-Amylase	480-481
Amyloglucosidase	484-486
Amylopectin	445
Amylose	445
Bovine serum albumin	21, 182-183, 199-200, 474, 705-706
N-Carboxyethylchitosan	138, 445
Carboxymethylcellulose	706
Carboxymethylcellulose-*g*-poly(*N,N*-dihexylacrylamide)	706
Carboxymethylstarch	706-707
Casein	445
Cellobiose	577, 581
Cellulose	445, 707
Cellulose acetate	445
Chitosan	599-600, 602
Chitosan-*g*-poly(*N*-isopropylacrylamide)	138, 445
Deca(ethylene glycol) mono(nonylphenyl) ether	138
Deca(ethylene glycol) monotridecyl ether	138
Dextran	21, 177-181, 183-184, 189-195, 446, 510, 571, 699, 707
Dextran-*g*-poly(*N*-isopropylacrylamide)	138
Ethylcellulose-*g*-[poly(ethylene glycol) methyl ether methacrylate]	138
Ethyl(hydroxyethyl)cellulose	138, 446-447
Ficoll	447
Gelatine	138, 447, 581
Guar gum	447, 600
Hepta(ethylene glycol) monododecyl ether	138, 447
Hepta(ethylene glycol) monohexadecyl ether	139
Hepta(ethylene glycol) monooctadecyl ether	139
Hepta(ethylene glycol) monotetradecyl ether	139, 448
Hexa(ethylene glycol) monodecyl ether	139, 448, 515

Polymer	**Page(s)**
Poly(acrylic acid)	23, 184-185, 256, 451, 510, 581, 604-605, 708
Poly(acrylic acid-*co*-*N*-isopropylacrylamide)	141
Poly(acrylonitrile-*co*-acrylic acid)	451
Poly(acrylonitrile-*co*-itaconic acid)	451
Poly(acrylonitrile-*co*-methyl acrylate-*co*-itaconic acid)	451
Poly(*N*-acryloylasparaginamide)	101
Poly(*N*-acryloyl-L-aspartic acid *N'*-propylamide)	207-208, 451
Poly(*N*-acryloyl-L-glutamic acid *N'*-ethylamide)	451
Poly(*N*-acryloyl-L-glutamic acid *N'*-methylamide)	451
Poly(*N*-acryloyl-L-glutamic acid *N'*-propylamide)	208, 451
Poly[*N*-(3-acryloyloxypropyl)pyrrolidone]	141
Poly(*N*-acryloyl-L-proline methyl ester-*b*-*N*-acryloyl-4-*trans*-hydroxy-L-proline)	141
Poly(*N*-acryloylpyrrolidine)	141
Poly(*N*-acryloylpyrrolidine-*b*-*N*-isopropylacrylamide)	142
Poly(*N*-acryloyl-L-valine *N'*-methylamide)	142
Poly(allylamine) hydrochloride	452
Poly[allylamine-*g*-poly(*N*-isopropylacrylamide)]	142
Poly(amidoamine) dendrimer	142
Poly(ammonium acrylate)	24, 55-56
Polyaspartamide copolymers	142
Poly(m-benzamide)	142
Poly(5,6-benzo-2-methylene-1,3-dioxepane-*co*-*N*-isopropylacrylamide)	142
Poly[1,3-bis(dimethylamino)propan-2-yl methacrylate]	142
Poly[1-(bis(3-(dimethylamino)propyl)amino)propan-2-yl methacrylate]	142
Poly{2-*tert*-butoxy-*N*-[2-(methacryloyloxy)ethyl]-*N,N*-dimethyl-2-oxoethanammonium chloride-*co*-*N*-isopropylacrylamide}	142
Poly(*N*-*tert*-butylacrylamide-*co*-acrylamide)	142, 452
Poly(*N*-*tert*-butylacrylamide-*co*-*N,N*-dimethylacrylamide)	143
Poly(*N*-*tert*-butylacrylamide-*co*-*N*-ethylacrylamide)	143
Poly(butyl acrylate-*b*-acrylic acid)	452
Poly(butyl acrylate-*co*-*N*-isopropylacrylamide)	143
Poly(*tert*-butyl acrylate)-*g*-poly[oligo(ethylene glycol) methyl ether methacrylate)]	143, 452
Poly(butylene oxide-*b*-ethylene oxide)	143
Poly(butylene oxide-*b*-ethylene oxide-*b*-butylene oxide)	143
Poly(*tert*-butyl glycidylether)-*b*-polyglycidol star polymer	143
Poly(4-*tert*-butyl styrene-*b*-*N*-isopropylacrylamide-*b*-4-*tert*-butyl styrene)	143
Poly(ε-caproamide)	510
Poly(ε-caprolactone)	587
Poly(ε-caprolactone-*b*-ethylene glycol-*b*-ε-caprolactone)	143
Poly(ε-caprolactone-*b*-*N*-isopropylacrylamide-*b*-ε-caprolactone)	143

Appendix 2 List of systems and properties in order of the polymers

Polymer	Solvent	Property	Page(s)
β-Lactoglobulin			
	water	A_2	708
Maltodextrin			
	water	A_2	708
	water	$\Delta_{dil}H^{12}$	577-578
	water	density	603
	water	VLE	22-23
Maltotriose			
	water	A_2	708
Methylguar			
	water	A_2	708
Methyl(hydroxypropyl)cellulose			
	water + potassium bromide	LLE	206
	water + potassium chloride	LLE	206
	water + sodium bromide	LLE	206
	water + sodium chloride	LLE	206
	water + sodium nitrate	LLE	207
	water + sodium phosphate	LLE	207
	water + sodium sulfate	LLE	207
Methyl(hydroxypropyl)guar			
	water	A_2	708
Penta(ethylene glycol) monodecyl ether			
	water	UCST/LCST	515
Poly(acrylamide)			
	water	A_2	708
Poly(acrylamide-*co*-hydroxypropyl acrylate)			
	water	UCST/LCST	515

Polymer	Solvent	Property	Page(s)
Poly(*N,N*-diethylacrylamide)			
	water	A_2	709
	water	UCST/LCST	515-516
Poly(*N,N*-diethylacrylamide-*co*-acrylic acid)			
	water	UCST/LCST	516
Poly(*N,N*-diethylacrylamide-*co*-*N*-ethylacrylamide)			
	water	UCST/LCST	516
Poly[di(ethylene glycol) methacrylate]			
	water	UCST/LCST	516
Poly(*N,N*-dimethylacrylamide)			
	water	A_2	709
Poly(*N,N*-dimethylacrylamide-*co*-allyl methacrylate)			
	water	LLE	101-102
Poly(*N,N*-dimethylacrylamide-*co*-*tert*-butylacrylamide)			
	water	VLE	25
	water + monoammonium phosphate	LLE	208
	water + monopotassium phosphate	LLE	208
	water + monosodium carbonate	LLE	208-209
	water + sodium chloride	LLE	209
Poly(*N,N*-dimethylacrylamide–*b*-*N*-isopropylacrylamide-*b*-*N,N*-dimethylacrylamide)			
	water	A_2	709

Polymer	Solvent	Property	Page(s)
Poly[2-(2-ethoxy)ethoxyethyl vinyl ether]			
	water	A_2	709
	water	LLE	105-106
Poly(ethoxyethyl glycidyl ether)			
	water	A_2	710
Poly[3-(2-ethoxyethyl)-1-vinyl-2-pyrrolidinone]			
	water	UCST/LCST	516
Poly(ethoxytriethylene glycol methacrylate)			
	water	LLE	106
Poly(*N*-ethylacrylamide)			
	water	UCST/LCST	516
	water + potassium chloride	LLE	209
	water + sodium dodecyl sulfate	LLE	209
Poly(*N*-ethylacrylamide-*co*-*N*-isopropylacrylamide)			
	water	UCST/LCST	517
Poly(ethylene-*co*-vinyl alcohol)			
	–	PVT	588-593
Poly(ethylene glycol)			
	water	A_2	710
	water	density	605-652
	water	V^E	611-612, 615-617
	water	VLE	26-30
	water + ammonium citrate	density	641-642
	water + ammonium citrate	LLE	210-211
	water + ammonium sulfate	LLE	212-216
	water + ammonium sulfate	VLE	59-60
	water + bovine serum albumin	LLE	182-183

Polymer	Solvent	Property	Page(s)
Poly(ethylene oxide-*b*-propylene oxide-*b*-ethylene oxide)			
	water + sodium chloride	LLE	383-384
	water + sodium chloride/ hexadecyltrimethylammonium bromide	LLE	500-501
	water + sodium chloride/ sodium dodecyl sulfate	LLE	501
	water + sodium citrate	LLE	384-386
	water + sodium hydroxide	LLE	386-387
	water + sodium nitrate	LLE	387-388
	water + sodium phosphate	LLE	388
	water + sodium phosphate/ hexadecyltrimethylammonium bromide	LLE	502
	water + sodium phosphate/ sodium dodecyl sulfate	LLE	502-503
	water + sodium sulfate	LLE	388-391
	water + sodium sulfate/ hexadecyltrimethylammonium bromide	LLE	503
	water + sodium sulfate/ sodium dodecyl sulfate	LLE	504
	water + L-threonine	LLE	197-198
	water + L-threonine/hexadecyltrimethylammonium bromide	LLE	504-505
	water + L-threonine/ sodium dodecyl sulfate	LLE	505
	water + L-valine	LLE	198
	water + L-valine/hexadecyltrimethylammonium bromide	LLE	506
	water + L-valine/sodium dodecyl sulfate	LLE	506-507
	water + zinc sulfate	LLE	391-392
Poly(*N*-ethylmethacrylamide)			
	water	UCST/LCST	517
Poly(*N*-ethyl-*N*-methylacrylamide)			
	water	UCST/LCST	517
	water + methanol	LLE	198-199
	water + 2-methyl-1-propanol	LLE	199
	water + 2-propanol	LLE	199

Polymer	Solvent	Property	Page(s)
Poly(*N*-isopropylacrylamide-*b*-*N,N*-dimethylacrylamide)			
	water	A_2	711
Poly(*N*-isopropylacrylamide-*co*-*N,N*-dimethylacrylamide)			
	water	LLE	119-120
	water	UCST/LCST	520
Poly(*N*-isopropylacrylamide-*b*-L-glutamic acid)			
	water	A_2	711
Poly(*N*-isopropylacrylamide-*co*-*N*-glycineacrylamide)			
	water	LLE	121
Poly(*N*-isopropylacrylamide-*co*-2-hydroxyethyl methacrylate)			
	water	LLE	121
Poly(*N*-isopropylacrylamide-*co*-2-hydroxyethyl methacrylate-*co*-acrylic acid)			
	water	UCST/LCST	520
Poly(*N*-isopropylacrylamide-*co*-2-hydroxyethyl methacrylate lactate-*co*-acrylic acid)			
	water	UCST/LCST	520
Poly[*N*-isopropylacrylamide-*co*-(2-hydroxyisopropyl)acrylamide]			
	water	UCST/LCST	520
Poly[*N*-isopropylacrylamide (isotactic)-*b*-*N*-isopropylacryl-amide-*b*-*N*-isopropylacryl-amide(isotactic)]			
	water	A_2	711

Polymer	Solvent	Property	Page(s)
Poly(*N*-isopropylacrylamide-*co*-sodium 2-acrylamido-2-methyl-1-propanesulfonate-*co*-*N*-*tert*-butylacrylamide)	water	LLE	127
Poly(*N*-isopropylacrylamide-*co*-p-vinylphenylboronic acid)	water	LLE	127-128
Poly(2-isopropyl-2-oxazoline)	water	UCST/LCST	521
Poly(2-isopropyl-2-oxazoline-*co*-2-butyl-2-oxazoline)	water	UCST/LCST	521
Poly(2-isopropyl-2-oxazoline-*co*-2-nonyl-2-oxazoline)	water	UCST/LCST	521
Poly(2-isopropyl-2-oxazoline-*co*-2-propyl-2-oxazoline)	water	UCST/LCST	521
Poly(DL-lactic acid)	water	VLE	40
Poly(L-lactic acid)	water	VLE	40-42
Poly(L-lactic acid-*co*-glycolic acid)	water	Henry	96
Poly(methoxydiethylene glycol methacrylate)	water	LLE	128

Polymer	Solvent	Property	Page(s)
Poly(sodium styrenesulfonate)			
	water	VLE	47
	water + sodium chloride	VLE	85
Poly(styrene-*b*-ethylene oxide)			
	water	A_2	712
Poly(styrene-*b*-*N*-isopropyl-acrylamide)			
	water	UCST/LCST	522
Poly(*N*-tetrahydrofurfuryl-acrylamide)			
	water	UCST/LCST	522
Poly(*N*-tetrahydrofurfuryl-methacrylamide)			
	water	UCST/LCST	522
Poly(vinyl acetate)			
	water	VLE	47
Poly(vinyl acetate-*co*-vinyl alcohol)			
	water	VLE	48
Poly(vinyl alcohol)			
	water	A_2	712
	water	density	682-683
	water	VLE	48-49
	water + dimethylsulfoxide	density	683
	water + nicotinic acid	density	685-688
	water + poly(ethylene glycol)	density	683-685
	–	PVT	596-597
Poly(vinyl alcohol-*co*-sodium acrylate)			
	water + ethanol	VLE	86
	water + 1-propanol	VLE	86-87
	water + 2-propanol	VLE	87

Polymer	Solvent	Property	Page(s)
Poly(vinyl chloride)			
	water	VLE	51
Poly(1-vinylimidazole)			
	water (NaCl)	A_2	712
Poly(vinyl methyl ether)			
	deuterium oxide	LLE	136
	water	LLE	136
Poly(vinyl methyl ether-*b*-vinyl isobutyl ether)			
	water	LLE	136-137
Poly(vinyl methyl ether-*b*-vinyl isobutyl ether-*b*-vinyl methyl ether)			
	water	LLE	137
Poly(1-vinyl-2-pyrrolidinone)			
	water	A_2	713
	water	density	688-691, 694-696
	water	$\Delta_{sol}H_B$	579
	water	VLE	51-54
	water + ammonium sulfate	VLE	88-89
	water + sodium chloride	VLE	89-90
	water + sodium citrate	VLE	91-92
	water + diammonium phosphate/monoammonium phosphate	LLE	426-427
	water + dimethylsulfoxide	density	696-697
	water + dipotassium oxalate	LLE	427-429
	water + dipotassium phosphate/monopotassium phosphate	LLE	429-431
	water + dipotassium tartrate	LLE	431-432
	water + disodium phosphate	LLE	432-434
	water + disodium succinate	LLE	434
	water + magnesium sulfate	LLE	434-435
	water + monosodium phosphate	VLE	92-93

Appendix 3 List of solvents in alphabetical order

Name	Formula	CAS-RN	Page(s)
acetic acid	$C_2H_4O_2$	64-19-7	177, 458
acetonitrile	C_2H_3N	75-05-8	461, 470, 574
2-aminoethanol	C_2H_7NO	141-43-5	470
1-butanol	$C_4H_{10}O$	71-36-3	446, 458, 462, 469, 512-513
tert-butanol	$C_4H_{10}O$	75-65-0	447, 453, 460, 462, 468, 470
carbon dioxide	CO_2	124-38-9	571-574
cyclohexane	C_6H_{12}	110-82-7	458
n-decane	$C_{10}H_{22}$	124-18-5	471
deuterium oxide	D_2O	7789-20-0	462, 465, 471, 520, 522-523, 569-570,
1,2-dichloroethane	$C_2H_4Cl_2$	107-06-2	459
3,4-dimethoxybenzaldehyde	$C_9H_{10}O_3$	120-14-9	462
N,N-dimethylacetamide	C_4H_9NO	127-19-5	178, 455, 471, 510, 514
1,4-dimethylbenzene	C_8H_{10}	106-42-3	574
N,N-dimethylformamide	C_3H_7NO	68-12-2	451, 453, 461-462, 471
dimethylsulfoxide	C_2H_6OS	67-68-5	451, 454-455, 462, 469, 571, 581, 683, 696
1,4-dioxane	$C_4H_8O_2$	123-91-1	450, 453, 462, 464, 466, 468
n-dodecane	$C_{12}H_{26}$	112-40-3	447-448, 450
1-dodecanol	$C_{12}H_{26}O$	112-53-8	450
ethanol	C_2H_6O	64-17-5	56-58, 61-63, 78, 86, 178, 446-447, 453-455, 459, 461-462, 465-468, 470-471, 512-514, 583
ethene	C_2H_4	74-85-1	574
2-(2-ethoxyethoxy)ethanol	$C_6H_{14}O_3$	111-90-0	449
formic acid	CH_2O_2	64-18-6	459
glycerol	$C_3H_8O_3$	56-81-5	462, 513
n-hexane	C_6H_{14}	110-54-3	513
1-hexanol	$C_6H_{14}O$	111-27-3	446, 459
isobutyric acid	$C_4H_8O_2$	79-31-2	452, 456-458, 460, 472
isopentanoic acid	$C_5H_{10}O_2$	503-74-2	472
2,6-lutidine	C_7H_9N	108-48-5	451

Appendix 4 List of solvents in order of their molecular formulas

Formula	Name	CAS-RN	Page(s)
CH_2O_2	formic acid	64-18-6	459
CH_4O	methanol	67-56-1	178-179, 198-199, 203, 446-447, 449, 452-453, 455, 459-462, 468-470
CO_2	carbon dioxide	124-38-9	571-574
$C_2HCl_3O_2$	trichloroacetic acid	76-03-9	472, 514
C_2H_3N	acetonitrile	75-05-8	461, 470, 574
C_2H_4	ethene	74-85-1	574
$C_2H_4Cl_2$	1,2-dichloroethane	107-06-2	459
$C_2H_4O_2$	acetic acid	64-19-7	177, 458
C_2H_6O	ethanol	64-17-5	56-58, 61-63, 78, 86, 178, 446-447, 453-455, 459, 461-462, 465-468, 470-471, 512-514, 583
C_2H_6OS	dimethylsulfoxide	67-68-5	451, 454-455, 462, 469, 571, 581, 683, 696
C_2H_7NO	2-aminoethanol	141-43-5	470
C_3H_6O	2-propanone	67-64-1	180, 204-205, 445, 453, 462, 465
C_3H_7NO	*N,N*-dimethylformamide	68-12-2	451, 453, 461-462, 471
C_3H_8O	1-propanol	71-23-8	86-87, 181, 446-447, 453, 459-460, 462, 466, 468-470, 512
C_3H_8O	2-propanol	67-63-0	65-67, 78, 87, 179, 199, 201-202, 204, 453, 462, 466, 468-470, 472
$C_3H_8O_3$	glycerol	56-81-5	462, 513
C_4H_8O	tetrahydrofuran	109-99-9	58, 181-182, 205, 452, 463, 470-471
$C_4H_8O_2$	1,4-dioxane	123-91-1	450, 453, 462, 464, 466, 468
$C_4H_8O_2$	isobutyric acid	79-31-2	452, 456-458, 460, 472
$C_4H_8O_2S$	sulfolane	126-33-0	463
C_4H_9NO	*N,N*-dimethylacetamide	127-19-5	178, 455, 471, 510, 514

INDEX